APPLICATIONS OF IONIC LIQUIDS IN SCIENCE AND TECHNOLOGY

Edited by **Scott T. Handy**

INTECHOPEN.COM

Applications of Ionic Liquids in Science and Technology
Edited by Scott T. Handy

Published by InTech

Janeza Trdine 9, 51000 Rijeka, Croatia

Publishing Process Manager Alenka Urbancic

Technical Editor Teodora Smiljanic

Cover Designer InTech Design Team

Additional hard copies can be obtained from orders@intechopen.com

Applications of Ionic Liquids in Science and Technology, Edited by Scott T. Handy
p. cm.
ISBN 978-953-307-605-8

For EU product safety concerns:
IN TECH d.o.o., Prolaz Marije Krucifikse Kozulić 3, 51000 Rijeka, Croatia,
info@intechopen.com or visit our website at intechopen.com.

Meet the editor

Dr. Scott Handy received his Ph.D. in Chemistry from Indiana University in 1996 under the direction of Professor Paul Grieco. Following an NIH postdoctoral fellowship at Stanford University in the labs of Professor Paul Wender, he joined the Chemistry Department at Binghamton University in 1999. In 2006 he moved to Middle Tennessee State University as an Associate Professor and was promoted to Professor in 2011. His research interests are in the areas of cross-coupling reactions (particularly regioselectivity in the couplings of polyhaloheteroaromatics) and ionic liquids (as recyclable solvents for transition metal-catalyzed reactions and electrochemistry).

Contents

Preface

Ionic liquids (more specifically, room temperature ionic liquids (RTIL)) have attracted considerable interest over the last several years. Although the specific definition of what is a RTIL varies from person to person, the most generally accepted definition is a salt with a melting point below 100 °C. Such a broad definition leaves considerable room for flexibility, which contributed to labeling RTILs as "designer solvents."

The history of ionic liquids (and the closely related molten salts) has a rather ill-defined beginning, although it is most commonly dated back to 1914 from the work of Walden on the use of alkylammonium nitrates. The next burst of interest occurred with the discovery of chloroaluminates formed by combining quaternary heterocyclic cations with aluminum chloride. These materials exhibited a great deal of potential for use in a variety of areas, but all suffered from extreme sensitivity to moisture.

A major step forward was made by Wilkes in the early 1990s, with the report of moisture stable ionic liquids created by replacing the aluminum chloride with other anions, such as tetrafluoroborate or hexafluorophosphate. Since that seminal report by Wilkes and co-workers, the family of RTILs has seen explosive growth. Starting with imidazolium cations, the cationic component has been varied to include pyridinium, ammonium, phosphonium, thiazolium, and triazolium species.

In general, these cations have been combined with weakly coordinating anions, although not all weakly coordinating anions result in RTILs (for example, the very weakly coordinating polyhedral borane anions of Reed afford salts with melting points between 45 and 156 °C for a series of imidazolium cations). Common examples include tetrafluoroborate, hexafluorophosphate, triflate, triflimide, and dicyanimide. Of these, the first two have been explored the most extensively, and must be treated with the greatest caution, as they are fairly readily hydrolyzed to boric acid and phosphate respectively. Indeed, various phosphate and phosphinate anions have been employed to some advantage in RTILs and the list of possible anionic components continues to grow at a rapid rate.

This volume, which is the second in a two volume set on ionic liquids, focuses on applications of ionic liquids in a growing range of areas. Throughout the 1990s, it seemed that the most attention in the area of ionic liquids applications was directed toward their use as solvents for organic and transition-metal-catalyzed reactions. This interest certainly continues to the present date, and is touched on in several chapters in volume 1 of the book, but the most innovative uses of ionic liquids span a much more diverse field than just synthesis.

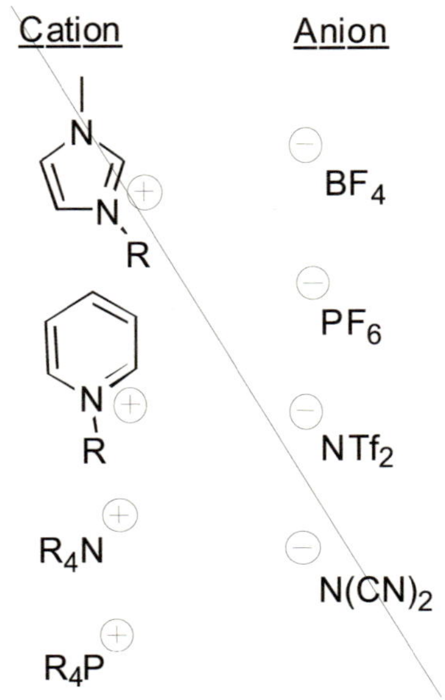

Fig. 1. Representative Ionic Liquid Components

For example, since ionic liquids are ions, they have been explored for use in various electronic applications, including electrolytes for batteries (Claude-Montigny, Stefan, and Violleau), capacitors and charge storage devices (Takaya, Shoko, and Takashi; Stenger-Smith, Chafin, Kline, Ostrom, and Quintana), as well as conducting and light emitting materials (Shao; and Luius, Garci-Verdugo, Burguete, Andrio, Molla, and Compañ).

There is also a list of ways to employ ionic liquids in the area of polymers: as grafted components (Maslowski and Zaborski), as solvents for polymerization (Tsubokawa), as modifiers of polymer morphology (Booker, Holdsworth, Bowyer, and McCluskey), and even as unusual components of polymers (Vidinha, Lourenco, Nunes, Duarte, and Barreiros). Similarly, ionic liquids are finding increasing use in the popular area of nanomaterials.

Much of this effort is directed toward using ionic liquids as either solvents for the synthesis of nanomaterials (Nan and Liebscher) or their stabilization (Kraynov and Mueller). Electrodeposition and recovery of metals is a related and fairly mature field (Anicai, Florea, and Visan).

A second, more mature area of application for ionic liquids is in the field of separations and spectroscopy (Flieger and Czajkowska-Belazko). For some time, ionic liquids have been explored as matrices for mass spectrometry (MALDI) (Fukuyama).

In addition to that, they have also been employed in liquid-liquid extraction (Yang, Xing, Yan, Bao, Su, Yang, and Ren), particularly for recovery of metal salts (Regel-Rosocka and Wisniewski).

Beyond these main areas of interest though, RTILs are finding their way into many other fields. One of timely international interest is utilization of biomass as non-petroleum-based fuels and as bio-renewable specialty chemicals (Kurata, Fujita, and Kishimoto; Hosmane).

Additional unusual applications are also displayed in this volume, including spectroscopy (Pawlak, Skrzypczak, and Bialek-Bylka), optical materials (Calixto, Rosete-Aguilar, Sanchez-Marin, Torres-Rocha, Prado, and Calixto-Solano), lubricants (Kondo), fuels for propulsion (Fahrat, Batonneau, Brohmi, and Kappenstein), and even refrigerants (Liang, Chen, Cheng, Guo, and Gui). As diverse as this list of topics is, it does not even begin to truly represent the wide array of areas impacted by RTILs. As more and more scientists become aware of the potential of RTILs, one can expect their applications to expand even further.

Sincerely,

Scott T. Handy,
Department of Chemistry
Middle Tennessee State University
Murfreesboro,
USA

References

[1] Freemantle, M. *Chem. Eng. News* 1998, 32-38.

[2] Walden, P. Bull. Acad. Imper. Sci. (St Petersburg), 1914, 1800.

[3] For an excellent review regarding the early history of ionic liquids, see Wilkes, J.S. *Green Chem.* 2002, *4*, 73.

[4] Wilkes, J.S.; Zaworotko, M.J. *J. Chem. Soc. Chem. Commun.* 1992, 965.

[5] Handy, S.T.; "Room Temperature Ionic Liquids: Different Classes and Physical Properties." *Curr. Org. Chem.* 2005, *9*, 959-989.

[6] Larsen, A.; Holbrey, J.D.; Tham, F.S.; Reed, C.A. "Designing Ionic Liquids: Imidazolium Melts with Inert Carborane Anions." *J. Am. Chem. Soc.* 2000, *122*, 7264-7272.

[7] Freire, M.G.; Neves, C.M.S.S.; Marrucho, I.M.; Coutinho, J.A.P.; Fernandes, A.M. "Hydrolysis of Tetrafluoroborate and Hexafluorophosphate Counter Ions in Imidazolium-Based Ionic Liquids." *J. Phys. Chem. A* 2010, *114*, 3744-3749.

[8] Lall, S.I.; Mancheno, D.; Castro, S.; Behaj, V.; Choen, J.I.; Engel, R. Polycations. Part X. LIPs, a new category of room temperature ionic liquid based on polyammonium salts." *Chem. Commun.* 2000, 2413-2414.

Part 1

Biomass Utilization

Ionic Liquids in Catalytic Biomass Transformation

Zhu Yinghuai[1,*], Algin Oh Biying[1], Xiao Siwei[1], Narayan S Hosmane[2] and John A. Maguire[3]
[1]*Institute of Chemical and Engineering Sciences, Jurong Island, Singapore*
[2]*Department of Chemistry and Chemical Biology, Northern Illinois University, DeKalb, Illinois*
[3]*Department of Chemistry, Southern Methodist University, Dallas, Texas*
[1]*Singapore*
[2,3]*USA*

1. Introduction

1.1 Alternatives to fossil fuels - biomass

For the past 150 years, coal, natural gas and petroleum functioned as our main sources of energy and chemicals. They provide an estimated 86% of energy and 96% of organic chemicals [1]. However, these natural resources will be depleted in the near future as studies has predicted that the global energy demand will grow over 50% by 2030 as shown in Figure 1 [2]. In addition, problem concerning global warming brought about from the emission of fossil fuels remains a major concern. Hence, there is an urgent need for us to seek alternative renewable and greener energy sources; the answer lies in biomass which possesses huge amounts of stored chemical energy. The stored chemical energy in biomass has the potential to be tapped to meet the rising energy demand.

Biomass is one of the most valuable products from living things where simple molecules such as H_2O, CO_2, N_2 are transformed into complex substances activated or catalyzed by biochemical processes (*e.g.* photosynthesis). Trees trap approximately 1% of solar energy and transform them into biomass. This energy is stored in the form of complex molecules such as lignin, carbohydrates, proteins, glycerides and others. Carbohydrates or saccharides (cellulose, hemicelluloses and starch) constitute 75% of the plant biomass, with 40% of the carbohydrate fraction of biomass being cellulose and 25% being hemicelluloses. Biomass energy is derived from five distinct energy sources: garbage, wood, waste, landfill gases, and alcohol fuels. Being the most abundant renewable resource as described in Figure 2, with approximately production of 1.0 x 10^{11} tons annually, biomass was recognized as the best candidate to replace fossil fuels [3].

The transformation of biomass into value added chemicals gives little or no harmful side products which address the pressing issue for greener alternatives to fossil fuels. Other than being renewable and abundant, biomass is a cost effective feedstock.

Cellulose, the major component of the carbohydrate fraction of biomass is hard to deal with. It is insoluble in most conventional organic solvent and water, which hinders the ease of

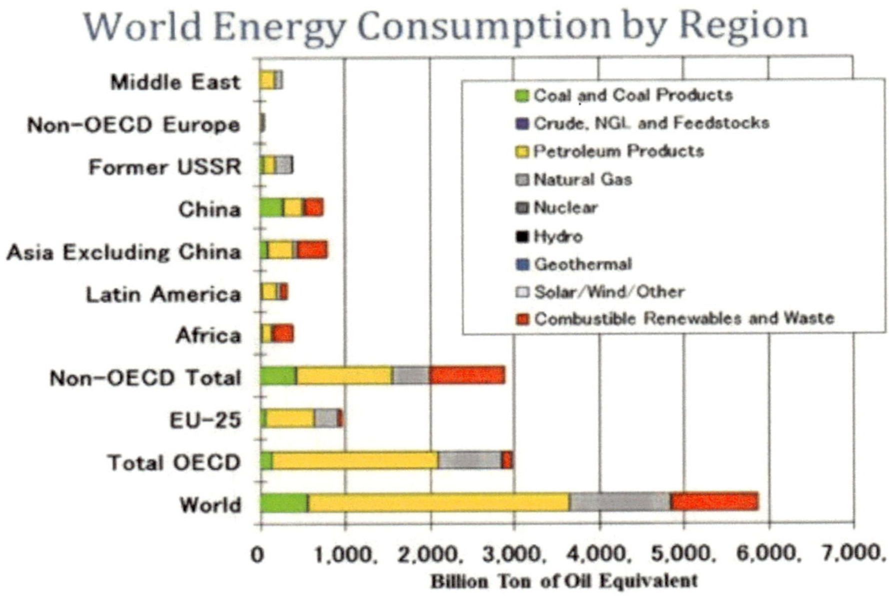

Fig. 1. World energy consumption by Region and will grow by 50% by 2030 (source: International Energy Agency (IEA) Non-Organization for Economic Co-operation and Development (OECD) Countries Energy Balance 2003).

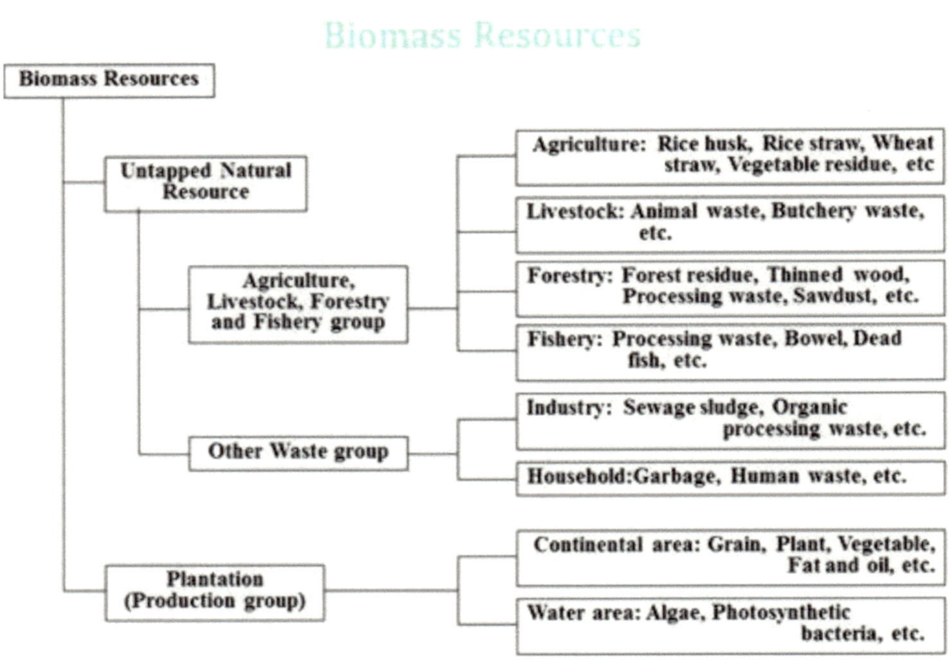

Fig. 2. Types of biomass resources.

transforming it into monosaccharides such as glucose. New solvents have to be sought which are able to dissolve the cellulose so as to disrupt the crystallinity structure of the cellulose and make it more susceptible to transformation into monosaccharides such as glucose.

Lignocellulosic cellulose refers to plant biomass which is made up primarily of cellulose, hemicellulose and lignin. Lignin acts as a strong adhesive holding the cellulose and hemicellulose together which makes lignocellulosic cellulose microbial and moisture resistant and imparts its rigid structure. Therefore, we need to devise a method to remove or reduce the lignin content in order to take advantage of the stored energy in lignocellulosic cellulose.

The problems of dissolving cellulose or reducing the lignin content have to be overcome. Ionic liquids (ILs) are a group of new organic salts that exist as liquids at relatively low temperatures can offer a solution to these problems. ILs can dissolve cellulose as well as remove or reduce the lignin content in lignocellulosic cellulose. ILs, based on polar organic solvents such as DMF, DMAc, DMI or DMSO and usually coupled with charged species such as LiCl, are frequently used for dissolving cellulose, while methyl methylsulfate imidazolium-based ILs show excellent results when dissolving softwood Kraft lignins.

1.2 Ionic liquids: Types and classfication

ILs is a group of new organic salts that exist in the liquid state at relatively low temperatures. An ideal IL for cellulose transformation should possess the following properties: (1) high dissolution capacity for cellulose; (2) low melting point; (3) good thermal stability; (4) non-volatile; (5) non-toxic; (6) chemically stable; (7) no cellulose decomposition; (8) easy cellulose regeneration and (9) low cost and simple process [4]. Imidazolium is one of the most commonly used cations of ILs, while anions include quite a number such as chloride, hexafluorophosphate, tetrafluroborate and others, as shown in Figure 3, and 4. To improve the solubility of cellulose and lignocellulose in ILs, it is important to optimize the suitable anions and cations in ILs, sometimes necessitating structural modifications.

1.3 Ionic liquids: Synthetic methods

Although a variety of methods have been reported for synthesizing ILs [5], the most commonly used procedures are summarized in Figure 5 and 6. (1) Path A: Metathetic exchange of anion. This is the most commonly used method to synthesize ILs. The downside of this method is that it produces harmful halide by-products (MX such as AgCl, etc.) which are difficult to remove by filtration, particularly for hydrophilic ILs. (2) Path B: neutralization of base with brØnsted acids and Path C: direct alkylation of alkylimidazole. Path B and C are more environmentally benign reactions as they do not produce halide by-products. However for path B, it can be hard to produce ILs with high purity; minute amounts of alkylimidazole or acid impurities could be present in the resulting ILs. For path C, it is restricted to the reactivity and availability of the alkylating agents. It is used to prepare sulfate, phosphate or sulfonate based ILs. (3) Path D is similar to Path A in that it involves a metathetic exchange of anions. This method uses dimethylcarbonate (DMC) as a clean methylating agent to substitute alkyl halides. It can prevent the formation of halide and other by-product that makes the ILs efficient [5]. The downside of this method is that it is restricted by the availability of the acid (HX) or NH_4^+ salts. (4) Non-conventional routes to synthesize ILs. Synthesis of ILs using non-conventional method such as microwaves and

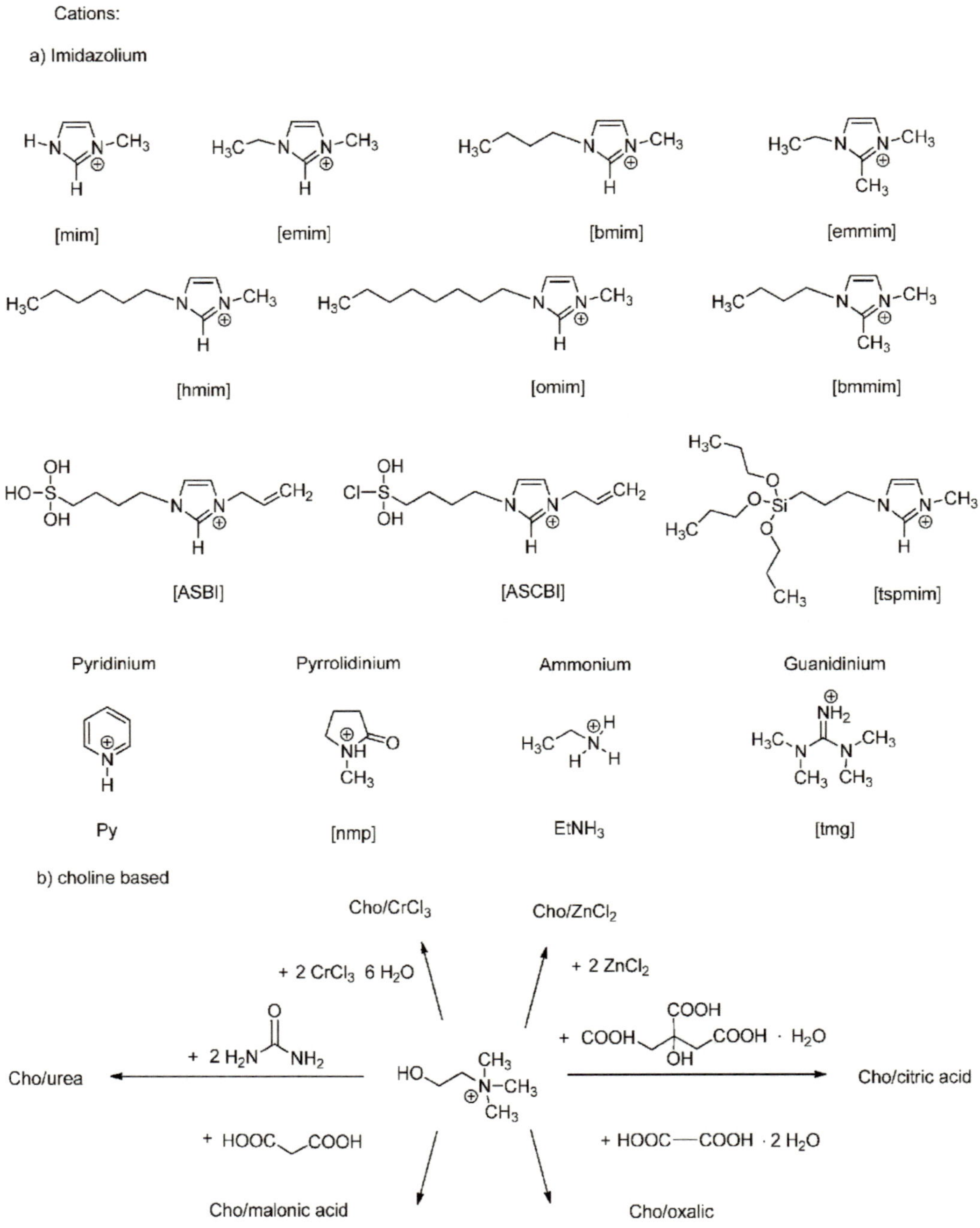

Fig. 3. (a) Most common cations of ILs. (b) Choline cations of ILs obtained by mixing of choline chloride based IL with the molecular donors (presented along the arrows).

ultrasounds are also possible. The disadvantage is that the ILs may decompose under sono-chemical conditions [6]. There are also other reports of synthesizing ILs using non-conventional methods. An example is [BMIM][BF_4] that has been synthesized *via* a one-pot-solvent-free route in a batch-mode reactor using microwave radiation (frequency of 5.8GHz), giving a high yield of 87% [7]. Successful synthesis of [BMIM][OH] in solution using an electro-dialysis set-up has also proven to be effective [8]. (5) Direct access to anion-functionalized ILs. The route to the direct access of anion-functionalized ILs is the one-step ring-opening reaction of sulfones (Figure 6). This route generates zwitterions which possess high melting points and can react with acids or $LiNTf_2$ to form new functionalized ILs [9]. This mixture, though very viscous, exists in liquid state at room temperature and is ionically conductive.

Anions:

Cl^-	PF_6^-	BF_4^-	$AlCl_4^-$	NO_3^-
chloride	hexafluorophosphate	tetrafluoroborate	tetrachloroaluminate	nitrate
$CH_3SO_3^-$	$CH_3CH_2OSO_3^-$	$CF_3SO_3^-$	CH_3COO^-	CF_3COO^-
methanesulfonate	ethylsulfate	trifluoromethanesulfonate	acetate	trifluoroacetate
$N(CN)_2^-$	$N(CF_3SO_2)_2^-$	$(CH_3O)_2PO_2^-$		
dicyanamide	bis(trifluoromethanesulfonyl)amide	dimethylphosphate		

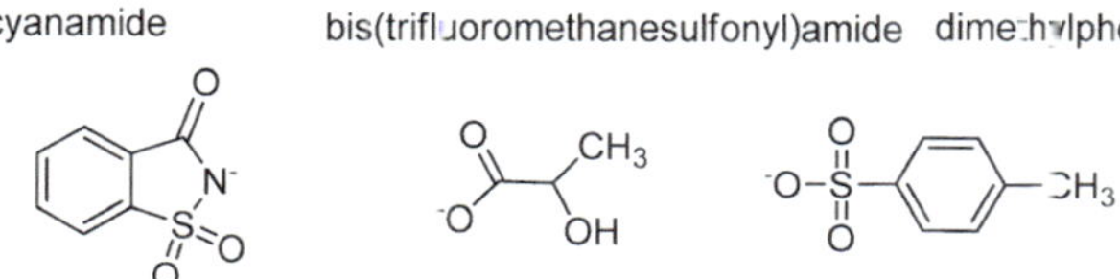

Fig. 4. Some anions of ILs [4].

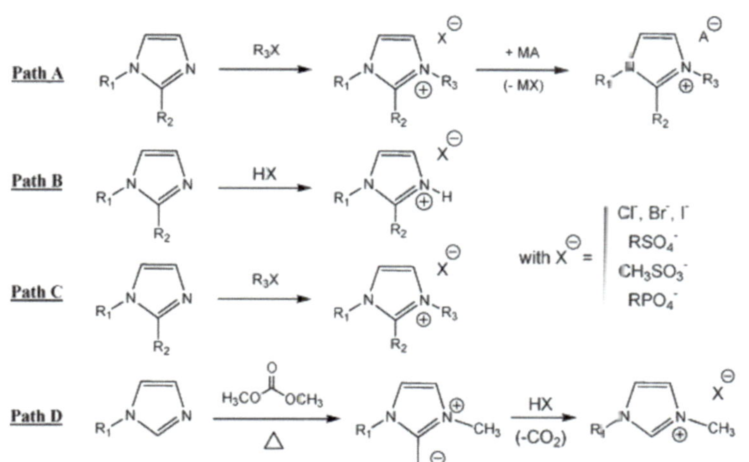

Fig. 5. General route for ILs synthesis. Path A: metathetic exchange of anion. Path B: neutralization of base with BrØnsted acids. Path C: direct alkylation of alkylimidazole. Path D: the carbonate method.

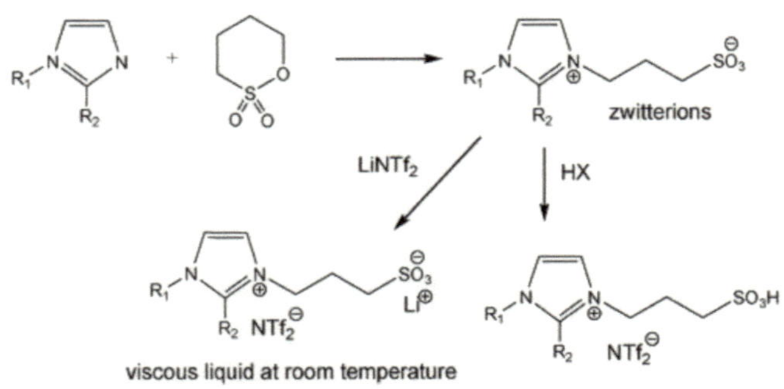

Fig. 6. Direct access to functionalized ILs.

2.Transformation of polysaccharides to monosaccharides

2.1 Solubility behavior of biomass in ionic liquids

Cellulose is made up of a linear chain with several β-(1→4) linked glucose repeating units (Figure 7). They are highly packed together with strong intra and inter hydrogen bonding and van der waals interactions between the cellulose fibrils. This accounts for its insolubility in water and conventional organic solvents except for concentrated phosphoric acid. Using concentrated phosphoric acid, however gives rise to environmental problems associated with the disposal of the solvent. Therefore, alternative solvents which are more benign have to be sought. Dissolving the cellulose can disrupt the highly crystalline structure in the cellulose to give amorphous cellulose which makes it more susceptible to transformation into monosaccharides such as glucose. Finding a suitable solvent to dissolve the cellulose is of utmost importance as it can affect its physical properties which include the degree of polymerization, its crystallinity and even the surface area of the substrate accessible in the case of further enzymatic hydrolysis. Some of the frequently used solvents in the dissolution of cellulose can be found in Table 1[10]. They are ionic liquids (ILs) based on polar organic solvents such as DMF, DMAc, DMI or DMSO usually coupled with charged species such as LiCl, etc.

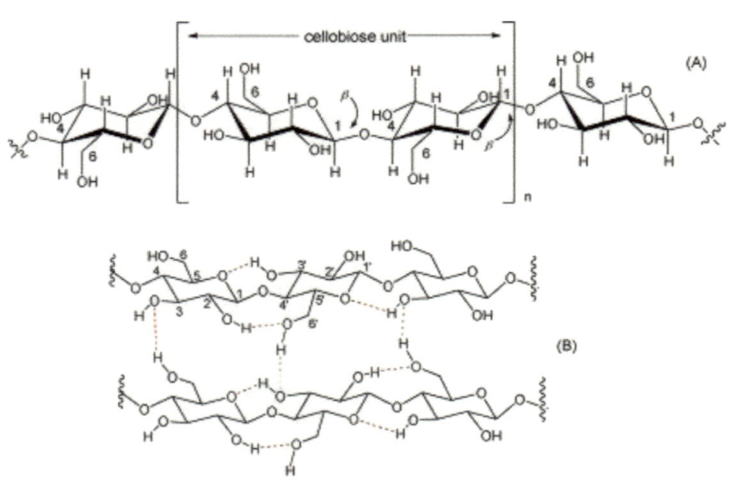

Fig. 7. The cellulose network (A: Cellulose chain and B: inter and intra H-bonds present in cellulose).

Acronym	Main systems used to dissolve cellulose
DMSO/TBAF	Dimethyl sulfoxide/tetrabutylammonium fluoride
LiCl/DMAc	Lithium chloride/dimethylacetamide
LiCl/DMI	Lithium chloride/dimethylimidazolidinone
LiCl/NMP	Lithium chloride/N-methyl pyrrolidine
N_2O_4/DMF	Dinitrogen tetraoxide/dimethylformamide
DMSO/CH_2O	Dimethyl sulfoxide/paraformaldehyde
NMMO	N-methylmorpholine-N-oxide monohydrate
	Aqueous solutions metal complexes

Table 1. Main solvent systems used in manufacturing cellulose and cellulose derivatives.

Dissolution of cellulose is highly dependent on the nature of the native cellulose (its degree of polymerization (DP) and its crystallinity), the operating conditions (temperature, reaction duration, initial concentration of cellulose in the IL, activation with microwaves) and presence of impurities such as water. Water can affect the dissolution of cellulose significantly. Optimal dissolution can be achieved with extremely dried ILs.
Both the cations and anions of ILs contribute to the dissolution of cellulose. The mechanism of the dissolution is that the anion of the IL acts as H-bond acceptor that interacts with the hydroxyl group of cellulose in a stoichiometric manner to generate an electron donor-electron acceptor (EDA) complex [4]. The cation, on the other hand, with its electron-rich aromatic π system acts as an electron acceptor center and prevents the cross linking of the cellulose. When the anion and the cation are in close proximity to each other, the formation of the EDA complexes between cellulose and ILs become feasible and the network of hydrogen bonds between the glycosidic monomers are disrupted, resulting in dissolution of the cellulose.

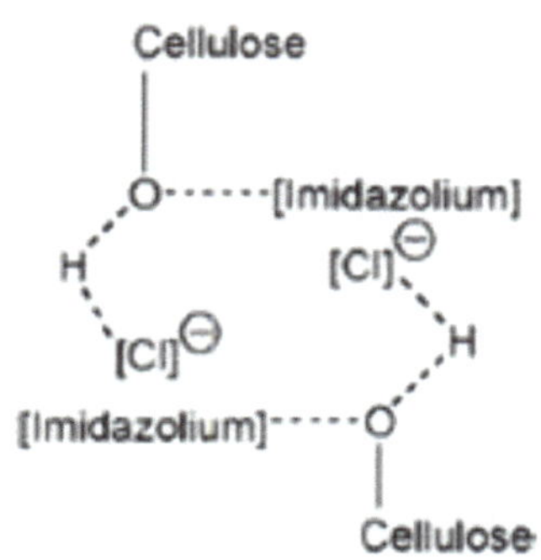

Fig. 8. Possible insertion of an imidazolium chloride in the cellulose matrix [4].

By adding a non-dissolution solvent such as water, ethanol or acetone, cellulose miscible in ILs can be precipitated from solution and then separated from the mixture of IL and non-dissolution solvent either by filtration or centrifugation. Due to non-volatile nature of IL, it can be recovered by distillation of the mixture, thus eliminating the non-dissolution solvent. The precipitated cellulose can appear in different forms such as monoliths, fibers and films and can have the same degree of polymerization and polydispersity as native cellulose but it depends largely on the operating conditions of the precipitation. Varying the condition of the precipitation can alter the degree of crystallinity of the precipitated cellulose. For example, cellulose precipitated after being dissolved in [AMIM][Cl] and [BMIM][Cl] had lower degree of crystallinity than native cellulose [11]. Results show that the precipitated cellulose is mostly amphorous with greater accessibility of the polysaccharides chains in the cellulose and exhibits enhanced enzymatic hydrolysis kinetics (Figure 9) [12-14].

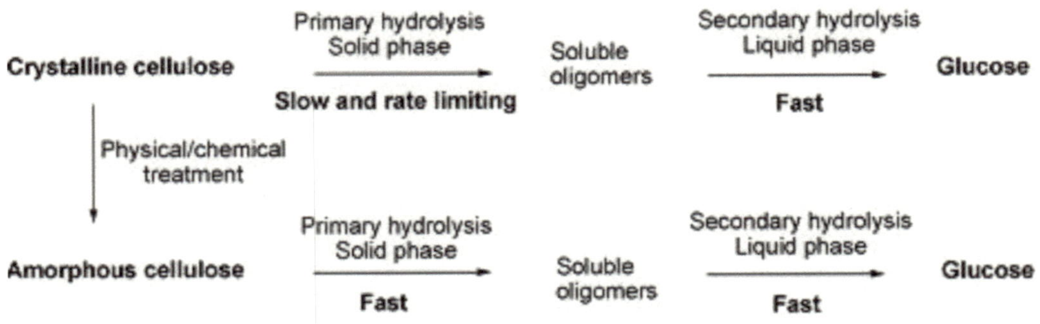

Fig. 9. Enzymatic hydrolysis of cellulose.

2.2 Lignocelluloses and challenges in their transformation

Lignocellulosic cellulose [4] refers to plants biomass which is made up primarily of cellulose, hemicellulose and lignin. Various challenges are faced when trying to convert lignocellulosic cellulose into fuels and other value added chemicals due to the existence of cellulose and lignin. Hemicellulose can be easily degraded but the highly crystalline structure cellulose makes it resistant to chemical and biological degradation. Lignin also poses a problem as it acts as an "adhesive" that holds hemicellulose and cellulose together, accounting for the microbial and moisture resistance and its rigid structure. Hence, it is necessary to devise methods to reduce or remove the lignin content and to disrupt the crystalline structure.

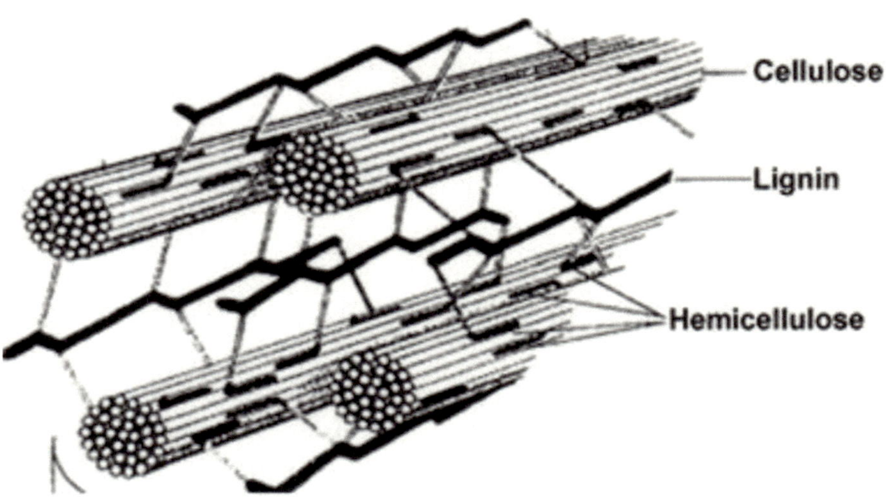

Fig. 10. Lignocelluloses network.

Various methods have been devised to reduce or remove the lignin content in lignocellulosic cellulose which includes physical (limited pyrolysis and mechanical disruption/comminution [15]), physiochemical (steam explosion, ammonia fiber explosion [16,17]), chemical (acid hydrolysis, alkaline hydrolysis, high temperature organic solvent pretreatment, oxidative delignification [18-20]), and biological (lignin degradation by white- and soft-rot fungi [21]) methods. The same problem arises from using these methods to extract lignin. After the removal of sufficient amount of lignin, the lignin experiences degradation and often results in the loss of fermentable sugar content in the polysaccharides.

The removal or reduction of lignin and disrupting the crystalline structure of lignocellulosic cellulose to make it more accessible to chemical and biological transformation remains a great challenge. Various works have been done to address these issues and the usage of ionic liquids (ILs) became a solution to this challenge. For example, research has shown that methylsulfate imidazolium-based ILs produce excellent results when dissolving softwood Kraft lignin [22]. Other research demonstrated that extraction of lignin from bagasse using the IL ethyl-methylimidazolium alkylbenzenesulfonate [EMIM][ABS] was successful giving a yield of >93% at atmospheric pressure. Although ILs proved to be effective in the dissolution of lignin, several problems such as the simplification of IL and the high operating temperature still exist [23].

Research has shown that when 40% of the lignin was extracted with a well chosen IL ([EMIM][Ac]), the cellulose crystallinity index dropped below 45 and the resulting hydrolysis of the wood flour cellulose was measured to be above 90%. This confirmed the close correlation between lignin extraction and residual cellulose crystallinity [11]. Another research group came to the conclusion that delignification alone was sufficient for effective hydrolysis over longer periods whereas for shorter hydrolysis times the combination of de-lignification and de-crystallization showed great benefits [24].

Pre-treatment of lignocelluloses leads to eventual fractionation of lignocelluloses which allow the disruption of the crystalline structure of the lignocellulosic cellulose to make it more accessible to chemical and biological transformation. An improved enzyme use in the cellulose hydrolysis step, with the possibility of enzyme recovery generates higher sugar yield with lower sugar degradation and less inhibitor formation. Generally, choosing a suitable pre-treatment will be the result of a compromise between hemicellulose and cellulose degradation and the ease of enzymatic hydrolysis of cellulosic substrate.

The two conditions needed for wood solubilization are wood particle size and the water content in the wood. Research has shown that water has a negative effect on the solubility of wood in ILs. The wood particle size can restrict the diffusion of IL into its interior causing poor dissolution of the wood particles. The relationship between lignin solubilization and wood dissolution is not clear; some ILs such as [MMIM][$MeSO_4$] can solubilize lignin without affecting the wood [11] The best solvent for lignocellulosic materials was found to be [BMIM][Cl] and for hardwood and softwood, it was [AMIM][Cl] [11,25]. The assistance of microwaves can greatly improve the dissolution of wood. A 5 wt% of six biomass types were rapidly and entirely dissolved in [BMIM][Cl] with the aid of microwave irradiation [26].

2.3 Transformation of polysaccharides to monosaccharides

Monosaccharides are the most basic units of biologically important carbohydrates and exist as the simplest forms of sugar. Polysaccharides are polymeric carbohydrate structures, formed of repeating units of monosaccharides or disaccharides joined together by glycosidic bonds.

2.3.1 Acid-catalyzed cellulose hydrolysis

The transformation of polysaccharides to monosaccharides can be achieved via acid-catalyzed cellulose hydrolysis. Although acid-catalyzed cellulose hydrolysis existed for many years, no alternate cost effective solutions can be offered for the transformation on a large-scale. Another downside is that, due to the involvement of acid, there is a need for non-corrosive materials and the acid must be removed or neutralized after the treatment, otherwise, the disposal of the acid will present be difficulties. The harsh conditions such as the high temperature and pressure for the hydrolysis can form degradation of products which lower the monosaccharides yield and prevent further fermentation.

Hence, acid-catalyzed cellulose hydrolysis remained a challenge prior to the introduction of ILs. When the acid-catalyzed cellulose hydrolysis was carried after cellulose was dissolved in ILs, the hydrolysis occurred at a lower temperature and with lower catalyst loading. The strength of the acid used also makes a major contribution to the successful hydrolysis. The optimized condition was determined to be a strong acid, water content between 5% and 10% (w/w), carbohydrate content less than 10% (w/w) and temperature range 80 and 150°C [4].

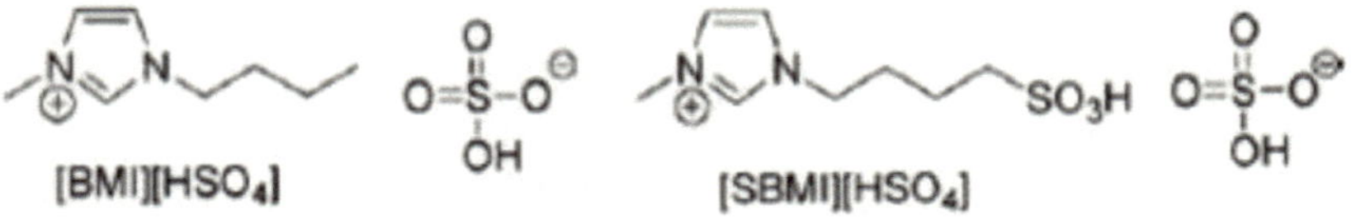

Fig. 11. ILs used for catalytic hydrolysis of lignocelluloses [4].

2.3.2 Enzyme-catalyzed cellulose hydrolysis

Enzyme-catalyzed cellulose hydrolysis is not very popular due to the high costs incurred and the lack of a feasible method for recovering the enzyme. Enzyme hydrolysis also suffers from low reaction rates that makes it undesirable. Dissolving cellulose in ILs generally involved interactions of the anions that create strong hydrogen bonding with the cellulose; this poses a serious problem of denaturing the enzyme. New types of ILs have been devised to counter the problem of inactivating enzymes [4].

2.3.3 Transition metal nanoparticle-catalyzed cellulose hydrolysis

Transition metal nanoparticle catalytic systems displayed excellent catalytic activities. Due to their small particle size and large surface area, they offer higher selectivity in comparison with conventional heterogeneous catalysts. The ILs used to dissolve the cellulose acts as a stabilizer for transition metal nanoparticles to maintain their small size, large surface area and to prevent nanoparticles from leaching. Boronic acids were predicted to disrupt the crystallinity of cellulose by binding reversibly with the numerous hydroxyl groups of cellulose, thereby improving the solubility and catalytic activity of cellulose, leading to the formation of 1 (Figure 12) [27]. Compound 1 serves as an excellent catalyst when coupled with the ILs-stabilized ruthenium nanoparticles for cellulose conversion. The catalyst has the advantage of being recycled which makes it desirable for cellulose hydrolysis.

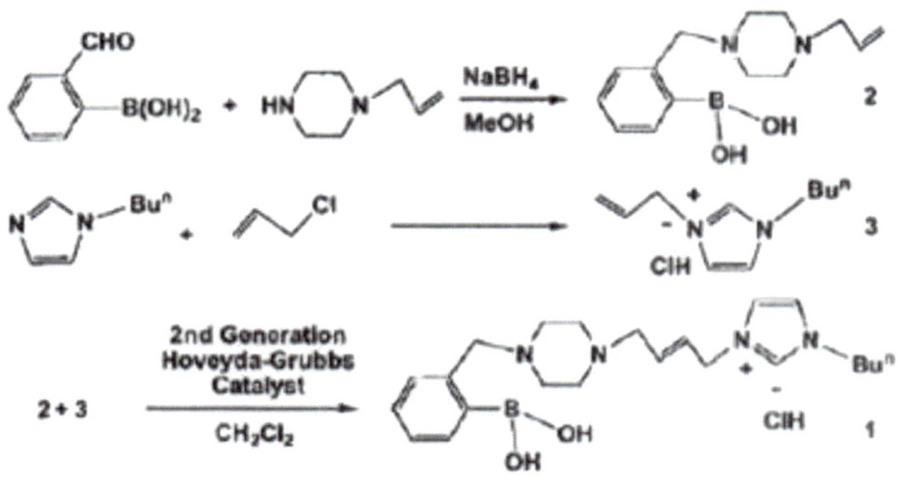

Fig. 12. Formation of boronic acid binding agent [27].

2.4 Conclusions

The use of ILs offers a good solution towards a more efficient transformation of polysaccharides to monosaccharide. But certain issues that arise with ILs must be addressed. ILs still remain expensive and the more work has to be done to reduce their costs as well as to improve on their recovery and reuse. After the pretreatment of lignocellulosic cellulose to remove lignin, methods of recovering the lignin should be explored to further tap its potential to be a combustible material and channeled for different usage.

3. Transformation of saccharides to 5-hydroxymethylfurfural (5-HMF) and other value added chemicals

When the ever declining fossil fuels meet with the ever increasing demands for energy, more and more people will be seeking renewable alternatives and product sustainability. Biomass has the potential to serve and excel as a sustainable source of energy and biomass resources are useful in many industrial applications. The basic chemical transformations include hydrolysis of cellulose to carbohydrates and the subsequent selective dehydration, hydrogenation, oxidation and condensation to yield the various useful chemicals. There are reports on the catalytic conversion of carbohydrates to liquid alkanes (C_3-C_{15}) and useful industrial chemicals such as 5-hydroxymethylfurfural (5-HMF), levulinic acid and furfural. Among the various possible types of biomass-derived chemicals, 5-hydroxymethylfurfural (5-HMF) seems to be the most promising building block that could substitute for the petrochemicals that are used in plastics, pharmaceuticals, fine chemicals and biofuels.

This part will discuss the use of ionic liquids and their role in the key chemical transformations of saccharides into value added chemicals such as 5-HMF. We also hope to provide some insights into the key mechanism of the transformation reactions; compile, discuss, elaborate, compare and contrast the different types of catalyst system, reaction conditions, reactors, solvents and novel techniques reported.

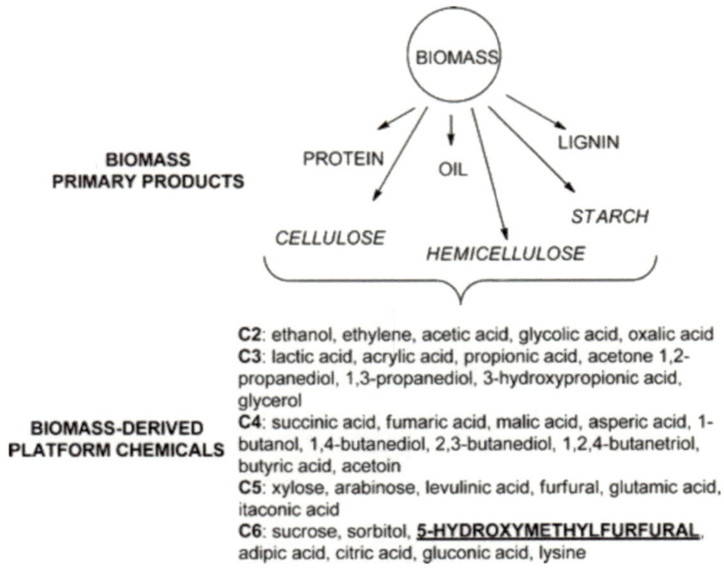

Fig. 13. Proposed biomass-derived platform chemicals.

3.1 5-hydroxymethylfurfural (5-HMF) and its applications

5-hydroxymethylfurfural (5-HMF) (Figure 14) is a furan based raw material that has been named one of the top building block materials obtained from biomass. It comprises an aromatic alcohol, aldehyde and furan ring system. Its expected usage is in the production of resins, fine chemicals, pharmaceuticals, polymers (polyester), solvents and liquid transportation fuels (2,4-dimethylfuran, a biofuel has 40% higher energy density than ethanol does [28]).

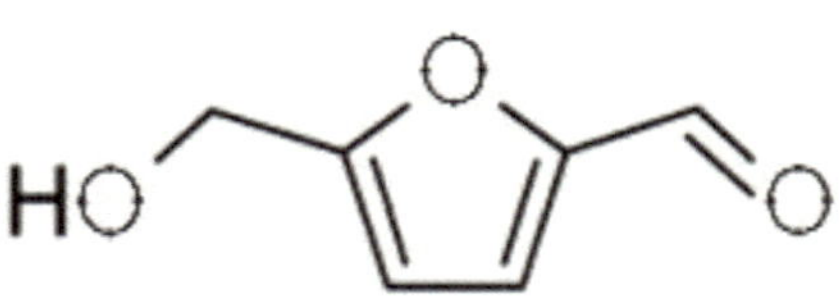

Fig. 14. Structure of 5-hydroxymethylfurfural (5-HMF)

Since it was first reported in the 19th century, 5-HMF has attracted much interest. Thereafter, much research and studies have been carried out to elucidate the structure, reaction mechanism, physical properties, chemical behaviours of 5-HMF. Antal et al. showed that 5-HMF was formed from hexoses via an acid catalyzed dehydration with the removal of three water molecules (Figure 15) [19].

About thirty years ago, van Dam and Cottier, et al. showed that an aqueous and non-aqueous process could produce 37% yield of 5-HMF. In the aqueous part of the system, 5-HMF takes up two molecules of water to form levulinic and formic acid while in the non-aqueous part, the hydrolysis of 5-HMF is suppressed. Nonetheless, cross-polymerization occurred under all circumstances leading to the formation of colored soluble polymers and insoluble brown precipitates. Thus, there is a need to come up with a suitable catalyst or a system for the selective formation of 5-HMF while not promoting the consecutive reactions leading to the formation of side products. Alternatively, continuous removal of 5-HMF from the reaction mixture would be an ideal process.

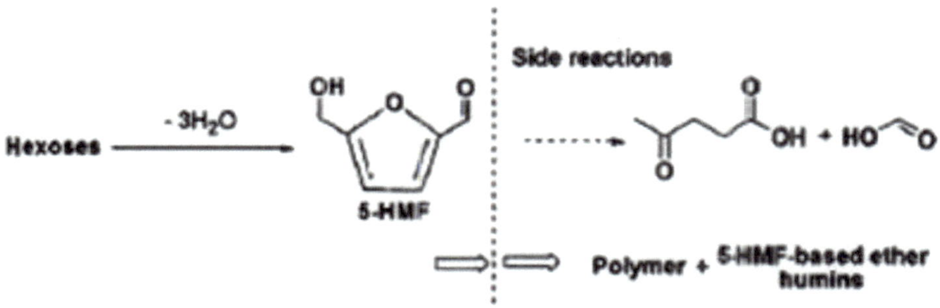

Fig. 15. Production of 5-HMF and the corresponding side reactions.

3.2 Dehydration of monosaccharides to 5-hydroxymethylfurfural (5-HMF) in ILs

1n 1983, the first report on the conversion of fructose to 5-HMF in the presence of pyridinium chloride with 70% yield under mild reaction conditions (30 min, 120 °C) was published [30]. This sparked an interest in investigating the dehydration of fructose over molten salts. Twenty years later, in 2003, Lansalot-matras et al. revisited the field and investigated the acid-catalyzed dehydration of fructose in commercially available ionic liquids, ([BMIM][BF_4]) and ([BMIM][PF_6]) with DMSO as co-solvent in the presence of Amberlyst-15 [31]. They demonstrated the advantages of using ILs as solvents and reported a yield of 80% for 5-HMF in 24 h and at a relatively low temperature of 80°C compared to that employed in the previous method. However, conventional methods require much higher temperatures of 100 to 300°C [31]. In an effort to further improve the dehydration reaction, the following reaction parameters were studied extensively with promising results:

3.2.1 Temperature and solvents

Qi et al. employed ionic liquid [BMIM][Cl] with different co-solvents such as DMSO, acetone, methanol, ethanol, ethyl acetate and supercritical carbon dioxide [32]. Fructose was first dissolved in ILs at 80°C and added Amberlyst-15, a strong acidic ion-exchange resin catalyst, and co-solvent was added to give a gel-like mixture after cooling. It was demonstrated that the reaction could proceed at an ambient temperature of 25°C with yields of 78% - 82% for 5-HMF [32].

3.2.2 Biphasic system

Zhang and co workers have demonstrated the first efficient room temperature catalytic system by using [BMIM][Cl] with tungsten chloride and a biphasic system composed of ILs and a modifier, tetrahydrofuran (THF) (see Figure 16) [33].

3.2.3 Metal halides as catalysts

Glucose and fructose are isomeric hexoses. Fructose is the direct and most efficient starting material for the formation of 5-HMF using old and conventional dehydration methods. For industrial application, glucose is converted to fructose using enzymatic hydrolysis.

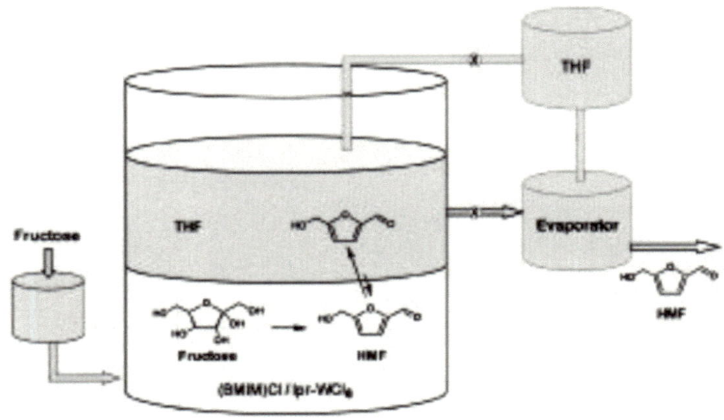

Fig. 16. Continuous batch process for the conversion of fructose to 5-HMF in a THF-([BMIM][Cl]) biphasic system [33].

Therefore, a number of investigations were made to examine whether the direct conversion of glucose to 5-HMF was viable. Zhao et al. made a major breakthrough by showing that $CrCl_2$ with IL [EMIM][Cl] are effective catalysts for the conversion of glucose to 5-HMF with a relatively good yield of 70% (Figure 17) [34]. This discovery, paved the way for the direct conversion of cellulose to 5-HMF in ILs under mild conditions (55% yield) [1], while microwave-assisted route gave 61% yield of the product [3,35].

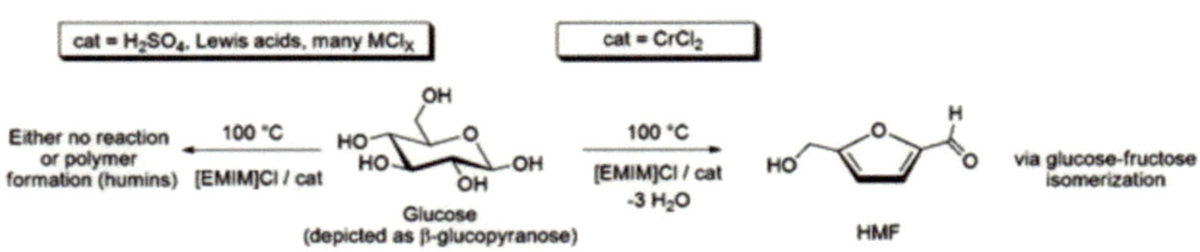

Fig. 17. Glucose conversion to 5-HMF at 100°C [34].

3.3 Mechanism

It has been proposed by Zhao et al. [34] that the reaction follows two possible pathways: (1) series of cyclic furan intermediates (A); (2) Open-chain pathway included formation of an enediol as an intermediate in the isomerization of glucose to fructose (B).

According to their report [34], complex between $CrCl_2$ and the IL interacts with the open chain glucose and helps in the isomerization to fructose and the direct conversion to 5-HMF (Figure 18). It was suggested that the mutarotation of the α-anomer to the β-anomer of glucose is the key step of the reaction

On the other hand, Binder et al. [1] proposed two variations on the mechanism previously suggested by Zhao et al (Figure 19). It was concluded that the fructofuranosyl cation undergoes attacks by chloride, bromide, or iodide. As bromide and iodide are better leaving groups than chloride, they were deemed as effective ionic additives. Chromium salts play an important role in the yield of 5-HMF. Research has shown that the yield of 5-HMF using chromium correlates with metal coordination. It was also proposed that the halide additives serve two roles: (1) as ligands for the chromium cation and (2) they facilitate the selective conversion of fructose.

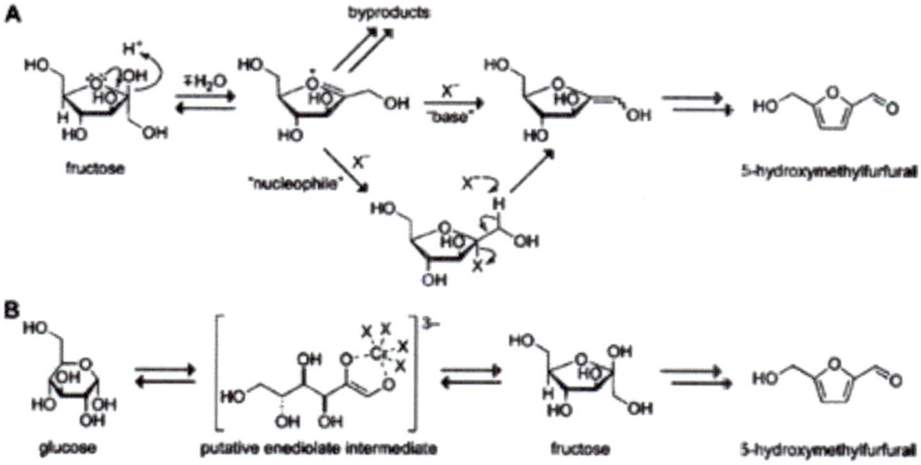

Fig. 18. Two possible mechanisms for the dehydration of fructose and isomerization of glucose.

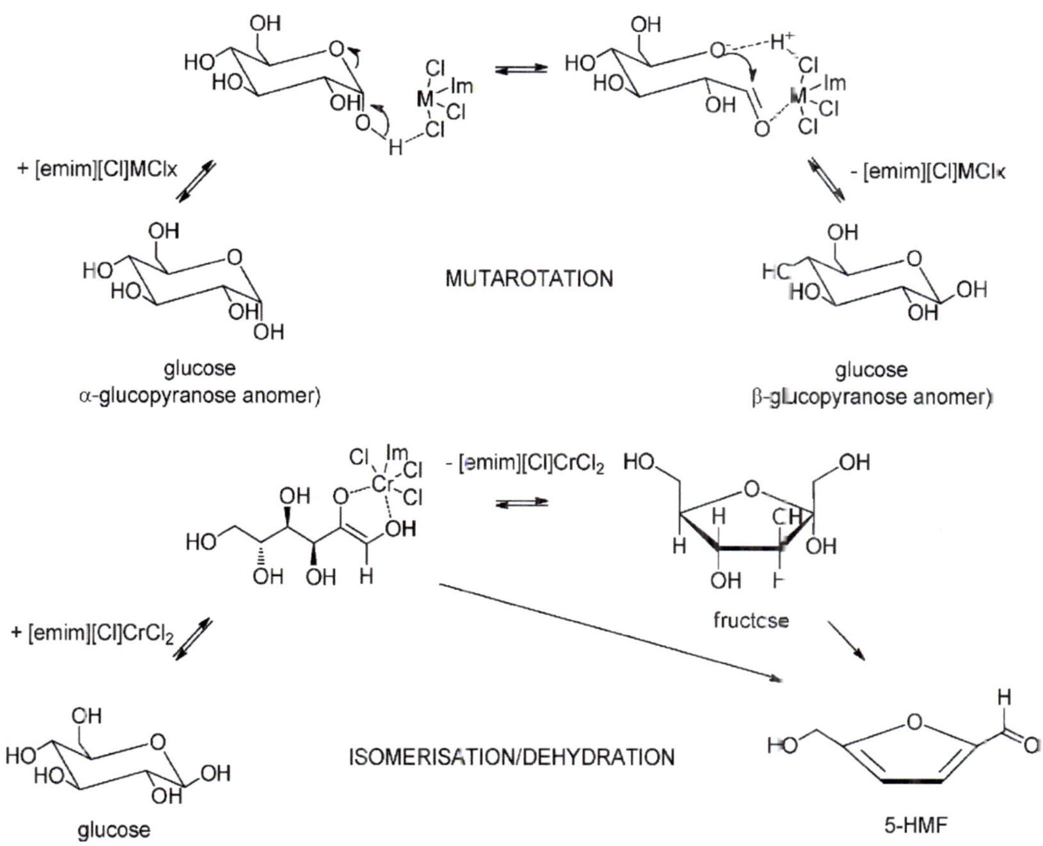

Fig. 19. Proposed metal halide interactions with glucose in [EMIM][Cl] [34].

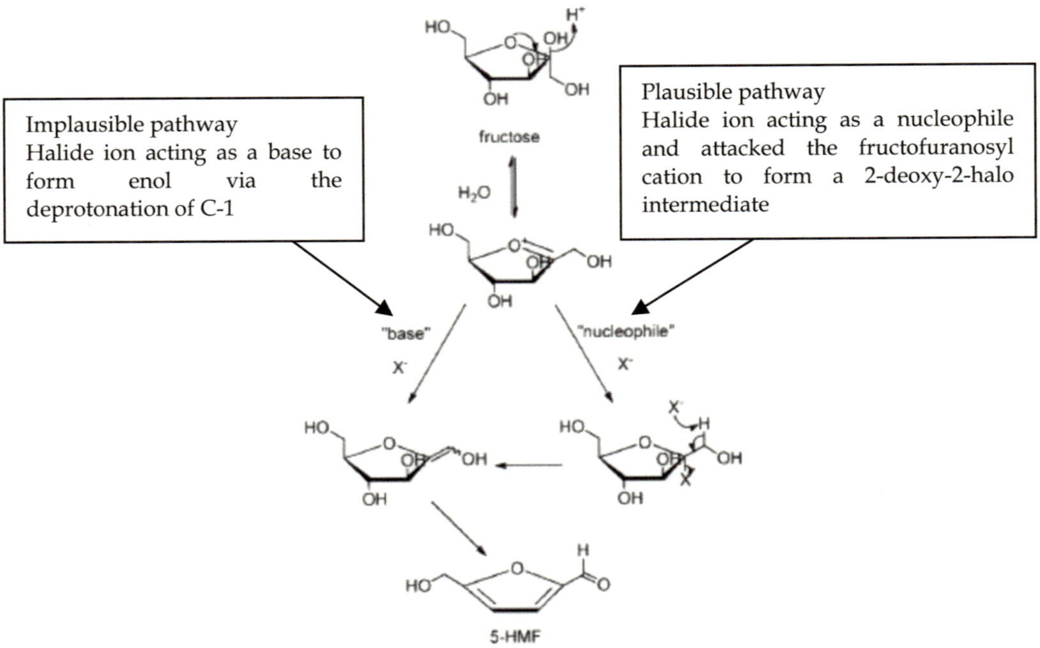

Fig. 20. Importance of halide in the mechanism (basic nucleophilic) of dehydration fructose to 5-HMF [1].

3.4 N-heterocyclic carbenes (NHCs)

It has been reported that there was no unusual observations in the catalytic activity for the previously described $CrCl_2$/[BMIM][Cl] catalytic system [36]. The chromium (II) chloride catalyst, modified with *N*-heterocyclic carbenes (NHCs), was investigated to show the stereochemical properties of the NHCs and not the oxidation states of the chromium that influence the catalytic properties. The best yields were achieved for the 1,3-bis(2,6-diisopropylphenyl)imidazolylidiene ligand that gave 96% and 81% yields of 5-HMF for fructose and glucose, respectively [36]. Interestingly, Zhao and coworkers used microwave irradiation and obtained the best results in ionic liquid for the conversion of glucose to 5-HMF with a yield of 91% [35]. Microwave-assisted reactions have great potential as they significantly reduce the amount of time required to complete the reactions.

3.5 Heterogeneous catalyst system

Solid catalyst, Amberlyst-15 sulfonic ion-exchange resin, proved to be the most effective solid catalyst that gave 82% yield of 5-HMF after 1 minute at 120°C [37]. Immobilization of homogeneous catalyst on solid support also demonstrated good activity. It was found that the immobilized ILs and acid modified silica gel were effective for the dehydration of fructose to 5-HMF. The advantages of using solid catalyst are in the area of product separation and catalyst recycling.

3.6 Dehydration of oligosaccharides to 5-HMF in ILs

Inulin, also known as fructan, is a carbohydrate that consists of fructose units with degrees of polymerization (DP) ranging from 2 to 60. The fructosyl units are linked by $\beta(2\rightarrow1)$

linkages with the glucose as the terminating unit of the polymer chain [38]. Inulins are found in plants such as artichoke tuber and chicory root and stored as energy sources. Since inulin-type fructans are resistant to hydrolysis by digestive enzymes in human intestine, they are classified as non-digestible carbohydrates as they [39]. Qi et al reported an effective one pot [40], two-step process for the production of 5-HMF using inulin in ionic liquids under mild conditions. Combining [BMIM][Cl] and a strong acidic cation exchange resin gave a 5-HMF yield of 55% in 20 min. Subsequently, a two-step in one pot system was reported in which two brϕnsted acidic ILs, [EMIM][HSO_4] and [BMIM][HSO_4], acted as both solvent and catalyst for the conversion of inulin with a 5-HMF yield of 82% at 80°C in 65 min. The advantage of the mixed ionic liquid system is that it is green and efficient system because recycling of the ILs is a simple process.

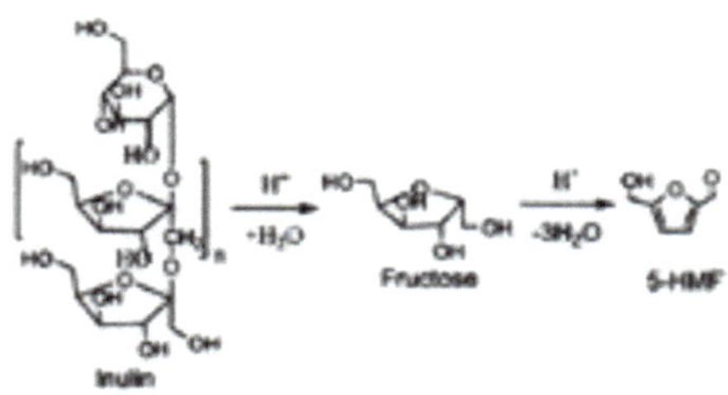

Fig. 21. The pathways for acid-catalyzed hydrolysis and dehydration of inulin to 5-HMF.

3.7 Dehydration of polysaccharides to 5-hydroxymethylfurfural (5-HMF) in ILs

For the synthesis of 5-HMF, *N,N*-dimethylacetamide (DMA) was used with lithium chloride (LiCl) in a single step from untreated lignocellulosic biomass [1], purified cellulose, glucose and fructose, since the conversion of cellulose to 5-HMF is unaffected by the presence of other biomass such as lignin and protein (Figure 22) [1].

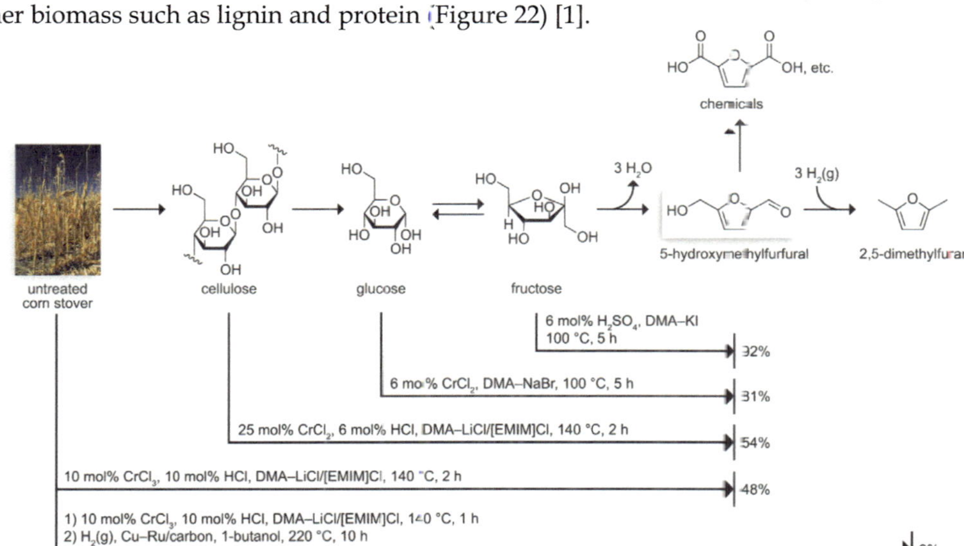

Fig. 22. Halide salts in DMA enable previously exclusive yields of bio-based chemicals from a variety of carbohydrates.

The production of 5-HMF and furfural from lignocellulosic biomass (corn stalk, rice straw and pine wood) in ILs, catalyzed by $CrCl_3$ under microwave irradiation, was reported with yields 45-52% and 23-31% respectively, from corn stalk, rice straw and pine wood in less than 3 min (Table 2) [3]. This method paved the way for energy-efficient and cost-effective conversion of biomass into biofuels and platform chemicals [3].

Entry	**Sample**	**Catalyst**	**Time (min)**	**HMF yield (%)**	**Furfural yield (%)**
1	Cellulose	$CrCl_3 \cdot 6H_2O$	2.5	62	nd
2	Xylan	$CrCl_3 \cdot 6H_2O$	2	nd	63
3	Xylan	–	2	nd	18
4	Cellulose + xylan	$CrCl_3 \cdot 6H_2O$	2	39	55
5	Cellulose + xylan	$CrCl_3 \cdot 6H_2O$	2.5	53	33
6	Corn stalk	$CrCl_3 \cdot 6H_2O$	3	45	23
7	Rice straw	$CrCl_3 \cdot 6H_2O$	3	47	25
8	Pine wood	$CrCl_3 \cdot 6H_2O$	3	52	31
9[b]	Pine wood	$CrCl_3 \cdot 6H_2O$	3	44	28
10[c]	Pine wood	HCl	60	2.1	4.4
11[d]	Pine wood	$CrCl_3 \cdot 6H_2O$	60	6.4	7.0
12[e]	Pine wood	$CrCl_3 \cdot 6H_2O$	6	35	18

nd = not detected

[a] Unless otherwise specified, reaction conditions were: substrate (100 mg) and $CrCl_3 \cdot 6H_2O$ (10 mg, 0.0375 mmol) were added to 2.0 g of [C_4mim]Cl, followed by MI at 400 W for the desired time. Yields of HMF from corn stalk, rice straw and pine wood were based on a hexose content of 36.1%, 37.5% and 54%, and yields for furfural were based on a pentose content of 21.4% , 21.2% and 7.6%, respectively.[b] [C_4mim]Br was used as a reaction medium.[c] Reaction conditions were: 20 mg of hydrochloric acid, 2.0 g of [C_4mim]Cl, 30 mg of H_2O, 0.1 g of pine wood, 100 °C with an oil-bath.[d] Reaction conditions were: 10 mg of $CrCl_3 \cdot 6H_2O$, 2.0 g of [C_4mim]Cl, 0.1 g of pine wood, 100 °C with an oil-bath.[e] Otherwise was the same as in Entry 11 except for the reaction temperature at 200 °C.

Table 2. Production of furan compounds under microwave irradiation.[a]

3.8 Challenges

In summary, one of the main challenges is picking the appropriate starting materials by considering the different aspects such as the availability, renewability, cost, viability, efficiency and so on. Currently, good and reasonable yields for the conversion of monosaccharides to 5-HMF can be achieved with the use of fructose and glucose. Since, glucose is a human food derived from starch and fructose is derived by an additional work-up from the enzymatic hydrolysis of glucose, the search for alternative starting materials that are renewable and from non-food sources presents challenges that must be overcome.

Another challenge would be found in up-scaling the reaction in ILs. The design and development of an efficient, simple, effective, high selectivity reactor and reaction system is necessary for the commercialization of this technology.

4. Transformation of oleate to value added chemicals

4.1 The oleochemical industry

Ever since the crude oil crisis in the late 1970s (Figure 23) [41], manufacturers have been switching from petrochemical to the oleohemicals that lead to a global expansion of the oleochemical industry. The oleochemical industry makes chemicals from renewable resources such as natural oils and fats in addition to waste or by-products from the food industry. Natural oils and fats are primarily composed of long chains of triglycerides with carboxylic acids. The primary olechemical products consist of glycerols, fatty acids, fatty alcohols and fatty acid methyl esters (FAME). These chemicals are useful raw materials for lubricants, fuels (biodiesels), paints, polymers (polyester), surfactants, detergents, soaps and cosmetics. Oleates are a class of fatty acid esters that are commonly synthesized from the reaction of natural fats and oils with methanol, catalyzed by alkali.

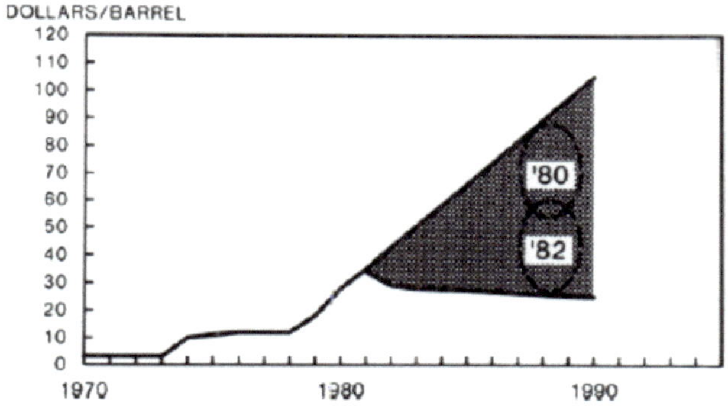

Fig. 23. Crude oil price projections, 1980 vs 1982. US refinery average acquisition costs, current dollars [41].

4.2 Transformation of oleates using olefin metathesis

After the first report on the $WCl_6/SnMe_4$ catalytic system for the metathesis of methyl oleate in 1972 [42], olefin metathesis has been considered as a useful catalytic reaction. It converts olefins into new and useful products by breaking and reforming the C=C bonds. The key step involves the formation of the [2+2] metallacyclobutane intermediate of the olefin and the transition metal alkylidene complex [43].

4.2.1 Homo-metathesis reaction of methyl oleate

In 1972, Boelhouwer and co-workers reported the selective transformation of methyl oleate to equimolar amounts of 9-octadecene and dimethyl 9-octadecene-1,18-dioate using $WCl_6/(CH_3)_4Sn$ as the catalyst (Figure 24) [42].

The successful metathesis of methyl oleate has led to the synthesis of many other unsaturated fatty acid esters with high selectivity. The development of homogeneous and heterogeneous catalyst systems and the metathesis of ω-unsaturated fatty acid esters such as methyl 10-undecenoate, that proceeds to completion via the continuous removal of the volatile co-product, ethylene, is an example of an industrial process based on these reactions (Figure 25).

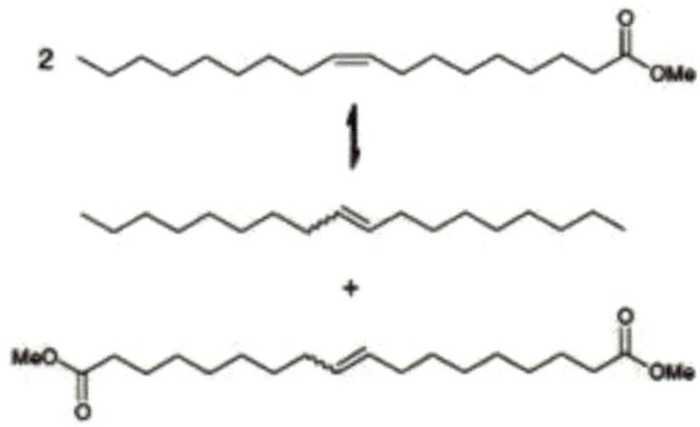

Fig. 24. Homo-metathesis of methyl oleate [42].

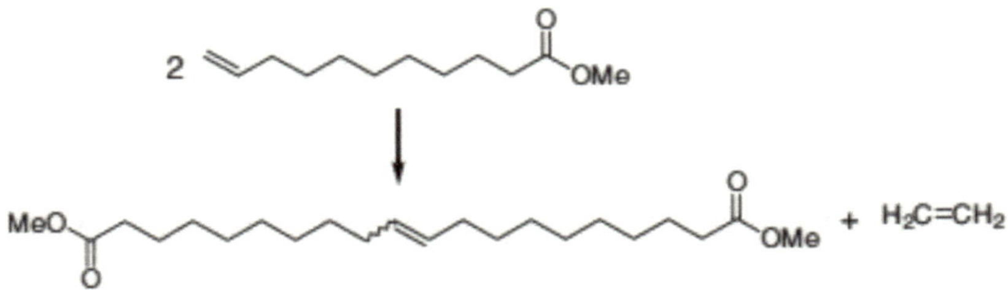

Fig. 25. Homo-metathesis reaction of methyl oleate.

4.2.2 Cross-metathesis reaction of unsaturated fatty acid esters

Cross-metathesis allows long chain unsaturated fatty acid esters to be shorten to produce highly desirable short chain esters such as detergent-range C_{10}-C_{14} esters (methyl 9-dodecenoate); the alkene co-products can be converted to C_{12}-C_{14} alcohols by hydroformylation. The cross-metathesis of unsaturated fatty acids with ethylene (Figure 26), or ethenolysis produce highly valuable short chain ω-unsaturated fatty acid esters. Various groups have reported on the ethenolysis of methyl oleate to form methyl 9-decenoate and 1-decene. Both methyl 9-decenoate and 1-decene are the key intermediates for the synthesis of chemical products (fragrances, lubricants [44]) and polymers (polyesters, polyamides), respectively.

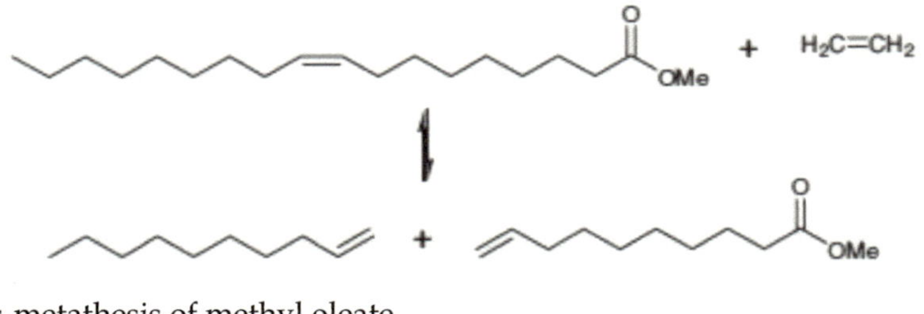

Fig. 26. Cross-metathesis of methyl oleate

4.2.3 Ethenolysis of methyl oleate in Ionic liquid [45]

In 2007, Bourbigou-Dixneuf and coworkers took the ethenolysis procedure one step further and reported the ethenolysis of methyl oleate using ionic liquids. They screened two imidazolium based ionic liquids of different counter anions and found that Hoveyda catalyst 8 (Figure 27) and [bdmim][NTf_2] ionic liquid gave the best conversion (95%) and high catalyst recyclability of 3 consecutive runs without significant loss in catalytic activity [45].

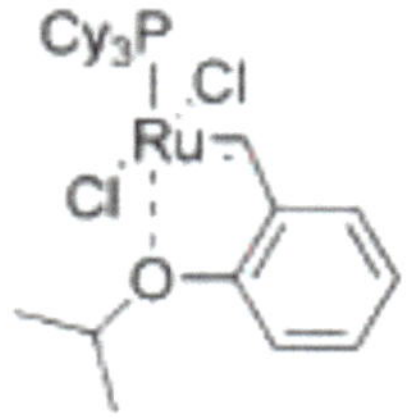

8

Fig. 27. Hoveyda catalyst.

4.3 Conclusions

The demand for inexpensive and renewable energy and chemical products will further spur the oleochemical industry. The metathesis of oleates offers new synthetic routes to valuable raw materials and feed stocks with high chemoselectivity. The reaction is catalytic and could be performed at ambient conditions. Moreover, solvents could be substituted with "green" ionic liquids that will further enhance the growth of production.

5. Conclusions and perspectives

Because of their high availability and low prices, the chemical industries in thelatter part of 20th and the early 21st century rely heavily on crude oil. However, this strong focus of raw materials now appears to be more precarious due to escalating international financial problems and unknown political factors.

At present, moving away from fossil fuel processes should be the main goal of industrial chemical development. This could be attained through the application of new technologies that utilized lignocellulosic biomass. Recent developments in ionic liquids show great potential in the utilization of lignocellulosic biomass processes; they have shown high efficacy and efficiency. One of the greatest shortcomings of ILs is their high cost; more research must be done to make them more cost-effective.

In spite of the possibilities that are opened up by using ionic liquids in biomass chemistry, both their recovery and reuse for continuous processes remain a formidable challenge for the industrialization of new technologies. Ionic liquids are relatively expensive compared to traditional solvents and the high cost of utilizing ionic liquids must be offset; this is mandatory for the development of commercially viable processes. In addition, the technology available now shows that ionic liquids can only dissolve about 10-15 wt % cellulose to give very thick and viscous solutions at these low concentrations.

The future of R&D for biomass processing utilizing ionic liquids relies on the development of novel, environmentally-friendly ionic liquids that allow high concentrations of dissolved cellulose and still possess good rheological properties. Furthermore, products should be readily isolated from ionic liquids for the efficient solvent recycling. Moreover, the separation of impurities from biomass has to be efficient otherwise the accumulation of impurities during the process may negatively affect the overall performance of the system.

6. Acknowledgements

This work was supported by the Institute of Chemical and Engineering Sciences (ICES), Singapore. NSH thanks the grants from the National Science Foundation (CHE-0906179), NIU Inaugural Board of Trustees Professorship, and the Research Prize for Senior US Scientists from the Alexander von Humboldt Foundation. JAM gratefully acknowledges the support from the Robert A. Welch Foundation (N-1322).

7. References

[1] Binder, J. B.; Raines, R. T., Simple Chemical Transformation of Lignocellulosic Biomass into Furans for Fuels and Chemicals. *J. Am. Chem. Soc.*, 2009, *131* (5), 1979-1985.

[2] Zakrzewska, M. E.; Ewa Bogel-Łukasik, E.; Bogel-Łukasik, R., Ionic Liquid-Mediated Formation of 5-Hydroxymethylfurfural — A Promising Biomass-Derived Building Block. *Chem. Rev.*, 2011, *111(2)*, 397-417.

[3] Zhang, Z.; Zhao, Z. K., Microwave-assisted conversion of lignocellulosic biomass into furans in ionic liquid. *Bioresource Tech.*, 2010, *101* (3), 1111-1114.

[4]Olivier-Bourbigou, H.; Magna, L.; Morvan, D., Ionic liquids and catalysis: Recent progress from knowledge to applications. *Appl. Catal. A: General* 2010, *373* (1-2), 1-56.

[5] Smiglak, M.; Holbrey, J. D.; Griffin, S. T.; Reichert, W. M.; Swatloski, R. P.; Katritzky, A. R.; Yang, H.; Zhang, D.; Kirichenko, K.; Rogers, R. D., Ionic liquids via reaction of the zwitterionic 1,3-dimethylimidazolium-2-carboxylate with protic acids. Overcoming synthetic limitations and establishing new halide free protocols for the formation of ILs. *Green Chem.*, 2007, *9* (1), 90-98.

[6] Oxley, J. D.; Prozorov, T.; Suslick, K. S., Sonochemistry and Sonoluminescence of Room-Temperature Ionic Liquids. *J. Am. Chem. Soc.*, 2003, *125* (37), 11138-11139.

[7] Horikoshi, S.; Hamamura, T.; Kajitani, M.; Yoshizawa-Fujita, M.; Serpone, N., Green Chemistry with a Novel 5.8-GHz Microwave Apparatus. Prompt One-Pot Solvent-Free Synthesis of a Major Ionic Liquid: The 1-Butyl-3-methylimidazolium Tetrafluoroborate System. *Org. Proc. Res. Develop.*, 2008, *12* (6), 1089-1093.

[8] Himmler, S.; Konig, A.; Wasserscheid, P., Synthesis of [EMIM]OH via bipolar membrane electrodialysis - precursor production for the combinatorial synthesis of [EMIM]-based ionic liquids. *Green Chem.*, 2007, *9* (9), 935-942.

[9] Ohno, H., Functional Design of Ionic Liquids. *Bull. Chem. Soc. Japan* 2006, *79* (11), 1665-1680.

[10] Heinze, T.; Liebert, T., Unconventional methods in cellulose functionalization. *Progress in Poly. Sci.*, 2001, *26* (9), 1689-1762.

[11] Lee, S. H.; Doherty, T. V.; Linhardt, R. J.; Dordick, J. S., Ionic liquid-mediated selective extraction of lignin from wood leading to enhanced enzymatic cellulose hydrolysis. *Biotechnology and Bioengineering* 2009, *102* (5), 1368-1376.

[12] Dadi, A. P.; Varanasi, S.; Schall, C. A., Enhancement of cellulose saccharification kinetics using an ionic liquid pretreatment step. *Biotechnology and Bioengineering* 2006, *95* (5), 904-910.

[13] Dadi, A.; Schall, C.; Varanasi, S., Mitigation of cellulose recalcitrance to enzymatic hydrolysis by ionic liquid pretreatment. *Applied Biochemistry and Biotechnology* 2007, *137-140* (1), 407-421.

[14] Liu, L.; Chen, H., Enzymatic hydrolysis of cellulose materials treated with ionic liquid [BMIM] Cl. *Chinese Science Bulletin* 2006, *51* (20), 2432-2436.

[15] Mosier, N.; Wyman, C.; Dale, B.; Elander, R.; Lee, Y. Y.; Holtzapple, M.; Ladisch, M., Features of promising technologies for pretreatment of lignocellulosic biomass. *Bioresource Technology* 2005, *96* (6), 673-686.

[16] Grous, W. R.; Converse, A. O.; Grethlein, H. E., Effect of steam explosion pretreatment on pore size and enzymatic hydrolysis of poplar. *Enzyme and Microbial Technology* 1986, *8* (5), 274-280.

[17] Mes-Hartree, M.; Dale, B. E.; Craig, W. K., Comparison of steam and ammonia pretreatment for enzymatic hydrolysis of cellulose. *Applied Microbiology and Biotechnology* 1988, *29* (5), 462-468.

[18] Chum, H. L.; Johnson, D. K.; Black, S.; Baker, J.; Grohmann, K.; Sarkanen, K. V.; Wallace, K.; Schroeder, H. A., Organosolv pretreatment for enzymatic hydrolysis of poplars: I. Enzyme hydrolysis of cellulosic residues. *Biotechnology and Bioengineering* 1988, *31* (7), 643-649.

[19] Zhang, Y.-H. P.; Ding, S.-Y.; Mielenz, J. R.; Cui, J.-B.; Elander, R. T.; Laser, M.; Himmel, M. E.; McMillan, J. R.; Lynd, L. R., Fractionating recalcitrant lignocellulose at modest reaction conditions. *Biotechnology and Bioengineering* 2007, *97* (2), 214-223.

[20] Gierer, J.; NorÃ©n, I., Oxidative Pretreatment of Pine Wood to Facilitate Delignification during Kraft Pulping. *Holzforschung* 1982, *36* (3), 123-130.

[21] Hatakka, A. I., Pretreatment of wheat straw by white-rot fungi for enzymic saccharification of cellulose. *Applied Microbiology and Biotechnology* 1983, *18* (6), 350-357.

[22] Pu, Y.; Jiang, N.; Ragauskas, A. J., Ionic Liquid as a Green Solvent for Lignin. *Journal of Wood Chemistry and Technology* 2007, *27* (1), 23-33.

[23] Tan, S. S. Y.; MacFarlane, D. R.; Upfal, J.; Edye, L. A.; Doherty, W. O. S.; Patti, A. F.; Pringle, J. M.; Scott, J. L., Extraction of lignin from lignocellulose at atmospheric pressure using alkylbenzenesulfonate ionic liquid. *Green Chemistry* 2009, *11* (3), 339-345.

[24] Zhu, L.; O'Dwyer, J. P.; Chang, V. S.; Granda, C. B.; Holtzapple, M. T., Structural features affecting biomass enzymatic digestibility. *Bioresource Technology* 2008, *99* (9), 3817-3828.

[25] Zavrel, M.; Bross, D.; Funke, M.; Büchs, J.; Spiess, A. C., High-throughput screening for ionic liquids dissolving (ligno-)cellulose. *Bioresource Technology* 2009, *100* (9), 2580-2587.

[26] Rayne, S. a. M., Giuseppe., Trichoderma reesei derived cellulase activity in three N,N-dimethylethanolammonium akylcarboxylate ionic liquids. *Available from Nature Precedings <http://hdl.handle.net/10101/npre.2007.632.1>* (2007).

[27] Zhu, Y.; Kong, Z. N.; Stubbs, L. P.; Lin, H.; Shen, S.; Anslyn, E. V.; Maguire, J. A., Conversion of Cellulose to Hexitols Catalyzed by Ionic Liquid-Stabilized Ruthenium Nanoparticles and a Reversible Binding Agent. *ChemSusChem* 2010, *3* (1), 67-70.

[28] Roman-Leshkov, Y.; Barrett, C. J.; Liu, Z. Y.; Dumesic, J. A., Production of dimethylfuran for liquid fuels from biomass-derived carbohydrates. *Nature* 2007, *447* (7147), 982-985.

[29] Antal, M. J.; Mok, W. S. L.; Richards, G. N., Mechanism of formation of 5-(hydroxymethyl)-2-furaldehyde from -fructose and sucrose. *Carbohydrate Research* 1990, *199* (1), 91-109.

[30] Fayet, C.; Gelas, J., Nouvelle méthode de préparation du 5-hydroxyméthyl-2-furaldéhyde par action de sels d'ammonium ou d'immonium sur les mono-, oligo- et poly-saccharides. Accès direct aux 5-halogénométhyl-2-furaldéhydes. *Carbohydrate Research* 1983, *122* (1), 59-68.

[31] Lansalot-Matras, C.; Moreau, C., Dehydration of fructose into 5-hydroxymethylfurfural in the presence of ionic liquids. *Catalysis Communications* 2003, *4* (10), 517-520.

[32] Qi, X.; Watanabe, M.; Aida, T. M.; Smith, R. L., Efficient Catalytic Conversion of Fructose into 5-Hydroxymethylfurfural in Ionic Liquids at Room Temperature. *ChemSusChem* 2009, *2* (10), 944-946.

[33] Chan, J. Y. G.; Zhang, Y., Selective Conversion of Fructose to 5-Hydroxymethylfurfural Catalyzed by Tungsten Salts at Low Temperatures. *ChemSusChem* 2009, *2* (8), 731-734.

[34] Zhao, H.; Holladay, J. E.; Brown, H.; Zhang, Z. C., Metal Chlorides in Ionic Liquid Solvents Convert Sugars to 5-Hydroxymethylfurfural. *Science* 2007, *316* (5831), 1597-1600.

[35] Li, C.; Zhang, Z.; Zhao, Z. K., Direct conversion of glucose and cellulose to 5-hydroxymethylfurfural in ionic liquid under microwave irradiation. *Tetrahedron Letters* 2009, *50* (38), 5403-5405.

[36] Yong, G.; Zhang, Y.; Ying, J. Y., Efficient Catalytic System for the Selective Production of 5-Hydroxymethylfurfural from Glucose and Fructose. *Angewandte Chemie International Edition* 2008, *47* (48), 9345-9348.

[37] Hu, S.; Zhang, Z.; Zhou, Y.; Han, B.; Fan, H.; Li, W.; Song, J.; Xie, Y., Conversion of fructose to 5-hydroxymethylfurfural using ionic liquids prepared from renewable materials. *Green Chemistry* 2008, *10* (12), 1280-1283.

[38] Sirisansaneeyakul, S.; Worawuthiyanan, N.; Vanichsriratana, W.; Srinophakun, P.; Chisti, Y., Production of fructose from inulin using mixed inulinases from <i>Aspergillus niger</i> and <i>Candida guilliermondii</i>. *World Journal of Microbiology and Biotechnology* 2007, *23* (4), 543-552.

[39] Verghese, M.; Walker, L. T.; Shackelford, L.; Chawan, C. B., Inhibitory effects of nondigestible carbohydrates of different chain lengths on azoxymethane-induced aberrant crypt foci in Fisher 344 rats. *Nutrition Research* 2005, *25* (9), 859-868.

[40] Qi, X.; Watanabe, M.; Aida, T. M.; Smith Jr, R. L., Efficient one-pot production of 5-hydroxymethylfurfural from inulin in ionic liquids. *Green Chemistry 12* (10), 1855-1860.

[41] Haupt, D.; Drinkard, G.; Pierce, H., Future of petrochemical raw materials in oleochemical markets. *Journal of the American Oil Chemists' Society* 1984, *61* (2), 276-281.

[42] van Dam, P. B.; Mittelmeijer, M. C.; Boelhouwer, C., Metathesis of unsaturated fatty acid esters by a homogeneous tungsten hexachloride-tetramethyltin catalyst. *Journal of the Chemical Society, Chemical Communications* 1972, (22), 1221-1222.

[43] Mol, J. C., Catalytic Metathesis of Unsaturated Fatty Acid Esters and Oils. *Topics in Catalysis* 2004, *27* (1), 97-104.

[44] Yadav, G. D.; Doshi, N. S., Development of a green process for poly-[small alpha]-olefin based lubricants. *Green Chemistry* 2002, *4* (6), 528-540.

[45] Thurier, C.; Fischmeister, C.; Bruneau, C.; Olivier-Bourbigou, H.; Dixneuf, P. H., Ethenolysis of Methyl Oleate in Room-Temperature Ionic Liquids. *ChemSusChem* 2008, *1* (1-2), 118-122.

2

Biotransformation of Underutilized Natural Resource to Valuable Compounds in Ionic Liquid: Enzymatic Synthesis of Caffeic Acid Phenethyl Ester Analogues from Immature Coffee Beans

Atsushi Kurata, Tokio Fujita, and Noriaki Kishimoto
Department of Applied Biological Chemistry, Kinki University
Nakamachi, Nara City, Nara
Japan

1. Introduction

Caffeic acid esters are widely distributed in plants and propolis (Tagashira & Ohtake, 1998; Ysrael & Nonato, 1999). Caffeic acid phenethyl ester (CAPE) especially has been found in propolis and has a broad spectrum of biological activities, including antimicrobial, anti-inflammatory, antioxidant, and antitumor activities (Bankova, 2005); it also has an inhibitory effect on HIV-1 integrase, cyclooxygenase, and lipooxygenase (Fesen et al., 1993; Michaluart et al., 1999; Nicklaus et al., 1997; Sud'ina et al., 1993). It has been reported that the ester part of CAPE is important for its antiproliferative effect on various human tumor cells (Nagaoka et al., 2003). Additionally, it was suggested that conversion of the phenyl group to a cyclohexyl group in a CAPE analogue enhanced the antiproliferative effect (Kadota & Tezuka, 2004).

Immature green coffee beans are not marketed as coffee because contamination of these beans negatively affects the flavor. However, they contain appreciable amounts of various caffeoylquinic acids, for example, 4.8–5.8 g of 5-caffeoylquinic acid/100 g of immature green coffee beans (Kishimoto et al., 2005a). These immature beans are notable among unused agricultural resources, and we are therefore currently investigating the enzymatic conversion of their caffeoylquinic acids to valuable products.

In recent work, we synthesized CAPE using 5-caffeoylquinic acid and 2-phenylethanol as substrates with chlorogenate hydrolase by a transesterification reaction in a biphasic aqueous-alcohol state and elucidated the antibacterial, antimutagenic, and anti-influenza virus activities of CAPE (Kishimoto et al., 2005b). The procedure using chlorogenate hydrolase provided various CAPE analogues, but the maximum conversion yield of CAPE was 50%. The insufficient yield was probably due to the hydrolysis of 5-caffeoylquinic acid by the enzyme to caffeic acid in the aqueous phase. Therefore, a new procedure for the synthesis of CAPE analogues superior to that method in terms of the conversion yield remained to be developed.

Ionic liquids (ILs), which are composed of a bulky asymmetric cation and a small anion, are easily modified with respect to the combination of cation and anion, and therefore, numerous IL compositions are possible (Sureshkumar & Lee, 2009). Unlike conventional organic solvents used for biocatalytic reactions, ILs are able to dissolve many compounds, have a wide temperature range for the liquid phase, and possess no vapor pressures. Thus, ILs have good properties for use as reaction solvents, and extensive studies of enzymatic synthesis using ILs as the solvents have been carried out (Moniruzzaman et al., 2010; van Rantwijk & Sheldon, 2007).

Hydrolases, especialy lipases, are noted for their tolerance of organic solvents, and are obvious candidates for the enzymatic synthesis in ILs (Sureshkumar & Lee, 2009). Indeed, lipases from *Candida antarctica*, *Burkholderia cepacia* (formerly *Pseudomonas cepacia*), and *Alcaligenes* sp. are catalytically active in ILs (Itoh et al., 2001; Nara et al., 2002). Additionally, lipases mediate transesterification reactions in these ILs with an efficiency comparable to that in tert-butyl alcohol, dioxane, or toluene (Lau et al., 2000; Nara et al., 2002; Park & Kazlauskas, 2001).

In this chapter, we describe alcoholysis with chlorogenate hydrolase (EC 3.1.1.42, Kikkoman, Chiba, Japan) from *Aspergillus japonicus* and transesterification with *C. antarctica* lipase B (Novozyme 435, Novozymes, Bagsvaerd, Denmark). Both reactions were performed with various ILs as the reaction solvent. Using chlorogenate hydrolase, various caffeoylquinic acids prepared from coffee beans were converted to methyl caffeate (Fig. 1). Using Novozyme 435, on the other hand, methyl caffeate was converted to CAPE analogues with various alcohols (Fig. 5). With consecutive reactions comprised of alcoholysis and transesterification, 5-caffeoylquinic acid from immature coffee beans was converted to a CAPE analogue, 3-cyclohexylpropyl caffeate (Fig. 9). Additionally, the CAPE analogues produced, namely, 2-cyclohexylethyl caffeate and 3-cyclohexylpropyl caffeate, were tested for their antiproliferative activities by MTT assay using four different human tumor cells, namely, colon carcinoma HT-29 cells, cervical carcinoma HeLa S3 cells, breast cancer MCF-7 cells, and chronic myeloid leukemia K-562 cells (Table 4).

2. Conversion of caffeoylquinic acids to methyl caffeate with chlorogenate hydrolase

The aim of this study was the development of conversion system of caffeoylquinic acids to valuable compounds. When an IL, 1-butyl-3-methylimidazolium bis(trifluoromethylsulfonyl)imide ([bmim][NTf_2]) was used as a reaction solvent, we found that immobilized chlorogenate hydrolase (Kikkoman) catalyzed the conversion of 5-caffeoylquinic acid to methyl caffeate with methanol (Fig. 1). The immobilized enzyme was prepared with chlorogenate hydrolase using quaternary ammonium sepabeads (Mitsubishi Chemical Co., Tokyo, Japan) (Kurata et al., 2011). To synthesize valuable compounds from caffeoylquinic acids, we attempted to develop a method for the conversion of caffeoylquinic acids to CAPE analogues *via* methyl caffeate. In section 2, we describe the properties of immobilised chlorogenate hydrolase in ILs. Using various caffeoyl quinic acid prepared from immature coffee beans, we developed a system to produce methyl caffeate.

2.1 Effect of aqueous solution on chlorogenate hydrolase in IL

When [bmim][NTf_2] was used as the reaction solvent, the enzymatic conversion with immobilized chlorogenate hydrolase proceeded under a non-aqueous condition (Kurata et al., 2011). In various enzymatic syntheses, water can shift the equilibrium in the direction of hydrolysis; therefore, in methods using lipase, the water produced as a by-product was

(A)

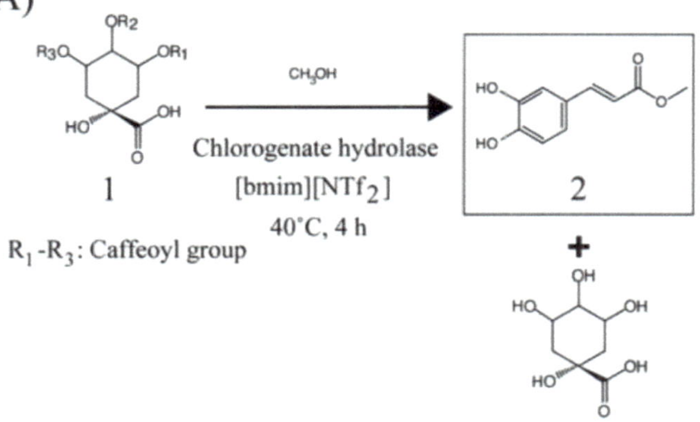

(B)

	Quinic acid		
	R1	R2	R3
3-Caffeoylquinic acid	Caffeoyl group	H	H
4-Caffeoylquinic acid	H	Caffeoyl group	H
5-Caffeoylquinic acid	H	H	Caffeoyl group
3,4-Dicaffeoylquinic acid	Caffeoyl group	Caffeoyl group	H
3,5-Dicaffeoylquinic acid	Caffeoyl group	H	Caffeoyl group
4,5-Dicaffeoylquinic acid	H	Caffeoyl group	Caffeoyl group

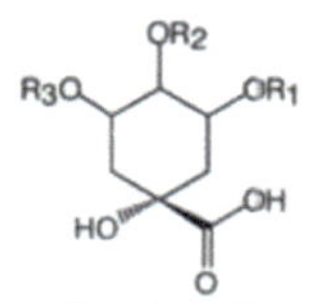

Quinic acid

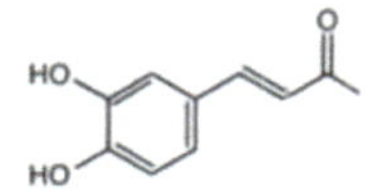

Caffeoyl group

Fig. 1. Conversion of caffeoylquinic acids to methyl caffeate by chlorogenate hydrolase. (A) Alcoholysis of various caffeoyl quinic aicds (compound 1) was catalysed by chlorogenate hydrolase with methanol and [bmim][NTf$_2$] as the reaction solvent, and methyl caffeate (compound 2) was produced. (B) Structures of substrates for synthesis of methyl caffeate with chlorogenate hydrolase are shown.

removed (Bélafi-Bakó et al., 2002; Gubicza et al., 2000). However, the catalytic activity in non-aqueous solvents depends on the amount of water associated with the enzyme conforming to its native conformation (Barahona et al., 2006). Thus, we examined the effect of the addition of an aqueous solution on the production of methyl caffeate in [bmim][NTf$_2$] by varying the additional volume of 50 mM sodium phosphate buffer (pH 6.5) from 0 to 5% (v/v) (Fig. 2).

When the concentration of the aqueous solution was 1% or lower, the production of methyl caffeate increased. However, the production of methyl caffeate was decreased in aqueous solutions greater than 2%. The production of caffeic acid was increased with addition of the buffer, indicating that the enzyme probably catalyzed hydrolysis of 5-caffeoylquinic acid to produce caffeic acid rather than alcoholysis to produce methyl caffeate. Thus, the result indicated that the addition of 1% aqueous solution was suitable for the production of methyl caffeate. It was suggested that ILs are able to maintain active structures of the enzymes with a monomolecular layer of water (Fehér et al., 2007). Thus, chlorogenate hydrolase would maintain the active structure with the layer of the buffer in [bmim][NTf$_2$].

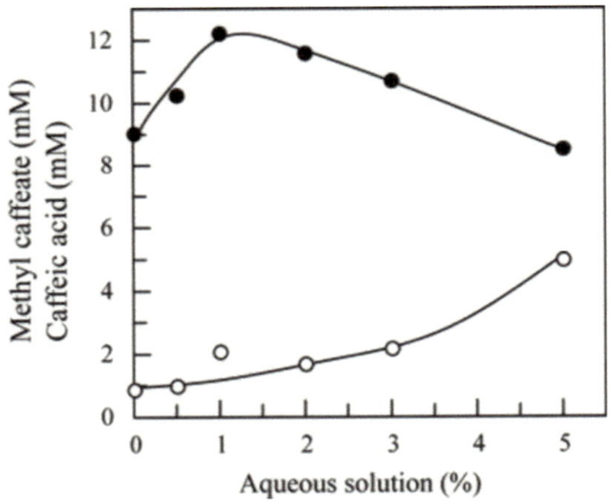

Fig. 2. Effects of water concentration on conversion of 5-caffeoylquinic acid to methyl caffeate by chlorogenate hydrolase (Kurata et al., 2011). The reaction was performed at 40°C using chlorogenate hydrolase and [bmim][NTF$_2$] as the reaction solvent with a 0–5% (v/v) aqueous solution of 50 mM sodium phosphate (pH 6.5). Each symbol indicates methyl caffeate (closed circle) and caffeic acid (open circle).

2.2 Selection of IL for chlorogenate hydrolase

In order to investigate the activities of chlorogenate hydrolase in various ILs, the alcoholysis of 5-caffeoylquinic acid with methanol was examined (Fig. 3).

The reactions were performed in five ILs, namely, [bmim][NTf$_2$], two [bmim] cation-containing ILs, 1-butyl-3-methylimidazolium tetrafluoroborate ([bmim][BF$_4$]) and 1-butyl-3-methylimidazolium trifluoromethanesulfonate ([bmim][CF$_3$SO$_3$]), and two [NTf$_2$] anion-containing ILs, *N*-methyl-*N*-propylpyrrolidinium bis(trifluoromethylsulfonyl)imide ([MPPro][NTf$_2$]) and *N*-methyl-*N*-propylpiperidinium bis(trifluoromethylsulfonyl)imide ([MPPip][NTf$_2$]).

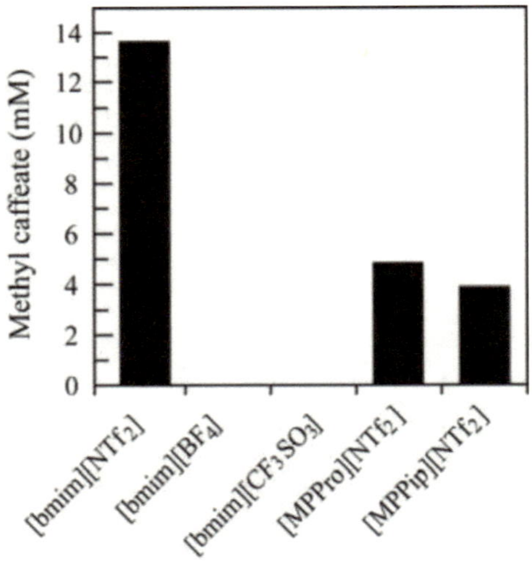

Fig. 3. Selection of IL for chlorogenate hydrolase. The reaction was performed with various ILs as the reaction solvent.

Although chlorogenate hydrolase catalyzed the alcoholysis reaction in [bmim][NTf_2] to produce 13.7 mM methyl caffeate, the reaction scarcely proceeded in [bmim][BF_4] and [bmim][CF_3SO_3]. Lipase from *Candida rugosa* was found to be active in [bmim][PF_6] for the transesterification, but inactive in ILs including [bmim][acetate] and [bmim][nitrate] (Kaar et al., 2003). Lipase from *Pseudomonas aeruginosa* was more stable in [bmim][PF_6] than in [bmim][BF_4] (Singh et al., 2009). Thus, the nature of the anion in IL plays a critical factor in determining the enzyme activity and stability. It was reported that the hydrogen-bond basicities of [bmim][BF_4] and [bmim][CF_3SO_3] are larger than that of [bmim][NTf_2] (Anderson et al., 2002; Kaar et al., 2003). Additionally, it was suggested that the [BF_4] and [CF_3SO_3] anions are more nucleophilic than the [NTf_2] anion, and that the [BF_4] and [CF_3SO_3] anions coordinate more strongly to positively charged sites in the structure of an enzyme. In consequence, the enzyme is deactivated by a conformation change in the enzyme structure due to these anions. As shown Figure 3, chlorogenate hydrolase was deactivated in [bmim][BF_4] and [bmim][CF_3SO_3]. Additionally, the ability to decrease the enzyme activity was in the order of cations: $[MPPip]^+ > [MPPro]^+ > [bmim]^+$. In the case of chlorogenate hydrolase, the enzyme activity was affected by the anions and cations. As shown in Figure 3, [bmim][NTf_2] was suitable for the alcoholysis of caffeoylquinic acids to produce methyl caffeate.

2.3 Effect of temperature

The effect of temperature on the chlorogenate hydrolase activity of the alcoholysis reaction with 5-caffeoylquinic acid and methanol was examined in [bmim][NTf_2] at temperatures from 20°C to 100°C (Fig. 4).

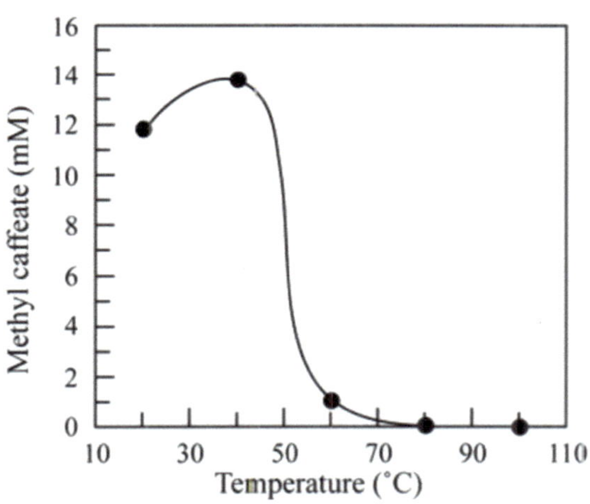

Fig. 4. Effect of temperature on conversion of 5-caffeoylquinic acid to methyl caffeate by chlorogenate hydrolase (Kurata et al., 2011). The reaction was performed at 20°C to 60°C for 4 h with immobilized chlorogenate hydrolase and [bmim][NTf_2] as the reaction solvent.

With increased temperature, the amounts of the product initially increased. However, at temperatures higher than 40°C, the amounts of the product decreased, indicating that 40°C is the optimum temperature for chlorogenate hydrolase. Additionally, the enzyme is deactivated at temperatures higher than 80°C.

2.4 Substrate specificity for chlorogenate hydrolase

We examined the production of methyl caffeate using the alcoholysis reaction by chlorogenate hydrolase in [bmim][NTf_2] (Table 1). Methyl caffeate was produced using various caffeoylquinic acids (15 μmol) and methanol (2200 μmol). Using 3-caffeoylquinic acid, 4-caffeoylquinic acid, 5-caffeoylquinic acid, 3,5-dicaffeoylquinic acid, and 4,5-dicaffeoylquinic acid, methyl caffeate was synthesized at concentrations of 9.0 mM (9.0 μmol), 9.2 mM (9.2 μmol), 12.9 mM (12.9 μmol), 13.9 mM (13.9 μmol), and 17.1 mM (17.1 μmol), respectively (Kurata et al., 2011).

	Quinic acid			
	R_1	R_2	R_3	Yield (%)
3-Caffeoylquinic acid	Caffeoyl group	H	H	60.0
4-Caffeoylquinic acid	H	Caffeoyl group	H	61.3
5-Caffeoylquinic acid	H	H	Caffeoyl group	86.0
3,5-Dicaffeoylquinic acid	Caffeoyl group	H	Caffeoyl group	92.7
4,5-Dicaffeoylquinic acid	H	Caffeoyl group	Caffeoyl group	114.0

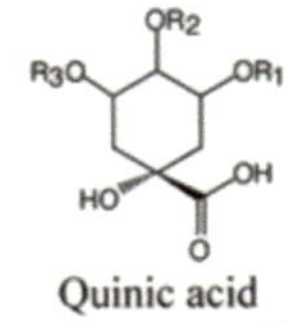

Quinic acid

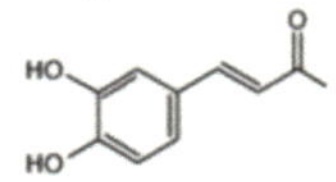

Caffeoyl group

Table 1. Conversion yields of caffeoylquinic acids to methyl caffeate.

Because dicaffeoylquinic acid and caffeoylquinic acid have two and one caffeoyl groups, respectively, the volumes of methyl caffeate prepared from 3,5-dicaffeoylquinic acid and 4,5-dicaffeoylquinic acid were greater than those of methyl caffeate prepared from 3-caffeoylquinic acid, 4-caffeoylquinic acid, and 5-caffeoylquinic acid. In the cases of 3,5-dicaffeoylquinic acid and 4,5-dicaffeoylquinic acid, both caffeoyl groups would be used for synthesis of methyl caffeate.

Additionally, using a mixture of 3,4-dicaffeoylquinic acid and 4,5-dicaffeoylquinic acid, which is a crude fraction prepared from coffee beans, HPLC analysis showed that all peaks of caffeoylquinic acids disappeared and that the peak of methyl caffeate occurred after a 4-h reaction with chlorogenate hydrolase in [bmim][NTf_2] (Kurata et al., 2011).

Chlorogenate hydrolase acted on caffeoylquinic acids and dicaffeoylquinic acids. Thus, methyl caffeate was produced from various caffeoylquinic acids prepared from coffee beans using this procedure.

3. Conversion of methyl caffeate to CAPE analogues with Novozyme 435

In section 2, we described success in converting various caffeoylquinic acids to methyl caffeate with good yields. Next, we tried to convert methyl caffeate to valuable compounds, namely, CAPE analogues, using IL as the reaction solvent (Fig. 5).

3.1 Selection of lipase for transesterification in IL

We initially selected methyl caffeate and 3-cyclohexyl-1-propanol as substrates for comparative study of the enzyme's performance in [bmim][NTf_2]. Four commercially available lipases, *C. antarctica* lipase B (Novozyme 435, Novozymes, 30 U mg^{-1}), *Rhizomucor miehei* lipase (RMIM, Novozymes, 1370 U mg^{-1}), *B. cepacia* lipases (PS-CI, Wako Pure Chemical, Osaka, Japan, 2560 U mg^{-1}), and *Thermomyces lanuginosus* lipase (TLIM, Novozymes, 1850 U mg^{-1}) were tested (Table 2).

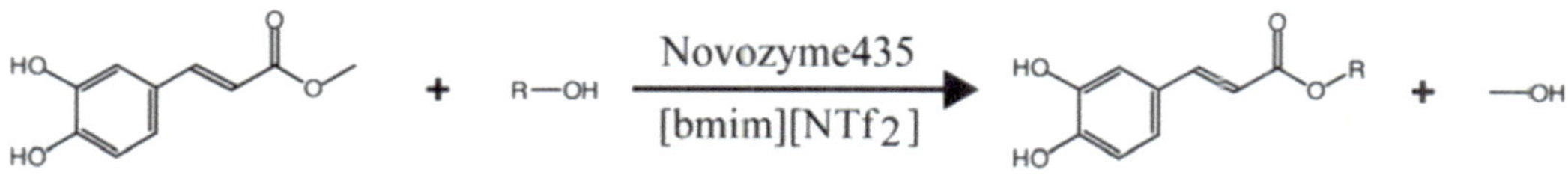

Compound	R
2-cyclohexylethyl caffeate	$-(CH_2)_2$Cyclohexyl
3-cyclohexylpropyl caffeate	$-(CH_2)_3$Cyclohexyl
4-phenylbutyl caffeate	$-(CH_2)_4$Phenyl
5-phenylpentyl caffeate	$-(CH_2)_5$Phenyl

Fig. 5. Convesion of methyl caffeate to CAPE analogues by Novozyme 435.

	3-Cyclohexylpropyl caffeate (mM)
Novozyme 435	30.8
RMIM	11.6
PS-CI	6.6
TLIM	1.8

Table 2. Selection of lipase for conversion of methyl caffeate (Kurata et al., 2010). The reaction was performed with 1,200,000 U of lipases.

Among these enzymes, Novozyme 435 was the most suitable for the procedure. PS-CI and TLIM hardly catalyzed the reaction, although RMIM catalyzed it relatively well. The difference in the recognition of the substrate by lipases is probably in accordance with the difference in the substrate orientation in the active center (Pleiss et al., 1998). With one possible exception, the substrates easily accessed the active center of Novozyme 435. Additionally, it is suggested that the substrates hardly accessed the active sites of RMIM, PS-CI, and TLIM.

3.2 Selection of IL for Novozyme 435

In order to investigate the activity of Novozyme 435 in various ILs, the transesterification of 2-cyclohexylethanol was examined (Fig. 6). The reactions were performed in eight ILs, namely, 1-butyl-3-methylimidazolium bis(trifluoromethylsulfonyl)imide ([bmim][NTf_2]), three [bmim] cation-containing ILs, 1-butyl-3-methylimidazolium hexafluorophosphate ([bmim][PF_6]), 1-butyl-3-methylimidazolium tetrafluoroborate ([bmim][BF_4]), and 1-butyl-3-methylimidazolium trifluoromethanesulfonate ([bmim][CF_3SO_3]), and four [NTf_2] anion-containing ILs, 1-propyl-2,3,5-trimethylpyrazolium bis(trifluoromethylsulfonyl)imide ([PMPra][NTf_2]), *N*-methyl-*N*-propylpyrrolidinium bis(trifluoromethylsulfonyl)imide ([MPPro][NTf_2]), *N*-methyl-*N*-propylpiperidinium bis(trifluoromethylsulfonyl)imide ([MPPip][NTf_2]), and *N*,*N*,*N*-trimethyl-*N*-propylammonium bis(trifluoromethylsulfonyl)imide ([TMPA][NTf_2]).

Although Novozyme 435 catalyzed the transesterification in [bmim][NTf_2] and [bmim][PF_6] to produce 2-cyclohexylethyl caffeate (40.1 mM and 22.9 mM, respectively), the transesterification scarcely proceeded in [bmim][CF_3SO_3] and [bmim][BF_4] (Fig. 6, closed bar). The amounts of 2-cyclohexylethyl caffeate produced in all [NTf_2] anion-containing ILs were smaller to that in [bmim][NTf_2], but the reactions proceeded effectively in these ILs (Fig. 6, open bars).

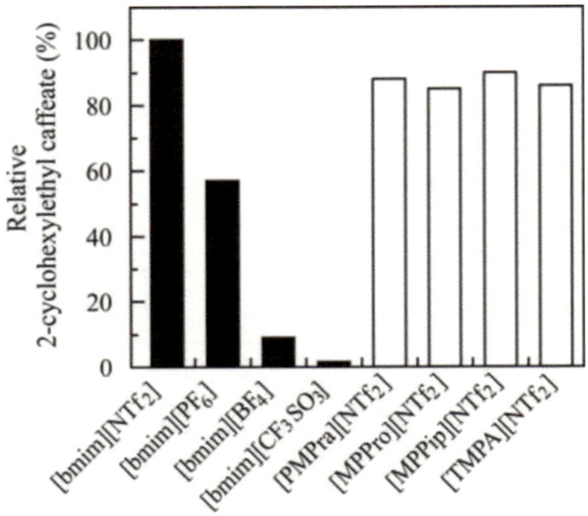

Fig. 6. Comparison of transesterifications in Ils. The reaction was performed with Novozyme 435 and [bmim][NTf_2], which was used as a reaction medium. The amount of 2-cyclohexylethyl caffeate (40.1 mM) in [bmim][NTf_2] was taken as 100%.

The results indicated that Novozyme 435 activity in ILs is anion dependent, like chlorogenate hydrolase (Fig. 3). As described in section 2. 2, chlorogenate hydrolase would be deactivated by conformation changes in the enzyme structure due to the [BF_4] and [CF_3SO_3] anions. Similary, Novozyme 435 was deactivated in [bmim][BF_4] and [bmim][CF_3SO_3] (Fig. 6). In addition, Novozyme 435 activity was higher in [bmim][NTf_2] than in [bmim][PF_6]. It was reported that the pH of the reaction mixture with [bmim][PF_6] was decreased by hydrogen fluoride, which was produced by the hydrolytic decomposition of the [PF_6] anion (Hernández-Fernández et al., 2007). Novozyme 435 activity in [bmim][PF_6] would be decreased with the decrease in pH value. Because the concentrations of 2-cyclohexylethyl caffeate produced in [bmim][NTf_2] and [bmim][PF_6] were 40.1 mM and 22.9 mM, respectively, [bmim][NTf_2] was suitable for the transesterification of methyl caffeate.

3.3 Effect of temperature

The effect of temperature on the activity of Novozyme 435 for the transesterification with 2-cyclohexylethanol and 3-cyclohexyl-1-propanol was examined in [bmim][NTf_2] by varying temperatures ranging from 60°C to 110°C (Fig. 7).

With increased temperature, the amounts of the product were initially increased. However, at higher than 90°C, the amounts of the product were decreased, indicating that 80°C is the optimum temperature for the transesterification.

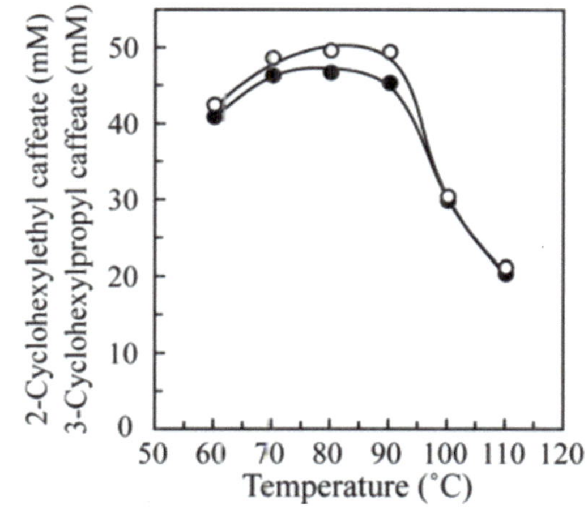

Fig. 7. Effect of temperature on productions of 2-cyclohexylethyl caffeate and 3-cyclohexylpropyl caffeate (Kurata et al., 2010). The reaction was performed at 60°C–110°C with Novozyme 435 and [bmim][NTf_2], which was used as a reaction medium.

3.4 Reusability of IL and lipase

We evaluated the repeated use of Novozyme 435 in [bmim][NTf_2] (Fig. 8). After the reaction, we added diethyl ether to the reaction mixture to form a diethyl ether–IL two-phase state. The great advantage of the state was the easy removal of the upper diethyl ether phase containing the product and unreacted substrate, while the lower phase containing the enzyme and IL could be used again.

After three cycles of repeated use, Novozyme 435 retained its ability to produce about 40 mM 2-cyclohexylethyl caffeate with a conversion yield of about 80% After four and five cycles, the amounts of 2-cyclohexylethyl caffeate were decreased to 28 7 mM and 27.4 mM, respectively.

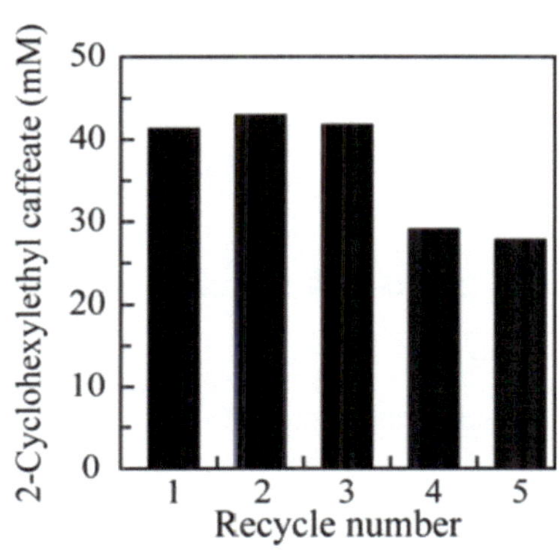

Fig. 8. Reuse of lipase in [bmim][NTf_2] solvent system (Kurata et al., 2010). The reaction mixture was comprised of Novozyme 435, methyl caffeate, and 2-cyclohexylethanol in [bmim][NTf_2]. The reaction was repeated five times.

3.5 Substrate specificity for Novozyme 435

We examined the production of CAPE analogues using transesterification with [bmim][NTf_2] (Kurata et al., 2010). CAPE analogues were produced using 50 mM methyl caffeate (50 μmol) and 300 mM alcohols (300 μmol). We synthesized 48.8 mM 2-cyclohexylethyl caffeate (48.8 μmol), 46.9 mM 3-cyclohexylpropyl caffeate (46.9 μmol), 49.4 mM 4-phenylbutyl caffeate (49.4 μmol), and 42.0 mM 5-phenylpentyl caffeate (42.0 μmol). The conversion yields are shown in Table 3.

In terms of the conversion yield, this procedure using Novozyme 435 is superior to the previously reported procedure affording CAPE with the maximum conversion yield of 50% (Kishimoto et al., 2005a). CAPE was obtained by transesterification catalyzed by Novozyme 435 using isooctane as the solvent with the conversion yield of 91.65% (Chen et al., 2010). The conversion yields of the CAPE analogues produced by the [bmim][NTf_2] system are comparable to that of CAPE produced by the isooctane system.

CAPE analogues		Yield (%)
2-cyclohexylethyl caffeate	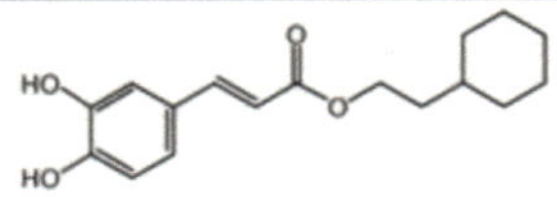	97.6
3-cyclohexylpropyl caffeate	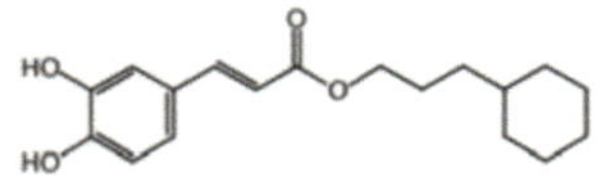	93.8
4-phenylbutyl caffeate	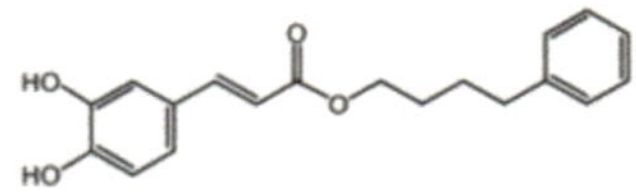	96.7
5-phenylpentyl caffeate	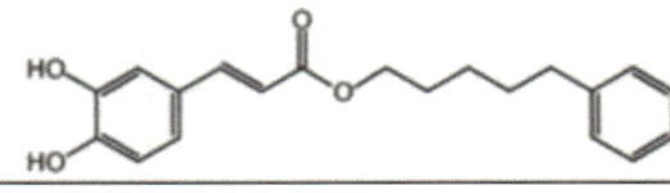	84.0

Table 3. Conversion yields of CAPE analogues. The reaction was performed with methyl caffeate, each of alcohol, Novozyme 435, and [bmim][NTf_2], which was used as a reaction medium. For production of 2-cyclohexylethyl caffeate, 3-cyclohexylpropyl caffeate, 4-phenylbutyl caffeate, and 5-phenylpentyl caffeate, alchols used were shown as follows: 2-cyclohexylethanol, 3-cyclohexyl-1-propanol, 4-phenyl-1-butanol, 5-phenyl-1-pentanol, respectively(Kurata et al., 2010).

4. Consecutive conversions of 5-caffeoylquinic acids to 3-cyclohexylpropyl caffeate

We described the alcoholysis of various caffeoylquinic acids with methanol to produce methyl caffeate in section 2, and the transesterification of methyl caffeate to various CAPE analogues in section 3. Both reactions was performed in the same IL, namely, [bmim][NTf_2], with good production yields. In section 4, we investigated a one-pot consecutive conversion of 5-caffeoylquinic acid to a CAPE analogue, 3-cyclohexylcaffeate, *via* methyl caffeate (Fig. 9). In the case of a one-pot two-step reaction, purification of the reaction intermediate is not required, so that the intended product was expected to be obtained with high conversion yield.

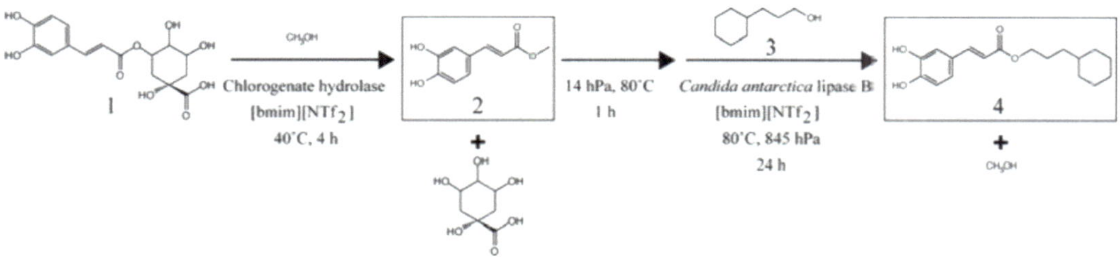

Fig. 9. Consecutive enzymatic reactions for synthesis of 3-cyclohexylpropyl caffeate from 5-caffeoylquinic acid. In all reaction steps, [bmim][NTf$_2$] was used as the solvent. Compound 1: 5-caffeoylquinic acid, 2: methyl caffeate, 3: 3-cyclohexyl-1-propanol, and 4: 3-cyclohexylpropyl caffeate.

4.1 Conversion of 5-caffeoylquinic acid with chlorogenate hydrolase

We initialy examined the production of methyl caffeate or 3-cyclohexylpropyl caffeate using immobilized chlorogenate hydrolase in [bmim][NTf$_2$] (Fig. 10).

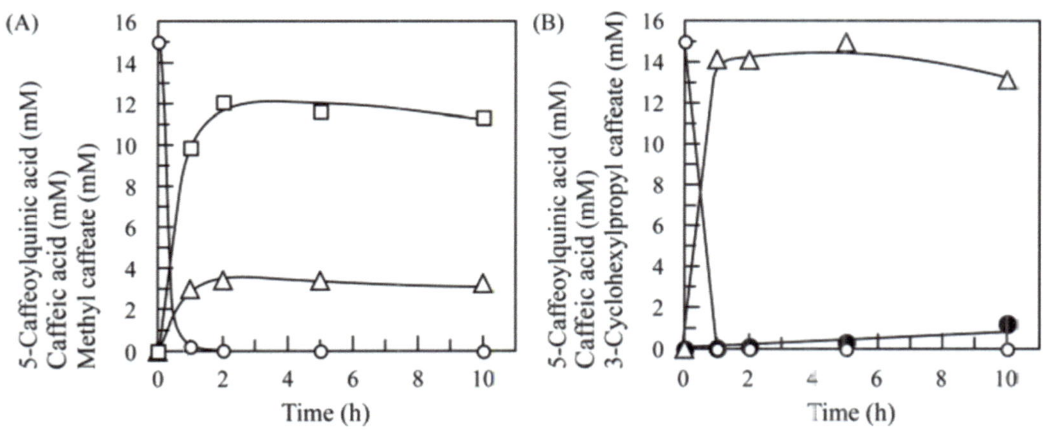

Fig. 10. Conversion of 5-caffeoylquinic acid to methyl caffeate and caffeic acid (A) and 3-cyclohexylpropyl caffeate and caffeic acid (B) by chlorogenate hydrolase (Kurata et al., 2011). The reaction was performed with 5-caffeoylquinic acid, methanol (A) or 3-cyclohexyl-1-propanol (B), chlorogenate hydrolase, and [bmim][NTF$_2$]. Symbols indicate 5-caffeoylquinic acid (open circle), caffeic acid (open triangle), methyl caffeate (open square), and 3-cyclohexylpropyl caffeate (closed circle).

The reaction mixture was comprised of 5-caffeoylquinic acid, alcohols, 50 mM sodium phosphate (pH 6.5), chlorogenate hydrolase, and [bmim][NTf$_2$] as the reaction solvent. After a 10-h reaction using 15 mM 5-caffeoylquinic acid and 2200 mM methanol, 11.3 mM methyl caffeate and 3.2 mM caffeic acid were produced with conversion yields of 75.4% and 21.5%, respectively (Fig. 10A), whereas 13.1 mM caffeic acid and 1.21 mM 3-cyclohexylpropyl caffeate were produced with conversion yields of 87.6% and 8.1%, respectively, using 15 mM 5-caffeoylquinic acid and 2200 mM 3-cyclohexyl-1-propanol (Fig. 10B). The result showed that methanol is a better substrate for chlorogenate hydrolase than 3-cyclohexyl-1-propanol to produce methyl caffeate with a high conversion yield. Additionally, Novozyme 435 did not catalyze the conversion of 5-caffeoylquinic acid to 3-cyclohexylpropyl caffeate or methyl caffeate (data not shown), but efficiently catalyzed the conversion of methyl caffeate to 3-cyclohexylpropyl caffeate with a conversion yield of 93.8%, as described in section 3.5. Thus, consecutive conversions of 5-caffeoylquinic acid to 3-cyclohexylpropyl caffeate *via* methyl caffeate using chlorogenate hydrolase and Novozyme 435 would be a more effective procedure than a single conversion of 5-caffeoylquinic acid to 3-cyclohexylpropyl caffeate.

4.2 Effect of methanol on transesterification by Novozyme 435

During conversion of 5-caffeoylquinic acid with methanol, the reaction mixture probably contained unreacted methanol. Additionally, when methyl caffeate prepared from caffeoylquinic acids was used as the acyl donor to produce 3-cyclohexylpropyl caffeate, methanol was produced as a byproduct.

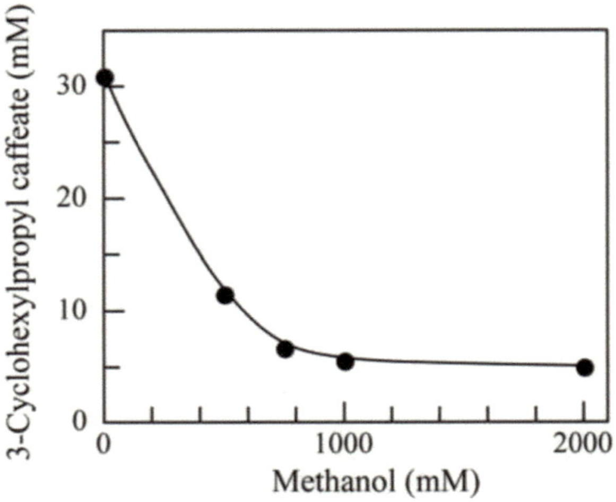

Fig. 11. Effect of methanol on conversion of methyl caffeate to 3-cyclohexylpropyl caffeate by Novozyme 435 (Kurata et al., 2011). Transesterification of methyl caffeate was performed with methyl caffeate, 3-cyclohexyl-1-propanol, Novozyme 435, 0–2200 mM methanol, and [bmim][NTf_2] as the reaction solvent.

In the conversion of methyl caffeate to 3-cyclohexylpropyl caffeate in the consecutive synthesis, therefore, the reverse reaction with methanol would occur easily to decrease the amount of the desired product. We tested the effect of methanol on Novozyme 435, and observed that methanol had an inhibitory effect on the production of 3-cyclohexylpropyl caffeate (Fig. 11).

For development of the two-step procedure, we attempted to avoid this difficulty by removing the unreacted methanol *in vacuo* (14 hPa) and performing the transesterification of methyl caffeate by Novozyme 435 under reduced pressure (845 hPa), so that methanol could be removed immediately from the reaction mixture (Itoh, 2007).

4.3 Consecutive enzymatic reactions for synthesis of 3-cyclohexylpropyl caffeate

For the development of a convenient procedure for synthesis of 3-cyclohexylpropyl caffeate from 5-caffeoylquinic acid *via* methyl caffeate, we tested the consecutive enzymatic reactions by chlorogenate hydrolase and Novozyme 435 using [bmim][NTf_2] as the solvent (Fig. 12).

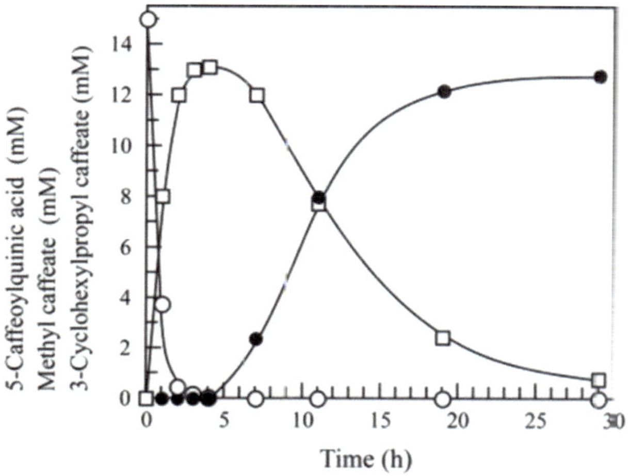

Fig. 12. Consecutive reactions for conversion of 5-caffeoylquinic acid to 3-cyclohexylpropyl caffeate *via* methyl caffeate (Kurata et al., 2011). The symbols indicate 5-caffeoylquinic acid (open circle), methyl caffeate (open square), and 3-cyclohexylproyl caffeate (closed circle).

As shown in Figure 9, the first alcoholysis reaction for 4 h at 40°C with methanol using chlorogenate hydrolase produced methyl caffeate from 5-caffeoylquinic acid. Next, the unreacted methanol was removed under reduced pressure (14 hPa) for 1 h at 80°C, and chlorogenate hydrolase was deactivated at 80°C. In the subsequent transesterification reaction by Novozyme 435 with 3-cyclohexyl-1-propanol, the methyl caffeate produced was converted to 3-cyclohexylpropyl caffeate under reduced pressure (845 hPa) to remove the by-product methanol.

In the consecutive synthesis, the methyl caffeate was transesterified efficiently to 3-cyclohexylpropyl caffeate by Novozyme 435 with deactivation of chlorogenate hydrolase by taking advantage of the difference between the optimum temperatures for chlorogenate hydrolase (40°C, Fig. 4) and Novozyme 435 (80°C, Fig. 7).

In the one-pot two-step synthesis, 15 mM 5-caffeoylquinic acid (15 µmol) was converted *via* 13.1 mM methyl caffeate (13.1 µmol, 87.3%) to 12.8 mM 3-cyclohexylpropyl caffeate (12.8 µmol, 97.7%). The net conversion yield of 3-cyclohexylpropyl caffeate toward 5-caffeoylquinic acid was 85.3%.

In the synthesis of CAPE analogues, the conversion yield of 3-cyclohexylpropyl caffeate toward 5-caffeoylquinic acid (85.3%) was superior to that of CAPE toward 5-caffeoylquinic acid (50%) (Kishimoto et al., 2005a). CAPE analogues were obtained by single transesterification systems catalyzed by Novozyme 435 using [bmim][NTf_2] or isooctane as the solvent with conversion yields of 93.8% and 91.65%, respectively (Chen et al., 2010; Kurata et al., 2010). The conversion yield of 3-cyclohexylpropyl caffeate produced by the consecutive conversion system was comparable to those of CAPE produced by single transesterification systems.

5. Antiproliferative effect against human tumor cells

In the study, we developed a convenient procedure for synthesis of CAPE analogues. As described in section 1, CAPE analogues have antimicrobial, anti-inflammatory, antioxidant, and antitumor activities (Kishimoto et al., 2005b). Finally, the CAPE analogues produced, namely, 2-cyclohexylethyl caffeate and 3-cyclohexylpropyl caffeate, were purified and tested for their antiproliferative activities against human tumor cells *in vitro*.

5.1 Antiproliferative effect

Cellular growth in the presence of 2-cyclohexylethyl caffeate or 3-cyclohexylpropyl caffeate produced in [bmim][NTf_2] was determined by an MTT assay (Carmichael et al., 1987). Rapidly growing human tumor cells, namely, colon carcinoma HT-29 cells, cervical carcinoma HeLa S3 cells, breast cancer MCF-7 cells, and chronic myeloid leukemia K-562 cells, were harvested, counted, and inoculated at appropriate concentrations (135 µl volumes) into a 96-well microtiter plate. After 24 h, 15 µl of CAPE analogues were applied to triplicate culture wells, and the culture was incubated at 37°C. After 72 h, 15 µl of 3-(4, 5-dimethylthiazol-2-yl)-2, 5-diphenyltetrazolium bromide (MTT, Nacalai Tesque, Kyoto, Japan) prepared at 5 mg/ml in PBS (Invitrogen, Carlsbad, CA) was added to microculture wells. After a 4-h incubation at 37°C, 200 µl of DMSO was added to the supernatant from each microculture well. After a thorough mixing with a mechanical plate mixer, absorbance at 540 nm was measured. IC_{50} values represent the compound concentration required to reduce the proliferation of the tumor cell by 50%.
Cell viability was calculated using the following equation:

$$\text{Cell viability (\%)} = [(A - B)/(C - B)] \times 100,$$

where A is the absorbance of the cells in the culture medium treated with drugs, B is the absorbance of the culture medium (blank), and C is the absorbance of static cells (control). The MTT assay was repeated two times.

	IC_{50} (µM)			
	HT-29	HeLa S3	MCF7	K562
2-Cyclohexylethyl caffeate	11.1	16.4	5.5	9.7
3-Cyclohexylpropyl caffeate	14.7	22.9	5.8	9.3
Caffeic acid	165.8	134.5	490.8	56.0
5-Fluorouracil	21.0	35.9	24.9	9.8

Table 4. Antiproliferative activities of CAPE analogues (Kurata et al., 2010).

Both 2-cyclohexylethyl caffeate and 3-cyclohexylpropyl caffeate halved the numbers of the four tumor cells at less than 25 µM. Although the IC_{50} value of caffeic acid for K562 cells was 56.0 µM, the values for other three tumor cells, HT-29, HeLa S3, and MCF-7, were 165.8 µM, 134.5 µM, and 490.8 µM, respectively. 2-Cyclohexylethyl caffeate and 3-cyclohexylpropyl caffeate exhibited stronger antiproliferative activities than caffeic acid, and the IC_{50} values of both compounds were comparable to that of 5-fluorouracil, which is a widely accepted anticancer drug (Yoshida et al., 2008).
It was also reported that CAPE inhibited the proliferation of human leukemic HL-60 cells and human colorectal cancer HCT116 cells and induced apoptosis (Chen et al., 2001; Wang et al., 2005). Thus, 2-cyclohexylethyl caffeate and 3-cyclohexylpropyl caffeate might induce apoptosis.

6. Conclusion

Generally, enzymes offer advantages in stereo and regioselectivity for synthesis of biologically active compounds (Luetz et al., 2008). Diminished yields, selectivity, and poor solubility of substrates in the aqueous phase may require that the enzymatic reactions be performed in the non-aqueous phase. In this regard, various solvents, such as organic solvents (Gubicza et al., 2000), ILs, and supercritical carbon dioxide (Oakes et al., 2001), were investigated to overcome the disadvantages of the aqueous solvent. Increasing environmental demands and legal restrictions currently necessitate the development of new types of green solvents and innovations for more sustainable enzymatic processes. ILs as solvents have at least four advantages for enzymatic reactions: 1) ILs are easily modified with cations and anions to dissolve polar and non-polar substrates, 2) the purification of products is very easy, 3) the enzymes can be used repeatedly with IL and 4) the thermal stability and non-volatility of ILs may promote the reaction to obtain the intended product.
In this study using various ILs as the reaction solvent, we examined the properties of chlorogenate hydolase from *A. japonicus* and Novozyme 435 (*C. antractica* lipase B). Additionally, we produced 2-cycylohexylethyl caffeate and 3-cyclohexylpropyl caffeate, and showed that they had antiproliferative effect on various human tumor cells.
We thus developed a convenient procedure for conversion of 5-caffeoylquinic acid to 3-cyclohexylpropyl caffeate *via* methyl caffeate. The procedure developed in this study would be useful for the exploitation of immature coffee beans that contain various caffeoylquinic acids (Fig. 13).

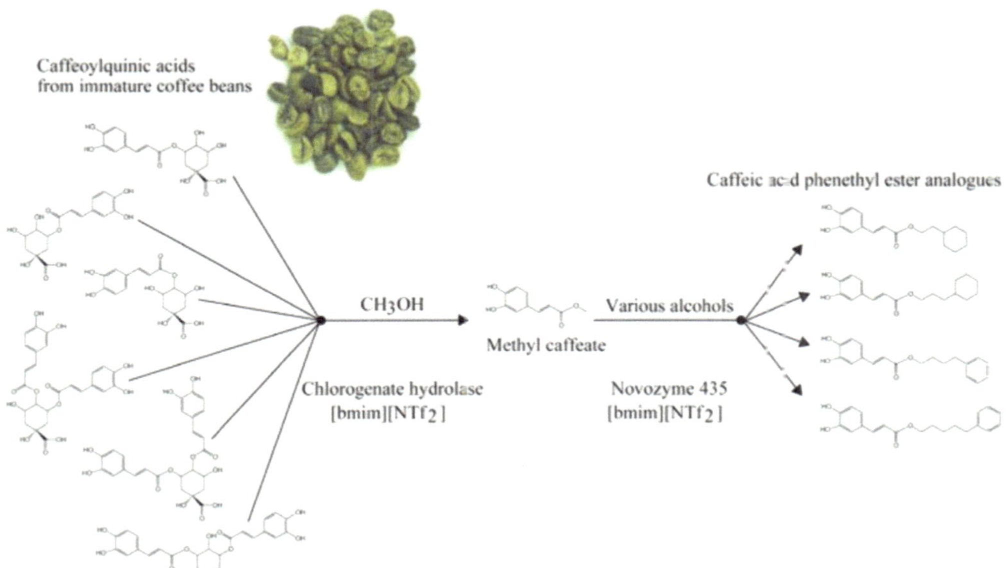

Fig. 13. One-pot two-step synthesis of various CAPE analogues from various caffeoylquinic acids prepared from coffee beans *via* methyl caffeate. Using chlorogenate hydrolase, Novozyme 435, and [bmim][NTf_2] as the reaction solvent, caffeoylquinic acids from coffee beans were converted to various CAPE analogues, which have antimicrobial, anti-inflammatory, antioxidant, and antitumor activities.

Using various enzymes with ILs as the solvent, an underutilized natural resource was converted to a valuable bioactive compound. Investigation of combinations of enzymes and ILs could lead to development of a novel bioconversion system.

7. Acknowledgment

We are grateful to Dr. Kazuya Iwai and Mr. Taiji Fukunaga at the R&D Center, UCC Ueshima Coffee Co., Ltd., for the donation of caffeoylquinic acids prepared from immature coffee beans and various ILs.

8. References

Anderson, J.L.; Ding, J.; Welton, T. & Armstrong, D.W. (2002). Characterizing ionic liquids on the basis of multiple solvation interactions. *J. Am. Chem. Soc.*, 124, 47, 14247-14254

Bankova, V. (2005). Recent trends and important developments in propolis research. *eCAM.*, 2, 1, 29-32

Barahona, D.; Pfromm, P.H. & Rezac, M.E. (2006). Effect of water activity on the lipase catalyzed esterification of geraniol in ionic liquid [bmim]pf6. *Biotechnol. Bioeng.*, 93, 2, 318-24

Bélafi-Bakó, K.; Dörmõ, N.; Ulbert, O. & Gubicza, L. (2002). Application of pervaporation for removal of water produced during enzymatic esterification in ionic liquids. *Desalination.*, 149, 1-3, 267-268

Carmichael, J.; DeGraff, W.G.; Gazdar, A.F.; Minna, J.D. & Mitchell, J.B. (1987). Evaluation of a tetrazolium-based semiautomated colorimetric assay: Assessment of radiosensitivity. *Cancer. Res.*, 47, 4, 943-946

Chen, H.C.; Ju, H.Y.; Twu, Y.K.; Chen, J.H.; Chang, C.M.; Liu, Y.C.; Chang, C. & Shieh, C.J. (2010). Optimized enzymatic synthesis of caffeic acid phenethyl ester by rsm. *N. Biotechnol.*, 27, 1, 89-93

Chen, Y.J.; Shiao, M.S.; Hsu, M.L.; Tsai, T.H. & Wang, S.Y. (2001). Effect of caffeic acid phenethyl ester, an antioxidant from propolis, on inducing apoptosis in human leukemic hl-60 cells. *J. Agric. Food. Chem.*, 49, 11, 5615-5619

Fehér, E.; Major, B.; Bélafi-Bakó, K. & Gubicza, L. (2007). On the background of enhanced stability and reusability of enzymes in ionic liquids. *Biochem. Soc. T.*, 35, 6, 1624-1627

Fesen, M.R.; Kohn, K.W.; Leteurtre, F. & Pommier, Y. (1993). Inhibitors of human immunodeficiency virus integrase. *Proc. Natl. Acad. Sci. U S A.*, 90, 6, 2399-2403

Gubicza, L.; Kabiri-Badr, A.; Keoves, E. & Bélafi-Bakó, K. (2000). Large-scale enzymatic production of natural flavour esters in organic solvent with continuous water removal. *J. Biotechnol.*, 84, 2, 193-196

Hernández-Fernández, F.J.; de los Rios, A.P.; Rubio, M.; Gomez, D. & Villora, G. (2007). Enhancement of activity and selectivity in lipase-catalyzed transesterification in ionic liquids by the use of additives. *J. Chem. Technol. Biotechnol.*, 82, 882–887

Itoh, T.; Akasaki, E.; Kudo, K. & Shirakami, S. (2001). Lipase-catalyzed enantioselective acylation in the ionic liquid solvent system: Reaction of enzyme anchored to the solvent. *Chem. Lett.*, 30, 3, 262

Itoh, T. (2007). Biotransformation in ionic liquid, In: *Future directions in biocatalysis*, Matsuda, T., 3-20, Elsevier Bioscience, The Netherlands

Kaar, J.L.; Jesionowski, A.M.; Berberich, J.A.; Moulton, R. & Russell, A.J. (2003). Impact of ionic liquid physical properties on lipase activity and stability. *J. Am. Chem. Soc.*, 125, 14, 4125-4131

Kadota, S. & Tezuka, Y. (2004). Study on antimetastatic activity of a propolis constituent, cape, and its analogues. *Honeybee. Sci.*, 25, 3, 107-112

Kishimoto, N.; Kakino, Y.; Iwai, K. & Fujita, T. (2005a). Chlorogenate hydrolase-catalyzed synthesis of hydroxycinnamic acid ester derivatives by transesterification, substitution of bromine, and condensation reactions. *Appl. Microbiol. Biotechnol.*, 68, 2, 198-202

Kishimoto, N.; Kakino, Y.; Iwai, K.; Mochida, K. & Fujita, T. (2005b). *In vitro* antibacterial, antimutagenic and anti-influenza virus activity of caffeic acid phenethyl esters. *Biocontrol. Sci.*, 10, 4, 155-161

Kurata, A.; Kitamura, Y.; Irie, S.; Takemoto, S.; Akai, Y.; Hirota, Y.; Fujita, T.; Iwai, K.; Furusawa, M. & Kishimoto, N. (2010). Enzymatic synthesis of caffeic acid phenethyl ester analogues in ionic liquid. *J. Biotechnol.*, 148, 2-3, 133-8

Kurata, A.; Takemoto, S.; Fujita, T.; Iwai, K.; Furusawa, M. & Kishimoto, N. (2011). Synthesis of 3-cyclohexylpropyl caffeate from 5-caffeoylquinic acid with consecutive enzymatic conversions in ionic liquid. *J. Mol. Catal. B: Enzyme.*, 69, 161-167

Lau, R.M.; van Rantwijk, F.; Seddon, K.R. & Sheldon, R.A. (2000). Lipase-catalyzed reactions in ionic liquids. *Org. Lett.*, 2, 26, 4189–4191

Luetz, S.; Giver, L. & Lalonde, J. (2008). Engineered enzymes for chemical production. *Biotechnol. Bioeng.*, 101, 4, 647-653

Michaluart, P.; Masferrer, J.L.; Carothers, A.M.; Subbaramaiah, K.; Zweifel, B.S.; Koboldt, C.; Mestre, J.R.; Grunberger, D.; Sacks, P.G.; Tanabe, T. & Dannenberg, A.J. (1999). Inhibitory effects of caffeic acid phenethyl ester on the activity and expression of cyclooxygenase-2 in human oral epithelial cells and in a rat model of inflammation. *Cancer. Res.*, 59, 10, 2347-2352

Moniruzzaman, M.; Kamiya, N. & Goto, M. (2010). Activation and stabilization of enzymes in ionic liquids. *Org. Biomol. Chem.*, 8, 13, 2887-2899

Nagaoka, T.; Banskota, A.H.; Tezuka, Y.; Midorikawa, K.; Matsushige, K. & Kadota, S. (2003). Caffeic acid phenethyl ester (cape) analogues: Potent nitric oxide inhibitors from the netherlands propolis. *Biol. Pharm. Bull.*, 26, 4, 487-491

Nara, S.J.; Harjani, J.R. & Salunkhe, M.M. (2002). Lipase-catalysed transesterification in ionic liquids and organic solvents: A comparative study. *Tetrahedron Lett.*, 43, 16, 2979-2982

Nicklaus, M.C.; Neamati, N.; Hong, H.; Mazumder, A.; Sunder, S.; Chen, J.; Milne, G.W. & Pommier, Y. (1997). Hiv-1 integrase pharmacophore: Discovery of inhibitors through three-dimensional database searching. *J. Med. Chem.*, 40, 6, 920-929

Oakes, R.S.; Clifford, A.A. & Rayner, C.M. (2001). The use of supercritical fluids in synthetic organic chemistry. *J. Chem. Soc. Perkin. Trans.* , 1, 917–941

Park, S. & Kazlauskas, R.J. (2001). Improved preparation and use of room-temperature ionic liquids in lipase-catalyzed enantio- and regioselective acylations. *J. Org. Chem.*, 66, 25, 8395-8401

Pleiss, J.; Fischer, M. & Schmid, R.D. (1998). Anatomy of lipase binding sites: The scissile fatty acid binding site. *Chem. Phys. Lipids.*, 93, 1-2, 67-80

Singh, M.; Singh, R.S. & Banerjee, U.C. (2009). Stereoselective synthesis of (*r*)-1-chloro-3(3,4-difluorophenoxy)-2-propanol using lipases from *pseudomonas aeruginosa* in ionic liquid-containing system. *J. Mol. Catal. B: Enzyme.*, 56, 4, 294-299

Sud'ina, G.F.; Mirzoeva, O.K.; Pushkareva, M.A.; Korshunova, G.A.; Sumbatyan, N.V. & Varfolomeev, S.D. (1993). Caffeic acid phenethyl ester as a lipoxygenase inhibitor with antioxidant properties. *FEBS. Lett.*, 329, 1-2, 21-24

Sureshkumar, M. & Lee, C.K. (2009). Biocatalytic reactions in hydrophobic ionic liquids. *J. Mol. Catal. B: Enzyme.*, 60, 1-2, 1-12

Tagashira, M. & Ohtake, Y. (1998). A new antioxidative 1,3-benzodioxole from *melissa officinalis*. *Planta. Med.*, 64, 6, 555-558

van Rantwijk, F. & Sheldon, R.A. (2007). Biocatalysis in ionic liquids. *Chem. Rev.*, 107, 6, 2757-2785

Wang, D.; Xiang, D.B.; He, Y.J.; Li, Z.P.; Wu, X.H.; Mou, J.H.; Xiao, H.L. & Zang, Q.H. (2005). Effect of caffeic acid phenethyl ester on proliferation and apoptosis of colorectal cancer cells *in vitro*. *World. J. Gastroenterol.*, 11, 26, 4008-4012

Yoshida, K.; Yamaguchi, K.; Osada, S.; Kawaguchi, Y.; Takahashi, T.; Sakashita, F. & Tanaka, Y. (2008). Challenge for a better combination with basic evidence. *Int. J. Clin. Oncol.*, 13, 3, 212–219

Ysrael, M.C. & Nonato, M.G. (1999). Biological activities of chemical constituents from *ipomoea muricata* [jacq] convolvulaceae. *Acta. Manilana. Ser. A.*, 47, 41-52

Part 2

Electronic Applications

Undeniable Contribution of Aprotic Room Temperature Ionic Liquids in the Security of Li-Ion Batteries

Claude-Montigny Bénédicte, Stefan Claudia Simona and Violleau David
Université François Rabelais, Laboratoire PCMB
France

1. Introduction

Nowadays, lithium-ion batteries are key components for portable electric devices for applications between 0°C and 50°C. Between the positive electrode and the negative one of the Li-ion battery, the transport of the Li^+ ions is insured by the electrolyte. Whatever the physical state of the electrolyte (liquid, polymer or gel), the direction of the transport stays the same: the Li^+ ion passes from the positive pole towards the negative one during the charging process, and conversely during the discharge. Up to now, electrolytes most commonly used in Li-ion batteries present in their formulation one or several aprotic and polar organic solvents, mixed with a lithium salt and some additives. Typical liquid electrolytes are based on a mixture of cyclic or linear alkylcarbonates. These solvents are perfectly adapted to the functioning of batteries Li-ion, but they are volatile and highly flammable (Aurbach et al. 2004; MacNeil et al., 1999, 2000; MacNeil & Dahn , 2001; Rasch et al. 1991; Richard & Dahn, 1999). If the temperature of the battery overtakes 60-70°C, the slightest warm-up or the internal short circuit can thus induce a thermal racing, leading to an inflammation (Balakrishnan el al., 2006; Zhang, 2006). It is well also known that over 50°C, cycle life and capacity of lithium-ion batteries based on usual alkylcarbonate mixtures may be drastically reduced. They are still not well adapted for other applications, such as security lighting, portable sensors, geothermal energy, drillings and explorations of potential hydrocarbon deposits, lithium batteries which have to operate at higher temperatures than 50°C. High energy density Li-ion batteries present usually a very short life over 60°C. Polymer electrolytes should be more convenient at high temperature either in the form of solid polymer electrolyte or polymer gel electrolyte. Nevertheless solid polymer Li-ion batteries have relatively low ionic conductivities as compared to gel polymer electrolytes (Kobayashi et al., 2007). Unfortunately gel electrolytes are unstable above 80°C (Kaneko et al., 2009; Shin & Cairns, 2008). Therefore, in order to enhance the conductivity of the electrolyte and simultaneously to get free of the flammability of organic solvents, ionic liquids (ILs) appear to be a good solution for use as electrolyte in large scale Li-ion batteries. ILs present a flame resistance, a low volatility (no OVCs), and a large electrochemical stability (Lewandowski & Galinski, 2004; Wasserscheid & Welton, 2008). In particular, room temperature ionic liquids (RTILs), which are liquid at ambient or sub-ambient temperature, look suitable for the replacement of organic solvents or their mixtures (Appetecchi et al.,

2008; Kobayashi et al., 2007, Lall-Ramnarine et al., 2008; Nokemann et al., 2008). N-alkyl-N-alkyl'pyrrolidinium bis(trifluoromethylsulfonyl)imide RTILs are chosen for this use. These RTIL compounds present a high thermal stability (up to 300°C), low flammability and low glass transition temperatures (Pan et al., 2010; Wasserscheid & Welton, 2008). All these characteristics make them good candidates as electrolytes for electrochemical devices as long as their conductivity remains high in the range of operating temperatures. The purpose of this chapter concerns the use of a series of N-alkyl-N-alkyl'pyrrolidinium bis(trifluoromethylsulfonyl)imide RTILs (P_{xy}-TFSI, presented in Table 1) as co-solvents in the formulation of a standard electrolyte (PC (propylene carbonate)/EC (ethylene carbonate)/DMC (dimethyl carbonate) (1:1:3 w/w) + $LiPF_6$ (lithium hexafluorophosphate) 1M + VC (vinylene carbonate) 1 %).

component	formula	abbreviation
N-methyl-N-propyl-pyrrolidinium	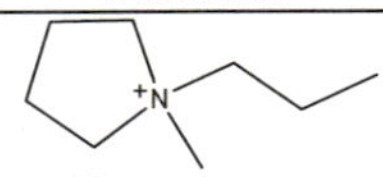	P_{13}^+
N-butyl-N-methyl-pyrrolidinium	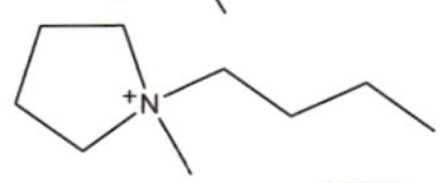	P_{14}^+
N-methyl-N-pentyl-pyrrolidinium	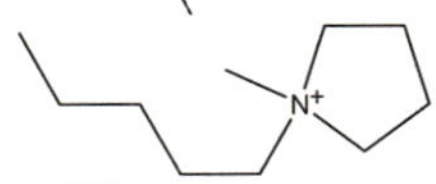	P_{15}^+
N-butyl-N-ethyl-pyrrolidinium	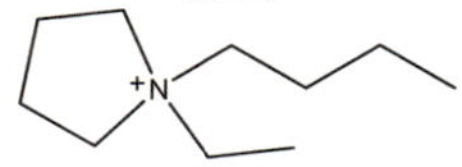	P_{24}^+
bis(trifluoromethanesulfonyl)imide	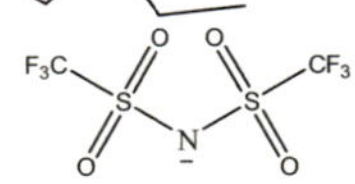	$TFSI^-$

Table 1. Names, formula and abbreviations of N-alkyl-N-alkyl'pyrrolidinium imide (P_{xy}-TFSI) under study.

Thanks to the exceptional fire break property of this class of RTILs (Wasserscheid & Welton, 2008), an improvement in the security of Li-ion batteries is expected, without changing the electrochemical performances of the well known $LiCoO_2$ cathode and a graphite negative electrode compared to the standard electrolyte. The first part of this chapter concerns the tests of thermal stability and the flammability of electrolytes based on N-alkyl-N-alkyl'pyrrolidinium bis(trifluoromethylsulfonyl)imide RTILs (from Solvionc, Toulouse, France), added to the standard electrolyte. A second part will be dedicated to the transport properties of electrolytes containing these RTILs as co-solvents. These properties are highly dependent on ionic conduction in these media. A third part will be devoted to the wettability of both electrodes and separators in the presence of these RTILs. Indeed, the wettability is another key factor for improving the cycling ability and the power of Li-ion batteries, especially when the temperature decreases. This chapter will end by an electrochemical study using these mixtures as electrolytes, leading to capacity determination.

2. Thermal stability of the mixed electrolytes: P_{xy}-TFSI / standard electrolyte

The study of the thermal stability of the mixed electrolytes (standard electrolyte/RTIL) is performed at sub ambient temperatures as a function of the alkyl chain length of the P_{xy}-

TFSI RTIL in the mixture. The effect of the RTIL content on the thermal stability is also studied. Finally this study is completed by the results of the flammability tests of the analyzed mixtures, according to the added RTIL content.

2.1 Thermal analysis at sub-ambient temperatures: effect of the P_{xy}-TFSI nature and of its content

Samples containing mixtures of standard electrolyte with P_{xy}-TFSI RTIL, which content is 20% (w/w), are analyzed by differential scanning calorimetry, DSC, between 20°C and 400°C at the fixed scanning rate of 5°C/min for both the heating and the cooling steps. The resulting thermograms are displayed in the Figure 1 and Table 2. The phenomenon of vaporization is well identified on each of them by the existence of an endothermic corresponding peak during the heating step at 139°C for 20% P_{13}-TFSI, 174°C for 20% P_{14}-TFSI and at 203°C for 20% P_{24}-TFSI. During the cooling step, no thermal phenomenon is observed, that means that vapours are not able to be condensed (capsules containing the analyzed products opened, letting escape the vapours of solvents). The results, obtained by DSC under the same analysing conditions, for samples containing 30% of RTIL in the standard electrolyte, are reported in Table 2. The DSC thermograms bring to light peaks of vaporisation at 140°C for 30% P_{13}-TFSI, at 195°C for 30% P_{14}-TFSI, at 224°C for 30% P_{15}-TFSI. These values are thus globally a little higher than those of the mixtures containing 20% of P_{xy}-TFSI. For both contents, the vaporisation temperatures are higher than the DMC boiling point reported in Table 3. Except for the P_{13}-TFSI, all the mixtures (RTIL/standard electrolyte) present higher vaporisation temperature than the EC/PC/3DMC mixture and the pure VC solvent.

	P_{13}-TFSI		P_{14}-TFSI		P_{15}-TFSI	
	20%	30%	20%	30%	20%	30%
$T_{vap.}$ / °C	139	140	174	195	203	224

Table 2. Temperatures of vaporization of various mixtures Pxy-TFSI/standard electrolyte

	PC	EC	DMC	VC	EC/PC/3DMC
Boiling point / °C	242	248	90	162	160

Table 3. DSC boiling points of the solvents composing the standard electrolyte.

These values show that the most vaporisable solvents are DMC and VC, even if they are already protected by EC and/or PC, they are better prevented from vaporisation in the presence of P_{xy}-TFSI whatever its content, except in the case of the P_{13}-TFSI.

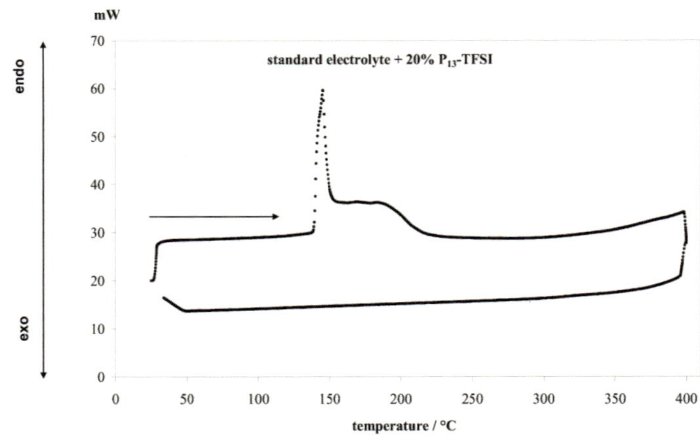

(a)

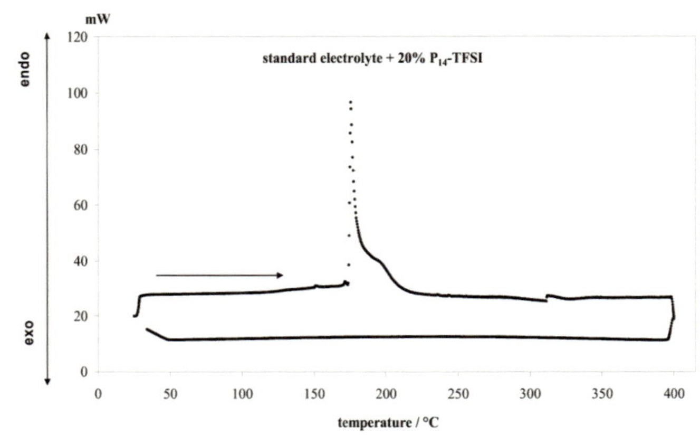

(b)

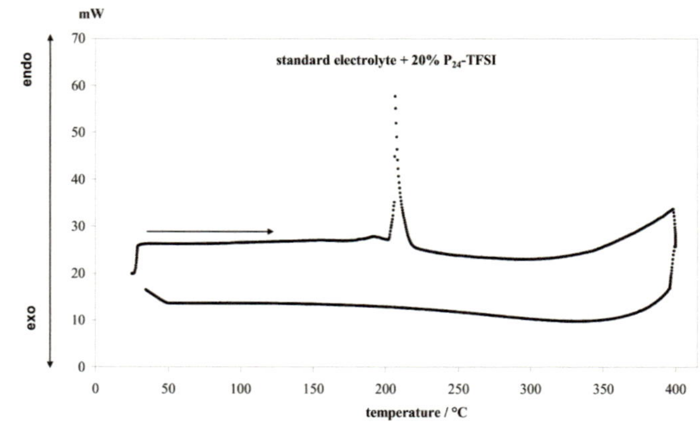

(c)

Fig. 1. DSC thermograms of mixtures of standard electrolyte with 20% (w/w) of P_{xy}-TFSI at 5°C/min.

These results clearly show that most of the studies P_{xy}-TFSI, added to the standard electrolyte as co-solvent, can contribute to the improvement of the electrolyte security.

2.2 Tests of flammability

In the event of a thermal incident affecting a Li-ion battery, the first risk is the ignition of the battery. Given the consequences which it can engender, it is essential to evaluate the behaviour of electrolytes in the contact with the flame. For this purpose, in order to investigate the flammability behaviour of P_{xy}-TFSI RTILs, tests are performed with strips of Manila paper soaked with the standard electrolyte containing or not various P_{xy}-TFSI contents from 3.1% to 90% (w/w). All the experimental flammability tests are photographed (or recorded) with a digital camera (Stefan et al., 2009). Some significant photos concerning the resulting tests are presented on Figures 2, 3 and 4. As shown on Figure 2, for the standard electrolyte, the inflammation of the soaked strip of paper is quickstriding (Figure 2a), leading to suffocating white smokes (Figure 2b). Strips soaked by the studied mixtures when the RTIL content is from 3.1% to 10 % present quite the same behaviour in the contact with the flame of the lighter (Figure 3a). Nevertheless, no white smoke is appearing, the electrolyte is gently burning (Figure 3b).

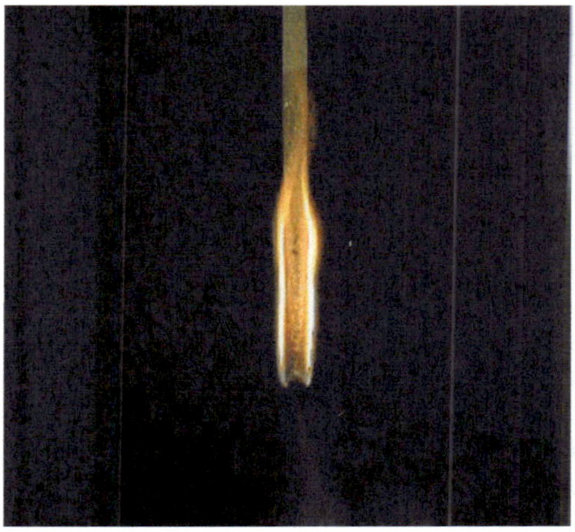

(a)

(b)

Fig. 2. Test of flammability of the standard electrolyte: (a) lighter flame in contact with the soaked paper, (b) inflammation of the strip.

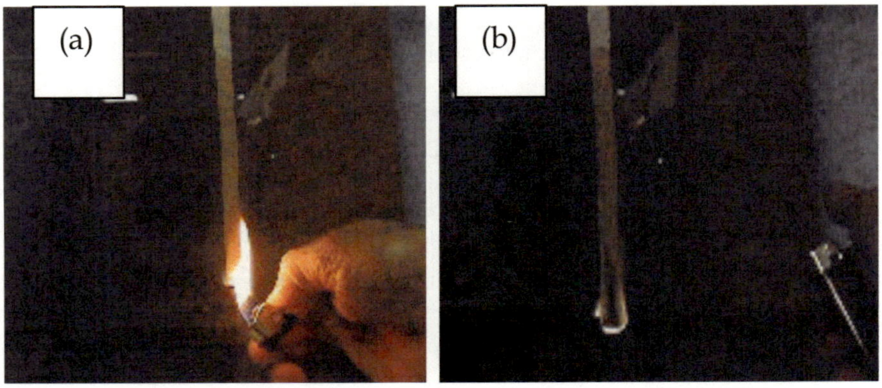

Fig. 3. Test of flammability of mixtures containing 3.1% to 10% of P_{xy}-TFSI: (a): lighter flame in contact with the soaked paper, (b): gentle burn.

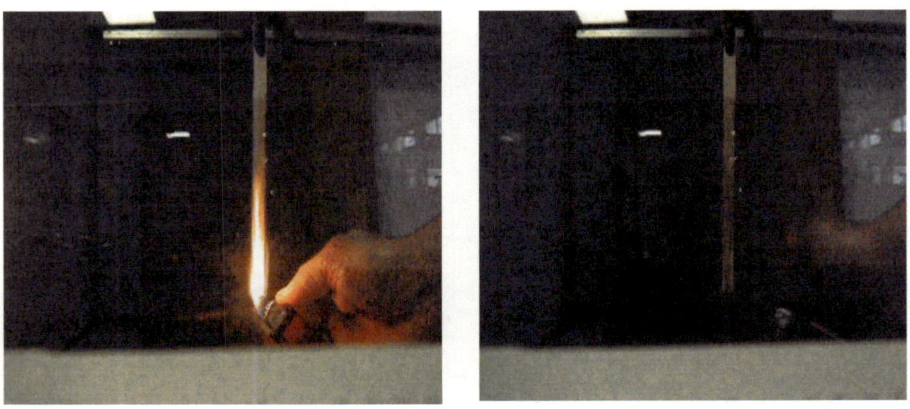

(a) (b)

Fig. 4. Test of flammability of mixtures containing 20% to 50% of P_{xy}-TFSI: (a): lighter flame in contact with the soaked paper, (b): flame self-extinguished.

In the case of a RTIL content in the range 20% to 50%, the soaked strips are very difficult to inflame (Figure 4a), then after a very brief inflammation, the tiny flame goes out. For mixtures of P_{xy}-TFSI content over 50 %, the soaked strips of paper present a remarkable resistance to the flame, like the pure P_{xy}-TFSI RTILs do, related to their thermal stability at high temperatures (Stefan et al., 2010). These tests show that under the influence of the heat of the flame of the gas lighter, the most volatile solvent (DMC) is partially vaporized and ignites briefly when the P_{xy}-TFSI RTIL content is over 10% in the mixture.

2.3 Conclusion on the thermal behaviour of mixtures containing Pxy-TFSI added as co-solvents in the standard electrolyte

Because of the thermal stability and the very low volatility of the pure P_{xy}-TFSI RTILs, when added as co-solvent to highly flammable organic solvents, such as DMC, PC, EC and VC, the obtained mixtures are more protected from a thermal incident than pure organic molecular solvents. This leads to a very good behaviour of these mixtures in the contact with the flame of a gas lighter. In particular, the flame extinguishes itself as soon as the content in P_{xy}-TFSI exceeds 20%. Thus P_{13}-TFSI, P_{14}-TFSI, P_{15}-TFSI or P_{25}-TFSI can be flame-retardant for the standard electrolyte, especially for contents over 20%. These very encouraging results for thermal stability purpose have to be confirmed with a study of transport properties of the mixtures under consideration, in order to be incorporated in a Li-ion battery device.

3. The study of the transport properties of electrolytes containing P_{xy}-TFSI added as co-solvents

The conductivity of electrolytes is a fundamental criterion for the electrochemical applications. This conductivity of ionic type is usually strongly correlated to the viscosity of the electrolyte. The two parameters are thus presented afterward in this chapter.

3.1 Study of the ionic conductivity of mixtures

The study of the conductivity of the mixed electrolytes based on standard electrolyte/P_{xy}-TFSI RTIL is performed as a function of the salt content: either the P_{xy}-TFSI RTIL or $LiPF_6$. The influence of the P_{xy}-TFSI nature on the ionic conductivity is also under consideration. $LiPF_6$, not directly soluble in P_{xy}-TFSI at ambient temperature, the necessary complementary quantity to keep the global content of 1M for the lithium salt, is added at first to the standard electrolyte, and then mixed with the tested P_{xy}-TFSI.

3.1.1 Influence of the P_{14}-TFSI co-solvent content (w/w) on the ionic conductivity

As all the P_{xy}-TFSI RTILs under study can be good flame retardant, P_{14}-TFSI is chosen to investigate the influence of its content (w/w) on the electrolyte ionic conductivity at 20°C only for availability purposes. P_{14}-TFSI content is varying from 0% to 100%. The corresponding conductivity values are presented on the Figure 5.

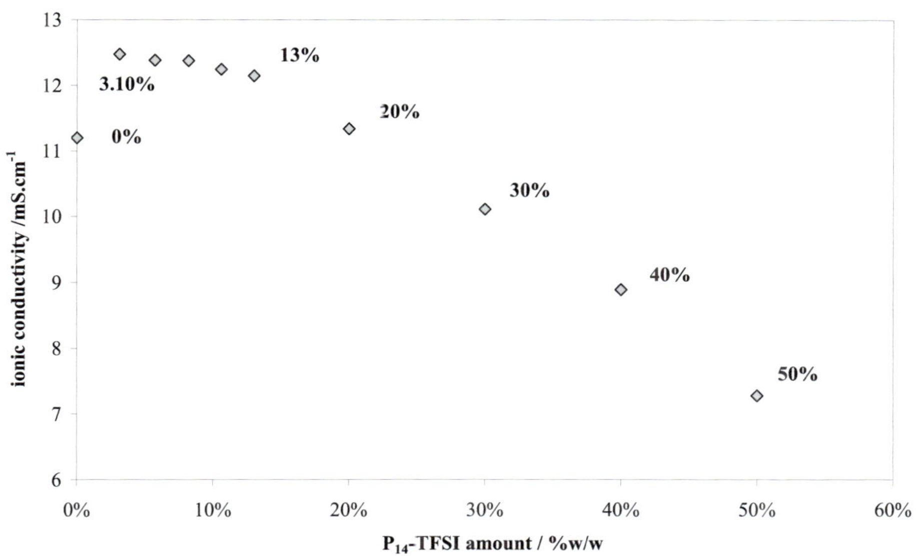

Fig. 5. Ionic conductivity at 20°C of standard electrolyte mixtures with P_{14}-TFSI RTIL of various contents.

On this graph, the conductivity in added co-solvent 0% corresponds to that of the standard electrolyte in 20°C, whereas that at 100% represents the conductivity of pure P_{14}-TFSI at 20°C. As seen on Figure 5, for a content of the co-solvent lower than approximately 20 %, the conductivity of mixtures exceeds that of the standard electrolyte. This range is characterized by the presence of the ionic conductivity maximum, obtained for the lowest content of P_{14}-TFSI, i.e. 3.1%. Then for further addition of P_{14}-TFSI up to 13%, the ionic conductivity decreases slightly leading to a loss of 0.5 $mS.cm^{-1}$. Between 13% and 20% of P_{14}-TFSI, the conductivity loss increases, it is around 1 $mS.cm^{-1}$, but the mixture ionic conductivity is still remaining over the standard electrolyte one up to around 30% of P_{14}-TFSI. Over 30%, the decreasing ionic conductivity of mixtures becomes lower than the standard electrolyte one, probably due to a growth of the electrolyte viscosity.

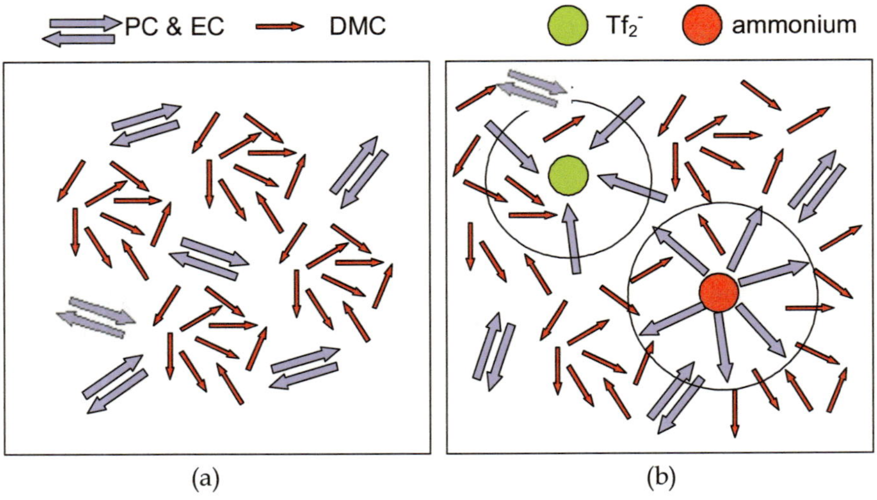

Fig. 6. Schematic arrangement of the solvent molecules, PC, EC, DMC respectively, with regard to the P_{xy}^+ and $TFSI^-$ (i.e. Tf_2^-) ions: (a) EC/PC/3DMC; (b) EC/PC/3DMC in the presence of P_{xy}-TFSI.

To better understand these phenomena, a schematic arrangement related to the structure is presented on the Figure 6. In this schematic layout, blue arrows symbolize the molecules of cyclic alkyl-carbonates, such as PC and EC, whereas the red arrows those of dimethylcarbonate. It is noticeable that PC and EC molecules are turning anti-parallel (Papoular et al., 2005), to favour a more important compactness. Within the framework of the carbonyl group present in the structure of these carbonate-based solvents, the carbon presents sp^2 hybrid orbital. Besides, the oxygen atom of higher electronegativity than the carbon is going to attract more the π electrons, the C=O bond is thus polarized according to:

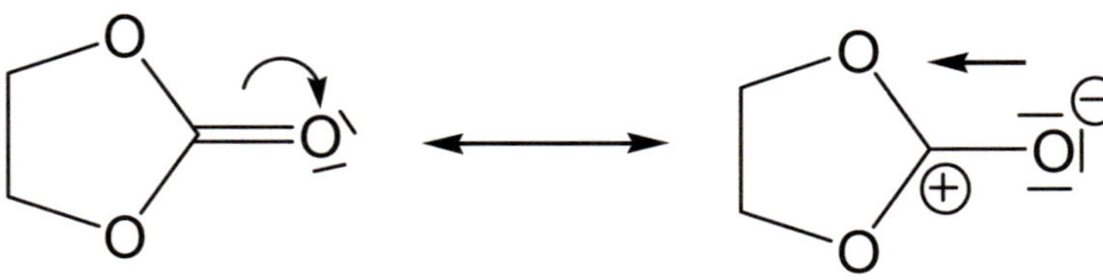

Consequently, the polarization of the carbonyl group is going to infer an electrophilic center on the carbon atom and a nucleophilic center on the oxygen atom. The direction of the blue arrows corresponds to the direction of dipole moments within the PC and EC molecules.
This structural property allows several phenomena to be explained, such as the decrease of the mixture density, according to the co-solvent content. Effectively, in the centrosymmetric electric field of the ions, the PC and EC molecules turn, taking up more space in the solution. Therefore, the molar volume of the free solvent is weaker than in the solvation layer, the density is thus decreasing with the addition of the RTIL in the electrolyte. Another consequence due to this phenomenon concerns the strong reduction of the DMC vapour pressure, which involves the decrease of the electrolyte flammability in the presence of P_{xy}-TFSI RTIL. The greater the content of the RTIL, the greater the mixture is flame resistant. Around of the density minimum, they are no more free molecules of solvent, because the ions are entirely solvated. Beyond this minimum, the solvation of the ions is incomplete, and then these charged species can interact by more intense Coulombic strengths, leading to an increase of the viscosity and a decrease of the ionic conductivity because of a drop in the ion mobility.

3.1.2 Influence of the temperature and of the P_{xy}-TFSI co-solvent type on the ionic conductivity

The temperature dependence of the ionic conductivity of mixtures is studied from 10°C to 65°C for various P_{xy}-TFSI RTILs, whose content is firstly fixed at 20% (w/w). The results presented on the Figure 7 show that the conductivity of these mixtures increases with the temperature according to the established order: 20%P_{15}-TFSI$<$standard electrolyte$<$20%P_{24}-TFSI$<$20% P_{14}-TFSI$<$20% P_{13}-TFSI. This order respects exactly that of the rise of the ionic conductivity obtained for the pure P_{xy}-TFSI RTILs (Mac Farlane et al., 1999; Wu et al. 2003a, 2003b). The experimental results for the ionic conductivity are well fitted by the VTF model as shown on Figure 7. The parameters used here are the temperatures of ideal glassy transition, T_{σ}°, already reported for the conductivity of the pure P_{xy}-TFSI RTILs (Mac Farlane et al., 1999; Wu et al. 2003a, 2003b).

3.1.3 Influence of the lithium salt concentration on the ionic conductivity of the mixtures containing the standard electrolyte and 30 % of P_{14}-TFSI

The influence of the temperature on the mixture ionic conductivity is studied according to the lithium salt concentration from 0.25 M and 1.25 M. For purpose of comparison with the standard electrolyte conductivity (cf. section 3.1.1), the P_{14}-TFSI content is fixed at 30%. The experimental results are depicted in the Figure 8. As expected, from the VTF model, the conductivity increases with the temperature. Furthermore at constant temperature, as the concentration of $LiPF_6$ increases, the conductivity decreases. This effect, often observed in the presence of lithium salts, is essentially attributed to an increase in the viscosity of the media (Taggougui, 2007). The standard electrolyte (Figure 7) presents a lower dependency towards the temperature than the studied mixtures. Over 30°C, mixtures present a higher ionic conductivity than the standard electrolyte, whatever the concentration in added $LiPF_6$ is. On the other hand between 10°C and 30°C, only mixtures of lithium concentration equal or lower than 0.75 M are more conducting than the standard electrolyte. This sensitivity to temperature variations can be quantified by the concept of fragility (Martinez &Angell, 2001; Wu et al. 2003a, 2003b).

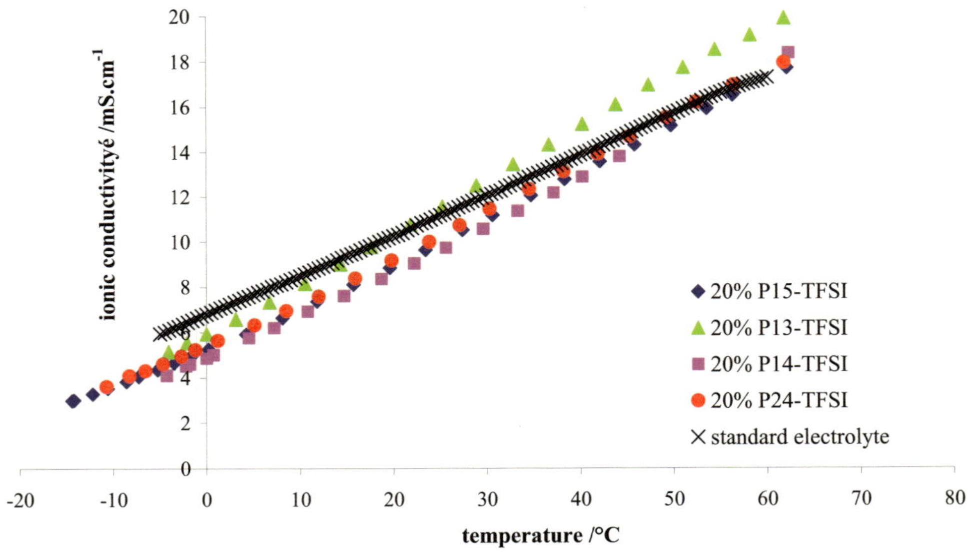

(a)

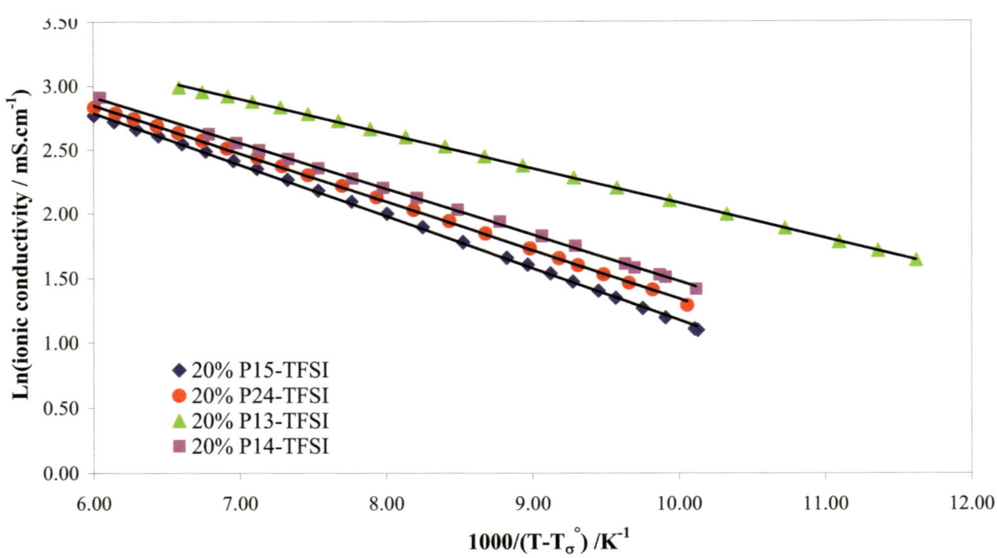

(b)

Fig. 7. Temperature dependence of the ionic conductivity of 20% P_{xy}-TFSI mixtures and of the standard electrolyte; (a) experimental results; (b) fits from the VTF model

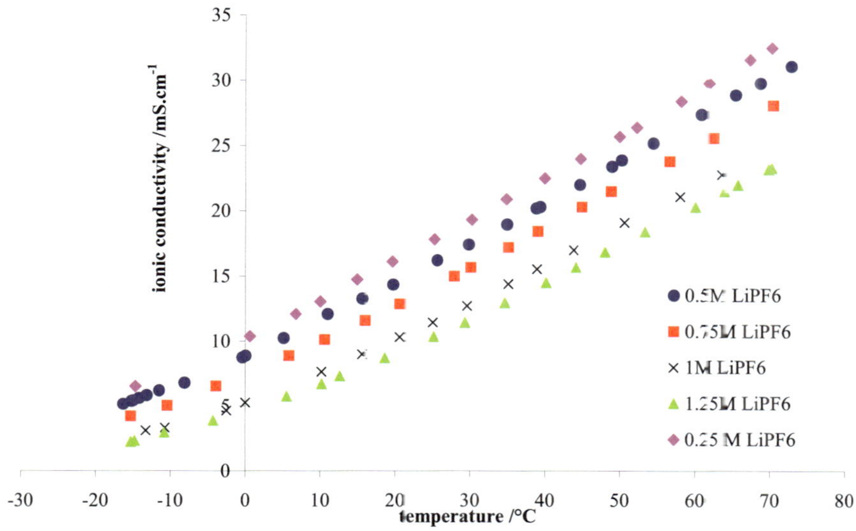

Fig. 8. Temperature dependence of the ionic conductivity of 30% P_{14}-TFSI mixtures as a function of $LiPF_6$ concentration.

3.2 Study of the dynamic viscosity of mixtures

The high viscosity of the P_{xy}-TFSI RTILS constitutes a real disadvantage for their applications in batteries. The only means to reduce the viscosity are an increase of the temperature or the addition of less viscous solvents. The study of the viscosity of the mixed electrolytes based on the mixture standard electrolyte/P_{xy}-TFSI RTIL is performed as a function of the salt content: either the P_{xy}-TFSI RTIL or $LiPF_6$. The influence of the P_{xy}-TFSI nature on the viscosity is also under consideration. The Figure 9 summarizes the evolution with the temperature of the viscosity according to the type of P_{xy}-TFSI added at 20%. Even if measures are performed at 2°C/min between 10°C and 90°C (insert in Figure 9), only the results from 10°C to 40°C are significant because of the vaporisation of one of the co-solvents (the DMC essentially, because of its high vapour pressure). Beyond 50°C, the viscosity of mixtures increases in a surprising way. As the used device (TA Instruments) is not equipped for airtight measurements, allowing it to work under an atmosphere saturated in vapour of the most volatile solvents, it is very likely that the observed variations of viscosity are connected to the modifications of the composition of the studied mixtures. Indeed the DMC solvent has already at ambient temperature a high vapour pressure (18 mmHg in 21.1°C (Weast, 1976). As a result after 40°C the electrolyte becomes impoverished in DMC, the solvent of lower viscosity, so that the dynamic viscosity of the mixtures increases beyond 50°C. Globally, electrolytes containing more RTILs are more viscous also. To bring to light the effect of the concentration of the lithium salt, $LiPF_6$, on the dynamic viscosity of mixtures containing 30% P_{15}-TFSI increasing amounts of $LiPF_6$ are added to mixtures from 0.25 M to 1.25 M. The corresponding results are reported in Figure 10 where they are compared to the dynamic viscosity of the standard electrolyte. For a 20% content, the viscosity decreases according to the order: P_{13}-TFSI> P_{24}-TFSI>standard electrolyte> P_{15}-TFSI> P_{14}-TFSI. As expected for the same amount of $LiPF_6$, the dynamic viscosity increases with the weight content of P_{xy}-TFSI.

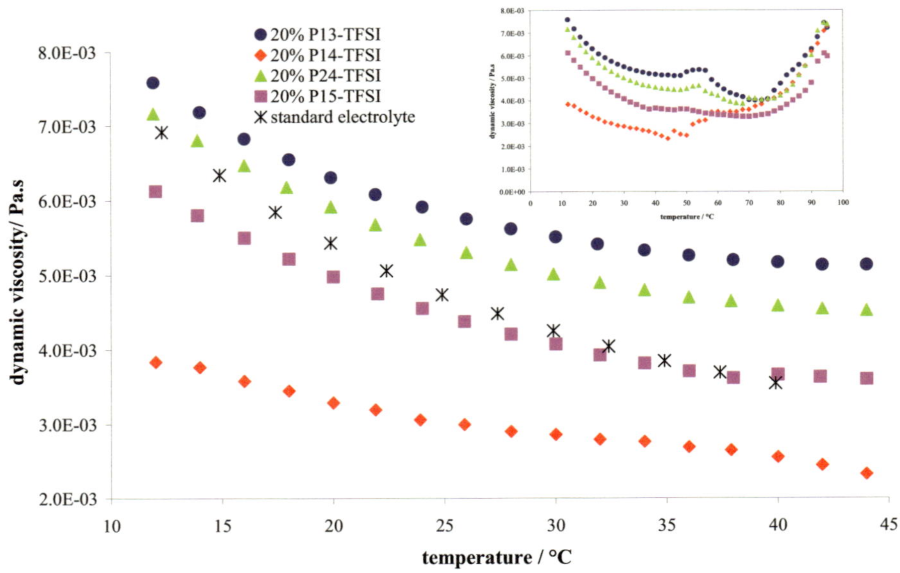

Fig. 9. Temperature dependence of the dynamic viscosity of 20% P_{xy}-TFSI mixtures, at 2°C/min (insert: full temperature range).

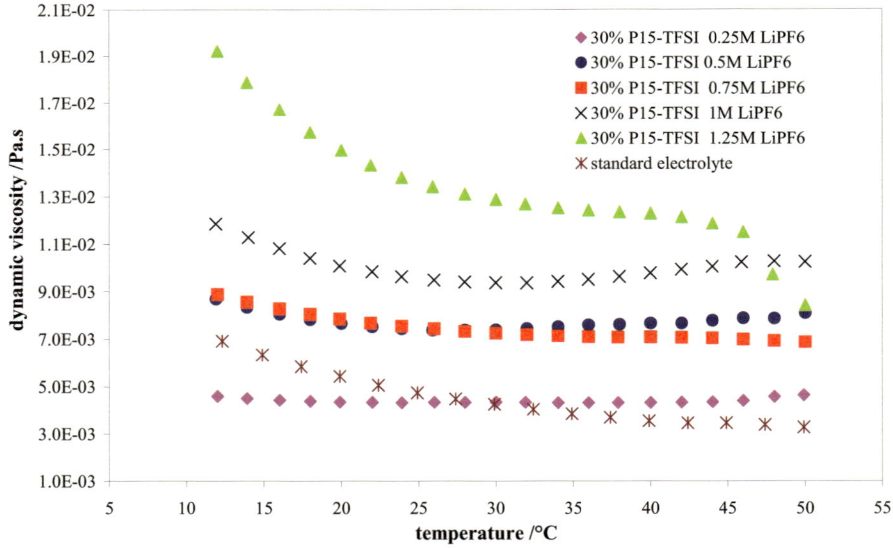

Fig. 10. Temperature dependence of the dynamic viscosity of 30% P_{15}-TFSI mixtures as a function of $LiPF_6$ concentration.

Globally these results confirm the fact that the viscosity increases with the lithium salt concentration. The addition of lithium salt in the electrolyte leads to a progressive increase of the mixture viscosity, which results in a decrease of the conductivity as mentioned previously (cf. section 3.1.3).

3.3 Conclusion on the transport properties of mixtures containing Pxy-TFSI added as co-solvents in the standard electrolyte

The weak ionic conductivity of the pure P_{xy}-TFSI RTILs can be improved if these RTILs are mixed with molecular solvents (or mixtures of molecular solvents). Mixtures containing up to 30% P_{xy}-TFSI present conductivity superior or equal to that of the standard electrolyte. The viscosity of the standard electrolyte does not vary in a significant way when P_{xy}-TFSI RTIL is added as co-solvent, for a content limited to 20% or 30% (w/w). All these results show that the moderate addition of P_{xy}-TFSI RTILs does not decrease the performances obtained in ionic conduction for the industrial standard electrolyte EC/PC/3DMC + $LiPF_6$ 1M + 1% VC. So, it is definite that the ionic transport in the studied mixtures does not constitute an obstacle to the use of co-solvent based on P_{xy}-TFSI RTILs for Li-ion batteries applications. But before their use in electrochemical devices, their compatibility with electrode and separator materials must be checked specifically for wettability purposes.

In order to better understand the effect of small cations, such as Li^+, and of molecular solvent addition on the conductivity and the dynamic viscosity of P_{xy}-TFSI, PGSE-NMR measurements are performed. The self-diffusion coefficients of lithium ^7Li (D(Li)) and hydrogen ^1H (D(H)) nuclei, are determined by PGSE-NMR. D(Li) is related to the diffusion of the Li^+ cation and D(H) is related to the diffusion of pyrrolidinium cation of the considered RTIL. P_{15}-TFSI is chosen as model IL for these experiments. In Figure 11(a), the variations of the self diffusion coefficient for ^1H and ^7Li are reported as a function of the concentration of LiTFSI in P_{15}-TFSI. Both self diffusion coefficient decrease linearly when the concentration in LiTFSI is increased. This is expected from the Stockes Einstein relation if the dynamic viscosity follows itself a linear dependence on the salt concentration as predicted by the semi-empirical relation (Jones & Dole, 1929; Kaminski, 1957): $\eta_r = \eta/\eta° = 1 + A\,C^{1/2} + BC$. If the term in $C^{1/2}$ is negligible, $\eta°$ is the viscosity of the pure RTIL, C the concentration in lithium salt, A can be calculated (Falkenhagen & Vernon, 1932) and B can be considered as an adjustable parameter linked to the molar volume of the added salt. It is noticeable that the RTIL self diffusion coefficient follow the same trend and hence verifies also the Stokes-Einstein relation. Nevertheless the slopes of the two lines are different and this means that the increase in viscosity has a different impact on the two diffusion coefficients, the Li salt being less affected than the RTIL. In order to test the influence of the anion of the Li salt, $LiPF_6$ is added to P_{15}-TFSI instead of LiTFSI. Results are presented as graphs in Figure 11(b). The self diffusion coefficient of the RTIL follows a linear variation with a slightly smaller slope but this is not the case of the Li^+ self diffusion coefficient which is quite constant between 0.25 M and 0.50 M, and then decreases.

As illustrated in Figure 12, the variation of the D(Li)/D(H) ratios is quite different for LiTFSI and $LiPF_6$ in the presence of P_{15}-TFSI: it increases with $LiPF_6$ and decreases with LiTFSI. This implies that Li^+ environment is changing when PF_6^- is added to the RTIL. D(Li) is thus raising more than D(H), which reflects the self-diffusion coefficient of P_{15}^+. According to the HSAB theory, $TFSI^-$ is a softer base than PF_6^-, because its negative charge is delocalized all

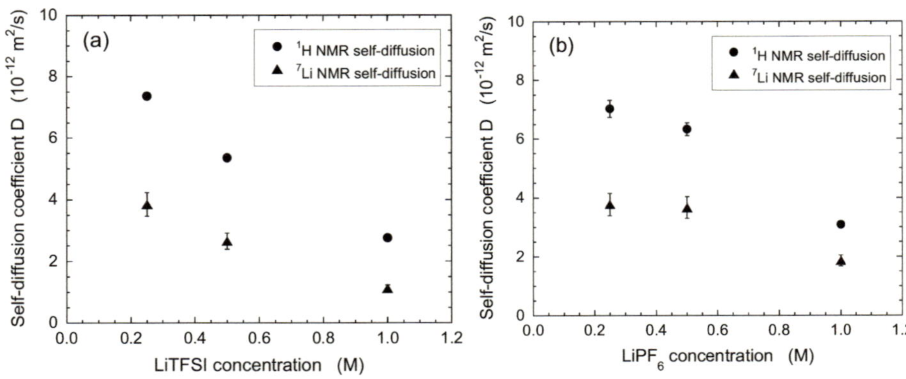

Fig. 11. Evolution of the self-diffusion coefficients at 25°C for the solvent molecules (^1H NMR) and the lithium cations (^7Li NMR) in the LiTFSI/P_{15}-TFSI mixture samples as a function of the LiTFSI concentration (a), and in the $LiPF_6$/P_{15}-TFSI mixture samples as a function of the $LiPF_6$ concentration (b).

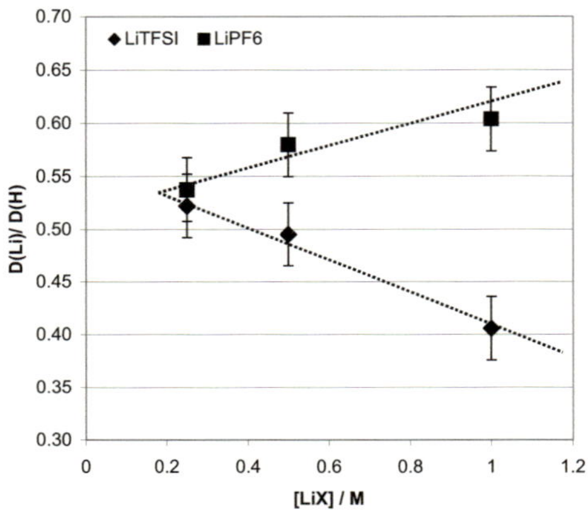

Fig. 12. Evolution of the ratio of lithium and hydrogen self-diffusion coefficients as a function of either the nature or the amount of the added salt to P_{15}-TFSI at 25°C.

over the anion structure, it is thus expected that preferential interactions occur between the hard acid (Li^+) and the hard base (PF_6^-) than the softer TFSI$^-$ base. Thus when $LiPF_6$ is added, the probability to get a tight Li^+-PF_6^- ion pair becomes more important, and especially over C=0.50 M in lithium salt. Because of its overall zero charge, the ion pair does not interact by Coulombic forces with the RTIL ions and this leads to an increase in the D(Li) to D(H) ratio. At the opposite, when LiTFSI is added, the Li^+ ion is intercalating within the free spaces of the RTIL ionic structure leading to the re-enforcement of the columbic interactions and to a decrease in D(Li)/D(H). In order to check the dynamic viscosity effect on D(Li) and D(H), P_{15}-TFSI is replaced by an alkylcarbonate solvent ternary mixture: EC/PC/3DMC (by weight). The corresponding results, reported in Table 4, show that, when the RTIL is replaced by the fluid alkyl carbonate mixture, the self-diffusion coefficients of both lithium and hydrogen atoms increase within a ratio of 100. Molecular solvents are preferentially solvating Li^+ and the Coulombic interaction with the RTIL ions are reduced leading to an enhanced mobility.

	P_{15}-TFSI + 1 M $LiPF_6$	EC/PC/3DMC + 1 M $LiPF_6$
from ^{7}Li NMR	1.9 10^{-12}	1.9 10^{-10}
from ^{1}H NMR	3.1 10^{-12}	4.9 10^{-10}

Table 4. Self-diffusion coefficients, in $m^2.s^{-1}$, for molecular solvents and P_{15}^+ cation (from ^{1}H NMR), as well as for the lithium cation (from ^{7}Li NMR), in the presence of 1 M $LiPF_6$.

4. Wettability of separators and electrodes by mixtures

The role of the separator is to electrically isolate the anode from the cathode while allowing the charged species to go through. The separator must be perfectly wet by the electrolyte to avoid any slowing down of the ionic transport. Characterization and prediction of wetting phenomenon by contact angle (CA) measurements, according to the sessile drop method using a G-11 goniometer (Krüss, Germany), and surface free energy (γ_L) calculations, from measurement with a tensiometer K10ST (Krüss, Germany) with the ring method, are powerful analysis tools widely used for many applications (Good, 1992). Thus, to comprehend the wettability of separators and electrodes commonly used in Li-Ion battery devices, it is also necessary to study the surface free energy of the P_{xy}-TFSI RTILs and to characterise the smooth polymeric materials as described elsewhere (Stefan et al., 2009).

4.1 Free surface energy of the mixtures

The values of free surface energy obtained at 25°C for mixtures based on 20 %. P_{xy}-TFSI RTIL and for the standard electrolyte are reported in Table 5.

γ_L / $mN.m^{-1}$	P_{13}-TFSI	P_{14}-TFSI	P_{15}-TFSI	P_{24}-TFSI	std.electrolyte.
pure P_{xy}-TFSI*	36.5± 0.1	34.5± 0.1	34.6± 0.1	35.2± 0.1	36.9 ± 0.1
20% P_{xy}-TFSI	37.7 ± 0.1	37.7 ± 0.2	28.4 ± 0.1	38.0 ± 0.1	(-)

Table 5. Surface free energy of the pure or mixed P_{xy} RTILs and of standard electrolyte (in $mJ.m^{-2}$) at 25°C; * from (Stefan et al., 2009).

Except the mixture containing the P_{15}-TFSI, the free surface energy of mixtures is close to that of the standard electrolyte and those of the pure P_{xy}-TFSI RTILs. This indicates that the surface layer probably consists mainly of the ionic liquid. The mixture based on 20% P_{15}-TSFI presents a free surface energy sharply lower than the others in connection with the longest substituted carbon chain, it should thus better wet the separator

4.2 Contact angle of mixtures on solid supports

The tests are done on smooth polymeric materials, i.e. polyethylene (PE) and polypropylene (PP) at -4°C. This choice of temperature is due to the fact that problems of non wettability generally occur at low temperatures. In Table 6 are reported the results contact angles (in degree) at the liquid/polymer/air interface, in the case of the standard electrolyte and the mixtures of various P_{14}-TFSI and P_{24}-TFSI contents. The CA values of liquids on PE are from

4 ° to 11 ° lower than those obtained on PP. PE is thus better wet than PP because of a higher free surface energy (for EP: 33 mJ.m-2, for PP: 30 mJ.m-2 (Cognard, 2000).

CA/	Std.	8%		21%		30%	
° (±1)	electrolyte	P_{14}-TFSI	P_{24}-TFSI	P_{14}-TFSI	P_{24}-TFSI	P_{14}-TFSI	P_{24}-TFSI
PE	38.3	41.1	41.5	42.3	48.3	42.7	47.6
PP	49.3	50.1	50.6	51.4	52.1	49.6	53.6

Table 6. Contact angles (in degree) at the liquid/polymer/air interface, for the standard electrolyte and the mixtures of various P_{14}-TFSI and P_{24}-TFSI contents, at -4°C.

The standard electrolyte presents the lower CA, whether it is on EP or on PP. But these differences between CA remain less 10°, the standard electrolyte wetting better the polyolefins than the mixtures containing P_{xy}-TFSI RTILs. The amount of P_{xy}-TFSI RTIL in the mixture does not affect the CA values; however the mixtures based on the P_{24}-TFSI RTIL present the highest values of CA, whether it is on EP or PP. In all cases, the results of Table 6 show that the CA on smooth polymeric materials are large enough up (>20-30°), that wetting problems could be engendered when the temperature is low on a porous support as the separators.

4.3 Wettability of Celgard® separators at 25°C

To estimate at the wettability of separators commonly used in Li-ion batteries, the contact angles are measured at 25°C on porous supports (i.e. Celgard® separator whose characteristics are reported in (Stefan et al., 2009) at the liquid/polymer/air interface, where the liquid is either the standard electrolyte, the ternary EC/PC/3DMC solvent or the pure P_{xy}-TFSI RTILs. The analysis of the results presented in Table 7 shows that the Celgard®2730 separator is the best wet, which seems evident because PE is the smooth polymeric material the best wet (Table 6). The addition of a salt (i.e. $LiPF_6$ or P_{xy}-TFSI) increases the CA values. The best wetting liquids are the ternary mixture EC/PC/3DMC and the P_{15}-TFSI RTIL, which already presents the relatively low free surface energy (Table 5). So, Celgard®2730/P_{15}-TFSI seems to be the best set for wettability purposes. The other P_{xy}-TFSI/Celgard® sets can not be ignored because the corresponding CA values are not too far from those obtained with the standard electrolyte. To study the influence of P_{xy}-TFSI RTIL content on the separator wettability, two mixtures based on of P_{14}-TFSI are chosen (i.e. 15% and 21%). The experimental results, obtained at 25°C, are gathered on Table 8.

CA/ °	Celgard®2730 PE 43% pososity	Celgard® 2320 PP-PE-PP 41% porosity	Celgard®2500 PP 55% porosity	Celgard® 2400 PP 37% porosity
Std. electrolyte	43.6 ± 1.0	50.6 ± 2.0	53.5 ± 0.9	52.6 ± 0.9
EC/PC/3DMC	42.0 ± 1.1	46.0 ± 1.5	45.6 ± 0.9	47.5 ± 0.8
P_{13}-TFSI	53.1 ± 0.7	60.9 ± 1.1	67.3 ± 0.8	60.9 ± 1.0
P_{14}-TFSI	46.9 ± 1.7	60.5 ± 0.7	58.5 ± 2.3	59.0 ± 1.3
P_{15}-TFSI	38.4 ± 1.4	49.0 ± 0.4	50.5 ± 1.4	52.6 ± 0.6
P_{24}-TFSI	46.4 ± 1.1	52.3 ± 0.7	53.9 ± 0.6	56.9 ± 1.0

Table 7. Contact angles (in degree) at the liquid/ Celgard® separators/air interface, for the standard electrolyte, the ternary molecular solvent mixture and the pure P_{xy}-TFSI RTILs, at 25°C.

Results from Table 8 show that the CA decreases when the amount of P_{14}-TFSI RTIL increases from 15% to 21%. In the case of the Celgard®2320 and Celgard®2400, CA for 21% of P_{14}-TFSI is lower than that measured for 100% of P_{14}-TFSI, but it is the opposite for the Celgard®2500 and Celgard®2730. All the presented results are not too far from those obtained with the standard electrolyte. Nevertheless, the weakest values of CA are obtained with the Celgard®2730, PE based, and with the Celgard®2320, a trilayered separator. Furthermore, the effect of the porosity (i.e. Celgard®2500 and Celgard®2400, both PP based) is not significant enough to be further considered.

CA / ° (±1)	Celgard®2320	Celgard®2400	Celgard®2500	Celgard®2730
15% P_{14}-TFSI	53.3	62.4	63.3	52.9
21% P_{14}-TFSI	49.3	57.6	59.0	48.3
100% P_{14}-TFSI	60.5	59.0	58.5	46.9

Table 8. Contact angles (in degree) at the liquid/ Celgard® separators/air interface dependence on the P_{14}-TFSI RTIL content, at 25°C.

4.4 Wettability of the membrane Separion® at 25°C

The wettability of the Celgard® separators based on PE or PP being insufficient for a practical use, especially at low temperatures in the presence of mixtures containing P_{xy}-TFSI, it looks interesting to investigate in a new generation of separator i.e. the Separion® separator made of porous alumina/silica ceramic coated by a polyethylenterephtalat (PET) polymer. For all the studied liquids (pure or mixed Pxy-TFSI RTILs, the ternary solvent mixture EC/PC/3DMC and the standard electrolyte), the contact angles are very difficult to measure on the Separion® separator as already mentioned (Stefan et al., 2009). All the initial CA values are systematically weaker than those observed for separators Celgard®. Thus, in the case of the ternary solvent EC/PC/3DMC and the standard electrolyte, the wetting is immediate leading to an initial contact angle value of 0°. In the case of the 20% and 30% P_{xy}-TFSI RTIL mixtures, the initial contact angle values are always lower than 10°.

4.5 Wettability of the positive and negative electrodes for Li-ion batteries at 25°C

The wettability is a key factor for improving the cycling ability and the power of Li-ion batteries. The wetting of electrodes commonly used for Li-ion battery devises is studied as a function of the P_{xy}-TFSI RTIL content. The experimental results are gathered in Table 9 and Table 10, where the CA values for pure P_{xy}-TFSI RTILs are mentioned for comparison purposes (Stefan et al., 2009).

CA/ ° (±1)	P_{13}-TFSI	P_{14}-TFSI	P_{15}-TFSI	P_{24}-TFSI
100% P_{xy}-TFSI	18.3	17.1	15.1	14.7
20% P_{xy}-TFSI	Total wetting			
30% P_{xy}-TFS	Good wetting: initial contact angle values are around 7°.			

Table 9. Contact angles (in degree) at the liquid/ $LiCoO_2$ electrode/air interface dependence on the P_{xy}-TFSI RTIL content and nature, at 25°C.

As seen in Table 9, the wettability of the composite positive electrode does not raise the problem; the penetration of liquids in the porosity of the electrode is very fast whatever the P_{xy}-TFSI RTIL used and its amount. Because oxides such as $LiCoO_2$ is a ceramic presenting a very high interfacial free energy, that facilitates the wetting (Stefan et al., 2009). For the graphite wetting, the angles of contact obtained with mixtures based on 20% and 30% P_{xy}-TFSI are lower than CA with the pure corresponding P_{xy}-TFSI RTIL, and, slightly superior to the CA obtained with the standard electrolyte (i.e. 10.5°). In every case, the wetting is good enough to let us suppose that the use of mixtures based on P_{xy}-TFSI RTIL should not raise problem of wetting even low-temperature.

4.6 Conclusion on wettability measurements

Results of the wettability of both the negative and the positive electrodes commonly used in Li-ion batteries allows the use of the electrolyte mixtures based on 20% or 30% of P_{xy}-TFSI RTILs. However all the Celgard® separators are insufficiently wet by these mixed electrolytes. This problem can be easily resolved by the replacement of this type of polyolefin separator by a ceramic-type separator, such as the Separion®. All these promising results let us trust the good electrochemical performances with the use of P_{xy}-TFSI/standard electrolyte mixtures.

CA/ ° (±1)	P_{13}-TFSI	P_{14}-TFSI	P_{15}-TFSI	P_{24}-TFSI
100% P_{xy}-TFSI	21.1	20.6	17.9	20.2
20% P_{xy}-TFSI	15.0	13.5	11.0	15.1
30% P_{xy}-TFSI	16.0	14.4	11.3	15.5

Table 10. Contact angles (in degree) at the liquid/graphite electrode/air interface dependence on the P_{xy}-TFSI RTIL content and nature, at 25°C.

5. Electrochemical studies on mixtures

To validate the P_{xy}-TFSI- RTIL mixtures as electrolytes for batteries Li-ion application, tests of cycling ability are required, after the determination of their electrochemical window. The quality of the passivating protective layer, and its stability at the graphite electrode, must be checked by mean of galvanostatic chronopotentiometric measurements. The P_{xy}-TFSI- RTIL contents lower than 20 % (w/w) are not of a real interest, in spite of an ionic conductivity over the standard electrolyte one, because the phenomenon of self-extinguished flame becomes striking only from 20 % of Pxy-TFSI. For between 20 % and 30 % P_{xy}-TFSI- RTIL, the performances are optimized from the point of view of the ionic transport, of the thermal stability and of the material wettability Beyond these contents, up to 50 %, the increase of the viscosity of mixtures can limit their applications at room temperature (mainly because of the decrease of the performances of ionic transport), but their use can be considered for higher temperature applications.

5.1 Electrochemical windows of the P_{xy}-TFSI RTILs

RTIL electrochemical windows are obtained from linear sweep voltamperometry performed under argon atmosphere. The rotating working disk electrode (1000 rpm) is made of glassy carbon. A Pt electrode is used as a counter electrode. An Ag/AgCl electrode, in the presence of saturated KCl in the RTIL, is used as a reference electrode (AgCl and KCl are sparingly soluble in these media). This electrode, especially designed for non-aqueous RTILs, is composed of salt-bridged compartments filled up with P_{14}-TFSI. The smallest one contains the silver wire and AgCl and KCl crystals are added up to saturation. This reference electrode is calibrated against the standard hydrogen electrode (SHE) using the ferrocene redox system (Fc/Fc$^+$). The Fc/Fc$^+$ (Snook et al., 2006) standard redox potential is taken as E°(Fc/Fc$^+$) = 0.45V vs. SHE whatever the solvent used. From Figure 13, the standard potential of Fc/Fc$^+$ system is 330 mV vs. the Ag/AgCl$_{(P_{14}\text{-TFSI})}$, KCl$_{(P_{14}\text{-TFSI})}$, reference electrode, leading to a potential of 120 mV vs. SHE or 3.14 V vs. Li/Li$^+$ which agrees well with 3.1 V elsewhere determined (Appetecchi et al., 2006).

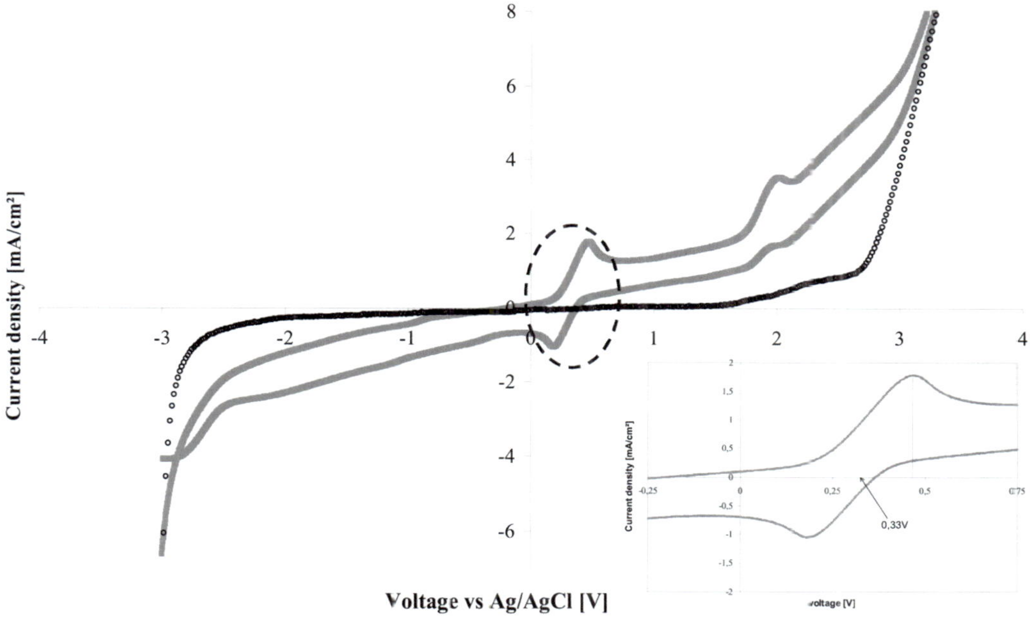

Fig. 13. Cyclic sweep voltammogram at 22°C of pure P_{14}-TFSI and P_{14}-TFSI + ferrocene; W.E.: glassy carbon; C.E.: Pt; R.E.: Ag/AgCl$_{(P14\text{-TFSI})}$, KCl$_{(P14\text{-TFSI})}$; at 100 mV.s^{-1} and 1000 rpm.

The electrochemical windows of the P_{xy}-TFSI RTILs, at the glassy carbon electrode, are displayed in Figure 14. The anodic limit is nearly the same for all P_{xy}-TFSI electrolytes. This suggests that it is the TFSI$^-$ ion which is first oxidized at 2.7 V vs. Ag/AgCl$_{(P_{14}\text{-TFSI})}$(KCl sat.) electrode (i.e. 5.84 V vs. Li/Li$^+$). The cathodic limit is nearly the same at -3 V (0.14 V vs. Li/Li$^+$) for P_{14}^+, P_{15}^+ and P_{24}^+, but clearly different for P_{13}^+ as in this case the reduction of the IL occurs at -2.4 V (0.74 V vs. Li/Li$^+$). This shows that the P_{13}-TFSI is less stable toward reduction at the glassy carbon electrode than the other RTILs. This compound will easily react to form a solid electrolyte interface (SEI) which is a passivating protective layer on metallic lithium or at a graphite electrode when the potential is driven to the intercalation potential of lithium ions in graphite.

5.2 Cycling ability study of graphite or $LiCoO_2$ electrodes in the presence of the mixtures

Button-type half cells composed of a metallic lithium anode and graphite or $LiCoO_2$ cathodes are used to test the behaviour of elaborated mixtures. The lithium metallic electrode plays the role of the reference electrode.

To ensure the formation of the passivating protective layer (SEI) on the graphite electrode, the first charge/discharge cycle is performed at 60°C, then after the following cycles at 25°C. For transport property and thermal behaviour purposes, the tested electrolytes are mixtures of 20% to 30% P_{xy}-TFSI RTILs, while the charge and discharge processes of the half cells are operated at the same rate C/20, D/20. Figure 15 and Figure 16 illustrate the resulting galvanostatic chronopotentiograms obtained respectively with the $LiCoO_2$ positive electrode and the graphite negative electrode at 25°C.

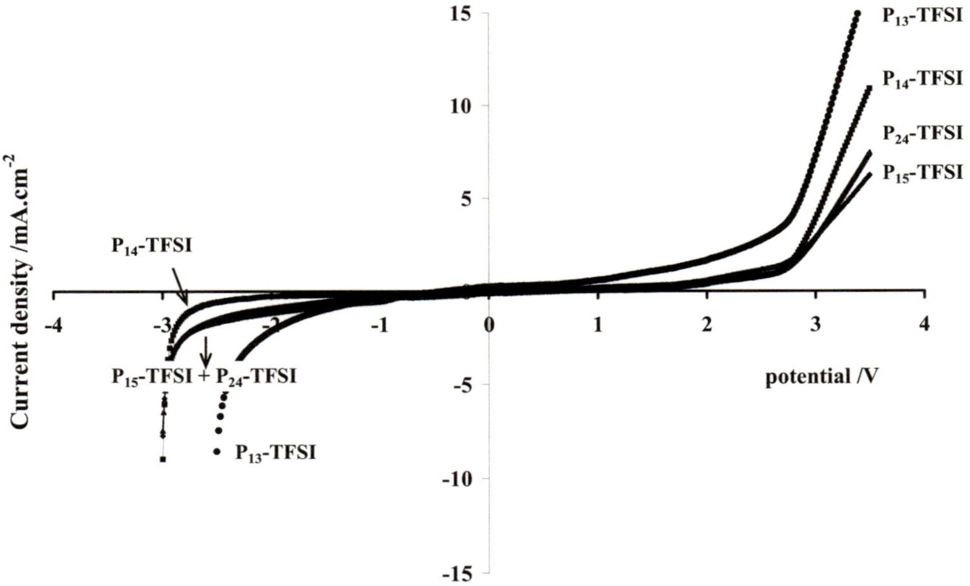

Fig. 14. Linear sweep voltammogram at 22°C of a several P_{xy}-TFSI RTILs tested in this study; W.E., glassy carbon; C.E., Pt; R.E., Ag/$AgCl_{(P14\text{-}TFSI)}$, $KCl_{(sat.\ in\ P14\text{-}TFSI)}$; scan rate, 100 $mV.s^{-1}$; at 1000 rpm.

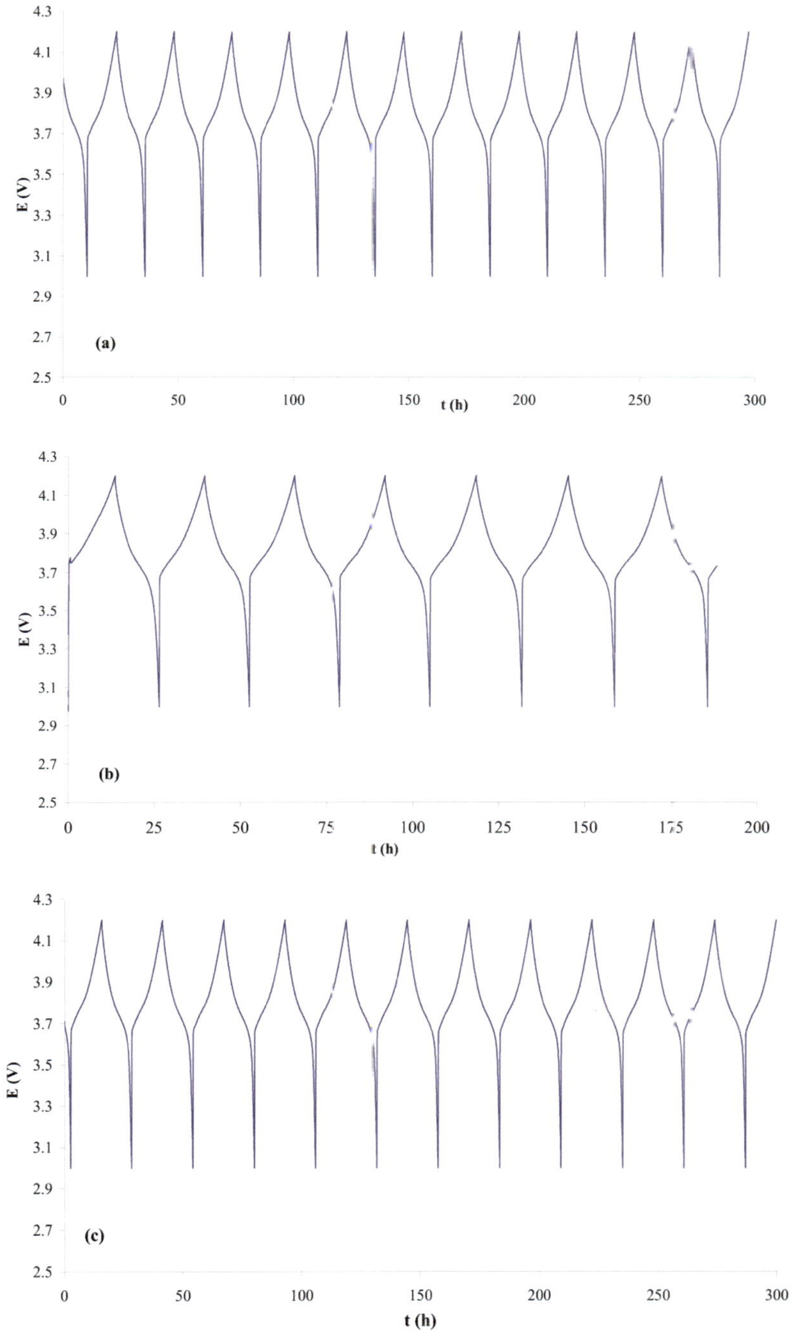

Fig. 15. Galvanostatic chronopotentiometry of a $LiCoO_2$ electrode, at C/20 and D/20 rates, at 25°C, with an electrolytic solution containing (a) 20% P_{14}-TFSI; (b) 30% P_{14}-TFSI; (c) 40% P_{14}-TFSI; reference electrode: Li/Li^+.

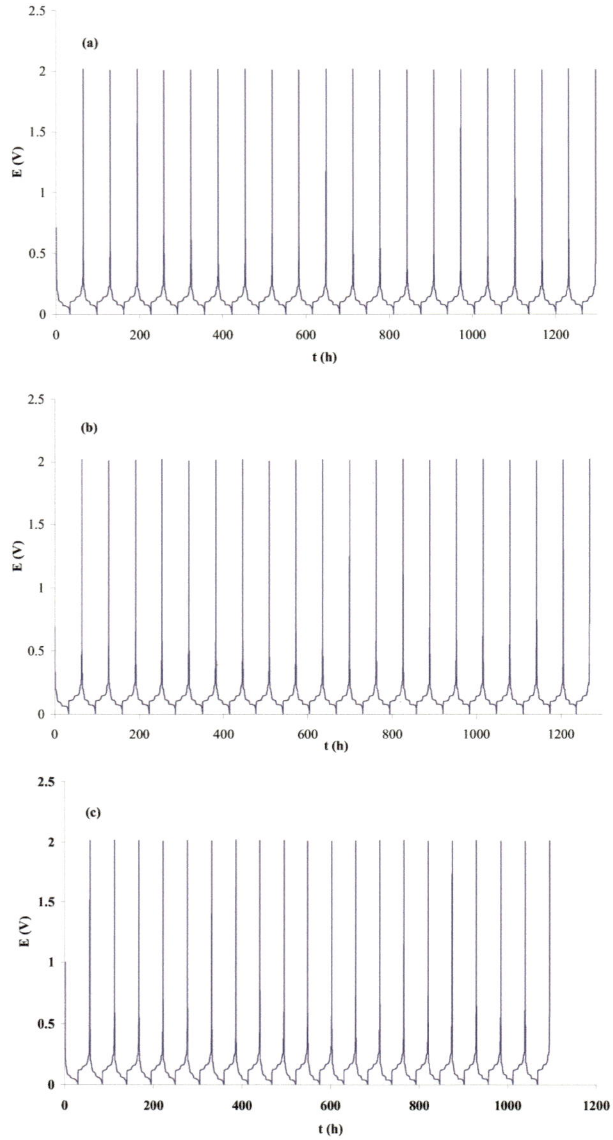

Fig. 16. Galvanostatic chronopotentiometry of a graphite electrode, with an electrolytic solution containing (a) 15 % P_{15}-TFSI, (b) 21 % P_{15}-TFSI and (c) 30 % P_{15}-TFSI; at C/20 and D/20 rates, at 25°C; reference electrode: Li/Li^+.

All the obtained results are gathered in Table 11. In general, all the tested mixtures give good cycling performances. However the analysis of the cycling of the positive electrode shows that the results depend on the P_{xy}-TFSI RTIL nature, even if no fading or polarisation phenomenon occurs. On one hand, whatever the weight content of P_{13}-TFSI or P_{14}-TFSI is, the resulting capacities do not degrade during the cycling and the standard deviation of the experimental values remains very low (except for the 30% P_{13}-TFSI mixture). On the other hand, for the mixtures of P_{15}-TFSI and P_{24}-TFSI, RTILs getting the same total number of carbon atoms on both alkyl chains substituting the nitrogen atom, the resulting capacities degrade during the cycling process. This leads to a bigger scattering of the capacity values which are always lower than those obtained with the P_{13}-TFSI or P_{14}-TFSI mixtures. For the graphite electrode, the presence of different plateus on the chronopotentiometric curves is indicative of good intercalation/de-intercalation processes of the lithium ions. The presence of an electrochemical process occurring the formation of intercalation compounds in the graphite is characteristics of the formation of the SEI layer on the graphite surface. No cycling fading or polarisation phenomenon is observed. Again the negative cycling results are depending on the P_{xy}-TFSI RTIL nature, P_{14}-TFSI giving the best performances and P_{13}-TFSI the worst because of its weaker resistance towards reduction process as expected from linear sweep voltammogram results (Figure 14).

	Positive cycling ability		Negative cycling ability	
	Mean capacity (mA.h.g^{-1})	ratio towards the theoretical value (%)	Reversible mean capacity (mA.h.g^{-1})	ratio towards the theoretical value (%)
20% P_{13}-TFSI	155	97 ± 1	337 ± 14	92 ± 4
20% P_{14}-TFSI	159	99 ± 1	354 ± 4	99 ± 1
20% P_{15}-TFSI	150	94 ± 3	344 ± 6	95 ± 2
20% P_{24}-TFSI	132	83 ± 9	340 ± 19	94 ± 5
30% P_{13}-TFSI	150	94 ± 3	268 ± 44	74 ± 12
30% P_{14}-TFSI	158	99 ± 1	352 ± 12	97 ± 3
30% P_{15}-TFSI	143	89 ± 9	317 ± 16	87 ± 4
30% P_{24}-TFSI	130	81 ± 3	309 ± 20	85 ± 5

Table 11. Reversible capacity values of LiCoO2 and graphite electrodes from cycling in the presence of P_{xy}-TFSI RTIL mixtures at 25°C at C/20 and D/20 rates.

5.3 Conclusion on the electrochemical study

The electrochemical window determination shows that all the studied pyrrolidinium imides can be incorporated in the electrolyte for Li-ion battery except the P_{13}-TFSI RTIL if a graphite electrode is used as negative electrode. These results are confirmed by the galvanostatic chronopotentiometric measurements to test the cycling ability with the electrolyte under consideration. The best electrochemical performances are achieved in the presence of the 20%or 30% P_{14}-TFSI RTIL mixtures

6. General conclusion

To improve the Li-Ion battery security, the use of a series of room temperature ionic liquids is considered. These compounds, abbreviated P_{xy}-TFSI RTILs, consist of a voluminous organic cation of pyrrolidinium-type (P_{xy}^+) and of the bis(trifluoromethanesulfonyl)imide anion (TFSI-). These compounds are not toxic and environmentally friendly because of their very low vapour pressure, which involves a weak volatility. To be compatible with any Li-ion battery applications, an electrolyte has to respect some specifications for thermal, chemical and electrochemical stabilities, for the ionic transport and for electrochemical ability.

The studied P_{xy}-TFSI/standard electrolyte mixtures present a wide temperature range of thermal stability, confirmed by the DSC study and the flammability tests. Even although the mixtures do not really respect the environment, because of the toxicity of the organic solvents and that of the salt of lithium used, the P_{xy}-TFSI RTILs in the mixtures contribute to the improvement of the security of the Li-ion batteries because of their flame-retardant effect. Indeed, the P_{xy}-TFSI co-solvent prevents the most volatile solvent (i.e. DMC) from vaporisation and inflammation, the better protection being achieved from 20% of P_{xy}-TFSI added to the standard electrolyte. The study of the ionic transport properties shows that the mixtures based on 20% of P_{xy}-TFSI are more conductive than the standard electrolyte (except the 20 % P_{15}-TFSI mixture). In the case of P_{14}-TFSI mixtures, when the content is lower than 30%, the resulting conductivity exceeds that of the standard electrolyte. When the content exceeds 30%, the ionic conductivity becomes lower than the standard electrolyte one. However, contents lower than 20% of P_{xy}-TFSI are not of real interest, because the phenomenon of self extinguished flame becomes striking only from 20% of P_{xy}-TFSI. Furthermore, the dynamic viscosity of the standard electrolyte does not vary in a significant way when P_{xy}-TFSI is added as co-solvent, for a content limited to 20% or 30%. Beyond these contents, from 30% to 50%, the increase of the viscosity of mixtures can be a disadvantage for the battery applications at room temperature. All the studied mixtures present a good chemical stability towards the electrode materials. No problem of electrode wetting by the mixtures under consideration is noticed. The use of these mixtures as electrolyte should not raise the problem of wetting, even at low-temperature. The tests of wettability of separators Celgard® (polyolefin type) show a wetting insufficiency. But this problem can be easily resolved by the use of a ceramic type separator such as the Separion®. In addition, the extent of the electrochemical window of the pure P_{xy}-TFSI RTILs is high enough (5.84 V vs. Li/Li^+), consequently, the electrochemical stability of the mixtures, to introduce them in a Li-ion battery device. In fact, a passivating protected layer is formed at the graphite surface before the lithium ion intercalation processes do occur, that prevents the graphite exfoliation. The galvanostatic chronopotentiometric study on graphite and $LiCoO_2$ in the presence of mixtures containing 20% or 30% of P_{xy}-TFSI RTILs in the standard electrolyte, gives fairly good results for the electrode cycling ability. However, values of reversible capacity are slightly smaller in the presence of P_{15}-TFSI or of P_{24}-TFSI, than those from tests using P_{13}-TFSI or P_{14}-TFSI RTIL. P_{14}-TFSI gives the best results.
Thus, this whole study shows the undeniable contribution a series of N-alkyl-N-alkyl'pyrrolidinium bis(trifluoromethylsulfonyl)imide RTILs in the security of Li-ion batteries, without changing their electrochemical performances.

7. Acknowledgement

We fully acknowledge the SAFT Company (Bordeaux, France) for its financial support and for having kindly supplied us with electrodes (graphite and $LiCoO_2$) and with the Separion® separator. We also thank Drs. M. Letellier and P. Porion (CRMD, CNRS-Université d'Orléans, France) for PGSE-NMR measurements, and Pr. D. Lemordant (PCMB, Université François Rabelais, Tours, France) for fruitful discussions.

8. References

Appetecchi, G. B.; Scaccia, S.; Tizzani, C.; Alessandrini, F. & Passerini, S. (2006). Synthesis of Hydrophobic Ionic Liquids for Electrochemical Applications. *Journal of The Electrochemical Society*, 153, 9, (2006), A1685-A1691

Appetecchi, G. B.; Kim, G.-T.; Montanino, M.; Alessandrini, F. & Passerini, S. (2008). Solvent-free, PYR14TFSI ionic liquids-based ternary polymer electrolyte systems. II. Battery tests. *ECS Transactions*, 11, 29, (2008, Rechargeable Lithium and Lithium Ion Batteries), 119-129

Aurbach, D.; Talyosef, Y.;. Markovsky, B; Markevich, E.; Zinigrad, E.; Asraf, L.; Gnanaraj, J.S. & Kim, H.J. (2004). Design of electrolyte solutions for Li and Li-ion batteries: a review. *Electrochimica Acta*, 50, 2-3, (2004), 247-254

Balakrishnan, P.G.; Ramesh, R. & Prem Kumar, T. (2006). Safety mechanisms in lithium-ion batteries. *Journal of Power Sources*, 155, 2, (2006), 401–414

Cognard, J. (2000) Sciences et Techniques du collage. PPUR, Lausanne, 2000, 26

Falkenhagen, H. & Vernon, E. L. (1932). The quantitative limiting law for the viscosity of simple strong electrolytes. *Physikalische Zeitschrift*, 33, (1932), 140

Good, R.J. (1992). Contact angle, wetting and adhesion: a critical review. *Journal of Adhesion Science Technololy*, 6, 12, (1992), 1269-1302

Jones, G. & Dole, M. (1929). The transference number of barium chloride as a function of the concentration. *Journal of the American Chemical Society*, 51, (1929), 1073-1091

Kaminsky M. (1957). The concentration and temperature dependence of the viscosity of aqueous solutions of strong electrolytes. III. KCl, K2SO4, MgCl2, BeSO4, and MgSO4 solutions. *Zeitschrift fuer Physikalische Chemie*, 12, (1957), 206-231

Kaneko, S.; Kono, Y.; Kobayashi, H.; Ishikawa, H. & Utsuki, K. (2009). Nonaqueous electrolyte solution, gel electrolyte, and secondary lithium battery using the electrolyte solution and the gel electrolyte. *Jpn. Kokai Tokkyo Koho* (2009), JP 2009140641 A 20090625

Kobayashi, Y.; Mita, Y.; Seki, S.; Ohno, Y.; Miyashiro, H. & Terada, N. (2007). Comparative study of lithium secondary batteries using non-volatil safety electrolytes. *Journal of the Electrochemical Society*, 154, 7, (2007), A677-A681

Lall-Ramnarine, S.; Castano, A.; Hatcher, J.; Kerr, K.; Li, X.; Munawar, A.; Parikh, A.; Ma, P.; McEntee, C. &Wishart, J. F. (2008). Manipulating the properties of ionic liquids by synthetic design. *Abstracts, 40th Middle Atlantic Regional Meeting of the American Chemical Society*, Queens, NY, United States, May 17-21, (2008). MRM-052

Lewandowski, A.& Galinski, M. (2004). Carbon-ionic liquid double-layer capacitors. *Journal of Physics and Chemistry of Solids*. 65, 2-3, (2004), 281-286

MacFarlane, D. R.; Meakin, P.; Sun, J.; A. N. & Forsyth, M. (1999). Pyrrolidinium Imides: A new family of molten salts and conductive plastic crystal phases. *Journal of Physical Chemistry B*, 103, 20, (1999), 4164-4170

MacNeil, D. D.; Larcher, D. & Dahn, J. R. (1999). Comparison of the reactivity of various carbon electrode materials with electrolyte at elevated temperature. *Journal of the Electrochemical Society*, 146, 10, (1999), 3596-3602

MacNeil, D. D.; Christensen, L.; Landucci, J.; Paulsen, J. M. & Dahn, J. R. (2000). An autocatalytic mechanism for the reaction of LixCoO2 in electrolyte at elevated temperature. *Journal of the Electrochemical Society*, 147, 3, (2000), 970-979

MacNeil, D. D. & Dahn, J. R. (2001). Test of reaction kinetics using both differential scanning and accelerating rate calorimetries as applied to the reaction of LixCoO2 in non-aqueous electrolyte. *Journal of Physical Chemistry A*, 105, 18, (2001), 4430-4439

Martinez, L. M. &. Angell, C. A (2001) A thermodynamic connection to the fragility of glass-forming liquids. *Nature*, 410, 6829, (2001), 663-667

Nockemann, P.; Thijs, B.; Parac-Vogt, T. N.; Van Hecke, K.; Van Meervelt, L.; Tinant, B.; Hartenbach, I.; Schleid, T.; Ngan, V. T. & Nguyen, M. T. (2008). Carboxyl-functionalized task-specific ionic liquids for solubilizing metal oxides. *Inorganic Chemistry*, 47, 21, (2008), 9987-9999

Pan, Y.; Boyd, L. E.; Kruplak, J. F.; Cleland, W. E., Jr.; Wilkes, J. S. & Hussey, C. L. (2010). Physical and transport properties of bis(trifluoromethylsulfonyl)imide-based room-temperature ionic liquids: Application to the diffusion of tris(2,2'-bipyridyl)ruthenium(II). *Journal of the Electrochemical Society*, 158, 1, (2010), F1-F9

Papoular, R. J.; Allouchi, H.; Chagnes, A.; Dzyabchenko, A.; Carré, B.; Lemordant, D. & V. Agafonov (2005). X-ray powder diffraction structure determination of γ-butyrolactone at 180 K: phase-problem solution from the lattice energy minimization with two independent molecules. *Acta Crystallographica, Section B:Structural Science*, B61, (2005), 312-320

Rasch, B.; Cattaneo, E.; Novak, P. & Vielstich, W. (1991). The influence of water on the oxidation of propylene carbonate on platinum. An electrochemical, in-situ FTIR and on-line MS study. *Electrochimica Acta*, 36, 9, (1991), 1397-1402

Richard, M. N. & Dahn, J. R. (1999). Accelerating rate calorimetry study on the thermal stability of lithium intercalated graphite in electrolyte. I. Experimental. *Journal of the Electrochemical Society*, 146, 6, (1999), 2068-2077

Shin, J. H. & Cairns, E. J. (2008). Characterization of N-methyl-N-butylpyrrolidinium bis(trifluoromethanesulfonyl)imide-LiTFSI-tetra(ethylene glycol) dimethyl ether mixtures as a Li metal cell electrolyte. *Journal of the Electrochemical Society*, 155, 5, (2008), A368-A373

Snook, G. A.; Best, A. S.; Pandolfo, A. G. & Hollenkamp, A. F. (2006). Evaluation of a Ag/Ag^+ reference electrode for use in room temperature ionic liquids. *Electrochemistry Communications*, 8, 9, (2006), 1405-1411

Stefan, C. S.; Lemordant, D.; Claude-Montigny, B. & Violleau, D. (2009). Are ionic liquids based on pyrrolidinium imide able to wet separators and electrodes used for Li-ion batteries? *Journal of Power Sources*, 189, 2, (2009), 1174-1178

Stefan, C. S.; Lemordant, D.; Biensan, P.; Siret, C. & Claude-Montigny, B. (2010). Thermal stability and crystallization of N-alkyl-N-alkyl'-pyrrolidinium imides. *Journal of Thermal Analysis and Calorimetry*, 102, 2, (2010), 685-693

Taggougui, M. (2007). Etudes d'additifs susceptibles de sécuriser les électrolytes des batteries lithium-ion du point de vue : contrôle de charge, de l'effet retardateur de flammes et de la mouillabilité. *Thèse d'Université de Tours*, (12 décembre 2007)

Wasserscheid, P. & Welton,T.(2008). Ionic Liquids in Synthesis. 2008 WILEY-VCH Verlag ISBN 978-3-527-31239-9

Weast, R. C., (1976). Handbook of Chemistry and Physics. 57th Edition (1976-1977) CRC Press

Xu, W.; Cooper, E. L. & C. Angell. A. (2003a). Ionic Liquidas: Ion Mobilities, Glass Temperatures, and Fragilities. *Journal of Physical Chemistry B*, 107, 25, (2003), 6170-6178

Xu, W.; Wang, L.-M.; Nieman, R. A. & Angell, C. A. (2003b). Ionic Liquids of Chelated Orthoborates as Model Ionic Glassformers. *Journal of Physical Chemistry B*, 107, 42, (2003), 11749-11756

Zhang, S. S., (2006). A review on electrolyte additives for lithium-ion batteries. *Journal of Power Sources*, 162, 2, (2006), 1379–1394

4

Ionic Liquids in Charge Storage Devices: Effect of Purification on Performance

Stenger-Smith, John D., Chafin, Andrew P., Kline Jr., Clare F., Ostrom, Gregory S. and Quintana, Roxanne L.
NAWCWD Chemistry Division, China Lake CA USA

1. Introduction

Ionic liquids have been the subject of study over the past several years and there are quite a few advantages to their use in charge storage devices such as extremely low volatility, high voltage window, and inherently high concentration of cations and anions for charge transport processes[1-8]. Several ionic liquids are also commercially available. Purity of ionic liquids has been the subject of several publications[9-11]. The impurities (chloride or water for example) discussed in these publications have a profound effect on the physical properties of the ionic liquids. There is a continuing discussion about the effect of ionic liquid purification on the electrochemical performance of electroactive charge storage devices.

In this chapter we examine the effect of column chromatography purification of ethyl methyl imidazolium bis(trifluoromethanesulfonylimide), (EMIBTI) (synthesized by our laboratory on the 100 gram scale) on the performance of a Type I electroactive polymer supercapacitor based upon poly(propylene dioxythiophene) (PProDOT).

The EMIBTI was purified on a 100-gram scale preparatory column using 20% acetonitrile and 80% chloroform as the eluting solvent.

A detailed analysis was performed on the ionic liquid and impurities such as water, lithium chloride, imidazolium, and other elements and compounds. Best efforts were made to quantify the amounts of these impurities. An attempt was made to correlate the presence or absence of an impurity with electrochemical growth and performance of the Type I electroactive supercapacitor.

2. Experimental

All electrochemical techniques were used as described previously[12,13] and all experiments were performed in a dry box. ProDOT and EMIBTI were synthesized according to published procedures[12]. Chromatography (as described in the Introduction) was performed on the 50 gram scale and approximately 100 mL fractions were collected until about 90% of the original material was recovered. There was a dark band at the origin that remained at the top of column.

2.1 Evaluation of capacitor performance

The 2-electrode devices were assembled according to previously published reports[12]. The devices were then switched from 0 to 1.0 Volts at 0.1, 1.0, and 10.0 Volts per second to determine the initial charge capacity and voltage performance. The devices were then cycled for 10,000 cycles at 1.0 Volt per second (approximately 70% depth of discharge). The devices were re-analyzed and the charge capacity and voltage performance was compared to initial values.

Another set of devices was prepared and the devices cycled to extreme over-voltage (2.5 Volts) for several hundred cycles to determine any differences.

2.2 Electron microscopy of electrodeposited films

Electron microscopy of the electrode ends employed a Zeiss EVO-50 SEM. Most images were taken using 6 kilo-volt (kv) accelerating voltage and a calculated beam current of 10 to 25 picoamperes (pA). Images were collected using either a Secondary Electron (SE) or Quadra-Pole Backscatter Detector (QBSD). The SE detector is the normal imaging detector (it shows surface topography) and the QBSD shows gray scale differences based on average atomic weight (the higher the atomic weight the brighter the image area). Semi-quantitative chemical analysis was done by Energy Dispersive Spectroscopy (EDS) using an EDAX Model Leo-EVO-50EP 132-10 Spectrometer. Working distance was approximately 9 mm. Locations on the electrode ends before coating and after coating were imaged. After coating, areas of thin coating as well as areas of thick coating were compared and contrasted.

2.3 Metals analysis of ionic liquids

Approximately 1 gram of each ionic liquid was weighed out to within 0.0001g into separate 50-mL, polypropylene digestion vials. Approximately 15 mL of concentrated nitric acid (trace metals grade) was added to the vials. The vials were covered with a polypropylene watch glass and allowed to reflux at ~90°C for 2 hours. The samples were cooled and brought to a final volume of 100 mL in polypropylene volumetric flasks. The solutions were then analyzed for metals.

The samples were analyzed on a TJA *IRIS Advantage* inductively coupled plasma atomic emission spectrometer (ICP-AES) interfaced to a COMPAQ Deskpro computer using ThermoSPEC software for data collection. The system was calibrated from 0 to 10 ppm using a certified calibration standard. The calibration curve was verified using a certified 1 ppm check standard. Each sample was aspirated into an argon plasma using an autosampler, a peristaltic pump, and a nebulizer. All elements were analyzed simultaneously. For each element, three replicate emission intensities were measured by the charge injection detector in the spectrometer, averaged, and converted to solution concentrations using the established calibration curve. A detection limit of 100 ppb (ug/L) was established for the analysis.

2.4 Anion analysis of ionic liquids

Approximately 1 gram of each ionic liquid was weighed out to within 0.0001g into separate 50-mL polyethylene centrifuge tubes. Exactly 20 mL of 18 Megaohm deionized water was pipetted into each the vial. The vials were capped, shaken vigorously for 1 minute, and then sonicated for 2 hours. The vials were then centrifuged for 1 hour at 5000 rpm. The supernatant liquids were analyzed for anions by ion chromatography.

A Dionex ICS-2500 High Performance Ion Chromatograph with Eluent Generator was used to perform the anion analyses. This was accomplished using a Dionex AS4A analytical column coupled to a Dionex AG4A guard column. A 2-mL/min flow rate of 30 mM hydroxide solution was produced electrolytically by the eluent generator using a KOH cartridge. Samples were injected onto the column using a 25 microliter sample loop. The background was suppressed electrolytically using an anionic self-regenerating suppressor (193 mA) and 18 Megaohm deionized water at a flow rate of 4 mL/min. Analyte peaks were detected using a conductivity detector. The chromatograph was interfaced to a Dell GX260 Pentium 4 computer using Windows XP and Chromeleon software (Version 6.50) for data acquisition and manipulation. Calibration curves for each analyte were generated using linear least squares best fit lines generated from the peak area of the standards.

2.5 Extraction of ionic liquids and analysis of purification column residue

The non-chromatographed and chromatographed EMIBTI were extracted with hexane. Methylene chloride, as well as other solvents were attempted, but all were miscible with the ionic liquid. Approximately 1 mL of each sample was added to approximately 1 mL of HPLC grade hexanes. The samples were thoroughly shaken and allowed to separate overnight (hexanes on top, ionic liquids on the bottom). The upper layer of hexanes was removed, and analyzed using GC/MS. Approximately 1 µL of the sample was injected into a gas chromatograph (Agilent 6890 with a Restek RTX-5MS 30-meter column) with a mass selective detector (Agilent 5973) for analysis. GC oven conditions were 3 minutes at 40°C and then increased to 350°C at 10°C/minute.

In another experiment, the resultant brown residue in the chromatography column (the top layer) was extracted with methanol to give about 0.01 grams of a tan solid. This material was analyzed using Mass Spectrometry and Nuclear Magnetic Resonance.

3. Results

The effect of the purity of the ionic liquid on the electrochemical growth of the PProDOT layers was compared using standard electrochemical techniques. Figure 1 shows the growth characteristics of PProDOT grown in the non-chromatographed ionic liquid. There were no discernable differences between growth characteristics of films grown using the non-chromatographed and the chromatographed ionic liquid.

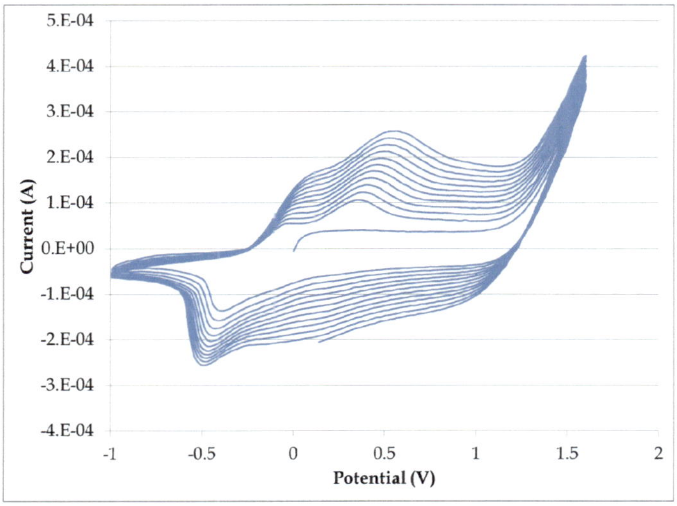

Fig. 1. Cyclic Voltammogram of PProDOT performed in unpurifed EMIBTI.

3.1 Capacitor performance

The results shown in Figure 2 indicate there appears to be very little, if any difference between the initial electrochemical behaviour of the devices that were grown and cycled in chromatographed EMIBTI and devices that were grown and cycled in non-chromatographed EMIBTI.

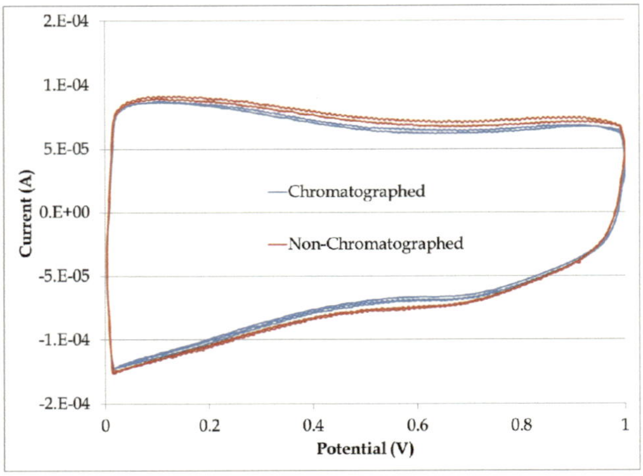

Fig. 2. Electrochemical behaviour comparison of supercapacitors using chromatographed (blue) and non-chromatographed (red) EMIBTI.

The long-term cycling performance of supercapacitors using the different grades of EMIBTI was then examined. As seen in Figure 3, devices grown and cycled using chromatographed EMIBTI improved long-term behaviour as far as charge retention and the ability to operate (store and release charge) at the maximum potential of 1 Volt.

Figure 3 shows that the electrochemical behaviour of the supercapacitor using chromatographed EMIBTI after 10,000 full cycles is nearly identical to that of the original (in Figure 2); whereas the electrochemical behaviour of the supercapacitor using the non-chromatographed sample is significantly degraded after 10,000 full cycles.

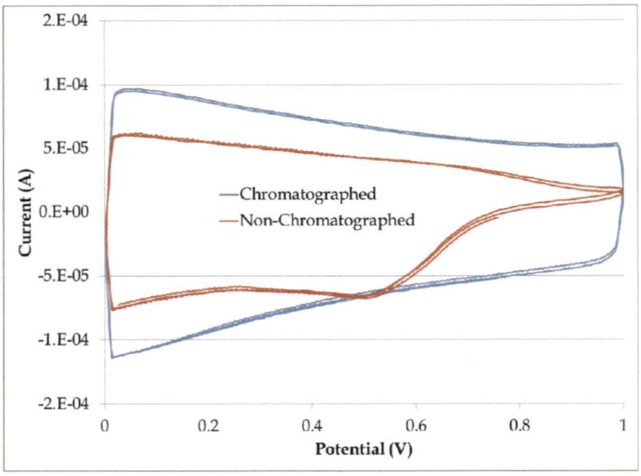

Fig. 3. Electrochemical behaviour comparison of supercapacitors after 10,000 cycles using chromatographed (blue) and non-chromatographed (red) EMIBTI.

In order to elucidate the effect of the growth versus the cycling electrolyte, the same sets of experiments and analysis were performed on capacitors in which the PProDOT was grown in the chromatographed and cycled in the non-chromatographed, and on capacitors in which the PProDOT was grown in the non-chromatographed and cycled in the chromatographed. Table 1 summarizes the results of all 4 possibilities.

Growth Medium	Cycle Medium	Initial Max Potential (V)	Potential After 10 K (V)	Normalized Capacity (10K)
Chromatographed	Chromatographed	1.0	1.0	0.93
Non-chromatographed	Non-chromatographed	1.0	0.7	0.53
Chromatographed	Non-chromatographed	0.7	0.65	0.88*
Non-chromatographed	Chromatographed	1.0	1.0	0.64

*Initial capacity was 30% lower

Table 1. Summary of dependence of performance on growth and switching electrolyte.

Interestingly, devices that were grown in the chromatographed EMIBTI and cycled in the non-chromatographed EMIBTI showed an immediate lack of performance at the full device potential, with 30% lower on average initial capacity, which is shown in Figure 4. The devices that were both grown and cycled in the non-chromatographed EMIBTI didn't lose the full cycle potential until after more than 2000 cycles.

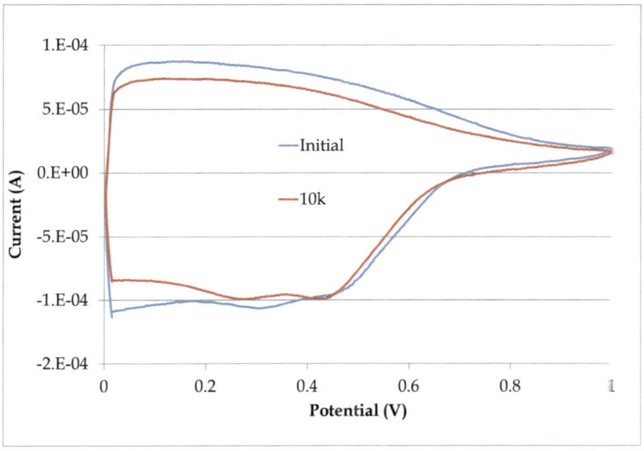

Fig. 4. Initial- (blue) and long-term (10k cycles red) electrochemical behaviour of supercapacitors using chromatographed EMIBTI as the growth electrolyte and non-chromatographed EMIBTI as the cycling electrolyte.

There was a slight decrease in the overall voltage, along with a 12% loss in initial capacity after 10,000 cycles. If the initial capacity loss is included, it can be qualitatively concluded that the overall capacity loss is about the same as those films cycled and grown in both the non-chromatographed EMIBTI.

Devices that were grown in the non-chromatographed EMIBTI then cycled in the chromatographed EMIBTI showed a 36% loss in capacity after 10,000 cycles but still retained the ability to move and store charge at 1.0 Volts.
The over-voltage behaviour of the devices using each grade of EMIBTI was compared. Devices were cycled to well past the potential where PProDOT stores and releases energy. Each system was subject to over-voltage at 2.5, 3.0 and 3.5 volts for 100 cycles at 90% depth of discharge. All samples, regardless of whether the EMIBTI was chromatographed, decayed rapidly, and there appeared to be no quantifiable difference between samples.

3.2 Electron microscopy of electrodeposited films

The properties of the films from each grade of EMIBTI were elucidated using Scanning Electron Microscopy to determine whether there are any significant and quantifiable differences between the architecture of the films. Figures 5 and 6 are typical Electron Micrographs of the PProDOT polymer films deposited shown at 15 and 500 times magnification. There was no discernable difference in the film topography by electron microscopy between films grown with the chromatographed and films grown with the non-chromatographed EMIBTI. Note that the polymer film is deposited in the circular area in the 15X magnification. The rest of the photo shows the poly(tetrafluoroethylene) casing of the electrode. The image quality was adversely affected since the intact, 158 mm long and 61.6 gram electrode needed to be in the chamber (a rather unique circumstance for SEM) to properly image the electrode end coated with the PProDOT film.

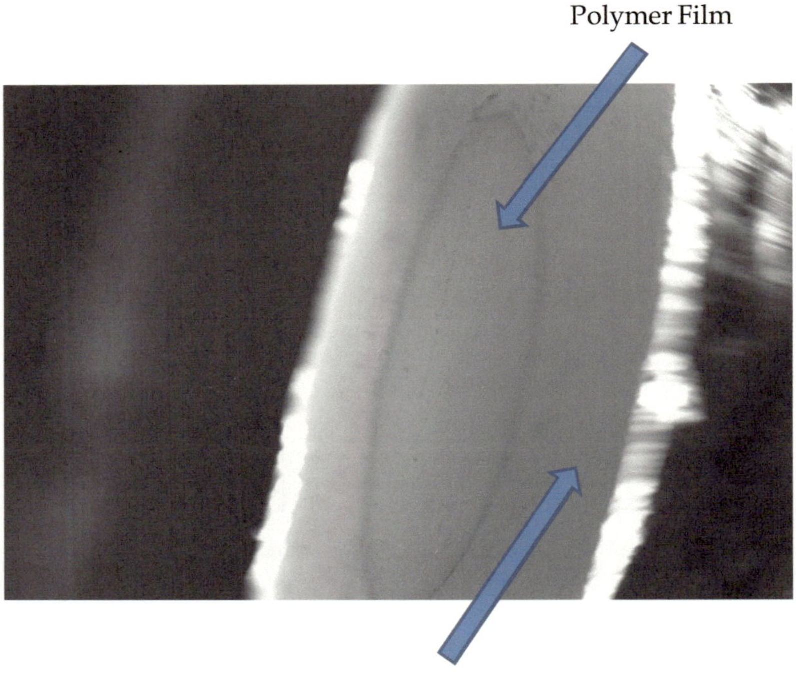

Fig. 5. Electron Micrograph of PProDOT film 15X (grown in non-chromatographed EMIBTI).

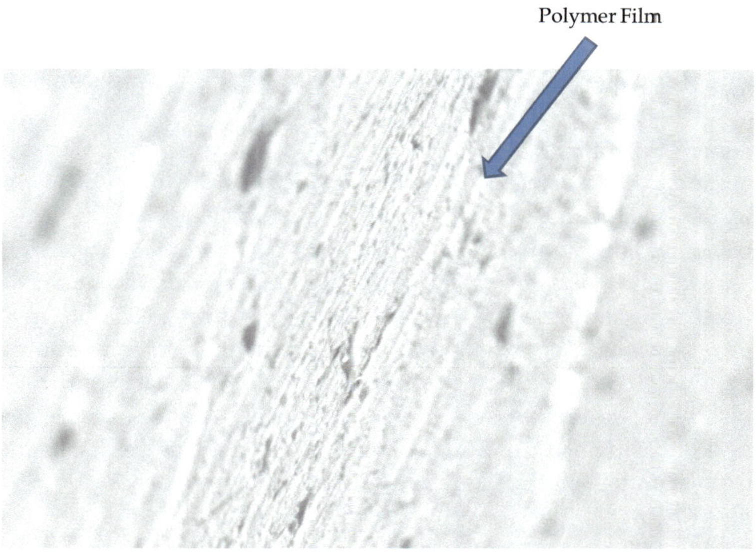

Fig. 6. Electron Micrograph of PProDOT film 500X (grown in non-chromatographed EMIBTI).

Analysis using EDS elemental analysis determined that there is no significant difference in the incorporation of key elements[13] (C, N, O, F and S for example) into the polymer layer as well as any traces of impurities such as Cl. At this semi-quantitative scale it could not be determined whether the amounts of impurities in the EMIBTI samples correlated with any impurities found in the polymer films. Figure 7 is an overlay of the EDS of polymer films grown in chromatographed and non-chromatographed EMIBTI, and shows that the elemental profiles of each film are for all practical purposes identical. These results support the hypothesis that the impurities in the non-chromatographed EMIBTI degrade the long-term switching of devices and not necessarily the short-term behaviour and/or the electrochemical growth of PProDOT.

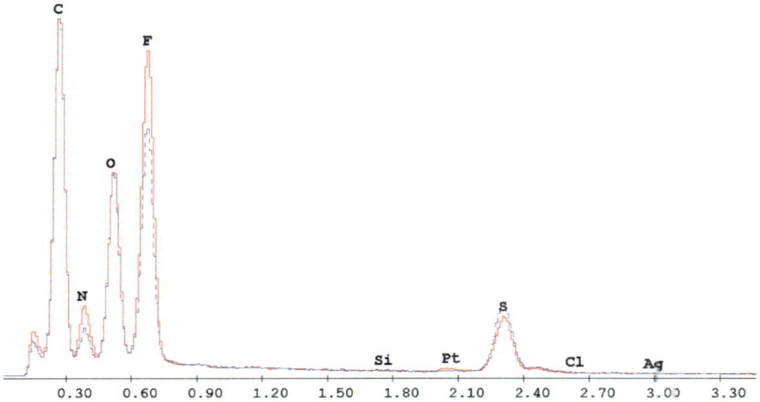

Fig. 7. Overlay of the EDS Spectra of polymer films chromatographed (blue) and non-chromatographed (red).

Neither film had any measureable amount of impurity such as chloride. Platinum was often detected over the area of measurement, but this was due to the thinness of the film on the platinum substrate and the x-ray interaction volume caused by the electron beam.

3.3 Metals analysis of ionic liquids

The samples were analyzed using the previous protocol for the following elements: Al, As, B, Ba, Be, Ca, Cd, Co, Cr, Cu, Fe, K, Li, Mg, Mn, Mo, Na, Ni, Pb, Sb, Se, Si, Sn, Sr, Ti, Tl, V and Zn. The detection limit for these elements is 100 ppb. Sodium, Na, was the only element that was detected: 210 ppm in the non-chromatographed sample and 290 ppm in the chromatographed sample. The propylene carbonate, used in the electrochemical growth of the polymer films, was also analyzed and found to have no detectable impurities at this reporting limit. These results for the most part are not particularly surprising since both EMIBTI samples were extracted many times with water. The lack of Li present in either sample is quite significant, as Li^+ is a major component of one of the starting materials, and could have an effect on the electrochemical properties.

3.4 Anion analysis of ionic liquids

Chloride and fluoride were detected above the threshold in each sample. This technique detected 1.25 ppm chloride and 2.23 ppm fluoride in the non-chromatographed sample and 0.265 ppm chloride and 0.401 ppm fluoride in the chromatographed sample. This result is noteworthy in that chloride is well-established as an undesirable impurity in ionic liquids and can affect many factors[10,11], not just electrochemistry.

3.5 Extraction of ionic liquids and analysis of purification column residue

The hexane extracts of both samples were clear and colorless. The extract of the chromatographed ionic liquid had a higher concentration of common components extracted (possibly contaminants from the glue in the vial caps) and visible in the gas chromatogram than the non-chromatographed sample. However, there were a few components present in the non-chromatographed sample that were not in the chromatographed sample. Since these compounds were most likely removed during the purification. The impurities (non-common components) detected by this technique appeared to be degradation products of imidazole.

The Mass Spectrometry analysis of the column residue was inconclusive. Only siloxanes (glassware grease and silica gel dissolved by the methanol) were detected, and only at very small levels.

Nuclear Magnetic Resonance Spectroscopy (^1H NMR) in d_6 acetone indicated the presence of an ethyl methyl imidazolium salt. This was verified by the 2 single peaks at 7.80 and 7.79 ppm, a single signature peak at 9.41 ppm, a single methyl peak at 4.0 ppm as well as the quartet-triplet signature peak of an ethyl group at 4.4 and 1.6 ppm respectively. Comparison of this spectrum to ethyl methyl imidazolium chloride, with a signature peak at 10.84 ppm and EMIBTI, with a signature peak at 9.02 leads to the conclusion that this imidazolium impurity is neither a pure chloride nor pure BTI salt of imidazolium, though it could be a mixture of chloride and BTI salts, or a completely different imidazolium compound.

NMR also indicated the presence of some type of hydrocarbon-based impurity which was difficult to identify.

4. Summary and conclusions

The purity of ionic liquids was found to be a critical factor in the performance of electroactive polymer-based energy storage devices. The chemical identity of the electrodeposited films is for all practical purposes identical as determined by semi-quantitative quantum backscatter experiments. Films grown and switched in the chromatographed ionic liquid retained the most activity at higher voltages and retained 93% of initial capacity after 10,000 cycles at 70% depth of discharge. Films grown and switched in the non-chromatographed ionic liquid retained the least activity at higher voltages and retained only 53% of initial capacity after 10,000 cycles at 70% depth of discharge. Films grown in the non-chromatographed ionic liquid and switched in the chromatographed ionic liquid retain the ability to store and transfer charge at the upper end of the voltage window, but retained only 64% of the initial capacity. Films grown in the chromatographed ionic liquid and switched in the non-chromatographed ionic liquid degrade more slowly as those grown and switched in the non-chromatographed liquid (88% capacity retained) but devices made from these films do not have the higher voltage (greater than 0.5 Volts), activity and the initial overall capacity of the films was consistently 30% lower.

Therefore, there must be some trace impurity in the non-chromatographed ionic liquid that is responsible for the performance degradation. This impurity affects the behaviour of the devices when it is used as a supporting electrolyte for the electrochemical growth, as well as an electrolyte for the device. Based upon the results using the chromatographed EMIBTI as the growth solvent and non-chromatographed EMIBTI as the cycling solvent, the incorporation of this impurity into the polymer film appears to significantly degrade the full device potential. Residual lithium ion can be ruled out for all practical purposes as it was well below 100 ppb for both samples. Devices using both types of EMIBTI degraded very rapidly at severe overvoltage, due to the irreversible oxidation of PProDOT being nearly independent of electrolyte impurity.

The impurities possibly responsible for performance issues have been narrowed down to the following: 1) The presence of 40% more sodium in the chromatographed sample may be beneficial to device performance. As unlikely as this is, this cannot be ruled out at this time. 2) The presence of 5.5-fold more fluoride and 4.7-fold more chloride in the non-chromatographed sample. This is a likely source of the performance degradation given the effect of chloride and fluoride on electrochemical performance[9]. 3) The presence of an organic impurity, namely an imidazolium salt of an anion that is neither chloride nor bis(trifluoromethyl sufonylimide) salt. This imidazolium impurity was detected from the material left behind on the chromatography column and must also be considered a very likely source of performance degradation. 4) An impurity that was undetected and/or unidentifiable by the analytical techniques that were used in this study.

The removal of these impurities using the chromatography described in this paper and/or other purification techniques[14] is recommended for electrochemical synthesis and analysis.

5. References

[1] Pringle, J. M., Forsyth, M., MacFarlane, D. R., Wagner, K., Hall, S. B., and Officer, D. L., (2005) *Polymer, 46,* 2047.

[2] Deepa, M., Awadhia, A., Bhandari, S., (2009) *Chem. Phys.* 11(27), 5674-5685.

[3] Ferrari, S.; Quartarone, E., Mustarelli, P., Magistris, A., Protti, S., Lazzaroni, S., Fagnoni, M., Albini, A. (2009) *J. Power Sources*, 194(1), 45-50

[4] Zakeeruddin, S. M., . Graetzel, M., *Adv. Funct. Mater.* 2009, 19(14), 2187-2202.

[5] Armand, M., Endres, F., MacFarlane, D. R., Ohno, H., and Scrosati, B. (2009) *Nature Mater.*, 8(8), 621-629.

[6] Fernicola, A., Scrosati, B., and Ohno, H. (2006) *Ionics* 12, 95-102.

[7] Ohno, H, (ed.), (2005) *Electrochemical Aspects of Ionic Liquids,* Wiley Interscience, Hoboken, NJ.

[8] Lu, W. Fadeev, A. G., Qi, B., Smela, E., Mattes, B. R., Ding, J., Spinks, G. M., Mazurkiewicz, J., Zhou, D., Wallace, G. G., MacFarlane, D. R., Forsyth, S. A., and Forsyth, M., (2002) *Science* 297, 983-987.

[9] Irvin, J. A. and Stenger-Smith, J. D., (2009) *Aldrich Material Matters™* Vol. 4, issue 4 on Alternative Energy

[10] Seddon, K. R., Stark, A., and Torres, M. J., (2000) *J. Pure Appl. Chem.* 72, 2275

[11] Zhang J., and Bond, A. M., (2005) *The Analyst*, 130, 1132

[12] Stenger-Smith, J., Webber C. K., Anderson N., Chafin, A.P., Zong, K and. Reynolds, J. R., (2002) *J. Electrochem. Soc.* 149(8), A973-A977

[13] Stenger-Smith, J. D., Guenthner, A., Cash, J., Irvin, J. A., and Irvin, D. J., (2010) *J. Electrochem. Soc.* 157 (3) A298

[14] Rosario Canales, M. R., Pravas, D., Therien, M. J., Gopu, P., Sotero-Esteva, J. O., Santiago-Aviles J. J., (2008) *Single-Wall Carbon Nanotube/Poly(p-napththalene ethynylene) Composites as Electtrode Materials for Supercapacitor Applications* http://xilonen.inaoep.mx/revista_electronica/volumen/descarga_articulo.php?articulo=c1-04-IB08.pdf&ida=cedebb6e872f539bef8c3f919874e9d7

5

Polymers with Ionic Liquid Fragments as Potential Conducting Materials for Advanced Applications

Santiago V. Luis[1], Eduardo García-Verdugo[1], M. Isabel Burguete[1], Andreu Andrio[2], Sergio Mollá[3] and Vicente Compañ[3]
[1]*Departamento de Química Orgánica, Universitat Jaume I., Castellón,*
[2]*Departamento de Física, Universitat Jaume I., Castellón,*
[3]*Departamento de Termodinámica Aplicada, Universidad Politécnica de Valencia, Valencia, Spain*

1. Introduction

During the last decades, the use of ionic liquids (ILs) has become one of the most useful techniques for the development of green chemistry tools.[1] The potential use of liquid salts based on delocalized organic cations (ammonium or phosphonium) as designer green solvents represents one of the most significant contributions to modern chemistry. Thus, the high modularity of the structures of the ILs based on the proper selection of the cation and anion moieties has allowed the development of an almost infinite number of compounds well suited for each specific application. Nevertheless, in recent years, the application of ILs also needs to confront some important challenges in order to be able to develop the expected practical applications.[2] Thus, some important drawbacks are limiting the general application of ILs in scientific and technological applications. First of all, the cost of ILs is clearly much higher than that for traditional solvents. Additionally, the benefits of the use of ILs as media for different chemical reactions and other applications is counterbalanced by the need of using traditional "non-green" solvents for the extraction of the desired products from the IL phase. Finally, recent studies have shown that, although ILs have been considered traditionally as environmentally friendly solvents because of the lack of any appreciable vapor pressure, some of the most usual ILs present some environmental concerns in particular in terms of their contact with aqueous media.[3] Many of those drawbacks can be drastically reduced by the use of supported ILs. The immobilization of ILs onto a solid support provides a simple way for reducing the amount of IL required for a given application, reducing accordingly the associated cost; facilitates their handling and manipulation decreasing the need of using traditional solvents in the corresponding process, and finally, greatly reduces the potential leaching to the environment of the ILs.[4]
In order to accomplish this target two main approaches have been studied:

1. Non-covalent support of IL phases on the surface of inorganic or organic supports (SILPs).
2. Covalent attachment of IL-like phases on the surface of inorganic or organic supports (SILLPs).

It seems clear to understand that the development of SILLPs presents some methodological advantages. In particular, it should be expected that the use of IL phases physically entrapped onto the surface of a given support would maintain the essential properties of the corresponding bulk IL, as no change in the structure is carried out through this process. On the contrary, the potential leaching of the supported IL phases in the presence of a solvent being able to partially solvate the IL structure represents the main limitation of this approach. The almost opposite reasoning is valid for the second approach. In this case, when the IL-like fragment is covalently attached to the support, the problem of leaching is totally eliminated. On the contrary, the transfer of the proper essential properties of the ILs to the solid phase needs to be guaranteed, and this is not a simple matter. However, it is reasonable to assume that if this goal of transferring IL properties to the surface of the solid is achieved, the resulting materials would present the advantages associated to ILs but would overcome the abovementioned drawbacks currently limiting many technological applications of bulk Ionic Liquids.

In recent years, most emphasis has been given to the application of ILs as selective and very specialized media for carrying out chemical transformations, including catalytic[5] and biocatalytic[6] applications or the extraction,[7] manipulation and transformation of natural products and biopolymers.[8] Nevertheless, it is important to bear in mind that the first applications of ILs were described in the field of the development of new electrolytes for advanced applications such as those involving new sensors, fuel cells, conducting materials or other applications in the general area of renewable energy sources.[9] In this regard, it is not surprising that the development of solid electrolytes based on the structures of ILs has received a great deal of attention, taking into account some of the advantages considered for SILPs and SILLPs.[10]

Thus, polymeric ILs represents an important field in the area of new advanced materials for energy applications.[11] Nevertheless, most applications in this field have been associated to the development of soluble polymers containing structural moieties related to ILs: imidazolium, phosphonium or ammonium groups.[12] The use of those polymers as conducting material can be easily rationalized in terms of the mobility of the polymeric chains carrying out the ionic species and giving place to the conduction phenomena.[13] Much less attention has been paid, however, to the study of insoluble polymers containing such structures. This is an interesting field because those materials can provide in full the advantages we have considered for supported ILs and because their study as conducting materials can provide some important insights on the conduction mechanisms in ILs. Here we will examine and discuss some essential properties of polymer-supported Ionic Liquid-like Phases (SILLPs), using crosslinked insoluble matrices as supports, and their behavior as conducting polymers. The importance of those materials has been highlighted by their application as "solid solvents" for the immobilization of different catalytic species that reveal that the corresponding polymers can advantageously substitute for the related bulk ILs for this purpose.[14]

2. Approaches for the preparation of polymeric-ILs

In general the preparation of SILPs in which the IL fragment is non-covalently attached onto the surface of a solid support, usually an inorganic support such as SiO_2, is carried out by the adsorption of a thin layer of a bulk IL onto the surface of the corresponding support (see Figure 1a). In some cases, however, it has been demonstrated that the previous modification of the structure of the support, in particular through the introduction of IL-like fragments provides an efficient way to improve the immobilization process and reduces the risk of leaching (see Figure 1b).

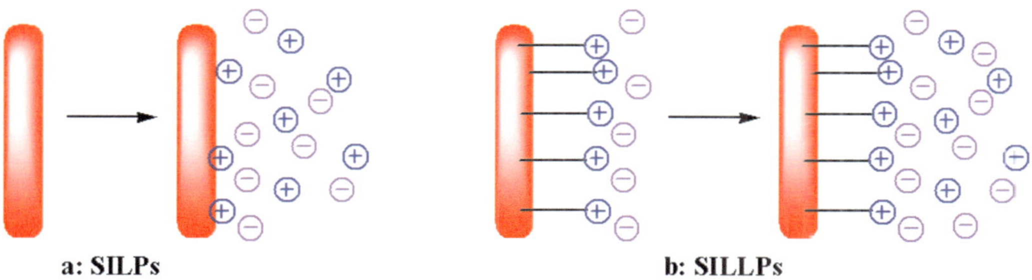

Fig. 1. General structures for SILPs (a) and SILLPs (b).

In the case of the preparation of SILLPs, it is considered that the modification of the structure of a support with the introduction of IL-like fragments such as imidazolium, phosphonium, ammonium or others will be able to transfer to this surface the properties of the related bulk IL. Two main approaches can be considered for this purpose in the case of organic polymers. The first one is the polymerization of the corresponding monomers containing IL fragments. This is exemplified in Figure 2 for the case of polystyrene-divinylbenzene backbones with imidazolium subunits. In the case of the use of monomeric mixtures just containing styrene (1) and functionalized styrene monomers (3) the resulting materials will be linear soluble polymers. When variable amounts of a crosslinking agent are added (i. e. divinylbenzene, DVB 2) the corresponding insoluble polymers can be obtained. When those polymers are prepared in the form of monoliths through a bulk polymerization process, the resulting materials can be easily obtained with different morphologies and shapes, according to the desired application. A more simple and versatile approach, however, for the preparation of those materials is the use of common polymeric precursors having the appropriate morphological and chemical properties. This is exemplified in Figure 3, in which chloromethylstyrene is used for the introduction of the precursor functionalities in the crosslinked polymeric backbone. The transformation of chloromethyl groups into the corresponding IL-like fragments (i.e. imidazolium groups) is very simple and can be carried out very efficiently. A quantitative transformation of the functional groups can be obtained after a few hours of heating of the corresponding polymer with the desired imidazole used as the reagent and the solvent. The progress of the process can be easily monitored through different methods that have been developed for analysing the transformation of chloromethyl groups. The advantage of this approach is that the polymerization of mixtures of styrene, divinylbenzene and chloromethylstryrene, in the presence of porogenic agents, is a well-known process and accordingly can be controlled, in terms of yields, conversions, morphology and so on, using standarised parameters. On the contrary, the introduction in the polymerization mixture of monomers such as 3 with very different structural and electronic properties can provide very different outcomes, according to the different polymerization rates and to the different monomers/oligomers/porogens solubility equilibria occurring in the global process. Finally, it must be noted that a simple metathesis procedure, as reported for bulk ILs, allows for the exchange of the initial chloride counteranion for any other anion of interest in the area of ILs. All those transformations can be easily monitored through the use of FT-IR spectroscopy and, very particularly in the case of the modification of the anions, with the use of FT-Raman spectroscopy.[14]

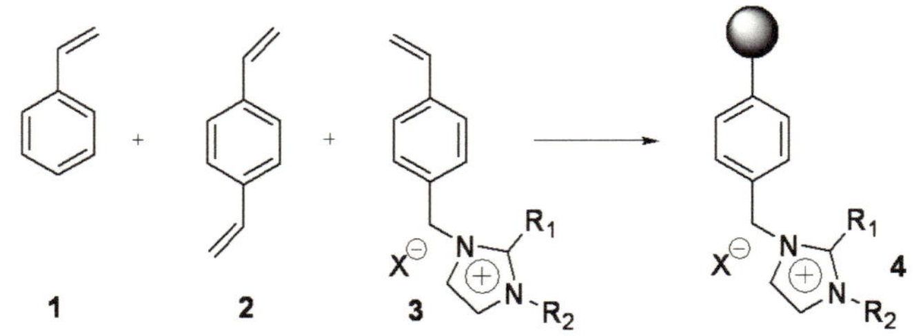

Fig. 2. Synthesis of SILLPs by polymerization of vinyl imidazolium monomers.

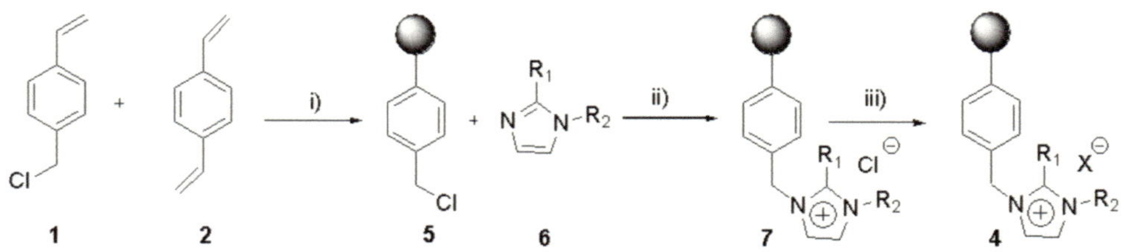

Fig. 3. Synthesis of SILLPs by modification of preformed polymers.

3. Properties of polymeric ionic liquids (SILLPs)

The potential application of the polymeric SILLPs prepared by any of the methodologies considered above is associated to the presence of some important properties according to the desired application. Some of them are associated to the morphology and the chemical properties of the polymeric backbone (thermal stability, porosity, surface…). In our case, however, the priority is to understand if the presence of the ionic liquid-like fragments on the surface of the polymer is able to transfer to this material the essential properties associated with the bulk ILs. In this regard, we have approached this subject through the analysis of a variety of properties characteristic of ILs on the corresponding polymeric SILLPs (4). As we will see below, we have been able to demonstrate that many of those properties are efficiently transferred to the polymer.[15] Moreover, we have checked that modifications on the structural parameters of the IL-like fragments provide a method to modulate the properties of the resulting material. This is an essential factor as much as some of the most important applications of ILs rely on their potential as "designer solvents" being able to provide the desired properties, almost in an unlimited way, through the proper modular modification of their structural parameters.

4. Thermal stability

For many applications, the thermal stability of ILs and SILLPs is a critical parameter.[16] In this regard we have carried out thermogravimetric (TGA) and differential scanning calorimetry (DSC) studies for the different SILLPs synthesised. Very interestingly, TGA data clearly show that the presence of imidazolium subunits provide an additional stabilisation of the SILLP when compared with the starting chloromethylated resin, which is rather surprising taking into account the presence of benzyl imidazolium subunits. This is, however, in good agreement with the studies on bulk ILs revealing the high thermal stability of those species. Some results from TGA analyses of SILLPs 4 are gathered on Figure 4 and demonstrate that those polymers follow the same trends observed for bulk ILs. Thus, the nature of the anion is the main parameter determining the thermal stability. This is higher for the less nucleophilic anions and lower for the most nucleophilic ones. The observed effects slightly increase with the loading of functional groups in the polymer. Thus the two extreme cases, in terms of the anion used, correspond, as we will see in all cases, to the chloride and the NTf_2^- anion. Although it is not represented in Figure 4, it is interesting to note that changes in the aliphatic chain of the imidazolium ring (i.e. changing from R_2 = CH_3 to R_2 = C_4H_9) have a much less pronounced effect on the thermal stability. The use of DSC techniques provides some additional information (Figure 5). DSC thermograms show the presence of small transitions on the SILLPs at temperatures well below the decomposition temperature. Those transitions can be associated to the movement of non-crosslinked polymeric chains. As can be seen in Figure 5, those transitions are very much affected by the nature of the counter anion. Thus, for instance, the exchange of Cl^- by NTf_2^- involves a reduction in the corresponding Tg from about 110ºC (4, R_2 = C_4H_9, R_1 = CH_3, X = Cl) to 50ºC (4, R_2 = C_4H_9, R_1 = CH_3, X=NTf_2). This suggests that the presence of the more coordinating anion provides a strong ordering of the structure through polymeric anion-cation interactions. It can be also seen that the presence of an acidic H atom at being able to participate in hydrogen bonding is accompanied by a small increase in this Tg. This is augmented in the case of the presence of the Cl^- anion. This indicates that intermolecular hydrogen bonding can play a role in determining intermolecular anion-cation interactions.

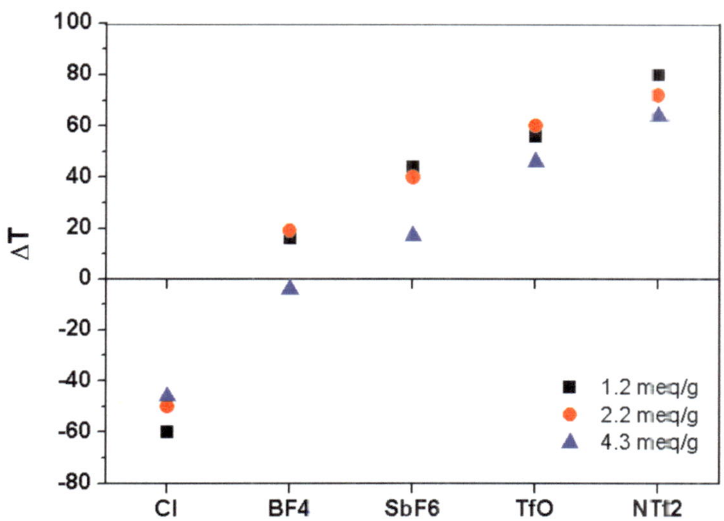

Fig. 4. Effect of loading and nature of the anion on the thermal stability of SILLPs as measured by TGA. ΔT is defined as the difference in thermal stability between the starting Merrifield's resin (5) and the polymer with the corresponding imidazolium subunits (4).

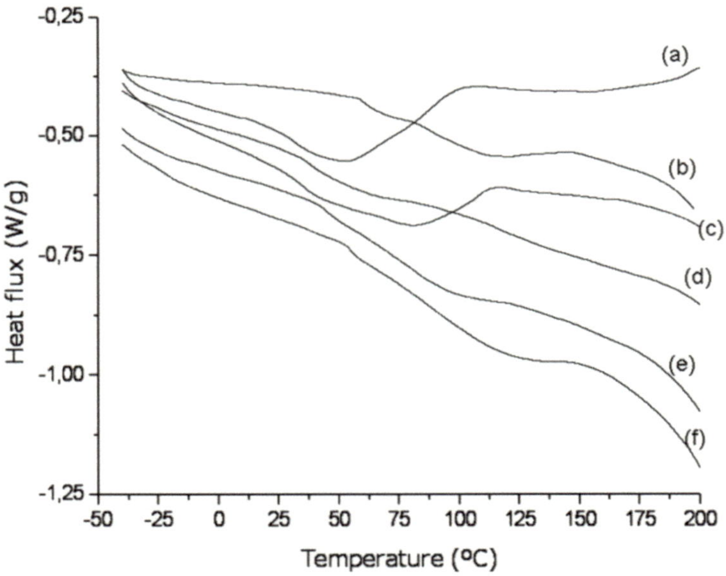

Fig. 5. DSC curves for different SILLPs 4: (a) KN10 ($R_1 = CH_3$, $R_2 = CH_3$, X = NTf_2), (b) KN5 (R_1 = H, $R_2 = CH_3$, X = Cl), (c) KN9 (R_1 = H, $R_2 = C_4H_9$, X = NTf_2), (d) KN6 (R_1 = H, $R_2 = CH_3$, X = NTf_2), (e) KN7 (R_1 = H, $R_2 = C_4H_9$, X = Cl) and (f) KN8 ($R_1 = CH_3$, $R_2 = CH_3$, X = Cl). Second run.

5. Assessment of polarity through solvatochromic probes

An important property of ILs is that significant changes in their polarity can be achieved through changes in their structure, either by modification of the imidazolium substituents or by modification of the counteranion.[17] In this regard polarity assessment through the use of solvatochromic probes is one of the most simple and accessible techniques for a rapid and semiquantitative analysis. In our case, we have selected two well known solvatochromic probes - pyrene[18] and Reichardt's probe[19] to analyze the properties of the SILLPs. Figure 6 includes some of the data obtained with the use of Reichardt's probe. As can be observed, values obtained for ET(N) in SILLPs are in good agreement with those reported for related bulk ILs (butyl imidazolium salts: BMIM).[20] In this case it is easy to observe that the introduction of larger aliphatic chains is accompanied by a decrease in polarity as described by the ET(N) parameter, although a clear pattern associated to the nature of the anion is not detected, according to a similar observation in solution. The effect of the anion is more visible, however, when pyrene is used as the fluorescent probe, using the I3/I1 ratio as a measure of polarity (see Figure 7).[21] In this case the more coordinating anion (Cl^-) is associated to a higher value of polarity (higher value of I3/I1), whereas the presence of the less coordinating NTf_2^- anion gives place to a lower value of I3/I1. Here, however, a clear pattern for the substitution of methyl by butyl in the cation is not detected. In any case the general values obtained are again in good agreement with the values obtained in solution for related solvents and materials and bulk ILs.[22] As clearly seen from the above data, as much as the former solvathochromic parameters are dependent on adsorption phenomena on a heterogeneous surface, the quantitative data obtained must be handled with caution and a more accurate method for polarity determination (see below) is required for properly determining this essential property.

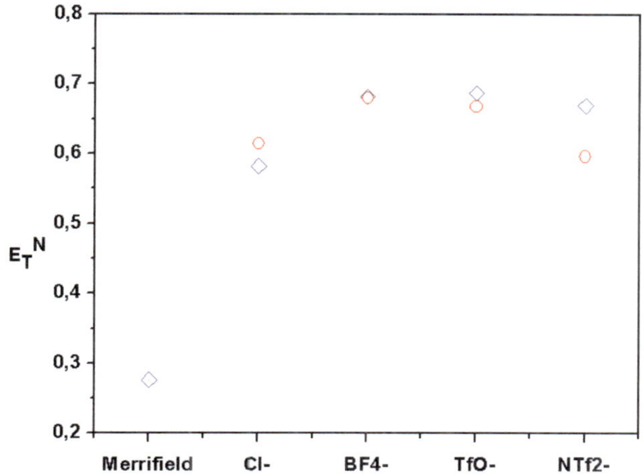

Fig. 6. Effect of the nature of the anion of the ET(N) solvathochromic parameter using the Reichart's probe. Comparison with related bulk ILs. Red circles indicate BMIMX bulk ILs; blue diamonds refer to the corresponding SILLPs.

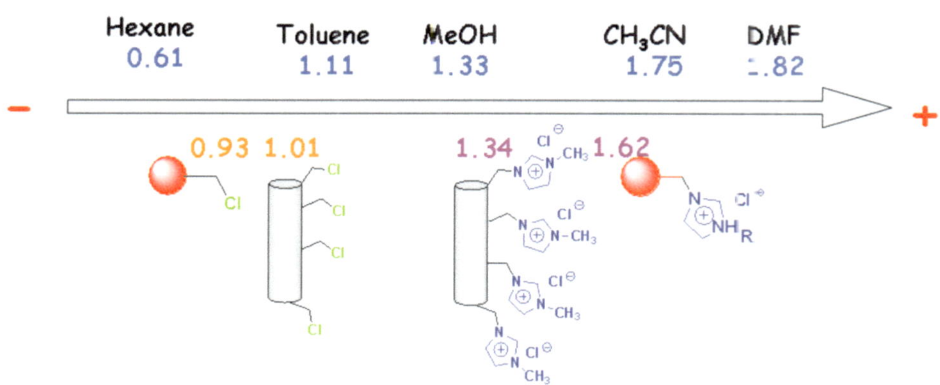

Fig. 7. Polarity determination of SILLPs using the I3/I1 ratio pyrene as the fluorescent probe.

6. Compatibility with polar solvents; swelling analysis

Gel-type polymers containing low levels of crosslinkers and prepared in the absence of porogenic agents are characterized by the absence of any permanent porosity in the dry state. Thus, the use of those polymers for applications requiring the accessibility of the functional sites on the polymeric backbone is always associated to the presence of a good solvent, compatible with the matrix, being able to expand the polymeric chains.[23] In general, PS-DVB polymers can be considered as hydrophobic materials, not being compatible with the use of polar solvents such as water or methanol. It can be expected that the introduction of imidazolium or other IL-like subunits will provide an important change in the polarity of the matrix, favouring the swelling of the polymer by polar solvents. In this regard, the swelling of a polymer with a polar solvent such as water (as determined through the change of the diameter from the dry to the swollen state) can be considered an indirect measurement of polarity. Although the corresponding values cannot be used for a quantitative assessment of polarity, they represent a very interesting parameter for practical applications as with the use of water (and other polar solvents). Figure 8 shows some data obtained for SILLPs prepared from a gel-type Merrifield resin with a high loading of functional groups (4.3 mmol Cl/g). This figure allows observing the expected trends associated to the changes in the structure of the IL-like fragment. Thus, when analysing the nature of the anion, it can be clearly seen that the presence of the Cl^- anion results in more polar resins. On the other hand, the substitution of the methyl substituent (R = CH_3) by a more lipophilic one (R = C_4H_9) produces an appreciable decrease of the polarity.

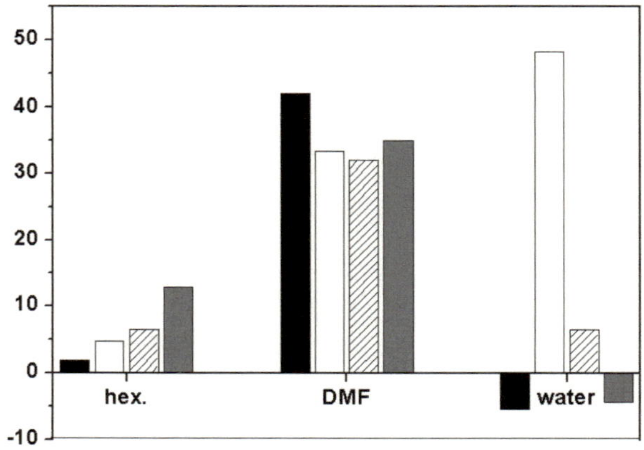

Fig. 8. Swelling effects for different solvents in SILLPs differing on the nature of the anion. Black bars: starting Merrifield's resin; white bars: SILLP with Cl^- anion; stripped bars: BF_4^- anion; grey bars: NTf_2^- anion.

7. Dielectric relaxation spectroscopy

The different crosslinked polystyrene resins containing ionic liquid fragments covalently attached (Supported Ionic Liquid-Like Phases: SILLPs) indicated above has been studied by means of dielectric impedance spectroscopy to obtain the conductivity. Impedance experiments were carried out on SILLPs samples at several temperatures lying in the range 298 K (25°C) to 398 K (125°C) and frequency window $10^{-2} < f < 10^6$ Hz. The measurements were performed with 100 mV amplitude, using a Novocontrol broadband dielectric spectrometer (Hundsangen, Germany) integrated by a SR 830 lock-in amplifier with an Alpha dielectric interface. The SILLP sample of interest was placed between two gold electrodes coupled to the impedance spectrometer. The area of the electrodes was 20 mm^2. The temperature was controlled by a nitrogen jet (QUATRO from Novocontrol) with a temperature error of 0.1 K during every single sweep in frequency.

8. Analysis from Bode and Nyquist diagrams

Under an alternating electric field, the response of dipoles associated with molecules or segments of chain molecules to the field as well as the ions movements depend on the physical state (glassy, liquid) of the system. The impedance response is usually measured with blocking electrode/polar system/blocking electrode configuration. Polarization processes arising from dipole motions in polar systems can empirically be represented by an equivalent circuit consisting of a resistance, R_p, accounting for the polarization of the system in parallel with a capacitor of capacity C that describes the interfacial double layer capacitance system-blocking electrode.[24] The complex impedance of the circuit is given by

$$Z^*(\omega)=\frac{R_p}{1+j\omega R_p C}=\frac{R_p}{1+j\omega\tau} \quad (1)$$

where R_pC, which has the dimensions of time, is replaced for τ, called relaxation time. Eq. 1 suggests that a plot of $Z''(\omega)$ against $Z'(\omega)$ should be a semicircle centered at $Z' = R_p/2$ that intersects the abscissa axis at the origin ($\omega \to \infty$) and $Z' = R_p$ ($\omega \to 0$). Relaxations defined by a single relaxation time are called Debye type processes. However, polarization processes in condensed matter are not defined by a single relaxation time, but by a distribution of relaxation times, in such a way that the complex impedance should be written as

$$Z^*(\omega)=\sum_i w_i \frac{R_{p;i}}{1+j\omega\tau_i} \quad (2)$$

where w_i is the fraction of relaxation mechanism i with polarization resistance R_{pi} and relaxation time τ_i. As in the case of Debye systems, the Z'' *vs* Z' plots are curves that intersect the abscissa axis at the origin ($\omega \to \infty$) and $Z' = \sum_{i=1}^{n} R_{pi}$ ($\omega \to 0$). The impedance behavior of these complex systems can be represented by an equivalent circuit consisting of a polarization resistance in parallel with a constant phase element, CPE, provided of an admittance $Y^* = Y_0\,(j\omega\tau)^n$, with n lying in the range $0 < n \leq 1$ and Y_0 given in Ω^{-1} units.[25] The impedance of the circuit is given by

$$Z^*(\omega)=\frac{R_p}{1+Y_1(j\omega\tau)^n} \quad (3)$$

Where, $Y_1 = Y_0R_p$. Notice that for $n = Y_1 = 1$, eq. 3 corresponds to the complex impedance of a Debye system.

For systems that combine polarization processes with a resistance independent on frequency or ohmic resistance, R_o, the electrical equivalent circuit is a parallel R-CPE circuit in series with an assembly of N circuits in series. Each circuit is made up of a resistance R_i that represent a polarization resistance in parallel with a constant phase element. The complex impedance of the circuit can be written as

$$Z^*(\omega) = R_0 + \sum_{i=1}^{N} \frac{R_i}{1 + R_i Y_{0i}(j\omega\tau_i)^{n_i}} \quad (4)$$

The real and imaginary parts of Z* are

$$Z'(\omega)=R_0+\sum_{i=1}^{N}\frac{R_i+Y_{0i}R_i^2(\omega\tau_i)^{n_i}\cos\frac{n_i\pi}{2}}{1+Y_{0i}R_i^2(\omega\tau_i)^{2n_i}+2Y_{0i}R_i(\omega\tau_i)^{n_i}\cos\frac{n_i\pi}{2}}$$

$$Z''(\omega)=-\sum_{i=1}^{N}\frac{Y_{0i}R_i^2(\omega\tau_i)^{n_i}\sin\frac{n_i\pi}{2}}{1+Y_{0i}R_i^2(\omega\tau_i)^{2n_i}+2Y_{0i}R_i(\omega\tau_i)^{n_i}\cos\frac{n_i\pi}{2}} \quad (5)$$

It is worth noting that the resistance R_0 is obtained from the Bode diagrams[26] in which the complex impedance modulus, $|Z^*(\omega)|=\left[Z'^2(\omega)+Z''^2(\omega)\right]^{1/2}$, is plotted against frequency. For R-CPE parallel circuits, $\lim_{\omega\to 0}|Z(\omega)|=R_p$ and $\lim_{\omega\to\infty}|Z(\omega)|=0$, whereas for circuits made up of a parallel R-CPE circuit in series with an ohmic resistor, $\lim_{\omega\to 0}|Z(\omega)|=R_p+R_o$ and $\lim_{\omega\to\infty}|Z(\omega)|=R_o$. In the latter case, the curve describing the dependence of the modulus of the impedance on frequency decreases from R_p+R_o, at low frequencies ($\omega\to 0$), where a plateau is reached until high frequencies ($\omega\to\infty$) at which $|Z^*(\omega)|$ tends to R_o. Moreover, the phase angle of the impedance $\phi=\tan^{-1}[-Z'(\omega)/Z''(\omega)]$ increases with frequency decrease from -90° at $\omega\to\infty$ to 0° at $\omega\to 0$. Then R_o is taken as the modulus of the impedance at the plateau coexisting with ϕ = 0. Bode diagrams in terms of the complex modulus of the conductivity obtained by means of $|\sigma^*(\omega)|=\frac{1}{|Z^*(\omega)|}$ are represented for KN8 and KN10 in the temperature range 298 K to 398 K are shown in Figures 9 and 10, respectively.

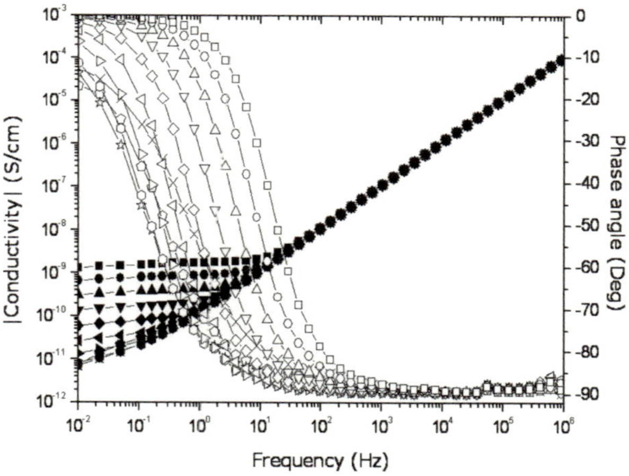

Fig. 9. Bode diagrams for KN8 (4, R_1 = CH_3, R_2 = CH_3, X = Cl) showing the logarithm of the conductivity and the phase angle at several temperatures: (square) 398K, (circle) 388K, (up triangle) 378K, (down triangle) 368K, (diamond) 358K, (left triangle) 348K, (right triangle) 338K, (hexagon) 328K, (star) 318K, (pentagon) 308K and (plus) 298K. Solid symbols for impedance and open symbols for phase angle.

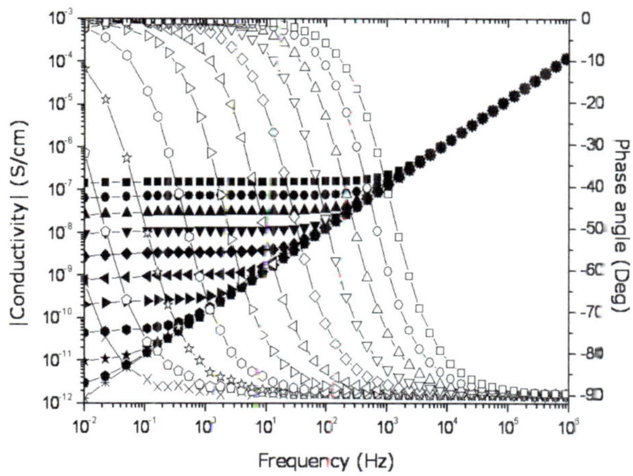

Fig. 10. Bode diagrams for KN10 (4, $R_1 = CH_3$, $R_2 = CH_3$, $X = NTf_2$) showing the logarithm of the conductivity and the phase angle at several temperatures: (square) 398K, (circle) 388K, (up triangle) 378K, (down triangle) 368K, (diamond) 358K, (left triangle) 348K, (right triangle) 338K, (hexagon) 328K, (star) 318K, (pentagon) 308K and (plus) 298K. Solid symbols for impedance and open symbols for phase angle.

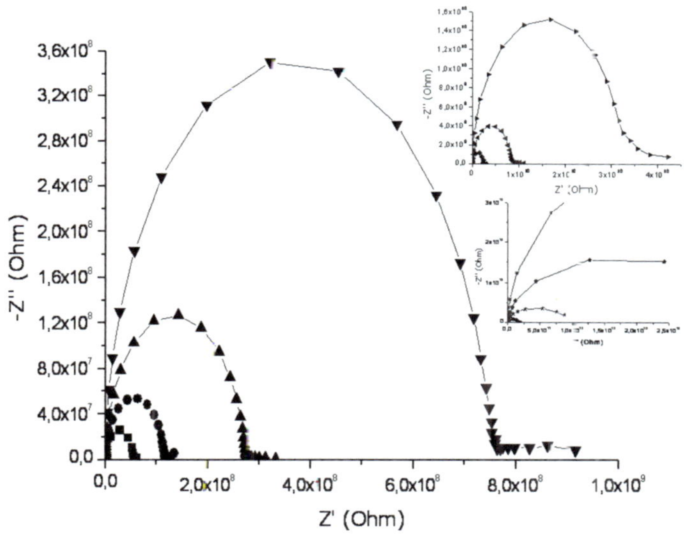

Fig. 11. Nyquist diagrams for KN10 (4, $R_1 = CH_3$, $R_2 = CH_3$, $X = NTf_2$) at all temperatures: (square) 398K, (circle) 388K, (up triangle) 378K, (down triangle) 368K, (diamond) 358K, (left triangle) 348K, (right triangle) 338K, (hexagon) 328K, (star) 318K, (pentagon) 308K and (plus) 298K.

At high temperatures, the isotherms are characterized by a plateau nearly independent on frequency, followed by a dispersive regime. Moreover, whereas the plateau for the KN10 isotherms coexists with $\phi = 0$ in a wide range of temperatures, this only occurs for the KN8 system at high temperatures. For low temperatures the values of $/\phi/$ at the frequencies studied are higher than zero, indicating that in these cases the ohmic resistance should be

slightly higher than the maximum value of $/Z^*(\omega)/$ at the apparent plateau for the conductivity $/\sigma\ ^*(\omega)/$. Then obtaining the whole Bode diagram would surely require measuring the impedance at frequencies significantly lower than those used in this study.
Similar results we can found from the study of Z" vs. Z' plot, called Nyquist diagram[27] such is observed in Figure 11 for KN10 at all temperatures. This Figure gives curves that intersect the abscissa axis at $Z'=R_0$ and $Z'=R_0+R_p$, when $w\rightarrow\infty$ and $w\rightarrow 0$, respectively. Departure from semicircles is observed in experimental Nyquist diagrams[27] as a result of polarization processes and other phenomena taking place in the membrane electrode interface.
From Figures 9 and 10 we can see that the length of the plateau in the Bode diagrams increases as temperature increases until a critical frequency f_c (= $\omega_c/2\pi$) dependent on temperature is reached, at which $\log/\sigma^*(\omega/$ collapses along a straight line with slope $d\log/Z^*(\omega/\ /d\log\ \omega = 1$. This is a typical behavior of a parallel R_0C circuit, where at high frequencies at which $\omega C >> 1$ and

$$\lim_{\omega\to\infty} \log|\sigma^*(\omega)| = \lim_{\omega\to\infty} \log\left|\frac{1+j\omega R_0 C}{R_0}\right| \cong \log(C\omega) \tag{6}$$

And for low frequencies at which $\omega C<<1$, then $\lim_{\omega\to 0}|\sigma^*(\omega)| = 1/R_0$.
Thus for frequencies $\omega \leq 1/(R_pC)$, the modulus of the complex impedance decreases from $R_o + R_p$ at low frequencies to R_o at high frequencies (i.e. the modulus of the complex conductivity increases from $1/(R_o + R_p)$ at low frequencies to $1/R_o$ at high frequencies). As indicated above, the detection of this drop in the modulus would require performing impedance experiments at very low frequencies. At the same time, the phase angle increases from -90° to 0°. However above the critical frequency $\omega_c \geq 1/(R_oC)$, the impedance is eminently capacitive and $|\sigma^*(\omega)| \propto \omega^n$ ($n\cong 1$), in agreement with the results shown in the Bode diagrams of Figures 9 and 10. The critical frequency was taken as that one corresponding to the inflexion point of the phase angle, i.e. $\phi = -45°$, at the temperature of interest. The plots of Figure 12 show that f_c follows Arrhenius behavior. The plots of Figure 12 show that f_c follows Arrhenius behavior with activation energies of (28.0±0.5), (26.4±0.4), (16.2±0.3), (20.5±1.0.3), (26.2±0.7) and (25.7±0.6) *kcal mol*$^{-1}$ for KN5, KN6, KN7, KN8, KN9 and KN10, respectively.
The conductivity patterns showed in Bode diagrams of Figures 9 and 10 shown a similar ac conductivity patterns with a frequency independent plateau in the low frequency region and exhibits dispersion at higher frequencies[28]. The effect of electrode polarization is evidenced small deviation from the modulus of the conductivity (plateau region) value in the conductivity spectrum. For the SILLPs studied, the geometric capacitors obtained from eq. (6) is independent on temperature. The values of these capacitors are 1.48, 1.67, 1.07, 1.55, 2.30, and 2.56 *pF* for KN5, KN6, KN7, KN8, KN9 and KN10, respectively.
The ionic conductivity thus estimated for KN5 and KN6 are 4.0×10^{-11} S/cm, at 358K (85°C), to ca. 1.1×10^{-9} S/cm at 398K (125°C) for KN5, and 3.3×10^{-9} S/cm to 1.4×10^{-7} S/cm for KN6 at the same temperatures. The conductivities of KN7 and KN9 are rather low increasing in the former material from 4.9×10^{-12} S/cm, at 328 K (55° C), to ca. 1.4×10^{-9} S/cm, at 398 K (125° C), S/cm, whereas in the SILLP KN9 the increase is from 8.5×10^{-10} to 5.3×10^{-7} S/cm, in the same range of temperature. Whereas the comparison of the ionic conductivity between KN8 and KN10 is estimated to change from 1.1×10^{-11} S/cm, at 328 K (55° C), to ca. 2.1×10^{-9}

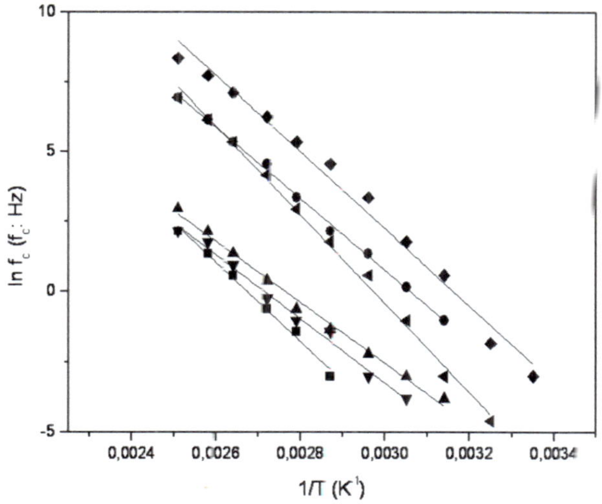

Fig. 12. Variation of the critical frequency f_c at which the capacitance governs the impedance behaviour of the SILLPs at high frequencies: KN5 (squares), KN6 (circles), KN7 (up triangle), KN8 (down triangle), KN9 (diamond) and KN10 (left triangle).

S/cm, at 398 K (125° C), S/cm for KN8, while for the material KN10 the increase is from 7.9×10^{-10} at 303 K (35°C) to 2.3×10^{-7} S/cm at 398 K (125° C). As shown in Figure 13, the temperature dependence of the conductivity is also a thermally activated process described by the Arrhenius equation.

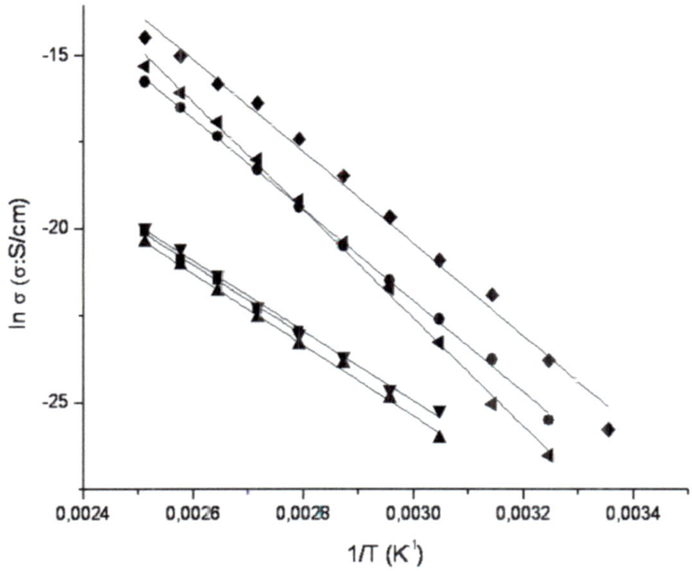

Fig. 13. Arrhenius Plot showing the temperature dependence of the conductivity of the SILLPs KN5 (squares), KN6 (circles), KN7 (up triangle), KN8 (down triangles), KN9 (diamonds) and KN10(left triangles).

The ionic conductivity of KN6, KN9 and KN10 is nearly two orders of magnitude higher than that of KN5, KN7 and KN8, respectively, whatever temperature as a basis of comparison is taken. In principle, one could think that ionic transport in the materials is carried out by the mobile anions. In this case, the results suggest that the mobility of NTf_2^- is higher than that of Cl^-, a result opposite to what one at first sight would expect if the anions volume were determinant in the conductive process. However, strong coulombic forces between cations and anions in absence of humidity, makes this transport mechanism unlikely. Moreover, charge transport by anions would be dependent on frequency. Therefore, to explain the conductivity of these materials it is necessary to postulate charge delocalization between cations and anions as mainly responsible for the conductivity independent of frequency that KN5, KN6, KN7, KN8, KN9 and KN10 exhibit. Owing to the high electronegativity of the chloride anion, charge delocalization is obviously easier for NTf_2^- than for Cl^- and hence the higher conductivity of KN6, KN9 and KN10. Similar behavior and results have been found for other supported ionic liquid-like phases prepared by anchoring imidazolium moieties onto highly crosslinked poly(p-chloromethylstyrene-codivinylbenzene) monolithic matrices such as polymers KN7 and KN10.[29] It is noteworthy that a similar effect is observed for both couples KN5/KN6, KN7/KN9 and KN8/KN10, the second one having a methyl substituent at the C2 position of the imidazolium ring. The change in the nature of the anion can be responsible for changing the conductivity observed by several orders of magnitude. On the contrary, changes on the structure of the cation are accompanied by observable modifications in conductivity, but always within the same order of magnitude. This suggests, for instance, that the interaction C2-H←X⁻ is not the main factor for determining the conductivity of the SILLPs considered for the SILLPs here considered. Nevertheless, for T > Tg, an increase in temperature increases the mobility of the chains, facilitating charge delocalization between neighboring cations and anions.

9. Analysis from dielectric complex permittivity

The current intensity across a capacitor produced by the alternating voltage $V=V_0\,\mathrm{Im}\exp(j\omega t)$ is $i=(dq/dt)=j\omega C_0\varepsilon^*(\omega)V$ where ω is the frequency of the field, where q is the positive charge at the electrodes, C_0 is the capacity of the capacitor in vacuum and ε^* is the relative complex dielectric permittivity of the material that acts as dielectric of the capacitor. Since the ratio i/V is a complex admittance or the reciprocal of the complex impedance, the complex permittivity of a material placed as dielectric between parallel arm plats is $\varepsilon^*(\omega)=l/(j\omega e_0 SZ^*)$ where e_0 is the dielectric permittivity in vacuum whereas S and l are, respectively, the area and the thickness of the dielectric placed between the electrodes. Illustrative experimental isotherms showing the components of the complex dielectric permittivity in the frequency domain for KN6 and KN10 are presented at several temperatures in Figures 14 and 15, respectively. For temperatures lower than 328 K (55 °C) for KN6 and 348 K (75°C) for KN10, ε' is only slightly dependent on frequency in the range $0.01 \leq f \leq 10^6$ Hz. Identical behavior has been observed for the comparison between the others SILLPs. At higher temperatures, ε' undergoes a significant increase as frequency decreases, the increase being higher with temperature. This augment of ε' is the result of the enhancement on the mobility of the chains occurring when the systems surpass the glass transition temperature. However, the plateau exhibited by the real component of ε^* at low frequencies in polar polymers is not reached in the SILLPs in the range of temperatures studied.[25,30]

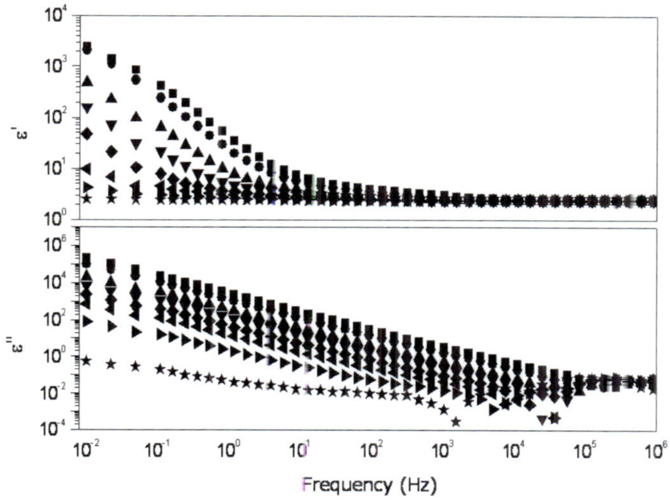

Fig. 14. Real (top) and loss (bottom) components of the complex dielectric permittivity for KN6 at several temperatures: (square) 398K, (circle) 388K, (up triangle) 378K, (down triangle) 368K, (diamond) 358K, (left triangle) 348K, (right triangle) 338K, (hexagon) 328K, (star) 318K, (pentagon) 308K and (plus) 298K.

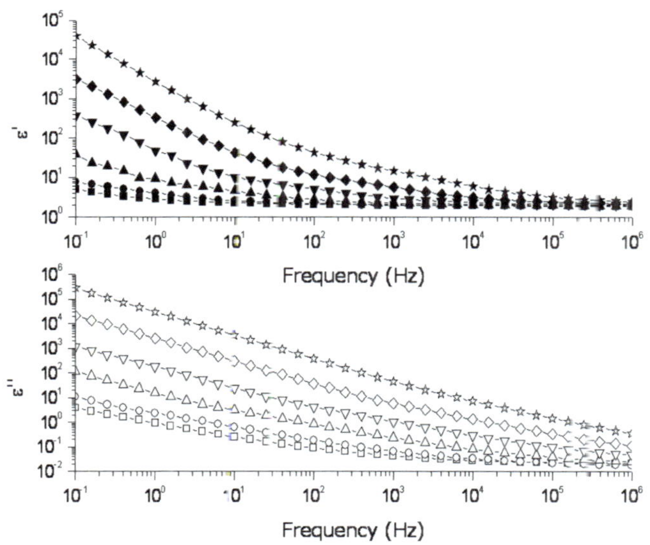

Fig. 15. Real (top) and loss (bottom) components of the complex dielectric permittivity for KN10 at several temperatures: (square) 398K, (circle) 388K, (up triangle) 378K, (down triangle) 368K, (diamond) 358K, (left triangle) 348K, (right triangle) 338K, (hexagon) 328K, (star) 318K, (pentagon) 308K and (plus) 298K.

As for the dielectric loss, the isotherms present a dielectric dispersion in the high frequency region of the spectra, only detectable in the isotherms obtained at high temperature. At moderate and low frequencies ε'' is nearly a linear increasing function of the reciprocal of the frequency and as a result dielectric dispersions associated with the dielectric loss are obscured by the conductivity.
The permittivity loss shows a straight line at low frequencies with a slope close to -1 at high temperatures. In these polymers the presence of secondary relaxation of in the range of high frequencies can be observed. This relaxation of very low intensity presumably is due to the fact tha, the conductive processes present in the SILLPs are very important. The relaxation strength of these polymers was very low, minor than 0.10 for all SILLPs for the range of temperatures between 95K and 125K in all cases. To characterize the dipolar absorption at high temperatures, $\varepsilon_d''(\omega)$, the following equation (7) was used:[31-33]

$$\varepsilon_d''(\omega) = \varepsilon''(\omega) - \left(\frac{\sigma}{\varepsilon_0 \omega}\right)^s \quad (7)$$

where σ and ε_0 are the specific conductivity and the dielectric permittivity in vacuum (=8.854 pFm^{-1}), respectively, and s≤1 is an exponent whose departure from the unit is caused by the interfacial electrode-sample process intervening in the α-relaxation. In our study, the value of s is compress between 0.80 and 0.99 depending on the temperature for all the SILLPs studied. The values of the conductivity obtained from the curves of permittivity loss by mean of equation (6) follow a Arrhenius behaviour. The values of the activation energies reach (27.1±0.4), (23.4±1.0), (21.5±0.2), (21.1±0.2), (24.3±0.7) and (30.3±0.4) kcal/mol, for KN5, KN6, KN7, KN8, KN9 and KN10 respectively. The main purpose of this chapter is not to characterize the dipolar relaxations that this SILLPs can have at high frequencies. We are essentially focused on the study of the conductivity of the SILLPs. Nevertheless, a few comments should be made before discussing the dielectric relaxation behaviour of the SILLPS. The high crosslinking degree of the supporting material, as well as the strong intermolecular interactions between the ionic groups should enhance the glass transition temperature of the SILLPS in such a way that, in principle, the magnitude of this parameter should be higher than 373 K, the Tg of polystyrene. Then the ostensible glass-rubber relaxation that uncrosslinked polar polymers present, should be absent in the dielectric relaxation spectra of the SILLPs. However, the abrupt changes occurring in ε' at temperatures slightly above 348 K for KN5, KN7 and KN8, and above of 328 K for KN6, KN9 and KN10, are the result of the glass-rubber relaxation taking place in the vicinity of these temperatures. The glass rubber relaxation in these apparently highly crosslinked systems may arise from the morphology of the crosslinked SILLPs. Owing to the high chemical reactivity of divinyl benzene, which is nearly four times that of styrene,[34] a gel with a nucleus of moderately crosslinked poly(dvinyl benzene) chains containing occasional chloromethyl styrene units may be firstly formed. As the crosslinking reaction proceeds, the Tg of the gel increases approaching to the reaction temperature, and the overall polymerization rate diminishes because the diffusivity of the reacting monomers to the growing radicals decreases. The remnant *p*-chloromethylstyrene may graft to the unreacted double bonds of divinyl benzene, forming dangling chains, responsible for the glass transition displayed by these systems at relatively low temperature.

10. Analysis of the conductivity from electric modulus formalism

When discussing the results presented in Figures 14 and 15, we have noted that the dielectric permittivity ε″ strongly increases with temperature in the low frequency region, which we attributed to a conduction process. In order to explore this phenomenon in detail, the electric modulus formalism M* has been widely used.[35-36]
Despite some doubts cast on the applicability of the method for this task,[37,38] a combined study of ionic motion using electrical relaxation and RMN relaxations suggests that the complex modulus is the most appropriate representation of dielectric data for analysis of ionic conductivity relaxation as in the present case. The points made by Elliot[37] and Roling[38] as well by others authors to discredit the dielectric modulus representation have been refuted and are summarily discussed in a review article, where the authors have answered all the unnecessary criticism on the use of the electric modulus.[39] The representation of the macroscopic data to describe the microscopic motion of the movement of the ions has been also well described, as in the present case, by Ngai et al.[40]
The electric modulus is the reciprocal of the permittivity M*=1/ε*. Generally, for a pure conduction process, a relaxation peak would be observed in the frequency spectra of the imaginary component M″ and no peak would take place in the corresponding plot of ε″, such is our case in all the SILLPs.
Isotherms at several temperatures showing the double logarithmic plots of the real and imaginary components of the complex dielectric modulus of KN6, are shown in the frequency domain in Figure 16. As occurs with the modulus of relaxation processes, *M′* increases as the frequency increases until a frequency is reached at which *M′* remains nearly constant. Moreover *M″* displays at low frequencies the characteristic absorption of conductive processes and a M″ peak in the plots M″ vs. frequency is observed in relation to conductivity relaxation of the SILLPs. In addition to the conduction peak, it can be seen in some cases at high temperatures a second peak which shift to higher frequency with a very small absorption.[29] These peaks indicate the transition from short range to long range mobility with decreasing frequency, where the low frequency side of the peak represents moving long distances, i.e., performing successful hopping from one site to the neighboring site, whereas, for the high frequency side, the ions are spatially confined to their potential wells and can execute only localized motion.[41,42]
The relaxation of the dielectric modulus can be mathematically represented equivalently as a function of time or of frequency. This approach developed for a single relaxation time for dielectric relaxations by Debye has been extended by Havriliak and Negami to a distribution of relaxation times.[43] The frequency representation for the complex modulus as found by the empirical Havriliak-Negami model function as[44]

$$M^*(\omega) = M_\infty + \frac{M_0 - M_\infty}{\left[1+\left(j\omega\tau_{HN}\right)^{\alpha}\right]^{\beta}} \quad (8)$$

where M_∞ and M_0 are the unrelaxed and relaxed moduli, ω is the excitation frequency and τ_{HN} is the average relaxation time. The parameter α ($0 < \alpha < 1$) broadens the width of the relaxation as it decreases in value; the parameter β ($0 < \beta < 1$) causes a skewing or asymmetry in the modulus in both real and imaginary parts as a function of frequency.

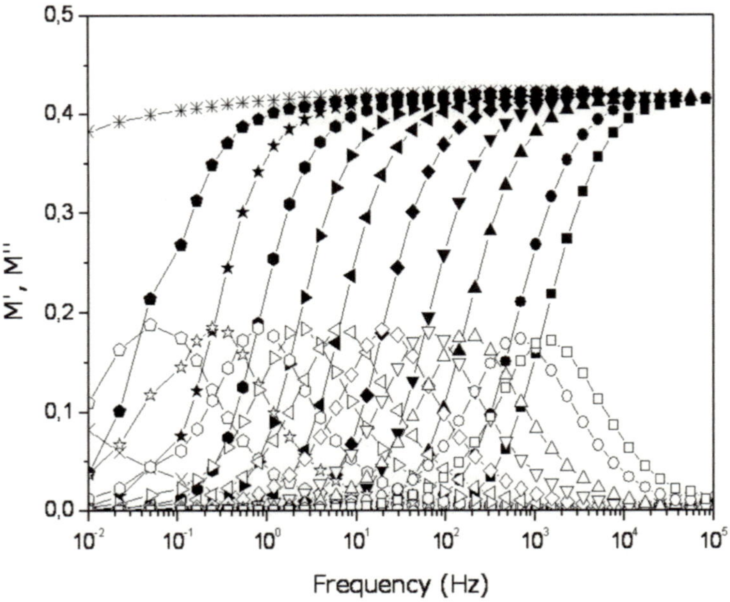

Fig. 16. Components of the complex dielectric modulus in the frequency domain for KN6 at several temperatures: (square) 398K, (circle) 388K, (up triangle) 378K, (down triangle) 368K, (diamond) 358K, (left triangle) 348K, (right triangle) 338K, (hexagon) 328K, (star) 318K, (pentagon) 308K and (plus) 298K; solid symbols for real dielectric modulus component and open symbols for loss dielectric modulus component.

In table 2 are given the parameters of the fitting the loss peak observed in Figure 16, for M″(w) spectra by mean of the HN equation for the KN6. In the same table, we also give the results obtained for the other SILLPs. The skewed form of the curves proceeds from ion-ion interactions and the high values of the β_{HN} parameter suggest the absence of those interactions because of the combined effect of the relatively low concentration of ionic species in those materials and the rigidity of the polymeric matrix that hinders ion-ion interactions. Actually, a second parameter that is in good agreement with this reasoning can be obtained from the full width at half maximum of the M″ peak in Figure 16. The value of this parameter for KN5, KN6, KN7, KN8, KN9 and KN10 is 1.4, 1.3, 1.3, 1.4, 1.2 and 1.2, respectively, and it is higher than the one corresponding to a single Debye peak (1.144) This suggest that the conductive peak includes not only pure ionic contributions but also electrodes polarization effects. It has been proven before that ion-ion interaction contributes to the width of the loss peak.[45]

The f_{max} is the frequency at which M″(f) reach the maximum value and $\tau_{max}=1/2\pi f_{max}$, where the ratio between τ_{max} and τ_{HN} is defined by the equation[46]

$$\tau_{max} = \tau_{HN}\left[\frac{\sin\left(\frac{\pi(\alpha_{HN})\beta_{HN}}{2(\beta_{HN}+1)}\right)}{\sin\left(\frac{\pi(\alpha_{HN})}{2(\beta_{HN}+1)}\right)}\right]^{\frac{1}{\alpha_{HN}}} \quad (9)$$

In Figure 17 we have plotted the dependence of $f_{max}=1/2\pi\tau_{max}$ (characteristic relaxation times of the modulus) obtained for KN5, KN6, KN7, KN8, KN9 and KN10 in the form of a Arrhenius plot. The variation of the dielectric loss modulus with temperature in the low frequency region obeys the Arrhenius behavior with activation energies of (22.8±0.7), (27.8±0.6), (20,4±0,4), (18,9±0,7), (25,1±0,5), and (28,5±0,5), kcal/mol for KN5, KN6, KN7, KN8, KN9 and KN10, respectively. A comparison between this values and the obtained from the conductivity calculated from the loss permittivity and from Bode diagram are in complete agreement, and on the other hand, these values are nearly similar to those obtained before in the analysis of the ionic conductivity by impedance spectroscopy indicating that ionic transport in the SILLPs is the origin of them. In the imaginary part of the modulus spectra a relaxation peak is observed for all the SILLPs for the conductivity processes of different intensity. The intensity of the modulus relaxation peak is 0.18, 0.19, 0.16, 0.23, 0.22 and 0.15, for KN5, KN6, KN7, KN8, KN9 and KN10, respectively; whereas no peak is observed in the dielectric spectra. Similar results have been reported by Fu et al.[47] for PPG-$LiCF_3SO_3$ system with higher salt concentration. So the conduction in polymer electrolytes takes place through charge migration of ions between coordinated sites of the polymer along with its segmental relaxation.

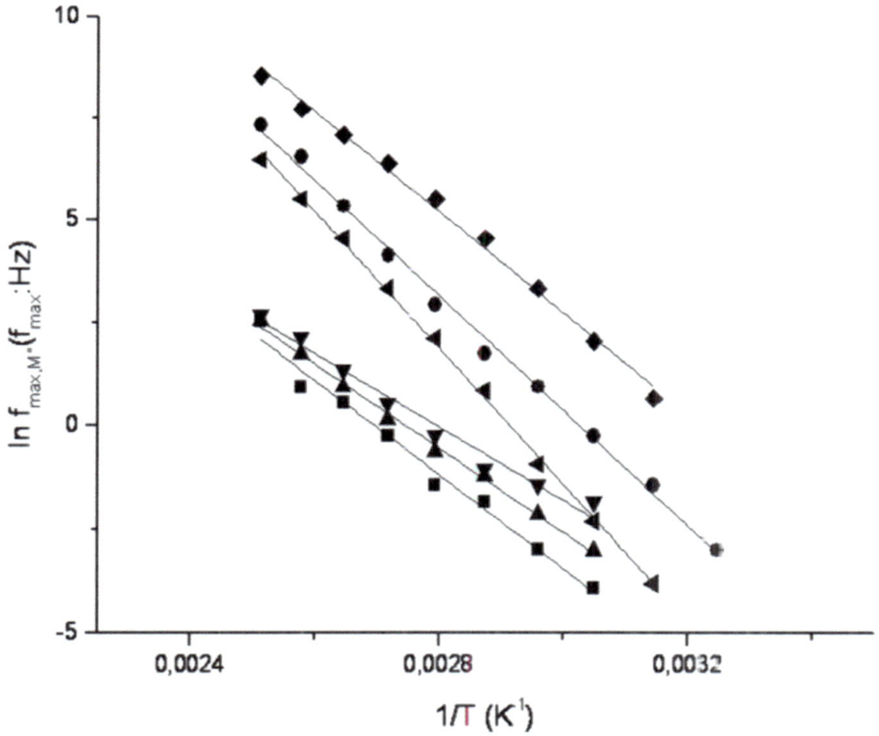

Fig. 17. Arrhenius behaviour of conduction relaxation times for the main dielectric mechanisms recorded in the SILLPs: KN5 (squares), KN6 (circles), KN7 (up triangles), KN8 (down triangles), KN9 (diamonds) and KN10 (left triangles).

For many ionically conducting materials, the M''(f) curves exhibit skewed shapes and, as a result, the complex dielectric modulus can be expressed in terms of the Laplace transform of the decay Kohlrausch-Williams-Watts (KWW) relaxation function.[48,49] The fit of the loss peak given at low frequencies for M"(w) spectra to the Fourier transform of the Kohlrausch-Williams-Watts (KWW) function[50] the values obtained for β_{KWW} of the samples KN5 and KN6 are respectively,[29] β_{KWW} = 0.77 and 0.78 at 398 K. The skewed form of the curves proceeds from ion-ion interactions and the high values of β_{KWW} parameter suggests absence of these interactions as consequence of the combined effect of the relatively low concentration of ionic species in these materials and the rigidity of the polymeric matrix that hinder ion-ion interactions. Also the crosslinks will inhomogeneously broaden the peak and to make it symmetrical. Actually another fact that can justify this can be taken from the full width at half maximum of the M" peak in Figure 18 for KN5 and KN6 it is about 1.5 decade at 338K (65°C), 368K (95°C) and 398K (125°C), respectively. The agreement between the experimental data and the fitting curves (solid lines) is very good. From the values showed in table 2 we can observe that α and β values from both frequency and time domains, for all the SILLPs, are not dependent on temperature. A similar behavior has been reported by Alvarez et al. for poly(hydroxyl ether of bisphenol-A) polymers.[51]

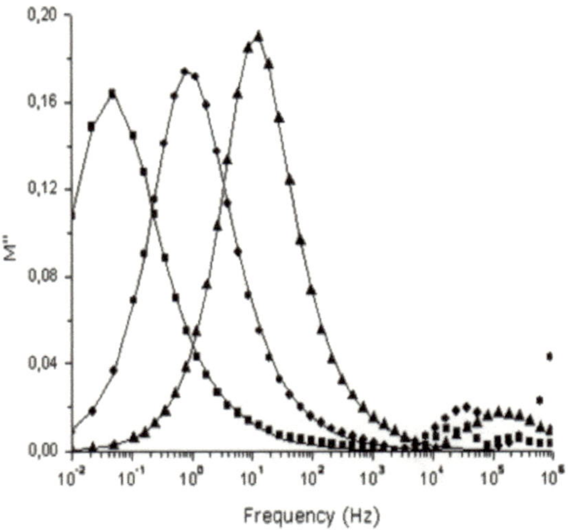

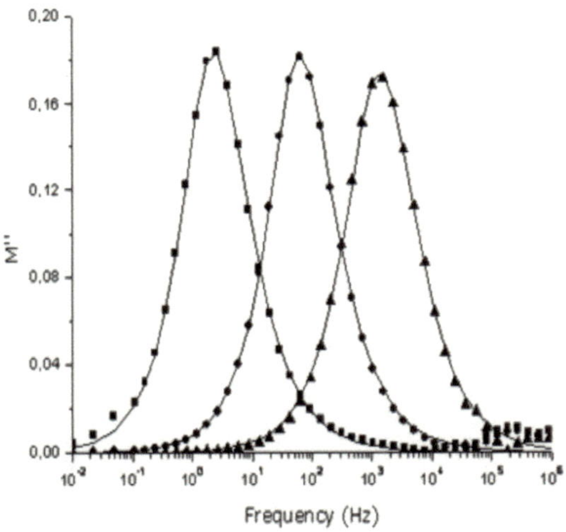

Fig. 18. Isothermal loss modulus peaks curves and their fits to the Havriliak-Negami expression at 338K (right triangles), 368K (down triangles) and 398 K (squares), respectively. Left: KN5; Rigth: KN6.

The lack of ion-ion interactions is also made evident in the fact that the conductivity of SILLPs are a thermally activated process, whereas in the case of supercooled ionic liquids with high ion-ion interactions the variation of the conductivity are governed by the volume.[52] At high frequencies, the secondary absorptions arising from local motions of moieties involving the imidazolium group which presumably arise from rotations about the C^{ar}-CH_2-N bonds of the ILs moieties, are affected by the nature of the counterion.

T	KN5	KN6	KN 7	KN8	KN9	KN10
(K)	$10^{10}\times\sigma$ S/cm	$10^{8}\times\sigma$ S/cm	$10^{10}\times\sigma$ S/cm	$10^{10}\times\sigma$ S/cm	$10^{8}\times\sigma$ S/cm	$10^{8}\times\sigma$ S/cm
398	19.0	14.5	14.0	21.3	53.0	22.8
388	8.6	7.0	7.4	11.8	30.8	10.6
378	4.8	3.0	3.5	5.6	13.8	4.6
368	2.1	1.1	1.6	2.2	8.0	1.5
358	1.1	0.39	0.74	1.0	2.8	0.47
348	0.16	0.13	0.43	0.52	0.95	0.14
338	-	0.05	0.16	0.20	0.29	0.038
328	-	0.016	0.05	0.11	0.085	0.079
318	-	0.005	-	-	0.031	0.013
308	-	0.0008	-	-	0.0047	0.0003
298	-	-	-	-	-	-
E_{act} Kcal/mol	20.2±0.4	26.2±0.4	17.2±0.2	17.0±0.2	25.5±0.5	29.4±0.6

Table 1. Conductivities and activation energies for the KN5, KN6, KN7, KN8, KN9 and KN10 SILLPs obtained from the plateau of the Bode diagrams.

SILLPs	Temperature (K)	ΔM	α	β	Max. Width	τ_{HN} (s)	τ_{max} (s)
KN5							
	328	0.41	1.0	0.58	1.48	11.5	7.4
	338	0.5	0.81	0.77	1.74	4.5	3.4
	348	0.50	0.80	0.80	1.73	1.5	1.2
	358	0.49	0.83	0.80	1.67	0.57	0.45
	368	0.49	0.86	0.78	1.60	0.23	0.18
	378	0.48	0.89	0.77	1.54	0.09	0.07
	388	0.48	0.99	0.76	1.31	0.06	0.05
	398	0.48	0.93	0.77	1.43	0.02	0.01
KN6							
	328	0.39	0.88	0.68	1.65	1.47	1.02
	338	0.51	0.73	0.89	1.91	1.34	1.16
	348	0.48	0.83	0.77	1.68	0.78	0.60
	358	0.47	0.88	0.71	1.59	0.37	0.27
	368	0.46	0.87	0.75	1.59	0.15	0.11
	378	0.46	0.90	0.71	1.5	0.07	0.05
	388	0.46	0.90	0,.72	1.54	0.03	0.02
	398	0.46	0.90	0.71	1.53	0.02	0.01

SILLPs	Temperature (K)	ΔM	α	β	Max. Width	τ_{HN} (s)	τ_{max} (s)
KN7							
	328	0.39	0.88	0.68	1.70	1.47	1.0200
	338	0.51	0.72	0.88	1.90	1.34	1.1562
	348	0.48	0.83	0.77	1.70	0.78	0.5954
	358	0.47	0.88	0.71	1.60	0.37	0.2712
	368	0.46	0.87	0.75	1.60	0.15	0.1133
	378	0.46	0.90	0.72	1.50	0.07	0.0530
	388	0.46	0.90	0.72	1.50	0.03	0.0246
	398	0.46	0.90	0.72	1.50	0.017	0.0125
KN8							
	328	0.67	0.84	0.77	1.4	1.4	0.135
	338	0.59	0.92	0.77	1.3	1.01	0.988
	348	0.54	0.94	0.77	1.25	0.55	0.52
	358	0.53	0.98	0.77	1.24	0.254	0.24
	368	0.52	1	0.75	1.24	0.119	0.108
	378	0.52	1	0.77	1.24	0.05	0.046
	388	0.52	1	0.76	1.25	0.025	0.023
	398	0.51	1	0.77	1.25	0.0128	0.012
KN9							
	328	0.44	1	0.88	1.21	0.36	0.33
	338	0.44	0.93	0.93	1.33	0.089	0.08
	348	0.48	1	0.82	1.25	0.025	0.02
	358	0.48	1	0.83	1.24	0.0067	0.0058
	368	0.48	1	0.82	1.25	0.002	0.0019
	378	0.48	1	0.84	1.23	0.00077	0.00068
	388	0.48	1	0.86	1.22	0.00031	0.00028
	398	0.48	1	0.82	1.25	0.00016	0.00014
	328	0.48	1	0.82	1.25	0.00008	0.000068
	338	0.48	1	0.84	1.24	0.00004	0.000037
KN10							
	328	0.33	0.98	0.82	1.29	0.460	0.39
	338	0.32	0.98	0.85	1.26	0.090	0.08
	348	0.32	0.99	0.85	1.25	0.024	0.021
	358	0.32	0.99	0.84	1.24	0.007	0.006

SILLPs	Temperature (K)	ΔM	α	β	Max. Width	τ_{HN} (s)	τ_{max} (s)
	368	0.32	0.99	0.85	1.24	0.002	0.0018
	378	0.32	0.99	0.86	1.24	0.0007	0.00065
	388	0.32	0.99	0.85	1.25	0.0003	0.00028
	398	0.32	0.99	0.85	1.25	0.00015	0.00013

Table 2. Parameters obtained from fit to the experimental data by the empirical Havriliak-Negami model function given by means of equation (8) for KN5, KN6, KN7, KN8, KN9 and KN10, respectively.

11. Conclusions

From the beginning, the study of bulk ILs has been associated with the development of new families of electrolytes. The reported needs for the preparation of new and efficient solid electrolytes was, on the other hand, at the origin of the preparation of polymeric species containing IL-like species. Up to now there has been a logical association between the preparation of linear polymeric SILLPs and their application as solid electrolytes. The high level of mobility that can be associated to those ionic materials, very often presenting low Tg values, makes reasonable this assumption. Nevertheless, the present results clearly highlight that SILLPs based on crosslinked polymeric matrices can also be of great interest in this field. The data obtained by different techniques reveal that preparation of polystyrene-divinylbenzene polymers containing imidazolium subunits is associated with the formation of polymeric surfaces to which most of the physico-chemical properties of the related bulk IL have been transferred. The modification of structural parameters in the cation, the methatesis of the anion or the variations in the loading of IL-like subunits are parameters that allow a simple modulation of the properties of the resulting SILLPs. Dielectric relaxation measurements have revealed the presence of an appreciable conductivity on those polymeric systems. A thorough theoretical analysis is possible from those experiments, being possible a complete characterization of those SILLPs as solid electrolytes. The analysis of the variation of the corresponding data with the structural modifications provides the potential for modeling the macroscopic structure for those functional polymeric surfaces, providing key information for many different applications considered for those materials. The proper understanding of the behavior of those crosslinked polymers as solid electrolytes opens the way for the design and preparation of new materials for electrical and electronic applications, in particular taking into consideration the porous morphology and the high level of structural variations available.

12. Acknowledgments

This work was supported by the Instituto de la Pequeña y Mediana Industria Valenciana (IMPIVA), Grant IMIDIC 2009/155. We thank the Spanish Ministerio de Ciencia y Tecnología (CTQ2008-04412/BQU and CTQ2008-04309/BQU) and Bancaixa-UJI (P1 1A2009-58) for financial support.

13. References

[1] P. Wasserscheid, T. Welton, *Ionic Liquids in Synthesis*, Wiley-VCH, Weinheim, 2007; b) R. D. Rogers, G. A. Voth, *Acc. Chem. Res.* 2007, 40, Ionic Liquids special issue.

[2] N. Sun, H. Rodríguez, M. Rahman, R. D. Rogers, *Chem. Commun.*, 2011, *47*, 1405; b) M. Palacio, B. Bhushan, *Tribol. Lett.*, 2010, *40*, 247; c) N. V. Plechkova, K.R. Seddon, *Chem. Soc. Rev.*, 2008, *37*, 123.

[3] R. F. M. Frade, C. A. M. Afonso, *Human and Experimental Toxicology*, 2010, *29*, 1038. b) T. P. Thuy Pham, C.-W. Cho, Y.-S. Yun, *Water Research*, 2010, *44*, 352. c) D. Coleman, N. Gathergood, *Chem. Soc. Rev.*, 2010, *39*, 600.

[4] C. P. Mehnert, *Chem. Eur. J.*, 2004, *11*, 50; b) A. Riisager, R. Fehrmann, S. Flicker, R. van Hal, M. Haumann, P. Wasserscheid, *Angew. Chem. Int. Ed.*, 2005, *44*, 815; c) Y. Gu, G. Li, *Adv. Synth. Catal.*, 2009, *351*, 817; d) C. Van Doorslaer, J. Wahlen, P. Mertens, K. Binnemans, D. De Vos, *Dalton Trans.*, 2010, *39*, 8377.

[5] H. Olivier-Bourbigou, L. Magna, D. Morvan, *Applied Catal. A*, 2010, *373*, 1; b) J. Dupont, R. F. De Souza, P. A. Z. Suarez, *Chem. Rev.*, 2002, *102*, 3667.

[6] a) P. Lozano, *Green Chem.*, 2010, *12*, 555; b) M. Moniruzzaman, K. Nakashima, N. Kamiya, M. Goto, *Biochem. Eng. J.*, 2010, *48*, 295.

[7] C. F. Poole, S. K. Poole, *J. Chrom.*, 2010, *1217*, 2268; b) H. Zhao, S. Xia, P. Ma, *J. Chem. Tech. Biotech.*, 2005, *80*, 1089.

[8] A. Stark, *Energy Env. Science*, 2011, *4*, 19; b) A., Pinkert, K. N., Marsh, S. Pang, M. P.Staiger, *Chem. Rev.*, 2009, *109*, 6712.

[9] M. Armand, F. Endres, D. R. MacFarlane, H. Ohno, B. Scrosati, *Nature Mat.*, 2009, *8*, 621.

[10] N. Srivastava, T. Tiwari, *E-Polymers*, 2009, art. no. 146.

[11] H. Ohno, *Bull. Chem. Soc. Jpn.*, 2006, *79*, 1665.

[12] J. Yuan, M. Antonietti, *Polymer*, 2011, *52*, 1469; b) M. D. Green, T. E. Long, *Pol. Rev.*, 2009, *49*, 291.

[13] K. Nakamura, T. Saiwaki, K. Fukao, *Macromolecules*, 2010, *43*, 6092; b) M. Lee, U. H. Choi, R. H. Colby, H. W. Gibson, *Chem. Mat.*, 2010, *22*, 5814.

[14] N. Karbass, V. Sans, E. García-Verdugo, M. I. Burguete, S. V. Luis, *Chem. Commun.*, 2006, 3095; b) P. Lozano, E. García-Verdugo, R. Piamtongkam, N. Karbass, T. DeDiego, M. I. Burguete, S. V. Luis, J. L. Iborra, *Adv. Synth. Catal.*, 2007, *349*, 1077; c) M. I. Burguete, H. Erythropel, E. García-Verdugo, S. V. Luis, V. Sans, *Green Chem.*, 2008, *10*, 401. d) M. I. Burguete, E. García-Verdugo, I. Garcia-Villar, F. Gelat, P. Licence, S. V. Luis, V. Sans, *J. Catal.*, 2010, *269*, 150; e) V. Sans, F. Gelat, N. Karbass, M. I. Burguete, E. García-Verdugo, S. V. Luis, *Adv. Synth. Catal.*, 2010, *352*, 3013.

[15] V. Sans, N. Karbass, M. I. Burguete, V. Compañ, E. García-Verdugo, S. V. Luis, M. Pawlak, *Chem. Eur. J.*, 2011, *17*, 1894.

[16] N. Meine, F. Benedito, R. Rinaldi, *Green Chem.*, 2010, *12*, 1711; b) H. L. Ngo, K. LeCompte, L. Hargens, A. B. McEwen, *Thermochim. Acta.*, 2000, *357-358*, 97; c) M. Koel, *Proc. Estonian Acad. Sci. Chem.*; 2000, *49*, 145. d) K. J. Baranyai,

G. B. Deacon, D. R. MacFarlane, J. M. Pringle, J. L. Scott, *Aust. J. Chem.* 2004, *57*, 145.
[17] C. Chiappe, M. Malvaldi, C. S. Pomelli, *Pure & Appl. Chem.*, 2009, *81*, 767.
[18] A. Nakajima, *Spectrochim. Acta,* 1974, *30A*, 360. b) D. C. Dong, M. A Winnik, *Photochem. Photobiol.*, 1982, *35*, 17. c) P. Lianos, S. Georghiou, *Photochem. Photobiol.*, 1979, *30*, 355.
[19] C. Reichardt, *Solvent Effects in Organic Chemistry*, Wiley-VCH, Weinheim, 3rd edn., 2003; b) C. Reichardt, *Chem. Soc. Rev.* 1992, *21*, 147, c) C. Reichardt, *Chem. Rev.* 1994, *94*, 2319.
[20] C. Reichardt, *Green Chem.*, 2005, *7*, 339; b) S. V. Dzyuba, R. A. Bartsch, *Tetrahedron Lett.*, 2002, *43*, 4657.
[21] D. C. Dong, M. A. Winnik, *Can. J. Chem.* 1984, *62*, 2560.
[22] K. A. Fletcher, I. A. Storey, A. E. Hendricks, S. Pandey, S. Pandey, *Green Chem.*, 2001, *3*, 210. b) D. S. Karpovich, G. J. Blanchard, *J. Phys. Chem.* 1995, *99*, 3951.
[23] D. C. Sherrington, *Chem Commun.* 1998, 2275
[24] E. Riande, R. Diaz-Calleja, *Electrical Properties of Polymers*; Marcel Dekker: New York, 2004; chapter 8.
[25] J.R. MacDonald, *Impedance Spectroscopy. Emphasizing solid materials and systems,* Chapter 2, pp. 87-88. Wiley-Interscience, 1987.
[26] W.W. Bode, *Network Analysis and Feedback Amplifier Design,* Van Nostrand, Princenton, N.J., 1956.
[27] H. Nyquist. *Phys. Rev.*, 1928, *32*, 10.
[28] A. K. Jonscher, *Nature*, 1977, *267*, 673.
[29] A. Garcia-Bernabé, V. Compañ, M. I. Burguete, E. Garcıa-Verdugo, N. Karbass, S. V. Luis, E. Riande, *J. Phys. Chem. C,* 2010, *114*, 7030.
[30] N. G. McCrum, B. E. Read, G. Williams, *Anelastic and Dielectric Effect in Polymeric Solids,* Dover Publications, New York, 1991.
[31] K. W. Wagner, *Arch. Electrotechnol.* 1914, *2*, 371.
[32] K. W. Wagner, *Arch. Electrotechnol.* 1914, *3*, 67.
[33] R. W. Sillars, *Proc. Inst. Electr. Eng. Lond.*, 1937, *80*, 378.
[34] R. H. Wiley, B. Davis, *J. Polym. Sci.*, 1960, *46*, 423.
[35] C. T. Moynihan, *J. Non-Cryst. Sol.*, 1994, *172-174*, 1395; *ibid* 1996, *203*, 359.
[36] C. T. Moynihan, *Solid State Ionics*, 1998, *105*, 175.
[37] S. R. Elliot, *J. Non-Cryst. Sol.*, 1994, *170*, 97.
[38] B. Roling, *Solid State Ionics*, 1998, *105*, 185.
[39] I. M. Hodege, K. L. Ngai, C. T. Moynihan, *J. Non-Cryst. Solids*, 2005, *351*, 104.
[40] K. L. Ngai, C. León, *Solid State Ionics,* 1999, *125*, 81.
[41] J. S. Kim, *J. Phys. Soc. Jpn.*, 2001, *70*, 3129.
[42] J. Liu, C.-G. Duan, W.-G. Yin, W. N. Mei, R. W. Smith, J. R. Hardy, *J. Chem. Phys.*, 2003, *119*, 2812.
[43] S. Havriliak, S. Negami, *Polymer*, 1967, *8*, 161.
[44] E. Neagu, P. Pissis, L;Apekis, J. L. Gomez-Ribelles, *J. Phys. D: Appl. Phys.*, 1997, *30*, 1551.
[45] K. L. Ngai, J. Habasaki, Y. Hiwatari, C. Leon. *J. Phys.: Condens. Matter.,* 2003, *15*, S1607.

[46] R. Diaz-Calleja, *Macromolecules,* 2000, 33, 8924.
[47] Y. Fu, K., Pathmanathan, J. R. Stevens, *J. Chem. Phys.*, 1991, *94*, 6323.
[48] G. Williams, D.C. Watts, *Trans. Faraday Soc.*, 1970, *66*, 80.
[49] R. Kohlrausch. *Prog. Ann. Phys.*, 1847, *12*, 393.
[50] R. Casalini, P. G. Santangelo, C. M., Roland. *J. Phys. Chem. B,* 2002, *106*, 11492.
[51] F. Alvarez, A. Alegría, J. Colmenero,. *Physical Review B*, 1991, *44*, 7306.
[52] H. Vogel, *Phys. Z.*, 1921, *22*, 645; b) V. G. Tammann, W. Hesse, *Z. Anorg. Allg. Chem.*, 1926, *156*, 245; c) G. S. Fulcher, *J. Am. Ceram. Soc.*, 1925, *8*, 339.

6

Ionic Liquids for the Electric Double Layer Capacitor Applications

Takaya Sato, Shoko Marukane and Takashi Morinaga
Tsuruoka National College of Technology
Japan

1. Introduction

The properties of ionic liquids (ILs) include non-volatility, non-flammability, and relatively high ionic conductivity (Rogers & Seddon, 2001). As novel green, reusable solvents that can substitute for organic solvents, ionic liquids have attracted much attention as good media in organic synthesis and other chemical processes (Zhao et al., 2009, 2011). At the same time, some studies have been reported which aim to improve the high-temperature safety and durability of such electrochemical devices as lithium rechargeable batteries (Sato et al., 2004), electric double layer capacitors (EDLC) (Sato et al., 2004), and titanium oxide dye-sensitized solar cells (Papageorgiou et al., 1996).

The electric double layer capacitor (EDLC) is an energy storage device based on the operating principle of the electric double-layer that is formed at the interface between an activated carbon material and an electrolyte. Various solvents and salts (solutes in otherword) are available, offering specific advantages such as high capacitance and low temperature performances. Generally, an organic electrolyte that is a solid quaternary ammonium salt, such as *N,N,N,N*-tetraethylammonium tetrafulucroborate (TEA-BF_4), dissolved in the high dielectric constant solvent propylene carbonate (PC) has been used for high voltage EDLCs of 2V or more. This device stores electricity physically, and lacks the chemical reactions found in rechargeable batteries during charging and discharging (Zheng et al., 1997). Therefore, compared to rechargeable batteries, the EDLC has a remarkably long cycle life and high power density. Such devices are now widely used in power electronics for peak power saving and back up memories, and in electronic power supplies for automated guided vehicle systems and construction equipment. One of their most promising applications is for use in transportation, especially in hybrid electric vehicles (HEVs). However, some issues in the development of EDLCs remain.

1. A lower energy density compared with lithium ion secondary batteries.
2. Flammable electrolytes raise safety concerns.
3. Low cycle durability in the high temperature region.
4. Poor charge-discharge properties at low temperatures.

In order to overcome these challenges, we are actively pursuing research and development in the use of ionic liquids for the electrolyte of EDLCs.

The energy *E* stored in an EDLC is proportional to its capacity, *C*, as well as the voltage *V* applied between the positive and the negative electrodes:

$$E = (CV^2) / 2 \tag{1}$$

In order to increase the energy density of EDLCs, it is necessary to increase *C* and *V*. A higher ionic concentration in the electrolyte is required for the improvement of the *C* value. And, since the energy density is proportional to the square of the cell voltage, increasing the cell voltage is a very important development target for EDLCs. Conventional electrolytes for EDLCs are solutions in which a certain type of solid ammonium salt is dissolved in an organic solvent. In this case, a high ionic concentration level is not achieved due to the restriction of the solubility of the ammonium salt; many kinds of solid ammonium salt have solubilities of only about 2M. On the contrary, ionic liquids are composed of dissociated cations and anions with a high concentration, and it is possible to use them alone as an electrolyte for EDLCs, that is to say, the ionic liquid electrolyte is both the solute and solvent. In this case, a high capacity can be expected because there are extremely high concentrations (3-5 M) of ion species that contribute to double layer formation in the ionic liquid. Furthermore, the non-flammability of the ionic liquid should be an attractive feature for the enhanced safety of EDLCs.

An ionic liquid with a wide electrochemical stability window is needed for the achievement of a high energy density EDLCs. Some cations of ionic liquids that have been the subjects of many previous studies are shown in Fig. 1.

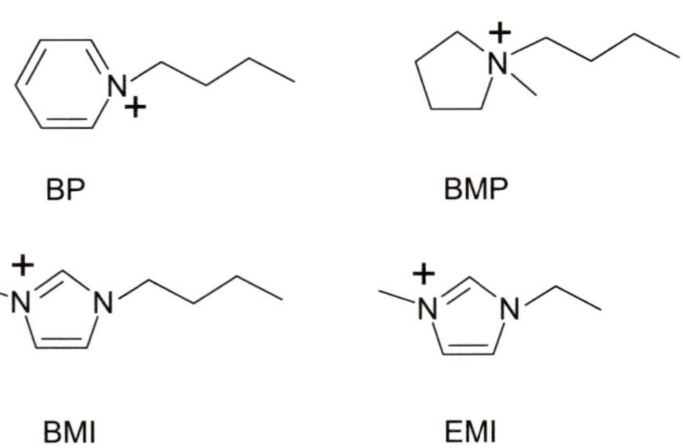

Fig. 1. Some cation structures of ionic liquids. BP: butylpyridinium, BMP: butylmethylpyrrolidinium, BMI: 1-butyl-3-methylimidazolium, EMI: 1-ethyl-3-methylimidazolium.

At first, many researchers focused on aromatic type ionic liquids, such as 1-ethyl-3-methylimidazolium tetrafluoroborate (EMI-BF_4), which has a relatively low viscosity and high ionic conductivity (Koch et al.,1995; Koch et al.,1994; McEwen et al.,1999; McEwen et al.,1997) for use in a variety of applications for electrochemical devices, including batteries and capacitors. However, since aromatic quaternary ammonium cations, such as imidazolium and pyridinium, have relatively low cathodic stability, electrochemical devices using these ionic liquid as electrolyte have not yet been shown to be practical (Ue et al.,2002). On the other hand, some papers have been published on the use of aliphatic quaternary ammonium-based ionic liquids, which would be expected to have a higher cathodic stability; this is due to the fact that the small size aliphatic quaternary ammonium-based ionic liquid has a relatively high melting temperature, compared with the aromatic ones. MacFarlane and co-workers reported that aliphatic quaternary ammonium cations which had relatively short (C1–C4) alkyl chains could not easily form an ionic liquid near room temperature (Sato, 2004, as cited in Sun et al.,1997,1998a,1998b & MacFarlane et al.,1999,2000). However, Angell and co-workers and Matsumoto et al reported that some asymmetric and short-chain aliphatic quaternary ammonium cations with a methoxyethyl or methoxymethyl group on the nitrogen atom formed ionic liquids below room temperature (Cooper & Angell,1986; Emanuel et al.,2003) (Matsumoto et al.,2000,2001,2002). Recently, the major advances in research and development of ionic liquid electrolytes have mainly concerned quaternary ammonium-based ionic liquids where the anion has a relatively high anodic stability, such as BF_4, bis(trifuluoromethanesulfonyl)imide ($^-N(SO_2CF_3)_2$: TFSI) and other fluorinated anions.

2. Neat ionic liquid electrolytes for EDLC application

2.1 Aliphatic quaternary ammonium type ionic liquids having a methoxyethyl group on the nitrogen atom

Generally, small sized aliphatic quaternary ammonium cations cannot easily form an ionic liquid, however, by attaching a methoxyethyl group to the nitrogen atom, many aliphatic quaternary ammonium salts can form ionic liquids with BF_4^- and TFSI anions. For instance, DEME-BF_4 and DEME-TFSI are novel ionic liquids whose liquid state covers a wide temperature range (Fig. 2) (Maruo et al.,2002).

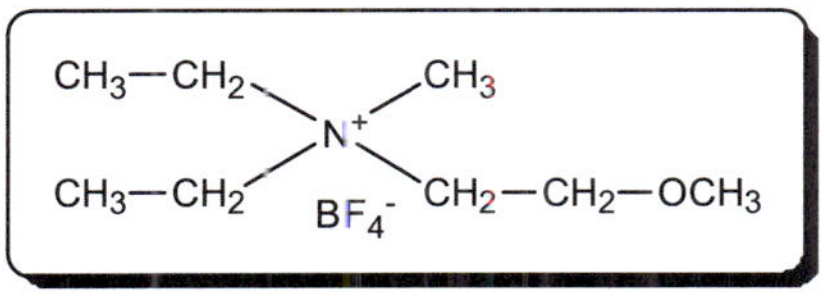

Fig. 2. Schematic illustration of the molecular structure of the DEME cation.

Since the electron donating feature of an oxygen atom in a methoxyethyl group weakens the cation's positive charge, the electrostatic binding between the ammonium cation and anion weakens, and an ionic liquid forms. The limiting reduction and oxidation potentials (E_{red} and E_{oxd}) on platinum of the ionic liquids were measured by cyclic voltammetry at room temperature as shown in Fig. 3.

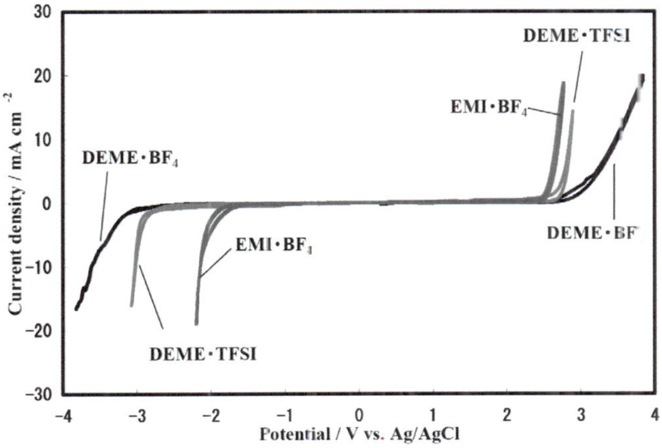

Fig. 3. Cyclic voltammogram of the ionic liquids based on the DEME cation and EMI-BF_4 at 25 °C. Scan rate: $1mVs^{-1}$, working and counter electrode: platinum, reference electrode: Ag/AgCl electrode.

The E_{red} and E_{oxd} were defined as the potential where the limiting current density reached $1mAcm^{-2}$. The potential window between the onset of E_{red} and E_{oxd} was 6.0V for the DEME-BF_4 and 4.5V for the EMI-BF_4, respectively. Since the E_{red} of the DEME-based ionic liquid was approximately 1V lower than that of the EMI-BF_4, we realized that DEME had a higher cathodic stability than aromatic type ionic liquids since it does not have a π-electron conjugated system. Compared with DEME-TFSI, DEME-BF_4 is a little more stable in the limiting reduction potential. Since DEME-BF_4 has the largest potential window so far reported, it should be suitable as an electrolyte for a high operating voltage EDLC.

2.2 The electric double layer capacitor using DEME-BF_4 as an electrolyte

The EDLC is an energy storage device based on charge storage between an activated carbon and an electrolyte solution. It may be possible to improve the safety and durability at high temperatures using an ionic liquid instead of an inflammable organic solvent. However, a typical ionic liquid, EMI-BF_4 cannot be used for an EDLC operating at a high voltage because of decomposition on the anode. We tried to prepare a test cell for the EDLC which would be stable at high temperature using DEME-BF_4 as an electrolyte. The cell configuration of a demonstration EDLC is shown in Fig.4.

Table 1 summarizes the capacitance *C* of the EDLC using various electrolytes at temperatures higher than room temperature. The capacity and the Coulombic efficiency between the charged and discharged capacity were obtained from the charge and discharge curves. As Table 1 shows, the EDLCs using ionic liquids showed higher discharge capacities than those using the standard electrolyte TEA-BF_4 in PC solution at all the temperatures tested. Within high ionic concentration electrolyte, the double layer might be specifically formed on the surface, while ion adsorption does not occur at a low ionic concentration, although the reason for this behavior is not clear.

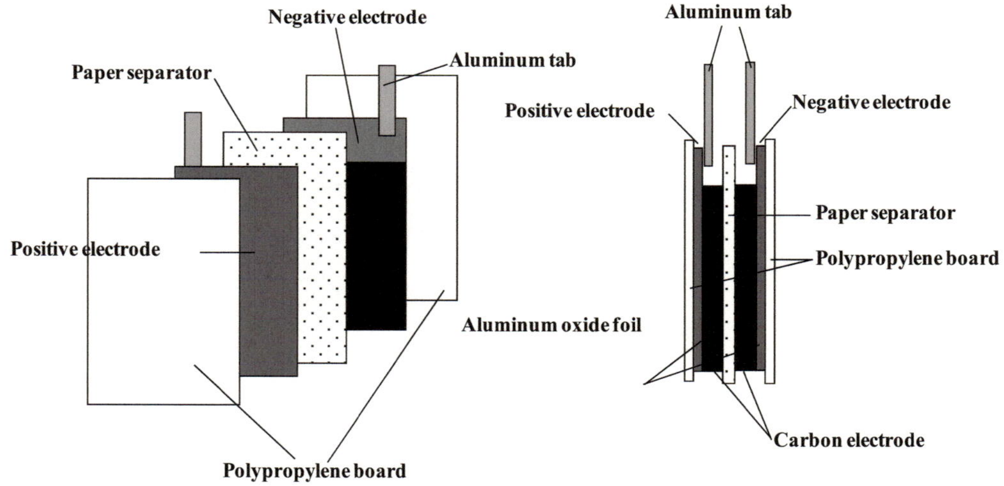

Fig. 4. EDLC test cell configuration.

Measurement temperature, T (℃)	Electrolytes		
	DEME-BF_4	EMI-BF_4	1M TEA-BF_4 in PC
	Discharge capacitance C^a (Fg^{-1}) and coulombic efficiency^b (%)		
25	25.4(98.5)	26.3(90.5)	21.3(98.3)
40	25.5(96.0)	26.6(84.7)	20.2(90.1)
70	25.7(88.9)	23.9(60.9)	20.1 (68.7)
100	23.7(78.5)	n.m.^{c} (n.m.)	20.1 (40.0)
130	21.3(58.7)	n.m. (n.m.)	n.m. (n.m.)
150	17.5(33.7)	n.m. (n.m.)	n.m. (n.m.)

[a] C was calculated by $C = 2WV^{-2}$. [b] Coulombic efficiency:[discharged capacity (mA h)/charged capacity (mA h)] × 100. Values of coulombic efficiency. [c] n.m. , not measurable due to unstable cell performance caused by gas evolution.

Table 1. Double layer capacitance of EDLC with various electrolytes above room temperature.

However, for the EDLC using EMI-BF_4 we observed both, the generation of gas from the decomposition of the ionic liquid and a remarkable drop in the capacity at 70°C. In the case of TEA-BF_4 in PC, the large drop in capacity was observed at 100 ◦C. On the other hand, the EDLC using DEME-BF_4 exhibited little gas generation and did not show such a large capacity drop at 100°C and capacity could be observed even at 150°C. These results indicate that the cell using DEME-BF_4 is more stable compared to the EDLC using the general organic electrolyte TEA-BF_4/PC. At high temperatures, the cell using DEME-BF_4 has the highest relative Coulombic efficiency of the tested electrolytes. Fig. 5 demonstrates that, even at 100 ◦C, the EDLC using DEME-BF_4 showed a practical good level of durability over cycles of charging and discharging. After 500 charge and discharge cycles, the capacity loss was just 15% of the initial discharge capacity. It is possible that an EDLC using DEME-BF_4 may be a practical energy storage device, which can stably charge and discharge at high temperature.

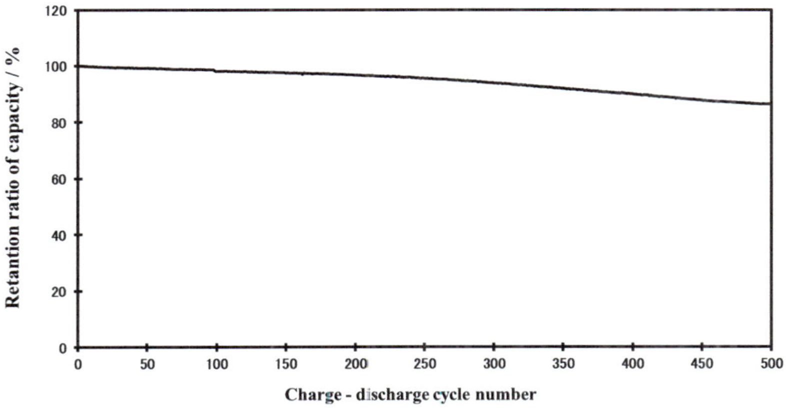

Fig. 5. Cycle test of an EDLC using DEME-BF_4 as an electrolyte at 100 ◦C The test cell was charged by a constant current–constant voltage mode from V = 0 to 2.5V at a constant current (CC) I = 15 mA and then charged to keep constant voltage (2.5 V) for 15 min and discharged to 0V by CC mode.

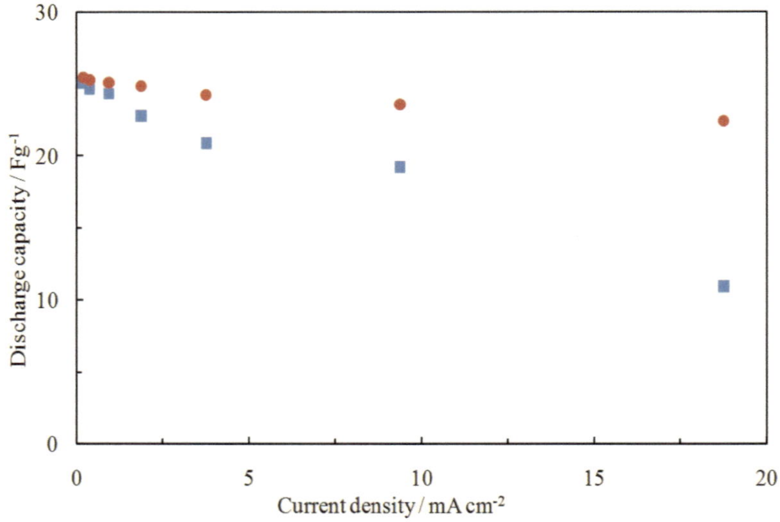

(a)

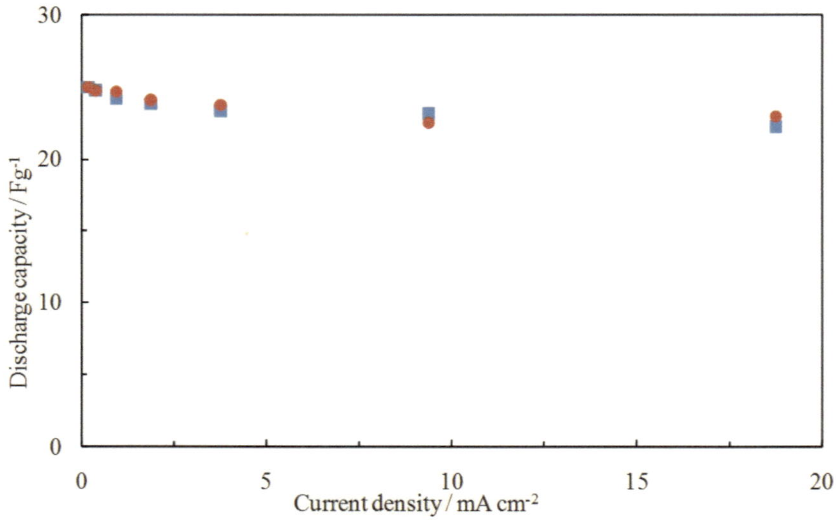

(b)

Fig. 6. The discharge capacity at various discharge currents for an EDLC using DEME-BF_4 (square) and 1M TEA-BF_4 (circle) as an electrolyte; (a) at 25 ◦C, (b) at 40 ◦C.

The relation between the capacity *C* and the discharge current *I* is shown in Fig. 6. At 25◦C the discharge capacity of the EDLC using DEME-BF_4 decreased significantly faster than that using TEA-BF_4 as the discharge current increased. It is probable that the larger decrease in *C* at discharge currents using DEME-BF_4 resulted from the large internal resistance of the EDLC cell when using this high viscosity electrolyte. In the case of an EDLC using an ordinary electrolyte the discharge capacity decreased only by 15% for a 100 times large current. The high viscosity of the ionic liquid decreases the mobility of ionic species. However, in operation at 40 ◦C shown in Fig. 6b, the capacity of the EDLC using DEME-BF_4

is remarkably improved. In this case the capacity of the ionic liquid cell is higher than that of a cell using TEA-BF_4/PC. A slight elevation of the temperature has decreased the viscosity of the ionic liquid, since the effect of a temperature change on the viscosity of ionic liquids is known to be large. We suggest that, at temperatures of 40 ◦C or above, an EDLC using DEME-BF_4 has a practical useable performance comparable to that of a conventional EDLC using an organic electrolyte. Furthermore, the durability at high temperatures of such an ionic liquid cell is higher than that of a conventional EDLC. However, below room temperature, the capacity of an EDLC using this ionic liquid was inferior to that of the conventional EDLCs using solid-ammonium salt / PC system due to the high viscosity of the ionic liquid and the fact that the charge and discharge operation of our demonstration cell might be compromised around 0°C.
On the other hand, Japanese researchers (Ishikawa et al., 2006 and Matsumoto et al., 2006) reported that ILs containing bis(fuluorosulfonyl)imide (FSI) had a quite low viscosity and high ionic conductivity when compared with those based on the TFSI anion, and various ILs containing the FSI anion had some suitable properties for use as the electrolyte of a lithium ion battery. Also, Handa et al., (2008) reported the application of an EMI-FSI ionic liquid as an electrolyte for an EDLC, and found that demonstration cell showed excellent capabilities comparable to one using a solid ammonium salt / PC electrolyte. Further, Senda et al. (2010) reported that fluorohydrogenate ionic liquids (FHILs) exhibit a low viscosity, high ionic conductivity and low melting point compared to other ionic liquids. Their EDLCs using some FHILs were operable even at -40 ◦C, exhibiting a capacitance higher than that of TEA-BF_4 / PC. The use of the ionic liquids with these new fluorine system anions as the electrolyte of various electrochemical devices will probably develop in the future, because it the structure of the anion has a big influence on the viscosity of the ionic liquid.

3. Practical use of ionic liquid as a solute for an EDLC electrolyte

3.1 Use of ionic liquids for EDLC performance improvement at low temperatures

Because a quaternary ammonium type ionic liquid has a higher solubility in the carbonate solvent than the previously investigated solid quaternary ammonium salts, it is possible to make an electrolyte with a high ion concentration, giving the EDLC a high capacitance. Some solid ammonium salts might be precipitated by recombination of dissociated ions the low temperature region due to their low solubility. As a result, an EDLC including a solid ammonium salt solute will suffer a large drop in capacitance at low temperatures. In many cases, an electrolyte including an ionic liquid that has a high solubility in propylene carbonate (PC) exhibit a high ionic conductivity, even at low temperatures, compared to a traditional solid ammonium salt electrolyte. Also, DEME-BF_4 can be homogeneously mixed with PC to produce a uniform electrolyte solution and it does not cause precipitation of the salt, even at temperatures below -40 ◦C. Figure 7 shows the relation between the ionic liquid concentration in PC and the solution viscosity at room temperature.
Many ionic liquids cause a remarkable viscosity decrement for just a small amount of solvent addition. It is the simplest method by which the problem of the high viscosity of the ionic liquid can be eased. Nisshinbo Industries Inc. in Japan manufactures successful large size EDLCs that have an extremely attractive high charge rate and discharge performance even at −40 ◦C by the use of an electrolyte including DEME-BF_4 diluted with PC (http:// www.nisshinbo.co.jp /english/r_d/capacitor/index. html).

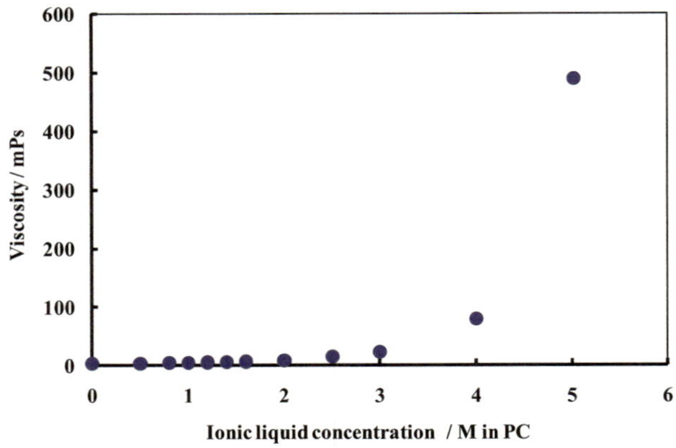

Fig. 7. Viscosity of ionic liquid / PC solution.

3.2 Ionic liquids containing the tetrafluoroborate anion have the best performance and stability for electric double layer capacitor applications

3.2.1 Ionic liquids having a methoxyalkyl group on the nitrogen atom as a solute

Generally, small sized aliphatic quaternary ammonium cations cannot easily form an ionic liquid; however, by attaching a methoxyalkyl group to the nitrogen atom, many aliphatic quaternary ammonium salts can form ionic liquids with BF_4^- and TFSI anions. In this section, we report on the performance and thermal stability of the EDLCs using various ionic liquids and some solid ammonium salts with methoxyethyl and methoxymethyl groups on the nitrogen atom as a solute in an electrolyte. The evaluation was performed using a large size cell (265F) with strict quality control at the industrial product manufacturing level. Of special interest is to determine which ionic liquid or ammonium salt with methoxyalkyl group shows the most attractive performance at low temperature and good thermal and electrochemical stability in a practical large size EDLC. We compare the direct current resistance of EDLCs at a relatively large current at low temperature, and the capacitance deterioration and the internal resistance increase when continuously charging at high temperature. It has been generally thought that a high viscosity of the electrolyte is detrimental to direct current resistance. However, we report that the direct current resistance of the EDLC depends on the size of the solute anion and is independent of the viscosity and the specific conductivity of the electrolyte. We prepared 14 kinds of ammonium salt that have 6 kinds of cation and 3 kinds of anion species as candidate electrolytes for EDLCs (Yuyama et al. 2006). Nine of these were ionic liquid in nature at 25 ◦C and some ammonium salts were not liquid at room temperature. The melting temperatures are summarized in Table 2. We prepared large size EDLCs using the 14 kinds of salt in Table 2. Because PF_6 and TFSI salts of *N,N*-diethyl-*N*-methoxymethyl-*N*-methylammonium (DEMM) and *N*-methoxymethyl-*N*-methylpyrrolidinium (MMMP) are expected to have a poor EDLC performance from the results of other experiments, they were not used.

3.2.2 Comparison of EDLC capacitance and internal resistance

We evaluated the cell performance and durability of EDLCs using all 14 types of ammonium salts, including 9 ionic liquids. The capacitance of the EDLCs using various ammonium salts with a methoxyalkyl group on nitrogen atom, and some imidazolium salts, in a 1M propylene carbonate solution was determined by charge-discharge cycling from 0 to 3.0V at 25 °C. Fig. 8 shows the variation of the cell capacitance in the 10th cycle as a function of the total molecular weight of the salts (a), and as a function of the molecular weight of the separate cations (b) and anions (c).

Code of cation	cation	Melting temperature / °C		
		BF_4	PF_6	TFSI
DEME	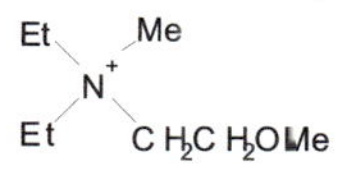	9 [a,b]	23 [a,c]	-91 [a,b,d]
MEMP	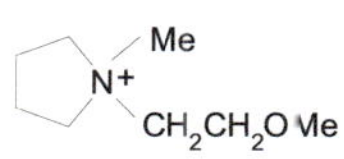	18 [a,b]	34 [a,c]	-95 [a,b,d]
DEMM	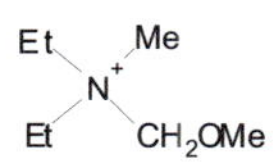	45 [a,b]	--	--
MMTM	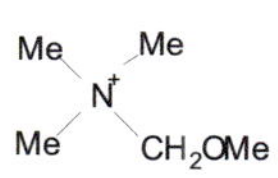	51 [a,b]	204 [a,c]	-2 [a,b]
MMMP	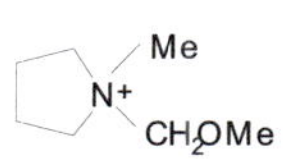	-5 [a,b]	--	--
EMI	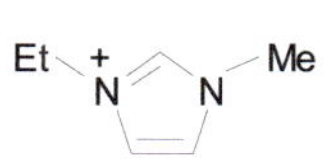	11 [a,b]	60 [a,c]	-16 [a,b]

a 1M of propylene carbonate solution was tested for EDLC electrolyte in this work.
b Melting temperature was measured by DSC.
c Melting temperature was measured by the microscope with heating plate for melting point measurement.
d Glass transition temperature was measured by DSC.

Table 2. Melting temperature of ammonium salts used in this work.

In general, it has been thought that an ion with a high molecular weight shows lower cell capacitance than a small size ion. However, we have previously observed that the EDLC using 1M DEME-BF_4, which a molecular weight of 233.05, has an approximately 10% higher capacitance than a cell using a 1M solution of a conventional ammonium salt,

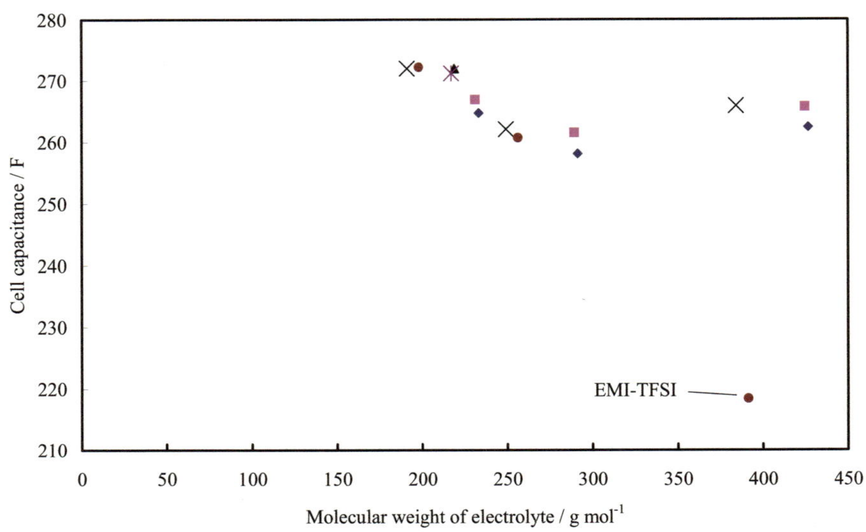
280
270
260
250
240
230
220
210
Cell capacitance / F
0
50
100
150
200
250
300
350
400
450
Molecular weight of electrolyte / g mol^{-1}
EMI-TFSI

(a)

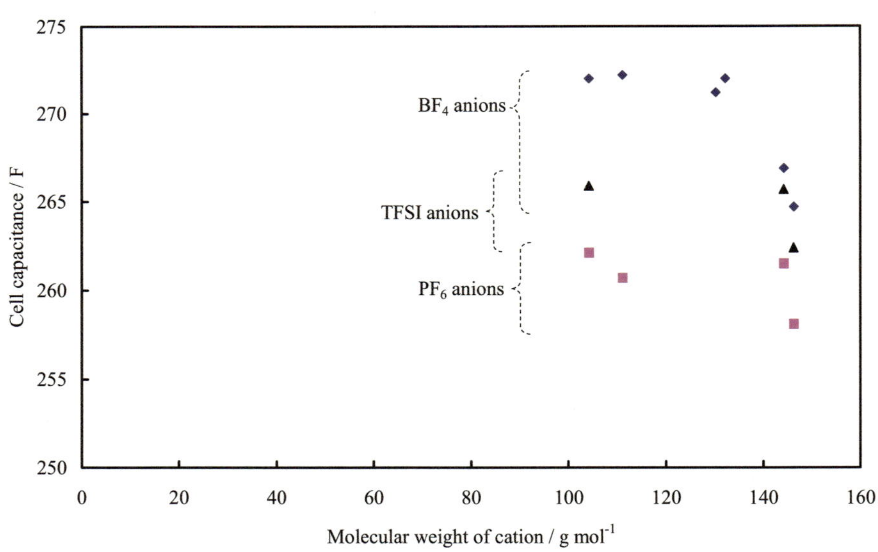
275
270
265
260
255
250
Cell capacitance / F
0
20
40
60
80
100
120
140
160
Molecular weight of cation / g mol^{-1}
BF_4 anions
TFSI anions
PF_6 anions

(b)

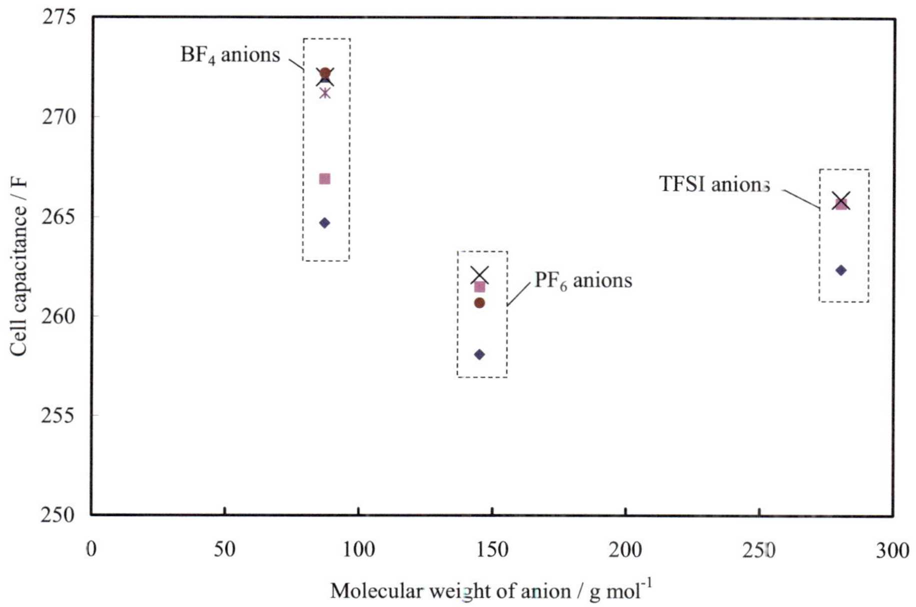

(c)

Fig. 8. The discharge capacitance of EDLCs containing various electrolytes. The test cell was discharged at a constant current I = 0.2 A from V = 3.0 to 0 V at 25 °C. Cell capacitance is plotted against the molecular weight of salts, cations and anions for (a), (b) and (c), respectively. The electrolyte concentration was 1M in PC: (♦) DEME-cation species; (■) MEMP-cation species; (▲) DEMM-cation species; (x) MMTM-cation species; (∗) MMMP-cation species; (●) EMI-cation species in figure (a) and (c). In figure (b): (♦) BF_4-anion species; (■), PF_6-anion species; (▲) TFSI-anion species. The prepared EDLC with 1M DEME-BF_4 in PC had 44 F g^{-1} of the specific capacitance per electrode weight.

tetraethylammonium tetrafluoroborate (TEA-BF_4), which has a molecular weight of 217.6 (Kim et al. 2005). However, in the results shown in Fig. 8(a), the cell capacitance with a 1M EMI-TFSI propylene carbonate solution exhibited an exceptionally very low value, and a clear correlation cannot be found between the molecular weights of the salts and cations and the cell capacitance based on the data in Fig. 8(a and b). We did not observe a large difference in the cell capacitance with the cation used even though the molecular weights of the cations varied by approximately 1.4 times. However, we realized from the data shown in Fig. 8(c) that the capacitance of the cell depends not on the total molecular weight of the salt or that of the cation, but on that of the anion. EDLCs using the BF_4 anion species had higher capacitances than those using other anion species. Because the deviation in the measured capacitance of each tested EDLC was only about 0.4%, the 2 F or more differences in capacitance found between different cells are significant. The ammonium salts with a methoxymethyl group show somewhat larger capacitance than ones with methoxyethyl group. In addition, it is undoubtedly the case that the state of the ammonium salt itself at room temperature, namely whether it is an ionic liquid or not, is irrelevant to the EDLC capacitance when used in PC solution.

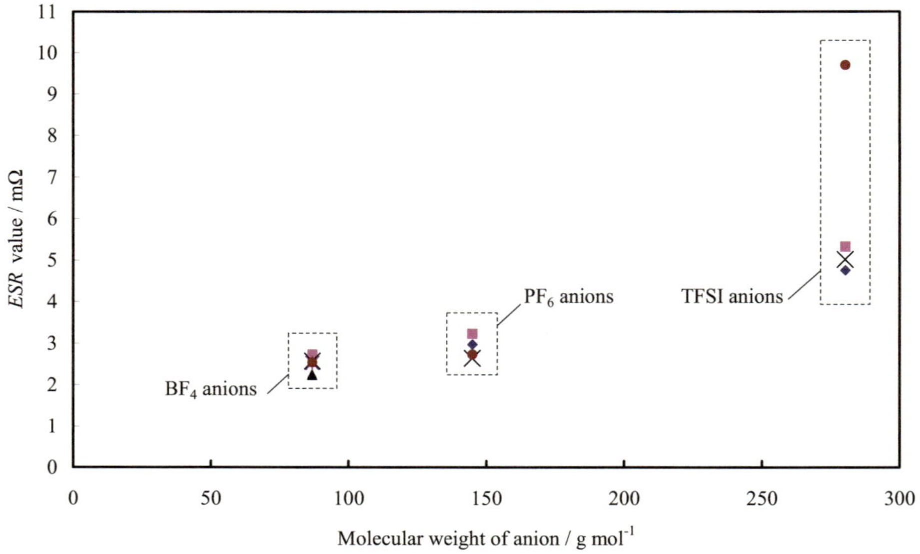

Fig. 9. The internal resistance (*ESR*) of EDLCs (265F) containing various electrolytes at 1kHz, 25 ◦C. The electrolyte concentration was 1M in PC: (♦) DEME-cation species; (■) MEMP-cation species; (▲) DEMM-cation species; (x) MMTM-cation species; (*) MMMP-cation species; (•) EMI-cation species.

We also understand that the kind of anion species used also influences the internal resistance (ESR) of an EDLC. Fig. 9 shows the ESR of EDLCs using various electrolytes at 1 kHz, 25 ◦C. Those EDLCs using the BF_4 anion in PC have the practicable value of ESR of 2.5mΩ. On the other hand, the ESR associated with the use of TFSI anions is twice or more than that of BF_4. The EDLC using the PF_6 anion showed a poor ESR value below a practical level. It seems that the ESR value of EDLCs significantly increased with an increase of the molecular weight of anion. As a result, we suggest that the use of solutes other than those containing BF_4 as electrolytes is unpractical for high voltage EDLCs.

3.2.3 Comparison of DCIR for EDLCs at low temperature

The major advantage of an EDLC versus a battery is long cycle life and high power density. Therefore, because a high power density is achieved by a low DCIR, an EDLC that has lower DCIR is desirable. It is obvious that cells using electrolytes including BF_4 have a significantly lower DCIR at room temperature than cells using other anion species. Cells using solutes with a large molecular weight anion have a high DCIR and ESR. The DCIR at the very low temperature of −30 ◦C increased between six to seven times compared to the room temperature value for BF_4 solutes, from ten to twelve times for PF_6 solutes, and for TFSI solutes by a factor of 30 or more. Fig. 10 shows the DCIR of EDLCs using various 1M solutions as an electrolyte at −30 ◦C. In the case of EDLCs using EMI-TFSI and DEME-TFSI, the cells could not discharge because the cell resistance was too large at low temperatures, and so they are not plotted in the figure.

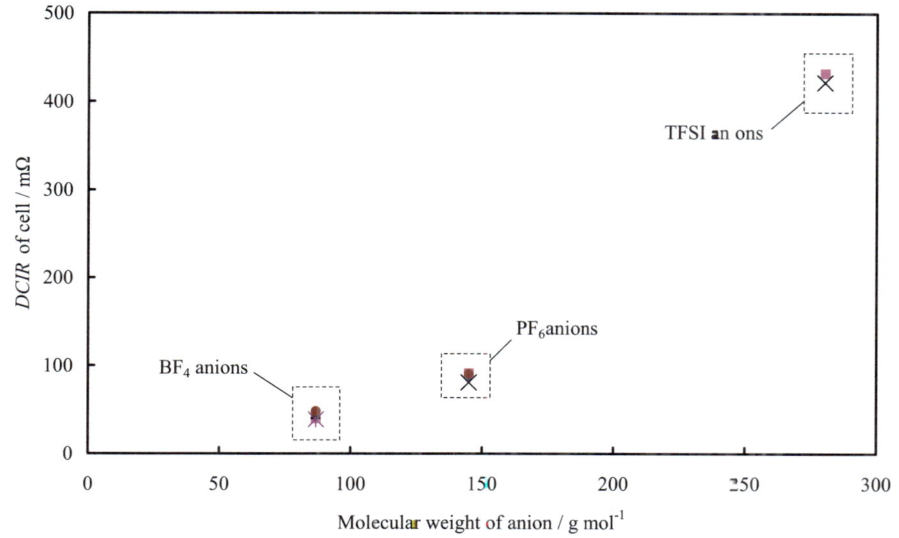

Fig. 10. The direct resistance (*DCIR*) of EDLCs containing various electrolytes at 5A discharge at -30 °C. The electrolyte concentration was 1M in PC: (♦) DEME-cation species: (■) MEMP-cation species; (▲) DEMM-cation species; (x) MMTM-cation species; (*) MMMP-cation species; (●) EMI-cation species.

It is generally considered that, since ionic liquids have high solubility in PC, an EDLC using an ionic liquid as a solute in the electrolyte should have a good performance compared to a cell using solid ammonium salts. However, based on our results, we suggest that not only should an ionic liquid be used as an electrolyte to obtain an EDLC with a high power density at low temperatures, but in particular, the ionic liquid should contain the BF_4 anion. However, MMTM-BF_4 in PC had a low DCIR at both temperatures, even though it was not an ionic liquid. Possibly, the very compact nature of the cation in this case was responsible for its good performance.
It is generally thought that the ionic conductivity of the electrolyte solution has an influence on the DCIR of an EDLC, namely, a high ionic conductive electrolyte gives a low DCIR value. The ionic conductivity (σ) for various electrolytes below 30 °C is shown in Fig. 11. Even though the nature of the anion and cation varied widely, the ionic conductivities of the various electrolytes were quite similar at a given temperature, over a range of temperatures from −30 °C to room temperature. There is essentially no difference among the six electrolytes' temperature dependence. The difference of ionic conductivity between BF_4 and TFSI at 25 °C was only 1.1 and 1.5mScm^{-1} in the MEMP and DEME series, respectively. It is not reasonable to suppose that a large difference in a DCIR value was caused by a small difference in ionic conductivity. It was surprising that at −30 °C, the DCIR difference between the BF_4 and TFSI anion combined with DEME or MEMP is ten times or more, yet the difference in ionic conductivity between BF_4 and TFSI was only 0.3mScm^{-1} at the same temperature.
The dynamic viscosity (η) of 1M solutions of DEME-BF_4, DEME-TFSI, MEMP-BF_4 and MEMP-TFSI in PC at various temperatures are displayed in Fig. 12. We focus attention first on the viscosity of DEME-BF_4 and its TFSI anion at −30 °C. If the viscosity of the electrolyte

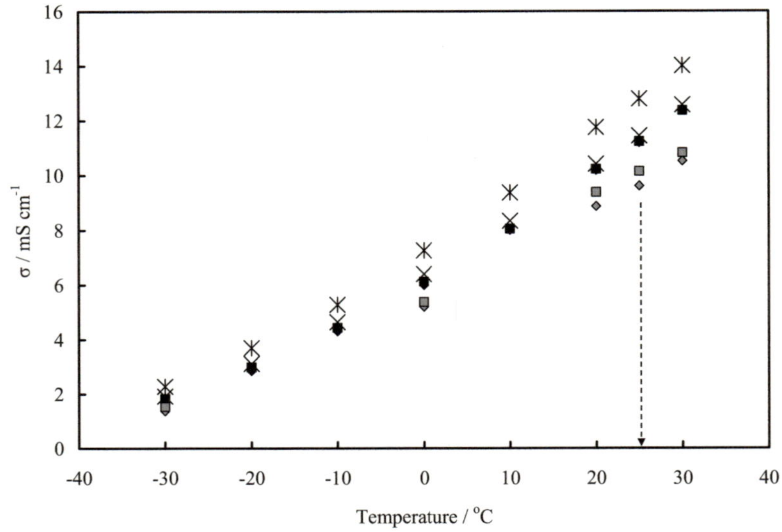

Fig. 11. The specific conductivity (σ) vs. temperature plots the various electrolytes containing a 1M concentration of ammonium salts. The electrolyte concentration was 1M in PC: (♦) DEME-BF_4; (♦, half-tone) DEME-TFSI; (■) MEMP- BF_4; (■, half-tone) MEMP-BF_4; (x) MMTM-BF_4; (*) MMMP-BF_4.

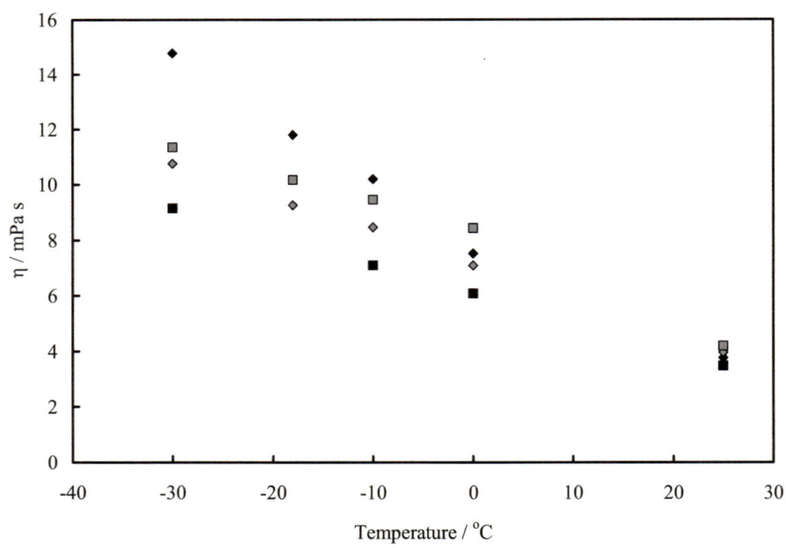

Fig. 12. Dynamic viscosity (η) as a function of temperature for the various electrolytes containing a 1M concentration of ammonium salts. The electrolyte concentration was 1M in PC: (◆) DEME-BF4; (◆ , half-tone) DEME-TFSI; (■) MEMP-BF4; (■, half-tone) MEMP-TFSI.

has a big influence on DCIR, due to a highly viscous electrolyte decreasing the mobility of the ionic species, then an electrolyte using DEME-BF_4 should have a large DCIR value compared to one using DEME-TFSI. However, an EDLC using DEME-BF_4 in PC had a remarkably small DCIR value compared with a DEME-TFSI cell. In fact, the EDLC using a DEME-TFSI electrolyte could not discharge as a result of having too large a value of DCIR at −30 ◦C. In the case of MEMP, the large difference of DCIR that existed between cells using BF_4 and TFSI anions seemed to overcome the difference of the viscosity. In a dilute PC system, the dynamic viscosity and the ionic conductivity of electrolyte did not much influence the DCIR of EDLCs with the same molar concentration. This suggests that the kind of anion species has the most significant effect on the DCIR at low temperatures. It seems that the molecular weight of the anion or the molecular size of the solvated anion most influences the DCIR value. In designing cells with such ions, it will be necessary to pay attention to the specific ion species used, especially for the anion, because the ease of ion adsorption and desorption in the confined space of porous activated carbon may influence the cell's ultimate DCIR. The compounds MEMP-BF_4, MMMP-BF_4 and DEMM-BF_4 in PC are more attractive candidates than DEME-BF_4 for use as an EDLC electrolyte from the view point of capacitance and DCIR.

3.2.4 Practical stability of EDLCs using various electrolytes

In contrast to batteries, the cycling test is less important for an EDLC, because deterioration mostly occurs at the maximum operating voltage. So, as a more useful life test we continuously operated the cell at 3.0V, 70 ◦C. Presumably, a good response to this test will indicate good durability at room temperature. Fig. 13 demonstrates that the EDLCs using MEMP-BF_4 and DEME-BF_4 showed a good practical level of durability after 1000 h of use. After 1000 h, the capacity loss was just 15 and 20% for EDLCs using MEMP-BF_4 and DEME-BF_4, respectively. On the other hand, MMMP-BF_4 and DEMM-BF_4 had good capacitance and DCIR value at low temperature, but their durability at high temperatures was inferior to that of DEME- and MEMP-BF_4 cells. Our conclusion is that ammonium salts with a methoxyethyl group have higher stability than those with a methoxymethyl group. Therefore, we recommend MEMP-BF_4 in PC as the preferred electrolyte for an EDLC in terms of capacitance, low temperature performance and thermal durability in practical use. The detailed ESR and maintenance ratio of the capacitance after 500 and 1000 h operation are summarized in literature (Yuyama et al. 2006).

We developed a new kinds of ammonium salt with a methoxyalkyl group on the nitrogen atom, including several kinds of ionic liquids, as electrolytes for an EDLC. A cell using an electrolyte containing the BF_4 anion had a higher capacitance at 25 ◦C, 3V than those including PF_6 and TFSI anions. The capacitance of an EDLC at room temperature depends on the nature of the anion, rather than the cation, species or whether the solute is an ionic liquid or is a solid itself at room temperature. The values of the resistance that are most relevant to the power density performance of an EDLC also differed greatly with different ionic species. At room (25 ◦C) and low (−30 ◦C) temperatures, both the resistance parameters, ESR and DCIR, followed the same trend that the cell resistance increased in the order of BF_4, PF_6 and TFSI. This order corresponds to the ranking of the molecular weights of the anion. Even in the ionic liquids, those including the PF_6 and TFSI anions had cell resistances that would make the practical performance of the cell such as to make it unusable. Of the ionic liquids tested, MEMP-BF_4 and DEME-BF_4, compounds that possess an aliphatic ammonium group including a methoxyethyl group, performed well at continuous charging at 70 ◦C. The aromatic type of ionic liquids of the EMI series were inferior to the aliphatic ones in terms of their practical long life stability. Our tests show that MEMP-BF_4 is the preferred ionic liquid for use as an electrolyte solute in an EDLC.

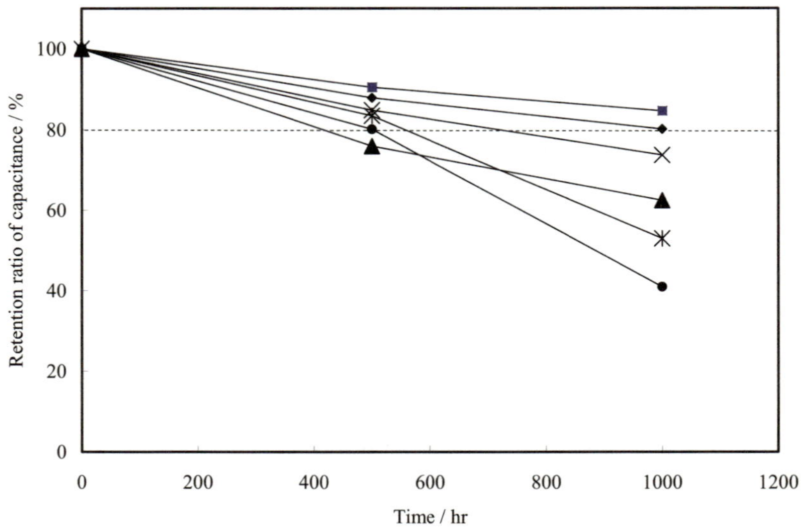

Fig. 13. Life test of EDLCs using various electrolytes including BF_4 anion. The cell was continuously charged at 3.0V, 70 ◦C for 1000 hrs. The electrolyte concentration was 1M in PC: (♦) DEME-BF_4; (■) MEMP-BF_4; (▲) DEMM-BF_4; (x) MMTM-BF_4; (*) MMMP-BF_4; (●) EMI-BF_4.

4. A thin layer including a carbon material improves the rate capability of an electric double layer capacitor

The discussion above showed that ionic liquids have very high potential as practicable EDLC electrolytes. However, the dilution of the ionic liquid with organic solvent is an indispensable step in order that such EDLCs may have a high power performance that is equal or better than cells with a conventional electrolyte such as the solid ammonium salt/PC solution system. While the addition of the organic solvent brings about a spectacular high power performance of the EDLC at low temperatures, at the same time it also has negative implications for the safety of the cells.

On the other hand, we develop a new method to improve the rate capability of an electric double layer capacitor (EDLC) using a thin polymer layer having a high concentration of carbon material on a current collector (CLC) (Sato et al., 2011). A novel thermocuring coating composed of a glycol-chitosan, a pyromellitic acid and a conductive carbon powder can form stable CLC on a metal foil current collector simply by spreading and curing at 160 ◦C for a couple of minutes. We compared the performance of some demonstration EDLC cells using three kinds of current collector: a conventional aluminum oxide foil for EDLC, an aluminum foil and an aluminum foil with CLC. The cell with the CLC had a much higher rate capability than the cell without CLC. Only the CLC cell was able to discharge at a current density of 500C. This cell shows a slight deterioration in capacity in a high temperature, continuous charging, life test, and the CLC has a suppressing effect on the internal resistance increase of EDLCs. The use of a CLC film current collector is one of the most effective and simple methods for the improvement of EDLC rate performance. In particular, a current collector consisting of aluminum foil coupled with a CLC promises to be a low cost alternative to the aluminum oxide foil commonly used in EDLCs. Also, these effects of a CLC have been confirmed in an EDLC with an undiluted ionic liquid as an electrolyte. In this connection, we note the use of a combination of undiluted aliphatic cyclic ammonium type ionic liquid, MEMP-BF_4 (Fig.14), which has a relatively low viscosity, with an aluminum collector, coupled with a CLC.

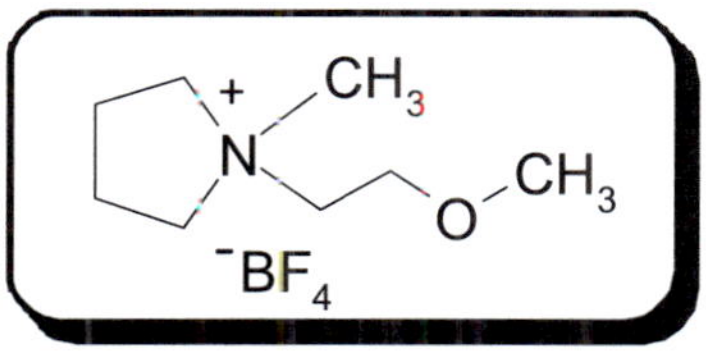

Fig. 14. Molecture structure of *N*-(2-methoxyethyl)-*N*-methylpyrrolidinium tetrafluoroborate (MEMP-BF_4).

4.1 Thermocuring coating on the current collector of an EDLC electrode

To improve rate performance, we have prepared cells containing a carbon layer on the current collector (CLC). This consists of a thin layer having a high concentration of conductive carbon material, situated between the activated carbon electrode layer and the current collector. We have found that a hydroxyalkylated chitosan (glycol-chitosan) derivative and 1,2,4,5-benzene-tetracarboxylic acid (pyromellitic acid) mixture acts as a thermally activated binder that adheres strongly to metal foil was very effective in improving the rate capability of EDLC cells.

Chitosan is a natural and low cost biopolymer prepared by the deacetylation of chitin, the most abundant polymer after cellulose, and is mainly obtained from crab shells (Fig. 15). Due to its unique physicochemical properties such as non-toxicity, chemical and thermal stability, hydrophilicity, remarkable affinity towards certain substances and film formation with relatively high physical strength, it has been extensively studied and is used in many fields (Rinaudo,2006; Mourya,2008) We found that we could form a stable coating layer on the metal substrate by spreading and heating an ink composed of a glycol-chitosan and a polycarboxylic acid compound. Because it is well known that the secondary amino group of the chitosan forms an imide linkage by reacting with two carboxylic acids, we propose that such cross-linkage between chitosan molecules can be formed by use of a tetracarboxylic compound. It is probable that a stable three-dimensional polymer network had been formed by the synergistic effect of combining a chitosan with a relatively rigid backbone with a chemically stable imide crosslinkage (Fig. 15). In addition the amino group of the chitosan has a strong affinity to carboxylic and hydroxyl groups on the surfaces of inorganic materials. Because the conductive carbon material used here, such as acetylene black and Ketjen black, has a lot of carboxylic acid and hydroxyl groups on the powder surface, it is likely that the chitosan binds strongly to them.

The chitosan dissolves easily in aqueous acids, though it does not dissolve in the NMP conventionally used as a solvent in electrode making. However, the chitosan can be converted into the amphipathic partially hydroxyalkylated derivative obtained on reacting the hydroxyl and amino groups in the glucosamine unit with epoxide compounds such as ethylene oxide, propylene oxide and butylene oxide. We developed a novel coating to use as a conductive layer on a metal current collector by reacting glycolchitosan and pyromellitic acid.

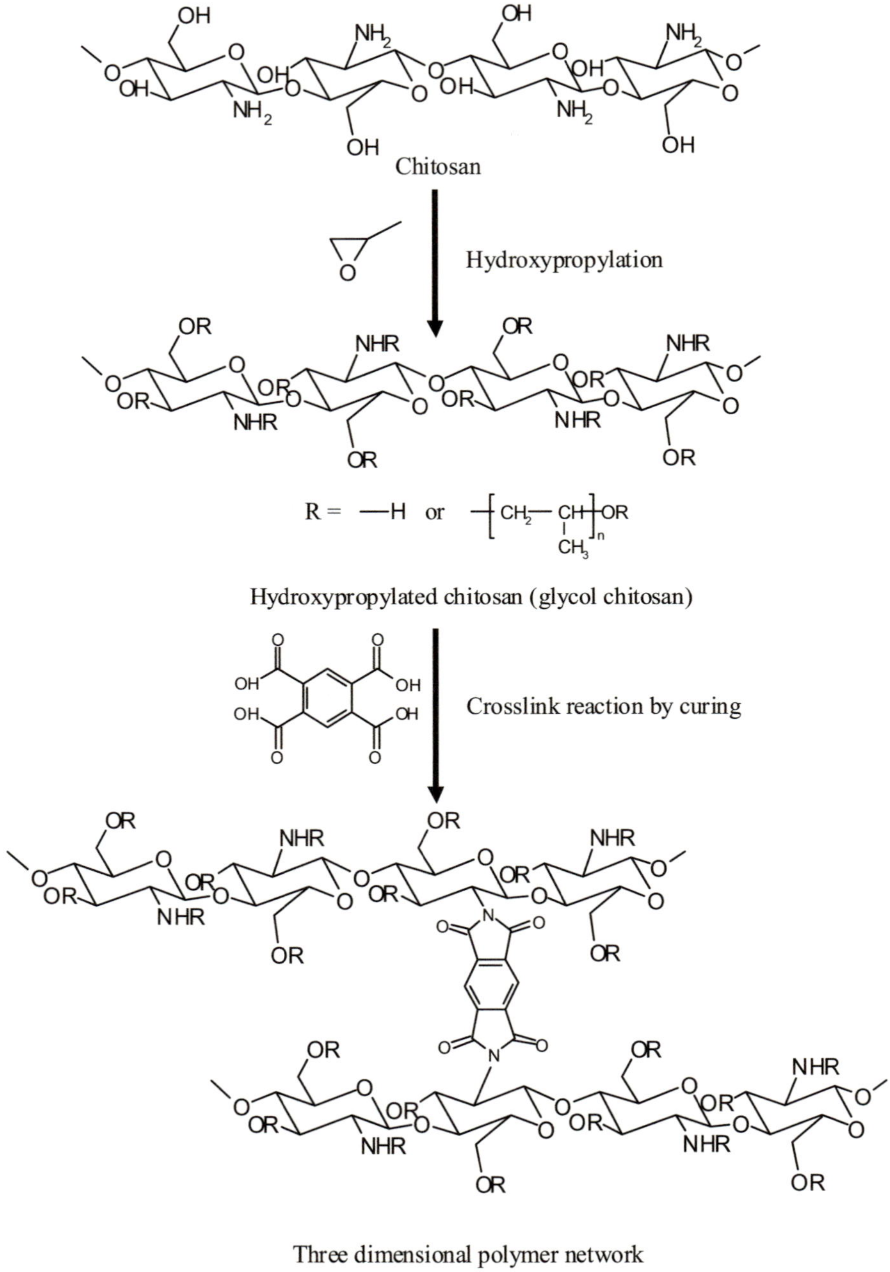

Fig. 15. Thermosetting chitosan / pyromellitic acid binder in this study.

4.2 CLC effect on EDLC cells

In this study, we assumed that the cell impedance could be decreased by decreasing the contact resistance between the activated carbon electrode layer and the current collector. The method we employed simply places stable, thin carbon layer composed of acetylene black and glycol-chitosan (CLC) between the activated carbon electrode and the current collector. The rate performance of demonstration EDLCs at 25 ◦C with and without, a CLC on the current collector and various electrolytes are summarized in Table 3.

The EDLC cell with an aluminum oxide current collector showed a slightly higher capacitance at low discharge current values. The pores present on the aluminum oxide foil surface might be contributing to a small capacity increase. The conventional aluminum foil cell has the lowest capacitance of the examined cells, resulting from the large internal resistance of the cell, even for low rate discharging, such as that at 1C. On the other hand, the cell with a thin carbon layer on both positive and negative conventional aluminum foil collectors (CLC) showed an excellent discharge rate character for an EDLC. Evidently, this cell has about the same capacitance as a common EDLC cell at 20C discharging, and at 50C, the discharging capacity of the CLC cell surpassed that of the common cell. The capacitance of the CLC cell at 300C discharging was five times larger than that of the common cell, and was thirty times for a 500C discharging process.

At 500C discharging, corresponding to approximately 94 mAcm^{-2} of current density to the electrode surface area, the EDLC with an aluminum oxide could not discharge. However, the CLC cell maintained 40% of the capacitance at 1C discharging. A remarkable improvement in the rate characteristic of an EDLC was also observed with a CLC using neat ionic liquid as an electrolyte. Such a cell showed almost the same output capacity as an aluminum oxide collector cell with diluted PC system as an electrolyte, although the viscosity at room temperature of the MEMP-BF_4 IL used is 72 mPa s, higher by a factor of 20 times or more compare with that of the conventional PC-diluted electrolyte. Fig. 16 shows discharging profiles of demonstration EDLCs at various discharge rates. In the discharge with a relatively large current, such as 300C and 500C, it is clear that the CLC cell has better rate character than the conventional EDLC. These results indicate that the installation of a CLC had an extremely positive effect on the rate performance of the cell.

Nyquist plots for a demonstration cell in the frequency range from 10mHz to 20 kHz are shown in Fig. 17. They consist of a semicircle at high frequency followed by an inclined line and a vertical line in the low frequency region. The intercept with the real axis at high frequency gives an estimate of the solution resistance (Rs). The diameter of semi-circle, namely the difference between the high frequency intercept (Rs) and low frequency intercept, indicates the interfacial resistance (Ri), which is attributed to the impedance at the interface between the current collector and carbon particles, as well as that between the carbon particles themselves.

The R_S value is different because of a difference in the ionic conductivity of the electrolyte. The most important conclusion is that the diameters of the semicircles were different in each cell. The cell with a CLC has a very small semicircle compared with AlCx cells. The cell with the largest semi-circle was that with a conventional aluminum current collector (not displayed in the figure). In the case of the cell with neat ionic liquid, the R_i value is small, similar to that of a CLC cell with diluted electrolyte. The activated carbon layer was prepared by the same methods and materials in each case, so the diameter of a semi-circle indicates the difference in the interfacial resistance between the activated carbon layer and current collector. The installation of a CLC has the effect of decreasing the interfacial

	Current collector [a]		Discharge capacity C (Fg^{-1}) in various discharging rates[b]					
Electrolyte	Positive electrode	Negative electlode	1C	20C	50C	100C	300C	500C
1M DEME-BF_4 in PC	AlOx	AlOx	25.2	21.3	16.9	11.6	2.4	0.3
1M DEME-BF4 in PC	Al	Al	21.3	14.3	7.9	3.4	0.1	0.2
1M DEME-BF4 in PC	Al+CLC	Al+CLC	22.5	20.8	19.1	17.6	12.3	9.1
Neat MEMP-BF_4	Al+CLC	Al+CLC	21.2	18.0	14.6	11.3	4.3	1.2

[a] Al + CLC, aluminum foil (20 μ m) with 2.5 μ m of CLC; Al, aluminum foil; AlOx, aluminum oxide foil (30 μ m). [b] Constant discharging current of 1.5mA corresponds to 1C and 0.19mAcm^{-2}.
*preparation of CLC: To form a CLC on a current collector, XSC-1N carbon ink suspension (Kyoritsu Chemical & Co., Ltd.) was spreaded by use of a gravure printing method. The thickness of the printed layer was approximately 5 μ m. After printing, the foil was dried at 160 ◦C in an oven to react chitosan and pyromellitic acid and evaporate the NMP solvent.

Table 3. Double layer capacitance of EDLCs with various current collectors at different discharging rate.

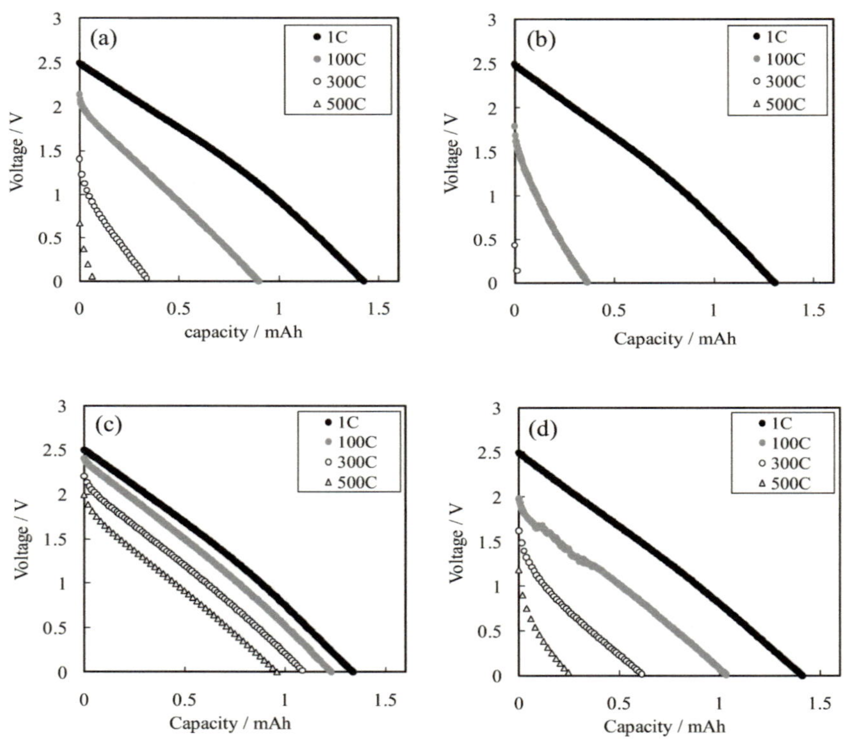

Fig. 16. Discharge curves of demonstration EDLCs using various current collectors at various current densities at 25 ◦C. (a) EDLC cell using aluminum oxide; (b) cell using aluminum foil; (c) cell using aluminum with CLC on both electrodes; (d) cell using aluminum with CLC on both electrodes and neat ionic liquid electrolyte. The discharging rate of 1C corresponds to 0.15mA of discharge current.

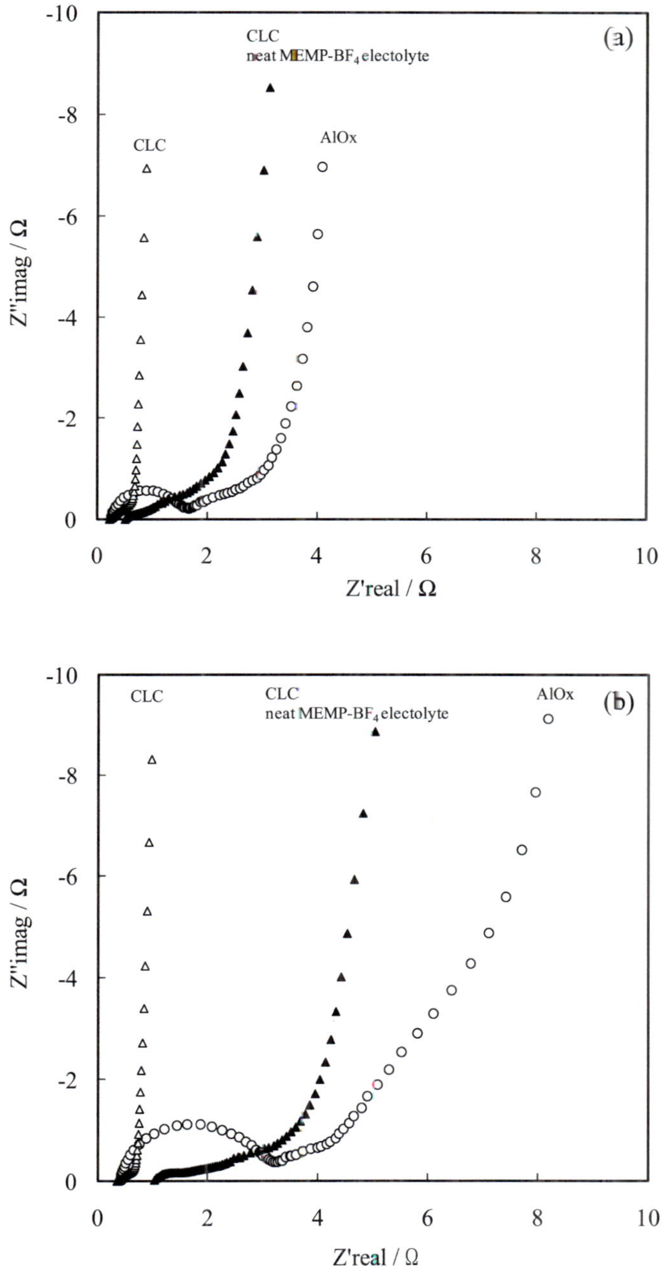

Fig. 17. Experimental Nyquist plots of demonstration EDLCs using various current collectors and electrolyte at 25 °C. (a) Fresh cell; (b) cell after life test, continuous charging at 2.5 V, 60 °C for 500 h. (CLC) = cell with aluminum foil with CLC on both electrodes; (AlOx) = cell with aluminum oxide foil on both electrodes. Unless otherwise indicated, 1M DEME-BF_4 in PC was used as an electrolyte.

resistance between the electrode layer including the active material and the collector. Of course, the viscosity of the ionic liquid does not decrease by employing a CLC, however, it is possible to improve the performance of a cell using neat ionic liquid to be similar to that of a conventional EDLC available on the market. It should be possible to develop nonflammable and high durability EDLCs by the combination of CLC technology and low viscosity ionic liquids including FSI or FH anions.

In contrast to batteries, the cycling test is less important for an EDLC, because deterioration mostly occurs at the maximum operating voltage. So, as a more useful life test we continuously operated the cell at 2.5 V, 60 ◦C. Presumably, a good response to this test will indicate good durability at room temperature. The maintenance of capacity of demonstration cells after 500 h operation is summarized in Fig. 18. All cells showed a good practical level of durability, the capacity loss being just 8%, regardless of the presence of a CLC in diluted electrolyte. In the case of the neat IL cell, the capacity loss from 100 to 500 hours was only 3%, though some deterioration was seen in the first 100 hours of operation. However, the interfacial impedance Ri of cells with neat ionic liquid cell and aluminum oxide collectors small increased after continuous charging at 60 ◦C, as shown in Fig.17(b).

We opened the demonstration CLC cell after the continuous charging examination and investigated the bonding between the electrode and current collector by rubbing the electrode surface with paper. The activated carbon layer of both electrodes with both the aluminum collector and aluminum oxide collector were peeled off easily by rubbing one or

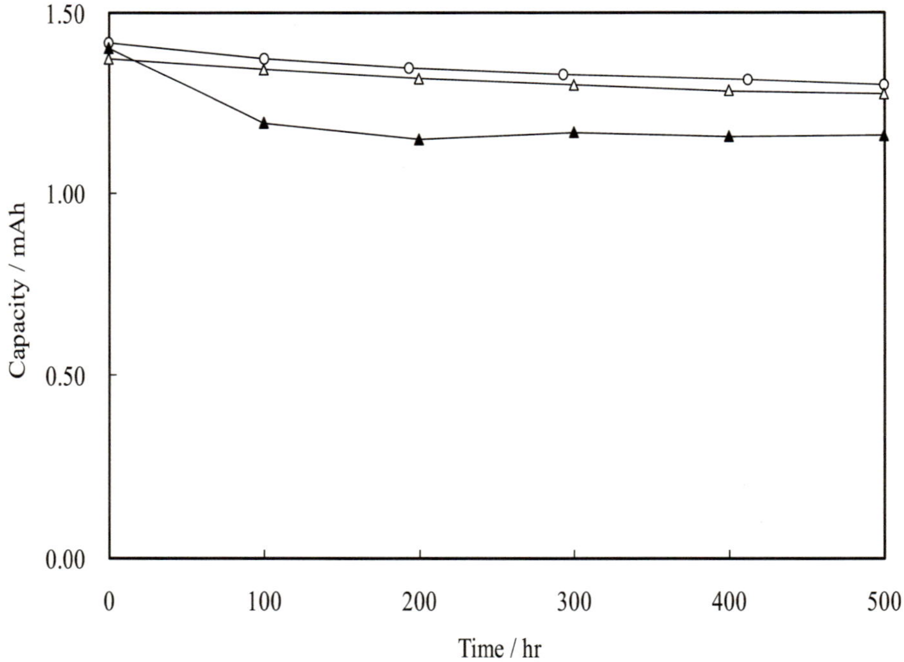

Fig. 18. Life test of demonstration EDLCs using various current collectors. The cell was continuously charged at 2.5 V, 60 ◦C for 500 h. The current collector used are: (open circle) aluminum oxide foil; (open triangle) aluminum foil with CLC; (filled triangle) aluminum foil with CLC and neat ionic liquid electrolyte.

two times. However, that on the CLC did not separate even after rubbing more than 20 times in both diluted and neat ionic liquid cells. Both the interfacial bonding between the CLC and aluminum foil, and between the CLC and the carbon electrode remained completely intact. The scanning electron micrograph observation data are summarized elsewhere (Sato. et.al 2011).We can conclude that the CLC acted as a long-lasting internal adhesive layer that improves durability.

5. Conclusion

1. The DEME-based ionic liquids have relatively high conductivities and remarkably wider potential windows compared to the other aromatic type ionic liquids that have been reported.
2. An EDLC using DEME-BF_4 as the electrolyte exhibits excellent stability and cycle durability even at temperatures over 100°C. Although below room temperature the capacity of an EDLC using this ionic liquid was inferior to that of a conventional EDLC using TEA-BF_4/PC due to the high viscosity of the ionic liquid, the ionic liquid EDLC showed a higher capacity for the discharge of a large current at temperatures of 40°C or above.
3. The major problem for the practical use of ionic liquids has been their high viscosity. Ionic liquids with relatively low viscosity with new type of anions, such as bis(trifuluorosulfonyl)imide and fluorohydrogenate have been developed recently, leading to the hope that the problem of the viscosity will be soon overcome.
4. The viscosity of many kinds of ionic liquids decreases markedly on adding an organic solvent. In this way, it is possible to make an electrolyte with a high ion concentration, giving the EDLC a high capacitance. Ionic liquids never precipitate and crystallize even at low temperatures. EDLCs made with PC diluted ionic liquid (DEME-BF_4) have an extremely attractive high charge rate and discharge performance even at −40 ◦C
5. We have evaluated various kinds of ammonium salt with a methoxyalkyl group on the nitrogen atom, including several kinds of ionic liquids, as solutes in PC for an EDLC A cell using an electrolyte containing the BF_4 anion had a higher capacitance at 25 ◦C, 3V than those including PF_6 and TFSI anions. The capacitance of an EDLC at room temperature depends on the nature of the anion, rather than the cation, species or whether the solute is an ionic liquid or is a solid itself at room temperature.
6. The value of the resistance that is most relevant to the power density performance of an EDLC, the cell resistance, increased in the order of BF_4, PF_6 and TFSI. Of the ionic liquids tested, MEMP-BF_4 and DEME-BF_4, both of which have a methoxyethyl group on nitrogen atom, showed the highest durability in a continuous charging life test. We conclude that ionic liquids containing the tetrafluoroborate anion have the best performance and stability for electric double layer capacitor applications.
7. We have developed a new method that improves the rate capability of EDLCs. It comprises a very simple method of creating a thin polymer film containing a high concentration of carbon material (CLC) on the current collector that makes high power discharging at 500C possible. It was revealed from impedance analysis that a CLC that was only 2.5 μm thick was effective in decreasing the interfacial impedance between the current collector and the electrode layer. The CLC technique was extremely effective

also for EDLCs with a neat ionic liquid. We anticipate that the use of a CLC will be one of the effective technologies to master the difficulties of using high viscosity ionic liquids.

6. Acknowledgments

The authors thank Mr. Hiroshi Yoshida, Mr. Gen Masuda and Ms. Kanako Yuyama at Nisshinbo Holdings Inc. and Dr. Taichi Uemura and Mr. Satoshi Yamazaki at Kyoritsu Chemical & Co., Ltd. Kisarazu R&D Center, for their valuable comments.

7. References

Rogers, R. D. & Seddon, K. R. (Eds.). (2001). *Ionic Liquids : Industrial Applications for Green Chemistry* (Acs Symposium Series), American Chemical Society, ISBN 0-8412-3789-1, Washington DC., USA.

Zhao, H.;Tang, Z.; Zhang, Q., You, J., Chen, Q. (2009). Inorganic synthesis in ionic liquids, *Progress in Chemistry,* Volume 21, Issue 10, pp. 2077-2083, ISSN: 10052 81X.

Zhao, H. ; Baker, G.A. ; Holmes, S. (2011). New eutectic ionic liquids for lipase activation and enzymatic preparation of biodiesel, *Organic and Biomolecular Chemistry,* Vol.9, Issue 6, pp. 1908-1916, ISSN: 14770520.

Sato, T.; Maruo, T.; Marukane, S. & Takagi, K. (2004). Ionic liquids containing carbonate solvent as electrolytes for lithium ion cells, *J. Power Sources,* 138, (2004). pp. 253-261.

Sato, T.; Masuda, G. & Takagi, K. (2004). Electrochemical properties of novel ionic lquids for electric double layer capacitor applications, *Electrochimica Acta,* 49, (2004), pp 3603-3611.

Papageorgiou, N.; Athanassov, Y.; Armand, M.; Bonhote, P.; Pettersson, A.; Azan, A. & Gratzel, M. *J. Electrochem. Soc,* 143, (1996), 3099.

Zheng, J.P.; Huang, J. & Jow, T.R. (1997). The Limitations of Energy Density for Electrochemical Capacitors, *J. Electrochem. Soc,* Vol.144, No.6, (June 1997), pp.2026-2031.

Koch, V.R.; Nanjundiah, C. & Goldman,J.L. (1995). International Seminar on Double Layer Capacitors and Similar Energy Storage Devices ,*Proceedings of the Fifth,* Florida Educational Seminars, 1995.

Koch, V.R.; Nanjundiah, C. & Carlin, R.T. (1994). Hydrophobic Ionic Liquid. U.S. Patent 5,827,602

McEwen,A.B.; Ngo, H.L.; LeCompte, K.; Goldman, J.L. (1999). Electrochemical Properties of Imidazolium Salt Electrolytes for Electrochemical Capacitor Applications, *J. Electrochem. Soc,* 146, (1999),1687.

McEwen, A.B.; McDevitt, S.F. & Koch, V.R. (1997). Nonaqueous Electrolytes for Electrochemical Capacitors: Imidazolium Cations and Inorganic Fluorides with Organic Carbonates, *J. Electrochem. Soc,* 144, (1997). L84.

Ue, M.; Takeda, M.; Takahashi, T. & Takehara, M. (2002). Anodic Stability of Several Anions Examined by *Ab Initio* Molecular Orbital and Density Functional Theories, *J. Electrochem. Soc,* 149, (2002). A1572.

Cooper, E.I. & Angell, C.A. (1986). Short-range order in a NASIGLAS sample by X-ray diffraction, *Solid State Ionics*, 18–19, (1986). 570.

Emanuel,W.X.; Cooper, I. & Angell, C.A. (2003). Ionic Liquids: Ion Mobilities, Glass temperatures and fragilities, *J. Phys. Chem*, 107, (2003). 6170.

Matsumoto, H.; Yanagida, M.; Tanimoto, K.; Nomura, M.; Kitagawa, Y. & Miyazaki, Y. (2000). Highly Conductive Room Temperature Molten Salts Based on Small Trimethylalkylammonium Cations and Bis(trifluoromethylsulfonyl)imide, *Chem. Lett*, (2000). 922.

Matsumoto, H.; Kageyama, H. & Miyazaki, Y. (2001). Room Temperature Molten Salts Based on Tetraalkylammonium Cations and Bis(trifluoromethylsulfonyl)imide, *Chem. Lett*, (2001). 182.

Matsumoto, H.; Kageyama, H. & Miyazaki, Y. (2002). Room temperature ionic liquids based on small aliphatic ammonium cations and asymmetric amide anions, *Chem. Commun*, (2002). 1726.

Maruo, T.; Sato, T.; Masuda, G. & Nozu, R. (2002). Ionic Liquids, Electrolyte Salts for Electrical Strage Devices, Liquid Electrolytes for Electrical Strage Devices, Electrical Double-Layer Capacitors and Secondary batterieis WO02076924.

Ishikawa, M.; Sugimoto, T.; Kikuta, M.; Ishiko, E. & Kono, M. (2006). A neat ionic liquid electrolyte based on FSI anion for electric double layer capacitor, *J. Power Sources*, 162, (2006). pp. 658–662.

Matsumoto, H.; Sakaebe, H.; Tatsumi, K.; Kikuta, M.; Ishiko,E. & Kono, M. (2006). Fast cycling of Li/$LiCoO_2$ cell with low-viscosity ionic liquids based on bis(fluorosulfonyl)imide $[FSI]^-$, *J. Power Sources*, 160, (2006). pp. 1308–1313.

Handa, N.; Sugimoto, T.; Yamagata, M.; Kikuta, M.; Kono, M. & Ishikawa, M. (2008). A neat ionic liquid electrolyte based on FSI anion for electric double layer capacitor, *Journal of Power Sources*, 185, (2008). pp. 1585–1588.

Senda, A.; Matsumoto, K.; Nohira, T. & Hagiwara, R. (2010). Effects of the cationic structures of fluorohydrogenate ionic liquid electrolytes on the electric double layer capacitance, *Journal of Power Sources*, 195, (2010). pp. 4414–4417

Nisshinbo Holdings Inc., *Nisshinbo Holdings Inc. Web page*, 03.03.2011, Available from: http://www.nisshinbo.co.jp/english/r_d/capacitor/index.html

Yuyama,K.; Masuda, G.; Yoshida, H. & Sato, T. (2006). Ionic liquids containing the tetrafluoroborate anion have the best performance and stability for electric double layer capacitor applications, *J. Power Sources*, 162, (2006). pp. 1401-1408.

Kim,Y.J.; Matsuzawa, Y.; Ozaki, S.; Park, K.C.; Kim, C.; Endo, M.; Yoshida,H.; Masuda, G.; Sato, T. & Dresselhaus, M.S. (2005). High Energy-Density Capacitor Based on Ammonium Salt Type Ionic Liquids and Their Mixing Effect by Propylene Carbonate, *J. Electrochem. Soc*, 152, (2005). A710.

Sato, T.; Marukane, S.; Morinaga, T.;Uemura, T.; Fukumoto, K.& Yamazaki, S. (2011). A thin layer including a carbon material improves the rate capability of an electric double layer capacitor, *J. Power Sources*, 196, (2011). pp.2835-2840.

Rinaudo, M. (2006). Chitin and chitosan: Properties and applications, *Prog. Polym. Sci*, 31, (2006). pp. 603–632.

Mourya, V.K. & Inamdar, N.N. (2008). Chitosan-modifications and applications: Opportunities galore, *React.Funct. Polym*, 68, (2008). pp. 1013–1051.

7

Ionic Liquid Used in Long-Lifetime Polymer Light-Emitting Electrochemical Cells

Yan Shao
Institute for Polymers and Organic Solids and Mitsubishi Chemical Center for Advanced Materials University of California, Santa Barbara Santa Barbara, CA USA

1. Introduction

Polymer light-emitting devices have been divided into two general types: polymer light-emitting diodes (PLEDs) and polymer light-emitting electrochemical cells (PLECs). (Burroughes et al., 1990; Pei et al., 1995; Pei et al., 1996; Gong et al., 2005; Liu et al., 2006;) The advantages for PLEDs include fast response and relatively long operating lifetime (with proper packaging). However, low work function cathodes and/or thin interfacial layers (e.g. LiF) between the metal and the emitting polymer layer are required. In contrast, PLECs have relatively low turn-on voltages (approximately equal to the band gap of the luminescent semiconducting polymer), and low work function metals are not required.

One of the serious disadvantages of PLECs, however, is the slow response time (time required for the mobile ions to diffuse during junction formation). A solution to this problem is to "freeze" the junction after ion redistribution. (Yu et al., 1998; Gao et al., 1999;) A frozen junction system that operates at room-temperature is necessary for practical use. A second limiting disadvantage of PLECs has been the short operating lifetimes compared with those of PLEDs. (Shin et al., 2005; Kervella et al., 2001)

We report here the results of an initial study of light emission from a luminescent polymer blended with a dilute concentration of an ionic liquid. Even with an aluminum cathode, the devices turn on at low voltage (approximately equal to the band gap of the luminescent semiconducting polymer. These ionic liquid containing LECs were operated continuously in the glove box (without packaging) for several days without significant degradation in brightness. After sealing with epoxy and a glass cover slide, the ionic liquid containing LECs were operated continuously in air for several weeks.

The major difference between PLEDs and PLECs is that the latter possess mobile ions inside the polymer; therefore, the selection of the mobile ions is one of the keys to fabricating high performance PLECs. Previously, the mobile ion systems that have been used fall into three categories. The first is polyethylene oxide (PEO) containing Li- salts. (Pei et al., 1995; Pei et al., 1996) Crown ethers (and derivatives) (Kervella et al., 2001; Cao et al., 1997) and other organic salts (Yang et al., 2003; Shin et al., 2006) have also been used in combination with metal salts. Finally, polymers with ionic side chains (polyelectrolyte conjugated polymers) and appropriate mobile counterions have been used. (Edman et al., 2005) For almost all PLECs, the additives comprise at least 5 weight percent. More important, these systems involve two-component phase separation with the emitting polymer in one phase and the mobile ions (e.g. dissolved in PEO) in a second phase. To create the p-i-n junction of the LEC, ions must move from one phase into the other; e.g. from the PEO into the luminescent polymer. This phase separation appears to degrade the device performance, especially the lifetime. (Cao et al., 1997) The phase separation results from the relatively poor compatibility of the ionic materials (hydrophilic) with host light-emitting polymers (hydrophobic). In order to reduce the phase separation, surfactants or bifunctional additives were introduced into the emitting layer and better performance was reported. (Cao et al., 1996) Single component PLEC polymers have been fabricated using luminescent polymers with ionic side chains (polyelectrolyte conjugated polymers), but the electroluminescence (EL) was weak and the operating lifetimes were poor. (Edman et al., 2005)

2. Long-lifetime polymer light-emitting electrochemical cells with ionic liquid

2.1 Material selection

In the devices described here, we utilized the simplest sandwich structure for the device configuration with poly(3,4-ethylenedioxythiophene)-poly(styrene sulfonate) (PEDOT-PSS) coated indium-tin-oxide (ITO) glass as the anode and aluminum as the cathode. The well-known soluble phenyl-substituted poly(para-phenylene vinylene) (PPV) copolymer ("superyellow" from Merck/Covion) (Spreitzer et al., 1998) was selected as our host light-emitting polymer and an organic ionic liquid, methyltrioctylammonium trifluoromethanesulfonate (MATS), was used to introduce a dilute concentration of mobile ions into the emitting polymer layer.

The molecular structure of MATS is shown in Fig.1 (a). The merits of MATS include its good solubility in common organic solvents, such as toluene, hexane, and acetonitrile, and its relatively high decomposition temperature (approximately 220 °C). Because MATS has a melting temperature of approximately 56 °C, frozen junction devices can be prepared for operation at room temperature.

2.2 Device fabrication

For device fabrication, the materials were used as received without further purification. MATS and superyellow were both dissolved into toluene at a weight ratio of 1:50 in a nitrogen-filled glove box (oxygen level under 3ppm); the total concentration of the solution was 6mg/mL. Solid thin films were prepared by spin casting from this solution in the glove box. Superyellow and superyellow with 2wt% MATS show strong photoluminescence (PL) both in solution and in the solid state with almost identical spectra. Fig. 2 shows the PL spectra of superyellow containing 2wt% MATS (solid lines) and pure superyellow (dash lines) both in solution (Fig. 2(a)) and as solid thin films (Fig. 2 (b)).

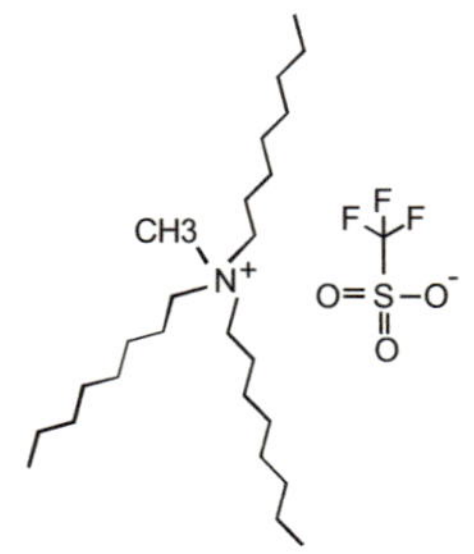

Fig. 1. (a)

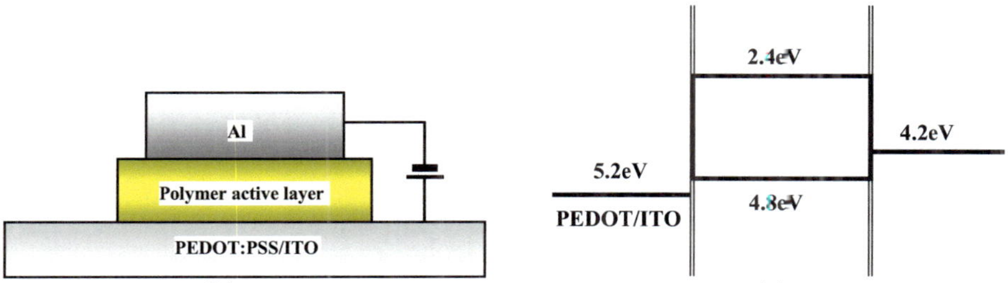

(b) (c)

Fig. 1. (a) Molecular structures of methyltrioctylammonium trifluoromethanesulfonate (MATS). (b) Device architecture. (c) Schematic energy level diagram for the open circuit status.

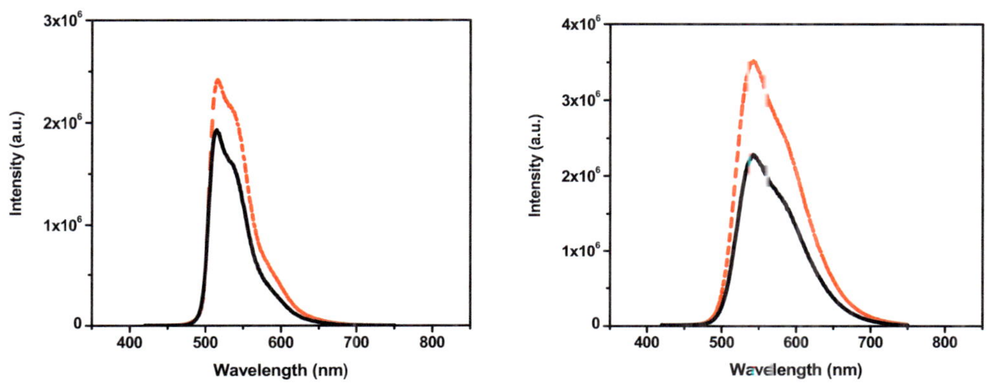

(a) (b)

Fig. 2. Photoluminescence spectra of 2wt% MATS in superyellow (solid lines) and pure superyellow (dash lines) in (a) 6mg/mL toluene solution and (b) solid thin films.

Polymer light-emitting devices were fabricated on patterned ITO-coated glass substrates, which had been cleaned by successive ultrasonic treatment in detergent, acetone, and isopropyl alcohol. The ITO glass was then subjected to UV-ozone treatment for about 30 minutes. A thin layer of PEDOT-PSS film was spin-cast onto the ITO glass substrate with a spin speed of 4000 rpm for 1 minute and then baked at 120 °C for 20 minutes in ambient. The polymer layers were then spin-cast from the solution containing 1:50 weight ratio of MATS and superyellow in toluene (spin-speed of 1500 rpm) for 1 minute in the nitrogen glove box. The Al cathode was evaporated through a shadow mask with an active area of approximately 14.8 mm^2 (vapor deposition of the aluminum cathode was carried out under a base pressure of ~ 1×10^{-6} Torr with deposition rates about 4 Å/s). Schematic diagrams of the device structure and the relevant energy levels of the various components (for open circuit conditions) are shown in Fig. 1(b) and (c).

2.3 Device measurement

All electrical measurements were performed under nitrogen in the glove box. The current-voltage (I-V) characteristics were recorded by a computer controlled Keithley 236 source-measure unit (SMU). When the devices were tested for the first time without any prior heat treatment, their behavior was just like that of a PLED with Al as cathode: the turn-on voltage was over 6 volts with relatively low brightness and low efficiency even at high

operating voltage. Under 4V forward bias, the current was in the range of a few μA/cm^2. The electrical behavior was asymmetric (characteristic of a diode). Light emission was not observed in reverse bias.
When the devices were heated to around 80°C under 4V forward bias, the current increased by approximately three orders of magnitude and reached 10mA/cm^2 in one minute. The current-rectification factor also improved by about two orders of magnitude and reached to about 10^4 (see Fig 3). The heating temperature was purposely chosen to be above the melting point of MATS and close to the glass transition temperature (T_g) of superyellow so that ions were generated and could move under the influence of the applied electrical field. The devices were subsequently cooled to room temperature under 4 V forward bias. After cooling, a frozen p-i-n junction was formed in the device. The devices demonstrated excellent performance: high brightness, high efficiency, long continuous operating time, and short response time. More important, as LECs, they demonstrated very low turn-on voltage even with the use of a stable cathode material.

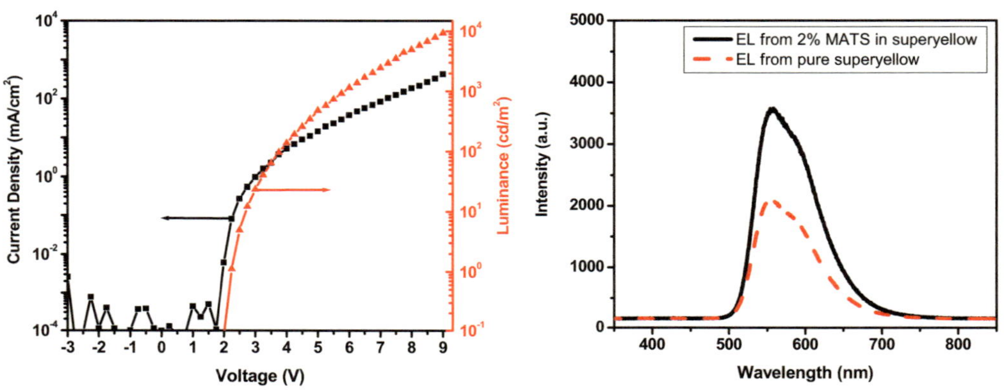

(a) (b)

Fig. 3. (a) Current-voltage (I-V) and brightness-voltage (B-V) curves of device ITO/PEDOT/2wt% MATS in superyellow/Al. (b) The EL spectrum of the ITO/PEDOT/2%MATS in superyellow/Al devices is shown as the solid line, and the EL spectrum of the reference device (ITO/PEDOT/superyellow/Al) is shown as the dash line.

Fig. 3(a) shows the I-V and brightness-voltage (B-V) curves. The turn-on voltages are as low as 2.2V (turn-on defined as the voltage required for achieving a brightness of 1cd/m^2); i.e. approximately equal to (or even a little lower than) the band gap of the superyellow semiconducting polymer. (deMello et al., 1998) The junction remained for more than 10 hours without application of an electrical field. Leakage currents below turn-on are quite small (10^{-7} A/cm^2). The brightness reached 10,000cd/m^2 at around 9V. As shown in Fig. 3(b), the light emission is from superyellow with almost identical EL spectrum as that obtained from the reference superyellow PLED with the following device structure: ITO/PEDOT/superyellow /Al. The device efficiency was approximately 3.3cd/A (at 5V with about 500cd/m^2 brightness), which is similar to the efficiency of our control PLED (maximum efficiency is about 4.5cd/A) with the same superyellow as emitting polymer and Ba as cathode.

2.4 Phase separation research in active layer

The use of chemical additives is known to play an important role in electron injection, carrier transport, and exciton energy confinement. (Gong et al., 2005; Hamada et al., 1999;

Shao and Yang, 2005; Kim et al., 2006) When two or more materials are used in a blend, the material compatibility or mutual solubility is one of the keys to high performance devices, especially for long time operation. (Bozano et al., 2003; Shao and Yang, 2005) Therefore, the proper selection of organic material components is an important issue.

We find that MATS shows excellent compatibility with superyellow. This was confirmed by atomic force microscope (AFM) studies as shown in Fig. 4(a). Fig. 4 (b) shows the AFM image of the control film of pure superyellow. Both were prepared on top of PEDOT-PSS coated ITO glass with a spin speed of about 1500rpm and then baked on hot plate at 80°C for about 25 minutes in the glove box. Then both films were stored in the glove box for about 12 days before measurement so that any possible phase separation can develop. Phase separation should be easily identified. In the AFM pictures, there is no evidence of phase separation. The two films show high surface quality and similar morphologies with similar uniformity, grain size, and height variation. The room-mean-square (RMS) roughness of both is about 0.7nm, which means they remain quite smooth after relatively long time storage. (Wenzl et al., 2004)

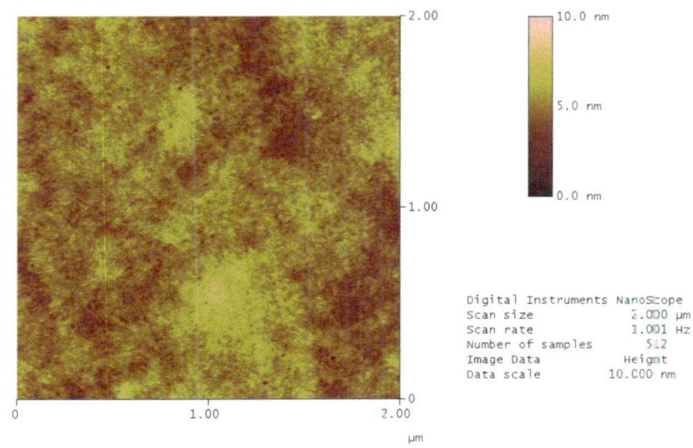

(a)

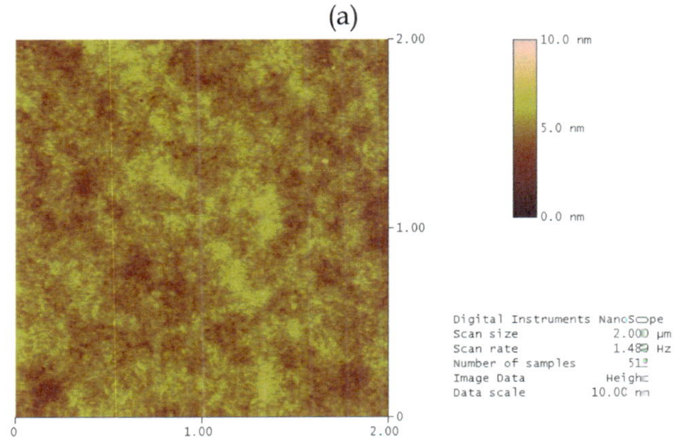

(b)

Fig. 4. Atomic force microscope (AFM) images for the thin films of (a) 2wt% MATS in superyellow and (b) pure superyellow.

The single-phase nature implies that the system is thermodynamically stable, one of the key characteristics of stable material systems. Phase separation exists in almost all the materials used previously in PLECs; including nanostructure multi-phase complexes, (Edman et al., 2004) interpenetrating networks, (Cao et al., 1996) needle-shaped fronts, (Shin et al., 2006) spherical aggregates, (Yang et al., 2003) and large-scale topographical separation. (Hu et al., 2006) In our emitting layer, the MATS comprises only 2 weight percent in the solid thin films. As a result of the good solubility of MATS in superyellow, the films exhibited properties almost identical to those of pure superyellow films. The single-phase stability of this material system with the ionic liquid additive is the most important reason for the long operating lifetime of the devices.

2.5 Origin of open circuit voltage

In order to investigate the origin of the diode junction inside the device during operation, photovoltaic effect measurements were performed and the open circuit voltages (V_{OC}), which provide information on the built-in potentials of the device, were recorded. Fig. 5 shows the I-V curves of the photovoltaic effects before and after device operation. The original V_{OC} of the devices is around 1.25V and there is almost no V_{OC} change after thermal treatment at 80 °C for about 1 minute. The I-V curves of the devices before and after heating are nearly identical. The formation of the junction is obvious in the data; when devices were heated to around 80 °C under 4V forward bias and cooled to room temperature, the V_{OC} increased by about 0.55V to 1.8 V. The measured V_{OC} of about 1.8V is comparable to the band gap of the emitting polymer. As a result of the electrochemical doping near the anode and cathode interfaces, there are very small barriers for both electron and hole injection and consequently device turn-on voltages are very close to the band gap of the semiconducting polymer. (Pei et al., 1995; Pei et al, 1996; Edman et al., 2004) This advantage reflects one of the most important characteristics of PLECs.

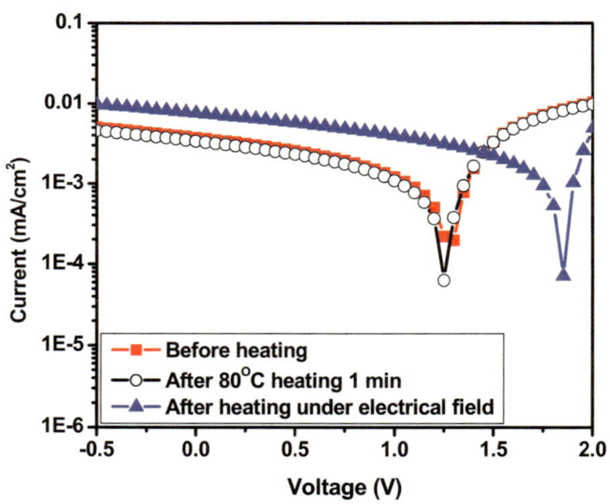

Fig. 5. Current-voltage (I-V) curves showing the photovoltaic effects of the devices before heating (solid square), after 80 °C heating for 1 minute (empty circle), and after 80 °C heating under 4V forward bias (solid triangle).

2.6 Device lifetime and response time

LECs fabricated with the MATS ionic liquid as the source of mobile ions exhibit long continuous operating lifetimes. Data showing the brightness vs time (at room temperature) are shown in Fig. 6(a) (constant current density at 6.76 mA/cm^2). The slow turn-on is indicative of the time required to form the p-i-n junction at room temperature (i.e. without pre-heating under bias). Independent measurements of the decay of the pre-formed p-i-n junction occurs at room temperature over the same time scale. Note, however, that the turn-on time (and junction formation time) is only approximately one minute at 80 °C. After forming the p-type-intrinsic-n-type (p-i-n) junction by pre-heating under bias, devices were operated continuously in the glove box (without packaging) for several days without significant degradation in brightness.

The operating lifetime can be further enhanced by introducing more stable components, for example, high T_g polymers, into this system. The frozen junction after redox and ion redistribution can also be further stabilized by the addition of high T_g components.

High molecular weight (M_W) polystyrene (PS) (MW~1×10^6) has a glass transition temperature of about 110°C. High molecular weight PS can be utilized to enhance the stability of LECS made with MATS without introducing phase separation because of the excellent compatibility of PS and superyellow. High molecular weight PS was dissolved into toluene and then blended with the previously described superyellow solution containing 2wt% MATS. The weight ratio of PS and superyellow was 1:4 and consequently the weight percent of MATS in the solution was diluted to 1.6wt%. Thin film formation and device fabrication processes were exactly same as described above. The brightness vs time for a superyellow:PS (1:4) device is shown in Fig 6(b). After 200 hours, the decay relative to the peak is less than 15wt%. After sealing with epoxy and a glass cover slide, LECS with MATS (and with PS) were operated continuously in air for several weeks.

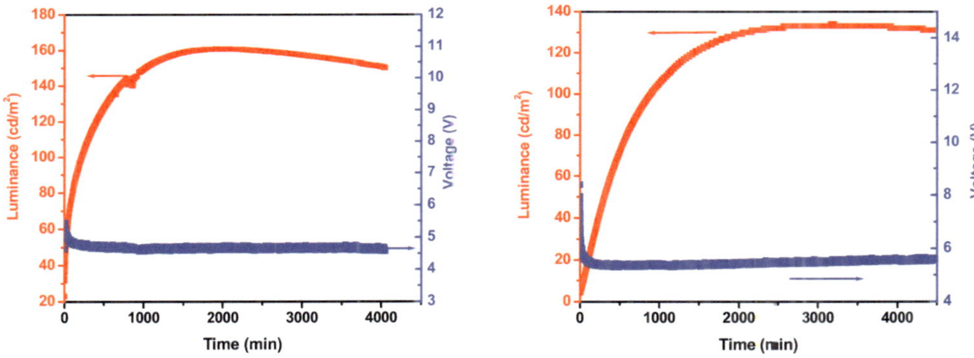

(a) (b)

Fig. 6. (a) Brightness vs time for the device ITO/PEDOT/2%MATS in superyellow/Al (continuous operation mode). (b) Brightness vs time for the device ITO/PEDOT/1.6%MATS + 20% PS in superyellow/Al (continuous operation mode).

Device response time is another important characteristic of frozen junction PLEC devices. To check the response time, devices were driven with a 6.3V pulse train at 152Hz with 43% duty cycle at room temperature, and the light output was measured with a photodetector (peak brightness approximately 1000cd/m^2). The response time was less than 2ms and consistent with the RC time constant of the device (data not shown).

2.7 Summary

In summary, polymer light-emitting devices with an ionic liquid blended into the semiconducting polymer have been demonstrated. After the frozen junction is formed, the devices exhibit all the good characteristics of both PLEDs and PLECs including excellent current-rectification (~10^4), fast response (<2ms), high brightness (>10,000 cd/m^2), high efficiency (~3.3cd/A), low turn-on voltage (2.2V), stable cathodes, and long operating lifetimes.

3. LED to LEC transition behavior in polymer light-emitting devices

3.1 Introduction for operation mechanism of polymer LED and LEC

Polymer light-emitting diodes (Braun & Heeger, 1991) and polymer light-emitting electrochemical cells (Pei et al., 1995; Pei et al., 1996) are of interest because of their potential applications in solid state lighting and in high information content displays. The two types of devices (LEDs and LECs) have different operating mechanisms. In PLECs, the anions and cations redistribute inside the active polymer layer under the influence of an applied electric field with associated electrochemical redox doping (n-type near the cathode and p-type near the anode). As a result of the electrochemical doping, a light-emitting junction with a built-in potential is formed. However, the ion redistribution is a relatively slow process compared with electron or hole transport in semiconducting polymers. Therefore, PLECs typically show a continuous (and slow) increase in emission after the electric field is applied. In PLEDs, the use of a low work function metal as the cathode material facilitates electron injection at the semiconductor interface. Thus, in PLEDs, electrons and holes are injected directly from the electrodes into the p*- and p-bands, respectively; PLEDs are inherently fast-response devices.

3.2 Device fabrication

In this part, polymer light-emitting devices with 2% MATS in superyellow were fabricated with thin films of barium (Ba) as the cathode material. The light-emitting devices were fabricated by spin-casting 6mg/mL superyellow with two weight percent MATS from solution in toluene onto PEDOT-PSS coated ITO glass (spin speed of 1500 rpm). After deposition of the polymer film, 5nm Ba and subsequently 100nm Al were deposited in a vacuum of about 10^{-6} torr. All the device fabrication steps were performed in a nitrogen glove box with oxygen level of about 3ppm. The thickness of the PEDOT-PSS and the active polymer layers were determined by AFM as 40nm and 50nm, respectively. For comparison, PLEDs without MATS were also fabricated under similar identical conditions.

3.3 Device measurement

All electrical measurements were performed under nitrogen in the glove box. The current-voltage characteristics were recorded by a computer controlled Keithley 236 source-measure unit.

Prior to electrochemical doping and p-i-n junction formation, the as-fabricated devices can be considered traditional PLEDs. After several forward scans from 0 to 8 V, the device was "charged" at room temperature by ion redistribution and electrochemical doping. Fig. 7 shows the device I-V curves before and after charging. After charging, the turn-on voltage was 2.18 V (@1cd/m^2), slightly lower than before charging (2.33V@1cd/m^2). As shown in Fig. 7, before charging the brightness at 4 V was only about 150 cd/m^2, whereas after charging the brightness at 4 V increased to nearly 600 cd/m^2. The injected currents also increased by a factor of 4 after charging. The device emission efficiency, therefore, remained at approximately 3 cd/A. As expected, there was no evidence of this kind of charging effect in regular PLEDs with pure superyellow as the active semiconductor polymer.

Fig. 8 shows the transition in operating mechanism from LED to LEC during continuous operation at room temperature. The device was operated in constant current mode at 6.76 mA/cm2 without prior charging or heating. The operating voltage was 4.2V at the beginning of the experiment and dropped to below 4V after about 5 minutes; i.e. lower than that of the control PLED with the same constant current (Fig. 9(b)). The brightness initially decreased from 200 cd/m2 to 128 cd/m2 during approximately 20 minutes. During the same period, the operating voltage dropped from 4.2 V to 3.7 V.

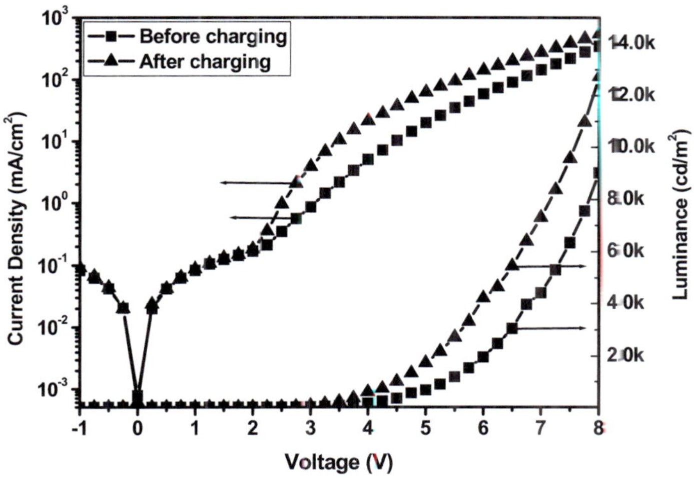

Fig. 7. The current-voltage (I-V) characteristics for the device ITO/PEDOT/2% MATS in superyellow/Ba/Al before and after charging.

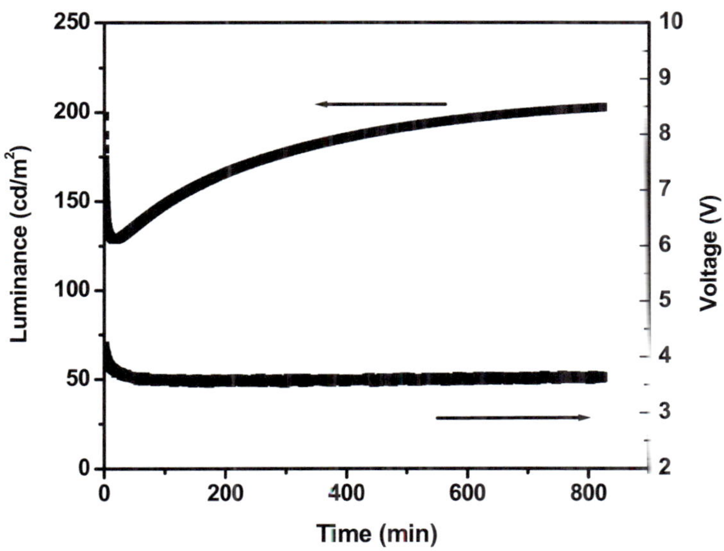

Fig. 8. The current-voltage-brightness (I-V-B) characteristics for the device ITO/PEDOT/2% MATS in superyellow/Ba/Al under continuous operation with constant current.

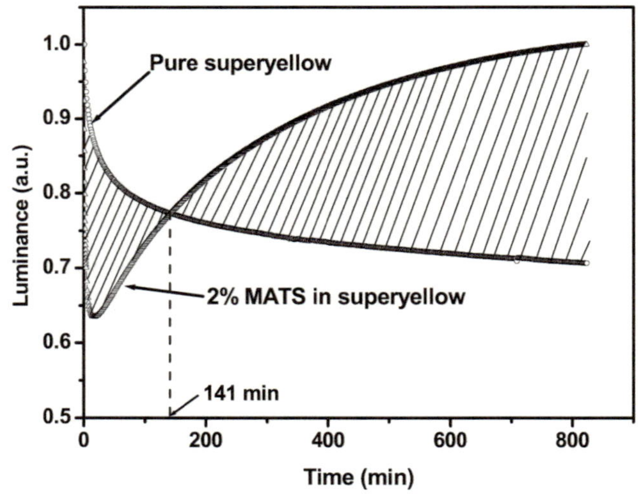

(a)

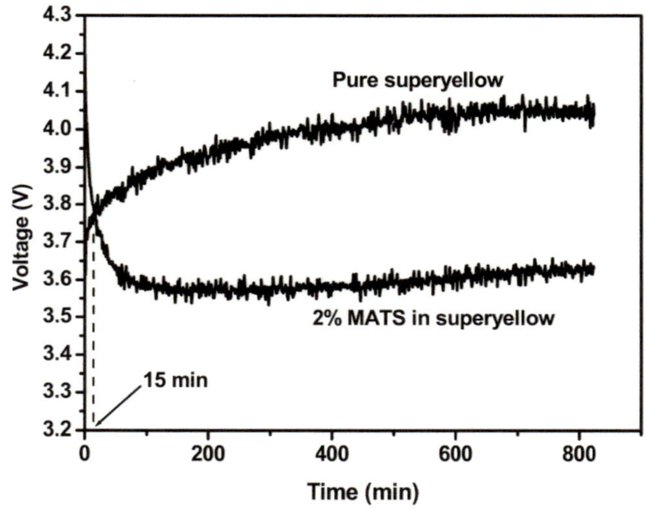

(b)

Fig. 9. (a) Comparison for the decay curves of regular PLED with pure superyellow and the device with 2% MATS in superyellow. (b) Comparison for the operational voltage curves of regular PLED with pure superyellow and the device with 2% MATS in superyellow.

3.4 Discussion

The initial rapid turn-on is characteristic of PLED behavior; the degradation of the low work function cathode resulted in a rapid initial decay in brightness. After a short time, however, ion motion and electrochemical redox doping were initiated. As a result of the redox doping, charge injection of both electrons and holes was enhanced.

After approximately 20 minutes continuous operation, the degradation of the PLED and the enhancement of the PLEC reached a balance point, and the brightness started to increase.

The initial brightness was fully recovered after 620 minutes continuous operation and continued to slowly increase, reaching the highest point after another two hundred minutes operation. Long term monitoring of the brightness indicated only approximately 10% decay after 10,000 minutes operation (measured with respect to the point of highest brightness); a time of order 1000 times longer than that of the control PLED (for the same 10% decay).
Fig. 9 shows the brightness decay and the increase in operating voltages for the control PLED with pure superyellow and the device with 2% MATS in superyellow. The brightness of device with 2% MATS in superyellow is larger than that of the control PLED after 141 minutes operation, and the corresponding operating voltage is less than that of the control after approximately 15 minutes. In Fig. 9(a), the shadowed area during the initial 141 minutes shows the loss of luminance and the shadowed area after 141 minutes shows the subsequent gain of luminance. As shown the operating voltage for the device with 2% MATS in superyellow is lower than that of the control PLED after 15 minutes continuous operation; see Fig. 9(b). At longer times, the operating voltage for the device with 2% MATS in superyellow remains approximately 0.4V below that of the control PLED.

3.5 Summary

In summary, hybrid polymer light-emitting devices with the combined features of LEDs and LECs were fabricated and investigated. The LED to LEC transition results from the formation of a built-in p-i-n junction. Because of the electrochemical doping (p-type near the anode and n-type near the cathode), both electron and hole injection are improved; the contact resistance and the operating voltage are correspondingly reduced. These hybrid polymer light-emitting devices exhibit fast turn-on with low turn-on voltage, low operating voltage, relatively long lifetime with brightness and efficiency comparable to PLEDs.

4. Long-lifetime polymer light-emitting electrochemical cells fabricated with crosslinked hole transport layers

4.1 Introduction for PLEC decay

The ions play an important role as counterions for the electrochemical doping of the semiconducting polymers in PLECs. This doping generates low-resistance contacts (ohmic contacts in many cases). The high doping levels, however, can lead to degradation of the polymers. (Yang et al., 2003; Hu et al., 2006; Dane and Gao, 2004) As a result, PLECs typically have relatively short device lifetimes, especially at high operating voltages. This lifetime issue remains the biggest obstacle for the applications of PLECs. Currently the doping degradation mechanism is not well understood. The heavy doping and/or related electrochemical side-reactions (Shin et al., 2006; Gao et al., 1997; Edman et al., 2004) in the doped polymer near the electrodes could affect the device stability.
At the interface between polymer layer and anode, hole injection is typically facilitated by casting a thin layer of PEDOT-PSS onto ITO glass. Generally in PLEDs, the injection barrier for holes is relatively lower than that for electrons, as confirmed by studies of "electron-only" and "hole-only" devices. (Parker et al., 1999; Blom et al., 1998) Therefore, in this case, it is not necessary to improve the hole injection by the double-layer/doping. Nevertheless, the redox doping near anode is inevitable since anions and cations are introduced into this system together in the form of neutral ionic materials. It has been suggested that the polymer layer might be degraded by holes from anode. (Blom et al., 1998)

4.2 Crosslinkable hole transport materials

In this part, a thin layer of a crosslinkable hole transport material is introduced into the PLEC structure and inserted between anode and active polymer layer. Therefore, there is no direct contact between anode and doped polymer. We find that the device stability is improved.

The two crosslinkable materials used here are polystyrene(PS) - N,N'-diphenyl-N,N'-bis(4-n-butylphenyl)-(1,1'-biphenyl)-4,4'-diamine(TPD)-perfluorocyclobutane(PFCB) (PS-TPD-PFCB) and 4,4',4''-tris(N-carbazoly) triphenylamine bis(vinlybenzylether) (VB-TCTA). (Niu et al., 2006; Liu et al., 2000; Zhao et al., 2006; Jiang et al., 2002; Gong et al., 2003) Fig. 10 shows their molecular structures and a schematic diagram of the device structure. The synthesis and characterisation of PS-TPD-PFCB and VB-TCTA were reported elsewhere. (Liu et al., 2000; Zhao et al., 2006; Niu et al., 2007) The advantages of using these two materials include solution-processability and simple thermal crosslinking with no side products involved. Both PS-TPD-PFCB and VB-TCTA were dissolved in 1,2-dichloroethane, and the 0.5 weight percent solutions were spin-cast at 3000rpm to form thin films on the PEDOT-PSS layer. The two-step heating process for crosslinking was conducted in a nitrogen glove box: 100 °C heating for 40 minutes and then 200 °C heating for 1 hour. Thicknesses were determined by atomic force microscopy (AFM); 12 nm for PS-TPD-PFCB layer and 6 nm for VB-TCTA layer. The room-mean-square (RMS) roughness values of the PS-TPD-PFCB and VB-TCTA layers were about 0.5 nm and 1.6 nm, respectively. (Fig. 11(a) and (b))

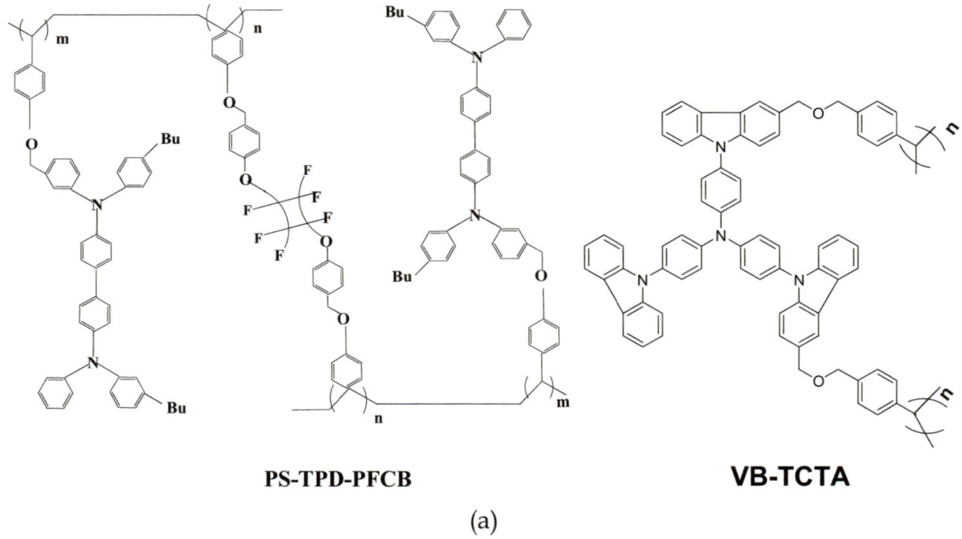

(a)

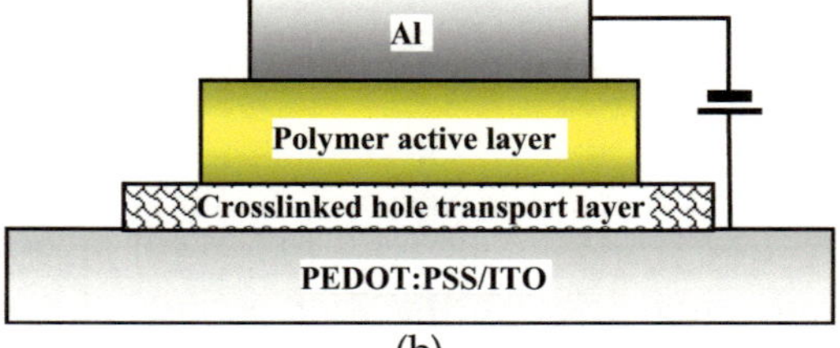

(b)

Fig. 10. (a) Molecular structures of PS-TPD and VB-TCTA and (b) Schematic device structure.

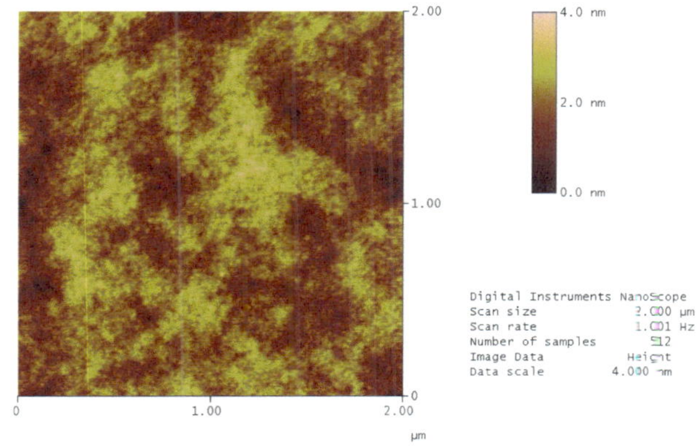

(a)

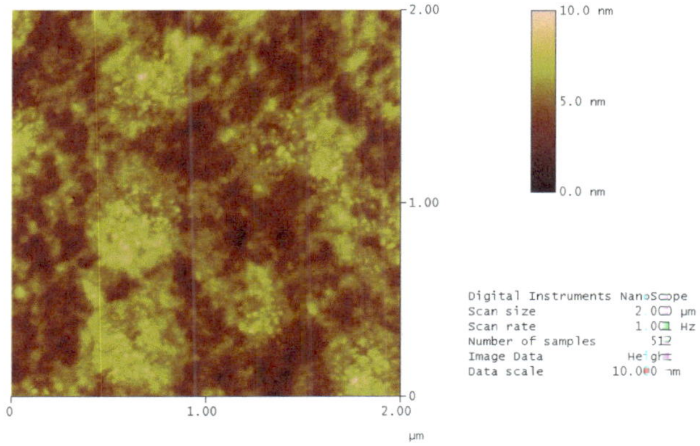

(b)

Fig. 11. Atomic force microscope image for (a) PS-TPD and (b) VB-TCTA layers on PEDOT-PSS.

4.3 Device fabrication on crosslinkable hole transport materials

After heating, the crosslinkable hole transport films were cooled to room temperature and 6mg/mL soluble phenyl-substituted superyellow from Merck/Covion with two weight percent MATS as ion source (both in solution in toluene) was spin-cast onto the crosslinked films with a spin speed of 1500 rpm. After annealing the superyellow film at of 80 °C for 30 minutes, 120nm Al was thermally deposited as the cathode under a vacuum of about 10^{-6} torr (1 torr =133Pa) through a shadow mask. The active device area was 14.8 mm^2. The thickness of the superyellow layer was about 50 nm (determined by AFM). In the following paragraphs, the device with ITO/PEDOT/PS-TPD-PFCB/superyellow:2%MATS/Al is labeled as device A, and the device with ITO/PEDOT/VB-TCTA/superyellow:2%MATS / Al is labeled device B.

4.4 Device measurement

Initially, without pre-bias and heating, the devices showed poor performance; before the ion redistribution, electron injection form Al was poor. After heating at 80 °C under 5V forward bias for about 1.5 minutes, the current increased approximately 1000 times. Then, the devices were cooled to room temperature to freeze in the p-i-n junction. In this process, the anions moved toward the ITO anode and cations moved toward the Al cathode. In our experiments, the p-i-n junction can survive for about 1 hour at room temperature without significant change in open circuit status after pre-bias. Generally, the device performance is measured within several minutes after the junction is formed and the junction can be considered almost unchanged. As a result, the effects of the double layer or redox doping (depending upon the voltage) improved the electron injection (ionic current only represented a very small part in the total current since ion source was very limited). Note, however, that the anions are blocked from the vicinity of the anode by the crosslinked hole transport layer. The frozen p-i-n junction was stable at room temperature for several hours without external electrical field since MATS has a melting point of 56 °C and superyellow possesses a T_g of around 80 °C.

The existence of the p-i-n junction was confirmed by measuring the built-in potentials of the devices. Photovoltaic effect measurements under AM 1.5 solar illumination at 100 mW/cm^2 (1 sun) were performed in a nitrogen glove box to obtain V_{OC} as a measure of the built-in potentials. The built-in potentials for the two devices changed significantly after the ion redistribution. Fig. 3 shows the changes of the built-in potentials for the two PLECs. After ion redistribution, the V_{OC} of device A changed from 1.30 V to 1.75 V (Fig. 12(a)) and the V_{OC} of device B changed from 1.15 V to 1.75 V (Fig. 12(b)).

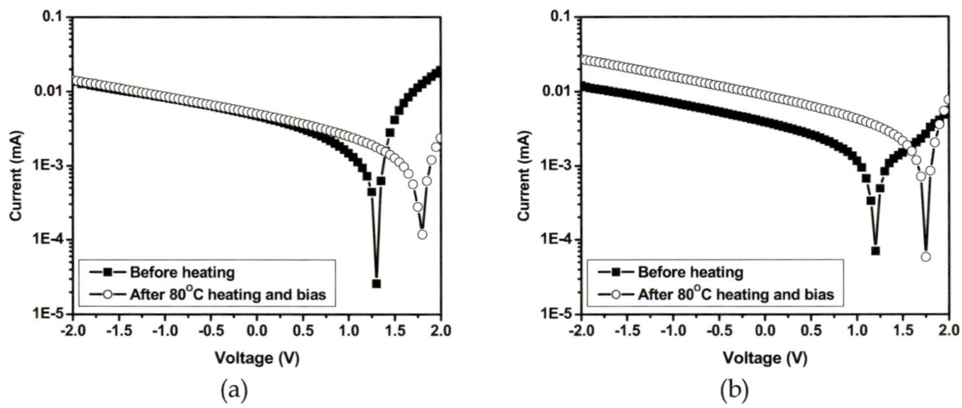

Fig. 12. Current-voltage (I-V) curves showing the photovoltaic effects of the devices before (solid square) and after 80 °C heating under 5V forward bias (circle) for (a) device A and (b) device B.

Fig. 13(a) and (b) show the voltage-current density-brightness curves for device A and device B, respectively. In device A, the turn-on voltage is approximately 3.0V (1 cd/m^2) and the maximum brightness is about 10,000 cd/m^2 at 11.5V. The current efficiency of the device changes from 1.5 cd/A at 4V to 2.5 cd/A at 11V. In device B, the turn-on voltage is approximately 2.5 V (1 cd/m^2) and the maximum brightness is about 9000 cd/m^2 at 9.0 V. The current efficiency again changes from 1.5 cd/A at 4 V to 2.5 cd/A at 9.5 V. In both cases, the current efficiencies increase when the current densities increase, a result which was not seen from traditional PLECs. Control PLECs without the crosslinked hole transport layers were also fabricated and measured. The maximum current efficiency occurred at low operating voltages and the device efficiency decreased slowly as the voltage and current density were increased. We speculate that undesired electrochemical reactions at the anodic interface will be inhibited after insertion of the crosslinked layer. Elimination of electrochemical side-reactions would be expected to stabilize the p-i-n junction. The crosslinked hole transport layers might also function as electron blocking layers which improve the device efficiency at relatively high voltages when the electron injection is very strong.

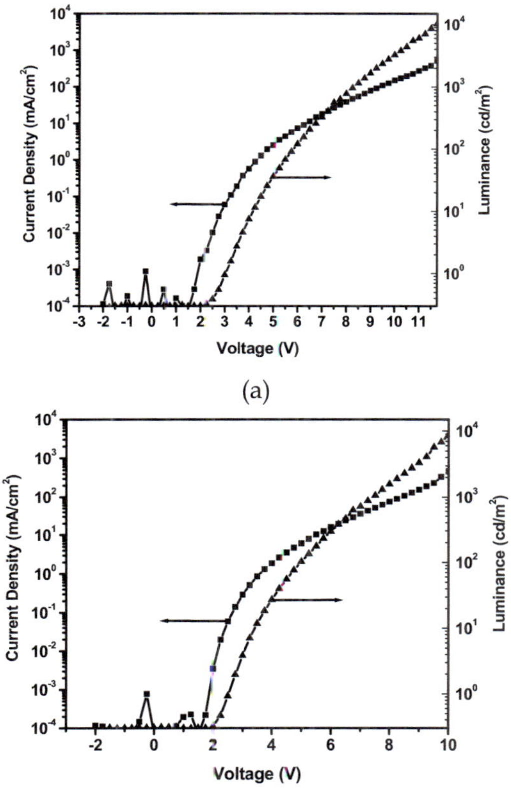

(a)

(b)

Fig. 13. Current-voltage (I-V) and brightness-voltage (B-V) curves of (a) device A and (b) device B.

4.5 Device lifetime enhancement and summary

More importantly, the crosslinkable hole transport layers further enhanced the PLEC operating lifetime. All the measurements were performed at room temperature in nitrogen glove box at a constant current density of 6.76 mA/cm^2. Fig. 14 shows the device decay trends for devices A and B, and the control device without the crosslinked layer. More than 5 devices have been fabricated for each case; Fig. 14 shows the typical results. For device B, after 500 hours the device had not yet reached the point of half brightness. Therefore, we can only conclude that Device B (with the cross-linked hole transport layer) had an even longer lifetime than Device A. All devices were measured from the original status without prior heating. The data clearly illustrate that the lifetimes are enhanced by the crosslinked hole transport layers. Note that Fig. 11 shows that the surfaces of the crosslinked hole transport layers were not very smooth. This surface roughness did not degrade the lifetime; i.e. the lifetime is apparently not very dependent on the surface roughness.

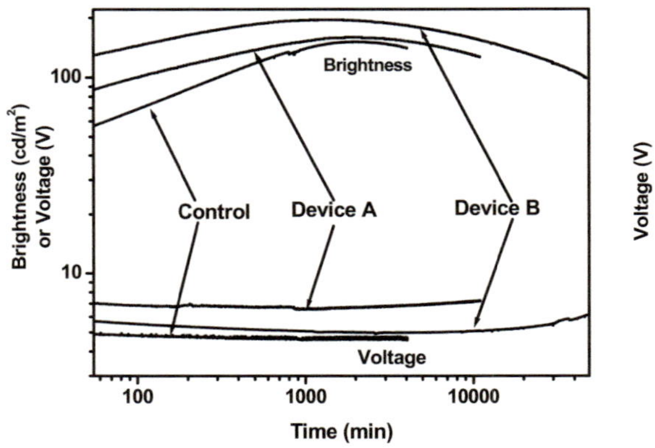

Fig. 14. Brightness vs time (continuous operation mode) for the device A, device B, and control device with structure ITO/PEDOT-PSS/2wt%MATS in superyellow/Al.

In summary, two kinds of crosslinkable hole transport materials have been introduced into the ionic liquid containing PLECs to enhance the performance. By separating the light-emitting layer from conducting PEDOT-PSS layer, the better lifetimes were obtained. The electron injection layer between emitting layer and Al can also be utilized to enhance device lifetime. Titanium sub-oxide sol-gel (TiOx) will be a good choice since it has good stability and hydrophilic property.

5. Conclusion

Ionic liquid has been used in polymer light-emitting electrochemical cells and high performance devices have been successfully demonstrated with long lifetime. With the excellent solubility, good thermal stability, and relatively wide chemical window, ionic liquids can be part of the electrochemical devices. It is believed that more and more ionic liquids and blend systems involving ionic liquids will be discovered.

6. Acknowledgment

This research was supported by a UC Discovery grant as part of the Mistsubushi Chemicals Center for Advamced Materials (MC-CAM) at UC Santa Barbara. The author thanks Dr. Daniel Moses and Nelson Coates for help in obtaining the response time of the frozen junction devices. Author appreciates Dr. M. Liu and A.K.-Y. Jen in Univeristy of Washington for providing the crosslinkable hole transport materials. Author also thanks Prof. G.C. Bazan, Dr. Xiong Gong, Dr. James Swensen and Dr. Jonathan D. Yuen for technical help and useful discussion. A special thank will be given to Prof. Alan J. Heeger, who always guide and encourage the author and support the research.

7. References

Blom, P.W.M., de Jong, M.J.M., Liedenbaum, C.T.H.F., (1998), Device Physics of Polymer Light-emitting Diodes. *Polym. Adv. Technol.* Vol.9, 390-401.

Bozano, L.D., Carter, K.R., Lee, V.Y., Miller, R.D., DiPietro, R., Scott, J.C., (2003) Electroluminescent Devices Based on Cross-linked Polymer Blends. *J. Appl. Phys.* Vol.94, 3061-3068.

Braun, D., Heeger, A.J., (1991), Visible Light Emission from Semiconducting Polymer Diodes. *Appl. Phys. Lett.* Vol.58, 1982-1984.

Burroughes, J.H., Bradley, D.D.C., Brown, A.R., Marks, R.N., Mackay, K., Friend, R.H., Burns, P.L., Holmes, A.B., (1990), Light-emitting Diodes Based on Conjugated Polymers. *Nature*, Vol.347, 539-541.

Cao, Y., Pei, Q.B., Andersson, M.R., Yu, G., Heeger, A.J., (1997)Light-Emitting Electrochemical Cells with Crown Ether as Solid Electrolyte. *J. Electrochem. Soc.* Vol.144, L317-L320.

Cao, Y., Yu, G., Heeger, A.J., Yang, C.Y., (1996), Efficient, Fast Response Light-emitting Electrochemical Cells: Electroluminescent and Solid Electrolyte Polymers with Interpenetrating Network Morphology. *Appl. Phys. Lett.* Vol.68, 3218-3220.

Dane J., Gao, J., (2004) Imaging the Degradation of Polymer Light-emitting Devices. *Appl. Phys. Lett.* Vol.85, 3905-3907.

deMello, J.C., Tessler, N., Graham, S.C., Friend, R.H., (1998), Ionic Space-charge Effects in Polymer Light-emitting Diodes. *Phys. Rev. B* Vol.57, 12951-12963.

Edman, L., Liu, B., Vehse, M., Swensen, J., Bazan, G.C., Heeger, A.J., (2005), Single-component Light-emitting Electrochemical Cell Fabricated from Cationic Polyfluorene: Effect of Film Morphology on Device Performance. *J. Appl. Phys.* Vol.98, 044502.

Edman, L., Pauchard, M., Moses, D., Heeger, A.J., (2004), Planar Polymer Light-emitting Device with Fast Kinetics at a Low Voltage. *J. Appl. Phys.* Vol.95, 4357-4361.

Edman, L., Summers, M.A., Buratto, S.K., Heeger, A.J., (2004), Polymer Light-emitting Electrochemical Cells: Doping, Luminescence, and Mobility. *Phys. Rev. B* Vol.70, 115212.

Gao, J. Li, Y.F., Yu, G., Heeger, A.J., (1999), Polymer Light-emitting Electrochemical Cells with Frozen Junctions. *J. Appl. Phys.* Vol.86, 4594-4599.

Gao, J., Yu, G., Heeger, A.J., (1997), Polymer Light-emitting Electrochemical Cells with Frozen p-i-n Junction. *Appl. Phys. Lett.* Vol.71, 1293-1295.

Gong, X., Moses, D., Heeger, A.J., Liu, S., Jen, A.K.-Y., (2003), High-performance Polymer Light-emitting Diodes Fabricated with a Polymer Hole Injection Layer. *Appl. Pyhs. Lett.* Vol.83, 183-185.

Gong, X., Wang, S., Moses, D., Bazan, G.C., Heeger, A.J., (2005), Multilayer Polymer Light-Emitting Diodes: White-Light Emission with High Efficiency. *Adv. Mater.* Vol.17, No.17, 2053-2058.

Hamada, Y., Kanno, H., Tsujioka, T., Takahashi, H., Usuki, T., (1999), Red Organic Light-emitting Diodes Using an Emitting Assist Dopant. *Appl. Phys. Lett.* Vol.75, 1682-1684.

Hu, Y.F., Tracy, C., Gao, J., (2006), High-resolution Imaging of Electrochemical Doping and Dedoping Processes in Luminescent Conjugated Polymers. *Appl. Phys. Lett.* Vol.88, 123507.

Jiang, X., Liu, S., Liu, M.S., Herguth, P., Jen, A.K.-Y., Fong, H., Sarikaya, M., (2002), Perfluorocyclobutane-Based Arylamine Hole-Transporting Materials for Organic and Polymer Light-Emitting Diodes. *Adv. Funct. Mater.* Vol.12, 745-751.

Kervella, Y., Armand, M., Stephan, O., (2001), Organic Light-Emitting Electrochemical Cells Based on Polyfluorene. Investigation of the Failure Modes. *J. Electronchem. Soc.* Vol.148, H155-H160.

Kim, T-H, Lee, H.K., Park, O.O., Chin, B.D., Lee, S-H, Kim, J.K., (2006), White-Light-Emitting Diodes Based on Iridium Complexes via Efficient Energy Transfer from a Conjugated Polymer. *Adv. Funct. Mater.* Vol.16, 611-617.

Liu, J., Zhou, Q.G., Cheng, Y.X., Geng, Y.H., Wang, L.X., Ma, D.G., Jing, X.B., Wang, F.S., (2006), White Electroluminescence from a Single-Polymer System with Simultaneous Two-Color Emission: Polyfluorene as the Blue Host and a 2,1,3-Benzothiadiazole Derivative as the Orange Dopant. *Adv. Funct. Mater.* Vol.16, 957-965.

Liu, S., Jiang, X., Ma, H., Liu, M.S., Jen, A.K.-Y., (2000), Triarylamine-Containing Poly(perfluorocyclobutane) as Hole-Transporting Material for Polymer Light-Emitting Diodes. *Macromolecules*, Vol.33, 3514-3517.

Niu, Y-H, Liu, M.S., Ka, J-W, Bardeker, J., Zin, M.T., Schofield, R., Chi, Y., and Jen, A.K.-Y., (2007), Crosslinkable Hole-Transport Layer on Conducting Polymer for High-Efficiency White Polymer Light-Emitting Diodes. *Adv. Mater.* Vol.19, 300-304.

Niu, Y-H, Liu, M.S., Ka, J-W, and Jen, A.K.-Y., (2006), Thermally Crosslinked Hole-transporting Layers for Cascade Hole-injection and Effective Electron-blocking/Exciton-confinement in Phosphorescent Polymer Light-emitting Diodes. *Appl. Phys. Lett.* Vol.88, 093505.

Parker, I.D., Cao, Y., Yang, C.Y., (1999), Lifetime and Degradation Effects in Polymer Light-emitting Diodes. *J. Appl. Phys.* Vol.85, 2441-2447.

Pei, Q.B., Yang, Y., Yu, G., Zhang, C., Heeger, A.J., (1996),Polymer Light-Emitting Electrochemical Cells: In Situ Formation of a Light-Emitting p−n Junction. *J. Am. Chem. Soc.* Vol.118, 3922-3929.

Pei, Q.B, Yu, G., Zhang, C., Yang, Y., Heeger, A.J., (1995),Polymer Light-Emitting Electrochemical Cells. *Science,* Vol.269, 1086-1088.

Shao, Y., Yang, Y., (2005), Organic Solid Solutions: Formation and Applications in Organic Light-Emitting Diodes. *Adv. Funct. Mater.* Vol.15, 1781-1786.

Shao, Y., Yang, Y., (2005), White Organic Light-emitting Diodes Prepared by a Fused Organic Solid Solution Method. *Appl. Phys. Lett.* Vol.86, 073510.

Shin, J-H, Xiao, S., Edman, L., (2006), Polymer Light-Emitting Electrochemical Cells: The Formation and Effects of Doping-Induced Micro Shorts. *Adv. Funct. Mater.* Vol.16, 949-956.

Shin, J-H, Xiao, S., Fransson, Å., Edman, L., (2005),Polymer Light-emitting Electrochemical Cells: Frozen-junction Operation of an "Ionic Liquid" Device. *Appl. Phys. Lett.* Vol.87, 043506.

Spreitzer, H., Becker, H., Kluge, E., Kreuter, W., Schenk, H., Schmidt, R. and Schoo, H., (1998), Soluble Phenyl-Substituted PPVs—New Materials for Highly Efficient Polymer LEDs. *Adv. Mater.* Vol.10, 1340-1343.

Wenzl, F.P., Pachler, P., Suess, C., Haase, A., List, E.J.W., Poelt, P., Somitsch, D., Knoll, P., Scherf, U., Leising, G., (2004), The Influence of the Phase Morphology on the Optoelectronic Properties of Light-Emitting Electrochemical Cells. *Adv. Funct. Mater.* Vol.14, 441-450.

Yang, C.H., Sun, Q.J., Qiao, J., Li, Y.F., (2003), Ionic Liquid Doped Polymer Light-Emitting Electrochemical Cells. *J. Phys. Chem. B* Vol.107, 12981-12988.

Yu, G., Cao,Y., Andersson, M., Gao, J., Heeger, A.J., (1998), Polymer Light-Emitting Electrochemical Cells with Frozen p-i-n Junction at Room Temperature. *Adv. Mater.* Vol.10, 385-388.

Zhao, J., Bardecker, J.A., Munro, A.M., Liu, M.S., Niu, Y., Ding, I-K, Luo, J., Chen, B., Jen, A.K.-Y., Ginger, D.S., (2006), Efficient CdSe/CdS Quantum Dot Light-Emitting Diodes Using a Thermally Polymerized Hole Transport Layer. *Nano. Lett.* Vol.6, 463-467.

Part 3

Polymers

Ionic Liquids Gelation with Polymeric Materials: The Ion Jelly Approach

Nuno M.T. Lourenço[1], Ana V.M. Nunes[2],
Catarina M.M. Duarte[3,4] and Pedro Vidinha[2]
[1]IBB/CEBQ Departamento de Bioengenharia, Instituto Superior Técnico, Universidade Técnica de Lisboa, Avenida Rovisco Pais, Lisboa,
[2]Requimte/CQFB, Departamento de Química, Faculdade de Ciências e Tecnologia Universidade Nova de Lisboa, Campus de Caparica, Caparica,
[3]Instituto de Tecnologia Química e Biológica, Universidade Nova de Lisboa, Avenida da Republica, Oeiras,
[4]Instituto de Biologia Experimental e Tecnológica, Apartado 12, Oeiras, Portugal

1. Introduction

The development of ionic liquids (ILs)-based materials is a promising field of research for the design of new advanced functional tailor made materials. In particular, ILs gelation induced by polymers originates quasi-solid materials commonly termed as ion gels which are very interesting, offering good mechanical strength and conductivity. (Torimoto et al., 2010) Indeed, ion gels hold both the processability and mechanical properties of polymers, but with added physico-chemical properties and although primarily developed as replacements for current solid-state polyelectrolytes (Delaney et al., 2010) several other applications are currently emerging as in biosensors and drug delivery applications.

2. From ILs towards new polymeric materials

Ionic liquids (ILs) are probably one of the most studied chemical compounds in the last decade. This is particularly motivated by ILs unique physical-chemical properties that enable their application in a broad range of scientific fields. ILs are comprised entirely by ions and most of them exhibit a negligible vapour pressure, ionic conductivity and a high thermal, chemical and electrochemical stability. (Fernicola et al., 2006; Galinski et al., 2006; Lu et al., 2002)

Nevertheless the tailor made design of ILs is probably the most fascinating and creative domain on ILs research. In fact the creative design of ILs has driven their application to completely different areas such as in physics or biology.

For the design, however is imperative to evaluate the fundamental physical-chemical properties before trying to evolve a given IL structure for a particular application. For instance, the type of molecular interaction between cation and anion is determinant for physical-chemical properties such as melting temperature, glass transition temperature or conductivity. (Yoshida et al., 2007) Thus, numerous works have published about the understating of fundamental properties of ILs. Most of these studies have been directed towards the elucidation of the impact of both anion and cation on ILs physical-chemical properties.

When we think about IL applications two fields emerge immediately: solvents and electrolytes. In the first case ILs can be designed in order to become green solvents. A very large number of works have been done at this level, namely on the use of ILs on different extraction processes or as reaction media. The industrial application is however very limited. Until now the only example of an industrial process using ionic liquid is the BASIL™ (Biphasic Acid Scavenging utilizing Ionic Liquids) process. (BASF, 2007) This process is used for the production of alkoxyphenylphosphines. The original process for the production of alkoxyphenylphosphines, triethylamine is used to scavenge the acid produced during the reaction. Nevertheless this reaction produces a solid that leads to the formation of an insoluble paste and consequently affects all the downstream purification processes. The idea was to replace triethylamine by 1-methyl imidazole which leads to the formation of 1-methylimidazolium chloride which is an ionic liquid. The substitution of triethylamine by 1-methylimidazole leads to the formation of a new phase on the reaction mixture which makes the purification a more straightforward process when compare with conventional techniques.

The great ability to dissolve organic compounds have put ILs on the edge of the chemicals development for commercial applications. More recently the design of biocompatible IL has lead to the application of ILs as drug solvents for drug delivery systems. Some authors have showed that ILs could in fact dissolve and stabilized different therapeutic agents and also assist the controlled drug release. (Dobbs et al., 2009; Moniruzzaman et al., 2010; Park et al., 2010; Stoimenovski et al., 2010; Zhang et al., 2009)

A good example was brought by Moniruzzaman et al. which have used ionic liquid-in-oil (IL/o) microemulsions to enhance the topical and transdermal delivery of acyclovir (ACV). The microemulsion was composed by a blend of nonionic surfactants, namely polyoxyethylene sorbitan monooleate (Tween-80) and sorbitan laurate (Span-20), isopropyl myristate (IPM) as an oil phase, and IL $[C_1mim](CH_3O)_2PO_2$ (dimethylimidazolium dimethylphosphate) as a pseudophase. The solubility of ACV on the microemulsion system significatively increased in the presence of IL, which act as a drug reservoir during the process of delivery. Moreover the transdermal delivery was only achieved when the IL was present in the microemulsion mixture.

Following the same line different authors have been using the IL chemistry to develop ionic hydrid delivery systems. Note that in spite of the success of these two examples this area is still in its infancy and with the recent examples of ILs based on active pharmaceutical ingredients (API) a world of creative developments will emerge to address some of the pains and limitations of pharmaceutical industry.

As salts the most natural application of ILs is of course as electrolytes. In fact ILs are known for their high conductivity (10^{-4} to 10^{-2} S cm^{-1}), high electrochemical stability (4-5.7 V) and thermostability (up to 300 °C). This set of properties together with the fact that most of ILs are nonvolatile and nonflammable, has driven their application as electrolytes for different electrochemical devices, such as dye synthesized solar cells, double layer capacitors, fuel cells, electrochemical windows and of course lithium secondary batteries.(Byrne et al., 2005; Fernicola et al., 2006; Galinski et al., 2006; Lu et al., 2002; Stephan, 2006)

In fact the actual trend in electrochemical devices points to ILs as the most promising approach to develop safety and highly conductive electrolytes.

Nevertheless the large scale fabrication of the above electrochemical devices is following the printing trend due to large scale production impositions. To address this issue different authors have tried to develop solid/polymeric/composites based ion liquids.(He et al., 2007; Leys et al., 2008; Winther-Jensen et al., 2009; Tiyapiboonchaiya et al., 2003) In fact some of these systems have been very competitive in terms of ionic conductivity.
The development of solid/polymeric/composites based ion liquids have of course a special interest to electrochemical applications. However as mentioned before these materials could also play an important role to develop different bioapplications such as the drug delivery.

2.1 IL-based polymer gels: the ion jelly approach

In general, the methods to prepare IL-based polymer gels can be classified into three major types: gelation of ILs within polymers/biopolymers, in situ polymerization of vinyl monomers in ILs, and polymerization of ILs containing polymerizable groups (e.g vinyl groups).(He et al., 2007; Le Bideau et al., 2011; Tiyapiboonchaiya et al., 2003)
The introduction of polymerizable groups into the ionic liquid structure has been presented as a very interesting way to obtain good ionic conductivity without liquid components. These polymerized ionic liquids (PILs) have been developed for battery electrolyte and for other solid electrolyte applications.(Jiang et al., 2006; Ohno et al., 2004; Washiro et al., 2004; Winther-Jensen et al., 2009)
PIL have been investigated to be used on polymeric lithium batteries. In these batteries besides a good conductivity it is necessary that specific molecules like the lithium ion could be easily transport through the electrolyte. Combining these two factors on PILs is not a simple task since the variables, conductivity and ion transport, are related with several issues. One of the major factors that can affect simultaneously the lithium transport and conductivity is the type of cation used in PIL.
Ohno and co-workers have shown that the type of cation could enhance the lithium transport trough the PIL. (Ogihara et al., 2006) They found that cations with the piperidinium salt structure could be an advantage for lithium ion conduction.
The most simple and efficient approach is based on gelation which is a simple method that allow a good compromise between the retention of the IL and its fluidity inside the polymeric network. These so called ion gels are simpler than solid polymer electrolytes and exhibit improved conductivities. In fact ion gels hold both the processability and mechanical properties of polymers, but with added physico-chemical properties and were primary developed as replacements for current solid-state polyelectrolytes in energy devices, such as dye-sensitized solar cells, supercapacitors, lithium ion batteries, and fuel cells. (Fernicola et al., 2006; Galinski et al., 2006; Le Bideau et al., 2011; Lu et al., 2002; Mazille et al., 2005; Stephan, 2006)
For instance, He and co-workers have shown the potential of an ion gel formed by gelation of poly(styrene-block-ethylene oxideblock-styrene) (SOS) triblock copolymer in 1-butyl-3-methylimidazolium hexafluorophosphate.(He et al., 2006) This system has shown interesting conductivity values at room temperature (above 10^{-3} S/cm^{-1}).
Another good example was described by McFarlane and co-workers. In this case the authors develop a novel self-polymerised ionic liquid (IL) gel prepared at room temperature (RT), without light or heat or addition of initiator, using choline formate (CF) and 2-hydroxyethyl methacrylate (HEMA). (Winther-Jensen et al., 2009)
This field of research has been developed in order to create ILs based materials that can work as electrolytes in different electrochemical devices and be used either as printer substrates or printable "inks".

Following this trend we recently reported Ion Jelly (IJ), a light flexible electrolyte that results from the combination of gelatin with an IL. This allows the production of gels that are extremely versatile conductive materials that can be molded into different shapes, using several techniques, and can be adapted to multiple surfaces. Moreover, on cooling, Ion Jelly can undergo a liquid-gel transition near room temperature (near 35°C) which makes a promising solution to develop electrolyte "inks" for printed electrochemical devices.(Vidinha et al., 2008)

Fig. 1. Ion jelly materials.

One of the main advantages of Ion Jelly is using gelatin which is a widely available, inexpensive and well studied biopolymer. Gelatin is prepared through the thermal denaturation of collagen, after acid or alkaline pre-treatment. Basically gelatin is a heterogeneous mixture of left-handed proline helix polypeptides and amino acid strands, with a typical sequence of -Ala-Gly-Pro-Arg-Gly-Glu-4Hyp-Gly-Pro. Dissolution of gelatin in water occurs at 30–35 °C, at which these polypeptide strands undergo a coil-helix transition.(Bigi et al., 2004) The presence of the hydroxyproline residue favors compactness of the quaternary structure due to the formation of hydrogen bonds between hydroxyl and carbonyl groups of the main chain helices. The physical properties of gelatin and collagen are influenced by this interaction.(Powel &Boyce, 2009)
Although inexpensive, biocompatible and biodegradable, the gelatin molecule only interacts strongly with molecules that are very soluble in aqueous media. The combination of gelatin with ionic liquids can tremendously increase the scope of the gelatin application, since ILs

can confer upon this polymer different physical-chemical properties and also change the gelatin microenvironment. The latter can in fact address one of gelatin's limitations which is related with the entrapment of poorly water soluble molecules.
The formation of an Ion Jelly should occur in much the same way as the formation of a water-based gelatin gel. Both gelatin and ILs have ionic character and this leads to strong interactions between the two species, and a high solubility of gelatin. During the renaturing or annealing process, the polypeptide strands will have a tendency to rearrange into the most thermodynamically favourable structure. X-ray diffraction experiments have shown important differences between the water-based gelatin and Ion Jelly films indicating a pronounced modification in the conformation of the gelatin left-hand helix. These results clearly show that Ion Jelly molecular structure is completely different from gelatin.
Due to the nature of ionic interaction between gelatin and ILs we also verify that hydrogen bonding played an important role on the mechanism of interaction between IL and gelatin. In fact Table 1 shows a selection of cation/anion combinations that provided the formation of either Ion Jelly or solid structures, 1-ethyl-3-methyl-imidazolium [emim], 1-butyl-3-methyl-imidazolium [bmim], 1-butyl-3-methyl-imidazolium [bdmim], 1-decyl-3-methyl-imidazolium [C_{10}mim] and tri-n-octyl-methylammonium (Aliquat336®).

Cation	Anion	Il:gelatin ratio (w/w)	Water-miscible	Type of material formed
[bmim]	[N $(CN)_2$]	1:1	YES	Solid Transparent films
		3:1	YES	Solid Transparent films
		6:1	YES	Liquid gel
[omim]	[N $(CN)_2$]	1:1	YES	Solid Transparent films
		1:3	YES	Solid Transparent films
Aliquat 336®	[N $(CN)_2$]	1:1	NO	No Ion Jelly formation
[Him]	[Cl]	1:1	YES	Solid Transparent films
[Him]	[Cl]	1:3	YES	Solid Transparent films
[Hmim]	[Cl]	1:1	YES	Solid Transparent films
		1:3	YES	Solid Transparent films
[bmim]	[Cl]	1:1	YES	Solid Transparent films
		1:3	YES	Solid Transparent films
[omim]	[Cl]	1:1	YES	Solid Transparent films.
		1:3	YES	Solid Transparent films.
[C_{10}mim]	[Cl]	1:1	YES	Solid Transparent films
[bdim]	[Cl]	1:1	YES	Solid Transparent films
		1:3	YES	Solid Transparent films
Aliquat 336®	[Cl]	1:1	NO	No Ion Jelly formation
[bmim]	[saccharin]	1:1	YES	Solid Transparent films
		1:3	YES	Solid Transparent films
Aliquat 336®	[saccharin]	1:1	NO	No Ion Jelly formation
[bmim]	[PF_6]	1:1	NO	No Ion Jelly formation

Table 1. Selected cation and anion combinations that led to the formation of an Ion Jelly.

The hydrophilicity of the IL correlates broadly with its ability to form Ion Jelly structures. This is well illustrated by the results obtained with the [bmim] series. The result for [bmim] BF_4, which is water-miscible but still does not form an Ion Jelly structure, may be due to the weak coordination ability of the BF_4 ion. More recently we have shown that 1-(2-hydroxyethyl)-3-methyl-imidazolium tetrafluoroborate ([C_2OHmim]BF_4) is able to form Ion Jelly materials. In fact Holbrey and co-workers found that the physico-chemical properties of ILs containing a hydroxyl group in the alkyl side-chain were substantially different from common N,N-dialkyimidazolium ILs. The presence of a hydroxyl group in the alkyl chain increases the polarity of the latter, leading to a stronger hydrogen bonding ability with surrounding groups. This reinforces the above idea that stronger hydrogen bonding between the ILs and gelatin is an essential requisite for forming Ion Jelly materials.
Moreover we also verify that the IL/gelatin ratio is an important feature to obtain IJ materials. For instance when using the ratio (6:1) we were unable to obtain IJ materials this may be related with the limited availability of sites on the gelatin molecules for hydrogen bonding with the IL, higher proportions of the latter decreasing the internal cohesion of the material formed.
With this we have tried to develop different applications for Ion Jelly. Our approach aims for scoping the advantages of gelatin and IL combination. Those applications were explored in three different fields: electrolytes, drug delivery and biosensing.

2.1.1 Electrolytes

Our primary approach was to develop Ion Jelly electrolyte materials. To this respect our first step was to evaluate the conductivity of Ion Jelly materials obtained with different ionic liquids in order to evaluate the impact of gelatin on the IL conductivity, Figure 2.
As we expected the conductivity of the Ion Jelly materials was affected by the type of IL used, since the type of interaction between IL and gelatin will be different in each case. Thus, in order to elucidate about the impact of gelatin on IL conductive properties we proceed to the physical chemical characterization of Ion Jelly materials. To accomplish this goal we have performed a dielectric relaxation spectroscopy characterization (DRS).
Basically, DRS spectra reproduce the set of molecular motions of all dipolar species present in the media. In ionic liquids these motions are highly correlated with the multiplicity of interaction between the different charged species present in the media, which makes it impossible to address a specific motion to a well defined dipole. In fact on ILs the molecular motions reflect the kinetics of the network rearrangement. (Aliaga et al., 2007; MacFarlane et al., 2001)
On the other hand, the Ion Jelly network is settled by the interaction between two polyelectrolyte molecules (gelatin and IL) creating a complex network with multiple interaction sites that can lead to a great variety of dipolar aggregates. Thus a comprehensive and detailed analysis of IL relaxation behavior inside gelatin matrix can result in important data about the crucial mechanisms implicated on the Ion Jelly conductivity. To conduct this study we have chosen the Ion Jelly materials based on 1-butyl-3-methyl imidazolium dicyanamide ([bmim]DCA). The dicyanamide (DCA) compounds are liquid at room temperature and characterized by their low viscosity, water miscibility, high thermo (over 100°C) and electrochemical stability (over 3.5 V). (Aliaga et al., 2007; MacFarlane et al., 2001)
Moreover, the dicyanamide ion is an anionic bridge ligand that has Lewis base attributes, which makes it particularly attractive to synthesize ionic liquids with very specific

properties. Compared to common anions such PF_6 or BF_4, DCA has a permanent dipole and thus facilitates the research of IL dynamics through dielectric spectroscopy.(Sangoro et al., 2008)

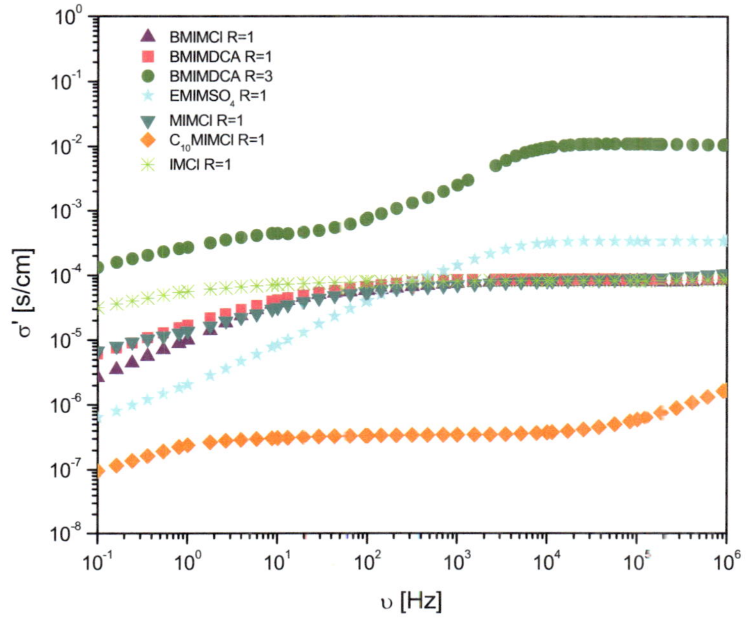

Fig. 2. Frequency dependent ionic conductivity was measured at 25°. C R= IL/gelatin (w/w).

Our aim in this study was to evaluate the impact of gelatin on IL diffusion. For that purpose we have used a correlation to separate the mobility, μ, and the effective number density of charge carriers, n, from conductivity (σ_0) obtained from the dielectric measurements, allowing also to estimate the diffusion coefficient of migrating charges, D, which is done by considering the following Einstein-Smoluchowski equation:

$$\sigma_0 = nq\mu = \frac{nq^2}{k_B T} D \qquad (1)$$

In figure 3 the obtained D and μ values are displayed for the ionic liquid and Ion Jelly; the inset shows the plot of the number density of charge carriers as a function of the reciprocal temperature.

Surprisingly, the diffusion coefficients and mobility of charge carriers in IJ3 are close to those in the pure [bmim]DCA. This means that the solid-like material retains a similar ability for charge transport as the ionic liquid. The same is not true in IJ1 due to the low ratio [bmim]DCA /gelatin. The difference in the temperature range where these quantities are able to be estimated is determined by the glass transition temperature that, which is nearly the same for IJ3 (174,3 K) and [bmim]DCA (174,6 K) and ~30K higher for IJ1 (204,1 K) .

Moreover we have obtained similar N values for all the three situations. Since N is the effective number density of charge carriers, this means that even in the Ion Jelly materials the number of charge carriers per volume is comparable to the one existing in the pure ionic liquid. This result clearly shows that the differences obtained in the transport properties do not arise from different magnitudes of N. These results were in fact quite promising since were able to obtain identical conductive properties between the pure IL and the Ion Jelly. In fact the diffusion and mobility of ionic species are identical on Ion Jelly and [bmim]DCA, meaning that the ionic conductivity is not significantly affected by the presence of gelatin. Thus Ion Jelly appears as a very promising solution to design different electrochemical devices.

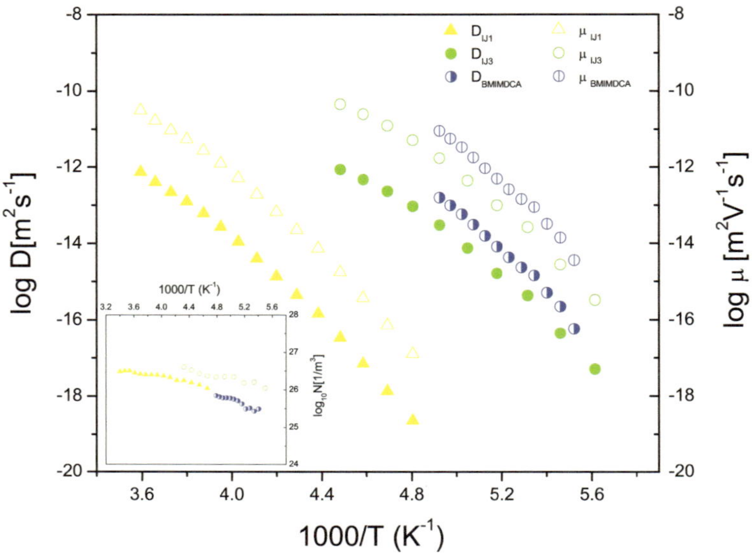

Fig. 3. Thermal activation plot for diffusion coefficients, D, and mobilities, μ,. (a) diffusion coeffiecents. (b) mobility. The inset shows the temperature dependence of N, the effective number of charge carriers. IJ1 – Ion Jelly with a IL/gelatin ratio (w/w) = 1 ; IJ3 – Ion Jelly with a IL/gelatin ratio (w/w) = 3.

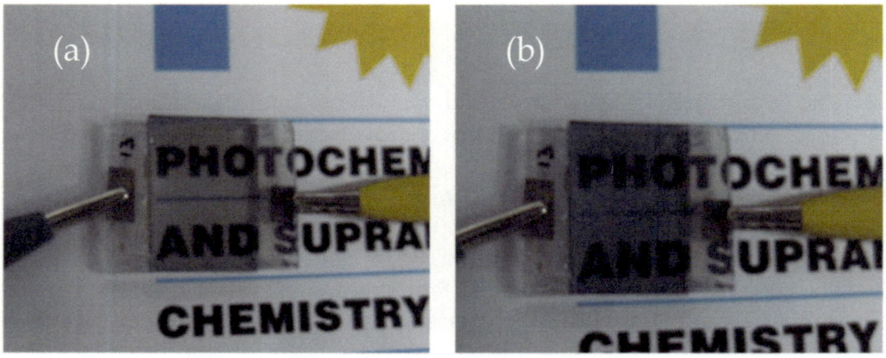

Fig. 4. Ion Jelly electrochromic window. Glass-ITO/PEDOT/Ion Jelly/PB/ITO-Glass.
(a)- Colored state; (b) – Bleached state.

In order to test this possibility, we built an electrochromic window based on Prussian blue (PB) and poly(3-4-ethylenedioxy thiophene (PEDOT) as electrochromic layers and we coated one of the electrode surface with a thin film of Ion Jelly based on [bmim]DCA (figure 4). This electrochromic window built performed reasonably well in what concerns contrast and stability (figure 5). Nevertheless further optimization of this concept should be performed in order to improve the switching velocity. (Vidinha et al., 2008)

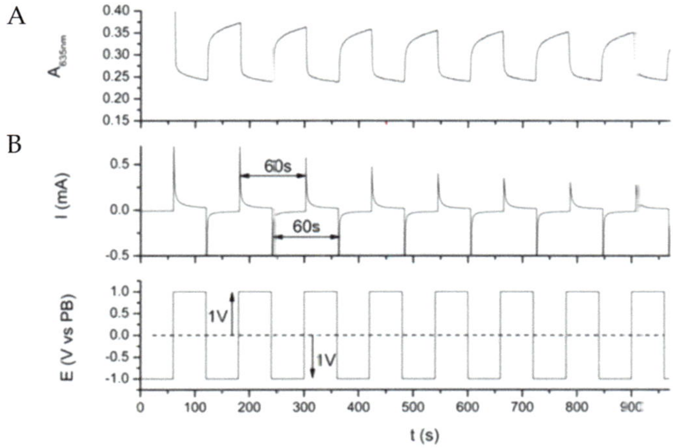

Fig. 5. In situ spectroelectrochemical cycling data for Glass-ITO/PEDOT/Ion Jelly/PB/ITO-Glass. A - Chronoabsorptometry recorded at 700nm; B - Square-wave switching between -1V (step duration 60 s) and +1, V (step duration 60 s) (vs. PB);

2.1.2 Biosensing

Another area that we have pursuing is biosensing. In this particular we are trying to develop biosensors based on the immobilization of different oxidoreductases in Ion Jelly matrices. This concept is based on the fact that enzyme exhibit an enhanced stability and activity in ILs.(Moniruzzaman et al., 2010) Thus, the possibility to tailor-make sensitive and specific enzyme-IL-polymeric materials is perceived as a great advantage, and is motivating extensive research.

Oxidoreductases, namely glucose oxidase (GOD) and horseradish peroxidase (HRP) have been used as model enzymes on the development of new enzyme-IL-polymeric materials. Most of the examples concern the use of these systems take advantage of the intrinsic IL electrochemical properties allowing a direct electron transfer. For example, the potential applications of several sol-gel–[bmim]BF_4 composite systems and the direct electrochemistry of HRP immobilized in such composites have been reported. (Liu et al., 2005) Another good example is the entrapment of GOD in nanogold-N,N-dimethylformamide-IL composite films on a glassy carbon electrode. The authors showed that ILs can affect the electron transfer through interactions with GOD. The influence of the IL on the thermal stability and catalytic activity of GOD entrapped in the composite was also demonstrated. (Li et al., 2007) HRP was also immobilized on a composite material based on the IL N-butylpyridinium hexafluorophosphate, as well as sodium alginate and graphite. The material allowed the determination of H_2O_2 with a detection limit of 0.5 μM.(Ding et al., 2008)

We have recently purpose the application of enzyme-IL-polymeric materials on biosensing, namely, colorimetric detection of glucose based on Ion Jelly materials. (Lourenço et al., 2011) Taking advantage of the preparation method of the Ion Jelly, which is a liquid material at temperatures above 35°C, it was possible to prepare glucose paper test strips by physical deposition of gelatin-[emim]$EtSO_4$ containing GOD and HRP as well as color-generating precursors - in this case, phenol-4-sulfonic acid (PSA) and 4-aminoantipyrine (4-AAP) as reducing substrates:

$$\text{Glucose} + O_2 \xrightarrow{GOD} \text{Gluconic acid} + H_2O_2$$

$$2\,H_2O_2 + \text{PSA} + \text{4-AAP} \xrightarrow{HRP} \text{quinoneimine dye} + 3\,H_2O + NaHSO_4$$

The GOD entrapment on the gelatin-[emim]EtSO4 polymeric material showed a activity decrease of about 16 times comparing with the free enzyme, however with excellent storage stability at 4°C for a period of two weeks, where it retains 70% of initial activity. Concerning HRP, this loses around 13 times activity but again with excellent storage stability, where it is retain 91% of initial activity. While free HRP loses activity rapidly above substrate concentrations of about 0,15 mM, the gelatin-[emim]$EtSO_4$ materials can protect the enzyme from immediate deactivation, making it possible to use much higher H_2O_2-concentration (110 mM), 1 thousand times higher concentration of H_2O_2 in relation to free enzyme.
The co-immobilization of phenol-4-sulfonic acid and 4-aminoantipyrine, as color-generating precursors, together with both enzymes (GOD, HRP) in gelatin-[emim]$EtSO_4$ demonstrate a positive effect on the enzymatic activity of both enzymes. (Figure 6)
Taking advantage of the observed positive effects on GOD-HRP activity and stability, a possible application of this enzyme-IL-polymeric materials was demonstrated on the colorimetric determination of glucose. In Figure 7 it is demonstrated the detection of glucose solutions at different concentrations 0.1- 15 g/L with a paper test strip that was prepared with the drop casting deposition of 110 µl of gelatin-[emim]$EtSO_4$ with the dye-components 4-AAP and PSA co-immobilized. The result of the glucose determination can be obtained in less than one minute.
The morphological and functional stability of these paper test strips was evaluated in a physiologic solution at 37 °C. These experiments reveal good stability for a period of one hour. For a longer period of time it is observed the dissolution of the gelatin-[emim]$EtSO_4$ functional polymers. The simplicity of paper test strip preparation allied to the straightforward detection can open new opportunities for new biosensing platforms or other deposition types, such as screen printing or even ink-jet printing.

2.1.3 Drug delivery and biomedical applications

Recent research has shown that IL-based polymer gel materials can be used as drug delivery systems which is envisaged as a very promising field of application. The search for new drug delivery systems is of major importance in drug development, and research studies in this area have grown enormously. (Gil et al., 2002) In order to minimize drug degradation and loss, to prevent harmful side-effects and to increase drug bioavailability, various drug delivery systems are currently under development (Kita & Dittrich 2011). Furthermore, as new therapeutic agents emerge, for example from studies in biotechnology, such as proteins and genes, there is an increasing need for the development of new technologies and new materials with improved properties (Ghandehari, 2008). Indeed, polymers are used in a wide range of medical applications which they were not specifically designed for, often

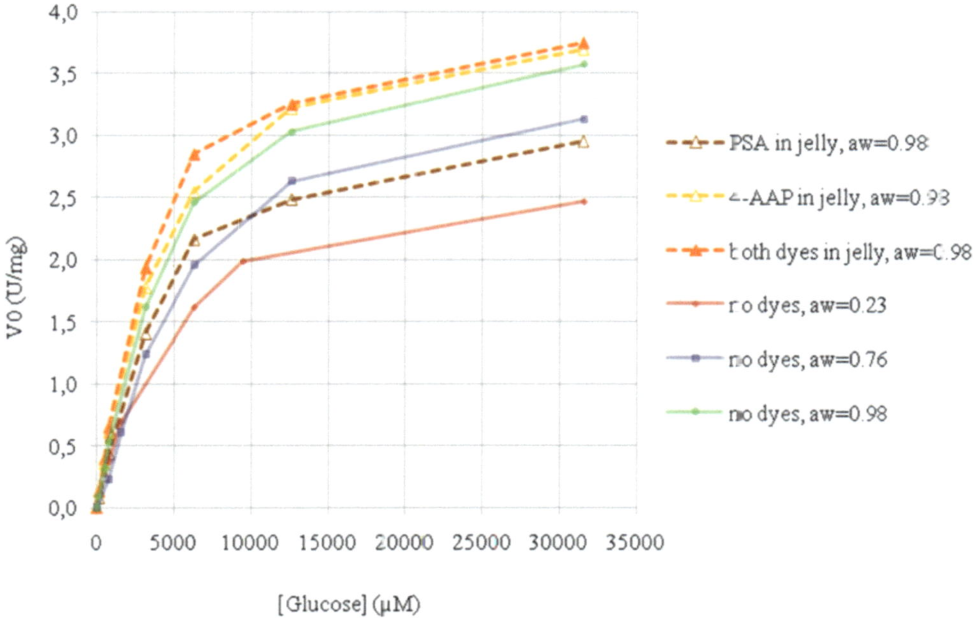

Fig. 6. Effect of the incorporation of the phenol-4-sulfonic acid and 4-aminoantipyrine, as color-generating precursors and water activity on the initial activity of GOD in gelatin-[emim]$EtSO_4$ functional polymers.

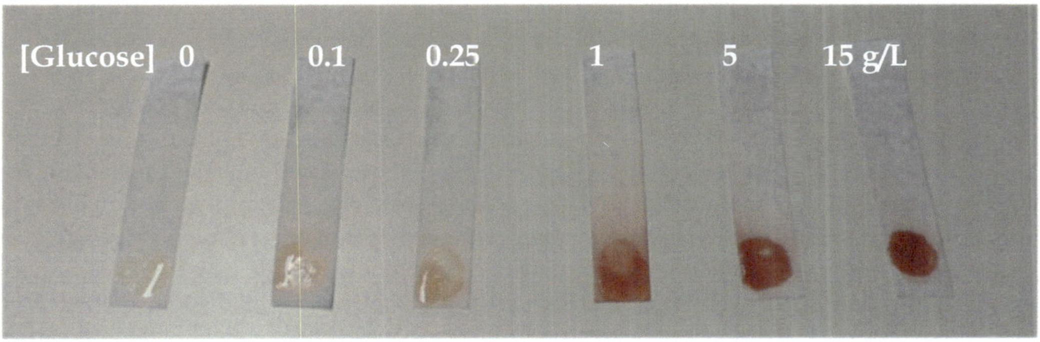

Fig. 7. GOD-HRP Paper Test Strip for glucose determination, from left to right: Control (50 μL MilliQ water) detection of 50μL of glucose solutions with different concentrations (0.1-15g/L).

presenting several problems such as "unsatisfactory entrapment efficiency"; poor stability; unacceptable delivery primarily to the liver and spleen; poor plasma pharmacokinetics and rapid dissociation of active agent. (Frokjaer & Otzen, 2005) What IL-based polymer gels offer is the opportunity of hybridized polymers properties with those of ILs, which can be further tuned by the appropriate choice of the anion-cation pair opening therefore millions of possibilities. These new hybrid materials are expected to present properties derived from both components, greatly increasing polymers performances and applicability.

In fact IL-based polymer gels hold several properties that make them adequate for drug delivery and other biomedical applications, for example due to IL tunable physico-chemical properties these materials can be tailored to respond to a number of stimuli e.g temperature, pH, electrical field. They can be design with different swelling capabilities, to be more or less stable and to have a controlled degradation rate. IL-based polymer gels may exhibit different bioadhesion degrees to facilitate drug targeting, especially through mucus membranes, delivering API for local or systemic effect and in this manner enhance bioavailability by avoiding or minimizing effects such as enzymatic or hepatic degradation. (Andrews et al., 2009) In addition, ion gels enjoy good mechanical strength and electrical conductivity which can be explored for more specialized applications as electrically stimulated controlled release devices (e.g iontophoresis) or artificial muscles as well as for cardiac and/or neuronal tissue engineering applications. (Guiseppi-Elie, 2010; Ravichandran et al., 2010) Finally, the versatility of sol-gel in shaping allows easy adaptations to a wide range of drug delivery applications. (Le Bideau et al., 2011)

Functionalization of IL-based polymer gels can be achieved using essentially two different strategies, by the incorporation of the active principle in the solid matrix or exploring the IL as the active principle ingredient, in which a specific biological activity is introduced through one or both of the ions. (Le Bideau et al., 2011) This latter strategy was recently reported by Vioux and co-workers in order to encapsulate ibuprofen in porous functionalized silica. (Viau et al., 2010) The authors synthesized a new ionic liquid based on ibuprofenate drug and imidazolium type cation and have further tested the ability of IL-silica gels to act as new drug delivery systems. The release kinetics were found to be slower with the gels than for both crystalline ibuprofen and pure ibuprofenate ionic liquid. The authors also reported that the release was governed by the inner and the outer surface of the IL-silica gel, as well as the chemical nature of the surface. They concluded that IL-silica gels act as drug reservoirs for controlled delivery and added that these gels are easily shaped and do not need any further processing to achieve a final pharmaceutical form.

For drug delivery purposes, the biocompatibility of the material and thus IL toxicity is a crucial issue to be addressed. Indeed the toxicity of ILs has been one of the most debated topics in the field, and several different approaches have been used in order to decrease the level of toxicology. A very interesting ion that can be used for this purpose is a quaternary ammonium ion, namely choline (Nockemann et al., 2007). Recently ILs composed solely of biomaterials were developed. In particular the combination of a choline cation with propionate, tiglate, hydrogen succinate and hydrogen maleate, yielded room temperature ionic liquids with strong hydrogen bonding characteristics (Fukaya et al., 2007). Another strategy was reported, based on the chemical modification of the alkyl side chain of ions that present a certain level of toxicity (Gathergood et al., 2004). Duarte and co-workers has performed toxicity measurements with different ionic liquids. Results showed that the presence of a carboxylic group at the end of the alkyl chain greatly decreases toxicity, and that by further changing the carboxylic group to an ester, an increase in the toxicity was observed. (Frade et al., 2007) This inherent tuneability was recently highlighted as a key feature that is entirely appropriate and applicable to the field of pharmaceuticals (Hough et al., 2007).

In this context, one good example of a biocompatible IL-based polymer gel was given by Sehgal and co-workers. (Vijayaraghavan et al., 2010) The authors have used choline salts, some of which can be described as ionic liquids, as crosslinking agents for collagen in order to replace commonly used glutaraldehyde, which is poorly biocompatible. In the reported work, the authors have hibridized collagen properties with those of ILs by actually synthesizing IL-collagen gels in order to improve collagen mechanical properties and above all to decrease collagen degradation rate, which is the main factor limiting it application for long-term biomedical implants. The materials prepared were found to be biocompatible and stable in water after 6 months of continuous immersion.

Along the same line, Ion Jelly is expected to improve gelatine properties and significantly increase it range of applications. An on-going research project supported by Fundação para a Ciência e Tecnologia, Portugal, is exploring the interesting features that result from the combination of the chemical versatility of an IL with the morphological flexibility, biocompatibility and bioavailabitity of gelatin, regarding an application in drug delivery systems. Gelatin is a very popular drug delivery vehicle, mainly because of its excellent biocompatibility and degradability to non-toxic products. The main limitations of gelatin are its extensive swelling, rapid dissolution and drug release. (Young et al., 2005) In this way, ILs can be used to replace chemical crosslinking in order to form relatively non-soluble networks and significantly expand gelatine applications. Furthermore also the possibility of introducing a determined biological activity in the immobilized IL phase will allow for the development of an entirely new concept of gelatine-based drug delivery systems

3. Conclusion

Ionic liquids are making in-roads in the development of new tailor-made and highly functional polymer-based materials. The hybridization between ILs and polymers allows taking advantage of ILs unique properties in the solid state, circumventing therefore some drawbacks related with liquids processability and opening for new ILs applications. On the other hand, as it was shown for the case of gelatine, widely known and already in use polymers can also see their applications greatly extended, as well as increase their performances by mixing with the right cation-anion pair. The multitude of possibilities, besides allowing for unlimited opportunities, may actually turn the choice of the best material to mix with the right IL a difficult task for researchers in the field. Close collaboration between engineers and chemists in this area will certainly open for new advancements and new materials with superior performances.

4. Acknowledgment

We acknowledge to Prof. Carlos Afonso, Prof. Susana Barreiros, Prof. Madalena Dionisio, Prof. Joaquim Cabral, Prof. Luis Fonseca, MSc Johannes Österreicher, MSc Tânia Carvalho, Prof. Jorge Parola and Dr. Carlos Pinheiro for their scientific contribution. Pedro Vidinha acknowledges its post-doc grant SFRH / BPD / 41546 / 2007, Nuno Lourenço acknowledges its post-doc grant SFRH / BPD / 41175 / 2007 and Ana Nunes its post-doc grant SFRH / BPD /74994/2010. This work was financed thought the following FCT projects: PTDC/EQU-EQU/104552/2008 and PTDC/EBB-EBI/099237/2008

5. References

Aliaga, C. & Baldelli, S. (2007). Sum frequency generation spectroscopy of dicyanamide based room-temperature ionic liquids. Orientation of the cation and the anion at the gas-liquid interface. Journal of Physical Chemistry B, Vol. 111, No. 33, pp. 9733-9740, ISSN 1520-6106.

Andrews, G. P.; Laverty, T. P. & Jones, D. S. (2009). Mucoadhesive polymeric platforms for controlled drug delivery. European Journal of Pharmaceutics and Biopharmaceutics, Vol. 71, No. 3, pp. 505-518, ISSN 0939-6411.

Bigi, A.P; Anzavolta, S. & Rubini, K. (2004). Relationship between triple-helix content and mechanical properties of gelatin films. Biomaterials, Vol. 25, No. 25, pp. 5675-5680, ISSN 0142-9612.

Byrne, N.; Howlett, P. C.; MacFarlane, D. R. & Forsyth, M. (2005). The zwitterion effect in ionic liquids: Towards practical rechargeable lithium-metal batteries. Advanced Materials, Vol. 17, No. 20, pp. 2497-2501, ISSN 0935-9648.

Delaney, J. T.; Liberski, A. R.; Perelaer, J. & Schubert, U. S. (2010). A Practical Approach to the Development of Inkjet Printable Functional Ionogels-Bendable, Foldable, Transparent, and Conductive Electrode Materials. Macromolecular Rapid Communications,Vol. 31, No. 22, pp. 1970-1976, ISSN

Delaney, J. T.; Liberski, A. R.; Perelaer, J. & Schubert, U. S. (2010). A Practical Approach to the Development of Inkjet Printable Functional Ionogels-Bendable, Foldable, Transparent, and Conductive Electrode Materials. Macromolecular Rapid Communications, Vol. 31, No. 22, pp. 1970-1976, ISSN 1022-1336.

Ding, C. F.; Zhang, M. L.; Zhao, F. & Zhang, S. S. (2008). Disposable biosensor and biocatalysis of horseradish peroxidase based on sodium alginate film and room temperature ionic liquid. Analytical Biochemistry, Vol. 378, No. 1, pp. 32-37, ISSN 0003-2697.

Dobbs, W.; Heinrich, B.; Bourgogne, C.; Donnio, B.; Terazzi, E.; Bonnet, M. E.; Stock, F.; Erbacher, P.; Bolcato-Bellemin, A. L. & Douce, L. (2009). Mesomorphic Imidazolium Salts: New Vectors for Efficient siRNA Transfection. Journal of the American Chemical Society, Vol. 131, No. 37, pp. 13338-13346, ISSN 0002-7863.

Fernicola, A.; Scrosati, B. & Ohno, H. (2006). Potentialities of ionic liquids as new electrolyte media in advanced electrochemical devices. Ionics, Vol. 12, No. 2, pp. 95-102, ISSN 0947-7047.

Frade, R. F. M.; Matias, A.; Branco, L. C.; Afonso, C. A. M. & Duarte, C. M. M. (2007). Effect of ionic liquids on human colon carcinoma HT-29 and CaCo-2 cell lines. Green Chemistry, Vol. 9, No. 8, pp. 873-877, ISSN 1463-9262.

Frokjaer, S. & Otzen, D. E. (2005). Protein drug stability: A formulation challenge. Nature Reviews Drug Discovery, Vol. 4, No. 4, pp. 298-306, ISSN 1474-1776.

Fukaya, Y.; Iizuka, Y.; Sekikawa, K. & Ohno, H. (2007). Bio ionic liquids: room temperature ionic liquids composed wholly of biomaterials. Green Chemistry, Vol. 9, pp. 1155-1157, ISSN 1463-9262.

Galinski, M.; Lewandowski, A. & Stepniak, I. (2006). Ionic liquids as electrolytes. Electrochimica Acta, Vol. 51, No. 26, pp. 5567-5580, ISSN 0013-4686.

Gathergood, N.; Garcia, M. T. & Scammells, P. J. (2004). Biodegradable ionic liquids: Part I. Concept, preliminary targets and evaluation. Green Chemistry, Vol. 6, No. 2, pp. 166-175, ISSN 1463-9262.

Ghandehari, H. (2008). Materials for advanced drug delivery in the 21st century: a focus area for Advanced Drug Delivery Reviews. Advanced Drug Delivery Reviews, Vol. 60, No. 9, pp. 956-956, ISSN 0169-409X.

Guiseppi-Elie, A. (2010). Electroconductive hydrogels: Synthesis, characterization and biomedical applications. Biomaterials, Vol. 31, No. 10, pp. 2701-2716, ISSN 0142-9612.

He, Y. Y.; Boswell, P. G.; Buhlmann, P. & Lodge, T. P. (2007). Ion gels by self-assembly of a triblock copolymer in an ionic liquid. Journal of Physical Chemistry B, Vol. 111, No. 18, pp. 4645-4652, ISSN 1520-6106.

Holbrey, J. D.; Turner, M. B.; Reichert, W. M. & Rogers, R. D. (2003). New ionic liquids containing an appended hydroxyl functionality from the atom-efficient, one-pot reaction of 1-methylimidazole and acid with propylene oxide. Green Chemistry, Vol. 5, No. 6, pp. 731-736, ISSN 1463-9262.

Hough, W. L.; Smiglak, M.; Rodriguez, H.; Swatloski, R. P.; Spear, S. K.; Daly, D. T.; Pernak, J.; Grisel, J. E.; Carliss, R. D.; Soutullo, M. D.; Davis, J. H. & Rogers, R. D. (2007). The third evolution of ionic liquids: active pharmaceutical ingredients. New Journal of Chemistry, Vol. 31, pp. 1429-1436, ISSN 1144-0546.

Jiang, J.; Gao, D. S.; Li, Z. H. & Su, G. Y. (2006). Gel polymer electrolytes prepared by in situ polymerization of vinyl monomers in room-temperature ionic liquids. Reactive & Functional Polymers, Vol. 66, No. 10, pp. 1141-1148, ISSN 1381-5148.

Kita, K. & Dittrich, C. (2011). Drug delivery vehicles with improved encapsulation efficiency: taking advantage of specific drug-carrier interactions. [Review]. Expert Opinion on Drug Delivery, Vol. 8, No. 3, pp. 329-342, ISSN 1742-5247.

Le Bideau, J.; Viau, L. & Vioux, A. (2011). Ionogels, ionic liquid based hybrid materials. Chemical Society Reviews, Vol. 40, No. 2, pp. 907-925, ISSN 0306-0012.

Leys, J.; Wubbenhorst, M.; Menon, C. P.; Rajesh, R.; Thoen, J.; Glorieux, C.; Nockemann, P.; Thijs, B.; Binnemans, K. & Longuemart, S. (2008). Temperature dependence of the electrical conductivity of imidazolium ionic liquids. Journal of Chemical Physics, Vol. 128, No. 6, ISSN 0021-9606.

Li, J. W.; Fan, C.; Xiao, F.; Yan, R.; Fan, S. S.; Zhao, F. Q. & Zeng, B. Z. (2007). Influence of ionic liquids on the direct electrochemistry of glucose oxidase entrapped in nanogold-N, N-dimethylformamide-ionic liquid composite film. Electrochimica Acta, Vol. 52, No. 20, pp. 6178-6185, ISSN 0013-4686.

Liu, Y.; Shi, L. H.; Wang, M. J.; Li, Z. Y.; Liu, H. T. & Li, J. H. (2005). A novel room temperature ionic liquid sol-gel matrix for amperometric biosensor application. Green Chemistry, Vol. 7, No. 9, pp. 655-658, ISSN 1463-9262.

Lodge, T. P. (2008). Materials science - A unique platform for materials design. Science, Vol. 321, No. 5885, pp. 50-51, ISSN 0036-8075.

Lourenco, N. M. T.; Osterreicher, J.; Vidinha, P.; Barreiros, S.; Afonso, C. A. M.; Cabral, J. M. S. & Fonseca, L. P. (2011). Effect of gelatin-ionic liquid functional polymers on glucose oxidase and horseradish peroxidase kinetics. Reactive & Functional Polymers, Vol. 71, No. 4, pp. 489-495, ISSN

Lu, J. M.; Yan, F. & Texter, J. (2009). Advanced applications of ionic liquids in polymer science. Progress in Polymer Science, Vol. 34, No. 5, pp. 431-448, ISSN 0079-6700.

Lu, W.; Fadeev, A. G.; Qi, B. H.; Smela, E.; Mattes, B. R.; Ding, J.; Spinks, G. M.; Mazurkiewicz, J.; Zhou, D. Z.; Wallace, G. G.; MacFarlane, D. R.; Forsyth, S. A. & Forsyth, M. (2002). Use of ionic liquids for pi-conjugated polymer electrochemical devices. Science, Vol. 297, No. 5583, pp. 983-987, ISSN 0036-8075.

Lu, Y.; Sun, Q. F.; Yu, H. P. & Liu, Y. X. (2010). Dissolution and Regeneration of Cellulose and Development in Processing Cellulose-Based Materials with Ionic Liquids. Chinese Journal of Organic Chemistry, Vol. 30, No. 10, pp. 1593-1602, ISSN 0253-2786.

MacFarlane, D. R.; Golding, J.; Forsyth, S.; Forsyth, M. & Deacon, G. B. (2001). Low viscosity ionic liquids based on organic salts of the dicyanamide anion. Chemical Communications, No. 16, pp. 1430-1431, ISSN 1359-7345.

Mazille, F.; Fei, Z. F.; Kuang, D. B.; Zhao, D. B.; Zakeeruddin, S. M.; Gratzel, M. & Dyson, P. J. (2006). Influence of ionic liquids bearing functional groups in dye-sensitized solar cells. Inorganic Chemistry, Vol. 45, No. 4, pp. 1585-1590, ISSN 0020-1669.

Moniruzzaman, M.; Kamiya, N. & Goto, M. (2010 a). Activation and stabilization of enzymes in ionic liquids. Organic & Biomolecular Chemistry, Vol. 8, No. 13, pp. 2887-2899, ISSN 1477-0520.

Moniruzzaman, M.; Kamiya, N. & Goto, M. (2010 b). Ionic liquid based microemulsion with pharmaceutically accepted components: Formulation and potential applications. Journal of Colloid and Interface Science, Vol. 352, No. 1, pp. 136-142, ISSN 0021-9797.

Nockemann, P.; Thijs, B.; Driesen, K.; Janssen, C. R.; Van Hecke, K.; Van Meervelt, L.; Kossmann, S.; Kirchner, B.; Binnemans, K. (2007). Choline saccharinate and choline acesulfamate: Ionic liquids with low toxicities. Journal of Physical Chemistry B, Vol. 111, No. 19, pp. 5254-5263, ISSN 1520-6106.

Ogihara, W.; Sun, J. Z.; Forsyth, M.; MacFarlane, D. R.; Yoshizawa, M. & Ohno, H. (2004). Ionic conductivity of polymer gels deriving from alkali metal ionic liquids and negatively charged polyelectrolytes. Electrochimica Acta, Vol. 49, No. 11, pp. 1797-1801, ISSN 0013-4686.

Ogihara, W.; Washiro, S.; Nakajima, H. & Ohno, H. (2006). Effect of cation structure on the electrochemical and thermal properties of ion conductive polymers obtained from polymerizable ionic liquids. Electrochimica Acta, Vol. 51, No. 13, pp. 2614-2619, ISSN 0013-4686.

Ohno, H.; Yoshizawa, M. & Ogihara, W. (2004). Development of new class of ion conductive polymers based on ionic liquids. Electrochimica Acta, Vol. 50, No. 2-3, pp. 255-261, ISSN 0013-4686.

Park, S. M.; Woo, J.; Jeon, E. & Kim, B. H. (2010). Synthesis of AZT-based cationic lipids and in vitro evaluation of siRNA delivery. Chemical Communications, Vol. 46, No. 9, pp. 1523-1525, ISSN 1359-7345.

Ravichandran, R.; Sundarrajan, S.; Venugopal, J. R.; Mukherjee, S. & Ramakrishna, S. (2010). Applications of conducting polymers and their issues in biomedical engineering. Journal of the Royal Society Interface, Vol. 7, pp. S559-S579, ISSN 1742-5689.

Sangoro, J. R.; Serghei, A.; Naumov, S.; Galvosas, P.; Karger, J.; Wespe, C.; Bordusa, F. & Kremer, F. (2008). Charge transport and mass transport in imidazolium-based ionic liquids. Physical Review E, Vol. 77, No. 5, ISSN 1539-3755.

Singh, T.; Trivedi, T. J. & Kumar, A. (2010). Dissolution, regeneration and ion-gel formation of agarose in room-temperature ionic liquids. Green Chemistry, Vol. 12, No. 6, pp. 1029-1035, ISSN 1463-9262.

Stephan, A. M. (2006). Review on gel polymer electrolytes for lithium batteries. European Polymer Journal, Vol. 42, No. 1, pp. 21-42, ISSN 0014-3057.

Stoimenovski, J.; MacFarlane, D. R.; Bica, K. & Rogers, R. D. (2010). Crystalline vs. Ionic Liquid Salt Forms of Active Pharmaceutical Ingredients: A Position Paper. Pharmaceutical Research, Vol. 27, No. 4, pp. 521-526, ISSN 0724-8741.

Tiyapiboonchaiya, C.; Pringle, J. M.; MacFarlane, D. R.; Forsyth, M. & Sun, J. Z. (2003). Polyelectrolyte-in-ionic-liquid electrolytes. Macromolecular Chemistry and Physics, Vol. 204, No. 17, pp. 2147-2154, ISSN 1022-1352.

Torimoto, T.; Tsuda, T.; Okazaki, K. & Kuwabata, S. (2010). New Frontiers in Materials Science Opened by Ionic Liquids. Advanced Materials, Vol. 22, No. 11, pp. 1196-1221, ISSN 0935-9648.

Viau, L.; Tourne-Peteilh, C.; Devoisselle, J. M. & Vioux, A. (2010). Ionogels as drug delivery system: one-step sol-gel synthesis using imidazolium ibuprofenate ionic liquid. Chemical Communications, Vol. 46, No. 2, pp. 228-230, ISSN 1359-7345.

Vidinha, P.; Lourenco, N. M. T.; Pinheiro, C.; Bras, A. R.; Carvalho, T.; Santos-Silva, T.; Mukhopadhyay, A.; Romao, M. J.; Parola, J.; Dionisio, M.; Cabral, J. M. S.; Afonso, C. A. M. & Barreiros, S. (2008). Ion jelly: a tailor-made conducting material for smart electrochemical devices. Chemical Communications, No. 44, pp. 5842-5844, ISSN 1359-7345.

Vijayaraghavan, R.; Thompson, B. C.; MacFarlane, D. R.; Kumar, R.; Surianarayanan, M.; Aishwarya, S. & Sehgal, P. K. (2010). Biocompatibility of choline salts as crosslinking agents for collagen based biomaterials. Chemical Communications, Vol. 46, No. 2, pp. 294-296, ISSN 1359-7345.

Washiro, S.; Yoshizawa, M.; Nakajima, H. & Ohno, H. (2004). Highly ion conductive flexible films composed of network polymers based on polymerizable ionic liquids. Polymer, Vol. 45, No. 5, pp. 1577-1582, ISSN 0032-3861.

Winther-Jensen, O.; Vijayaraghavan, R.; Sun, J. Z.; Winther-Jensen, B. & MacFarlane, D. R. (2009). Self polymerising ionic liquid gel. Chemical Communications, No. 21, pp. 3041-3043, ISSN 1359-7345.

Yoshida, Y.; Baba, O. & Saito, G. (2007). Ionic liquids based on dicyanamide anion: Influence of structural variations in cationic structures on ionic conductivity. Journal of Physical Chemistry B, Vol. 111, No. 18, pp. 4742-4749, ISSN 1520-6106.

Young, S.; Wong, M.; Tabata, Y. & Mikos, A. G. (2005). Gelatin as a delivery vehicle for the controlled release of bioactive molecules. Journal of Controlled Release, Vol. 109, No. 1-3, pp. 256-274, ISSN 0168-3659.

Zhang, Y. Y.; Chen, X.; Lan, J. B.; You, J. S. & Chen, L. J. (2009). Synthesis and Biological Applications of Imidazolium-Based Polymerized Ionic Liquid as a Gene Delivery Vector. Chemical Biology & Drug Design, Vol. 74, No. 3, pp. 282-288, ISSN 1747-0277.

9

Environmentally Friendly Synthesis of Polymer-Grafted Nanoparticles

Norio Tsubokawa
Graduate School of Science and Technology, Niigata University
Japan

1. Introduction

Inorganic nanoparticles such as silica, and titanium oxide, are widely used industrially as fillers and pigments for polymer materials, because inorganic nanoparticles have excellent properties such as chemical, heat and weather resistance, lightweight thermal conductivity and low thermal expansion.

On the other hand, nanocarbons, such as carbon black, nanodiamond, and carbon nanotube, are well known as one of the industrially important carbon materials. Carbon materials also have outstanding properties such as electro-conductivity, heat-resistance, biocompatibility, and chemical-resistance. Carbon nanotubes in particular have attracted attention as nanotechnology related materials.

In general, dispersing inorganic nanoparticle and nanocarbons uniformly into a polymer or an organic solvent is very difficult because of aggregation. In addition, the mechanical properties of polymer composite from nanoparticles and nanocarbons are considered to depend on not only the mechanical properties of the polymer matrix but also on the properties of interfacial regions between surface of nanoparticles and matrix polymers.

The chemical and physical modifications of inorganic nanoparticle and nanocarbon surfaces, therefore, have been extensively studied. The chemical modification of surfaces is permanent, but physical modification is temporary. We have pointed out that the dispersibility of silica nanoparticles and nanocarbons is extremely improved by surface grafting of polymers, namely, chemical binding of polymers, onto nanoparticle and nanocarbon surfaces (Tsubokawa, 1999; Tsubokawa, 2002; Tsubokawa, 2007).

In addition grafting of polymers onto these surfaces interests us for designing new functional composite materials which have the excellent properties both of nanoparticles as mentioned above and of grafted polymers, such as photosensitivity, biorepellent activity, antibacterial activity, and pharmacological activity (Tsubokawa, 2007).

We have succeeded in the grafting of various polymers such as vinyl polymer (Tsubokawa et al., 1988a; Tsubokawa et al., 1990; Fujiki et al., 1990; Tsubokawa et al., 1992a), polyester (Tsubokawa et al., 1982; Tsubokawa et al., 1983), polyether (Tsubokawa et al., 1986; Tsubokawa et al., 1988b), poly(organophosphazene) (Tsubokawa et al., 1992b), and poly(dimethysiloxane) (Tsubokawa et al., 1992c) onto silica nanoparticle and carbon black surfaces using surface functional groups as grafting sites. Furthermore, many experimental attempts by other researchers also have been made to graft polymers onto silica nanoparticle and carbon black surfaces.

For example, it has been reported that controlled grafting of polymers, so called "polymer brush"), from nanoparticles was successfully achieved by atom transfer radical polymerization (ATRP) initiated by a system consisting of surface functional groups and transition metal complexes (Pyun & Matyjaszewski, 2001; Ohno et al., 2002; Lui et al., 2003; Harrak, et al., 2005; Lui et al., 2005). In situ radical transfer addition polymerization (Lui et. al., 2004) [19] and emulsion polymerization from silica nanoparticles also has been reported (Ding et al., 2004). Wang et al. have reported the synthesis of a well-defied organic/inorganic nano-composite *via* reverse ATRP (Wang et. al., 2005a; Wang et al., 2005b).

However, scale-up synthesis of polymer-grafted nanoparticles was hardly achieved, because complicated reaction processes, such as centrifugation, filtration, and solvent extraction, are required for the synthesis and isolation of polymer-grafted nanoparticles, and a lot of waste solvent is generated.

We have reported the scale-up synthesis of polymer-grafted silica nanoparticle in a solvent-free dry-system. In solvent-free dry-system, the isolation and purification after the graft polymerization was easily achieved, because untreated monomer can be removed under high vacuum (Tsubokawa, 2007).

In this Chapter, for the purpose of the prevention of the environmental pollution and the simplification of surface grafting process, the radical graft polymerization of vinyl monomers onto silica nanoparticle and carbon black surfaces by "grafting onto" and "grafting from" process in an ionic liquid will be reviewed. In addition, the grafting of hyperbranched poly(amidoamine) (PAMAM) onto silica nanoparticle surface by dendrimer synthesis methodology in an ionic liquid will be discussed.

2. Advantages of polymer grafting onto nanoparticles in an ionic liquid

Recently, polymer synthesis in an ionic liquid as a solvent for the polymerization has been reported. It is well known that ionic liquids are non-volatile, non-flammable, and thermally stable solvents. These properties promise replacements for the traditional volatile organic solvents for polymer synthesis. Therefore, the number of studies on ionic liquids as polymerization media have increased markedly (Wilkkers, 2002; Harrisson et al., 2002; Carmichael & Haddleton, 2003).

For example, the polymerization of vinyl monomers initiated by benzoyl peroxide (BPO) and 2,2'-azobisisobutyronitrile (AIBN) in ionic liquids has been reported. It is interesting to note that the rate of the radical polymerization and molecular weight of resulting polymer in ionic liquids were much larger than those in a conventional organic solvent.

In addition, Cu^{I}-*N*-propyl-2–pyridylmethanimine mediated "living" radical polymerization of vinyl monomers by use of 1-butyl-3-methylimidazolium hexafluorophosphate as solvent has been reported (Carmichael et al., 2000). It has been pointed out that the rate of polymerization was enhanced in comparison to other polar/coordinating solvents. Moreover, the polymerization product was made copper free by a simple solvent wash, which avoids the contamination of the polymer product by the catalyst. Other atom transfer radical polymerizations in ionic liquids have recently been reported (Sarbu & Matyjaszewski, 2001; Biendron & Kubisa, 2001).

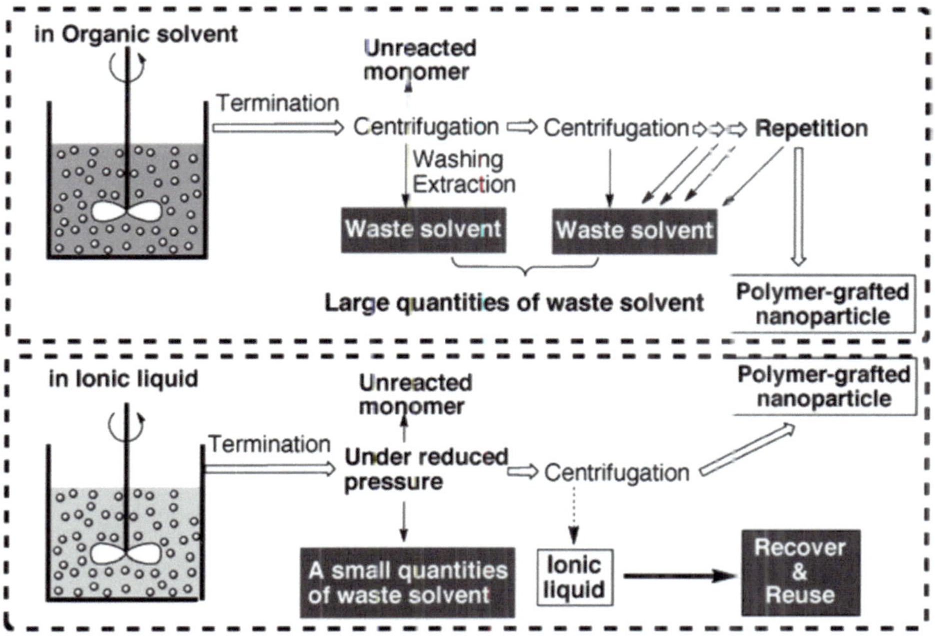

Fig. 1. Graft polymerization in organic solvent system *vs.* in ionic liquid.

Figure 1 shows the comparison of procedures of graft polymerization of various monomers onto nanoparticles, such as inorganic nanoparticles, carbon blacks, and carbon nanotubes, in conventional organic solvents with those in ionic liquids.

In an organic solvent system, after the reaction, the purification and isolation of the resulting nanoparticles was achieved by troublesome procedures, such as filtration and centrifugation as mentioned above. Therefore, the scale-up synthesis of polymer-grafted nanoparticles was hardly achieved and large quantities of waste solvent result.

On the contrary, in ionic liquids, the isolation and purification after the graft polymerization was easily achieved, because untreated monomer can be removed under high vacuum as well as solvent-free dry-system. Therefore, it is expected that we can achieve scale-up and environmentally friendly synthesis of polymer-grafted nanoparticles by use of ionic liquid.

3. Methodology of surface grafting of polymers onto nanoparticles

Several methodologies for the preparation of a variety of graft and block copolymers were established. These methodologies can be applied to the preparation of polymer-grafted nanoparticles. The methodology of surface grafting of polymers onto nanoparticle surfaces is summarized in Table 1. We can apply one of the following principles for the surface grafting of various polymers onto inorganic nanoparticles and nanocarbons (Tsubokawa, 1999; Tsubokawa, 2002; Tsubokawa, 2007).

By process (1), although we can obtaine polymer-grafted nanoparticles, the percentage of grafting (weight percent of grafted polymer onto nanoparticles) onto nanoparticles is less than 10%, because of the preferential formation of ungrafted polymers. We can obtaine polymer-grafted nanoparticles by termination of living polymers with well-defined molecular weight and narrow molecular weight distribution.

	Grafting process	Methodology
1	Grafting onto	Termination of growing polymer radical, cation, and anion, formed during the polymerization of various monomers initiated by conventional initiator in the presence of nanoparticles and the deactivation of living polymer radical, cation, and anion with functional groups on nanoparticle surface
2	Grafting from	Initiation of graft polymerization of various monomers from radical, cationic and anionic initiating groups previously introduced onto nanoparticle surfaces
3	Polymer reaction	Reaction of surface functional groups on nanoparticles with polymers having functional groups, such as hydroxyl, carboxyl, and amino groups
4	Stepwise growth	The growth of hyperbranched polymer chains from surface functional groups on nanoparticles by repeated reaction of low molecular weight compounds by dendrimer synthesis methodology

Table 1. Methodology of surface grafting of polymers onto nanoparticles.

Process (2) is one the most favorable for the preparation of polymer-grafted nanoparticles with a higher percentage of grafting. We can control the molecular weight and number of grafted polymer chains by use of surface initiated living polymerization as mentioned above.

An important characteristic of process (3) is that not only are the molecular weight and the number of grafted chains on the nanoparticle surfaces easily controlled, but also commercially available polymers having a well-defined structure can be grafted. But the number of grafted polymer chains on nanoparticle surface decreases with increasing molecular weight of polymer because of steric hindrance. We can include the process (2) with "grafting onto" process.

By process (4), although dendron with theoretical structure was not easily grafted, hyperbranched polymers having a large number of terminal functional groups can be grafted onto nanoparticle surfaces.

4. Radical grafting of vinyl polymers onto nanoparticles by "grafting onto" process in an ionic liquid

We have pointed out that the radical polymerizations of vinyl monomers initiated by benzoyl peroxide (BPO) and 2,2′-azobisisobutyronitrile (AIBN) were remarkably retarded in the presence of carbon black. It is considered that the competitive reaction of initiator radicals with low molecular weight growing polymer radicals against the carbon black surface proceeds during the radical polymerization (Scheme 1). As a result, a part of polymer formed was grafted onto the surface based on the termination (trapping) of growing polymer radicals by carbon black (Ohkita et al., 1972; Ohkita et al., 1976); but the percentage of grafting was less than 10%. This may be due to preferential trapping of the low molecular weight initiator radicals by the carbon black surface.

Therefore, we have investigated the effect of 1-butyl-3-methylimdazolium hexafluorophosphate as an ionic liquid on the competitive reaction during the radical polymerization in the presence of carbon black (Ueda et al., 2008).

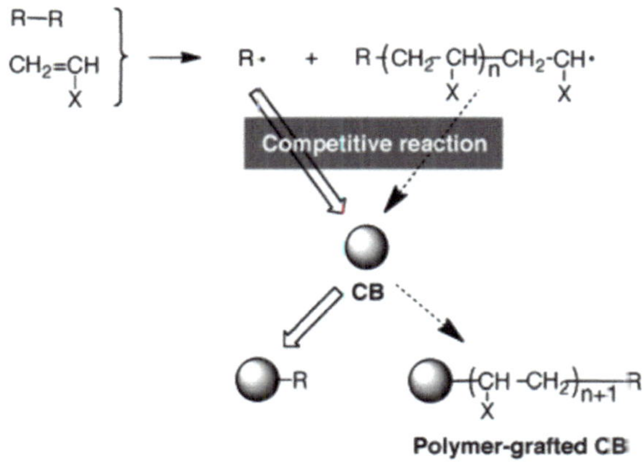

Scheme 1. Competitive reaction of radicals against carbon black surface.

4.1 Radical polymerization of vinyl monomers initiated by BPO in the presence of carbon black

Figures 2 and 3 show the effect of ionic liquid on the radical polymerization of styrene (St) and methyl methacrylate (MMA), respectively, initiated by BPO in the presence of carbon black. As shown in Figures 2 and 3, in toluene as solvent, the polymerization was remarkably retarded. On the contrary, the retardation of polymerization observed in the presence of carbon black was almost eliminated in the initial stage of the polymerization in ionic liquid.

Figures 4 and 5 show the relationship between the reaction time and percentage of polySt and polyMMA grafting, respectively, onto carbon black surface. When toluene was used as solvent, the percentage of polySt grafting onto the carbon black surface was less than 1%. On the contrary, in ionic liquid, the percentage of polySt grafting onto carbon black increased with increasing reaction time and reached to 4.5%.

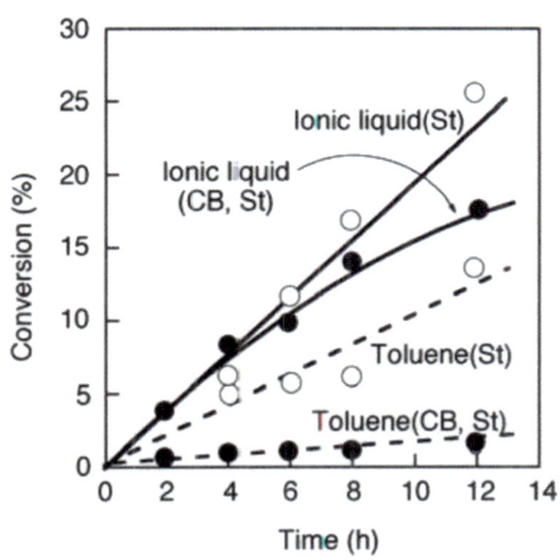

Fig. 2. Effect of ionic solvent on the polymerization of St initiated by BPO in the presence of carbon black. Carbon black, 0.10 g; BPO, 0.10 g; St, 5.0 mL; solvent, 10.0 mL; 60 °C.

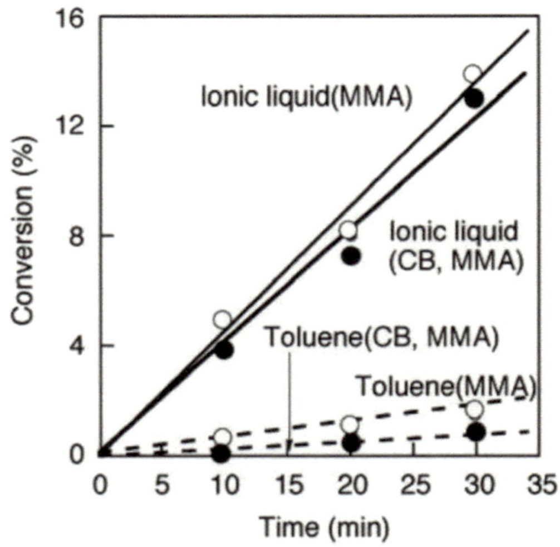

Fig. 3. Effect of ionic liquid on the polymerization of MMA initiated by BPO in the presence of carbon black. Carbon black, 0.10 g; BPO, 0.10 g; MMA, 5.0 mL; solvent, 10.0 mL; 60 °C.

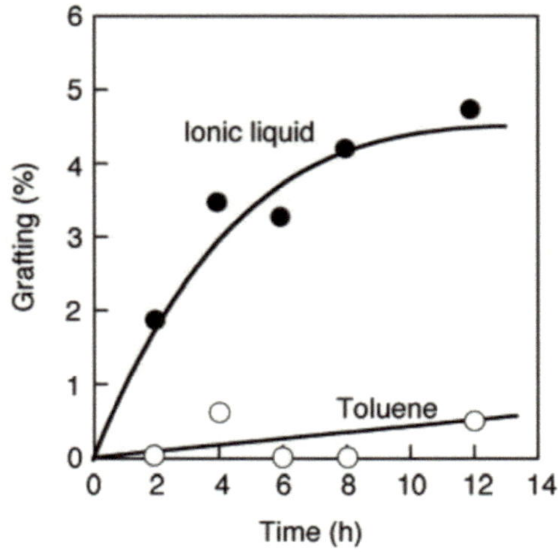

Fig. 4. Relationship between polySt grafting and reaction time. Polymerization conditions are given in Figure 2.

In addition, it is interesting to note that the effective grafting of polyMMA onto carbon black was achieved in ionic liquid: the percentage of grafting reached 35% after 30 h.

Figure 6 shows relationship between conversion and grafting efficiency [percentage of grafted polymer to total polymer (ungrafted polymer + grafted polymer) formed] during the polymerization shown in Figure 2. It is interesting to note that the grafting efficiency in ionic liquid was larger than that in toluene. These results indicate the formation of ungrafted polymer is inhibited in ionic liquid in comparison with toluene in the initial stage of the polymerization. However, the grafting efficiency in ionic liquid also remarkably decreased with increasing conversion as well as in toluene.

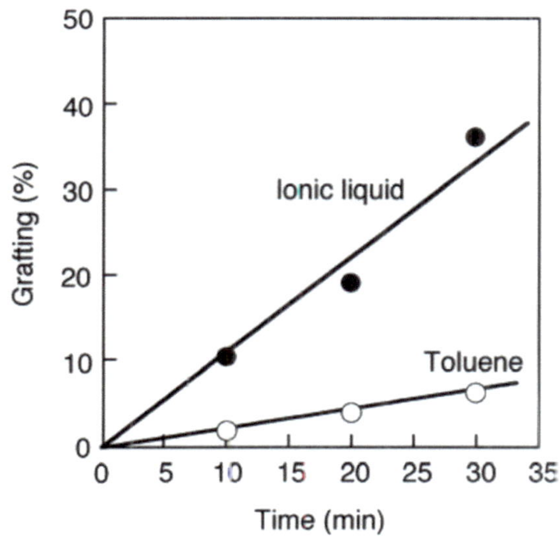

Fig. 5. Relationship between polyMMA grafting and reaction time. Polymerization conditions are given in Figure 3.

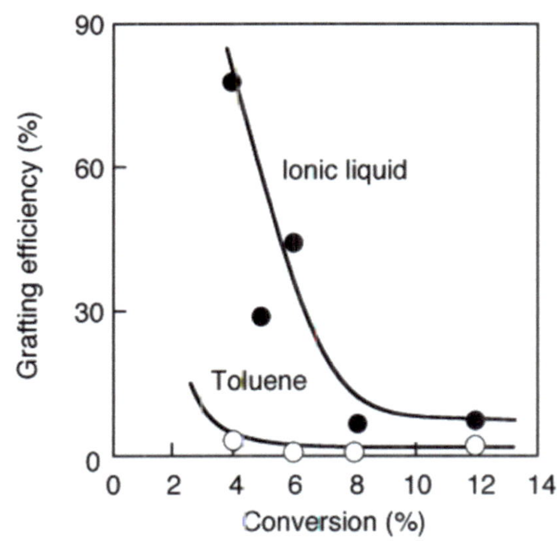

Fig. 6. Relationship between conversion and grafting efficiency of polySt in ionic liquid. Polymerization conditions are given in Figure 2.

The grafting of polySt and polyMMA onto carbon black surface was confirmed by FT-IR and thermal decomposition GC-MS.

These results may be explained as follows: in toluene, initiating radicals (benzoyloxy radicals) were preferentially trapped by carbon black surface and the polymerization is retarded. During the polymerization, a part of growing polySt radicals was trapped by the carbon black surface, the percentage of grafting was very small. This indicates very low molecular weight of polySt was grafted onto the surface in toluene.

On the contrary, in ionic liquid, growing polySt radicals with relatively high molecular weight was trapped by the carbon black surface because of stabilization of polymer radicals in ionic liquid, resulting in higher the percentage of grafting (Scheme 1).

4.2 Radical polymerization of vinyl monomers initiated by AIBN in the presence of carbon black

Figure 7 shows the results of the radical polymerization of MMA in the presence of carbon black initiated by AIBN in ionic liquid and in toluene. It was found that the retardation of the radical polymerization of MMA in the presence of carbon black was also remarkably reduced in ionic liquid. The polymerization behavior of MMA initiated by AIBN shows almost the same tendency as that initiated by BPO in the presence of carbon black.

Figure 8 shows the relationship between reaction time and percentage of polyMMA grafting. As shown in Figure 8, the percentage of grafting increased with increasing reaction time. The percentage of grafting of polyMMA onto carbon black in ionic liquid was also much larger than that in toluene: the percentage of polyMMA grafting was determined to be 40% after 30 h.

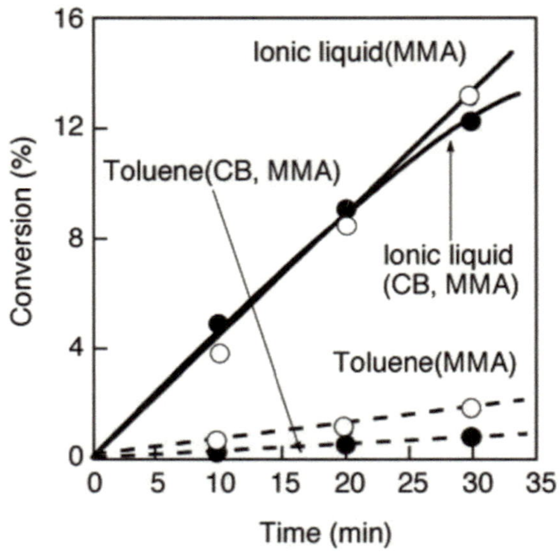

Fig. 7. Effect of ionic liquid on the polymerization of MMA initiated by AIBN in the presence of carbon black. Carbon black, 0.10 g; AIBN, 0.10 g; MMA, 5.0 mL; solvent, 10.0 mL; 60 °C.

5. Radical grafting of vinyl polymers onto nanoparticles by the "grafting from" process in an ionic liquid

As mentioned above, the radial graft polymerization of vinyl monomers onto silica nanoparticle and carbon black surface was initiated by surface azo groups previously introduced onto the surface, and we can obtain the corresponding polymer-grafted silica nanoparticle and carbon black by the "grafting from" process (Tsubokawa et al., 1990; Fujiki et al., 1990; Tsubokawa et al., 1992a).

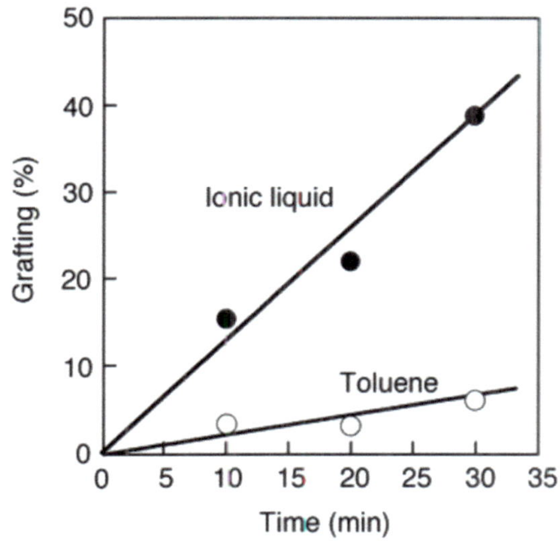

Fig. 8. Relationship between polyMMA grafting and reaction time. Polymerization conditions are given in Figure 7.

Therefore, the effect of ionic liquid on the radical grafting of vinyl polymers onto silica nanoparticle and carbon black by the "grafting from" process was investigated (Ueda et al., 2008).

5.1 Radical graft polymerization of vinyl monomers initiated by azo groups introduced onto silica nanoparticle and carbon black surface in ionic liquid

5.1.1 Radical graft polymerization initiated by CB-Azo

(1) Introduction of azo groups onto the carbon black surface

The introduction of azo groups onto the carbon black surface was readily achieved by the reaction of 4,4'-azobis(4-cyanopentanoic acid) with surface isocyanate groups, which were introduced by the reaction of phenolic hydroxyl and carboxyl groups on the carbon black surface with tolylene-2,4-diisocyanate (TDI) as shown in Scheme 2 (Tsubokawa et al., 1990).

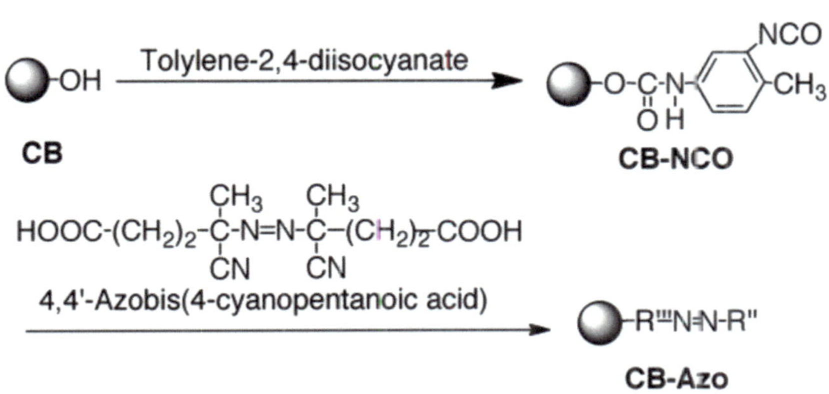

Scheme 2. Introduction of azo groups onto carbon black surface.

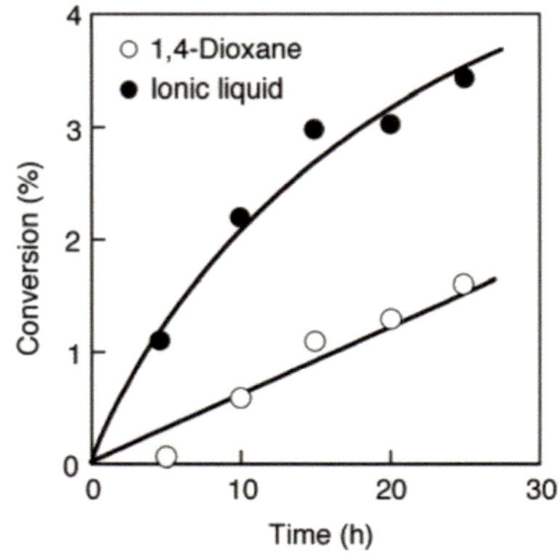

Fig. 9. Effect of ionic liquid on the graft polymerization of MMA initiated by CB-Azo. CB-Azo, 0.15 g; MMA, 10.0 mL; solvent, 5.0 mL; 70 °C.

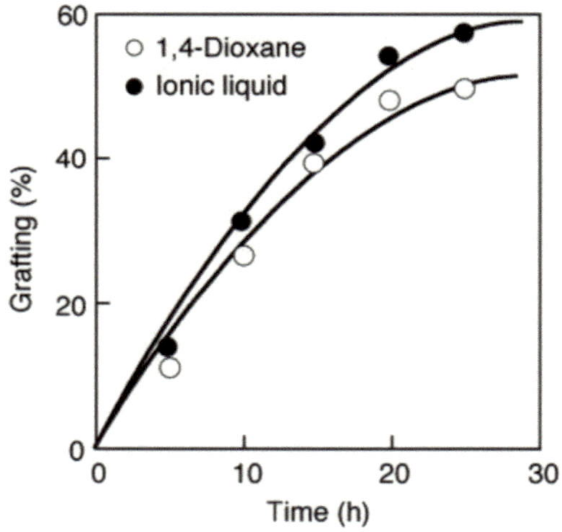

Fig. 10. Relationship between polyMMA grafting and reaction time. Polymerization conditions are shown in Figure 9.

(2) Effect of ionic liquid on the graft polymerization of MMA initiated by CB-Azo

The effect of ionic liquid as solvent on the radical graft polymerization of MMA initiated by CB-Azo was investigated. Figure 9 shows the relationship between reaction time and conversion in ionic liquid and 1,4-dioxane. It was found that the rate of the polymerization in ionic liquid was considerably larger than that in 1,4-dioxane.

Figure 10 shows the effect of ionic liquid on the polyMMA grafting onto the carbon black surface obtained from the graft polymerization shown in Figure 9. It was found that the percentage of grafting increased with increasing reaction time in both solvents and the percentage of grafting in ionic liquid was larger than that in 1,4-dioxane.

5.1.2 Radical graft polymerization of St and MMA initiated by azo groups introduced onto silica nanoparticle surface in ionic liquid

(1) Introduction of azo groups onto silica nanoparticle surface

The introduction of azo groups onto silica nanoparticle surface was achieved by the reaction of surface amino groups, which were previously introduced onto the surface by the treatment with 3-aminopropyltriethoxysilane, with 4,4′-azobis(4-cyanopentanoic acid) dichloride (ACPC) as shown in Scheme 3

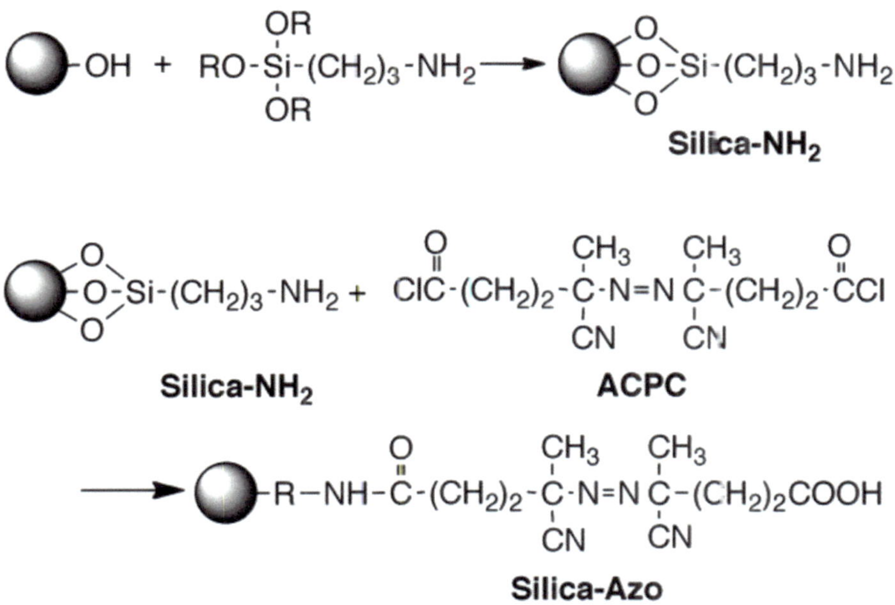

Scheme 3. Introduction azo groups onto silica nanoparticle surface.

(2) Effect of ionic liquid on the graft polymerization of St and MMA initiated by Silica-Azo

The effect of ionic liquid as solvent on the radical graft polymerization of St and MMA initiated by Silica-Azo was investigated. Figures 11 and 12 show the effect of ionic liquid on the rate of polymerization of St and MMA initiated by Silica-Azo, respectively. It was found that the rate of the polymerization of both monomers initiated by Silica-Azo in ionic liquid was considerably larger than those in 1,4-dioxane.

Figures 13 and 14 shows the effect of ionic liquid on the grafting of polySt and polyMMA onto the silica nanoparticle surface obtained from the graft polymerization shown in Figures 11 and 12, respectively. It was found that the percentage of grafting onto the silica surface increased with increasing reaction time in both solvents. It is interesting to note that the percentage of grafting in ionic liquid was much larger than those in 1,4-dioxane: the percentage of polySt and polyMMA grafting in ionic liquid were 180% and 300%, respectively.

This may be due to the fact that the lifetime of radical is prolonged because of the high viscosity of the ionic liquid (Wilkkers, 2002; Harrisson et al., 2002; Carmichael & Haddleton, 2003).

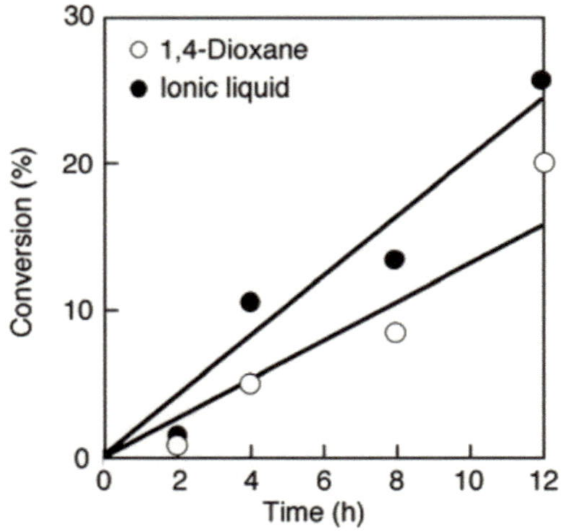

Fig. 11. Effect of ionic liquid on the graft polymerization of St initiated by Silica-Azo. Silica-Azo, 0.20 g; St, 5.0 mL; solvent, 10.0 mL; 70 °C.

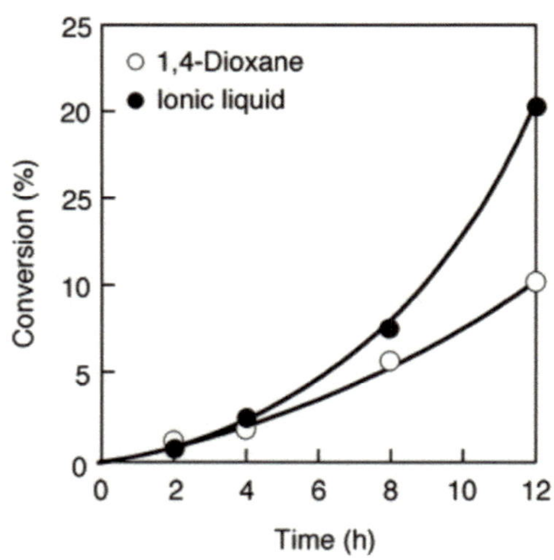

Fig. 12. Effect of ionic liquid on the graft polymerization of MMA initiated by Silica-Azo. Silica-Azo, 0.20 g; MMA, 5.0 mL; solvent, 10.0 mL; 70 °C.

Both surface radicals on the nanoparticle surface and fragment radicals formed by the thermal decomposition of surface azo groups initiate the polymerization in the first stage of the polymerization: the surface radicals produce grafted polymer, but the fragment radicals produce ungrafted polymer as shown in Scheme 4. It is considered that at the last stage of the graft polymerization, the latter reaction preferentially proceeds to give ungrafted polymer, which decreases the grafting efficiency.

On the contrary, it is suggested that in ionic liquid, the propagation of grafted chains from surface radicals formed by the thermal decomposition of azo groups effectively proceeded, because of stabilization of surface radicals and depression of chain transfer reaction.

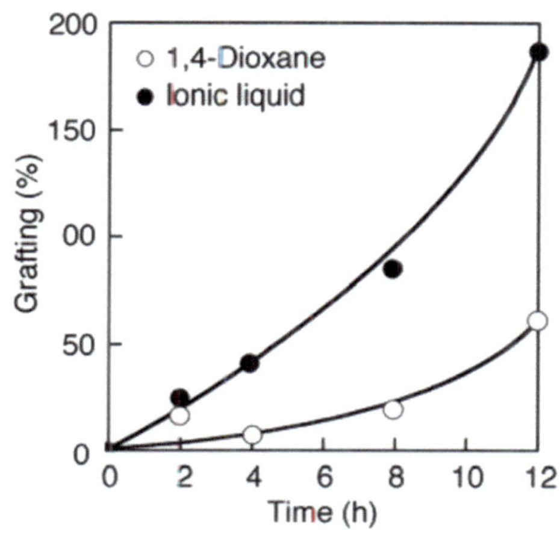

Fig. 13. Relationship between polySt grafting and reaction time. Polymerization conditions are shown in Figure 11.

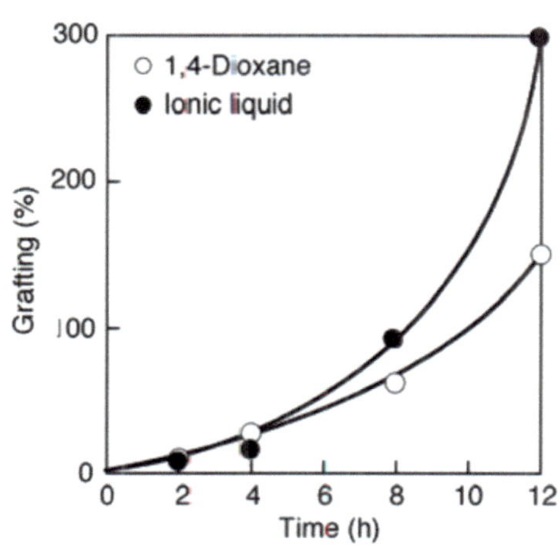

Fig. 14. Relationship between polyMMA grafting and reaction time. Polymerization conditions are shown in Figure 12.

5.1.3 Effect of ionic liquid on the molecular weight of grafted polymer

PolySt grafted onto silica nanoparticle was isolated by the alkali treatment of polySt-grafted silica nanoparticle. Table 2 shows the molecular weight (*Mw*) of grafted polySt on silica nanoparticle surface and ungrafted polySt obtained from the graft polymerization in ionic liquid and 1,4-dioxane. It was found that molecular weight and molecular weight distribution in ionic liquid were almost equal to those in 1,4-dioxane.

Based on the above result, the number of the grafted polymers on the silica surface was calculated. The results are shown in Table 3. The number of grafted polySt on silica obtained from the polymerization in ionic liquid was about 5-times larger than that in 1,4- dioxane as schematically shown in Figure 15. The proportion used for the grafting site to azo group on the surface (*R*) was about 5 % in 1,4-dioxane, but about 25 % in ionic liquid.

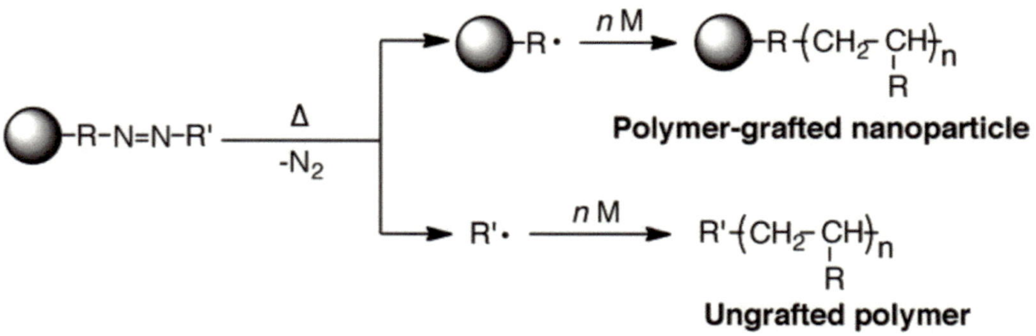

Scheme 4. Graft polymerization initiated by surface azo groups on silica nanoparticle.

Solvent	*Mw* x10^{-4}		*Mw/Mn*	
	Ungrafted	Grafted	Ungrafted	Grafted
1,4-Dioxane	5.16	5.45	2.01	1.95
Ionic liquid	5.17	5.47	2.08	2.12

Table 2. Molecular weight of polySt grafted onto silica nanoparticle surface.

Solvent	Number of grafted polymer	*R* (%)[b]
1,4-Dioxane	2.1 x 10^{18}	5.5
Ionic liquid	9.6 x 10^{18}	25

[a]Silica-Azo, 0.20 g; styrene, 5.0 mL; solvent, 10 mL; 70 °C; 8 h.
[b]*R* =[Grafted polymer (mmol/g)/Azo group (mmol/g)] x 100.

Table 3. Number of grafted polySt chain grafted onto silica nanoparticle surface[a].

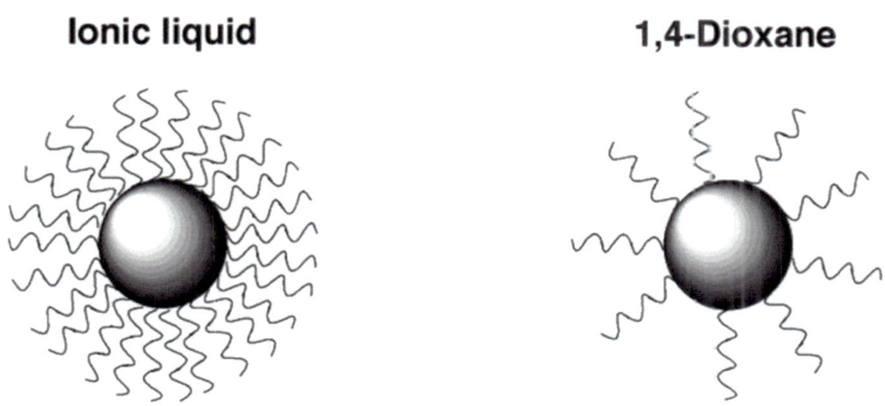

Fig. 15. Polymer chains grafted onto silica nanoparticle surface in ionic liquid and 1,4-dioxane.

The results indicate that in 1,4-dioxane, the growth of the grafted polymer chain from the silica nanoparticle surface is inhibited by the blocking effect of surface radical by grafted polymer chain, but the blocking effect is reduced, because of stabilization of growing polymer radical in ionic liquids.

5.2 Recycle of ionic liquid after the radical graft polymerization

Ungrafted polymer precipitated because the polymer is insoluble in ionic liquid. After the graft polymerization, polymerization was terminated by the addition of a small amount of methanol. Therefore, ionic liquid contains polymer-grafted nanoparticles, unreacted monomer, ungrafted polymer, and methanol. The precipitate, containing polymer-grafted nanoparticle and ungrafted polymer, was removed by filtration.
Filtrate contains unreacted monomer and methanol. Methanol and monomer was removed from the filtrate under high vacuum at 70 °C and the resulting ionic liquid could be reused after washing with water followed by drying under high vacuum at 50 °C.
Therefore, the reduction of waste solvent and environmentally friendly scale-up synthesis of polymer-grafted nanoparticles can be achieved by use of ionic liquid as a reaction solvent

6. Grafting of polymers onto MWCNT in an ionic liquid

1,1-Bis(*t*-butyldioxy)cyclohexane (BDOC) thermally decomposes near 70 °C to give radicals having a peroxyester moiety and a *t*-butoxy radical (Scheme 5). We introduced peroxyester groups onto carbon black surface and MWCNT by the trapping of the radicals having peroxyester moiety. And the radical graft polymerization of vinyl monomers initiated by the peroxyester groups on the surface was achieved (Hayashi et al., 1996).

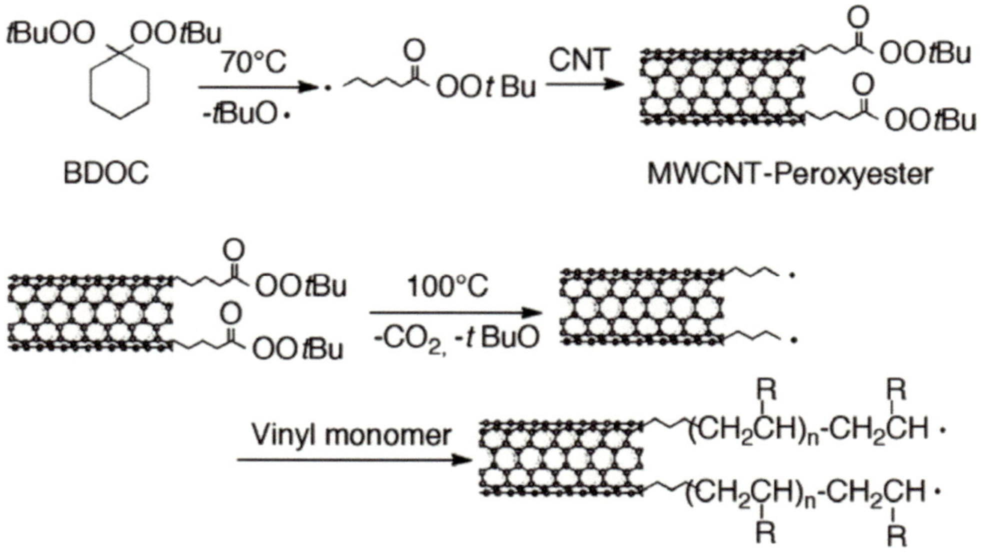

Scheme 5. Introduction of peroxyester groups onto MWCNT surface by the reaction with BDOC.

6.1 Introduction of peroxyester groups onto MWCNT by radical trapping

The effect of ionic liquid as reaction solvent on the trapping of radicals having peroxyester moiety by MWCNT was investigated. The results are shown in Table 4. The amount of peroxyester groups introduced onto MWCNT by the reaction with BDOC in ionic liquid was much larger than that in hexane, because of the stabilization of free radicals in ionic liquid as mentioned above.

Solvent	Time (h)	Peroxyester group (μmol/g)
Ionic liquid	8	30.1
Ionic liquid	10	34.9
Ionic liquid	12	38.3
Hexane	12	9.8

Table 4. Introduction of peroxyester groups onto MWCNT by the reaction with BDOC. MWCNT, 0.02 g; BDOC, 10.0 mL; solvent 10.0 mL; 70°C.

6.2 Grafting of polymers onto MWCNT initiated by peroxyester groups on the surface

Radical graft polymerization of MMA was successfully initiated by surface radicals formed by the thermal decomposition of peroxyester groups introduced onto MWCNT and grafted chains propagated from the surface to give polyMMA-grafted MWCNT: the percentage of grafting was determined to be 60.2% after 4 h at 100°C.

PolyMMA-grafted MWCNT was found to give stable dispersions in a good solvent for polyMMA: untreated MWCNT precipitated completely in THF within 1 h, but no precipitation of polyMMA-grafted MWCNT in THF was observed even after 1 week.

7. Grafting of hyperbranched poly(amidoamine) onto silica nanoparticles by dendrimer synthesis methodology in an ionic liquid

A great interest has been recently focused on "Dendrimer", because dendrimers have the fundamental building blocks, controlled molecular weight, controlled branching and versatility in modification of terminal groups (Tomalia et al., 1985; Tomalia et al., 1987). We have reported that hyperbranched poly(amidoamine) (PAMAM) can be grown from an amino group on silica nanoparticle, chitosan powder, and carbon black surface using dendrimer synthesis methodology in methanol solvent (Tsubokawa et al., 1988c; Tsubokawa et al., 2000; Fujiki et al., 2000; Tsubokawa et al., 2001).

Grafting of hyperbranched PAMAM onto silica nanoparticle surface was achieved by repeating two steps, (1) Michael addition of methyl acrylate (MA) to amino group on the surface and (2) amidation of terminal methyl ester group with ethylenediamine (EDA) as shown in Scheme 6.

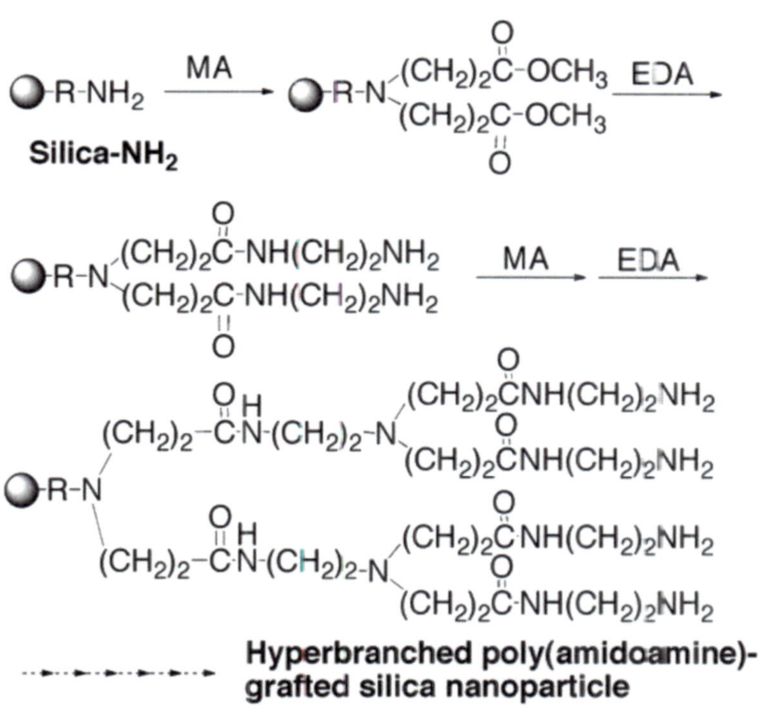

Scheme 6. Grafting of PAMAM onto silica nanoparticle surface by dendrimer synthesis methodology.

The hyperbranched PAMAM-grafted nanoparticle has the possibility of to be utilized as a catalyst support (Hagiwara et al., 2009), biorepellent capsaicin (Yamauchi et al., 2010), and a curing agent of epoxy resin (Ukaji et al., 2008), because hyperbranched PAMAM-grafted nanoparticle has many terminal amino groups.

Therefore, we designed the scale-up synthesis of hyperbranched PAMAM-grafted silica nanoparticle in a solvent-free dry-system (Murota et al., 2002). The preparation of PAMAM-grafted silica in solvent-free dry-system was conducted as follows. MA was sprayed onto silica having amino groups and the silica was agitated at 300 rpm at 50°C. After the reaction, unreacted MA was removed under high vacuum. Then EDA was sprayed and the reaction was conducted at 50°C with agitation. After the reaction, unreacted EDA was also removed under high vacuum at 50°C and MA was sprayed again. The procedures were repeated to grow PAMAM from the surface. The percentage of PAMAM grafting onto the surface was determined to be 141% with repeated reaction cycles of 8-times.

It is considered that by use of ionic solvent as reaction solvent, unreacted MA and EDA can be removed under high vacuum after the reaction as well as solvent-free dry-system. Therefore, we investigated the preparation of hyperbranched PAMAM-grafted silica in ionic solvent.

7.1 Procedures for the grafting of PAMAM onto silica in an ionic liquid

Into a 200-mL four-necked flask, that contained 50 mL of ionic liquid and 5.0 g of silica nanoparticle having amino group, 1.0 g of MA (11.6 mmol; excess of surface amino groups) was added and the reaction mixture was agitated at 300 rpm at 50 °C under argon gas. After 21 h, unreacted MA was removed under high vacuum at 50 °C. After 2 hr, into the flask, 1.0 g of EDA (16.7 mmol; excess of surface ester group) was added and the reaction was conducted at 50 °C with agitation. After 21 h, unreacted EDA was also removed under vacuum at 50 °C for 3 h and MA was added again. Both the Michael addition with MA and the amidation with EDA were repeated to grow PAMAM from the silica surface. After the reaction, PAMAM-grafted silica was isolated by centrifugation.

In order to remove ungrafted PAMAM from PAMAM-grafted silica, the product was dispersed in methanol and centrifuged at 10^4 rpm. The supernatant solution containing ungrafted PAMAM was removed and PAMAM-grafted silica was dispersed in methanol and centrifuged. The procedures were repeated for several times, until no more ungrafted PAMAM was detected in the supernatant solution.

7.2 Grafting of PAMAM in an ionic liquid

The treatment of silica nanoparticle having amino group with MA and EDA in ionic liquid was repeated for *n*-times to obtain PAMAM-grafted silica nanoparticle as shown in Scheme 6 (Murota, et al., 2002).

Table 4 shows the amino group content of silica nanoparticle and the percentage of PAMAM grafting onto silica nanoparticle surface after the grafting reaction. As shown in Table 4, the amino group content and percentage of PAMAM grafting of the resulting silica nanoparticle increased with increasing repeated reaction cycles.

On the other hand, when untreated silica nanoparticle was used, no increase of surface amino group and no grafting of PAMAM onto the surface were observed even with repeated reaction cycles of 8-times.

The percentage of grafting and amino group content of PAMAM-grafted silica obtained in ionic liquid was slightly smaller than those in methanol.
During the grafting reaction, the formation of a small amount of ungrafted polymer was observed: methanol soluble part (major product) was hyperbranched PAMAM and THF soluble part (minor product) was a mixture of polyMA and unknown viscous materials. This indicates that the removal of unreacted monomer in each step was incomplete.

Cycle of repeated reactions	Amino group (mmol/g)		Grafting (%)		R^b
	Found	Calcud.	Found	Calcud.	
2-times	1.5	1.3	13.5	22.6	0.60
6-times	6.8	21.1	90.1	474.0	0.19
8-times	7.7	84.5	99.8	1918.6	0.05

[a]Amino group content of grafting site, 0.33 mmol/g
[b]*R* =Experimental value/Calculated value

Table 4. Grafting of hyperbranched PAMAM onto silica nanoparticle surface in ionic liquid[a]

The amino group content of silica nanoparticle and percentage of PAMAM grafting, however, were considerably smaller than those of the calculated value. The ratio of experimental value for calculated value, *R*, remarkably decreased with increasing repeated reaction cycles. The same tendency was observed in the grafting of PAMAM onto silica nanoparticle in methanol solvent system (Tsubokawa et al., 1988; Tsubokawa et al., 2000; Fujiki et al., 2000; Tsubokawa et al., 2001).

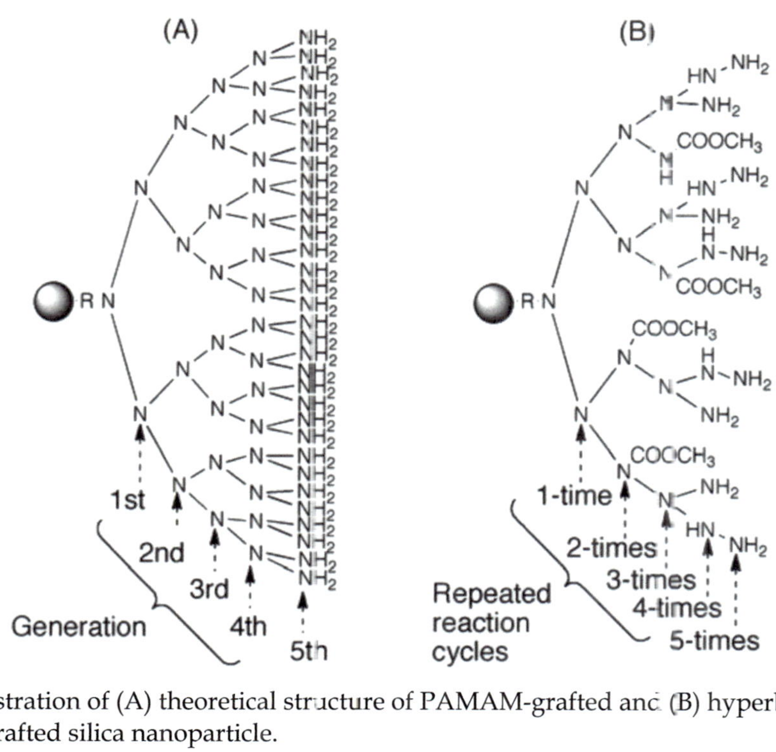

Fig. 16. Illustration of (A) theoretical structure of PAMAM-grafted and (B) hyperbranched PAMAM-grafted silica nanoparticle.

Based on the above results it is concluded that the theoretical propagation of PAMAM from silica nanoparticle surface was hardly achieved and hyperbranched PAMAM was grafted onto the surface: Figure 16 shows illustration of (A) theoretical PAMAM-grafted silica nanoparticle and (B) hyperbranched PAMAM-grafted silica nanoparticle.
This may be due to the fact that (1) complete Michael addition and the amidation with surface amino and ester group hardly proceeded because of aggregation of silica nanoparticles and (2) the grafted chains on the silica surface interfere with the propagation of PAMAM from the surface because of steric hindrance (Tsubokawa et al., 1988; Tsubokawa et al., 2000; Fujiki et al., 2000; Tsubokawa et al., 2001).

8. Conclusion

1. The effect of ionic liquid on the radical polymerization of St and MMA in the presence of carbon black initiated by BPO and AIBN was investigated. When a conventional organic solvent was used as solvent, the polymerization was remarkably retarded by carbon black. On the contrary, the retardation of polymerization was almost eliminated in the initial stage of the polymerization in ionic liquid. It is interesting to note that the percentage of grafting of polySt and polyMMA onto carbon black in ionic liquid was much larger than those in organic solvent.
2. The radical graft polymerizations of vinyl monomers initiated by azo groups introduced onto silica nanoparticle and carbon black surfaces in ionic liquid were investigated. The percentage of grafting in ionic liquid was mach larger than those in 1,4-dioxane. The molecular weight of polySt grafted onto the silica nanoparticle surface in ionic liquid was almost equal that in 1,4-dioxane. The result indicates that the number of grafted polySt in ionic liquid is 5-times that in 1,4-dioxane. This may be due to the fact that lifetime of the surface radical formed by the thermal decomposition of azo groups is prolonged because of the high viscosity of ionic liquid. Therefore, the surface azo groups were effectively used as initiating sites for the graft polymerization.
3. Peroxyester groups were successfully introduced onto MWCNT surface by the trapping of radicals containing peroxyester moieties formed by the thermal decomposition of BDOC. The number of peroxyester groups introduced onto MWCNT in ionic liquid was much larger than those in conventional organic solvent. The radical graft polymerization of MMA was initiated by peroxyester groups on the surface to give polyMMA-grafted MWCNT.
4. Grafting of hyperbranched PAMAM onto silica nanoparticle surface was achieved by repeating two steps in ionic liquid: (1) Michael addition of MA to amino group on the surface and (2) amidation of terminal methyl ester group with EDA. The reaction procedures are considerably simplified by use of ionic liquid as solvent, because unreacted MA and EDA can be removed under high vacuum.
5. The reduction of waste solvent was achieved by use of ionic liquid as reaction solvent, because unreacted monomer could be removed under high vacuum after the reaction and the reuse of ionic liquid was easily achieved. Therefore, the graft polymerization onto nanoparticle surface in ionic liquid enables environmentally friendly scale-up synthesis of polymer-grafted nanoparticles.

9. Acknowledgment

This study was partly supported by a Grant in Aid for Scientific Research from the Ministry of Education, Culture, Sport, Science and Technology of Japan (No. 21560711), which is gratefully acknowledged.

10. References

Biendron, T. & Kubisa, P. (2001). Atom-Transfer Radical Polymerization of Acrylates in an Ionic Liquid. *Macromolecular Rapid Communications*, 22: 1237-1242

Carmichael, A. J.; Haddleton, D. M.; Bon, S. A. F. & Seddon, K. R. (2000). Copper(I) Mediated Living Radical Polymerization in an Ionic Liquid. *Chemical Communications*, 14, 1237-1238

Carmichael, A. J. & Haddleton, D. M. (2003). Polymer Synthesis in Ionic Liquids. In: *Ionic Liquids in Synthesis*, Wasserscheid, P. & Welton, T. (Editors), Ionic Liquids in Synthesis. 319-335, Wiley-VHC Verlag, Weinheim

Ding, X.; Zhao, J.; Liu, Y.; Zhang, H. & Wang, Z. (2004). Silica Nanoparticles Encapsulated by Polystyrene via Surface Grafting and in situ Emulsion Polymerization. *Material Letters*, 58, 3126-3130

Fujiki, K.; Tsubokawa, N. & Sone, Y. (1990). Radical Grafting from Carbon Black. Graft Polymerization of Vinyl Monomers Initiated by Azo Groups Introduced onto Carbon Black Surface. *Polymer Journal*, 22, 661-670

Fujiki, K.; Sakamoto, M.; Satoh.; T. & Tsubokawa, N. (2000). Postgrafting of Hyperbranched Dendritic Polymer from Terminal Amino Groups of Polymer Chains Grafted onto Silica Surface. *Journal of Macromolecular Science, Pure and Applied Chemistry*, A37, 357-377

Hagiwara, H.; Sekifuji, M.; Tsubokawa, N; Hoshi, T. & Suzuki, T. (2009). Nano-silica PAMAM Dendrimer as a Novel Catalyst for Knoevenagel Reactions. *Chemistry Letters*, 38, 926-927

Harrak, A. E.; Carrot, G.; Oberdisse, J.; Jestin, J. & Bou, F. (2005). Atom Transfer Radical Polymerization from Silica Nanoparticles using Grafting from Method and Structural Study via Small Angle Neutron Scattering. *Polymer*, 46, 1095-1104

Harrisson, S.; Mackenzie, S. R. & Huddleton, D. M. (2002). Unprecedented Solvent-induced Acceleration of Free-radical Propagation of Methyl Methacrylate in Ionic Liquids. *Chemical Communications*, 23, 2850-2851

Hayashi, S; Handa, S. & Tsubokawa, N. (1996). Introduction of Peroxide Groups onto Carbon Black Surface by Radical Trapping and Radical Graft Polymerization of Vinyl Monomers Initiated by the Surface Peroxide Groups. *Journal of Polymer Science: Part A: Polymer Chemistry*, 34, 1589-1595

Liu, P.; Liu, W. M. & Xue, Q. J. (2004). In situ Radical Transfer Addition Polymerization of Styrene from Silica Nanoparticles. *European Polymer Journal*, 40, 267-271

Lui, T.; Jia, S.; Kowalewski, T.; Matyjaszewski, K.; Casadio-Portilla, R. & Belmont, J. (2003). Grafting Poly(*n*-butyl acrylate) from a Functionalized Carbon Black surface by Atom Transfer Radical Polymerization. *Langmuir*, 19, 6342-6345

Lui, T.; Casadio-Portilla, R.; Belmont, J. & Matyjaszewski, K. (2005). ATRP of Butyl Acrylates from Functionalized Carbon Back. *Journal Polymer Science: Part A: Polymer Chemistry*, 43, 4695-4709

Murota, M.; Sato, S. & Tsubokawa, N. (2002). Scale-up Synthesis of Hyperbranched Poly(amidoamine)-grafted Ultrafine Silica using Dendrimer Synthesis Methodology in Solvent-free Dry system, *Polymers for Advanced Technology*, 13, 144-150

Ohkita, K.; Tsubokawa, N.; Kadoi, H. & Suneya, Y. (1972). Polymerization Reaction of Vinyl Monomers in the Presence of Carbon Black. *Nippon Gomu Kyokaishi*, 45, 1074-1080

Ohkita, K.; Tsubokawa, N. & Takashina, N. (1976). The Free Radical Polymerization of Vinyl Monomers in the Presence of Carbon Black. *Nippon Gomu Kyokaishi*, 49, 223-230

Ohno, K.; Koh, K.; Tsuji, Y. & Fukuda, T. (2002). Synthesis of Gold Nanoparticles Coated with Well-defined High-density Polymer Brushes by Surface-initiated Living Radical Polymerization. *Macromolecules*, 35, 8989-8993

Pyun, J. & Matyjaszewski, K. (2001). Synthesis of Nanocomposite Organic/Inorganic Hybrid Materials using Controlled Living Radical Polymerization. *Chemistry Materials*, 13, 3436-3448

Sarbu, T. & Matyjaszewski, K. (2001). ATRP of Methyl Methacrylate in the Presence of Ionic Liquids with Ferrous and Cuprous Anions. *Macromolecular Chemistry and Physics*, 202, 3379-3391

Tomalia, D. A.; Baker, H.; Dewald, J.; Hall, M.; Kallos, G.; Martin, S.; Roeck, J.; Ryder, J. & Smith, P. (1985). A New Class of Polymers: Starburst-dendritic Macromolecules. *Polymer Journal*, 17, 117-132

Tomalia, D. A.; Berry, V.; Hall, M. & Hedstrand, D. M. (1987). Starburst Dendrimers. 4. Covalently Fixed Unimolecular Assemblages Reminiscent of Spheroidal Micelles. *Macromolecules*, 20, 1164-1167

Tsubokawa, N.; Funaki, A.; Hada, Y. & Sone, Y. (1982). Grafting onto Carbon Black: Graft Polymerization of β-Propiolactone onto Carbon Black Surface. *Journal Polymer Science, Polymer Chemistry Edition*, 20, 3297-3304

Tsubokawa, N.; Yamada, A. & Sone, Y. (1983). Grafting of Polyester onto Carbon Black. 4. Copolymerization of Epoxide with Phthalic Anhydride Initiated by COOK Groups on Carbon Black Surface. *Polymer Bulletin*, 10, 63-69

Tsubokawa, N.; Nunokawa, H. & Sone, Y. (1986). Cationic Grafting from Carbon Black. 4. Grafting of Poly(oxytetramethylene) to Carbon Black by Ring-opening Polymerization of Tetrahydrofuran Initiated by $CO^+ClO_4^-$ Groups on Carbon Black. *Journal of Macromolecular Science, Pure Applied Chemistry*, A23, 105-115

Tsubokawa, N.; Fujiki, K. & Sone, Y. (1988a). Graft Polymerization of Vinyl Monomers onto Carbon Black Surface Initiated by Peroxyester Groups. *Polymer Journal*, 20, 213-220

Tsubokawa, N.; Jian, Y. & Sone, Y. (1988b). Cationic Grafting From Carbon Black. VIII. Cationic Ring-opening Polymerization and Copolymerization of Oxepane Initiated by Acylium Perchlorate Groups on Carbon Black. *Journal of Polymer Science, Polymer Chemistry Edition*, 26, 2715-2724

Tsubokawa, N.; Ichioka, H.; Satoh, T.; Hayashi, S. & Fujiki, K (1998c). Grafting of Dendrimer onto Ultrafine Silica Surface. *Reactive and Functional Polymer*, 37, 75-82

Tsubokawa, N.; Kogure, A.; Maruyama, K.; Sone, Y. & Shimomura, M. (1990). Graft Polymerization of Vinyl Monomers from Inorganic Ultrafine Particles Initiated by Azo Groups Introduced onto the Surface. *Polymer Journal*, 22, 827

Tsubokawa, N. & Ishida, H. (1992a). Graft Polymerization of Vinyl Monomers by Peroxyester Groups Introduced onto the Surface of Inorganic Ultrafine Particles. *Polymer Journal*, 24, 809-816

Tsubokawa, N. & Tsuchida, H. (1992b). Heat-resistant Polymer-grafted Carbon Black: Grafting of Poly(organophosphazenes) onto Carbon Black Surface. *Journal of Macromolecular Science, Pure Applied Chemistry*, A29, 311-321

Tsubokawa, N. & Yanadori, K. (1992c). Reaction of Polymer Radicals formed by the Decomposition of Azopolymer with Carbon Black Surface. *Kobunshi Ronbunshu*, 49, 865-870

Tsubokawa, N. (1999). Modification of Inorganic particles by Grafting of Functional Polymers. In: *Fundamental and Applied Aspects of Chemically Modified Surface*, Blits, J. P & Little, C. B (Eds.), 36-51, The Royal Society of Chemistry, London

Tsubokawa, N. & Takayama, T. (2000). Surface Modification of Chitosan Powder by Grafting of "Dendrimer-like" Hyperbranched Polymer onto the Surface. *Reactive and Functional Polymer*, 43, 341-350

Tsubokawa, N.; Satoh, T.; Murota, M.; Sato, S. & Shimizu, H. (2001). Grafting of Hyperbranched Poly(amidoamine) onto Carbon Black Surface using Dendrimer Synthesis Methodology. *Polymers for Advanced Technology*, 12, 569-602

Tsubokawa, N. (2002). Functionalization of Carbon Materials by Surface Grafting of Polymers. *Bulletin of Chemical Society of Japan*, 75, 2115-2136

Tsubokawa, N. (2007). Surface Grafting of Polymers onto Nanoparticles in a Solvent-free Dry-system and Applications of Polymer-grafted Nanoparticles as Novel Functional Hybrid Materials. *Polymer Journal*, 39, 983-1000

Ueda, J.; Yamaguchi. H.; Yamauchi, T. & Tsubokawa, N. (2008). Radical Graft Polymerization of Vinyl Monomers onto Nanoparticles in Room Temperature Ionic Liquid Initiated by Azo Groups Introduced onto the Surface. *Journal of Applied Polymer Science*, 107, 3300-3305

Ukaji, M.; Takamura, M.; Shirai, K.; Yamauchi, T. & Tsubokawa, N. (2008). Curing of Epoxy Resin by Hyperbranched Poly(amidoamine)-grafted Silica Nanoparticles and Their Properties. *Polymer Journal*, 40, 607-613

Wilkes, J. S. (2002). A Short History of Ionic Liquids-from Molten Salts to Neoteric Solvents. *Green Chemistry*, 4, 73-80

Yamauchi, T.; Saitoh, T.; Shirai, K.: Fujiki, K. & Tsubokawa, N. (2010). Immobilization of Capsaicin onto Silica Nanoparticle Surface and Stimulus Properties of the Capsaicin-immobilized Silica. *Journal of Polymer Science: Part A: Polymer Chemistry*, 48, 1800-1805

10

Ionic Liquids as Porogens in the Synthesis of Molecularly Imprinted Polymers

Katherine M Booker[1], Clovia I Holdsworth[1],
Michael C Bowyer[2] and Adam McCluskey[1]
[1]*Chemistry Building, School of Environmental & Life Sciences,*
[2]*Discipline of Applied Sciences, Central Coast Campus,*
The University of Newcastle,
Australia

1. Introduction

Although the information regarding room temperature ionic liquids (RTILs) as polymerisation solvents is extensive, there has been very little research published using RTILs as solvents for the synthesis of crosslinked polymers in general (Cooper 2004; Pavlova 2006) and molecularly imprinted polymers (MIPs) in particular (Booker et al 2006, 2007; Wang 2006, 2008; He 2008). Herein we examine a model system, cocaine, and will review the properties and performance of imprinted polymers prepared in volatile organic compounds (VOCs) with those prepared in RTILs, and the experimental parameters such as polymerisation temperature, solvent volume, rebinding conditions and template-RTIL combination, which may have a role to play in these systems.

1.1 The design and synthesis of molecularly imprinted polymers

No discussion on the effect of RTILs on the efficacy of MIPs would be complete without a brief discourse on MIP design and synthesis. MIPs are a specialty class of polymers that possess the capacity to selectively sequester a target species from a solution matrix. Selectivity arises through the inclusion of the molecular target in the pre-polymer formulation, which leads to the generation of tailored molecular cavities within the final polymer. MIP synthesis requires four basic ingredients: a template (T); a functional monomer (FM); a crosslinking agent to impart stability and cavity rigidity to the resultant polymer; and a porogen to generate a pore structure within the polymer to aid mass transfer during template rebinding. The typical process for MIP formation is outlined in Figure 1.

With thousands to millions of available template binding pockets, MIPs possess the ability to recognize and bind specific target molecules. Whilst they are the synthetic counterparts to biological receptors and are robust, insoluble materials exhibiting high stability in most media, they generally lack the natural homogeneity of active sites associated with biological receptors. The population of binding sites generated in MIPs typically presents a highly heterogeneous profile because of the influence of the equilibria that govern the T-FM complex formation and the stability and dynamic nature of the growing polymer chains.

Binding site homogeneity and selectivity is a product of the stability of the T-FM cluster during polymerization. Two approaches to cluster formation have been explored; one based on covalent interactions and the other centred around self-assembly (utilising non-covalent binding interactions). With the former, post polymerization template removal requires the destruction of covalent T-FM linkages. The stable nature of this system yields a highly defined cavity (binding site) that complements the size, shape and electronic properties of the template. Non-covalent or self-assembly of the T-FM cluster in the pre-polymer mix relies on combinations of hydrogen bond interactions, electrostatic attraction and associated weak interactions to form, resulting in a distribution of binding site profiles reflecting the stability of the T-FM cluster. Post-polymerisation, the template is removed from the cavity via exhaustive extraction.

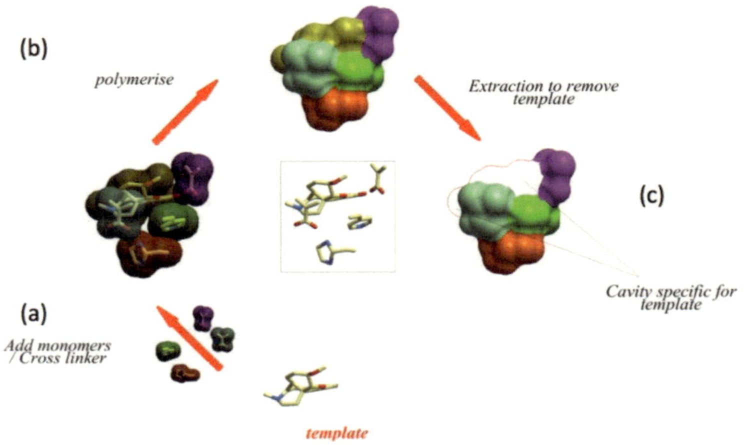

Fig. 1. Development of MIPs. (a) Addition of functional monomers and crosslinker to a solution of the template, followed by prearrangement association of template with functional monomers; (b) resultant preassociation functional monomer / cross linker / template cluster subjected to free radical polymerisation conditions results in cavity formation; and (c) template removal to leave template specific cavity.

MIP technology has been applied in a myriad of areas including, but not limited to, separation and isolation, antibody and receptor mimics, and biosensor style devices (McCluskey et al., 2007). Their shelf stability, robustness and reusability mean that they are highly usable and flexible materials. The variety of molecules 'imprinted' is impressive in both breadth of template and also diversity, highlighting the utility of MIPs. The general MIP area has been extensively reviewed over the past decade. More recently, efforts have focused on garnering a greater understanding of the influence of the template on MIP morphology and function, as well as the role of the porogen in the initial MIP synthesis. In this latter regard, we are one of the few groups who have explored the potential use of room temperature ionic liquids as porogen (Booker et al 2006, 2007; Wang 2006, 2008; He 2008).
The tunable solvating properties of ionic liquids hold considerable potential to facilitate an increase in favorable T-FM interactions, whilst limiting those associated with non-specific binding. An additional attraction of RTILs as porogens is the considerable body of literature pertaining to polymerization rate enhancements (which allows precipitation polymerization

approaches to be applied in a short time frame), and our earlier studies that showed an enhancement of MIP selectivity relative to the identical MIP formulation manufactured in a traditional VOC porogen (Booker et al 2006, 2007).

2. Scope of this study

Prior studies conducted by our group using cocaine as a template molecule have produced data showing that polymers selective for this template could be generated in both a VOC ($CHCl_3$) and RTILs ([bmim]BF_4 and [bmim]PF_6). These results provide a good foundation for the current study (Booker et al 2007). Previous research conducted by our group indicated that a change in the anion of the RTIL used as polymerisation solvent (from BF_4^- to PF_6^-) influences polymer selectivity for the template for both *trans*-aconitic acid and cocaine-imprinted polymers (Booker et al, 2005, 2006, 2008) when prepared under otherwise identical conditions.

Porogen	Structure	Viscosity at 25°C (mPa s)
Chloroform	$CHCl_3$	0.54
[bmim]BF_4	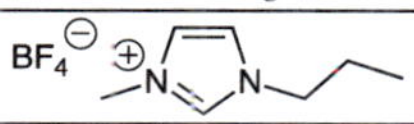	104.2
[bmim]PF_6	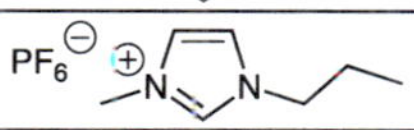	195.9
[bmim]HSO_4	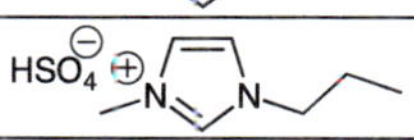	900

Table 1. Chemical structure of RTILs examined in this study and the viscosity of porogens used in the preparation of MIPs and the control polymers, i.e. the non-imprinted polymers (NIPs).

This work considers the RTILs [bmim]BF_4, [bmim]PF_6 and [bmim]HSO_4 and compares MIP efficacy to chloroform. Of the three selected RTILs, only [bmim]PF_6 and [bmim]BF_4 have previously been used as polymerisation solvents (Kubisa 2004), [bmim]HSO_4 to the best of our knowledge, has not. Both [bmim]PF_6 and [bmim]BF_4 have been used in the production of MIPs (Booker et al 2006, 2007; Wang et al 2006, 2008; He et al 2008). The work reported herein aims to identify RTILs that reproducibly create well-performing MIPs, and assess factors contributing to variations in polymer performance. Of particular interest is the effect of anion and RTIL viscosity and how they affect polymer morphology. The selected RTILs have viscosities of 104.2, 195.9 and 900 mPa s; substantially more viscous than the VOC chloroform at 0.51 mPa s (Crabtree and O'Brien 1991) (Table 1). Other features that affect polymer morphology, and hence will be evaluated as part of this study, include:

(1) *Polymerisation temperature (0 °C and 60 °C)* as it is has been reported that MIP preparation at low temperature, photoinitiated by UV light, allows for better 'freezing' of the interaction between the template and T-FM pre-polymerisation cluster, leading to an enhanced imprinted framework and improved MIP selectivity (Lu et al 2004). The increased polymerisation rates in RTILs are advantageous here (Andrejewska et al 2009), with some polymerisations failing to proceed in VOCs at 0-5 °C (Booker 2005).

(2) *Polymerisation volume (5 mL and 25 mL),* as this typically differentiates between two of the main imprinting formats, i.e. bulk or precipitation. Bulk (or monolith) polymers are prepared using minimal porogen volumes, where the polymer particles coalesce during

polymerisation to create a monolithic polymer structure (Venn and Goody 1999). By contrast, precipitation polymers are prepared in a large porogen volume (usually >95% of the polymerisation mixture by volume), allowing the polymer to form in solution as discrete particles. In the case of VOCs, minimal amounts of porogen (5 mL) produce monolithic polymers, whereas high porogen volume (25 mL) promotes the formation of polymer nanoparticles (Castell et al 2006). The effect of porogen volume on MIP morphology and selectivity has been documented in both VOCs and RTILs (Booker et al. 2006; Kotrotsiou et al 2009).

In terms of pore size, high porogen volumes increase the pore volume of MIPs, which, in turn, increase the rebinding capacity of the polymers (Kotrotsiou et al 2009). The effect of RTIL porogen volume on MIP morphology, however, is less distinct. Even at 5 mL porogen volume, clusters of irregular shaped nanoparticles have been found to form. Increasing the volume to 25 mL resulted in only minor visible changes, although differences in selectivity were observed over the various template/solvent/polymerisation temperature combinations studied (Booker 2005). Only a limited number of characterisation studies have been conducted on the polymers and much remains to be understood about the precise nature of the polymers and how this leads to the observed variations in template selectivity.

3. Experimental section

3.1 Synthesis of 1-butyl-3-methylimadazolium chloride ([bmim]Cl)

1-Methylimidazole (76mL, 0.9mol) was refluxed for 72 hours under nitrogen with a mixture of chlorobutane (110mL, 1mol) and ethyl acetate (50mL, 0.5mol) following the procedure of Whitehead (Whitehead et al. 2004).

3.2 Synthesis of 1-butyl-3-methylimadazolium tetrafluoroborate ([bmim]BF_4)

[Bmim]Cl (60 g, 0.28 mol) was dissolved in water (60 mL) and cooled in an ice bath. To this chilled solution HBF_4 (60 mL, 0.9 mol) was added over 15 minutes. After complete addition the solution was stirred at room temperature overnight. The aqueous layer was extracted with dichloromethane (3x 50mL), the combined extracts dried over $MgSO_4$ and the solvent removed by rotary evaporation (Whitehead et al., 2004).

3.3 Synthesis of 1-butyl-3-methylimidazolium hexafluorophosphate ([bmim]PF_6)

[Bmim]Cl (120 g, 0.56 mol) was dissolved in distilled water (100 mL) and HPF_6 (68 mL, 60% aqueous solution, 0.56 mol) slowly added over 20 minutes. The aqueous upper phase was decanted and water added (50mL). The mixture was stirred vigorously, allowed to settle and the upper phase decanted. This procedure was repeated until the pH of the upper phase was ~7. The water was removed by rotary evaporation (Whitehead et al., 2004).

3.4 Synthesis of 1-butyl-3-methylimidazolium hydrogen sulfate ([bmim]HSO_4)

[Bmim]Cl (30 g, 0.14 mol) was dissolved in water (30 mL) and conc. H_2SO_4 (96 mL) added. The solution was refluxed for 4 hours at 100 °C then heated at 120 °C until the vapour given off was no longer acidic. The remaining water was removed by rotary evaporation (Whitehead et al., 2004).

3.5 Cocaine-imprinted polymers: preparation and rebinding of MIPs

MIPs were prepared following the procedure of Holdsworth with cocaine base (0.14 mmol, 42.6 mg), MAA (0.28 mmol, 24.1 mg) and EGDMA (1.4 mmol, 280 mg), in desired amount of

solvent (Holdsworth et al, 2005). The reaction mixture was degassed with N_2 before AIBN (10 mg) was added and the solution was heated at 60 °C in a Syncore Polyvap Platform (Buchi). Photoinitiated polymerisation was done at 0 ˚C for 3 hours in an ice bath using a UV probe (Acticure). The porogens evaluated were $CHCl_3$ (control), [bmim]BF_4, [bmim]PF_6 and [bmim]HSO_4. Non-imprinted control polymers (NIPs) were prepared using the same method without the addition of the cocaine base.
Template extraction was by exhaustive washing with methanol until no cocaine peak was registered by GCMS analysis. The polymer was then filtered and dried under vacuum. Rebinding was carried out by suspension of 10 mg of polymer in 2 mL of 25 μM cocaine solution in various porogens for 1 hour. The resulting solution was filtered using a 2 μm PTFE membrane filter and analysed using GCMS. The amount of cocaine bound by the polymer was calculated from the difference in solution concentration before and after rebinding. The total selective binding of the polymer (ΔB) was calculated as the difference between the MIP binding and the NIP binding (B_{MIP}-B_{NIP}). Imprinting factors (*I*) were calculated as B_{MIP}/B_{NIP}.

4. Results and discussion

In our hands, the MIP preparation commences with combinatory molecular modelling – NMR titration studies which identified the most favourable FM interactions of cocaine (1) to be with methacrylic acid (MAA, **2**) in a 1:2 (T-FM) ratio (Holdsworth et al, 2005). . Thus, MIP preparation was conducted with a 1:2:10 ratio of cocaine : MAA : EGDMA (ethylene glycol dimethacrylate (EDGMA, **3**)) (Figure 2).

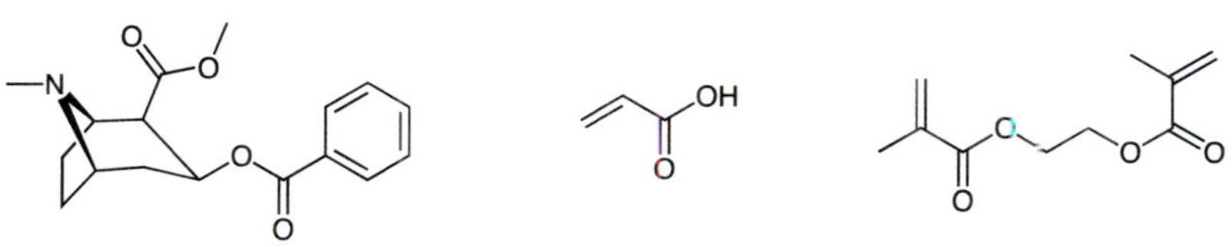

1 **2** **3**

Fig. 2. Chemical structures of cocaine (**1**), methacrylic acid (**2**) and ethylene glycol dimethyl acrylate (**3**).

MIPs corresponding to each RTIL and the VOC ($CHCl_3$) were produced at 60 ºC and 0 ºC and at 5 mL and 25 mL porogen volume. This resulted in the synthesis of fourteen MIPs (Table 2). The corresponding NIPs were also synthesised (data not shown).
With $CHCl_3$, MIPs were produced at 60 ºC and 5 mL and 25 mL porogen volume. MIP preparation in $CHCl_3$ at 0 ºC failed. This is the first RTIL / VOC point of differentiation with the rapid polymerisation rates inherent with RTILs facilitating the production of sufficient quantities of MIP for further studies when the equivalent VOC preparation failed. Additionally, the use of RTILs showed marked advantages over $CHCl_3$ in terms of production efficiency. Yields were increased by up to 50% for some RTIL-mediated polymerisations, for example with $BF_{4\text{-}60\text{-}5}$ where a 94% yield was obtained, compared to 46% in $CHCl_{3\text{-}60\text{-}5}$. Reaction times were also markedly reduced in RTILs, with particularly fast reaction times observed in the low temperature photoinitiated polymerisations. At 60 °C in 5 mL solvent, reaction time was reduced from 6 hours in $CHCl_3$ to 2 hours in the RTILs and in 25 mL, from 18 hours in $CHCl_3$ to 8 hours in RTILs. At 0 °C, using UV-initiated

polymerisation, a high polymer yield was obtained in RTIL polymerisations within 30-45 minutes, while there was no polymer formed in $CHCl_3$ (after a reaction time of 6 hours). The effect of increased reaction rate in RTIL polymerisations has been attributed to a combination of an increased rate of propagation coupled with decreased rates of termination, a function of the high viscosity of the RTILs (Schmidt-Naake et al 2009).

Polymer Code*	Porogen	Temperature (°C)	Porogen Volume (mL)	Reaction Time (h)	Polymer Yield (%)[b]
$CHCl_{3\text{-}60\text{-}5}$	$CHCl_3$	60	5	6	46
$CHCl_{3\text{-}60\text{-}25}$	$CHCl_3$	60	25	18	40
$BF_{4\text{-}60\text{-}5}$	[bmim]BF_4	60	5	2	94
$BF_{4\text{-}60\text{-}25}$	[bmim]BF_4	60	25	8	86
$BF_{4\text{-}0\text{-}5}$	[bmim]BF_4	0	5	0.75	78
$BF_{4\text{-}0\text{-}25}$	[bmim]BF_4	0	25	2	81
$PF_{6\text{-}60\text{-}5}$	[bmim]PF_6	60	5	2	80
$PF_{6\text{-}60\text{-}25}$	[bmim]PF_6	60	25	8	84
$PF_{6\text{-}0\text{-}5}$	[bmim]PF_6	0	5	0.5	70
$PF_{6\text{-}0\text{-}25}$	[bmim]PF_6	0	25	2	73
$HSO_{4\text{-}60\text{-}5}$	[bmim]HSO_4	60	5	2	69
$HSO_{4\text{-}60\text{-}25}$	[bmim]HSO_4	60	25	8	52
$HSO_{4\text{-}0\text{-}5}$	[bmim]HSO_4	0	5	0.5	55
$HSO_{4\text{-}0\text{-}25}$	[bmim]HSO_4	0	25	2	60

*Polymer codes relate to the conditions used to synthesise each polymer: porogen-temp-volume, thus $CHCl_3$-60-5 identifies the MIP prepared in $CHCl_3$ at 60 °C and 5 mL of porogen. [b] Average of MIP and Values. Yield was determined gravimetrically.

Table 2. Cocaine-imprinted polymer synthesis conditions, reaction time and yield. All polymers were prepared using MAA as functional monomer and EGDMA as crosslinker.

4.1 MIP rebinding

Having successfully prepared the required MIPs (and NIPs), our attention turned to determining the efficacy of cocaine rebinding in each MIP. This was conducted via batch rebinding assays (in triplicate, see Experimental - section 3.5.1.). Typically, rebinding is carried out in the original MIP porogen, however high viscosity of the RTILs means that this is not practical due to mass transport issues significantly hampering ingress and egress of the template to the cocaine specific cavities. The optimal rebinding time was determined to be 60 minutes (data not shown), and as such all analyses were conducted at this time point (see Experimental – section 3.5.1.). Determination of the amount of cocaine rebound to both MIP and NIP allowed the calculation of the *imprinting factor (I)*, where $I = B_{MIP}/B_{NIP}$. These values are shown in bold in Figure 3 below.

A cursory examination of the data presented in Figure 3 reveals a broad range of imprinting values from $I_{HSO4\text{-}60\text{-}25} = 0.4$ (Fig. 3c) to $I_{BF4\text{-}0\text{-}25} = 2.2$ (Fig. 3d). An $I < 1.0$ indicates preferential binding of the template (cocaine) to the control polymer. The $I_{BF4\text{-}0\text{-}25} = 2.2$ value compares very favourably than the best I value noted for $CHCl_3$ (VOC) of $I_{CHCl3\text{-}60\text{-}5} = 1.6$ (Fig. 3a). Interestingly, while none of the $RTIL_{60\text{-}5}$ preparations afforded a selectivity enhancement relative to the $CHCl_3$ preparations, these RTIL-MIPs (and NIPs) displayed a considerably higher binding capacity (B% up to ca 45%) in all cases. This suggests the formation of a greater number of low specificity binding sites at higher temperature (in keeping with MIP theory that suggests lower temperature favour high specificity binding sites). Regardless, the specificity of these $RTIL_{60\text{-}5}$ MIPs is poor at best with a maximum $I = 1.4$. Also of note is the HSO_4 systems that returned $I = 1.0$, that is, no selectivity for the template relative to the NIP. The HSO_4 RTIL displays, by a considerable margin, the highest viscosity of all the systems examined (Table 2), as such the poor specificity was unexpected as the high viscosity should retain the FM and T in close proximity, allowing for a longer interaction lifetime, and thus an enhanced I value. However the lack of specificity may be an artefact of the poor mass transport associated with high viscosity and the short pre-association time (20 minutes) used in the preparation of these MIPs. In the initial MIP pre-association (pre-polymerisation phase), to generate high specificity binding sites, the template and functional monomer must participate in favourable interactions which invariably requires a co-localisation of the FM and T units. The high viscosity of the HSO_4 RTIL may have prevented this association and thus lead to no or little T-FM interactions, in turn leading to no observable specificity.

The 0 °C, 5 mL MIP preparation (Figure 3c) were effectively non-selective. We do note, however, the highest levels of cocaine re-binding across the board in these systems. Given that this is a non-specific effect, it is most likely due to an increased surface area / surface binding rather than the development of template specific cavities.

These findings appear to contradict the conventional MIP theory, which stipulates that the template selectivity of MIPs is enhanced by low temperature polymerisation due to lower system internal energies. At present, with the limited number of RTIL-MIP studies, there appears to be no fixed correlation between polymerisation temperature and MIP efficacy. It is important to note here that these findings are limited to the $RTIL_{0\text{-}5}$ systems. These systems strongly favour the formation of a bulk (monolithic) MIP. Under these conditions we believe that polymer precipitation induces a high level of template to co-precipitate, effectively destroying the T-FM association crucial for high specificity MIPs. Thus, increasing the lifetime of a homogeneous system would be expected to result in the preparation of a higher specificity MIP (see below). Another possible explanation for this is that the generation of selective sites is limited by the fast polymerisation kinetics at low temperatures in the RTILs. Fast reaction kinetics make the overall polymerisation times very short compared to the polymers prepared at 60 °C. Although this is generally seen as an advantage of the RTIL-based system (due to vastly improved efficiency), short polymerisation times have been linked with decreased selectivity of polymers for the template (Piletska et al 2009). Shorter polymerisation times limit the thermally driven equilibrium between functional monomers and template required for the generation of selective sites. These reduced polymerisation times will also reduce the level of crosslinking in the polymer which inturn may reduce cavity specificity.

The rebinding performance for the 60 °C, 25 mL polymer systems (Fig. 3c) showed retention of selectivity with $CHCl_{3\text{-}60\text{-}25}$ and $PF_{6\text{-}60\text{-}25}$ ($I = 1.5$), with enhanced specificity for $BF_{4\text{-}60\text{-}25}$ ($I = 1.9$). Selectivity has been derived at the cost of binding capacity with a decrease in binding noted for $BF_{4\text{-}60\text{-}25}$ (B = 24%; c.f. $BF_{4\text{-}60\text{-}5}$ B = 39%). The VOC system, $CHCl_{3\text{-}60\text{-}25}$ showed increased levels of total binding (B = 28%) relative to the corresponding $CHCl_{3\text{-}60\text{-}5}$ (B = 16%) with no change in specificity. Again, the HSO_4-prepared polymer system showed different behaviour to other RTILs; in $HSO_{4\text{-}60\text{-}25}$, NIP binding was higher (B = 33%) and MIP binding was lower (B = 13%) than in $HSO_{4\text{-}60\text{-}5}$. In the $MIP_{0\text{-}25}$ systems (Fig. 3d), there was increased specificity with the BF_4 ($I = 2.2$) and HSO_4 ($I = 1.9$) systems. This was due to both an increase in MIP binding and a reduction in NIP binding compared to both other systems. The selectivity of $PF_{6\text{-}0\text{-}25}$, however was adversely affected with $I = 0.8$ and total binding levels significantly lower than in all other [bmim]PF_6 systems.

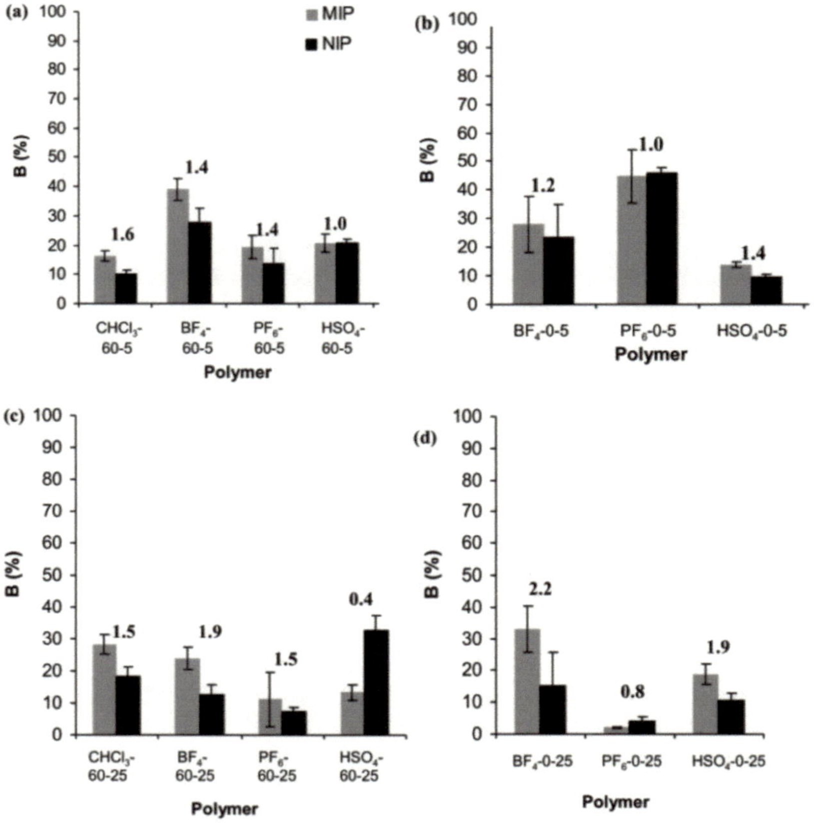

Fig. 3. Rebinding data (percentage cocaine rebound MIP and NIP, and *I* values shown in bold over the MIP/NIP graph columns) for cocaine-imprinted polymers prepared in $CHCl_3$, [bmim]BF_4, [bmim]PF_6 and [bmim]HSO_4. MIPs were prepared at temperatures and porogen volumes of: (a) 60 °C, 5 mL; (b) 0 °C, 5 mL; (c) 60 °C, 25 mL; and (d) 0°C, 25 mL. All rebinding studies were conducted in $CHCl_3$. The error bars represent the standard deviation after 3 measurements.

This rebinding data highlights the complex nature of the RTIL effects on MIP systems. Overall, the best performing RTIL-MIP system was [bmim]BF_4. The [bmim]BF_4-MIPs retained and at times showed enhanced selectivity across all temperature and porogen volumes studied. The highest imprinting factor was achieved in $BF_{4\text{-}0\text{-}25}$ with an imprinting factor of $I = 2.2$. This represents an improvement over the VOC-$CHCl_{3\text{-}60\text{-}5}$ with $I = 1.6$. While MIPs have been prepared using RTILs that show selectivity comparable to, and in some cases better than, polymers prepared in $CHCl_3$, this has been shown to be dependent on the RTIL composition and the synthesis conditions. These effects are not directly linked to RTIL viscosity. Various factors related to the physical characteristics of the polymers, such as swelling and surface charge, were found to be contributing factors driving this selectivity (Rampey et al 2004).

4.2 Polymer morphology

The MIPs herein have been generated as monolithic (5 mL porogen) and precipitation polymers (25 mL porogen), with the precipitation approach displaying the highest *I* values. This suggests that the final polymer morphology affects the selectivity of template rebinding. Our initial study of the physical characteristics of MIPs commenced with evaluation of the polymer surfaces by scanning electron microscopy (SEM). SEM images of cocaine-imprinted polymers prepared in both $CHCl_3$ and RTILs are shown in Fig. 4 below.

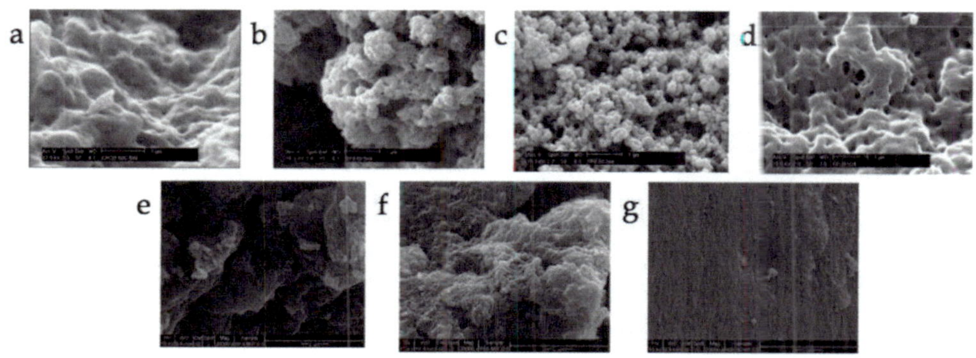

Fig. 4. SEM images of cocaine-imprinted polymers prepared in 5 mL porogen at 60 and 0 °C. (a) $CHCl_{3\text{-}60\text{-}5}$; (b) [bmim]$BF_{4\text{-}60\text{-}5}$; (c) [bmim]$PF_{6\text{-}60\text{-}5}$; (d) [bmim]$HSO_{4\text{-}60\text{-}5}$; (e) [bmim]$BF_{4\text{-}0\text{-}5}$; (f) [bmim]$PF_{6\text{-}0\text{-}5}$; (g) [bmim]$HSO_{4\text{-}0\text{-}5}$.

There are marked differences between the MIP-$CHCl_3$, which is clearly amorphous monolithic in nature, typical of polymers synthesised under similar conditions (Sun & Fung, 2006). However the RTIL preparations show much greater structural definition and are typically composed of nano-particulate clusters (< 300 nm), which result in greatly enhanced surface area. This morphology difference is attributed to increased polymer solubility in the RTIL, promoting a much later phase separation, yielding the observed cluster formation. The $HSO_{4\text{-}60\text{-}5}$ system forms a space filled gel (according to Stover's naming system (Goh & Stover, 2002)) (Fig. 4d).

Examination of these polymers now at 0 °C and 5 mL porogen, leads to a marked decrease in the visible surface area (Fig. 4e-g). Presumably this results from an increase in RTIL viscosity with lower temperatures (note that the $CHCl_3$ 0 °C preparation failed), decreasing polymer solvation, decreasing diffusion resulting in an increasing monolithic polymer characterisation.

Typically, the MIPs prepared at lower temperature displayed much smoother surfaces. This was anticipated, as it is known that polymerisation temperature has a direct influence on the polymer particle size and porosity (Lu et al., 2004; Kotrotsiou et al., 2009; Li et al., 2007; Rajaram & Hudson, 1996). Interestingly, no fixed correlation between polymer surface area and porosity has been reported for polymers formed in $CHCl_3$, with character dependent upon the nature of the template, functional monomer and cross linker used (Li et al., 2007; Kotrotsiou et al., 2009).

The effect of increasing porogen volume (5 mL to 25 mL) was next examined. Of particular note is the now particulate character (~300 nm) of the $CHCl_3$ MIP (cf. Fig. 4a ($CHCl_{3\text{-}60\text{-}5}$) and Fig. 5a ($CHCl_{3\text{-}60\text{-}25}$)). In all instances, there was little difference noted between the 5 mL and 25 mL RTIL preparations. Both the BF_4 and PF_6 systems afforded nano-particulate clusters (Fig. 5 a and b), with the HSO_4 system, again, the outlier with a gel structure lacking any obvious structural features (Fig. 5d). Reducing the polymerisation temperature to 0 °C, resulted in a similar reduction in the observable surface area (as with the 60-5 and 0-5 mL preparations above). All RTIL polymers adopt an amorphous surface morphology. As before, this is most likely due to high RTIL viscosity and low polymer solvation.

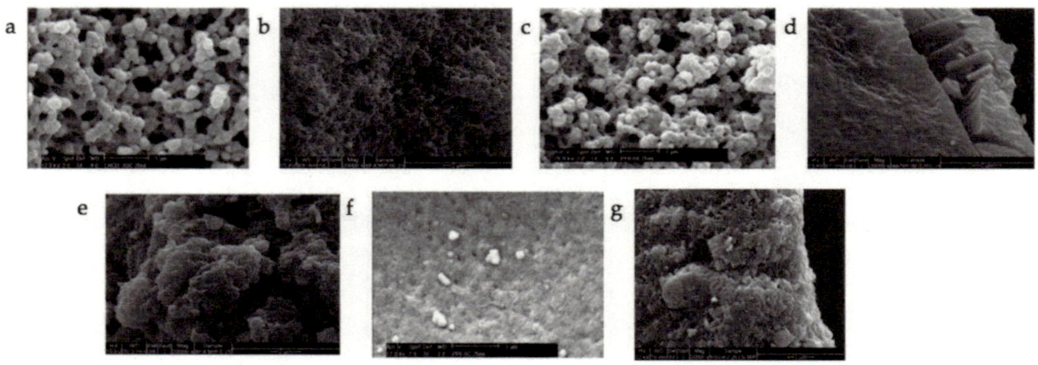

Fig. 5. SEM images of cocaine-imprinted polymers prepared in 25 mL porogen at 60 and 0 °C. (a) $CHCl_{3\text{-}60\text{-}25}$; (b) [bmim]$BF_{4\text{-}60\text{-}25}$; (c) [bmim]$PF_{6\text{-}60\text{-}25}$; (d) [bmim]$HSO_{4\text{-}60\text{-}25}$; (e) [bmim]$BF_{4\text{-}0\text{-}25}$; (f) [bmim]$PF_{6\text{-}0\text{-}25}$; (g) [bmim]$HSO_{4\text{-}0\text{-}25}$.

4.3 MIP swelling

Swelling behaviour is known to impact on MIP performance in two ways. Some degree of swelling facilitates template access to the specific binding cavity within the MIP. However, high levels of swelling can lead to cavity deformation reducing the efficacy of the template-cavity interaction (Spivak, 2005). Typically polymers prepared in solvating porogens will experience a greater dynamic swelling range due to a decreased crosslinking network (Sellergren, 2001). This results from later phase separation of the growing polymer chains, resulting in a lower degree of aggregation of the polymer particles. Generally, RTIL prepared MIPs and NIPs have been observed to swell less (Booker et al., 2006). It is important to note here though, that the absence of polymer swelling does not preclude the development of high specificity, high capacity MIPs.

MIP	60 °C 5 mL	0 °C 5 mL	60 °C 25 mL	0 °C 25 mL
$CHCl_3$	100%	-	250%	-
[bmim]BF_4	20%	33%	100%	170%
[bmim]PF_6	70%	160%	285%	80%
[bmim]HSO_4	35%	50%	35%	10%

Table 2. Degree of swelling exhibited by MIPs prepared at 60 °C, 5 mL porogen; 0 °C, 5 mL porogen; 60 °C, 25 mL porogen; and 0 °C 25 mL porogen, in $CHCl_3$.

The degree of MIP swelling was determined in $CHCl_3$, the batch rebinding solvent. Minimal differences were noted between MIPs and NIPs. In the 60 °C, 5 mL polymer preparation $CHCl_{3\text{-}60\text{-}5}$ MIP showed approximately 100% swelling, significantly higher than the $BF_{4\text{-}60\text{-}5}$ (20%), $PF_{6\text{-}60\text{-}5}$ (70%), and $HSO_{4\text{-}60\text{-}5}$ (35%) MIPs (Table 2). The MIPs prepared at 0 °C in 5 mL solvent showed a similar pattern of swelling behaviour to the 60 °C preparation across the solvent range, with an increase in MIP swelling compared to the 60 °C, 5 mL polymer preparation was observed in most cases: $BF_{4\text{-}0\text{-}5}$ (33%), $PF_{6\text{-}0\text{-}5}$ (160%) and $HSO_{4\text{-}0\text{-}5}$ (50%). This is most likely attributable to the decreased crosslinking density when prepared under photoinitiated conditions (Sellergren, 2001).

Short polymerisation times may also result in lower degrees of crosslinking, which limits the formation of well-defined rebinding cavities, which in turn can reduce selectivity due to decreased polymer rigidity. This could also serve as an explanation of the increased swelling and template uptake, as a more flexible polymer will facilitate mass transfer through the polymer and allow for more binding to take place. The associated cavity deformation has a substantial impact on the reproducibility of results and on selectivity for the template (Spivak, 2005).

The porogen volume in the initial synthesis also influenced the polymer 'swellability'. Polymers prepared at 60 °C in 25 mL porogen (Table 2) showed enhanced swelling compared to their 5 mL counterparts for most polymer preparations; $CHCl_{3\text{-}60\text{-}25}$ (250%), $BF_{4\text{-}60\text{-}25}$ (100%) and $PF_{6\text{-}60\text{-}25}$ (285%). This is not the case in $HSO_{4\text{-}60\text{-}25}$ where the degree of swelling remained constant. Again, this effect may be attributed to a general decrease in crosslinking density as a result of the dilution of the polymerisation components. This had no effect on the [bmim]HSO_4-prepared polymer where the high viscosity (900 mPa s) promotes high levels of crosslinking (due to very slow diffusion). When prepared at 0 °C in 25 mL porogen, a different trend was observed. The $BF_{4\text{-}0\text{-}25}$ prepared polymer showed enhanced swelling (170%) compared to the 0 °C, 5 mL preparation (33%), while in $PF_{6\text{-}0\text{-}25}$ (80%) and $HSO_{4\text{-}0\text{-}25}$, swelling was reduced (10%). We speculate, and are currently investigating, that the lower viscosity of these solvents, which will be increased at lower temperatures, prevents polymer diffusion, thereby increasing crosslink density.

The fact that MIP and NIP polymers do not vary greatly indicates that it is the preparation solvent, not the presence of the template, which has the greatest effect on the polymer swelling behaviour (data not shown).

4.4 MIP zeta potential

Zeta potential analysis can give valuable data regarding the surface charge of the polymers. This indicates whether or not the polymer is colloidally stable in solution whilst also giving some indication as to the nature and relative density of functional groups that may be present at the polymer surface. The zeta potential of polymers of this type can be related to the number of surface accessible MAA carboxylic acid groups (Perez-Moral & Mayes, 2004). These carboxylic acid groups on the polymer surface will influence the degree of surface binding, i.e. a larger zeta value would be expected to correlate with higher levels of surface binding, assuming uniform distribution of MAA units.

While all the RTIL samples examined by SEM were nano-particulate in nature, no sample showed high levels of dispersion, instead showing a high degree of polymer aggregation (Fig. 4 and 5). Pilot studies showed that sonication time had no effect on the measured zeta potential (data not shown). With polymers prepared at 60°C, 5mL, most measurements fell within zeta values of -18 mV and -22 mV, with readings obtained that were within the same error range across the various solvent preparations. The major outliers were clearly $BF_{4\text{-}60\text{-}5}$ MIP (-28mV) and $HSO_{4\text{-}60\text{-}5}$ NIP (-14 mV). The increased surface charge of $BF_{4\text{-}60\text{-}5}$ indicates a higher concentration of MAA present on the surface of the polymer. This could have an impact on the rebinding behaviour of the polymer as it would increase the levels of template

rebound to the polymer surface (Perez-Moral & Mayes, 2004). The decrease in zeta potential observed in the $HSO_{4\text{-}60\text{-}5}$ NIP further emphasises the difference in the nature of this polymer, as the TGA indicated (see section 4.5).
As the SEM data showed a significant decrease in the visible surface area of the polymers at low polymerisation temperatures, it follows that there may be some differences in zeta potential. Image **b**, Figure 3.6 shows the zeta potential data for the polymers prepared at 0 °C in 5 mL solvent. Overall, the polymers showed a slight decrease in the average zeta values compared to the 60 °C-prepared system (approximately 2mV lower in most cases), corresponding to the decrease in visible surface area observed via SEM. Again, the effect of solvent choice for the polymerisation process appeared to have little effect on the observed zeta values. The $HSO_{4\text{-}0\text{-}5}$, however, showed a significant drop in the zeta potential of the NIP (-12 mV compared to -18 mV in the MIP), as was observed for this RTIL in the 60 °C system.

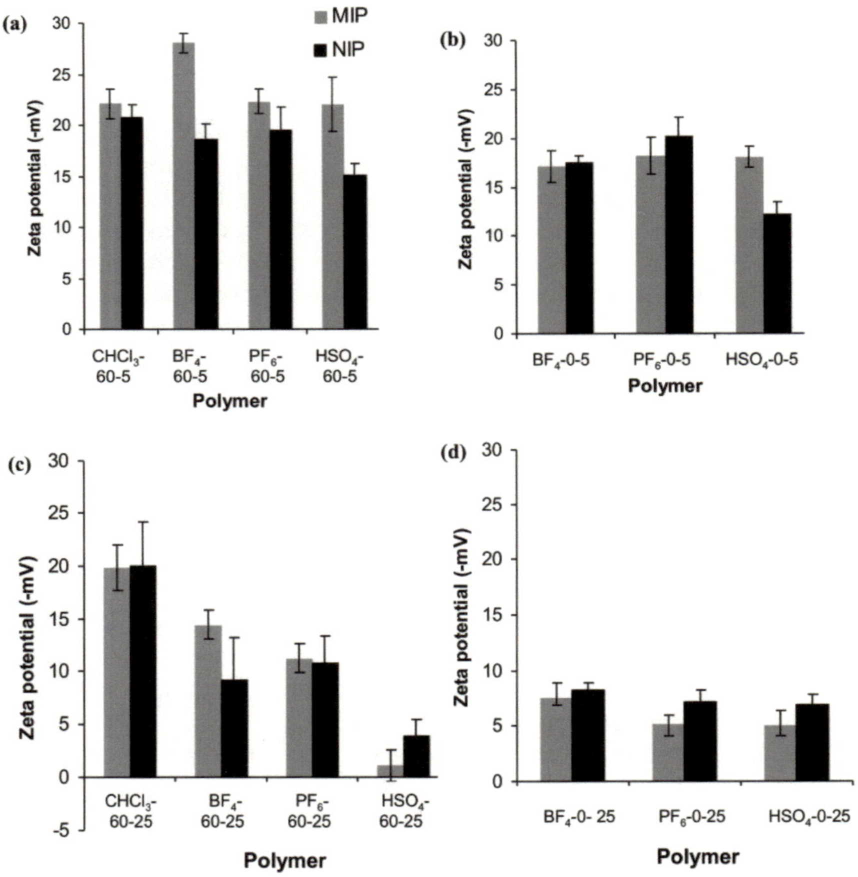

Fig. 6. Zeta potential data for cocaine-imprinted polymers prepared in various solvents. Polymers were prepared at temperatures and solvent volumes of (a) 60 °C, 5 mL, (b) 0 °C, 5 mL, (c) 60 °C, 25 mL and (d) 0 °C, 25 mL. The error bars represent standard deviation after 10 measurements.

4.5 MIP stability

All MIPs showed good thermal stability up to 260 °C, as did the corresponding NIPs (data not shown). In the 0 °C polymer preparations (Figure 7), the polymers were again observed to be stable up to 260 °C. The TGA traces show little variation between the polymers, with the exception of $[bmim]PF_6$ which showed enhanced thermal stability. There is one very distinct decomposition phase, indicating a more homogeneous composition than the 60 °C-prepared polymers.

MIPs prepared under precipitation polymerisation conditions (25 mL progen) showed similar levels of thermal stability to those prepared in 5 mL porogen (data not shown), with most polymers stable to 260°C. The exception to this was polymer $HSO_{4\text{-}60\text{-}25}$, which showed substantial mass losses (>20%) at temperatures below 260 °C and high char yields, again indicating that the ionic liquid may not have been completely removed from solution. Polymers prepared at low temperature again showed a more distinct decomposition phase, indicating a more homogeneous composition. This decomposition data shows that a good level of thermal stability can be achieved in polymers prepared in RTILs, which is, in most cases, comparable to polymers prepared in VOCs. These results are in line with literature for linear polymers prepared in RTILs, which showed similar thermal stability to that of VOC-prepared polymers (Cheng et al., 2004).

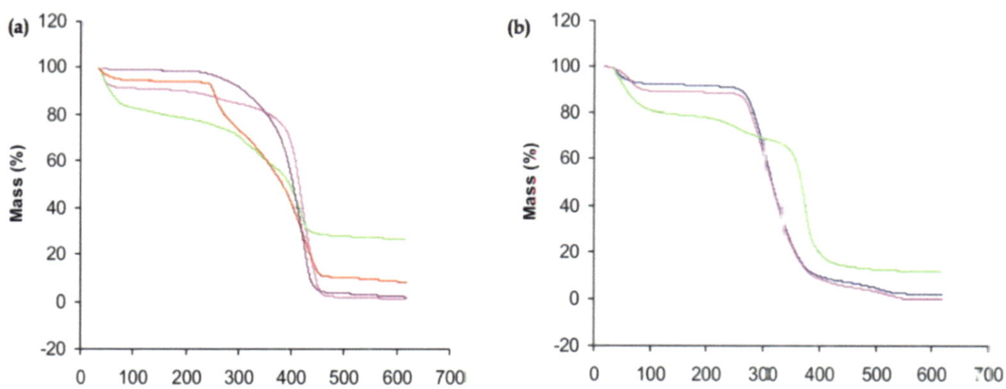

Fig. 7. TGA spectra for cocaine-imprinted MIPs prepared in: —— $CHCl_3$; —— $[bmim]PF_6$; —— $[bmim]BF_4$; —— $[bmim]HSO_4$ at (a) 60 °C, 25 mL and (b) 0 °C, 25 mL.

5. Conclusions

This chapter has highlighted the many different factors related to using RTILs as polymerisation solvents, which can influence polymer characteristics and rebinding performance. In depth analysis of the previously studied cocaine-imprinted system showed that the porogen used for polymerisation, temperature and solvent volume all have significant effects on polymer properties. Some general trends could be observed, such as the increased binding capacity of polymers when prepared at lower temperatures, due to a decrease in crosslinking density. It showed that extremely high levels of crosslinking, whilst producing very thermally stable polymers, results in decreased uptake of the template and a loss of selectivity for both VOC and RTIL-prepared polymers. Overall, using $[bmim]BF_4$ as polymerisation solvent gave the best results in terms of polymer selectivity from all the RTILs studied. Imprinting factors >1 were achieved across all temperature ranges and solvent volumes with better imprinting factors achieved than in the VOC-prepared polymers (the highest *I* value in [bmim]BF_4 was 2.2 compared to 1.6 in the $CHCl_3$).

The RTIL-prepared polymers show promise as VOC alternatives in MIP systems, with numerous advantages (such as improved reaction efficiency, the elimination of the need to grind the polymers, the ability to polymerise at low temperatures and the improved selectivity for the template in some instances). There is also the possibility of recycling and reusing the RTIL for numerous reactions, although this was not examined in the current study. However, the data presented thus far has highlighted one of the problems associated with MIP synthesis, which was the lack of reproducibility observed in rebinding studies. This is presumably a result of heterogenous binding sites present in the polymers. In order to achieve good selectivity for the template, it is necessary to create the right balance between the different synthesis and rebinding conditions. The fact that this is further affected by the choice of RTIL makes the process even more complex. The prediction of polymer properties in any given RTIL-MIP system has been shown to be challenging, which places limitations on the utility of these systems. A more comprehensive study is required, to include analysis such as molecular modelling and NMR studies right through to rebinding isotherms on a well studied, 'model' MIP template. This may assist in furthering our understanding of the role of RTILs in MIP polymerisations and enhance the scope of RTILs in MIP applications.

6. References

Andrzejewska, E.; Podgorska-Golubskaa, M.; Stepniaka, I.; Andrzejewski, M. Photoinitiated polymerization in ionic liquids: Kinetics and viscosity effects. *Polymer* 2009, *50*, 2040-2047.

Booker, K. Molecularly Imprinted Polymers and Room Temperature Ionic Liquids: Evaluation of Applicability to the Detection of Illicit Drugs. University of Newcastle, Newcastle, 2005.

Booker, K.; Holdsworth, C.; Bowyer, M.; Mc Cluskey, A. Efficient preparation and improved sensitivity of molecularly imprinted polymers using room temperature ionic liquids. *Chemical Communications* 2006, *16*, 1730-1732.

Booker, K.; Holdsworth, C.; Bowyer, M.; Lennard, C.; Mc Cluskey, A. Molecularly imprinted polymers and room temperature ionic liquids: Impact of template on polymer morphology. *Australian Journal of Chemistry* 2007, *60*, 51-56.

Castell, O.; Allender, C.; Barrow, D. Novel biphasic separations utilising highly selective molecularly imprinted polymers as biorecognition solvent extraction agents. *Biosensors and Bioelectronics* 2006, *22*, 526-533.

Liang, C.; Yumei, Z.; Tingting, Z.; Huaping, W. Free radical polymerization of acrylonitrile in green ionic liquids. *Macromolecular Symposia* 2004, *216*, 9-16.

Crabtree, A.; O'Brien, J. Excess viscosities of binary mixtures of chloroform and alcohols *Journal of Chemical and Engineering Data* 1991, *36*, 140-142.

Cooper, A.; Scott, K.; Sneddon, P.; Winterton, N. Cross-linked polymers in ionic liquids: Ionic liquids as porogens. *Polymer Preprints* 2004, *45*, 291-292.

Goh, E.; Stover, H. Cross-linked poly(methacrylic acid-co-poly(ethylene oxide)methyl ether methacrylate) microspheres and microgels prepared by precipitation polymerisation: a morphology study. *Macromolecules* 2002, *35*, 9983-9989.

He, C.; Long, Y.; Pan, J.; Li, K.; Liu, F. Molecularly imprinted silica prepared with immiscible ionic liquid as solvent and porogen for selective recognition of testosterone. *Talanta* 2008, *74*, 1126-1131.

Holdsworth, C. I.; Bowyer, M. C.; Lennard, C.; McCluskey, A. Formulation of cocaine-imprinted polymers utilizing molecular modelling and NMR analysis. *Australian Journal of Chemistry* 2005, *58*, 315-320.

Kotrotsiou, O.; Chaitidou, S.; Kiparissides, C. Boc-*L*-tryptophan imprinted polymeric microparticles for bioanalytical applications. *Materials Science and Engineering C* 2009, *29*, 2141-2146.

Li, X.; Li, H.; Huang, M. Productive synthesis and properties of polydiaminoanthraquinone and its pure self-stabilised nanoparticles with widely adjustable electroconductivity. *Chemistry: A European Journal* 2007, *13*, 1884-1896.

Lu, Y.; Li, C.; Wang, X.; Sun, P.; Xing, X. Influence of polymerization temperature on the molecular recognition of polymers. *Journal of Chromotography B* 2004, *804*, 53-59.

McCluskey, A.; Holdsworth, C. I.; Bowyer, M. C. Molecularly imprinted polymers (MIPs): sensing, an explosive new opportunity? *Organic and Biomolecular Chemistry*, 2007, *5(20)*, 3233-3244.

Pavlova, L.; Pavlov, M.; Davankov, V. The first representatives of hypercrosslinked hydrophilic networks:alkylation and polymerisation of 4-vinylpyridine in an ionic liquid. *Daklady Akademii Nauk* 2006, *406* (2), 200-202.

Perez-Moral, N.; Mayes, A. Noncovalent Imprinting in the shell of core-shell nanoparticles. *Langmuir* 2004, *20*, 3775-3779.

Piletska, E.; Guerreiro, A.; Whitcombe, M.; Piletsky, S. Influence of the polymerisation conditions on the performance of molecularly imprinted polymers. *Macromolecules* 2009, *42*, 4291-4298.

Rajaram, C.; Hudson, S. Effect of polymerisation temperature on the morphology and electrooptic properties of polymer-stabilised liquid crystals. *Chemical Materials* 1996, *8*, 2451-2460.

Schmidt-Naake, G.; Woecht, I.; Schmalfus, A.; Gluck, T. Free radical polymerisation in ionic liquids- solvents of a new dimension. *Macromolecular symposia* 2009, *275-276*, 204-218.

Sellergren, B., *Molecularly Imprinted Polymers - Man-made mimics of antibodies and their applications in analytical chemistry*. Elsevier: 2001.

Spivak, D. Optimisation, evaluation and characterisation of molecularly imprinted polymers. *Advanced Drug Delivery Reviews* 2005, *57*, 1779-1794.

Venn, R.; Goody, R. Synthesis and properties of molecular imprints of darifenacin: the potential of molecular imprinting for bioanalysis. *Chromatographia* 1999, *50*, 407.

Wang, H.; Zhu, Y.; Yan, X.; Gao, R.; Zheng, J. A room temperature ionic liquid (RTIL)-mediated, non-hydrolytic sol-gel methodology to prepare molecularly imprinted, silica-based hybrid monoliths for chiral separation. *Advanced Materials* 2006, *18*, 3266-3270.

Wang, H.; Zhu, Y.; Yan, X.; Lin, J., Fabrication of molecularly imprinted hybrid monoliths via a room temperature ionic liquid-mediated nonhydrolytic sol-gel route for chiral separation of zolmitriptan by capillary electrochromotography. *Electrophoresis* 2008, *29*, 952-959.

11

Magnetorheological Elastomers Containing Ionic Liquids

Marcin Masłowski and Marian Zaborski
Technical University of Lodz; Institute of Polymer and Dye Technology
Poland

1. Introduction

The term ionic liquid (IL) refers to a class of liquids that are composed solely of ions[1]. It is a synonym for molten salt. IL in a narrow sense often indicates room-temperature ionic liquid (RIL) that exists as a liquid at room temperature.

Ionic liquids (ILs), which have been widely promoted as "green solvents", are attracting much attention for applications in many fields of chemistry and industry due to their chemical stability, thermal stability, low vapour pressure and high ionic conductivity properties[2]. In recent years, ILs have been used in polymer science, mainly as polymerisation media in several types of polymerisation processes, including conventional free radical polymerisation[3], living/controlling radical polymerisations (such as atom transfer radical polymerisations (ATRP), reversible addition-fragmentation transfer (RAFT))[4], and ionic and coordination polymerisations[5]. When radical polymerizations are conducted in an IL, a significant increase of the kp/kt ratio is normally observed compared radical polymerizations conducted in other polar/coordinating solvents. As solvents for ATRP and RAFT, ILs facilitate the separation of the polymer from residual catalyst and reduce the extent of side-reactions. Applications of ILs as solvents for polymerisation processes have been reviewed by Kubisa[6] and Shen and Deng[7]. However, application of ILs in polymer science are not limited to traditional polymerisation media. ILs have also been investigated as components of polymeric matrixes (such as polymer gels), as templates for porous polymers and as novel electrolytes for electrochemical polymerisations. This review focuses on recent developments and applications of ILs in the preparation of functional polymers.

There is much current interest in ionic liquids. The much lower melting points of ionic liquids compared to those of inorganic salts can be partially attributed to the bulky cationic groups, i.e., the low charge density and incompatibility of the Coulombic attraction forces with steric hindrance. Ionic liquids have very low vapour pressures, although it was recently shown that they are distillable[8]. Therefore, they do not produce hazardous vapors (in contrast to many conventional organic solvents). Most ionic liquids have high ignition points, and they do not generate explosive air-vapour mixtures. They can act as solvents for chemical reactions, including catalytic reactions[9].

As a result of their interesting physical-chemical features ionic liquids have been extensively evaluated as environmentally friendly or "green" alternatives[10] to conventional organic solvents for a broad range of organic synthetic applications. In addition, ionic liquids have been used, *e.g.*, as catalysts[11] in organic synthesis, in compositions for stabilising and/or isolating nucleic acids in or from micro-organisms[12], as process aids for the synthesis of polynucleotides[13], as lubricants[14] in the preparation and stabilisation of nanoparticles[15] and in the synthesis and stabilisation of metal[16] and metal oxide nanoparticles[17].

It is common practice to use dispersing agents to homogeneously stabilise fillers and pigments in liquid media and to obtain storage stable pigment pastes, paints and lacquers. According to the state of the art, dispersing agents have to fulfil the following requirements:

- Easy incorporation of fillers and pigments, which leads to shorter dispersing times and, allows for the use of simpler and cheaper dispersing aggregates (a dissolver instead of ball mills).
- Reduction of the pigment paste viscosity, which allows for high pigment loading and economical tinting systems.
- Development of optimal colour strength, *i.e.*, of the optimal hiding power when using opaque pigments.
- Avoidance of sedimentation in pigment pastes when stored over long periods of time or under extreme climatic conditions.
- Avoidance of flocculation for good reproducibility of colours, which is becoming particularly important when using mixing machines for the tinting of white base paints and lacquers.

These requirements have to be independently fulfilled regardless of the liquid medium into which a filler or a pigment must be dispersed. Hence, there has been no lack of attempts to develop polymeric dispersants[18] with truly universal applicability. Additionally, a good dispersant wets and stabilises not only inorganic but also organic pigments exhibiting both hydrophilic and hydrophobic surfaces. Most modern polymeric dispersants are universal for pigments, but not for the surrounding medium. Generally, a pigment in a solvent-based liquid medium (hydrophobic) is only sterically stabilized, whereas a pigment in a water-based medium (hydrophilic) can be both sterically and electronically stabilized, depending on the presence of the appropriate functional groups in the polymeric dispersant (electro-steric stabilisation)[19].

Magnetorheological fluids (MRFs) are dispersions of micrometre-sized (from 1 to 20 lm) magnetic particles in a carrier fluid, whose rheological behaviour can be controlled by means of a magnetic field[20]. Thus, MRFs can change from a liquid to a solid-like state and vice versa almost instantaneously. MRFs have been used in various technological applications since their first preparation[21]. Current fundamental research on MRFs focuses mainly on the settling of dispersed magnetic particles (and redispersion phenomena), which may restrict their use in specific applications[22]. To overcome the problem of sedimentation, several strategies have been proposed (e.g., the addition of thixotropic agents, surfactants, and nanoparticles; the use of viscoplastic media as carriers; and polymeric core/shell-structured magnetic particles)[23]. The sedimentation problem in MRFs is of such importance that it has even been investigated (together with other properties) under microgravity conditions in outer space[24]. A very recent study reports on the rheological behaviour of suspensions of hematite nanoparticles in an IL[25]. On one hand, it was found that concentrated suspensions of nanoparticles show non-Newtonian characteristics, including shear thinning and shear thickening, which probably originate from particle–particle interactions. On the other hand, suspensions with a low content of nanoparticles show Newtonian behaviour similar to that shown by pure ILs. However, this study does not provide any information about the magnetorheological behaviour of the suspensions and/or the influence of the structure of the ILs on the stability of the suspensions against sedimentation. It has also been reported recently that ILs can be used as stabilising agents in different heterogeneous systems[26]. For these reasons, it is thought that the use of ILs as carriers of MRFs may lead to magnetic dispersions that are colloidally stable (against flocculation) and which offer an improved stability against sedimentation (in this report the term 'stability' is used in reference to sedimentation unless otherwise indicated).

Magnetorhological elastomers (MREs) are smart materials whose rheological properties can be changed by applying external magnetic field[27]. They are solid analogues of magnetorheological fluids (MRFs). MRE and MRF consist of micro- or nano- sized particles of iron compounds dispersed in an elastomer matrix[28]. The size of magnetically polarisable particles has a great influence on the stress of magnetorheological materials. It is well known from the literature[29] that the aggregation of magnetic particles results in local anisotropy and dispersion irregularity, which drastically diminishes the magnetic and magnetorheological effects of materials. Therefore, it is evident that the best magnethoreological effect can be achieved by improving the dispersion stability. As in the case of MR fluids, magnetoactive particles attempt to arrange themselves in the direction of an applied magnetic field[30]. Ferromagnetic composites with an elastomer matrix are characterised by unique properties that are not characteristic of bulk magnetic materials. These properties, among others, are a high susceptibility to elastic strains in the magnetic field and the dependence of the magnetic permeability on stress. An interesting and useful property of the composites described is the change of their elasticity coefficients and other material parameters in a sufficiently high magnetic field. Due to their properties, ferromagnetic composites with an elastomer matrix have already found many applications, for example, in sensors, converters and controlled vibration dampers.

2. Materials and methods

2.1 Materials

Acrylonitrile-butadiene rubber ((NBR) – Europrene N3960, acrylonitrile unit content: 28 %) was purchased from Lanxess, and ethylene-propylene rubber ((EPM) Dutral CO 054, propylene unit content: 40 %) was obtained from Montedison Ferrara. Crosslinking agents: dicumyl peroxide DCP (Aldrich), triallyl-1,3,5-triazine-2,4,6(1H,3H,5H) TAC (Aldrich), zinc oxide ZnO (Huta Będzin), mercaptobenzothiazole MBT (Pneumax MBT) and sulphur S_8 (Z.P.S. Siarkopol).

Fillers: magnetite (Fe_3O_4), both micro- and nano- sized, were obtained from Sigma Aldrich, and carbonyl iron powder CIP was provided by BASF.

Ionic liquids: 1-ethyl-3-methylimidazolium diethylphosphate EMIMDEP, 1-butyl-3-methylimidazolium hexafluorophosphate $BMIMPF_6$, 1-hexyl-3-methylimidazolium chloride HMIMCl, 1-butyl-3-methylimidazolium trifluoromethanesulfonate BMIM OTf, 1-butyl-3-methylimidazolium tetrafluoroborate $BMIMBF_4$, trihexyltetradecylphosphonium chloride HPCl, and 1-butyl-3-methylimidazolium chloride BMIMCl were supplied by Alrdich.

Composition of a typical elastomer mixture: rubber - 100 phr, DCP - 2 phr, TAC – 0.5 phr, ZnO - 5 phr, MBT - 2 phr, S - 2 phr, Ionic liquids – 3 phr, filler 20-100 phr.

2.2 Methods

Elastomer mixtures, based on EPM or NBR filled with micro- and nano- sized iron oxides and carbonyl iron powder were prepared by common manufacturing procedures for rubber.

Magnetoactive particles were mixed with the rubber and vulcanisation system using two - roll mills at 30 °C. The injection and stirring of ionic liquids into the elastomer composites were performed in a Brabender measuring mixer N50 (temp. 50 °C, speed range 40 min^{-1}, time of the process 10 min). Then, the mixtures were vulcanised at 160 °C and, 15 MPa pressure for 30 min. The vulcanisates were produced in two different ways: under the influence of a magnetic field - to align the magnetic particles in the elastomer matrix and without the presence of a magnetic field.
The particle size of the aggregates in water and parafin oil and the zeta potential of the dispersions were measured with a Zetasizernano S90 and Zetasizernano 2000 respectively.
The mechanical properties were examined with a Zwick static materials testing machine, according to ISO 37 standards. The magnetic properties of the samples were measured with a vibrating sample magnetometer VSM LakeShore 7410, with a the field of 960 kA/m. The magnetorheological properties were studied with an Ares Rheometer (plate-plate system, plate diameter, 20 mm; gap 2 mm; magnetic field range, 0-300 mT). The cross-linking density of the vulcanisates was calculated based on rapid solvent-swelling measurements (in toluene and in toluene with ammonium vapours) using the Flory-Rehner equation. Microstructural observations were made using scanning electron microscopy (SEM). A heat ageing process was performed with an FD series dryer (Binder) according to PN-82/C-04216 standards. An ultraviolet ageing process was performed with UV2000 equipment (Atlas).

3. Influence of ionic liquids on the characteristic of magnetoactive fillers

3.1 Aggregate size of magnetoactive fillers

The particle size of ferromagnetic particles is a main parameter that has a great influence on the properties of magnetorheological composites. A reduction in the particle size results in an increase in the specific surface area of magnetoactive particles, providing better contact between the crosslinking agent particles and the elastomer chains. Moreover, the morphology of the filler particles determines the size of the interphase between the magnetoactice fillers and the elastomer. The sizes of the nanometric and micrometric Fe_3O_4 and carbonyl iron powder aggregates in water and in paraffin oil with and without the addition of ionic liquids are presented in Tab. 1-3. The morphologies of the magnetorheological elastomer composites are presented in Section 4.4 "Scanning Electron Microscopy Images".
The sizes of the aggregates of the fillers were measured in water and in paraffin oil. Elastomers have a hydrophobic nature; therefore, the size of the magnetic particles was measured in a liquid hydrophobic medium-paraffin oil, which was chosen as the model for an elastomer matrix. The aim of this study was to estimate the tendency of particles to agglomerate in the elastomer. In all cases, the magnetoactive fillers particles exhibited a higher ability for aggregation or agglomeration in paraffin oil than in water. As expected nanometric Fe_3O_4 presented the smallest aggregate size (695 nm) measured in water in contrast to carbonyl iron powder, which showed the greatest aggregate size (3151 nm). Nearly the same situation was observed in paraffin oil. According to the assumptions of the work on the stabilisation of particles, ionic liquids were added together with ferroparticles to the water and oil dispersions, which led to decrease of the aggregate size no matter what type of the filler was used, *e.g.*, micrometric magnetite aggregates in water decreased from 695 nm (without IL) to 213 nm (with IL), nanometric magnetite aggregates from 1188 nm to 632 nm and CIP from 3151 nm to 429 nm. Magnetic filler aggregates size measured in paraffin oil were even more spectacular *e.g.*, for CIP, it decreased from 11210 nm to 3283 nm.

Ionic liquid	Aggregate size in water [nm]	
	Micrometric Fe_3O_4	Nanometric Fe_3O_4
-------	1188	695
BMIMBF$_4$	940	812
BMIM OTf	1035	625
BMIMCl	632	552
EMIMDEP	790	543
HMIMCl	799	606
HPCl	690	213

Table 1. The micro and nanometric Fe_3O_4 aggregate sizes measured in water containing ionic liquids.

Ionic liquid	Aggregates size in paraffine oil [nm]	
	Micrometric Fe_3O_4	Nanometric Fe_3O_4
-------	4411	4916
BMIMBF$_4$	1141	2247
BMIM OTf	5061	4249
BMIMCl	4080	2678
EMIMDEP	5438	5007
HMIMCl	2622	2453
HPCl	6744	7340

Table 2. The micro and nanometric Fe_3O_4 aggregate sizes measured in paraffin oil containing ionic liquids.

Ionic liquid	Aggregates size in water [nm]	Aggregates size in paraffine oil [nm]
	Carbonyl Iron Powder	
----------	3151	11210
BMIMBF$_4$	1106	3283
BMIM OTf	4122	4907
BMIMCl	2695	10100
EMIMDEP	2801	3247
HMIMCl	3031	3432
HPCl	429	9615

Table 3. The carbonyl iron powder aggregate size measured in paraffin oil and water containing ionic liquids.

3.2 The electrokinetic potential (ζ - zeta) for water magnetoactive fillers suspensions

The zeta potential is an electrokinetic potential measured on the surface of a particle in solution. A charged surface results in the formation of an electric double layer, and the zeta potential is the potential between the charged surface and the electrolyte solution[31]. The zeta potentials of water-magnetoactive filler (nanometric Fe_3O_4, micrometric Fe_3O_4, carbonyl iron powder) suspensions containing ionic liquids were examined versus pH (Fig. 1-3). From the zeta potential measurements, the isoelectric points (IEP) of water dispersions were determined. The IEP is the pH value at which the zeta potential is zero. It is not a description of the absolute basicity or acidity of solid surface but a description of their relative strength. A high IEP value indicates that the surface shows more basic functionality compared with its acidic functionality. On the contrary, a low IEP surface reveals less basic functionality compared with its acidic functionality[32]. IEP is mainly related to the Brönsted acid and base definitions. The higher IEP corresponds to stronger affinity to protons. However, IEP is also a measure of the Lewis acidity or basicity because, when the oxide surface adsorbs a proton, the electron (or electron density) is transferred from the oxygen to the proton. Therefore, a strong proton acceptor is also a strong electron donor[32].

The isoelectric point (IEP) of micrometric iron oxide (Fe_3O_4) was determined at pH 4.5, of nanometric iron oxide (Fe_3O_4) at pH 5.0 and of CIP at pH 3.8, whereas the IEP for ionic liquids and micrometric iron oxide (Fe_3O_4) dispersions was determined within a pH range from 4.0 (for BMIMBF4) to 5.7 (for BMIMCl), and, for HPCl, there was no IEP. The IEP for ionic liquids and nanometric iron oxide (Fe_3O_4) dispersions was determined within the pH range from 5.0 (for EMIMDEP) to 6.0 (for HMIMCl and BMIMCl), and, for HPCl, there was no IEP. The IEP for ionic liquids and carbonyl iron powder dispersions was determined within the pH range from 3.2 (for $BMIMBF_4$) to 4.1 (for HMIMCl) and, at pH 8.3 for BMIMCl, and, for HPCl, there was no IEP. Thus, the properties of the surfaces of ionic liquids containing magnetic filler were determined to be acidic (in almost all dispersions), similar to pure micrometric and nanometric iron oxide and CIP (the isoelectric points are in the range of pH from 3.8 to 5.0; a negative zeta potential was determined over almost the whole range of measured pH), and the properties of micro- and nano- sized magnetite and CIP containing HPCl were basic (a positive zeta potential at the whole pH range).

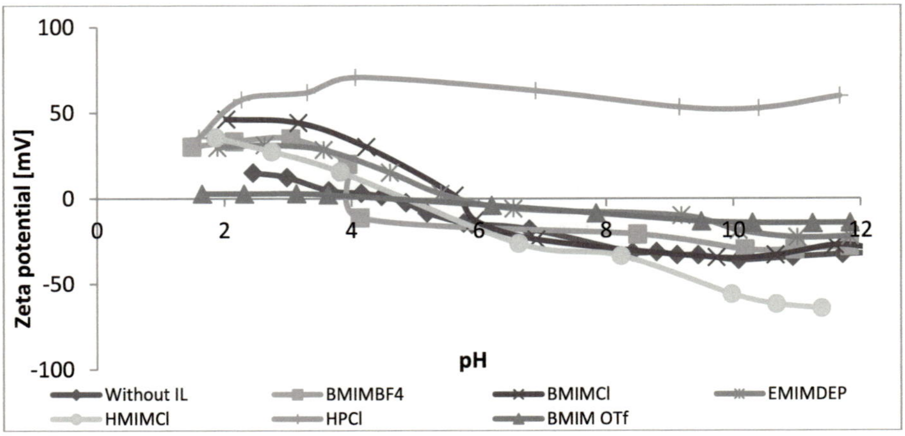

Fig. 1. The zeta potential of micrometric iron oxide (Fe_3O_4) dispersions containing ionic liquids.

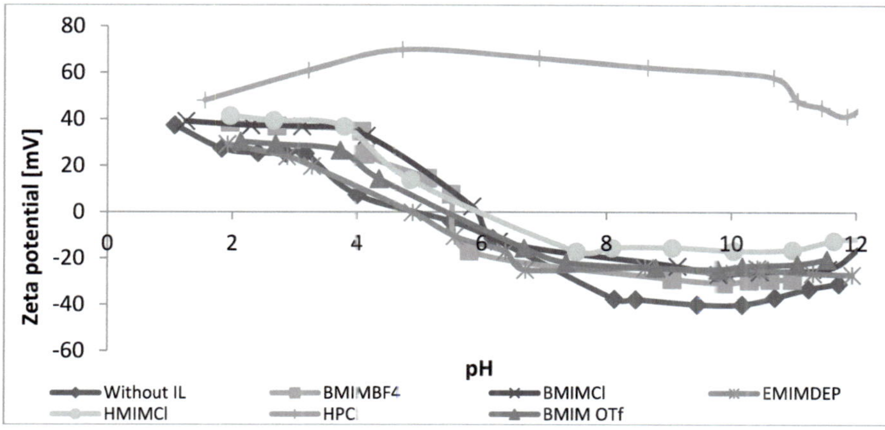

Fig. 2. The zeta potential of nanometric iron oxide (Fe_3O_4) dispersions containing ionic liquids.

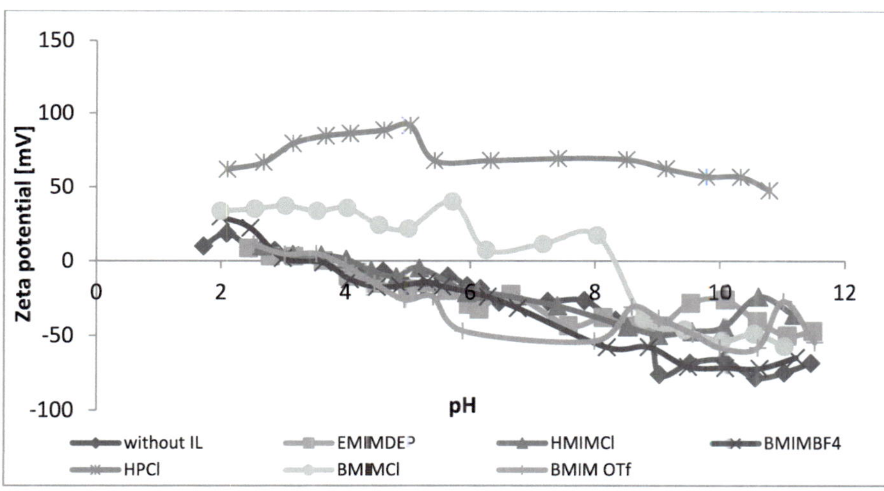

Fig. 3. The zeta potential of carbonyl iron powder (CIP) dispersions containing ionic liquids.

4. Specific properties of MRE containing ionic liquids

4.1 Mechanical properties, cross-linking density and ageing processes

The influence of ionic liquids on the ability of carbonyl iron powder particles in the crosslinking of EPM and NBR elastomers was estimated based on the tensile properties and crosslink density of the vulcanisates. The results are given in Tables 4 and 5. The application of ionic liquids into CIP particles increased the tensile strength of the composites significantly (especially in NBR vulcanisates) compared to those produced without ionic liquids as dispersing agents. Moreover, the stress at a relative elongation of 100% increased, and the elongation at break decreased, due to an increase in the crosslink density of

vulcanisates (Tab. 4 and 5). To increase the tensile strength of composites, the most effective ionic liquid appeared to be EMIMDEP apart from the type of used elastomer.
The crosslink density was determined by the equilibrium swelling of the vulcanisates in toluene. The influence of the addition of different ionic liquid on the crosslink density of NBR and EPM vulcanisates was analysed. Applying ILs as dispersing agents of carbonyl iron powder magnetic particles had a detrimental effect on the crosslink density. The calculated values confirmed that almost all of the applied ionic liquids increased in cross-linking density, and the most effective dispersing agent was HPCl, which did not depend on what type of vulcanisate was considered (EPM/NBR).
To determine the influence of ionic liquids on the mechanical properties of the magnetorheological elastomer composites, another inference occurred. Generally metals with variable oxidation state can influence the mechanical properties of materials after ageing. However, the conditions that were applied herein for ageing (thermal or UV), maintained the values of tensile strength and elongation at break of the NBR vulcanisates. Tables 6 and 7 show that, ageing factors measured for samples containing ionic liquids with micro or nanosized Fe_3O_4 were closer to unity than those of vulcanisates without dispersing agents. The application of selected ionic liquids in the elastomer matrix could effectively protect vulcanisates from thermal and UV ageing processes.

Ionic liquid	v_e*10^{-5} [mol/cm^3]	SE_{100} [MPa]	T.S [MPa]	E.B [%]
-	5,37	1,01	2,43	537
$BMIMBF_4$	6,11	1,16	2,31	384
BMIM OTf	6,93	1,55	3,14	233
BMIMCl	6,00	1,03	3,2	455
EMIMDEP	6,15	1,14	3,47	434
HMIMCl	6,48	1,12	3,08	321
HPCl	6,90	1,28	2,86	326

Table 4. Cross-link density (v_e), tensile strength - T.S, elongation at break - E.B and strength at elongation 100% - SE_{100} of NBR vulcanisates containing carbonyl iron powder and ionic liquids.

Ionic liquid	v_e*10^{-5} [mol/cm^3]	SE_{100} [MPa]	T.S [MPa]	E.B [%]
-	5,53	0,72	2,01	632
$BMIMBF_4$	6,48	0,75	1,96	558
BMIM OTf	3,56	0,74	2,08	866
BMIMCl	5,93	0,81	2,18	570
EMIMDEP	3,97	0,71	2,74	820
HMIMCl	4,09	0,72	2,63	738
HPCl	7,54	0,89	1,69	557

Table 5. Cross-link density (v_e), tensile strength - T.S, elongation at break - E.B and strength at elongation 100% - SE_{100} of EPM vulcanisates containing carbonyl iron powder and ionic liquids.

Ionic liquid	Heat ageing factors		UV ageing factors	
	$T.S_{ag}/T.S$	$E.B_{ag}/E.B$	$T.S_{ag}/T.S$	$E.B_{ag}/E.B$
-	0,78	0,82	0,73	0,83
$BMIMBF_4$	0,87	0,93	0,83	0,86
BMIM OTf	0,80	1,02	0,64	0,87
BMIMCl	0,89	1,00	0,87	1,00
EMIMDEP	0,89	1,07	0,74	0,85
HMIMCl	0,52	0,87	0,59	0,91
HPCl	0,95	1,10	0,88	0,94

Table 6. Ageing factors (T.S – tensile strength, E.B – elongation at break, $T.S_{ag}$ – tensile strength measured after ageing, and $E.B_{ag}$ – elongation at break measured after ageing) of NBR vulcanisates containing micro-sized Fe_3O_4 and ionic liquids.

Ionic liquid	Heat ageing factors		UV ageing factors	
	$T.S_{ag}/T.S$	$E.B_{ag}/E.B$	$T.S_{ag}/T.S$	$E.B_{ag}/E.B$
-	0,78	0,82	0,73	0,83
$BMIMBF_4$	0,87	0,93	0,83	0,86
BMIM OTf	0,80	1,02	0,64	0,87
BMIMCl	0,89	1,00	0,87	1,00
EMIMDEP	0,89	1,07	0,74	0,85
HMIMCl	0,52	0,87	0,59	0,91
HPCl	0,95	1,10	0,88	0,94

Table 7. Ageing factors (T.S – tensile strength, E.B – elongation at break, $T.S_{ag}$ – tensile strength measured after ageing, and $E.B_{ag}$ – elongation at break measured after ageing) of NBR vulcanisates containing nano-sized Fe_3O_4 and ionic liquids.

4.2 Magnetic properties

The orientation of particles and their arrangement were investigated by VSM. Studies of the magnetic properties of the MREs were conducted parallel to the sample long axis, corresponding to the magnetic field direction during curing. Four types of composites were investigated: NBR/microsized magnetite, EPM/micro-sized magnetite, NBR/CIP and EPM/CIP. Significant differences between the characteristic magnetic values recorded for vulcanisates containing different magnetoactive fillers can be seen in Tables 8-11. Higher values of coercivity and retentivity and lower values of magnetisation are characteristic of composites containing micrometre Fe_3O_4 in comparison to NBR/CIP and EPM/CIP compositions, which is certainly due to the specific magnetic properties of the different filler. Carbonyl iron powder presented a higher magnetisation saturation than micrometric magnetite. Coercivity values obtained for the MREs with micrometre Fe_3O_4 and CIP particles varying from 15 to 140 G indicate magnetically soft materials.

Additionally, to improve the dispersion of the applied filler properties, ionic liquids were added during preparation. No matter what kind of IL was admixed, all composites demostrated magnetic properties regardless of the type of elastomer matrix, applied filler or ionic liquid. Moreover, an increase of magnetisation values was observed for magnetorheological NBR composites with micro-sized magnetite containing ionic liquids compared to vulcanisates where no ionic liquid was applied.

Ionic liquid	Coercivity (Hci) [G]	Magnetisation (Ms) [emu/g]	Retentivity (Mr) [emu/g]
----------	140,45	25,392	3,8768
$BMIMBF_4$	137,71	27,179	3,4498
BMIM OTf	138,65	26,792	3,1009
BMIMCl	139,89	26,159	2,8098
EMIMDEP	140,13	28,336	3,2073
HMIMCl	135,39	27,509	2,798
HPCl	136,23	27,514	2,9412

Table 8. Magnetic properties of NBR vulcanisates containing micro- sized magnetite (60 phr) and ionic liquids.

Ionic liquid	Coercivity (Hci) [G]	Magnetisation (Ms) [emu/g]	Retentivity (Mr) [emu/g]
----------	144,64	29,132	4,38
$BMIMBF_4$	145,06	29,160	4,294
BMIM OTf	140,44	29,054	4,5374
BMIMCl	143,18	28,715	4,4912
EMIMDEP	144,25	28,313	4,327
HMIMCl	142,92	28,755	4,4267
HPCl	144,14	28,535	4,1544

Table 9. Magnetic properties of EPM vulcanizates containing micro- sized magnetite (60 phr) and ionic liquids.

Ionic liquid	Coercivity (Hci) [G]	Magnetisation (Ms) [emu/g]	Retentivity (Mr) [emu/g]
---------	22,06	73,27	0,64
$BMIMBF_4$	23,07	70,16	0,54
BMIM OTf	23,23	69,86	0,46
BMIMCl	22,92	70,18	0,51
EMIMDEP	22,51	70,16	0,53
HMIMCl	22,60	76,64	0,41
HPCl	22,47	70,28	0,47

Table 10. Magnetic properties of NBR vulcanisates containing carbonyl iron powder (60 phr) and ionic liquids.

Ionic liquid	Coercivity (Hci) [G]	Magnetisation (Ms) [emu/g]	Retentivity (Mr) [emu/g]
--------	22,11	75,27	0,75
$BMIMBF_4$	15,77	72,81	0,56
BMIM OTf	22,45	72,87	0,71
BMIMCl	22,09	73,76	0,71
EMIMDEP	22,81	74,86	0,74
HMIMCl	22,66	72,43	0,72
HPCl	21,84	74,51	0,74

Table 11. Magnetic properties of EPM vulcanisates containing carbonyl iron powder (60 phr) and ionic liquids.

4.3 Magnetorheological properties

MR elastomers are normally operated with small deformations in the pre-yield regime of the linear viscoelastic region. MREs are intended to be used as structural materials in applications where the load is often of a dynamic type. In cyclic dynamic loading, the material deforms and returns to its original form over one cycle. The oscillating force is varied periodically, usually with a sinusoidal amplitude at the angular frequency x. In viscoelastic materials, some of the deformation energy input is stored and recovered during each cycle, and some is dissipated as heat. The storage modulus G′ represents the ability of the viscoelastic material to store the energy of deformation, which contributes to the material stiffness. The loss modulus G″ represents the ability of the material to dissipate the energy of deformation. The elastic (storage) modulus G′, loss modulus G″ and loss tangent (tanδ) of NBR composites containing different types of magnetic fillers were measured with an ARES rheometer (TA Instruments) as a function of angular frequency x under different magnetic fields (Fig. 4-9). Experiments were conducted at 25 °C on samples with various particle contents. Thin slices were cut to a thickness of 2 mm and were 20 mm in diameter. The slope of the particle chains was equal to 45°.

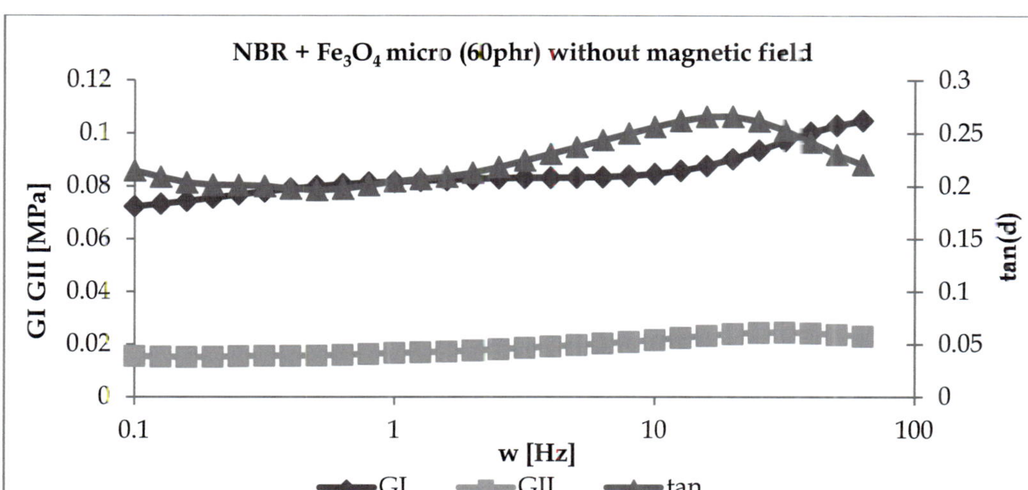

Fig. 4. Dependence of the elastic (storage) modulus (G′ = GI), loss modulus (G″ = GII) and loss tangent (tanδ) of MRE samples containing micro Fe_3O_4 on the strain frequency.

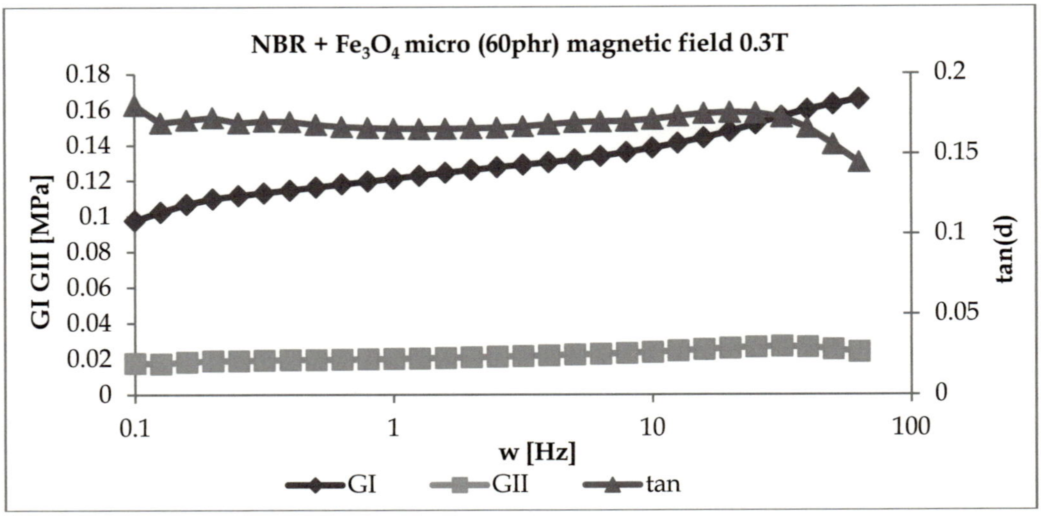

Fig. 5. Dependence of the elastic (storage) modulus (G′ = GI), loss modulus (G′′ = GII) and loss tangent (tanδ) of MRE samples containing micro Fe_3O_4 on the strain frequency.

The elastic modulus increased with the strain frequency for all filler types. In all cases, the application of an external magnetic field led to a significant increase in the elastic modulus (G′) and loss modulus (G′′). Higher initial values of the storage modulus for composites containing micrometre-sized Fe_3O_4 (Fig. 5) than nanoometre-sized Fe_3O_4 or CIP (Fig. 7 and 9) show the appropriate alignment in the elastomer matrix and larger magnetic susceptibility.

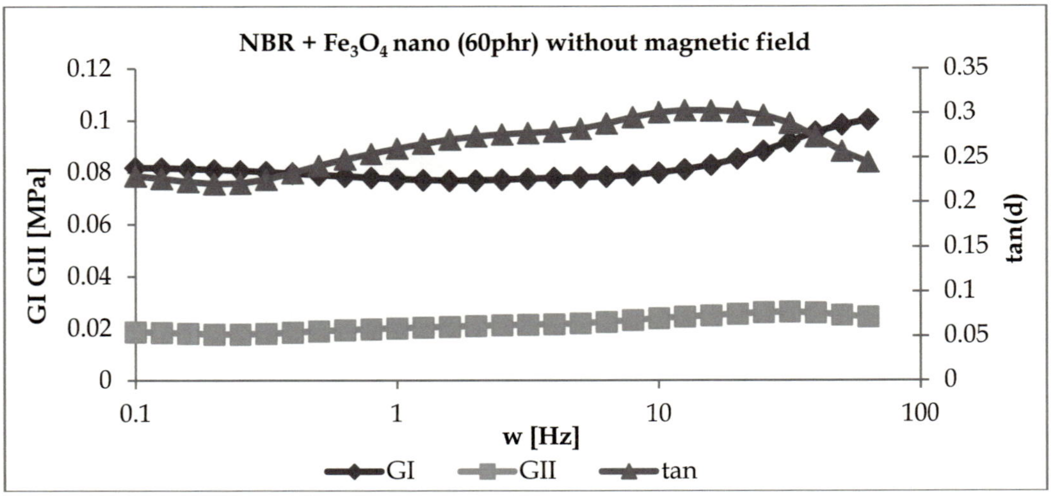

Fig. 6. Dependence of the elastic (storage) modulus (G′ = GI), loss modulus (G′′ = GII) and loss tangent (tanδ) of MRE samples containing nano Fe_3O_4 on the strain frequency.

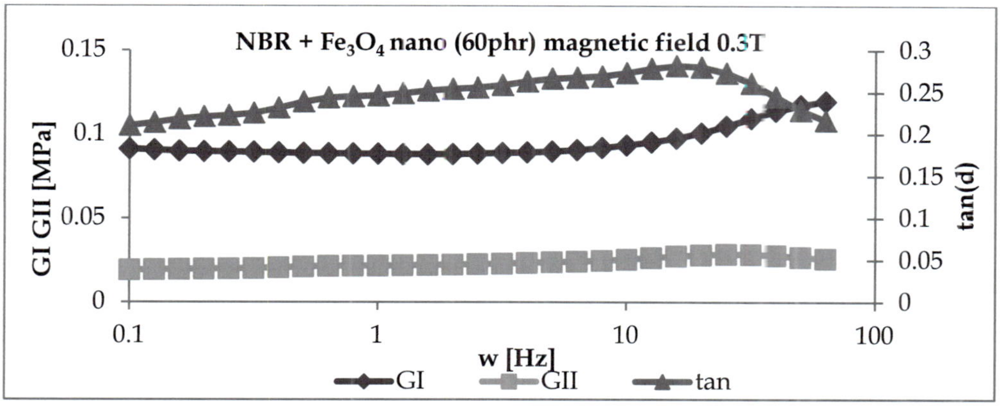

Fig. 7. Dependence of the elastic (storage) modulus (G′ = GI), loss modulus (G′′ = GII) and loss tangent (tanδ) of MRE samples containing nano Fe_3O_4 on the strain frequency.

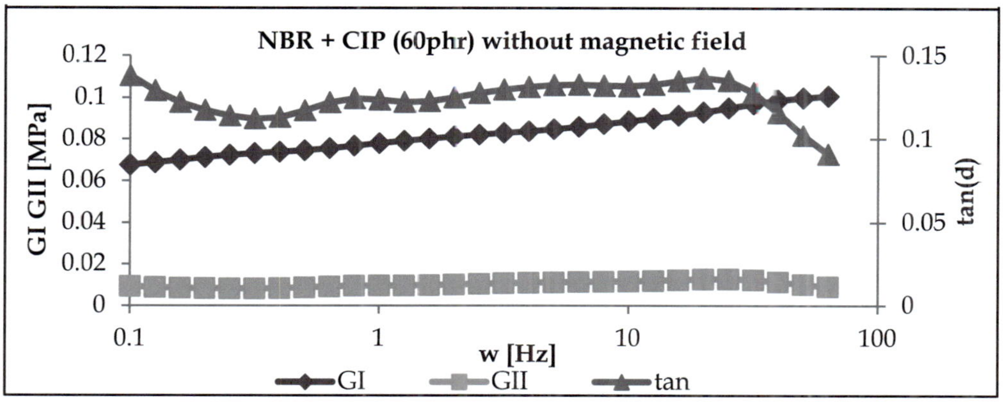

Fig. 8. Dependence of the elastic (storage) modulus (G′ = GI), loss modulus (G′′ = GII) and loss tangent (tanδ) of MRE samples containing CIP on the strain frequency.

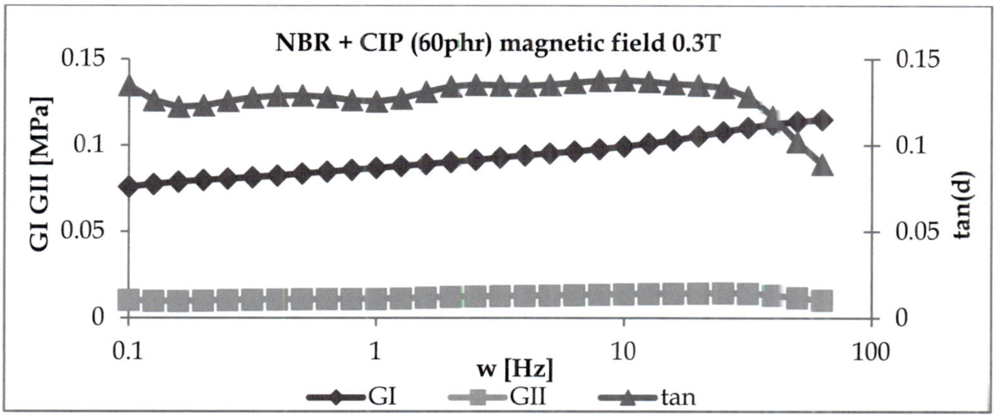

Fig. 9. Dependence of the elastic (storage) modulus (G′ = GI), loss modulus (G′′ = GII) and loss tangent (tanδ) of MRE samples containing CIP on the strain frequency.

4.4 Scanning Electron Microscopy Images – dispersion of magnetoactive filler particles in elastomer matrixes

The dispersion of magnetoactive filler particles in the elastomer has a significant effect on the activity of micrometre and nanometre iron oxides and carbonyl iron powder and their influence on vulcanisate properties. The dispersion of particles in the elastomers (EPM and NBR) was estimated based on the SEM images of the vulcanisate surfaces (Fig. 10–19). Micro-sized Fe_3O_4 (Fig. 10 and 12), nano-sized Fe_3O_4 (Fig. 12) and carbonyl iron powder (Fig. 16 and 18) particles were poorly dispersed in the elastomer matrix (EPM and NBR) and therefore were not homogeneously distributed. They created clusters of particles (agglomerates) with complex structures. The tendency of ferromagnetic particles to agglomerate in the elastomer matrix was confirmed by particle size measurement in water and paraffin oil (model of elastomer matrix), as presented in Section 3.1.

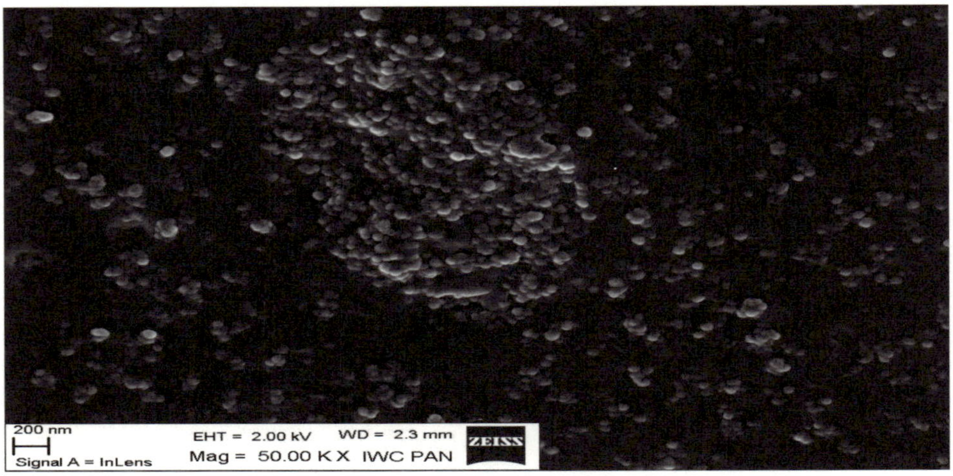

Fig. 10. SEM image of NBR vulcanisate filled with micro-sized 60 phr Fe_3O_4.

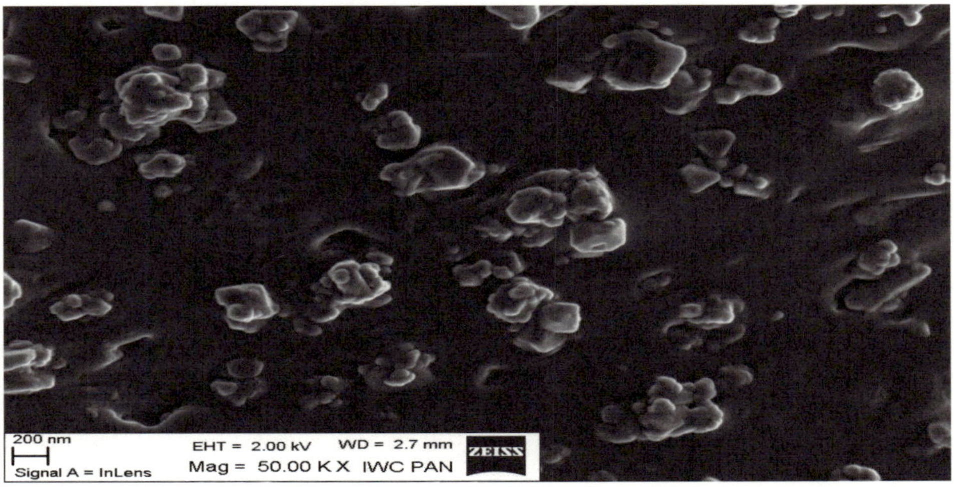

Fig. 11. SEM image of NBR vulcanisate filled with micro-sized 60 phr Fe_3O_4 and 1-butyl-3-methylimidazolium trifluormethanesulphonate.

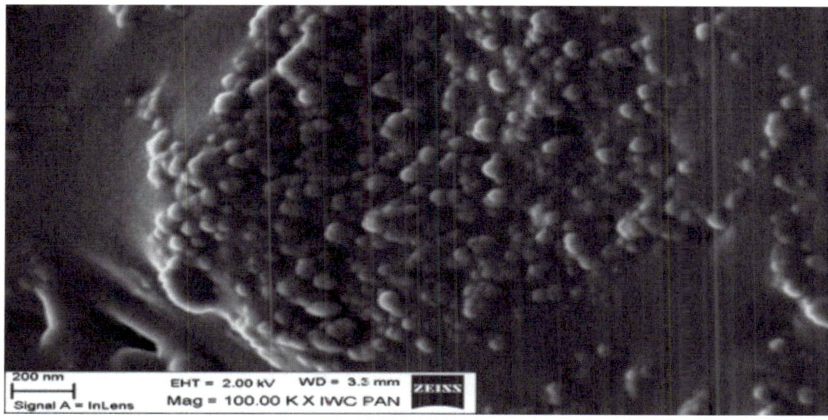

Fig. 12. SEM image of NBR vulcanisate filled with nano-sized 60 phr Fe_3O_4.

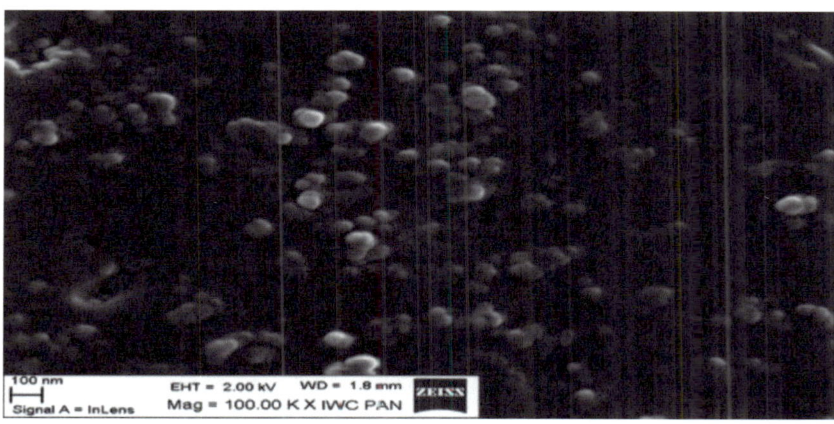

Fig. 13. SEM image of NBR vulcanisate filled with nano-sized 60 phr Fe_3O_4 and 1-butyl-3-methylimidazolium tetrafluoroborate.

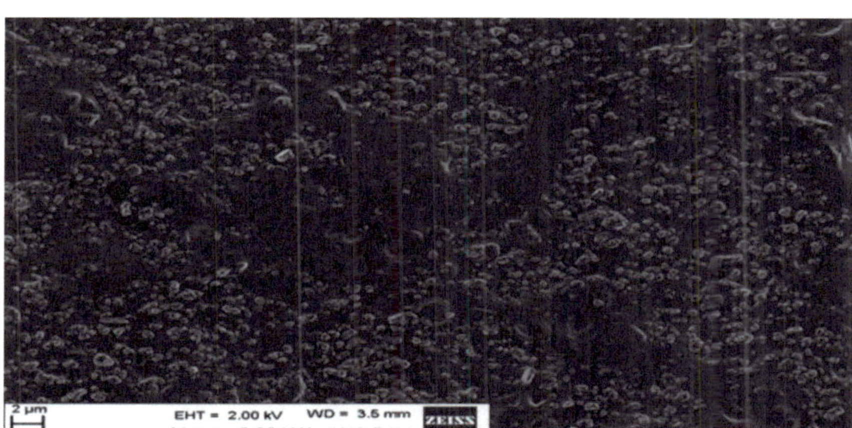

Fig. 14. SEM image of EPM vulcanisate filled with micro-sized 60 phr Fe_3O_4.

Fig. 15. SEM image of EPM vulcanisate filled with micro-sized 60 phr Fe_3O_4 and 1-butyl-3-methylimidazolium chloride.

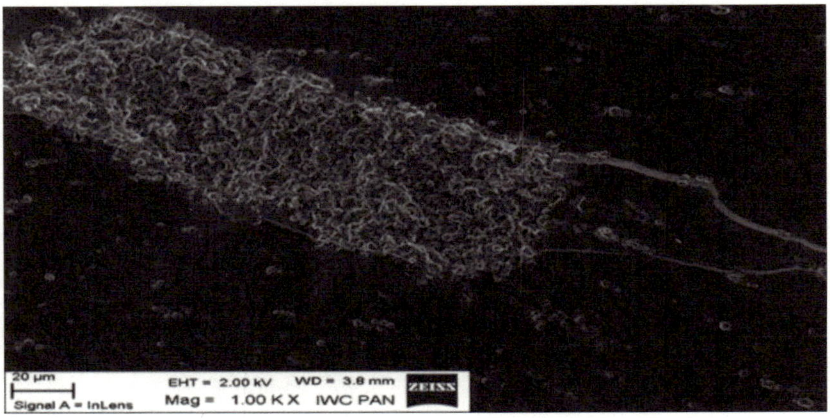

Fig. 16. SEM image of EPM vulcanisate filled with 60 phr CIP.

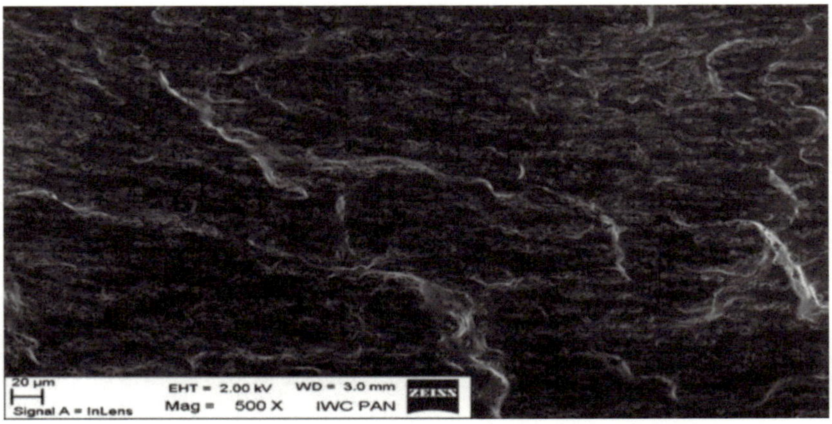

Fig. 17. SEM image of EPM vulcanisate filled with 60 phr CIP and 1-butyl-3-methylimidazolium tetrafluoroborate.

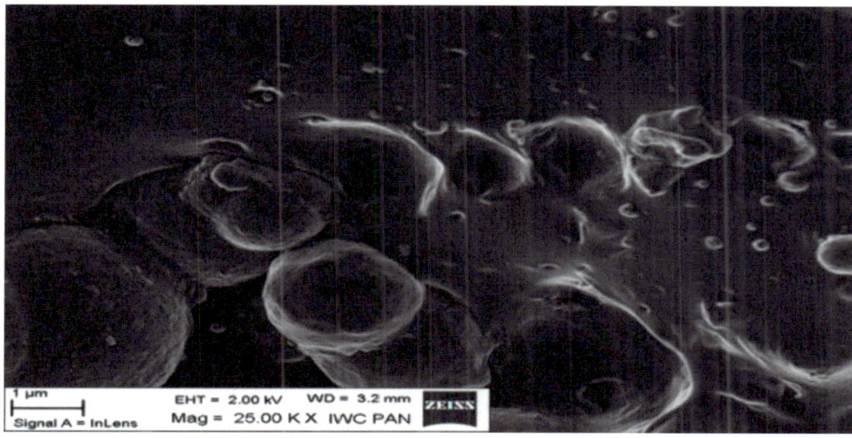

Fig. 18. SEM image of NBR vulcanisate filled with 60 phr CIP.

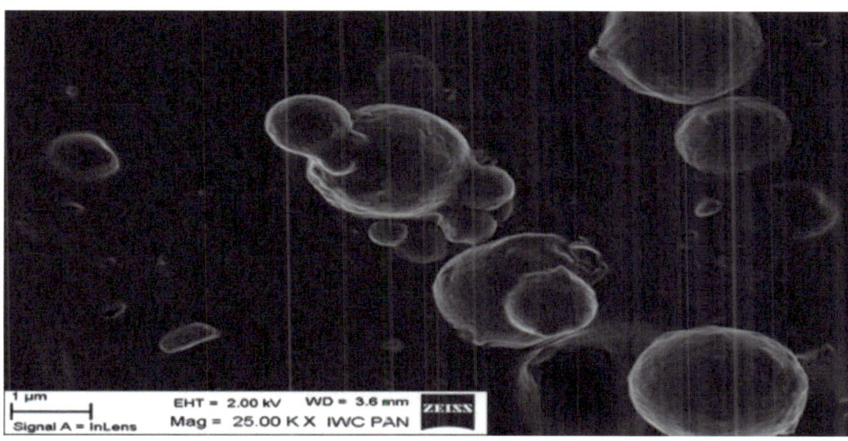

Fig. 19. SEM image of NBR vulcanisate filled with 60 phr CIP and trihexyltetradecylphosphonium chloride.

The addition of selected ionic liquids improved the dispersion of applied magnetic particle fillers for all types of elastomers (Fig. 11, 13, 15, 17, 19). The influence of ionic liquids as dispersing agents in magnetorheological composites included higher tensile strength and crosslink density in these vulcanisates, which was presented in Section 4.1.

Magnetorheological materials can be fluids, gels or even solids such as an elastomer. Magnetorheological materials have magnetically polarisable colloidal particles suspended in some functional suspension, *i.e.*, a viscous fluid (*e.g.*, silicone oil) or an elastomer matrix (*e.g.*, silicone rubber). A magnetorheological fluid operates on the principle that the magnetic particles are randomly distributed in the liquid when no magnetic field is applied, but the particles acquire a magnetic polarisation and form chains in the presence of a magnetic field of sufficient strength. If the elastomer, with suspended ferromagnetic particles, is cured in the presence of a magnetic field, the magnetisable particles will form chains along the direction of the magnetic field prior to the elastomer cross linking process (curing), and a MRE is produced (Fig. 16-19). However, if the mixture is not cured in the presence of a magnetic field and the particles are hence left randomly distributed, an elastomer ferromagnet composite (EFC) is produced (Fig. (10-15).

5. Conclusion

MRE composites were prepared according to commonly used methods. Micro- and nano-sized iron oxides and carbonyl iron powder were the active fillers of acrylonitrile – butadiene and ethylene – propylene rubber. It was found that micro- and nano- sized iron oxides and carbonyl iron powder improved mechanical properties of elastomers. They also changed their magnetic properties and reinforced the magnetorheological effects of the composites.
The dispersion of nano- and micro- sized Fe3O4 or CIP in the elastomer matrix can be improved by using ionic liquids as dispersing agents. The addition of ionic liquids decreased the aggregates size independently of the type of magnetic filler used. Scanning electron microscopy images used to observe magnetorheological elastomer microstructures proved that the improvement in dispersion had a significant effect on the composites properties. The application of ionic liquids with magnetoactive particles increased the tensile strength of the composites significantly. Moreover, the stress at a relative elongation of 100% increased, and the elongation at break decreased, due to an increase in the crosslink density of the vulcanisates. Finally, the application of selected ionic liquids in the elastomer matrix could effectively protect the vulcanisates from thermal and UV aging processes.

6. Acknowledgment

These studies were financially supported with the project no. N N507 514838 funded by the Polish Ministry of Science and High Education.

7. References

[1] Welton, T. (1999). Room-temperature ionic liquids. Solvents for synthesis and catalysis. *Chem. Rev.*, vol. 99, pp. 2071–2083.

[2] Dupont J, de Souza RF, Suarez PA. (2002). Ionic liquid (molten salt) phase organometallic catalysis. *Chem. Rev.*, vol. 102, pp. 3667–92.

[3] Zhang H, Hong K, Mays JW. (2002). Synthesis of block copolymers of styrene and methyl methacrylate by conventional free radical polymerization in room temperature ionic liquids. *Macromolecules* vol. 35, pp. 5738–41.

[4] Perrier S, Davis TP, Carmichael AJ, Haddleton DM. (2002). First report of reversible addition-fragmentation chain transfer (RAFT) polymerization in room temperature ionic liquids. *Chem Commun.*, vol. 19, pp. 2226–7.

[5] Chiefari J, Chong YK, Ercole F, Krstina J, Jeffery L, Mayadunne RTA. (1998). Living free-radical polymerization by reversible additionfragmentation chain transfer: the RAFT process. *Macromolecules.*, vol. 31, pp. 5559–62.

[6] Kubisa P. (2004). Application of ionic liquids as solvents for polymerization processes. *Prog. Polym. Sci.*, vol. 29, pp. 3–12.

[7] Shen Y, Ding S. (2004). Catalyst separation in atom transfer radical polymerization. *Prog. Polym. Sci.*, vol. 29, pp. 1053–78.

[8] Wasserscheid, P. (2006). Volatile times for ionic liquids. *Nature.*, vol. *439*, pp. 797-797.

[9] Wasserscheid P., Welton, T. (2002). *Ionic Liquids in Synthesis*; Wiley-VCH: Weinheim, Germany.

[10] Rogers RD., Seddon KR. (2003). Ionic Liquids--Solvents of the Future? *Science*, vol 302, pp. 792.

[11] Laas HJ., Halpaap R., Richter F., Köcher J. , Ionic liquids (as catalysts for the oligomerization of isocyanates), Germ. Pat. Appl., DE 102 19 227, 30.4.2002, Bayer AG

[12] Oelmüller U., Wille T., Use of compositions consisting of cationic compounds and proton donors for stabilizing and/or isolating nucleic acids in or from micro-organisms such as prokaryots, fungi, protozoa or algae, PCT. Pat. Appl., WO 02/00600, 20.6.2001, Quiagen GmbH.

[13] Myerson J., Perobost MGM., Dellinger DJ., Dellinger GF., *Method of synthesizing polynucleotides using ionic liquids, US Pat. Appl.*, US 03/0083489, 31.10.2001, Agilsent Technologies, Inc.

[14] Reich RA., Stewart PA., Bohaychick J., Urbanski JA. (2003). Base oil properties of ionic liquid *Lubr. Eng.* Vol. 59, pp. 16.

[15] Zhou Y., Antonietti M. (2003). Communication Synthesis of Very Small TiO_2 Nanocrystals in a Room-Temperature Ionic Liquid and Their Self-Assembly toward Mesoporous Spherical Aggregates. *J. Am. Chem. Soc.*, vol. 125, pp. 14960

[16] Schmidt FG., Petrat FM., Pawlik A., Haeger H., Weyershausen B., Polymeric compositions containing ionic liquids as plasticizers, PCT Int. Appl., WO04/005391, 15.1.2004, Creavis GmbH.

[17] Price KN., Hartshorn RT., Rohrbaugh RH., Scheper WM., Showell MS., Baker KH., Sivik MR., Scheibel JJ., Gardner RR., Reddy PK., Aiken III JD,. Addison MC., *Ionic liquid based products and method of using the same, PCT Int. Appl.*, WO04/003120, 8.1.2004, The Procter & Gamble Company.

[18] Kirchmeyer S., König K., Mazanek J., Wilmes O., Polyaddition products for dispersing pigments in water-based paint, Eur. Pat. Appl., EP0731148, 11.9.1996, Bayer AG; Merten G., Zöller J., Urbano E., Hydrophilic polyurethane-polyureas and their use as a dispersant for synthetic resins, Germ. Pat. Appl., DE4416336, Hoechst AG, H. L. Jakubauskas, Acrylic polymer dispersant for aqueous acrylic coating compositions, US Pat. Appl., US3980602, 14.9.1976, Du Pont; I.-C. Chu, Fryd M., Lynch LE., Aqueous graft copolymer pigment dispersants, PCT Pat. Appl., WO94/21701, 29.9.1994, Du Pont: Denker DH., Pons DA., Wilhelmus HJ., Water-soluble and air-drying resin, PCT Pat. *Appl.*, WO94/18260, 18.8.1994, DSM; Quednau PH., Wulff WA., *Dispersant, Eur. Pat.*, EP0311157, 12.04.1989, assigned to EFKA Chemicals.

[19] Weyershausen B., Lehmann K., (2007). Industrial application of ionic liquids as performance additive, *Green Chem.* Vol. 7, pp. 15-19 DOI:10.1039/B411357H.

[20] Chaudhuri A., Wereley NM., Kotha S., Radhakrishnan R., Sudarshan TS., (2005) Viscometric characterization of cobalt nanoparticle-based magnetorheological fluids using genetic algorithms, *J. Magn. Magn. Mater.* Vol. 293, pp. 206.

[21] Winslow WN., (1949). Induced Fibration of Suspensions *J. Appl. Phys.* Vol. 20, pp. 1137.

[22] Odenbach S., (2002) in Magnetoviscous Effects in Ferrofluids, Lecture notes in Physics: Monograph 71, Springer, Berlin, Germany , Ch. 5.

[23] López-López MT., Zugaldia A., González-Caballero F., Durán JDG., (2006). Sedimentation and redispersion phenomena in iron-based magnetorheological fluids, *J. Rheol.* Vol. *50*, pp. 543.

[24] Häfeli UO., Pauer GJ., (1999). Lung Perfusion Imaging with Monosized Biodegradable Microspheres *J. Magn. Magn. Mater.* Vol. 194, pp. 76.

[25] Altin E, Gradl J., Peukert W., (2006). First Studies on the Rheological Behavior of Suspensions in Ionic Liquids *Chem. Eng. Technol.* Vol. 29, pp. 1347.

[26] Clavel G., Larionova J., Guari Y., Guérin C., (2006). Synthesis of Cyano-Bridged Magnetic Nanoparticles Using Room-Temperature Ionic Liquids, *Chem. Eur. J.* vol. 12, pp. 3798.

[27] Munoz BC, Jolly MR, (2001) Performance of Plastics. Carl Haser Verlag, Munich.

[28] Jolly MR., Carlson JD., Munoz BC., (1996). A model of the behaviour of magnetorheological materials" *Smart Mater. Struct.* vol. 5, pp. 607.

[29] Lokander M, Stenberg B, (2003) Improving the magnetorheological effect in isotropic magnetorheological rubber materials, *Polym Test* vol. 22, pp. 677.

[30] Farshad M, Benine A, (2004) Magnetoactive elastomer composites, *Polym. Test.* Vol. 23, pp. 347.

[31] Sun C., Berg J. C. (2003). A review of the different techniques for solid surface acid-base characterization. *Advances in Colloid and Interface Science*, vol. 105, pp. 151–175.

[32] Ohm R. F.: (1997). Rubber chemicals. in 'Kirk-othmer encyclopedia of chemical technology' (ed.: Howe-Grant M.) John Wiley and Sons, New York, Vol 21, 460–481.

Part 4

Nanomaterials and Metals

12

Concepts for the Stabilization of Metal Nanoparticles in Ionic Liquids

Alexander Kraynov and Thomas E. Müller
CAT Catalytic Center, ITMC, RWTH Aachen University
Germany

1. Introduction

Nanoparticles - assemblies of hundreds to thousands of atoms and a size in the range of 1-50 nm - can be considered at first approximation as a state of matter intermediate between single atoms or molecules and bulk bodies. While the properties of atoms and molecules can be described *via* quantum mechanics, the properties of bulk bodies are described by solid-state physics. Using quantum chemical models for describing nanoparticles is demanding because of the huge number of strongly interacting atoms, which have to be taken into account. On the other hand, methods of solid state physics cannot always be applied as nanoparticles often demonstrate size dependent quantum effects,[1] *e.g.*, in surface plasmon resonance.[2, 3] Magnetic,[4] thermodynamic,[5] catalytic [6] and other properties of nanoparticles can also depend on their size.

Due to their small size, nanoparticles have a very high specific surface area and dispersion, the latter being defined as the ratio of the number of surface atoms to the total number of atoms in the particle. Platinum particles with 2 nm diameter, *e.g.*, have a surface area of 140 m^2/g and a dispersion of 50 %. Note that also the relative concentration of atoms located at corners, edges, and faces is strongly size dependent (Fig. 1).[7] These surface atoms are not equivalent to each other and frequently play different roles in catalysis. In the following, we focus on metal nanoparticles,[8-10] as one can find good reviews about metal oxide [11-16] and semiconductor [17-23] nanoparticles elsewhere.

One of the most attractive features of unsupported metal nanoparticles is the possibility to apply physicochemical methods to investigate them and to catalyze chemical reactions by exactly the same material. Thus, unsupported metal nanoparticles are frequently used as a model system for conventional heterogeneous catalysts as many physicochemical methods cannot be applied directly to conventional catalysts due to their heterogeneous nature, interfering support effects, and low transparency for electromagnetic radiation. In other words, nanoparticles can serve as models for ideal surfaces and are good candidates to fill the so-called pressure material gap.[24, 25]

It is of great interest to study nanoparticles in ionic liquids, primarily for applications in electrochemistry and catalysis, as they display advantages in comparison to polymer stabilized nanoparticles. These advantages lie, *e.g.*, in the stabilization mechanism, the relative ease of adjusting the properties of the nanoparticle-ionic liquid system, as well as aspects associated with green chemistry. The applicability of nanoparticle suspensions depends on the strategy employed for stabilization (*vide infra*), which determines how strongly the nanoparticles are stabilized against agglomeration. In the following, we will focus on the use of ionic liquids as a unique medium and highly effective stabilizer. We will describe the concepts of electrostatic, steric, and diffusional stability of nanoparticles as well as discuss the different preparation strategies for metal nanoparticles in ionic liquids. Although we will describe the boundary conditions for the design of the nanoparticle-ionic liquid system for catalytic applications, we will not discuss the use of nanoparticles as a catalyst for specific reactions.

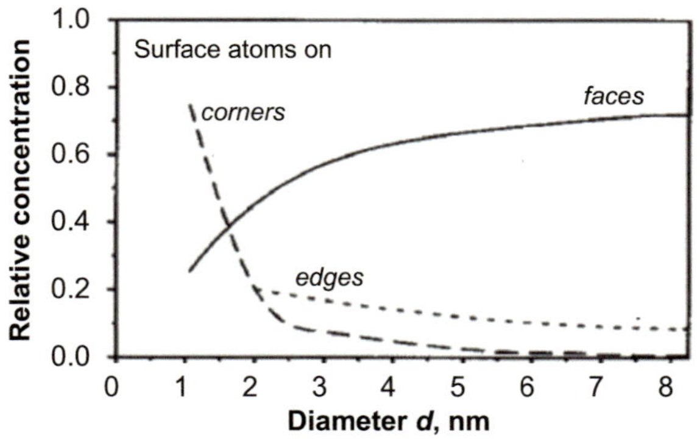

Fig. 1. The relative concentration of the different types of surface atoms n^S_i/n^S in dependence of the size of, here, a silver nanoparticle.[7]

2. Stabilization of metal nanoparticles

Dealing with phenomena associated with surfaces, it is useful to specify the term "clean surface". Atomically clean metal surfaces can be made and maintained only under ultra high vacuum, but do not exist in a chemical flask. As soon as the surface gets into contact with matter, it adsorbs it (as this process generally results in a decrease of the total energy of the interface), even when this material is an inert gas or an alkane. The adsorption mode (physi- or chemisorption) determines how strong the interaction is. Generally, we call adsorption to be strong, when the adsorption energy ΔG_{ads} is higher than ~ 30-40 kJ/mol. This implies that the probability that the sorbate desorbs from the surface at room temperature is very small (RT = 2.48 kJ/mol at 298 K). The residence time τ_a, the time passed between adsorption of a sorbate molecule and subsequent desorption from the surface, is an exponential function $\tau_a = \tau_0 e^{\Delta G_{ads}/RT}$ of the adsorption energy ΔG_{ads} and the reverse of the temperature T, where τ_0 is the characteristic time of surface atom vibrations (typically 10^{-12} 1/s).[26] The residence time quickly decreases with increasing temperature. At a given temperature, the residence time can be decreased, when the system (sorbate on surface) is exposed to an external influence, such as an external electric field, *e.g.*, induced by the presence of polar solvent molecules (*vide supra*). In further, we call a surface to be clean, when no molecules are strongly adsorbed, while it may be covered with weakly interacting molecules (molecules of dissolved gas, alkanes, *etc.*).
Nanoparticles have a particular tendency to lower their very high surface energy, which is the origin of their thermodynamic instability. Bare nanoparticles tend to stabilise themselves either by sorption of molecules from the surroundings or by lowering the surface area

through coagulation and agglomeration. In order to avoid the later, nanoparticles have to be (kinetically [26]) stabilized. The three conceptions of *electrostatic*, *steric*, and their combination *electrosteric* repulsive forces are generally discussed in literature and are summarized below. The concept of stabilising nanoparticles in viscous media, where the diffusion constants are low, will be introduced new.

2.1 Electrostatic stabilization

The origin of electrostatic stabilization is the repulsive electrostatic force, which nanoparticles experience, when they are surrounded by a double layer of electric charges. The Derjaguin, Landau, Verwey and Overbeek (DLVO) theory [27 28] considers initially charged colloidal particles whereby the electric charges are uniformly distributed over their surface. The total energy potential V_T of the interaction between two particles is then described as the sum of attractive (van der Waals) contributions and repulsive forces (due to a double layer of counter ions). The height of the overall potential barrier V_T determines, whether the particles are stable (the kinetic energy E_k of particle motion is less than V_T) or not ($E_k > V_T$). The average kinetic energy is defined through the quadratic mean velocity or temperature of the system according to formula (1).

$$E_k = \frac{m < V^2 >}{2} = \frac{3}{2}kT \tag{1}$$

One of the central parts of the DLVO theory is the quantitative prediction of particle coagulation, when electrolytes are added to the system.[29] The formation of virtual charges, however, is neglected. Later is important for conducting (metal) spheres in contrast to non-conducting (metal oxide) spheres. Further, the stability or instability of metal nanoparticles cannot be described based on electrostatic stabilization only.[30] To the best of our knowledge, there is no example in the literature, where colloidal metal nanoparticles are stabilized over an extended period of time only by charges induced by the presence of small ions. This instability is probably a consequence of the very high specific surface energy of metal surfaces (1000-2000 mJ/m^2) [31-35] in comparison with other organic and inorganic materials (~20 mJ/m^2 for teflon and 462 mJ/m^2 for silica) [36] as well as additional attractive dipole-dipole interactions between the metal nanoparticles.

Charge polarization of two interacting metal spheres as origin of virtual charges was included by Levine and Dube in 1940.[37] They postulated that the interaction energy results from the undistorted penetration of the double layers as well as induced dipole and quadrupole moments. Extending this description to metal nanoparticles requires exhaustive mathematical treatment even in the approximation of classic electrostatic theory, which is well beyond the scope of this chapter. Below, we will introduce a simple, well established model,[38, 39] where the redistribution of electron density within the metal nanoparticle is taken into account. The resulting electric multipoles determine the strength of the electrostatic interaction between the nanoparticles and, thus, their stability.

Let us consider now, how a neutral metal particle interacts with one or several approaching ions assuming that chemical interactions are absent and then investigate how this single particle interacts with another similar particle.

It is generally assumed that the approach of negatively charged anions (rather than positive cations) to a metal sphere induces a partial positive charge (δ^+) on the surface (Table 1 A).[40]

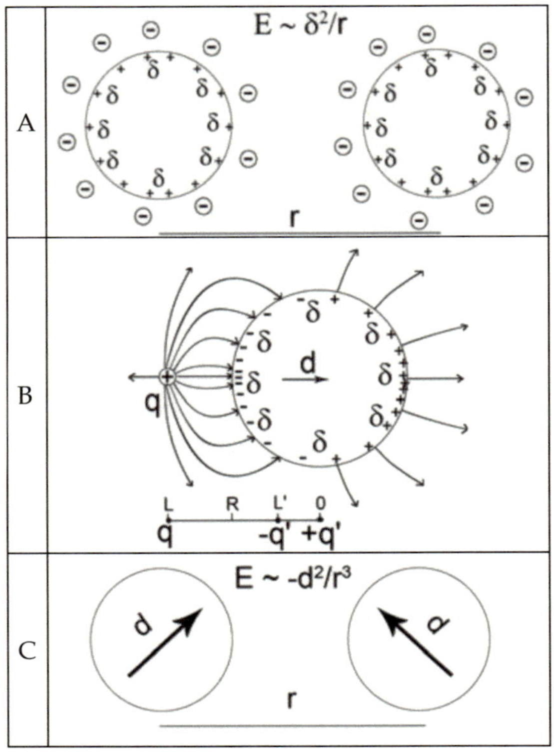

Table 1. *A* – Coulomb repulsion between partly charged particles as the origin of electrostatic stabilization. *B* – Distribution of the surface charge and geometry of the electric field of the neutral metal sphere of radius R , when the adsorbate is a single external charge q in the distance L from the centre. The electric potential of this system is equivalent to the superposition of the potential of an external point charge q and the induced dipole moment d. *C* – Attractive dipole-dipole interaction between electrically neutral metal particles.

Equally "charged" particles repel each other, which is the basis of general electrostatic stabilization.[30, 40] However, this description does not fully consider that the electron charge density on the metal sphere is redistributed. Let us assume that an external charge q approaches the surface of a neutral non-grounded metal sphere (Table 1 B). Note that this implies that the overall electric charge of this sphere is zero and remains so. Close to the approaching charge, an excess of surface charges (with opposite sign) accumulates, whereas excess charges of opposite sign appear on the other side of the sphere.[38, 39] Table 1 B illustrates schematically the induced surface charge and the geometry of the electric field. Based on classic electrostatic theory, it can be shown that the total electrostatic potential (and electric field) is a superposition of the potential of the external charge q and an induced dipole moment d (Eq. 2), where q' and $-q'$ are virtual charges located in the centre and at a distance L' from the centre of the sphere). The interaction energy (E) between the external charge q and the metal sphere is half of the sum of the interaction energy between the charge pairs q, $-q'$ and q, $+q'$ (Eq. 3). Since the second term is larger than the first, the total interaction energy is negative, thus, resulting in an attractive force.

$$d = -q'L' = qR^3 / L^2 \text{ , where } q' = -\frac{R}{L}q \text{ , } -q' = \frac{R}{L}q \text{ , } L' = \frac{R^2}{L} \tag{2}$$

$$E = \frac{qq'}{2}\left(\frac{1}{L} - \frac{1}{L-L'}\right) \tag{3}$$

It is important to note that the metal particle stays electrically neutral (rather than partly charged), carrying an electric dipole moment. The later is a function of the magnitude of the external charge q, its distance $(L-R)$ to the approaching particle and the radius R of the metal particle. Thus, the closer the external charge to the metal surface and the smaller the size of the particle is, the larger the dipole moment and, thus, the stronger the attractive dipole-dipole interaction between two particles becomes. Another important fact is that attractive dipole-dipole interactions are significantly weaker and more dependent on the distance ($\sim 1/r^3$) than the repulsive interaction between monopoles ($\sim 1/r^2$).

It is useful to consider this attractive electrostatic energy quantitatively, assuming one ion with a charge e^- at a distance of 0.2 nm (close to the ionic radius of Br^-) from a non-grounded neutral metal sphere with a radius of 3 nm. The electrostatic energy of interaction is then -147 kJ/mol. Note that an approaching electrically neutral molecule would not lead to this type of electrostatic interaction. However, weak interaction due to dipole or higher multipole moments is possible.

The electrostatic interaction energy depends on the number of ions around the nanoparticle (for simplicity we assume equal distance from the sphere and equal charge). This energy increases with growing number of ions (N) and becomes positive for $N \geq 9$ (see Table 2). The electrostatic attraction energy is very high, when only few charges approach the nanoparticles and in magnitude is comparable with the energy of chemisorption. With increasing number of approaching charges, the energy becomes significantly smaller, which is particularly important in steric conception of nanoparticle stabilization as described below.

When nanoparticles are approached by more than one molecule (or ion), the system becomes more complex. For example, from basic vector algebra, the overall dipole moment is reduced to zero in case of two adsorbates of the same charge approaching opposite sides of the nanoparticle and enhanced, if they are of opposite charge. Collectively, cations and anions approaching the metal particles (assuming here equal probability to approach) induce many image charges δ^+ and δ^- on the surface (Fig. 2) and the particle obtains a multipole moment. As cations and anions often have a different chemical nature resulting in distinctly different interactions with the metal sphere, the metal surface is frequently surrounded predominately with one sort of ions and a shell of counter ions.

Thus, a single ion interacts very strongly with a non-charged non-grounded metal particle, as long as the number of equal ions around the nanoparticle is small. When the interaction of the metal particle with cations and anions is different, a double layer of oppositely charged ions can be formed. This double layer leads to repulsive forces between two nanoparticles in vicinity,[27, 28] whereas induced electric dipole moments (or higher order multipole moments) promote the approach of metal nanoparticles. It is interesting to note that changes in the magnitude of the dipole moments located close to the surface of the metal nanoparticle can be utilized for controlled agglomeration (when the dipole moment is high) and re-dispersion (when the dipole moment low) of the corresponding nanoparticles. This concept has been applied in the design of "smart" nanomaterials.[41-43] It is important to note that we assumed the nanoparticles to be ideal metal spheres without considering the influence of quantum effects on the electronic structure of the nanoparticles. These effects may be particularly relevant to particles smaller than 1 nm. For a more detailed description of quantum properties of metal nanoparticles in a dielectric environment, we refer the reader to the Effective Medium Theory [44-46] and Green's Function based approach.[47]

Number of charges	Geometry	Electrostatic interaction energy, kJ/mol
1		-147
2		-126
3	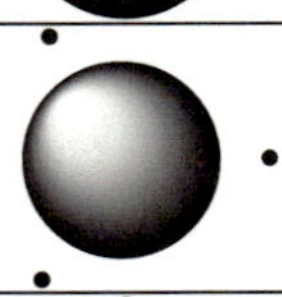	-106
4	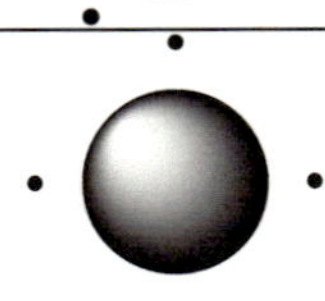	-86
6	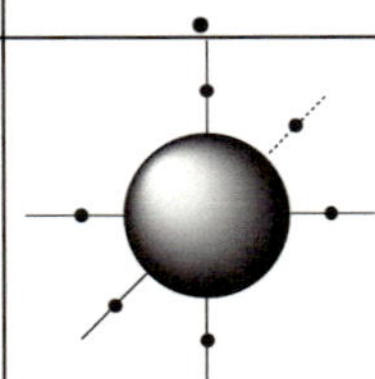	-50
8	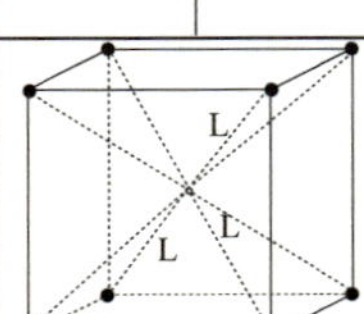	-16
10	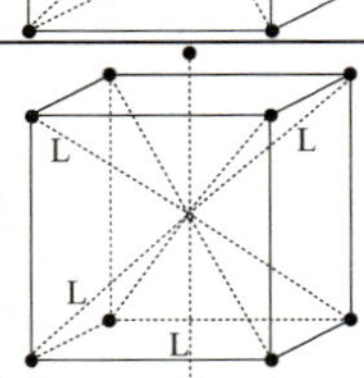	+24

Table 2. Idealised geometry of ion(s) with charge *e* approaching a neutral non-grounded metal sphere and corresponding electrostatic interaction energies between one of the ions and the metal sphere. *L* is the distance between the approaching ions and the centre of the sphere. For $N \geq 8$, only the distances (L) to the external charges are shown, while the metal sphere is omitted. For the assumptions refer to the text.

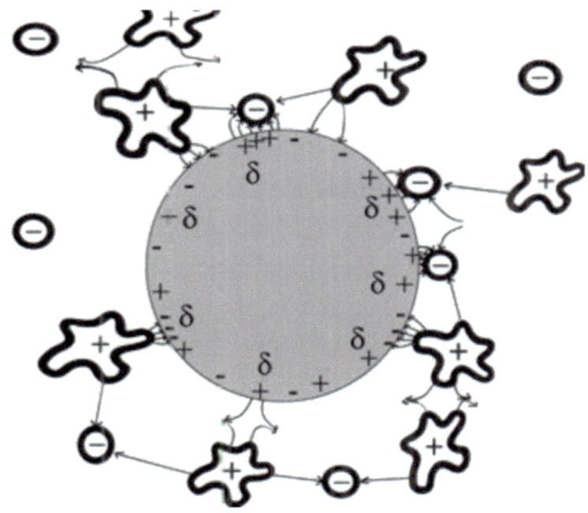

Fig. 2. Schematic representation of a metal particle surrounded by interacting cations and anions, leading to an alternating surface charge distribution.

2.2 Steric stabilization

Another stabilization mechanism is based on the steric repulsion between molecules or ions adsorbed on neighbouring particles. Size and chemical nature of these molecules determine the degree of stabilization. Due to geometric constraints around nanoparticles, large, bulky molecules provide a particularly effective stabilization and an elongated or conical geometry is advantageous to keep the approaching nanoparticles apart (Table 3 A).

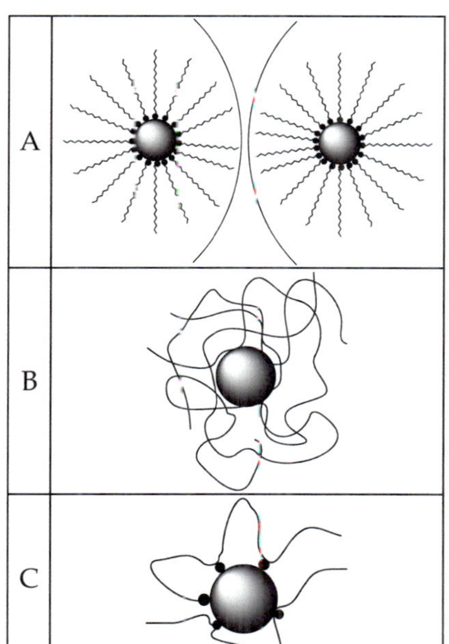

Table 3. Schematic illustration of steric stabilization: *A* – Elongated or conical molecules adsorbed *via* anchoring centres (small black dot) hinder nanoparticles from close contact. *B* – Long polymer threads encapsulate a nanoparticle. *C* – Chelate effect when the stabilizer is adsorbed *via* more than one anchoring centre (small black dots).

When the length of the stabilizer is significantly longer than the characteristic size of the nanoparticles, a sphere can be formed encapsulating the nanoparticle (Table 3 B). Because of that, high molecular weight polymers are often employed as stabilizers for nanoparticles.
Another important requirement is that the stabilizer has to be adsorbed strongly enough on the surface of the nanoparticles to provide long residence time and to prevent its spontaneous desorption. When a stabilizer provides more than one adsorption centre, the chelating effect can increase the probability that the stabilizer remains adsorbed (Table 3 C). Frequently, chemisorption is the driving force for strong binding between adsorbate and metal surface. Metals with more valence orbitals than valence electrons have an "electron-deficient" surface. Thus, molecules readily "donating" electron density (*i.e.* with chemical groups associated with free electron lone pair, such as divalent sulphur, trivalent phosphorus and trivalent nitrogen moieties or molecules with π-electrons, *e.g.*, aromatic systems) often adsorb very strongly on metal surfaces (although the opposite examples are also known [48, 49]). Strongly adsorbing, large molecules are prime candidates for stabilization of nanoparticles.
The concept of steric stabilization plays a very important role in the successful synthesis of nanoparticles. It is not restricted to metals, and can be applied on its own.

2.3 Stabilization by ionic liquid

The idea to employ bulky and highly charged adsorbents in the stabilization of nanoparticles rather than a neutral stabilizer results in the so-called electrosteric stabilization. The best stabilizers for nanoparticles are those, which are ionic, adsorb strongly enough and meet the conceptions of steric and electrostatic stabilization simultaneously.
Ionic liquids, consisting of cations and anions only, provide a huge excess of ions favouring coordination also of less strongly coordinating ions. Note that strong coordination is good for physical applications of the nanoparticles, *e.g.*, as quantum dots, but is not suitable for applications, such as catalysis, where access of the substrate molecules to the metal surface is required. It is worth to mention that nanoparticles with strongly adsorbing ligands can be isolated from excess stabilizer and isolated as powder, whereby the layer(s) of stabilizer on the surface is maintained.[50, 51] This test can be used as criterion for a particularly good stabilization of the nanoparticles.[30]
Due to the high ion density, ionic liquids enable particularly strong electrostatic stabilization, while at the same time they can be selected in such a way that either the cations or the anions are strongly chemisorbed, and either the cations or the anions have the appropriate bulky geometry. The bromide anions of tetraoctylammonium bromide ($[N_{8,8,8,8}^+][Br^-]$), for example, adsorb very strongly on many metal surfaces, while they have a relatively small size. The tetraoctylammonium cation on the other hand provides the bulky geometry necessary for steric repulsion. In fact, many examples for $[N_{8,8,8,8}^+][Br^-]$ stabilized metal nanoparticles are reported in the literature.[50] Charged polymers functioning as polyelectrolytes [52] are other well known stabilizers of metal nanoparticles [53-57] as well as metal oxide nanoparticles [58] and organic micelles.[59-61] They meet the conceptions of steric and electrostatic stabilization simultaneously, *i.e.*, they provide electrostatic stabilization.
Metal nanoparticles stabilized by tetraalkylammonium halides were reported for the first time in 1979.[62] Later, these nanoparticles were prepared *via* chemical reduction in the liquid phase [63] or electrochemically.[64] Thus obtained nanoparticles are stabilized by electrochemical and steric conceptions, as halogen ions adsorb very strongly on the metal

surface, while the surrounding tetraalkylammonium cations constrain close contact between the nanoparticles as illustrated in Fig. 3. Although tetrahexyl-, tetrabutyl- and tetraoctyl-ammonium bromide (and chloride) have a melting point around 100°C and above (and are therefore classified as molten salts), they provide the first examples describing the ability of ionic liquids to stabilize nanoparticles.[63, 64] Note that the use of imidazolium based ionic liquids for the stabilization of Ir nanoparticles was reported much later in 2002.[65]

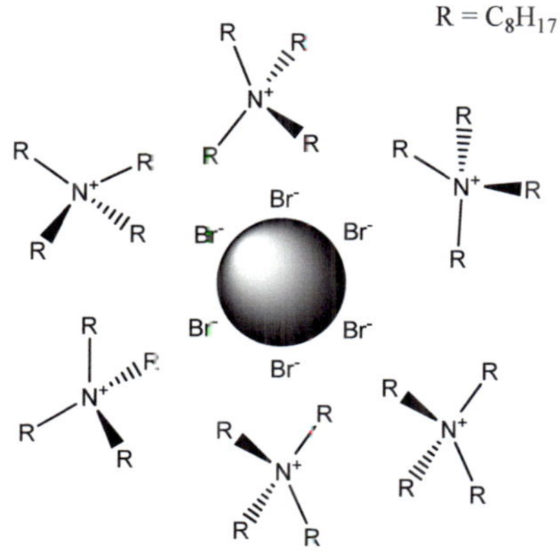

Fig. 3. Schematic illustration of the electric double layer around a tetrahexylammonium bromide stabilized metal nanoparticle as an example of electrosteric (combined electrostatic and steric) stabilization.

Another aspect of ionic liquids is that the thermal movement of the colloidal nanoparticles is suppressed due to the high viscosity of the surrounding medium minimizing the probability of close contacts. Let us analyse this aspect in further detail. Let us assume that nanoparticles with size $a = 3$ nm have just formed and are dispersed at room temperature (20°C) in a medium with viscosity η at a volume fraction φ of 0.01. To assess qualitatively their half-life time, we assume here a very simple model of rapid random coagulation where every collision of two nanoparticles immediately leads to coagulation and, in consequence, agglomeration. The number of collisions ν that one nanoparticle experiences per time unit with other nanoparticles can be expressed by eq. 4,[66] which can be obtained based on the Einstein–Smoluchowski [67, 68] formalism of Brownian motion of colloidal particles.

$$\nu = \frac{kT}{\pi \eta a^3}\varphi \tag{4}$$

The inverted number of collisions per unit of time (eq. 5) is the time constant τ for rapid coagulation. This half-life time is a measure for the decrease of the concentration $c(t)$ of nanoparticles (eq. 6) with time.[66]

$$\tau \sim \frac{1}{\nu} \tag{5}$$

$$\frac{c(t)}{c(t=0)} = \frac{1}{1+t/\tau} \tag{6}$$

Typical collision frequencies ν and time constants τ for such rapid coagulation for 3 nm sized particles in dependence of the viscosity of the surrounding medium are collected in Table 4. Since the viscosity of a typical ionic liquid is a factor of hundred or thousand higher than that of commonly used organic solvents, such as tetrahydrofuran (THF), the collision frequency becomes hundred and thousand times smaller. This results in a longer half-life time of the nanoparticles. Thus, ionic liquids are characterised by very low self diffusion coefficients [69-71] of approximately 10^{-12}–10^{-11} m^2/s compared to 10^{-9} m^2/s for classic solvents with viscosities similar to water (1 mPa·s).[72] Although suppressed diffusion in media with high viscosity, cannot be the only mechanism for stabilization of small particles (the absolute values of half-life are still very small ~10^{-4}-10^{-3} s), the effect may contribute considerably to the overall stabilization of nanoparticles in ionic liquids, as is discussed below.

Medium	Viscosity, mPa·s	Diffusion constant of nanoparticles, m^2/s	Collision frequency, 1/s	Half-life time, s
THF	0.46	$2\cdot10^{-10}$	10^{6}	10^{-6}
$[C_4MIm^+][Tf_2N^-]$	60 [73]	10^{-12}	10^{4}	10^{-4}
$[C_4MIm^+][PF_6^-]$	376 [73]	$2\cdot10^{-13}$	10^{3}	10^{-3}
Solid body	~10^{15}	10^{-25}	$5\cdot10^{-10}$	10^{9}

Table 4. Diffusion constant and estimated half-life time of 3 nm sized particles in environments with different viscosity assuming rapid coagulation at room temperature (see text).

The half-life time of the particles will be longer, if we assume the presence of surface species, which act as stabilizer (a coagulation probability of $1/10^6$ may be reasonable for collisions between two nanoparticles). In this case, a typical half-life time of colloidal particles calculates to half an hour in a highly viscous medium, whereas in classic solvents it would be in the range of seconds. The situation changes drastically, when the temperature of the surrounding medium is below its melting point. In this case, the ionic liquid is in a solid or quasi-solid,[74] *i.e.*, amorphous glassy state [75] with very high viscosities in the range of ~ 10^{15} mPa·s. Embedded nanoparticles experience practically no Brownian movement and the half-life time is in the order of magnitude of 10^3 years. This effect may be applied to storage of nanoparticles for very long times in solid matrices of ionic liquids, until they are being used. Another potential application lies in the field of electrochemistry or photochemistry, when semiconductor nanoparticles serve as quantum dots or photoactive elements for solar cells.

In summary, suppressed diffusion of nanoparticles in highly viscous media, such as ionic liquids, leads to an increase in the lifetime by a factor of 10-1000 compared to classic low viscous solvents, contributing significantly to the stabilization of colloidal nano-sized particles. As mentioned above, ionic liquids can also serve as electrostatic and steric stabilizers. The physicochemical properties of ionic liquids and, thus, the properties of ionic liquid stabilized nanoparticles can be readily adjusted by changing cations and anions. Compared to polymers, ionic liquids as stabilizers for nanoparticles provide some particular advantages. Non-ionic polymers (in contrast to polyelectrolytes) have no ionic nature (only steric stabilization applies) and their ability to dissolve various compounds, for instance, metal precursors or substrates is limited (*vide infra*). Thus, often an organic solvent is required, where both polymer, metal precursor and the substrate are soluble (*e.g.*, in case of applications in catalysis). The use of organic solvents can be avoided, when catalytic reactions are performed in neat ionic liquids (with dispersed nanoparticles). Reactions can also be conducted in neat substrate, where small amount of ionic liquid stabilized nanoparticles are added.

As we have mentioned above, strongly adsorbed species are needed for good stabilization of nanoparticles. Many ions generally employed in ionic liquids, like the cations tetraalkylammonium ($N_{k,l,m,n}^+$), tetraalkylphosphonium ($P_{k,l,m,r}^+$), and 1,1-dialkyl-pyrrolidinium ($C_mC_nPyr^+$) and the anions bis(trifluoromethanesulfonyl)amide (Tf_2N^-), trifluoromethanesulfonate ($CF_3SO_3^-$), alkylsulfate ($C_nOSO_3^-$), tetrafluoroborate (BF_4^-), hexafluorophosphate (PF_6^-) are weakly coordinating and, thus, need to be complemented with an appropriate counter ion to stabilize nanoparticles efficiently.

On the other hand, there are reports on the ability of BF_4^- to coordinate to metallic Ir and act as a stabilizer for Ir nanoparticles.[76, 77] Addition of 1,8-*bis*(dimethylamino)naphthalene (PS) provided improved stability, yet transmission electron microscopy (TEM) showed assemblies typical for the early stages of agglomeration. The PS may serve as a proton scavenger or the two tertiary amino groups of PS act as an anchor for strong adsorption. Also the bulky geometry of PS may be beneficial for steric stabilization. After addition of excess [N_{4444}^+][BF_4^-] (melting point 155°C), the life time of the nanoparticles increased significantly,[77] most likely, due to the increased viscosity of the surrounding medium and, thus, reduced Brownian diffusion.

On the contrary, Ir nanoparticles prepared in [C_4MIm^+][PF_6^-], [C_4MIm^+][BF_4^-] and [C_4MIm^+][$CF_3SO_3^-$],[76] were found to be very stable possibly due to the formation of strongly coordinating *N*-heterocyclic carbenes (*vide infra*).[78]

Various physical properties of ionic liquids, as viscosity and molar concentration, are directly dependent on the sum of the ionic volumes of cation and anion.[72] In this light, it is interesting to consider the effect of the choice of the ionic liquid. The size of the nanoparticles frequently increases with increasing molecular volume (or size) of the (weakly) coordinating anion (BF_4^-, PF_6^- and Tf_2N^-), *e.g.*, for Ag nanoparticles from 2.5 to 27 nm [79] and for W nanoparticles from 1.5 to 30 nm.[80]

The size of the ion determines the distance from the surface of the nanoparticle and the number of ions around it. This influences directly the total electrostatic energy of the system. In the case of weakly coordinating ions, the interaction energy with the metal particle is almost only electrostatic in nature and its value depends on the particle size and the distance between particle and the ion and can reach up to 100 kJ/mol. The size of the ion determines also the maximum number of ions, which can be accommodated around a nanoparticle. In result, the thickness of the electric double layer is influenced and, thus, the stability of nanoparticles in terms of the electrostatic conception.

Halogens (Cl^-, Br^-), dicyanimide ($(CN)_2N^-$), carboxylates ($RCOO^-$), *etc.* adsorb strongly on metal surfaces and, thus, provide good stabilization. It is important to note that even traces of halogens, as present in many commercially available ionic liquids in concentrations of several tens to hundred ppm, may contribute to the stabilization of the metal nanoparticles.[30] The contribution of the cation to the stabilization of the nanoparticles also has to be taken into the account.[80] Bulky imidazolium cations, such as 1-alkyl-3-methylimidazolium (C_nMIm^+) as well as ions, such as 1-alkylpyridinium (C_nPy^+) may contribute to varying degree to the stabilization of the nanoparticles. Functionalized ionic liquids [81] (often, but not necessarily, imidazolium based) with a specific coordinating group attached to the structure of cation or/and anion can be used for enhanced stabilisation. Alkylnitrile pyridinium [$N{\equiv}CC_nPy^+$] and 1-alkylnitrile-3-methylimidazolium [$N{\equiv}CC_3MIm^+$] based ionic liquids, *e.g.*, provide better stabilization of Pd nanoparticles [82, 83] than alkyl pyridinium ionic liquids [C_nPy^+] and 1-alkyl-3-methylimidazolium [C_nMIm^+] based ionic liquids. Pt and Ru nanoparticles were also stabilized successfully in 4,5-dicyano triazolium based ionic liquid.[81] Thiol functionalized imidazolium ionic liquid, specifically 1-mercaptoalkyl-3-methylimidazolium, was found to be an excellent stabilizer for, *e.g.*, Au nanoparticles.[84] The strong adsorption of carboxylic groups on metal surfaces [85, 86] makes ions carrying one or a few carboxylic groups to be very attractive for the stabilization of metal nanoparticles. Pivalic (Piv^-), tartaric (TA^{2-}), Glycinate (Gly^{2-}) as well as malic acid (MA^{2-}) based ionic liquids, combined with long-chain tetraalkylammonium cations are good to excellent stabilizers for, *e.g.*, Pt, Pd, Rh, Ru and Ni nanoparticles.[87, 88]

It is interesting to note that even non-functionalized imidazolium-based ionic liquids with weakly coordinating anions are stabilizers for various transition metal nanoparticles.[76, 89-92] Explanations may be that (*i*) the imidazolium cation can, in principle, adsorb on the metal surface *via* its π-electron system and (*ii*) imidazolium cations, which are not substituted in 2-position, can form *N*-heterocyclic carbenes, which are strongly coordinating to many metals.[93-96] On the other hand, the strong adsorption of carbenes results in poisoning of the metal surface. This type of inhibition was demonstrated for the hydrogenation of acetone over Ir nanoparticles in imidazolium-based ionic liquids.[78]

Additives can be used to further stabilize nanoparticles in imidazolium based ionic liquids. Stabilization with polyvinylpyrrolidone (PVP) is necessary to obtain stable mono- and bimetallic Pt/Au nanoparticles immobilized in $[C_4MIm^+][PF_6^-]$.[97] Also, ionic liquid-like copolymers can be used to stabilize Rh [98] and Pt [99] nanoparticles, where $[C_4MIm^+][BF_4^-]$ itself can not stabilize the platinum nanoparticles sufficiently.[99]

2.4 Stability against oxidation

When metals are exposed to air, dioxygen is adsorbed on their surface, often accompanied by dissociation of the dioxygen. This is the first step to oxidation of metals with low standard electrode potential (*e.g.*, Ni, Fe, Cu, Co). Monooxygen on the surface of metals with high standard electrode potential (*e.g.*, Au) does not lead to noticeable oxidation of the metal at normal conditions. The oxidation of metals is a highly complex process,[100] whereby dissociative adsorption of oxygen on the metal surface constitutes a key step. A straightforward idea of protecting the metal from oxidation is to prevent the adsorption of oxygen. In other words, one can modify metal surfaces with protective agents before exposing them to air. In the best case, agents stabilizing against agglomeration act at the same time as a protective agent against oxidation.[101] Generally, the adsorption energy of the stabilizer has to be significantly higher than the adsorption energy of monooxygen on the same adsorption site. However, it is difficult to choose a proper stabilizer *a priori*. It is easier to prepare nanoparticles with a chosen stabilizer and then check, if it protects the surface of the nanoparticles from oxidation. Ni nanoparticles protected (and stabilized) with $[N_{8,8,8,8}^+][Br^-]$ prepared *via* the wet chemical way (*vide infra*) [63] are stable in air and keep the dark-brown colour specific for metal colloids for more than 2 years. In contrast, $[N_{8,8,8,8}^+][Br^-]$ stabilized Ni nanoparticles prepared *via* electrochemical reduction [64, 102] were found to be extremely oxygen sensitive. The stability of metal nanoparticles against oxidation can also be size dependent. Under oxygen plasma conditions, 1.4 nm sized Au nanoparticles showed better oxidation resistance compared to larger particles.[103] A similar effect was predicted for Pt nanoparticles.[104]

3. Design of metal nanoparticles in ionic liquids for specific applications

The task for which nanoparticles are to be used often determines the preparation strategy. In the case, when access to the surface of the nanoparticles is not needed, one can choose a stabilizer, which adsorbs very strongly on the surface of nanoparticles (*vide supra*). Au nanoparticles, *e.g.*, are often used as quantum dots in applications, where their optical properties are important (*e.g.*, in cancer cell imaging).[105] In these cases, stabilizers such as tetraalkylammonium bromide [105, 106] or bulky, divalent sulphur containing molecules [107-109] are commonly employed.

For applications of nanoparticles in catalysis, a balance between the adsorption strength of the stabilizer and the substrate has to be found. Stabilisation with strongly coordinating adsorbents results in a reduced or completely suppressed activity. Besides the well-know surface poisoning effect induced by divalent sulphur,[110, 111] the presence of halogens can poison the metal surface.[86] In fact, $[N_{8,8,8,8}^+][Br^-]$ stabilized Ni nanoparticles were found to be completely inactive in the hydrogenation of, *e.g.*, aliphatic ketones and β-ketoesters, while PVP stabilized Ni nanoparticles demonstrate high catalytic activity in these reactions.[93] As stabilizer and substrate molecules compete for coordination to the metal surface, the poisoning effect may vary in dependence of the substrate molecule. Rh nanoparticles, *e.g.*, catalyze the hydrogenation of benzene, while they show no activity in the hydrogenation of aliphatic ketones. Similarly, Ru nanoparticles stabilised by $[N_{8888}^+]_2[TA^{2-}]$ or $[N_{8888}^+]_2[Gly^{2-}]$, catalyze the hydrogenation of nitrobenzene to cyclohexylamine, while $[N_{8888}^+]_2[Me_2Gly^{2-}]$ stabilised Ru nanoparticles provide aniline.[88] Selective surface poisoning can be utilized to direct selectivity, as is illustrated by the example of enantioselective hydrogenation of β-ketoesters, when a small amount of Br^- ions added to tartaric acid modified Ni particles resulted in higher enantioselectivity.[86] Functional groups can be introduced into the ionic liquid to combine the stabilization effect with additional properties, such as a co-catalytic group, a proton buffer or chirality.

4. Preparation concepts for ionic liquid stabilised metal nanoparticles

Methods for the synthesis of nanoparticles have a more than 150 year old history dating back to the work of Faraday in 1857.[112] It appears logical to employ these methods also in the synthesis of ionic liquid stabilized nanoparticles. Herein, we give an overview on strategies to obtain nanoparticles in general and ionic liquid stabilised particles in particular. For an overview on metal nanoparticles and stabilizers we refer the reader to the corresponding review articles [8, 14, 30, 101, 113-127] and for more detailed procedures to the original literature.

In general, the preparation methods for nanoparticles are classified as "top-down" and "bottom-up" approaches.[101, 128] Herein, we do not consider classic "top-down" methods, which deal with mechanic grinding of bulk metals to the nano-sized scale. We will rather focus on "bottom-up" (starting from molecular metal precursors) and a combination of "top-down" and "bottom-up" strategies. The basic idea is to obtain metal atoms (from bulk material) and to control the nucleation and growth to nano-sized scale. For clarity, we call "bottom-up" methods *chemical*, due to chemical reduction of metal precursors to neutral atoms. Methods being combinations of "top-down" and "bottom-up" we call *physical*, as the formation of metal atoms (from bulk metals) often involves various physical techniques.

4.1 Chemical methods

Typically, a precursor to the metal of interest (such as an inorganic salt or an organometallic complex) is reduced (or decomposed) to generate neutral metal atoms. In the absence of stabilizers, the atoms quickly agglomerate to microscopic crystallites. Coagulation of these crystallites leads to precipitation of the metal as "metal black". It is very difficult to directly measure the size of nanoparticles, while they are growing.[129] An indirect method for following the kinetics of nanoparticle formation deals with the hydrogenation of cyclohexene as probe reaction.[30] A recent *in situ* electron microscopy study describes the dynamics of the growth of Pt nanoparticles under typical synthesis conditions.[130] Along with the conventional growth mechanism by incorporation of metal atoms from solution, frequent coalescence events between the particles were observed.[130]

In the presence of an appropriate stabilizer, it is possible to halt the nucleation at a certain stage and to obtain colloidal nanoparticles. Besides the appropriate choice of the stabilizer, careful selection of the metal precursor, the reducing agent, and the solvent are also important. Stabilizer and reducing agent have to be soluble in the solvent employed, while it is not necessary that the metal precursor is soluble.[131]

The associated processes during formation of nanoparticles are summarized in Fig. 4. Typically, metal atoms are formed by reduction (1) or decomposition of the metal precursor (2A). Free atoms are highly unstable and nucleate (3) quickly, until this process is prevented by a stabilizer (4). The reducing agent may also act as a stabilizer.[101] The range of suitable metal precursors can be expanded by employing strongly reducing agents (or harsh reaction conditions), while decomposition of the reducing agent or reaction with the stabiliser has to be avoided. By-products formed from the metal precursor (2B) can adsorb on bare or stabilized nanoparticles, which can lead to reduced stability or/and surface poisoning.[110] Some by-products (*e.g.* strong acids and bases) can also lead to decomposition of the stabilizer. In order to avoid these problems metal precursors, reducing agent, and stabilizer have to be carefully chosen, especially when ionic liquids are employed as stabilizers.

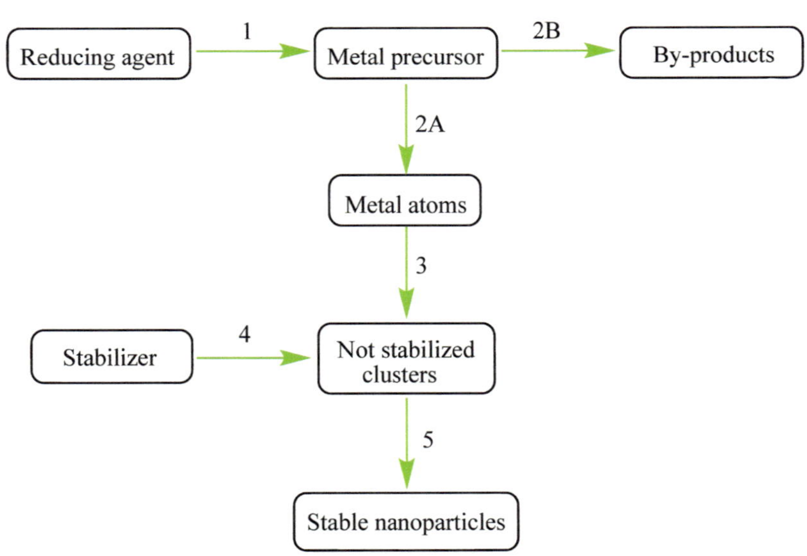

Fig. 4. Schematic sequence of the formation of metal nanoparticles.

The chemical stability of ionic liquids has to be taken into the account, when they are used as stabilizers for nanoparticles. Ionic liquids often react with strong bases, acids, some reducing agents (*e.g.* $NaBH_4$, $LiAlH_4$, $Al(alkyl)_3$, *etc.*) or may decompose at relatively high temperatures. Preferentially, mild reducing agents, like hydrogen or alcohols should be used. This requirement limits the selection of appropriate metal precursors. Preferably, hydrogen or reducing agents, which give non-coordinating, volatile, non-reactive by-products, are used at temperatures below the decomposition temperature of ionic liquids. An elegant way to avoid the decomposition of ionic liquids by acids HX (X = NO_3^-, BF_4^-, PF_6^- or $CF_3SO_3^-$) produced during reduction of AgX is the addition of an imidazole scavenger, which resulted in increased stability, narrowed size distribution and decreased average size of the Ag nanoparticles formed.[79]

Interestingly, if the metal precursor is soluble in ionic liquids (sometimes this generally slow process can be speeded up by stirring, heating, ultrasonic treatment or addition of a few drops of a suitable solvent), the use of additional organic solvents is no longer needed. For instance, the Ru precursor $[Ru(MeAllyl)_2(COT)]$ (COT = cyclooctatetraen) is soluble in $[N_{8,8,8,8}^+][Piv^-]$ and can be reduced easily with hydrogen, while this ionic liquid successfully stabilizes the Ru nanoparticles and protects the Ru surface from oxidation for a long time. Other examples describing the preparation of nanoparticles in ionic liquids *via* the chemical route can be found elsewhere.[8, 30, 89, 91, 121, 122, 132-134] The preparation steps of this so-called direct route are schematically illustrated in Fig. 5.

Fig. 5. Preparation of nanoparticles in ionic liquids *via* direct method. In the first step a suitable metal precursor is dissolved in the ionic liquid (IL), followed by reduction with a suitable reducing agent or heating the mixture to decompose the metal precursor.

A major advantage of the direct method is its simplicity. However, it cannot always be applied, as only few metal precursors meet the above-mentioned requirements. In case of difficulties in preparing nanoparticles directly in ionic liquids, the following alternative approach (indirect method) may be considered. The nanoparticles of interest are prepared using a convenient and suitable combination of stabilizer, metal precursor and reducing agent in an organic solvent and then transferred into the required ionic liquid (preferably after isolation of the nanoparticles or removal of the by-products, excess reducing agent, stabilizer and solvent); the indirect method is illustrated in Fig. 6. Note that the stabilizer has to be well soluble in the solvent in order to obtain a good dispersion of thus prepared nanoparticles.

Tartaric acid based ionic liquids $[N_{6,6,6,6}^+][TA^-]$, *e.g.*, are not stable against strong reducing agents as, *e.g.*, $NaBH_4$. Therefore, the reduction of metal precursors in the presence of such ionic liquids needs to be avoided. Nonetheless, $NiCl_2$, *e.g.*, can be reduced with $NaBH_4$ in a classic solvent in the presence of PVP (polyvinylpyrrolidone) as stabilizer. Then, thus prepared nanoparticles can be isolated and transferred into the ionic liquid of interest.[see, e.g., 135]

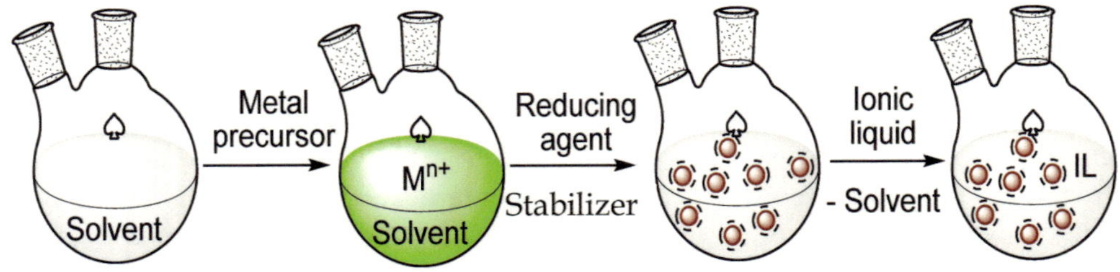

Fig. 6. Indirect method for the synthesis of nanoparticles in ionic liquids. After dissolution (not always needed) of the metal precursor in an organic solvent, the former is reduced in the presence of an external stabilizer, then the ionic liquid (IL) is added and solvent (and any by-products) are removed.

Another example is the precursor [Ni(acetylacetonate)$_2$], which can be reduced readily with Al(^{i}Bu)$_2$H in THF solution. Then, after addition of ionic liquid and removal of solvent very stable and air insensitive Ni nanoparticles of ~ 2-3 nm size (Fig. 7) are obtained and can be used, *e.g.*, in the hydrogenation of ketones and ketoesters.[87]

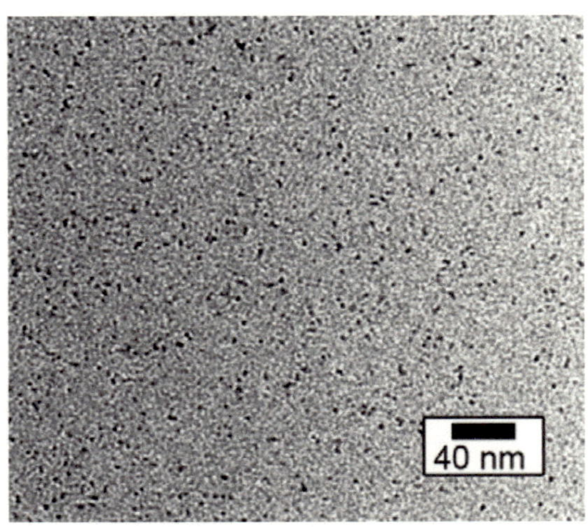

Fig. 7. TEM image of Ni nanoparticles immobilized in [$N_{6,6,6,6}^+$][TA^-] prepared *via* the indirect method, see text for details.

4.2 Physical methods

As mentioned above, nanoparticles can be prepared *via* a combination of "top-down" and "bottom-up" routes. First, free metal atoms are obtained *via* a chosen physical method and their subsequent agglomeration is controlled by a stabilizer. Below we give a short overview of some of these physical methods, which found attention in the past or recent literature.

4.2.1 Solvated metal atoms

If one would think about methods to obtain free metal atoms, the first idea, which might come to mind, is to transfer solid bulk metal to the gas phase through strong heating in ultra high vacuum. In fact, condensation of the metal vapour together with a weakly stabilizing organic solvent (pentane, acetone, THF) at low temperature (-196 °C) (so called metal-vapour synthesis) results in solvent pre-stabilized metal clusters, consisting of a few atoms (5-10).[136] These are stable at low temperatures (typically below -50 °C), while a precipitate of a finely dispersed metal powder forms rapidly at ambient temperatures. In order to avoid agglomeration, additional stabilizers have to be used. In fact, mono-dispersed 4-5 nm sized Au nanoparticles were obtained *via* metal vapour synthesis, quick reaction with dodecanethiol after warming up and "digestive ripening" by refluxing in toluene.[137] This method can be applied to other metals and stabilizers,[137-145] including ionic liquids. Note that this method can be scaled up easily. In a slightly different approach, Au and Ag can be sputter deposited onto $[C_4MIm^+][PF_6^-]$ to obtain monometallic and bimetallic Au and Ag particles with 3-5 nm diameter.[146, 147]

4.2.2 Laser ablation

Nanoparticles of various metals (including bimetallic particles [148-155] and metal oxides [156-160]) can be made by laser ablation methods. Instead of using chemical precursors, the laser ablation method deals with precursors in form of macroscopic metal plates or relatively "large" nanoparticles (100 nm in size and larger), which are submerged in a solvent. When focused on a very small spot of the metal surface, laser ray induces quick heating and subsequent transport of metal atoms to the liquid medium. If the later contains a proper stabilizer, the atoms nucleate to stable metal nanoparticles. Au nanoparticles can be prepared *via* laser ablation from Au foil in neat ionic liquids.[161] It is interesting to note that $[C_8MIm^+][BF_4^-]$ provided better stabilities compared to ionic liquids with shorter alkyl-chains, such as $[C_4MIm^+][BF_4^-]$ and $[C_2MIm^+][BF_4^-]$.

Laser irradiation can also fragment existing nanoparticles.[162] The efficiency of nanoparticle formation depends on the frequency of the laser pulses applied, the pulse duration time, pulse power and the wavelength of the laser. It is interesting to note that the nanoparticles are most efficiently heated, when the wavelength of laser irradiation matches the frequency of the plasmon absorption band.[163]

Laser ablation [164] is a relatively new, not fully investigated, and not widely used (in comparison with numerous chemical methods) method for the synthesis of nanoparticles. Reports on catalytic application of nanoparticles made by laser ablation are limited.[155] A particular disadvantage is the slow preparation, although the laser beam can be split into several beams. Typically, only 4-5 mg of Pt nanoparticles is produced after 1 hour of laser work. This can be accepted in the case of model catalysts, where the material demonstrates excellent catalytic properties or high metal loadings are not needed.

4.2.3 Electrochemical synthesis

In an electrochemical way of making metal nanoparticles, a solution of stabilizer (*e.g.*, tetraalkylammonium bromide in THF) is electrolyzed using an anode made of the metal of interest and an inert cathode.[64, 166] Under the conditions of the electrolysis, dissolution of the anode material takes place *via* oxidation. The metal cations transfer to the cathode, where reduction, nucleation and finally stabilization occur. The size of thus-prepared nanoparticles can be easily controlled by the current density. The method is applicable to many transition metals. Moreover, systems with two anodes of different metals result in the formation of bimetallic nanoparticles. When oxidation of the anode metal is difficult, the metal of interest can be introduced as inorganic salt.[167] Pd nanoparticles, *e.g.*, prepared *via* this method were found to be catalytically active in the Heck reaction [168] and the hydrogenation of olefins [169] both as colloid (in propylene carbonate as solvent) and supported on alumina. Although the electrochemical reduction is very simple, it is not yet used for routine synthesis of metal nanoparticles. The approach can be adapted readily to the synthesis of nanoparticles in ionic liquids, whereby a metal precursor dissolved in ionic liquid is reduced *via* an electric current. Thus, Pd nanoparticles can be electrodeposited in pyrrolidinium based ionic liquids [170] and Au and Ag nanoparticles in imidazolium based ionic liquids.[171] The size, shape and morphology of the obtained nano-sized Au and Ag structures were found to be strongly dependent of the physical state of the reaction medium (solid or liquid).[171]

4.2.4 Plasma method

An interesting approach for preparation of nanoparticles in ionic liquids was recently found *via* plasma induced reduction of metal salts.[172-174] Ag,[175] Cu [176, 177] and Al [177] nanoparticles, *e.g.*, can be prepared by treating a solution of the corresponding salt in an ionic liquid with a low temperature ionized Ar plasma, which results in electron transfer into the liquid phase (close to the interface) and subsequent reduction of metal cations.

4.2.5 Electron beam or γ–irradiation

The method of reducing of metal cations employing an electron beam or γ-irradiation for making mono- or bimetallic nanoparticles (as well as semiconductor nanoparticles) is known since the late seventies.[178-182] and [183-186] Recently, this method found a practical application in nanolithography [187] for optical and electronic data transfer. High molecular weight polymers like polyvinyl alcohol (PVA) or poly(acrylamide) (PAM) were employed to stabilize the nanoparticles. Now it becomes logical to explore the use of ionic liquids acting as a stabilizer instead of a polymer. Thus, Au nanoparticles, *e.g.*, can be prepared in ionic liquid by electron beam irradiation.[188] In $[C_4MIm^+][Tf_2N^-]$, relatively large nanoparticles with average diameter ~120 nm were obtained.[188]

4.2.6 Photochemical method

Certain metal precursors can be reduced under UV irradiation. This process is often assisted by semiconductors,[189-192] which provide electrons generated after light induced excitation. Reduction of alkylpyridinium Au^{3+} complexes under UV irradiation provides Au particles, whereby size, shape, and morphology are strongly dependent on the physical state of the complex.[193] Thus, reduction in the solid state resulted in the formation of hexagonal platelets with a size of 10-20 μm. Small spherical particles in the range of 50-100 nm were obtained close and above the melting point.

4.2.7 Comparison of chemical and physical methods

A strong advantage of the physical methods for making nanoparticles is that no chemical reducing agent is involved, as laser ablation and metal vapour deposition methods do not require reduction, while metal cations are reduced by electric current in case of the electrochemical synthesis. The absence of chemical precursors and reducing agents significantly decreases the limitations in the synthesis of nanoparticles. Involving less chemical substances, nanoparticles with particularly high purity can be obtained that may be important in catalysis. In particular, the metal, which is employed in form of the bulk metal, *e.g.*, as a thin foil, can be used in very high purity. However, physical methods generally require a specific apparatus (laser, solvated metal atom dispersion (SMAD) reactor, source of γ-rays, *etc.*). Such equipment is available on the market and can be adapted to the synthesis of nanoparticles.

A comprehensive comparison of the catalytic properties of nanoparticles obtained *via* chemical and physical routes is difficult, because of the scarce information on nanoparticles prepared *via* physical methods. Catalytic applications were described for Pd nanoparticles obtained *via* electrochemical route,[168, 169] and metal clusters obtained in a SMAD reactor.[138] The differences in the catalytic behaviour relative to analogous conventional catalysts obtained *via* chemical route [138] concerning activity and selectivity in, *e.g.*, the hydrogenation of 1-heptene or 1,4-butadiene lie not only in the size and shape of the particles, but also in specific interactions with the solvent. Thus, it is claimed that Pd nanoparticles, *e.g.*, react with polar solvent molecules and acquire a negative charge.[136] The absence of other chemical species, which can interact with the metal and the small size of the nanoparticles make them in some degree comparable with gas phase metal clusters and their unique reactivity.[194]

5. Conclusions

Ionic liquid modified nanoparticles are a system with many degrees of freedom and particularly interesting for applications in catalysis. As the physical properties (viscosity, conductivity, melting point, *etc.*) of ionic liquids can be tuned readily by changing the structure of cation and anion, the physical behaviour (solubility, dispersion, accessibility, *etc.*) of nano-sized catalysts can be tailored to specific applications. Functional groups can be introduced into the cation or anion in order to increase the stability of the nanoparticles and to tailor the catalytic properties, in particular, activity and selectivity. Note that simple mixtures of two (or more) ionic liquids with one or two sorts of nanoparticles will result in new physicochemical properties, which are different from those of the single components. Extending the preparation and use of monometallic nanoparticles immobilized in ionic liquid to bimetallic or semiconductor nanoparticles appears particularly interesting for further exploration and application. There are numerous reports about excellent catalytic properties of nanoparticles in ionic liquids. Particularly interesting are those cases, where the substrate is readily soluble in the ionic liquid utilizing the quasi-homogeneous nature of the suspended nanoparticles, and where the product separates readily from the ionic liquid phase. To extract the product, the reaction may be performed in biphasic systems (for example with supercritical carbon dioxide *sc*CO_2 as mobile phase) [195] or the ionic liquid with embedded nanoparticles is immobilized on a heterogeneous support.[135, 196] A conceptual review on molecular homogeneous and nano-sized catalysts immobilized in a thin film of ionic liquids supported on a heterogeneous carrier will be published in due course.[197]

6. Acknowledgment

Financial support of Bayer MaterialScience AG and RWTH Aachen University is gratefully acknowledged. The authors thank Dr. Sergey Subach from Jülich Research Centrum for discussions concerning XPS and surface science in general, as well as Dr. Carsten Sievers from Georgia Institute of Technology for comments and suggestions.

7. References

[1] G. Schmid, *Nanoparticles. From Theory to Applications*, Wiley-VCH, Weinheim, 2003.

[2] K. L. Kelly, E. Coronado, L. L. Zhao, G. C. Schatz, *Journal of Physical Chemistry B* 2003, *107*, 668.

[3] C. J. Murphy, T. K. San, A. M. Gole, C. J. Orendorff, J. X. Gao, L. Gou, S. E. Hunyadi, T. Li, *Journal of Physical Chemistry B* 2005, *109*, 13857.

[4] D. Zitoun, M. Respaud, M. C. Fromen, M. J. Casanove, P. Lecante, C. Amiens, B. Chaudret, *Physical Review Letters* 2002, *89*, 4.

[5] Z. L. Wang, J. M. Petroski, T. C. Green, M. A. El-Sayed, *Journal of Physical Chemistry B* 1998, *102*, 6145.

[6] F. Klasovsky, P. Claus, in *Metal nanoclusters in catalysis and material science: the issue of size control* (Eds.: B. Corain, G. Schmid, N. Toshima), Elsevier, Amsterdam, 2008, pp. 167.

[7] P. Claus, H. Hofmeister, *Journal of Physical Chemistry B* 1999, *103*, 2766.

[8] B. Chaudret, K. Philippot, *Oil & Gas Science and Technology - Revue de l'IFP* 2007, *62*, 799.

[9] B. Corain, G. Schmidt, N. Toshima, *Metal nanoclusters in catalysis and material science: the issue of size control*, Elsevier, Amsterdam, 2008.

[10] M. A. Neouze, U. Schubert, *Monatshefte für Chemie* 2008, *139*, 183.

[11] J. F. Hochepied, *European Journal of Inorganic Chemistry* 2008, 835.

[12] H. Itoh, S. Utamapanya, J. V. Stark, K. J. Klabunde, J. R. Schlup, *Chemistry of Materials* 1993, *5*, 71.

[13] R. V. Kumar, Y. Diamant, A. Gedanken, *Chemistry of Materials* 2000, *12*, 2301.

[14] M. A. Neouze, U. Schubert, *Monatshefte für Chemie* 2008, *139*, 183.

[15] J. A. Rodríguez, M. Fernández-García, *Synthesis, Properties, and Applications of Oxide Nanomaterials*, John Wiley and Sons, New York, 2006.

[16] M. Niederberger, G. Garnweitner, *Chemistry - A European Journal* 2006, *12*, 7282.

[17] N. A. Anderson, T. Q. Lian, *Annual Review of Physical Chemistry* 2005, *56*, 491.

[18] E. Galoppini, *Coordination Chemistry Reviews* 2004, *248*, 1283.

[19] K. Grieve, P. Mulvaney, F. Grieser, *Current Opinion in Colloid & Interface Science* 2000, *5*, 168.

[20] N. Herron, D. L. Thorn, *Advanced Materials* 1998, *10*, 1173.

[21] V. I. Klimov, *Semiconductor and Metal Nanocrystals: Synthesis and Electronic and Optical Properties*, Marcel Dekker, New York, 2004.

[22] S. Sapra, J. Nanda, D. D. Sarma, in *Encyclopedia of Nanoscience and Nanotechnology, Vol. 3* (Ed.: H. S. Nalwa), American Scientific Publishers, 2004, pp. 181.

[23] T. Trindade, P. O'Brien, N. L. Pickett, *Chemistry of Materials* 2001, *13*, 3843.

[24] G. A. Somorjai, R. L. York, D. Butcher, J. Y. Park, *Physical Chemistry Chemical Physics* 2007, *9*, 3500.

[25] A. Kraynov, A. Suchopar, L. D'Souza, R. Richards, *Physical Chemistry Chemical Physics* 2006, *8*, 1321.

[26] J. D. Aiken III, R. G. Finke, *Journal of Molecular Catalysis A - Chemical* 1999, *145*, 1.

[27] B. Derjaguin, L. D. Landau, *Zhurnal Eksperimentalnoi I Teoreticheskoi Fiziki* 1945, *15*, 663.

[28] E. J. Verwey, J. T. G. Overbeek, *Theory of the Stability of Lyophobic Colloids*, Elsevier, Amsterdam, 1948.

[29] J. T. G. Overbeek, *Pure and Applied Chemistry* 1980, *52*, 1151.

[30] L. S. Ott, R. G. Finke, *Coordination Chemistry Reviews* 2007, *251*, 1075.

[31] M. I. Baskes, *Physical Review B* 1992, *46*, 2727.
[32] C. L. Liu, J. M. Cohen, J. B. Adams, A. F. Voter, *Surface Science* 1991, *253*, 334.
[33] Y. N. Wen, H. M. Zhang, *Solid State Communications* 2007, *144*, 163.
[34] J. M. Zhang, M. Fei, K. W. Xu, *Chinese Physics* 2004, *13*, 1082.
[35] J. M. Zhang, F. Ma, K. W. Xu, *Applied Surface Science* 2004, *229*, 34.
[36] D. Myers, *Surfaces, interfaces, and colloids: principles and applications*, second ed., Wiley-VCH, New York, 1999.
[37] S. Levine, G. P. Dube, *Trans Faraday Soc* 1940, *35*, 1141.
[38] L. D. Landau, E. M. Lifshitz, *Elektrodinamika Sploshnykh Sred (Electrodynamics of Continuous Media), Vol. 8*, Nauka, Moscow, 1982.
[39] D. B. Sivukhin, *Obshchii Kurs Fiziki (Course in General Physics) Vol. 3 Elektrichestvo (Electricity), Vol. Vol. 3 Elektrichestvo (Electricity)*, Fizmatlit - Izd. MFTI, Moscow, 2002.
[40] S. Özkar, R. G. Finke, *Journal of the American Chemical Society* 2002, *124*, 5796.
[41] A. M. Kalsin, M. Fialkowski, M. Paszewski, S. K. Smoukov, K. J. M. Bishop, B. A. Grzybowski, *Science* 2006, *312*, 420.
[42] R. Klajn, K. J. M. Bishop, M. Fialkowski, M. Paszewski, C. J. Campbell, T. P. Gray, B. A. Grzybowski, *Science* 2007, *316*, 261.
[43] R. Klajn, P. J. Wesson, K. J. M. Bishop, B. A. Grzybowski, *Angewandte Chemie-International Edition in English* 2009.
[44] S. Hussain, R. K. Roy, A. K. Pal, *Materials Chemistry and Physics* 2006, *99*, 375.
[45] E. Lidorikis, S. Egusa, J. D. Joannopoulos, *Journal of Applied Physics* 2007, *101*.
[46] H. C. Weissker, J. Furthmuller, F. Bechstedt, *Physical Review B* 2002, *65*.
[47] V. A. Markel, M. R. Geller, *Journal of Physics-Condensed Matter* 2000, *12*, 7569.
[48] S. K. M. Henze, O. Bauer, T. L. Lee, M. Sokolowski, F. S. Tautz, *Surface Science* 2007, *601*, 1566.
[49] R. Temirov, S. Soubatch, A. Luican, F. S. Tautz, *Nature* 2006, *444*, 350.
[50] H. Bönnemann, W. Brijoux, *Active Metals*, VCH, Weinheim, 1996.
[51] A. Kraynov, R. Richards, *Physical Chemistry Chemical Physics* 2007, *9*, 884.
[52] J. Ruhe, M. Ballauff, M. Biesalski, P. Dziezok, F. Grohn, D. Johannsmann, N. Houbenov, N. Hugenberg, R. Konradi, S. Minko, M. Motornov, R. R. Netz, M. Schmidt, C. Seidel, M. Stamm, T. Stephan, D. Usov, H. N. Zhang, *Polyelectrolytes with Defined Molecular Architecture I* 2004, *165*, 79.
[53] M. Ballauff, G. Sharma, R. Kempe, T. Irrgang, Y. Talmon, S. Proch, *Abstracts of Papers of the American Chemical Society* 2005, *230*, U4116.
[54] Y. Mei, Y. Lu, F. Polzer, M. Ballauff, M. Drechsler, *Chemistry of Materials* 2007, *19*, 1062.
[55] Y. Mei, G. Sharma, Y. Lu, M. Ballauff, M. Drechsler, T. Irrgang, R. Kempe, *Langmuir* 2005, *21*, 12229.
[56] S. Proch, Y. Mei, J. M. R. Villanueva, Y. Lu, A. Karpov, M. Ballauff, R. Kempe, *Advanced Synthesis & Catalysis* 2008, *350*, 493.
[57] G. Sharma, M. Ballauff, *Macromolecular Rapid Communications* 2004, *25*, 547.
[58] S. Creutz, R. Jerome, *Langmuir* 1999, *15*, 7145.
[59] H. Mori, A. H. E. Muller, J. E. Klee, *Journal of the American Chemical Society* 2003, *125*, 3712.
[60] H. Mori, M. G. Lanzendorfer, A. H. E. Muller, J. E. Klee, *Langmuir* 2004, *20*, 1934.

[61] S. Forster, V. Abetz, A. H. E. Muller, in *Polyelectrolytes with Defined Molecular Architecture II*, M. Schmidt, (Ed.), Advances in Polymer Science, 2004, Vol. 166, pp 173.
[62] J. Kiwi, M. Grätzel, *Journal of the American Chemical Society* 1979, *101*, 7214.
[63] H. Bönnemann, W. Brijoux, in *Active Metals: Preparation Characterization Applications* (Ed.: A. Fürstner), Wiley-VCH, Weinheim, 1996, pp. 339.
[64] M. T. Reetz, W. Helbig, S. A. Quaiser, in *Active Metals: Preparation Characterization Applications* (Ed.: A. Fürstner), VCH, Weinheim, 1996, pp. 279.
[65] J. Dupont, *Abstracts of Papers of the American Chemical Society* 2002, *224*, U622.
[66] J. Lyklema, *Fundamentals of Interface and Colloid Science: Vol. IV Particulate colloids, Vol. IV Particulate colloids*, Elsevier, Amsterdam, 2005.
[67] A. Einstein, *Annalen der Physik* 1905, *17*, 549.
[68] M. Smoluchowski, *Annalen der Physik* 1906, *21*, 756.
[69] M. H. Kowsari, S. Alavi, M. Ashrafizaadeh, B. Najafi, *Journal of Chemical Physics* 2008, *129*, 224508
[70] M. H. Kowsari, S. Alavi, M. Ashrafizaadeh, B. Najafi, *Journal of Chemical Physics* 2009, *130*, 014703
[71] P. Wasserscheid, T. Welton, *Ionic Liquids in Synthesis*, Wiley-VCH, Weinheim, 2002.
[72] J. M. Slattery, C. Daguenet, P. J. Dyson, T. J. S. Schubert, I. Krossing, *Angewandte Chemie-International Edition* 2007, *46*, 5384.
[73] J. Jacquemin, P. Husson, A. A. H. Padua, V. Majer, *Green Chemistry* 2006, *8*, 172.
[74] V. Cimpeanu, M. Kocevar, V. Parvulescu, W. Leitner, *Angewandte Chemie, International Edition* 2009, *48*, 1085.
[75] A. Triolo, A. Mandanici, O. Russina, V. Rodriguez-Mora, M. Cutroni, C. Hardacre, M. Nieuwenhuyzen, H. J. Bleif, L. Keller, M. A. Ramos, *Journal of Physical Chemistry B* 2006, *110*, 21357.
[76] G. S. Fonseca, G. Machado, S. R. Teixeira, G. H. Fecher, J. Morais, M. C. M. Alves, J. Dupont, *Journal of Colloid and Interface Science* 2006, *301*, 193.
[77] L. S. Ott, R. G. Finke, *Inorganic Chemistry* 2006, *45*, 8382.
[78] L. S. Ott, S. Campbell, K. R. Seddon, R. G. Finke, *Inorganic Chemistry* 2007, *46*, 10335.
[79] E. Redel, R. Thomann, C. Janiak, *Inorganic Chemistry* 2008, *47*, 14.
[80] E. Redel, R. Thomann, C. Janiak, *Chemical Communications* 2008, 1789.
[81] Z. F. Fei, T. J. Geldbach, D. B. Zhao, P. J. Dyson, *Chemistry-a European Journal* 2006, *12*, 2123.
[82] D. B. Zhao, Z. F. Fei, T. J. Geldbach, R. Scopelliti, P. J. Dyson, *Journal of the American Chemical Society* 2004, *126*, 15876.
[83] Z. F. Fei, D. B. Zhao, D. Pieraccini, W. H. Ang, T. J. Geldbach, R. Scopelliti, C. Chiappe, P. J. Dyson, *Organometallics* 2007, *26*, 1588.
[84] H. Itoh, K. Naka, Y. Chujo, *Journal of the American Chemical Society* 2004, *126*, 3026.
[85] V. Humblot, S. Haq, C. Muryn, R. Raval, *Journal of Catalysis* 2004, *228*, 130.
[86] T. Osawa, T. Harada, O. Takayasu, *Current Organic Chemistry* 2006, *10*, 1513.
[87] A. Kraynov, D. Chen, V. Cimpeanu, J. Klankermayer, W. Leitner, in *Green Solvents - Progress in Science and Application*, Lake Constance, Friedrichshafen, Germany, 2008.
[88] A. Kraynov, E. Gebauer, W. Leitner, T. E. Müller, *International Scientific Journal for Alternative Energy and Ecology (ISJAEE)* 2010, *84*, 37.
[89] J. Dupont, G. S. Fonseca, A. P. Umpierre, P. F. P. Fichtner, S. R. Teixeira, *Journal of the American Chemical Society* 2002, *124*, 4228.

[90] G. S. Fonseca, A. P. Umpierre, P. F. P. Fichtner, S. R. Teixeira, J. Dupont, *Chemistry-a European Journal* 2003, *9*, 3263.
[91] C. W. Scheeren, G. Machado, J. Dupont, P. F. P. Fichtner, S. R. Texeira, *Inorganic Chemistry* 2003, *42*, 4738.
[92] E. T. Silveira, A. P. Umpierre, L. M. Rossi, G. Machado, J. Morais, G. V. Soares, I. L. R. Baumvol, S. R. Teixeira, P. F. P. Fichtner, J. Dupont, *Chemistry-a European Journal* 2004, *10*, 3734.
[93] D. S. McGuinness, K. J. Cavell, *Organometallics* 2000, *19*, 741.
[94] D. S. McGuinness, W. Mueller, P. Wasserscheid, K. J. Cavell, B. W. Skelton, A. H. White, U. Englert, *Organometallics* 2002, *21*, 175.
[95] L. S. Ott, M. L. Cline, M. Deetlefs, K. R. Seddon, R. G. Finke, *Journal of the American Chemical Society* 2005, *127*, 5758.
[96] J. Julis, M. Hölscher, W. Leitner, *Green Chemistry* 2010, *12*, 1634.
[97] P. Dash, N. A. Dehm, R. W. J. Scott, *Journal of Molecular Catalysis A-Chemical* 2008, *286*, 114.
[98] C. Zhao, H. Z. Wang, N. Yan, C. X. Xiao, X. D. Mu, P. J. Dyson, Y. Kou, *Journal of Catalysis* 2007, *250*, 33.
[99] C. X. Mao, H. Z. Wang, X. D. Mu, Y. Kou, *Journal of Catalysis* 2007, *250*, 25.
[100] P. A. Chernavskii, N. V. Peskov, A. V. Mugtasimov, V. V. Lunin, *Russian Journal of Physical Chemistry B* 2007, *1*, 394.
[101] H. Bönnemann, R. M. Richards, *European Journal of Inorganic Chemistry* 2001, 2455.
[102] M. T. Reetz, W. Helbig, S. A. Quaiser, Studiengesellschaft Kohle mbH, Mülheim an der Ruhr, Germany, 1996.
[103] H. G. Boyen, G. Kastle, F. Weigl, B. Koslowski, C. Dietrich, P. Ziemann, J. P. Spatz, S. Riethmuller, C. Hartmann, M. Moller, G. Schmid, M. G. Garnier, P. Oelhafen, *Science* 2002, *297*, 1533.
[104] Y. Xu, W. A. Shelton, W. F. Schneider, *Journal of Physical Chemistry A* 2006, *110*, 5839.
[105] X. H. Huang, I. H. El-Sayed, W. Qian, M. A. El-Sayed, *Journal of the American Chemical Society* 2006, *128*, 2115.
[106] D. K. Smith, B. A. Korgel, *Langmuir* 2008, *24*, 644.
[107] C. Gentilini, F. Evangelista, P. Rudolf, P. Franchi, M. Lucarini, L. Pasquato, *Journal of the American Chemical Society* 2008, *130*, 15678.
[108] A. B. R. Mayer, J. E. Mark, *European Polymer Journal* 1998, *34*, 103.
[109] X. Q. Zou, H. F. Bao, H. W. Guo, L. Zhang, Q. Li, J. G. Jiang, L. Niu, S. J. Dong, *Journal of Colloid and Interface Science* 2006, *295*, 401.
[110] C. H. Bartholomew, *Applied Catalysis A-General* 2001, *212*, 17.
[111] J. Oudar, *Catalysis Reviews-Science and Engineering* 1980, *22*, 171.
[112] M. Faraday, *Philosophical Transactions of the Royal Society of London* 1857, *147*, 145.
[113] M. Antonietti, D. B. Kuang, B. Smarsly, Z. Yong, *Angewandte Chemie-International Edition* 2004, *43*, 4988.
[114] M. Antonietti, E. Wenz, L. Bronstein, M. Seregina, *Advanced Materials* 1995, *7*, 1000.
[115] S. Forster, M. Antonietti, *Advanced Materials* 1998, *10*, 195.
[116] Z. G. Li, Z. Jia, Y. X. Luan, T. C. Mu, *Current Opinion in Solid State & Materials Science* 2009, *12*, 1.
[117] S. N. Sidorov, L. M. Bronstein, P. M. Valetsky, J. Hartmann, H. Colfen, H. Schnablegger, M. Antonietti, *Journal of Colloid and Interface Science* 1999, *212*, 197.

[118] A. Taubert, *Acta Chimica Slovenica* 2005, *52*, 183.
[119] A. Taubert, Z. Li, *Dalton Transactions* 2007, 723.
[120] M. Willert, R. Rothe, K. Landfester, M. Antonietti, *Chemistry of Materials* 2001, *13*, 4681.
[121] P. Migowski, J. Dupont, *Chemistry-a European Journal* 2007, *13*, 32.
[122] D. Astruc, F. Lu, J. R. Aranzaes, *Angewandte Chemie-International Edition* 2005, *44*, 7852.
[123] D. V. Goia, E. Matijevic, *New Journal of Chemistry* 1998, *22*, 1203.
[124] H. Bonnemann, K. S. Nagabhushana, *Journal of New Materials for Electrochemical Systems* 2004, *7*, 93.
[125] C. Burda, X. B. Chen, R. Narayanan, M. A. El-Sayed, *Chemical Reviews* 2005, *105*, 1025.
[126] A. Roucoux, J. Schulz, H. Patin, *Chemical Reviews* 2002, *102*, 3757.
[127] L. D. Pachón, G. Rothenberg, *Applied Organometallic Chemistry* 2008, *22*, 288.
[128] J. D. Aiken III, R. G. Finke, *Journal of Molecular Catalysis A-Chemical* 1999, *145*, 1.
[129] L. D'Souza, A. Suchopar, R. M. Richards, *Journal of Colloid and Interface Science* 2004, *279*, 458.
[130] H. Zheng, R. K. Smith, Y. Jun, C. Kisielowski, U. Dahmen, A. P. Alivisatos, *Science* 2009, *324*, 1309.
[131] H. Bönnemann, W. Brijoux, R. Brinkmann, E. Dinjus, T. Joussen, B. Korall, *Angewandte Chemie-International Edition in English* 1991, *30*, 1312.
[132] M. Moreno-Manas, R. Pleixats, *Accounts of Chemical Research* 2003, *36*, 638.
[133] X. D. Mu, D. G. Evans, Y. A. Kou, *Catalysis Letters* 2004, *97*, 151.
[134] D. B. Zhao, Z. F. Fei, W. H. Ang, P. J. Dyson, *Small* 2006, *2*, 879.
[135] R. Knapp, S. A. Wyrzgol, M. Reichelt, T. Hammer, H. Morgner, T. E. Müller, J. A. Lercher, *Journal of Physical Chemistry C* 2010, *114*, 13722.
[136] K. J. Klabunde, G. Cardenas-Trivino, in *Active Metals: Preparation Characterization Applications* (Ed.: A. Fürstner), VCH, Weinheim, 1996, pp. 237.
[137] S. Stoeva, K. J. Klabunde, C. M. Sorensen, I. Dragieva, *Journal of the American Chemical Society* 2002, *124*, 2305.
[138] K. J. Klabunde, Y. X. Li, B. J. Tan, *Chemistry of Materials* 1991, *3*, 30.
[139] K. Matsuo, K. J. Klabunde, *Journal of Organic Chemistry* 1982, *47*, 843.
[140] N. Panziera, P. Pertici, L. Barazzone, A. M. Caporusso, G. Vitulli, P. Salvadori, S. Borsacchi, M. Geppi, C. A. Veracini, G. Martra, L. Bertinetti, *Journal of Catalysis* 2007, *246*, 351.
[141] A. A. Ponce, K. J. Klabunde, *Journal of Molecular Catalysis A-Chemical* 2005, *225*, 1.
[142] A. B. Smetana, K. J. Klabunde, C. M. Sorensen, *Journal of Colloid and Interface Science* 2005, *284*, 521.
[143] S. I. Stoeva, B. L. V. Prasad, S. Uma, P. K. Stoimenov, V. Zaikovski, C. M. Sorensen, K. J. Klabunde, *Journal of Physical Chemistry B* 2003, *107*, 7441.
[144] S. I. Stoeva, A. B. Smetana, C. M. Sorensen, K. J. Klabunde, *Journal of Colloid and Interface Science* 2007, *309*, 94.
[145] M. Yoshinaga, H. Takahashi, K. Yamamoto, A. Muramatsu, T. Morikawa, *Journal of Colloid and Interface Science* 2007, *309*, 149.
[146] K. I. Okazaki, T. Kiyama, K. Hirahara, N. Tanaka, S. Kuwabata, T. Torimoto, *Chemical Communications* 2008, 691.
[147] T. Torimoto, K. Okazaki, T. Kiyama, K. Hirahara, N. Tanaka, S. Kuwabata, *Applied Physics Letters* 2006, *89*.

[148] N. Hajdukova, M. Prochazka, J. Stepanek, M. Spirkova, *Colloids and Surfaces A-Physicochemical and Engineering Aspects* 2007, *301*, 264.
[149] Y. Ishikawa, K. Kawaguchi, Y. Shimizu, T. Sasaki, N. Koshizaki, *Chemical Physics Letters* 2006, *428*, 426.
[150] J. Lee, D. K. Kim, W. Kang, *Bulletin of the Korean Chemical Society* 2006, *27*, 1869.
[151] H. Masuhara, T. Asahi, Y. Hosokawa, *Pure and Applied Chemistry* 2006, *78*, 2205.
[152] T. Tsuji, K. Iryo, Y. Nishimura, M. Tsuji, *Journal of Photochemistry and Photobiology A-Chemistry* 2001, *145*, 201.
[153] T. Tsuji, K. Iryo, H. Ohta, Y. Nishimura, *Japanese Journal of Applied Physics Part 2-Letters* 2000, *39*, L981.
[154] T. Tsuji, T. Kakita, M. Tsuji, *Applied Surface Science* 2003, *206*, 314.
[155] L. Yu, J. Yao, Z. W. Fu, *Acta Physico-Chimica Sinica* 2007, *23*, 945.
[156] C. He, T. Sasaki, Y. Shimizu, N. Koshizaki, *Applied Surface Science* 2008, *254*, 2196.
[157] K. Kawaguchi, J. Jaworski, Y. Ishikawa, T. Sasaki, N. Koshizaki, *IEEE Transactions on Magnetics* 2006, *42*, 3620.
[158] C. H. Liang, Y. Shimizu, T. Sasaki, N. Koshizaki, *Applied Physics A-Materials Science & Processing* 2005, *80*, 819.
[159] T. Sasaki, Y. Shimizu, N. Koshizaki, *Journal of Photochemistry and Photobiology A-Chemistry* 2006, *182*, 335.
[160] L. Zbroniec, T. Sasaki, N. Koshizaki, *Journal of Ceramic Processing Research* 2005, *6*, 134.
[161] Y. Kimura, H. Takata, M. Terazima, T. Ogawa, S. Isoda, *Chemistry Letters* 2007, *36*, 1130.
[162] M. A. Gelesky, A. P. Umpierre, G. Machado, R. R. B. Correia, W. C. Magno, J. Morais, G. Ebeling, J. Dupont, *Journal of the American Chemical Society* 2005, *127*, 4588.
[163] M. Rele, S. Kapoor, D. K. Palit, T. Mukherjee, *Research on Chemical Intermediates* 2004, *30*, 847.
[164] V. Amendola, M. Meneghetti, *Physical Chemistry Chemical Physics* 2009, *11*, 3805.
[165] S. Kim, B. K. Yoo, K. Chun, W. Kang, J. Choo, M. S. Gong, S. W. Joo, *Journal of Molecular Catalysis A-Chemical* 2005, *226*, 231.
[166] M. T. Reetz, W. Helbig, *Journal of the American Chemical Society* 1994, *116*, 7401.
[167] M. T. Reetz, S. A. Quaiser, *Angewandte Chemie-International Edition in English* 1995, *34*, 2240.
[168] M. T. Reetz, G. Lohmer, *Chemical Communications* 1996, 1921.
[169] M. T. Reetz, S. A. Quaiser, R. Breinbauer, B. Tesche, *Angewandte Chemie-International Edition in English* 1996, *34*, 2728.
[170] Y. Katayama, Y. Bando, T. Miura, *Transactions of the Institute of Metal Finishing* 2008, *86*, 205.
[171] W. Dobbs, J. M. Suisse, L. Douce, R. Welter, *Angewandte Chemie-International Edition* 2006, *45*, 4179.
[172] I. G. Koo, M. S. Lee, J. H. Shim, J. H. Ahn, W. M. Lee, *Journal of Materials Chemistry* 2005, *15*, 4125.
[173] J. Shim, K. Y. Joung, J. H. Ahn, W. M. Lee, *Journal of the Electrochemical Society* 2007, *154*, B165.
[174] J. J. Zou, Y. P. Zhang, C. J. Liu, *Langmuir* 2006, *22*, 11388.
[175] S. A. Meiss, M. Rohnke, L. Kienle, S. Z. El Abedin, F. Endres, J. Janek, *ChemPhysChem* 2007, *8*, 50.

[176] M. Brettholle, O. Höfft, L. Klarhöfer, S. Mathes, W. Maus-Friedrichs, S. Zein El Abedin, S. Krischok, J. Janek, F. Endres, *Physical Chemistry Chemical Physics* 2009, *12*, 1750.
[177] S. Z. El Abedin, M. Polleth, S. A. Meiss, J. Janek, F. Endres, *Green Chemistry* 2007, *9*, 549.
[178] A. Henglein, *Topics in Current Chemistry* 1988, *143*, 113.
[179] A. Henglein, *Berichte der Bunsen-Gesellschaft-Physical Chemistry Chemical Physics* 1995, *99*, 903.
[180] A. Henglein, *Chemistry of Materials* 1998, *10*, 444.
[181] A. Henglein, B. G. Ershov, M. Malow, *Journal of Physical Chemistry* 1995, *99*, 14129.
[182] A. Henglein, D. Meisel, *Langmuir* 1998, *14*, 7392.
[183] J. Belloni, M. Mostafavi, H. Remita, J. L. Marignier, M. O. Delcourt, *New Journal of Chemistry* 1998, *22*, 1239.
[184] M. O. Delcourt, J. Belloni, J. L. Marignier, C. Mory, C. Colliex, *Radiation Physics and Chemistry* 1984, *23*, 485.
[185] E. Gachard, H. Remita, J. Khatouri, B. Keita, L. Nadjo, J. Belloni, *New Journal of Chemistry* 1998, *22*, 1257.
[186] J. L. Marignier, J. Belloni, M. O. Delcourt, J. P. Chevalier, *Nature* 1985, *317*, 344.
[187] R. Abargues, J. Marques-Hueso, J. Canet-Ferrer, E. Pedrueza, J. L. Valdes, E. Jimenez, J. P. Martinez-Pastor, *Nanotechnology* 2008, *19*.
[188] A. Imanishi, M. Tumura, S. Kuwabata, *Chemical Communications* 2009, 1775
[189] C. Y. Wang, C. Y. Liu, X. Zheng, J. Chen, T. Shen, *Colloids and Surfaces A-Physicochemical and Engineering Aspects* 1998, *131*, 271.
[190] C. Y. Wang, C. P. Liu, J. Chen, T. Shen, *Journal of Colloid and Interface Science* 1997, *191*, 464.
[191] T. Alammar, A. V. Mudring, *Journal of Materials Science* 2009, *44*, 3218.
[192] M. Malenovska, M. A. Neouze, U. Schubert, A. Peled, J. P. Lellouche, *Dalton Transactions* 2008, 4647.
[193] A. Taubert, I. Arbell, A. Mecke, P. Graf, *Gold Bulletin* 2006, *39*, 205.
[194] K. Eller, H. Schwarz, *Chemical Reviews* 1991, *91*, 1121.
[195] S. K. Kumar, L. Greiner, A. Kraynov, T. E. Müller, B. Niemeijer, W. Leitner, *Chemical Communications* 2010, *46*, 6705.
[196] T. E. Müller, *The chemical engineer (tce)* 2009, *813*, 42.
[197] T. E. Müller in *Supported Ionic Liquids. Fundamentals and Applications*, R. Fehrmann, A. Riisager, M. Haumann (Eds.), Wiley-VCH, Weinheim, 2011.
[198] T. Osawa, T. Harada, O. Takayasu, *Current Organic Chemistry* 2006, *10*, 1513.

13

Studies Regarding the Nickel Electrodeposition from Choline Chloride Based Ionic Liquids

Liana Anicai[1], Andreea Florea[1] and Teodor Visan[2]
[1]*Direction of Research, PSV Company SA, Bucharest*
[2]*Department of Physical Chemistry and Applied Electrochemistry, University POLITEHNICA of Bucharest, Romania*

1. Introduction

The electrochemical technologies are extensively applied in a large range of industrial fields, from decorative/protective coatings in electrical engineering, electronics, automotive or machine building domains towards the manufacturing of microelectronic components, of micro-electro-mechanical systems (the so-called MEMS) or of the high precision micro- and nanostructures, including biological ones (Data & Landolt, 2000; Gurunathan et al., 1999). Due to the fact that electrochemical deposition is an atomic/molecular level process, the formed layer entirely takes the three-dimensional shape of the substrate with a very high accuracy.

As a result of the last years scientific progress, the theoretical aspects of electrochemical engineering principles applicable for electrochemical micro-processing have significantly developed so that, associated with performance techniques and equipment lead to the new and innovative materials development having controlled micro- or nanostructures.

In spite of a long history and experience, the metals surface finishing industry still has not been successful in extending the range of the metals able to be electrodeposited towards some of them with a promising technical potential, such as Al, W, Mo, Ti. Their capability to form stable oxides represents an obstacle against their application as coating layers, especially from classical aqueous electrolytes.

Moreover, in the last decade sustained efforts have been made to preserve the environment and to minimize the toxic effects of certain electrochemical technologies on human health. Thus, certain metallic coatings have been and still are subjected to some use restrictions such as Cr, Ni, but Cu, Ag and Au as well, due to the involvement of aqueous cyanide electrolytes.

Taking into account the above mentioned considerations, in the last 10 years investigations to develop and promote alternative more environmentally friendly electrochemical media have been reported, including the class of the so-called “ionic liquids”, defined as “ionic materials in liquid state for temperatures lower than 100°C” (Endres et al., 2008; Wasserscheid & Welton, 2007).

The ionic liquids have some properties which make them adequate as metal electrodeposition electrolytes, such as:

- wide potential window;
- high solubility of metallic salts;
- minimization/absence of secondary reactions due to the water presence;
- higher conductivity as compared with non-aqueous/organic solvents.

These types of ionic media thus may offer the possibility to obtain metallic coatings that are hard or impossible to be obtained in classical aqueous solutions or to apply these coating layers with a suitable adherence on water sensitive metallic substrates such as Al, Mg, Ti and their alloys, stainless steels, other alloys containing high contents of refractory or rare earth metals (Endres et al., 2008; Endres, 2002; Mukhopadhyay et al., 2005).

Recently it has been shown the possibility of formation of ionic liquids from eutectic mixtures of quaternary ammonium salt such as choline chloride (2-hydroxy-ethyl-trimethyl ammonium chloride) with a hydrogen bond donor species such as amides, glycols or carboxylic acids (Abbott et al., 2003; Endres, 2002; Endres et al., 2008; Mukhopadhyay et al., 2005; Wasserscheid & Welton, 2007). These media, also known as "deep eutectic solvents", have been further used to electrodeposit a large range of metals and alloys including Zn and Zn alloys, Cr, Sn and Sn alloys, Cu and Ag (Abbott et al., 2003; Abbott et al., 2006; Abbott et al., 2007; Cojocaru et al., 2009; Endres et al., 2008; Florea et al., 2010).

Having in mind all of the above, the chapter will present some general information regarding the preparation and characterization of choline chloride based ionic liquids, as well as their use as plating electrolytes. Afterwards, several experimental results obtained by our group are discussed, regarding the electrochemical Ni deposition which allowed us to get more consistent information on the technological parameters influence during plating onto various metallic substrates (e.g. Cu, Al, Mg). $NiCl_2$ and for the first time $NiSO_4$ have been involved as precursors for the metallic cations.

2. General aspects regarding the synthesis and characterization of choline chloride based ionic liquids

An ionic medium with interesting perspectives in metals electrodeposition is that based on choline chloride (2-hydroxy-ethyl-trimethyl ammonium chloride) mixed with urea. They proved to be very attractive in the field of metal and alloy electrodeposition. The ionic liquid is formed due to the hydrogen bonds formed between the ammonium quaternary salt and amide.

These new ionic media are potentially recyclable, biodegradable and with no proven adverse effects on human health. Thus they show good potential for a "green alternative" for metal and alloy electrodeposition as well as for other various chemical processes. Currently, the interest is focused towards those kinds of ionic liquids formed by the mixture of choline chloride with a metal salt, alcohol, amide or organic acid (Abbott et al., 2001; Abbott et al., 2003; Abbott et al., 2003; Abbott et al., 2004; Abbott et al., 2006; Endres 2002). This electrochemical medium is also characterized by good air and water stability with no additional precautions.

Abbott et al.(Abbott et al., 2004) have investigated in detail the possibility of formation of various ionic liquids based on choline chloride which usually are characterized by a freezing point of up to 100°C, but more usual around 60°C, through the reaction between a quaternary ammonium salt (compound A) having the general formula:

$$R^1R^2R^3R^4\,N^+X^- \qquad (1)$$

and an organic compound B which is in a solid form at 20°C and is able to form hydrogen bonds with X^-.
R^1, R^2, R^3 and R^4 can be:

- H,
- a C_1 to C_5 alkyl group or
- a C_6 to C_{10} cycloalkyl group or alkylene group or
- a C_6 to C_{12} alkyl or cycloalkyl group, substituted with various functional groups such as: OH^-, Cl, Br, F, I, NH_2, etc.

X^- represents a suitable counter ion.
The organic compound B may usually be urea, aniline, hydroxyaliphatic acids, dicarboxylic acids, phenol, ethylene glycol or citric acid.
A and B compounds are mixed in molar ratios between 1:1.5 up to 1:2.5. Generally, the preferred quaternary ammonium salts are those exhibiting asymmetry and R^1, R^2, R^3 and R^4 are recommended not to be identical.
Choline chloride mixtures with various types of B compounds may be considered as eutectic ones, having the melting point much lower as compared with that of the individual components (Abbott et al., 2004).
Choline chloride – carboxylic acids eutectic mixtures are easy to prepare, they don't react with water and are usually biodegradable. The components toxicological characteristics are already well known. These mixtures have led to the identification of a new class of solvents whose physical properties are similar to the "traditional" ionic liquids and are correlated with the number of functional groups, of alkyl/aryl substituents and with the mixture composition. These "eutectic solvents" dissolve various metallic oxides so that they may be considered as potential suitable media for the metal extraction (Abbott et al., 2004; Abbott et al., 2004).
The deep eutectic solvents based ionic liquids may be thus described by the general formula:

$$R_1R_2R_3R_4N^+X.zY \qquad (2)$$

and they have been characterized into three types depending on the complexing agent, Y:
Type I Y = MCl_x, M = Zn, Sn, Fe, Al, Ge
TypeII Y = MCl_x yH_2O, M = Cr, Cu, Co, Ni, Fe
Type III Y = RZ, Z = $CONH_2$, COOH, OH
X is usually a halide anion (mostly Cl^-). (Abbott et al., 2001; Abbott et al., 2004; Abbott & McKenzie, 2006).
In the case of Type I and II eutectics the potential window is limited at high potentials by chlorine gas evolution and at low potentials by the metal ion reduction with metal deposition from the melt. Type I eutectics have been prepared using Zn, Sn, Fe, Al, Ge and Cu chlorides. Their reduction potential is shifted towards more electronegative values as the metal halide is closer by Lewis acid. Because the reduction potential is associated with Lewis acidity, the corresponding proportions of metal and quaternary ammonium salts affect the potential window. Type II eutectics have been developed in order to extend the range of metals able to be electrodeposited from ionic liquids and Cr electrodeposition with good characteristics has been reported (Abbott et al., 2004; Abbott et al., 2004; Benaben & Sottil, 2006). Hydration water plays a significant role on the stability and fluidity of choline chloride based ionic liquids. In this case water behaviour is different compared to the case of aqueous electrolytes and the potential window is limited rather by the metallic species

reduction than the water-related reduction processes. The metal can be reduced with a high cathodic efficiency and the reaction is not affected by a max. 10% water content; this phenomenon suggests the water is associated either with chloride anions or with the central metal.

The eutectic based ionic liquids are considerably less sensitive to water addition and sometimes its presence has a positive effect on the deposit morphology.

In the case of ionic liquids containing high amounts of halides, the chemical attack of the metallic oxides at the electrode surface occurs; that means that only few metals are really inert in these media and consequently the anodic potential limit is governed rather by the metal oxidation. Even metals such as Pt, Au, Al and Ti may be oxidized in ionic liquids and their oxides can be dissolved especially in eutectics containing chloride ions as ligands for metallic ions dissolution. This phenomenon allows the use of soluble anodes during electrodeposition procedure.

The process of metals electrodissolution in ionic liquids has been applied to develop several electropolishing procedures suitable for stainless steel, when a Type III eutectic has been involved containing ethylene glycol as the hydrogen bond donor (Abbott et al., 2004; Abbott et al., 2006) The anodic dissolution occurs in the same manner as in the case of the aqueous phosphoric and sulphuric acids electrolytes but the current efficiency was significantly improved up to about 90% and the ionic medium is much less toxic and corrosive.

Abbott et al (Abbott et al., 2004; Abbott et al., 2004; Abbott et al., 2004; Benaben & Sottil, 2006) have reported several investigations regarding the formation of choline chloride based ionic liquids with various kinds of hydrated metallic salts, potentially able to be applied for electrodeposition and electropolishing. The involved metallic compounds have chlorides or nitrates as anions and the freezing point is up to 50°C. When other types of metallic salts have been employed, such as acetates, sulphates, the resulted ionic compound doesn't become solid at temperatures over 50°C but the liquid state is maintained on a narrower domain.

The use of ionic liquids to perform various metal surface treatments allows the replacement of several toxic chemical reagents such as CrO_3 in the case of chromium plating from aqueous electrolytes and KCN involved in Ag, Cu and Zn electrodeposition.

When ionic liquid systems are intended to be applied for electrodeposition their behaviour has to be assessed as compared with the case of aqueous electrolytes. The main factors which affect the overall electrochemical process include viscosity, conductivity, the potential window, the ionic medium chemistry as well as the structure of the electrical double layer and redox potentials. All these parameters will influence the diffusion rate of metallic ions at the electrode surface as well as the thermodynamics and kinetics of the reduction process. Consequently, the nucleation/growth mechanisms and the deposit morphology will be affected, too. More detailed discussions on this topic may be found in (Abbott et al., 2004; Abbott et al., 2004; Abbott & McKenzie, 2006; Abbott et al., 2007; Endres et al., 2008 and included references).

3. Metal and alloy electrodeposition from choline chloride based ionic liquids

The electrodeposition of metals and alloys in this type of ionic liquid appeared attractive because they are air and moisture stable. Moreover, some of these systems are very good candidates as replacement technologies for aqueous processes that often require strong inorganic acids or toxic co-ligands. Abbott and his group had a significant contribution on this topic through reporting a large range of experimental results regarding metal and alloy plating including Zn and Zn alloys, Cr, Sn and Sn alloys, Cu and Ag (Abbott et al., 2004; Abbott et al., 2007; Abbott et al., 2007; Abbott et al., 2008; Abbott et al., 2009; Abbott et al., 2009; Benaben & Sottil, 2006).

Zn and Zn alloys electrodeposition has attracted a great interest due to their large applications as anticorrosive layers in a wide range of industrial fields. Moreover, they represent an environmentally friendly alternative to Cd depositions as well. Regarding the electrodeposition process, Abbott et al. (Abbott et al., 2009) found that nucleation in the urea based liquid is fast but bulk growth is slow, whereas nucleation in the glycol based liquid is slow but bulk growth is relatively fast. Critical surface coverage necessary for the transition from nucleation to bulk growth is similar for both liquids. Whitehead et al.(Whitehead et al., 2010) have investigated the zinc deposition process from a basic choline chloride/ethylene glycol deep eutectic solvent containing $ZnCl_2$ at 30°C. On sweeping or stepping the potential to −0.5 to −0.8 V vs. Zn/Zn(II), only little deposition takes place initially, with a more rapid deposition being observed when the potential was subsequently raised to −0.4 to −0.2 V. The role of choline chloride was also studied by comparing with a choline-free electrolyte, which exhibited a more conventional voltammetric response. The formation of a dissolved, intermediate species Z on the cathodic sweep was proposed to account for the observed deposition behavior in the deep eutectic. Furthermore, an observation of the electrodeposition behavior with the addition of sodium ethoxide supported the suggestion that Z is a complex of Zn^{2+} and deprotonated components of the solvent. To improve the appearance of Zn coatings from ionic liquids, several brighteners have been investigated, such as acetonitrile, ethylene diamine and ammonia (Abbott et al. 2010). Ammonia was initially chosen as previous studies have determined that Zn is effectively complexed by this ligand in aqueous sulphate electrolytes. Acetonitrile and ethylene diamine were chosen to represent a weaker and a stronger nitrogen containing ligand respectively. Ammonia and ethylene diamine have shown to be effective brighteners for Zn deposition. EXAFS investigation proved however that this was not because of the Zn ions complexation but rather due to the different hydrogen bond strengths of the additives.

Some experimental investigations dealing with Cr electrodeposition from choline chloride based ionic liquids have been reported in (Abbott et al., 2004; Abbott et al., 2004; Abbott et al., 2004; Benaben & Sottil, 2006), involving a Type II eutectic with $CrCl_3 \cdot 6H_2O$. The obtained results have shown that thick, adherent, crack free deposits of chromium black can be electrodeposited from an ionic liquid containing Cr (III). The morphology of the deposits can be made nanocrystalline by the incorporation of LiCl into the ionic liquid. The chromium containing species present in the liquid change following the addition of LiCl and the species $[Cr(H_2O)_2Cl_4]^-$has been identified for the first time. The chromium films showed corrosion resistance following a post-treatment electrolysis in aqueous potassium nitrate solutions.

Tin and tin alloys with Zn (Abbott et al., 2007) or Ni (Cojocaru et al., 2009; Cojocaru et al., 2009) have been also electrodeposited from deep eutectic solvents. It has been found that the nature of hydrogen bond donor significantly influences the ionic liquid electrochemistry and the morphology of the deposit. In the case of ZnSn alloy the morphology and composition can be changed by judicious choice of the ionic liquid. It was proposed that metal speciation is a cause of metal reduction thermodynamics.

Ni-Sn alloy morphology was also found to be dependent on the ionic liquid composition and on operating parameters, but the chemical composition was found to be quite constant, regardless the value of the applied current density (Florea et al., 2010).

Abbott et. Al (Abbott et al., 2009) described the electrolytic deposition of copper and copper composites from a solution of the metal chloride salt in either urea–choline chloride, or ethylene glycol–choline chloride based eutectics. It has been found that the deposition kinetics and thermodynamics are quite unlike those in aqueous solution under comparable conditions and that the copper ion complexation is also different. The mechanism of copper nucleation was investigated using chronoamperometry and it was shown that progressive nucleation leads to a bright nano-structured deposit. In contrast, instantaneous nucleation, at lower concentrations of copper ions, leads to a dull deposit. The electrochemical quartz crystal microbalance (EQCM) has been involved to monitor both current efficiency and the inclusion of inert particulates into the copper coatings. It was found that the composition of composite material was strongly dependent on the amount of species suspended in solution. The majority of material was dragged onto the surface rather than settling on to it. The distribution of the composite material was found to be even throughout the coating. This technology may be important because it facilitates deposition of bright copper coatings without co-ligands such as cyanide.

Abbott et al (Abbott et al., 2007; Abbott et al. 2008) have reported for the first time the electroless Ag deposition onto Cu substrates from Ag ions containing choline chloride - ethylene glycol ionic liquids, which represents a technological procedure with a reduced impact on environment. The authors have shown that Ag deposition on Cu takes place due to the thermodynamic reduction of Ag^+ and to the Cu^+ dissolution which are determined by the redox potential difference of the two involved metallic ions. In these eutectic mixtures, the hydrogen bond donor species is ethylene glycol. The obtained experimental results have supported a deposition mechanism involving a rapid nucleation followed by a slower growth step that suggests a limitation due to the mass transport. The Ag layer adherence is adequate to prevent Cu substrate corrosion and welding procedures on printed circuit tracks have been successfully tested. Florea et al. (Florea et al., 2010) showed some preliminary results regarding electrochemical Ag plating involving choline chloride and triethanolamine based ionic liquids which allowed successful deposition of Ag layers of a good quality and adherence to the Cu substrate. Based on XRD measurements, the average size of crystallites was about 28-35 nm for pure Ag deposits.

Abbott et al. (Abbott et al., 2008) have reported several preliminary investigations regarding Ni coatings from nickel chloride dissolved separately in either a urea or ethylene glycol/choline chloride based ionic liquid. From cyclic voltammograms the shape of the stripping peaks depended upon sweep rates, illustrating that the process is kinetically slow. The morphology and composition can be changed by the judicious choice of IL and the addition of complexing agents such as ethylenediamine and acetylacetonate as brighteners. Both additives suppress UPD of Ni and lead to smaller particulate deposits.

More detailed investigations performed by our group dealing both with the synthesis and characterization of the formed nickel salts (chloride and for the first time sulphate) containing ionic liquids and the electrochemical Ni deposition onto various metallic substrates to get more consistent information on the technological parameters, will be presented in the next sections.

4. Nickel electrodeposition from choline chloride based ionic liquids

4.1 Synthesis and characterization of ionic liquids systems containing nickel salts

Various ionic liquids based on choline chloride (noted as ChCl) - urea (symbolized IL) and choline chloride - ethylene glycol (symbolized ILEG) (1:2 molar ratio) eutectics have been synthesized involving various Ni salts, respectively $NiCl_2.6H_2O$ and $NiSO_4.7H_2O$ with

concentrations between 0-1.68M. They have been further characterized with regard to the preparation conditions, appearance, fluidity against applied temperature, time stability. All chemical reagents have been of analytical grade. Choline chloride (Aldrich) has been used as received. The dependence of electrical conductivity against the temperature has been determined for several systems, using a WTW (Germany) 340i model conductometer with the cell constant K=0,469 cm^{-1}.
In the case of IL - *$NiCl_2$ containing mixtures*, all of them have been liquids at room temperature (25°C) and the freezing point attained 5°C. Their color varied from green to dark green depending on the Ni (II) salt content, as it can be also seen in Figure 1.

Fig. 1. Ionic liquids containing $NiCl_2$ in various concentrations

Usually, the electrical conductivity takes values between 1-14 mS/cm for a temperature interval between 25-80°C. Slightly higher values are evidenced in the case of $NiCl_2$ low content in the mixture, as it can be also noticed in Figure.2. At a 0.5M Ni(II) the conductivity exhibited a minimum. This behaviour might be due to the possible formation of several metallic complexes, $M^{x+}Cl_{x+}1$, through the dissolution of $M^{x+}Cl_x$ metallic halide in this eutectic solvent, whose speciation has not yet explained (Abbott et al., 2008).
The conductivity of ionic liquids σ varies with temperature according to equation (3):

$$\ln\sigma = \ln\sigma_o - \frac{E_\Lambda}{R.T} \qquad (3)$$

where E_Λ is the activation energy for conduction, T is the absolute temperature and σ_o is a constant. Consequently, from semi logarithmic Arrhenius type plots (Abbott et al., 2004; Florea, 2010) activation energies between 25-28 kJ/mol have been determined (r > 0.988), which are slightly higher as compared to those of high temperature molten salts.
For the above presented systems, viscosities (η) in the range of 50-700 cP have been measured for temperatures between 25°C – 70°C. These values are significantly lower as compared with other choline chloride–metallic salts systems, respectively ChCl- $CrCl_3.6H_2O$ (Abbott et al., 2004) or ChCl-$ZnCl_2$ (Abbott et al., 2004). In the case of ionic fluids, the conductivity is usually governed by the charge carriers mobility. Consequently, the σ- $1/\eta$ representations should be approximately linear for a certain charge carrier. This situation might be more complex in the case of ionic liquids containing metallic halides (in our case $NiCl_2$) due to the possibility of various complex compounds formation. Figure 3 shows this

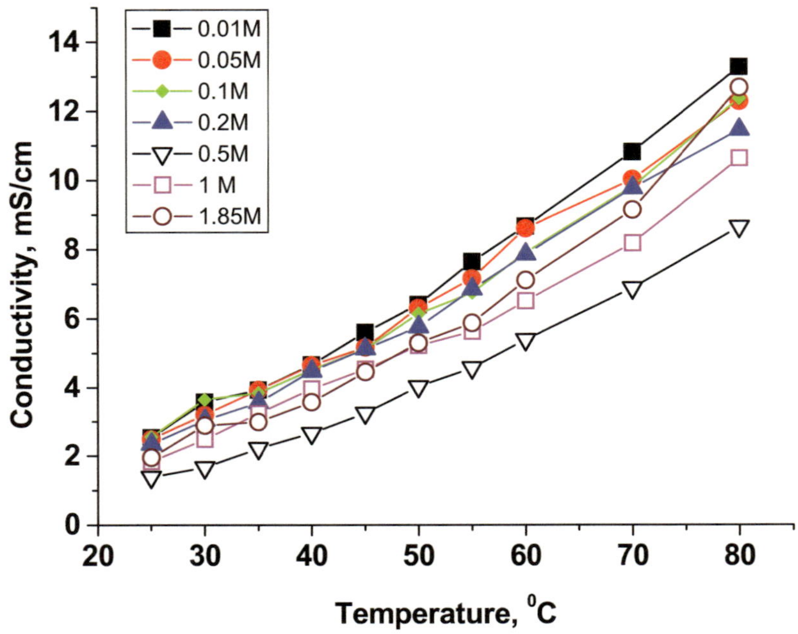

Fig. 2. Dependence of conductivity *vs.* temperature for IL systems containing various nickel chloride concentrations.

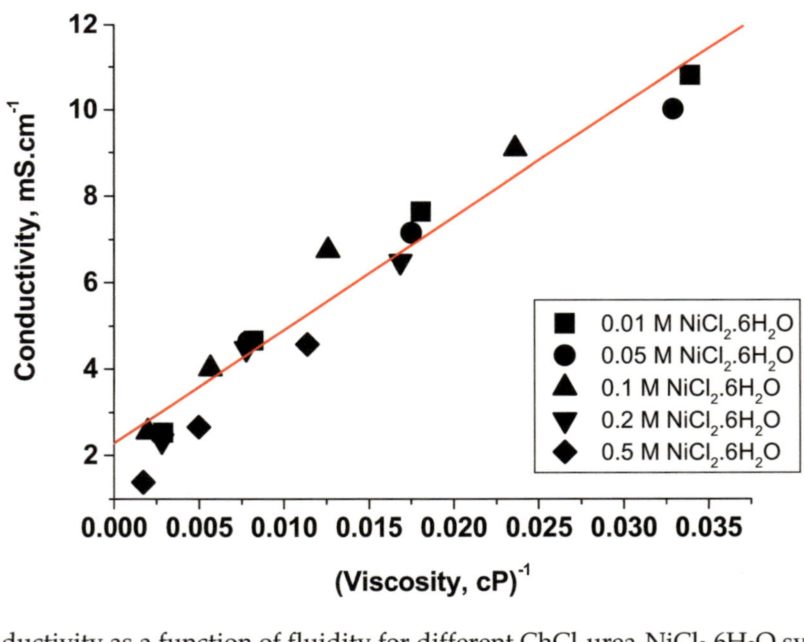

Fig. 3. Conductivity as a function of fluidity for different ChCl-urea-$NiCl_2.6H_2O$ systems.

dependence for several concentrations of the dissolved metallic salt in the ChCl-urea eutectic mixture. It can be evidenced that the ionic mobility is controlled by the ionic liquid conductivity and the slope is almost constant for all analyzed compositions, thus suggesting the presence of a single charge carrier species. These results are in agreement with other literature data (Abbott et al., 2004; Abbott et al., 2004).

The possibility of the synthesis of choline chloride based systems but where the metallic cation source was $NiSO_4$ was also investigated, since according to the environmental legislation and to REACH requirements (**R**egistration, **E**valuation and **A**uthorisation of **Ch**emicals - the REACH system), there are discussions dealing with the restricted use of $NiCl_2$ due to its potential risk on human health.

In the case of *ionic liquids based on Ni (II) sulfate*, the maximum salt concentration allowing a liquid state was found to be 1M, higher values leading to the formation of a solid mixture. Typically, the electrical conductivity is situated in the range of 1-12 mS/cm for a temperature domain between 20-80°C.Slightly higher values were measured for lower Ni(II) concentrations. Figure 4 presents examples of the semi logarithmic Arrhenius type plots in the case of ChCl-urea-$NiSO_4.7H_2O$ systems. The determined values of activation energies (E_A) as shown in Table 1, are relatively larger than those characteristic to high temperature molten salts.

The same investigations were performed but using ILEG as solvent. According to the experimental results (not shown here), green, stable liquids at room temperature (25°C) were obtained regardless of the anion of the Ni salt, having conductivities in the range of 2 - 22 mS/cm for a temperature range between 20-80°C.

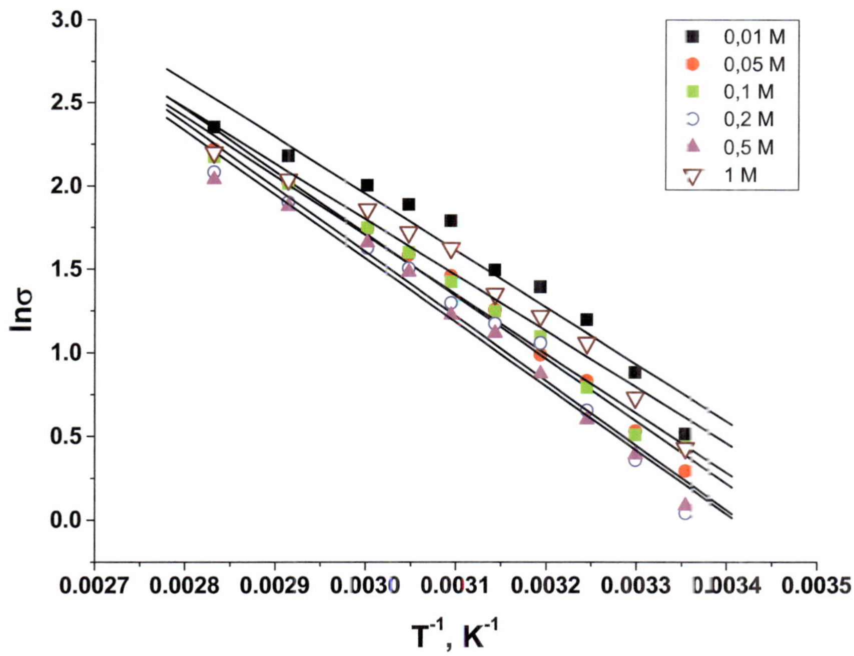

Fig. 4. Plots of log conductivity vs. reciprocal of temperature for IL systems containing various nickel sulfate concentrations.

Molar concentration of $NiSO_4.7H_2O$ in ChCl:urea IL as solvent, Mol/L	E_Λ, kJ/mol	r
0.01	28.3	0.975
0.05	31.04	0.993
0.1	29.5	0.991
0.2	32.07	0.981
0.5	31.75	0.989
1	27.8	0.981

Table 1. Activation energies for conductivity (E_Λ) against ionic liquid composition (Coefficients of correlation (r) are also given).

The activation energies were found to be between 11 - 22 kJ/mol for ILEG-$NiCl_2$ systems and between 12-24 kJ/mol for ILEG-$NiSO_4$ ones. It should be noted that the use of ChCl-ethylene glycol eutectic as solvent led to slightly lower values of activation energy as compared with the case of ChCl-urea mixture, probably due to a relatively different speciation (Abbott et al., 2008).

4.2 Electrochemical characterization of processes in Ni(II) compounds based ionic liquids through cyclic voltammetry

To obtain more information on the electrochemical processes occurring in choline chloride based systems containing various amounts of nickel salts (chloride and sulfate), cyclic voltammetry experiments have been performed using an Autolab PGSTAT 12 potentiostat controlled with GPES software. A three electrode system consisting of a Pt working electrode with a constant geometrical surface of 0.19625 cm^2, a graphite counter electrode and a silver wire quasi-reference electrode was used. The working electrode surface has been mechanically prepared using 1200 abrasive paper, rinsed and dried before each measurement. All voltammograms have been recorded in stationary conditions, at a potential sweep rate of 10 mV/s. The Ni^{2+} concentration has been in the range of 0-1.85M for metallic chlorides and of 0-1M for sulfates.

Several examples of the recorded voltammograms at different applied working temperatures are presented in Figures 5, 6 and 7, for the case of choline chloride-urea based ionic liquids where the cation source has been provided as chloride.

As it can be seen from the figures, a quasi-reversible behaviour is evidenced on the investigated range of potential. The Ni^{2+} reduction process on the cathodic branch occurs at around -0.81 ÷ -1.3 V/Ag quasi-ref. at relatively low temperatures between 30-40°C, which shifts towards more electropositive values of -0.55 ÷ -0.85 V/Ag quasi-ref. as temperature increased to 80°C.

Increasing the Ni^{2+} concentration from 0.01M to 1.85M shifts the cathodic potential towards more electronegative values especially at higher values of the applied temperature. When the metal content is higher than 0.2M and the working temperature is in the range of 40°C-60°C a small cathodic limiting current plateau is evidenced of about 2.5 - 9.8 mA/cm^2, higher as the metal content and temperature increased. It should be noted that the reduction process in this case is not so well defined compared to the electrochemical behaviour of other metals in the same ionic medium, such as Cu and Ag (Abbott et al., 2005; Abbott et al., 2008; Abbott et al., 2009; Anicai et al., 2008).

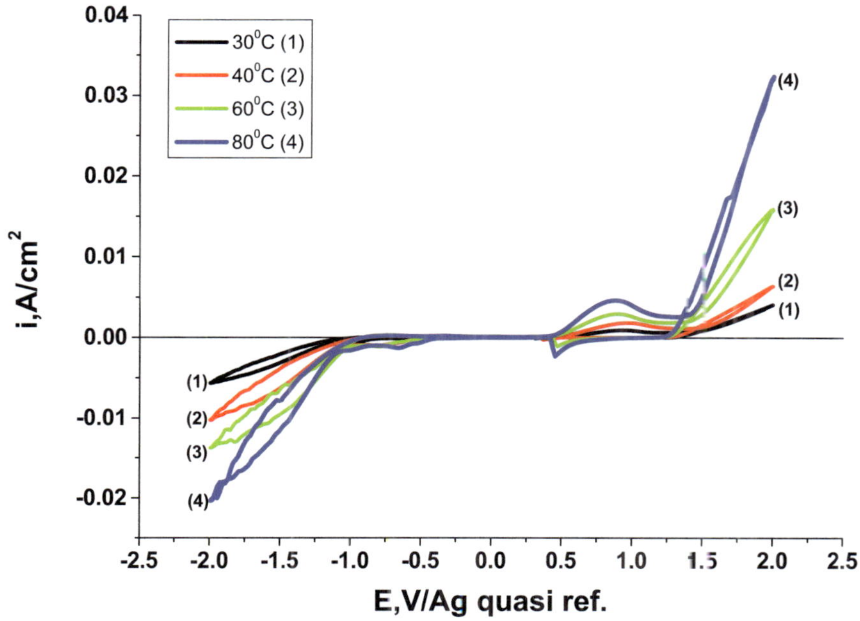

Fig. 5. Cyclic voltammograms for choline chloride-urea-$NiCl_2$ systems containing 0.1M Ni^{2+} at various temperature values.

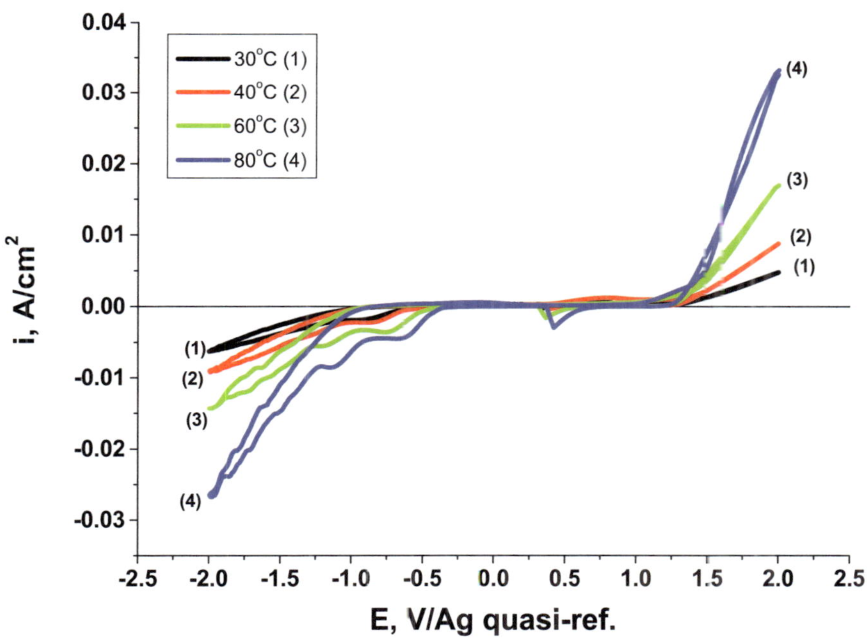

Fig. 6. Cyclic voltammograms for choline chloride-urea-$NiCl_2$ systems containing 0.5M Ni^{2+} at various temperature values.

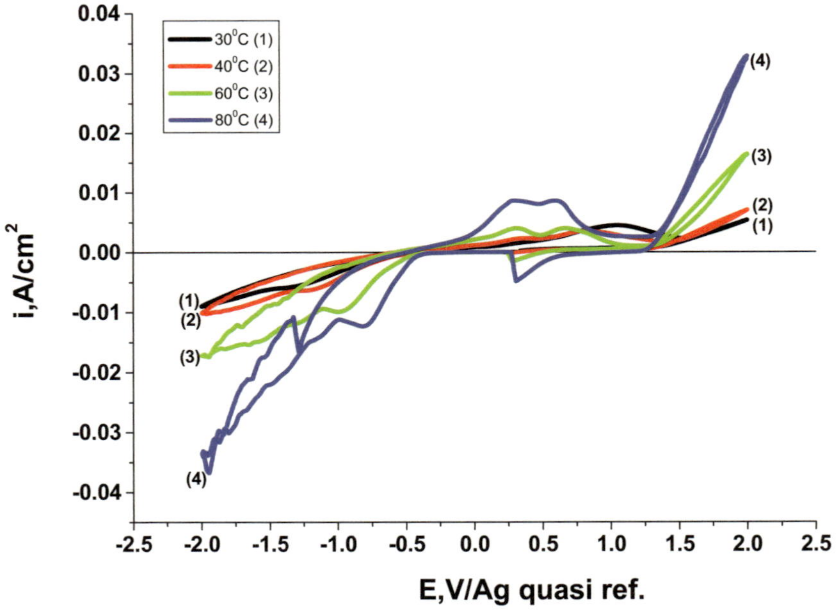

Fig. 7. Cyclic voltammograms for choline chloride-urea-$NiCl_2$ systems containing 1M Ni^{2+} at various temperature values.

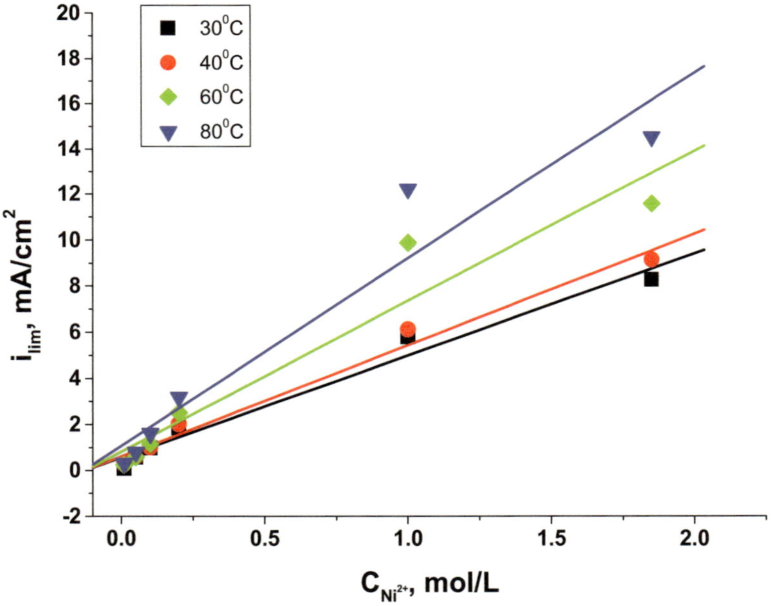

Fig. 8. Plots of cathodic limiting current *vs.* Ni^{2+} concentration for IL-$NiCl_2$ systems at different temperatures.

On the anodic scan the dissolution peak is observed at around 0.85 - 1.05 V/Ag quasi-ref., and shifts to more negative values as the operating temperature increases.
If the dependence of the cathodic limiting current i_{lim} against Ni^{2+} concentration $-c_{N(II)}$ is analysed one can find a linear relationship for the investigated temperature domain, as is also shown in Figure 8. The following empirical equation may be written:

$$i_{\lim} = A + B \cdot c_{Ni^{2+}} \tag{4}$$

where A and B are y-intercept and slope of the lines from Figure 8, respectively, which are temperature dependent. A and B values against temperature are shown in Table 2.

T, °C	A	B
30	0.56253	4.41926
40	0.60644	4.82317
60	0.81778	5.53998
80	1.07909	8.1366

Table 2. Values of A and B at different temperatures.

The dependencies of A and B against 1/T as are shown in Figures 9 and 10 are linear, too, so that the equations (5) and (6) can be written, respectively:

$$\text{Log } A = a + b \bullet 1/T \tag{5}$$

$$B = \alpha + \beta \bullet 1/T \tag{6}$$

Thus a mathematical correlation between the cathodic limiting current, metallic cation concentration in the ionic liquid and the temperature may be proposed, as follows:

$$i_{\lim} = 10^{(a+b/T)} + (\alpha + \beta\, 1/T) \bullet c_{Ni^{2+}} \tag{7}$$

$$i_{\lim} = 10^{(1.78-622.7/T)} + (31.2-8186/T) \bullet c_{Ni^{2+}} \tag{8}$$

Thus, relationship (8) allows the calculation of the cathodic limiting current as a function of Ni(II) concentration in the ionic medium at the applied temperature. This behaviour proves the diffusive character of the electrode process control.

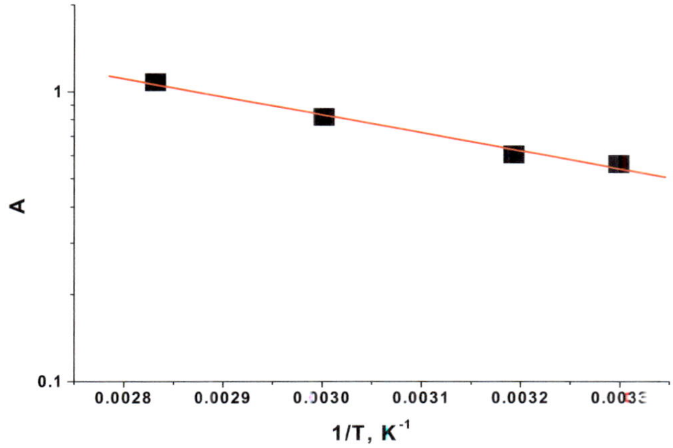

Fig. 9. Plot of log A against 1/T.

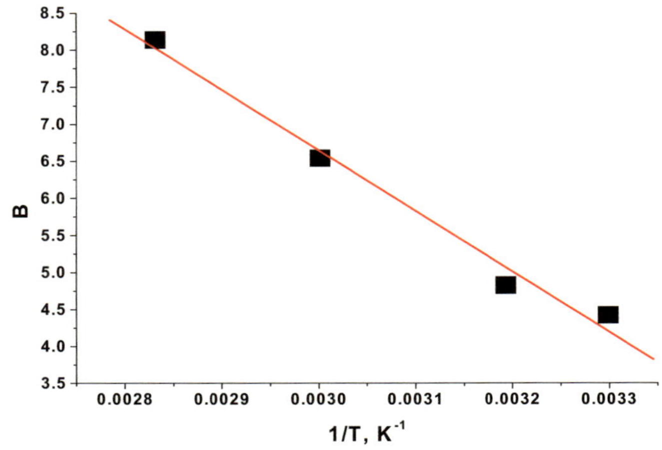

Fig. 10. Plot of B versus 1/T.

Figures 11 and 12 show examples of the recorded voltammograms at different applied working temperatures for the case of choline chloride-urea based ionic liquids where the cation source has been provided as sulfate. The reduction potential of Ni^{2+} cationic species is situated in the range of -0.650 ÷ -0.730 V/Ag quasi ref. and it is shifted with about 50-100 mV towards more electropositive values as the temperature increased. For ionic liquids systems containing metallic cation species concentrations higher than 0.2M a second cathodic peak was evidenced at -1.05 ÷ -1.1 V/Ag quasi ref., for temperatures above 60°C.

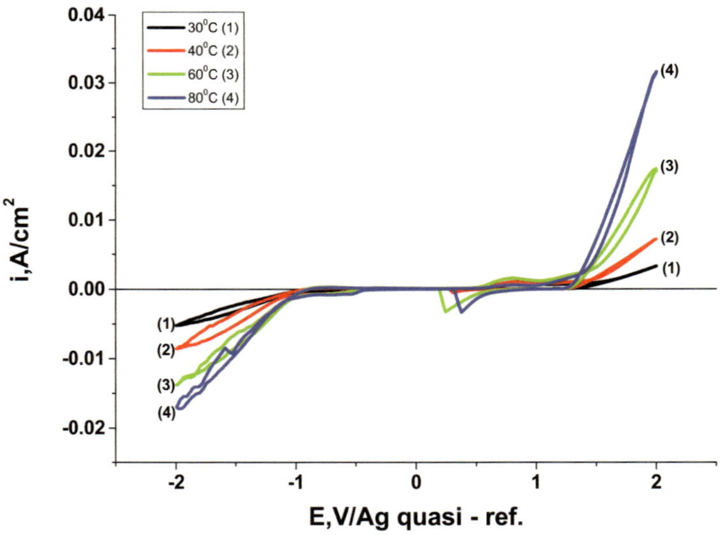

Fig. 11. Cyclic voltammograms for choline chloride-urea-$NiSO_4$ systems containing 0.1M Ni^{2+} at various temperature values.

This behaviour may suggest a sequence of reduction processes for some possible nickel complex compounds in the presence of an ionic medium containing chloride anions and ammonium cations (from the eutectic composition acting as a solvent) (Ji & Cooper, 1996; Zheng et al., 2005).

The same linear dependence in agreement with equation (4), between the cathodic limiting current i_{lim} against Ni^{2+} concentration $-c_{Ni(II)}$ has been also found for IL-$NiSO_4$ system for the investigated domain, as shown in Figure 13. Following the same processing procedure as in the previous system, a similar mathematical correlation between the cathodic limiting current, metallic cation concentration (introduced as sulfate) in the ionic liquid and the temperature was proposed:

$$i_{lim} = 10^{(92.9/T - 0.9)} + (35.1 - 10177.8/T) \bullet c_{Ni^{2+}} \quad (9)$$

Based on the relationships (8) and (9) one can conclude that the reduction process of the metallic cations from chloride salts occurs with relatively higher rates as compared with the use of sulfates, materialized by higher values of the cathodic limiting currents.

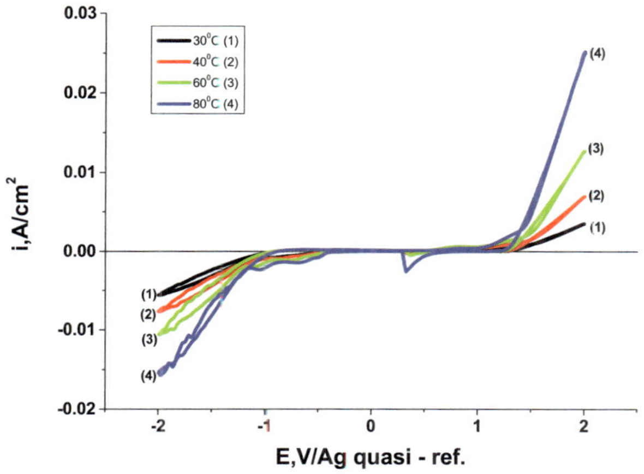

Fig. 12. Cyclic voltammograms for choline chloride-urea-$NiSO_4$ systems containing 0.2M Ni^{2+} at various temperature values.

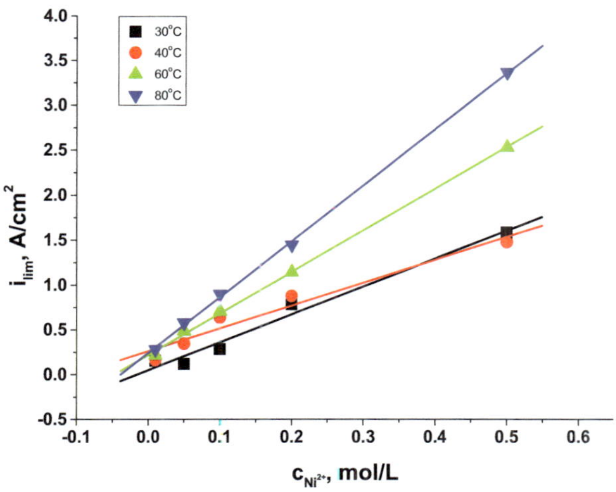

Fig. 13. Plots of cathodic limiting current *vs.* Ni^{2+} concentration for IL-$NiSO_4$ systems at different temperatures.

This fact has been experimentally confirmed. For example for an ionic medium based on 0.2M Ni^{2+} in ChCl-urea eutectic mixture, a cathodic current of 0.881 mA/cm^2 has been determined when the salt has been added as sulfate as compared with a value of 2.012 mA/cm^2 in the case of metallic chloride utilization, for a working temperature of 40°C. The diffusive character of the electrode process control evidenced by the linear dependence between the cathodic limiting current and cation concentration was found to be the same regardless the metallic salt anion and regardless the type of the eutectic mixture (Florea, 2010).

4.3 Ni layers electrodeposition – technological aspects

To evaluate the quality of the electrochemical coatings obtained from choline chloride based ionic liquids, constant current electrolyses were carried out under stationary conditions and in open air (the electrolyte has been in contact with atmospheric air and humidity). Different ionic liquids systems have been involved, as shown in Table 3. A two-electrode configuration has been employed, using various metallic substrates, respectively of Cu, Al and Mg as metallic substrates, of 70x35 mm sizes, with a constant geometrical area of 24.5 cm^2. Table 4 briefly presents the surface preparation of the metallic electrode before electrodeposition, as well as the electrolysis conditions.

No.	System type	Solvent composition (molar ratio)	Metallic compound concentration (Mole/L)
1	IL-Ni-chloride	Choline chloride-urea 1:2	0.42 - 1.68 M $NiCl_2.6H_2O$
2	IL-Ni-sulfate	Choline chloride-urea 1:2	0.71 M $NiSO_4.7H_2O$
3	IL-Ni	Choline chloride-urea -$NiCl_2.6H_2O$ 1:2:0.5 (molar ratio)	

Table 3. Ionic liquids systems involved in Ni electrodeposition.

Metallic substrate type	Surface preparation	Electrolysis conditions
Cu sheet, 99.99%, 70x35 mm;	I.Pickling in 1:1 HNO_3:H_2O for 10-20 s; II. Rinsing; III.Drying;	Duration: 10-60 min. Temperature: 30-80°C Current density: 1-18 A/dm^2 Stationary conditions Anode: Graphite/Nickel
Al strip, 99.5% purity (thermally treated), 70x35 mm;	I. Alkaline degreasing in: 5-10 g/L NaOH+ 30-40 g/L Na_2CO_3.10 H_2O + 30-40 g/L Na_3PO_4.10 H_2O, at 60-70oC, for 15-20 min; II. Rinsing; III.Drying	
Mg foil, 99.5% purity, 50x3 mm	I. Mechanical preparation with abrasive paper of 1000 and 2000; II. Pickling in 1:4 (volumic ratio) HF:H_2O for 30-60 s; III. Rinsing; IV. Drying.	

Table 4. Surface preparation of the involved metallic substrates and Ni electrodeposition conditions from the ionic systems shown in Table 3.

The obtained electrodeposits have been characterized with respect to appearance, adherence and layer thickness (deposition rate). In order to obtain information about the composition and the structure of the deposit, XRD investigations were performed, too. Morphological investigations have been performed involving AFM (Quesant Scope 350 equipment), SEM associated with EDX analysis (Zeiss EVO 50 equipment provided with Bruker–EDX probe) and optical microscopy involving OPTIKA B-253 equipment with a digital EverFocus Model EQ250 video color camera.

4.3.1 Nickel electrodeposition from choline chloride based ionic liquids containing chlorides

Generally, the electrodepositions have been characterized by a very good adherence on Cu metallic substrates. The use of working temperatures of min. 60°C and relatively high current densities in the range of 3.5-8.5 A/cm^2 determined the formation of light grey, uniform deposits, relatively similar to the semi bright Ni coatings obtained in aqueous electrolytes.

Fig. 14. Photograph of a Ni coating onto Cu substrate using IL-Ni-chloride system as electrolyte (i=.35 A/dm^2, 70°C, 60 min.).

Figure 14 shows a photograph of a Ni deposit onto Cu substrate and Figure 15 presents two examples of the coating morphology from IL-Ni electrolyte against a Ni anode, through AFM images.

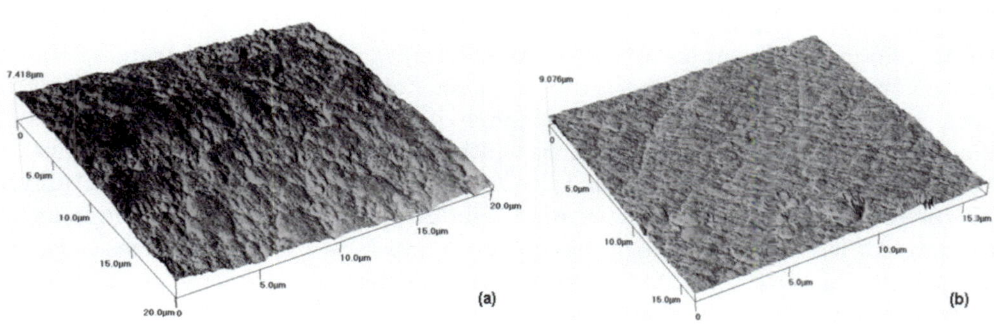

Fig. 15. AFM images of Ni electrodeposited on Cu substrate from IL-Ni-chloride system containing 0.5M $NiCl_2$ for various operating conditions: (a) 1 A/dm^2, 50°C, 30 min. and (b) 0.65 A/dm^2, 30°C, 30 min., against a Ni anode.

The presence of some cracks within the layer were noticed, less pronounced when a Ni anode has been involved, maybe due to the still present hydrogen discharge reaction that may be facilitated by the water content, of about 5-10% in the ionic liquid. Mechanical stresses that might be developed during deposition are not excluded, due to the relative high content of chlorides, which may also be responsible for cracks.

The use of soluble Ni anode facilitated an easier operation of IL-Ni electrolyte as compared with a graphite one, materialized by a better control of nickel content in the bath and by formation of a brighter coating layer.

According to XRD investigations and using the Scherrer's formula ($d = 0,9\lambda/(\beta\cos\theta)$, where λ is X-ray wavelength, θ represents the diffraction angle and β is the half width at half height for the diffraction peak in radians), average sizes of the crystallites have been of 9.2 - 14.4 nm, higher for increased applied current densities, maybe due to the temperature increase during electrodeposition. An amorphous phase is also observed, as it can be seen in Figure 16.

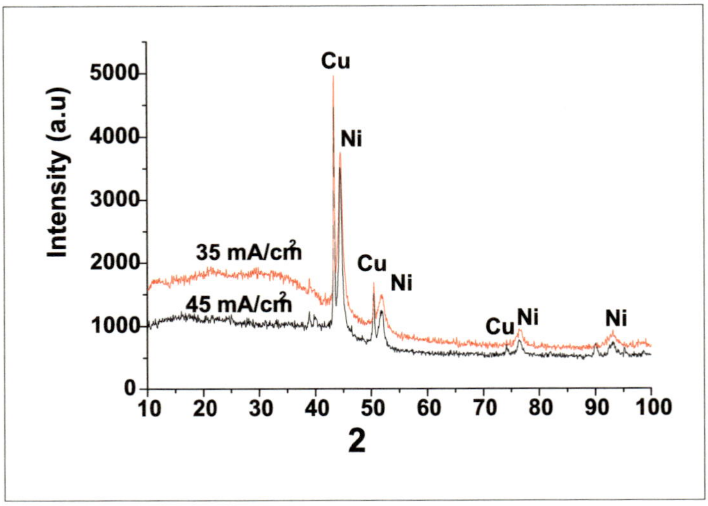

Fig. 16. X-ray diffractograms for electrodeposited Ni from an IL containing 1.84M $NiCl_2.6H_2O$ (70°C, 60 min.).

The deposition rate at various applied current densities was determined using gravimetric method. Thus, it has been founded that the ionic liquid electrolyte allows relatively low deposition rates, in the range of 0.1-0.4 μm/min. for the investigated current densities domain, as compared to aqueous solutions which show typical values of 0.5-1 μm/min.

When the IL-Ni system (as noted in Table3) has been involved, consisting of ChCl-urea-$NiCl_2.6H_2O$ mixture in a 1:2:0.5 molar ratio, light grey, semi bright, uniform deposits have been formed, with an excellent adhesion onto Cu substrate. Cathodic efficiencies between 85-99% have been determined when the operation temperature was in the range of 60-70°C, for electrolysis periods of 20-40 minutes.

This system was then used to electrodeposit Ni layers onto Al and Mg substrates. Adherent, mat black deposits onto Al have been obtained at current densities of about 1-1.1 A/dm² and relatively high temperatures of 80°C for deposition durations of minimum 30 minutes. Adherent, mat grey-black and slightly porous layers have been evidenced onto Mg substrate for the same operation conditions. Figures 17 and 18 present examples of optical micrographies of Ni deposits onto Al, respectively onto Mg substrates from IL-Ni electrolyte, at 1 A/dm², 80°C, for 30 minutes.

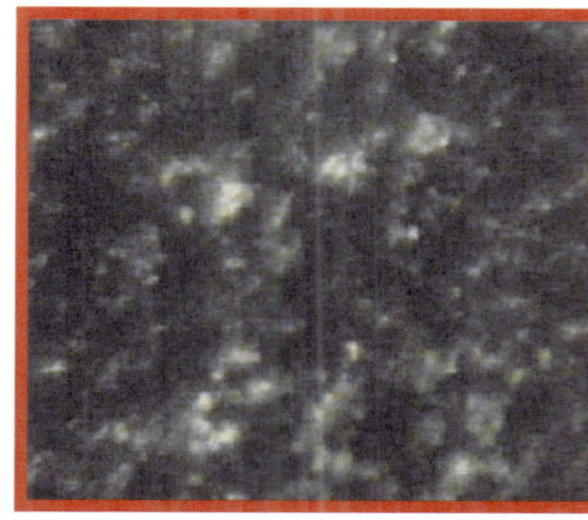

Fig. 17. Optical micrographies of Ni deposits onto Al substrate from IL-Ni system, 1 A/dm^2, 80°C, 30 minutes. Left: magnification: X40; Right: magnification : X 400.

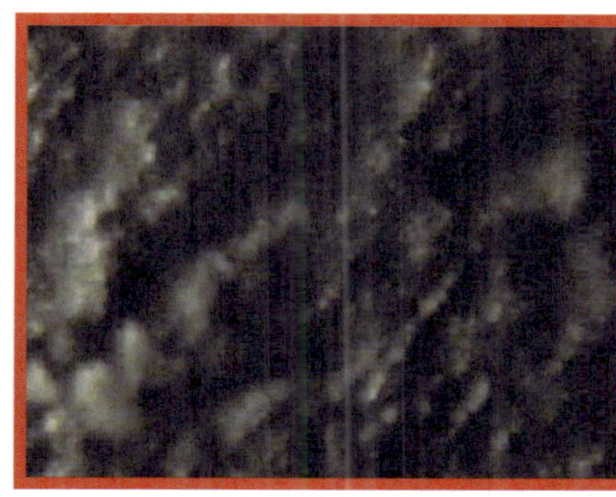

Fig. 18. Optical micrographies of Ni deposits onto Mg substrate from IL-Ni system, 1.1A/dm^2, 80°C, 30 minutes. Left: magnification: X40; Right: magnification : X 400.

4.3.2 Nickel electrodeposition from choline chloride based ionic liquids containing sulphates

The Ni electrodeposition from choline chloride based ionic liquids containing the metallic cation as sulfate (see Table 3) was also investigated, onto the same previously mentioned metallic substrates, respectively Cu, Al and Mg ones. The electrolysis was performed using a Ni anode for coating durations between 30-60 minutes.

Suitable coatings with a very good adherence were deposited onto Cu substrate. Temperatures of minimum 70°C and current densities of 1 - 1.5 A/dm^2 afford optimum coatings. The deposit entirely covers the metallic copper substrate, with a very good uniformity. Usually metallic grey layers are formed, with a micro/nano-crystalline structure.

When Ni electrodeposition using IL-Ni-sulfate system is applied onto Al substrate, uniform, dull, dark grey up to black layers are obtained for current densities between 10-15 mA/cm^2, at 80°C. The coating is adherent to the substrate in the most of the cases.

However, if the specimen is bent at 90°, a partial detaching was noticed (see Figure 19) suggesting the presence of a thin oxide layer that hinders the direct contact between Al and Ni deposit.

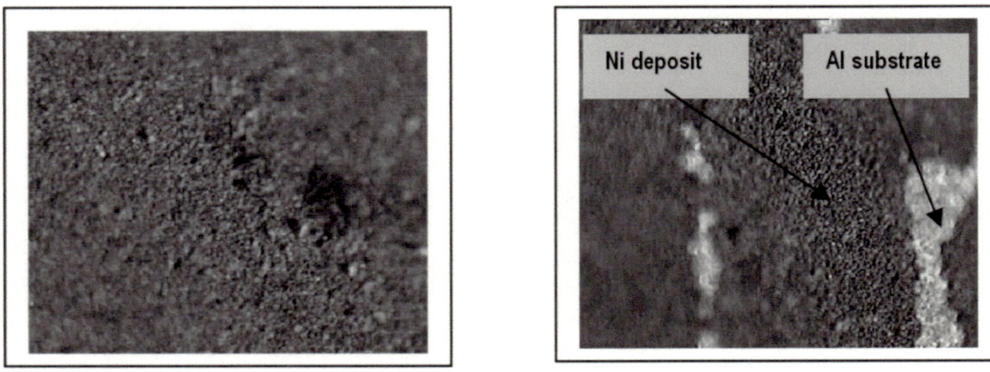

Fig. 19. Optical micrography of Ni electrodeposited onto Al substrate from IL-Ni-sulfate containing 0.71M $NiSO_4$ 1 A/dm^2, 80°C, 30 min. (left) and coating cracking after adherence test (right). Magnification : X 100.

Figure 20 presents am AFM image of the Ni deposit on Al from IL-Ni-sulfate system. The presence of spherical, cluster like formations can be clearly observed. A coating thickness of about 4 μm and a roughness around of 430 nm have been estimated, probably due to the initial non-homogeneities of the Al metallic substrate.

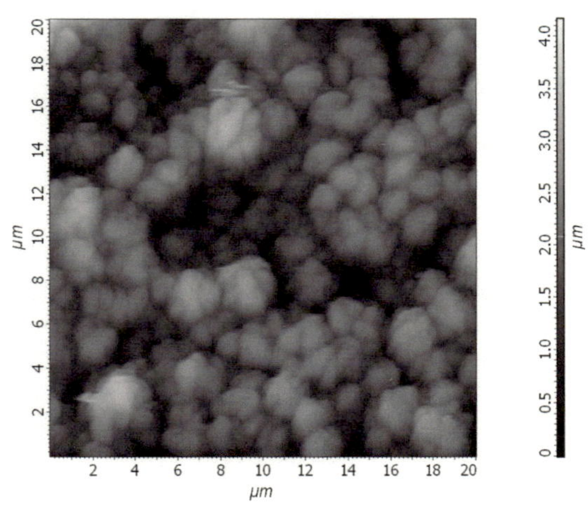

Fig. 20. AFM image of Ni electrodeposited onto Al from IL-Ni-sulfate system at 2.5 mA/cm^2, 80°C, 60 min.

Ni coatings were successfully electrodeposited directly onto Mg substrate. The formed layers in this case are dull grey, uniform on the entire surface, with a good adherence on the substrate and slightly porous (Florea et al., 2010) .

4.3.3 Corrosion behaviour of Ni coatings obtained from choline chloride based ionic liquids

In order to get some information on the corrosion protection performance of the electrodeposited Ni coatings from choline chloride based ionic liquids, several accelerated corrosion tests were performed: (i) continuous immersion in 0.5M NaCl for 336 hours with intermediary visual examinations and recording of corrosion potential; (ii) potentiodynamic polarization curves in aerated 0.5M NaCl against a Ag/AgCl reference electrode and Pt mesh counterelectrode, with a sweep rate of 5 mV/s and (iii) impedance spectra at open circuit potential, in a frequency range between 10^5 Hz - 0.1 Hz. in the same three electrodes cell. For both electrochemical investigations the geometrical surface of working electrode was 0.63585 cm^2. Three Cu specimens of 35x70 mm have been electrochemically coated with Ni deposits using IL-Ni system (see Table 3) of 10-15 µm and subjected to the corrosion tests. Figure 21 shows an example of the recorded impedance spectra for Ni deposit as Nyquist plot.

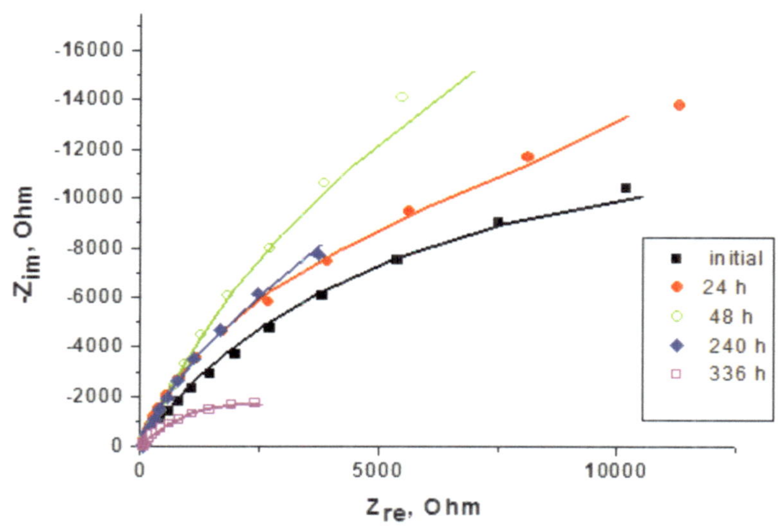

Fig. 21. Nyquist plot in 0.5 M NaCl at open circuit potential for Ni coating initially and after various periods of continuous immersion, representing both experimental and fitted (solid lines) points

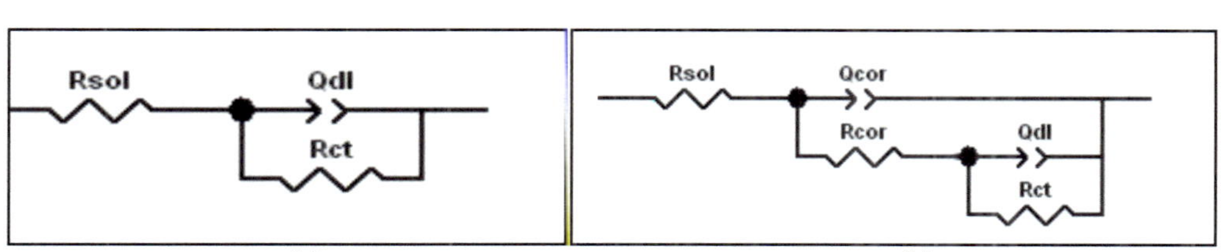

(a) (b)

Fig. 22. The proposed equivalent circuits to fit the experimental data from Figure 21

To fit experimental data, a simple equivalent circuit model has been used (as shown in Figure 22a), comprising the solution resistance (R_{Sol}), in series with a combination between a constant phase element corresponding to the double layer capacitance (Qdl) in parallel with charge transfer resistance (Rct). After 96 hours of immersion, a second time constant can be noticed on Bode plots which may be associated with corrosion processes determined by the pitting initiation at the coating surface. In this case, a different circuit model has been used by addition of two more elements, Q cor and Rcor (see Figure 22b). This model considers that the passive film doesn't entirely cover the metallic surface. As it can be seen from Figure 21, at the initial moment of immersion the charge transfer resistance is about 30 kΩ comparable to those obtained in Ni aqueous electrolytes. As the immersion period increases, the Rct increases towards 70 kΩ after 48 hours, due to the formation of a thin passive film on the metallic surface, which may enhance protective characteristics. This is a common occurence in the case of Ni immersion in aerated solutions. On visual examination of the metallic specimens, no significant surface modifications were found and no pits were observed.

5. Conclusions

The performed investigations showed the possibility of ILs preparation with quite large concentrations of Ni (II) compounds (0-1.68M $NiCl_2.6H_20$, respectively 0-1M $NiSO_4.7H_2O$) using deep eutectic solvents involving choline chloride. The formed low temperature melts are characterized by a good stability in time and are liquid at 25-30°C.

The cyclic voltamogramms evidenced the Ni (II) reduction at about -1.05 V/Ag cvsi-ref. for low working temperatures (30°C) which shifts towards more positive values with temperature increase, towards -0.75÷-0.88 V/Ag cvasi-ref at 80°C. Usually the cathodic limiting current increases with temperature. The anion nature does not significantly influence the cathodic process.

The obtained Ni deposits are adherent and uniform onto Cu substrate and the morphology evidenced the presence of some cracks that further may influence the corrosion behaviour. According to XRD investigations, average sizes of the crystallites have been of 9.2 - 14.4 nm, higher for increased applied current densities, maybe due to the temperature increase during electrodeposition. Also an amorphous phase has been evidenced. The direct deposition of nickel on magnesium and aluminum was noticed from ILs containing either $NiCl_2$ or $NiSO_4$ as a source of metallic cations. Operating temperatures in the range of 60-80°C and applied current densities in the range of 40 – 150 mA/cm^2 are recommendable, to allow deposition rates of 0.1-0.4 µm/min.

Ni alloys coatings formed in ILs showed a moderate corrosion protection, materialized in polarization resistances of 30-40 kΩ, probably due to the cracked morphology. However, after 240 hours of continuous immersion in chloride containing aggresive aqueous medium the exposed specimens did not exhibited any major surface modification and no pits have been evidenced.

Further investigations will be performed, for a better understanding of metal deposition mechanism associated with obtained morphology as an important factor to influence the final performance and to establish industrially compliant technological processes.

6. Acknowledgment

Part of this work was supported by the Romanian Ministry of Education and Research, PNCDI II Program, under Research Contract No.31066/1-2007.

7. References

Abbott, A.P.; Capper, G.; Davies, D.L.; Munro, H.L.; Rasheed, R.K. & Tambyrajah, V. (2001), Preparation of novel, moisture stable, Lewis-acidic ionic liquids containing quaternary ammonium salts with functional side chains, *Chem .Commun.*, No.19, 2001, pp. 2010-2011, ISSN 1359-7345

Abbott, A.P.; Capper, G.; Davies, D.L.; Rasheed, R.K. & Tambyrajah, V. (2003), Novel solvent properties of choline chloride/urea mixtures, *Chem. Commun.*, No.1, 2003, pp. 70-71, ISSN 1359-7345

Abbott, P.; Capper, G.; Davies, D.L.; Munro, H.; Rasheed, R.K. & Tambyrajah, V. (2003). Electrochemical Studies of Ambient Temperature Ionic Liquids Based on Choline Chloride, In: *Ionic liquids as green solvents: progress and prospects*, Rogers, R.D. & Seddom, K.R. (Ed)., pp. 439-452, An American Chemical Society Publication, ISBN: 9780841238565. Washington, DC

Abbott, A.P.; Capper, G.; Davies, D.L. & Rasheed, R.K. (2004), Ionic liquids based upon metal halide/substituted quaternary ammonium salt mixtures, *Inorg. Chem.*, Vol.43, No.11, May 2004, pp. 3447-3452, ISSN 0020-1669

Abbott, A.P.; Davies, D.L.; Capper, G.; Rasheed, R.K. & Tambyrajah, V, USA Patent 2004/0097755

Abbott, A.P.; Boothby, D.; Capper, G.; Davies, D.L. & Rasheed, R.K. (2004), Deep eutectic solvents formed between choline chloride and carboxylic acids: versatile alternatives to ionic liquids, *J. Am. Chem. Soc.*, Vol.126, No.29, July 2004, pp. 9142-9147, ISSN 0002-7863

Abbott, A.P.; Capper, G.; Davies, D.L. & Rasheed, R.K. (2004), Ionic liquid analogues formed from hydrated metal salts, *Chem. Eur. J.*, Vol.10, No.15, August 2004, pp. 3769-3774, ISSN 1521-3765

Abbott, A.P.; Capper, G.; Davies, D.L.; Rasheed, R.K.; Archer, J. & John, C. (2004), Electrodeposition of chromium black from ionic liquids, *Trans. Inst. Met. Finish.*, Vol.82, No.1-2, 2004, pp. 14-17, ISSN 0020-2967

Abbott, A.; Davies, D.L.; Capper, G.; Rasheed, R.K. & Tambyrajah, V., Ionic liquids and their use, Patent SUA nr.US 2004/0054231

Abbott, A.P.; Capper, G.; Davies, D.L.; Rasheed, R.K. & Shikotra, P. (2005), Selective extraction of metals from mixed oxides matrixes using choline – based ionic liquids, *Inorg. Chem.*, Vol.44, No.19, September 2005, pp. 6497-6499, ISSN 0020-1669

Abbott, A.P.; Capper, G.; McKenzie, K.J. & Ryder, K.S. (2006), Voltammetric and impedance studies of the electropolishing of type 316 stainless steel in a choline chloride based ionic liquid, *Electrochim. Acta*, Vol.51, No. 21, June 2006, pp. 4420-4425, ISSN 0013-4686

Abbott, A.P. & McKenzie, K.J. (2006), Application of ionic liquids to the electrodeposition of metals, *Phys. Chem. Chem. Phys.*, Vol.8, No.37, 2006, pp. 4265-4279, ISSN 1463-9076

Abbott, A.P.; Capper, G.; McKenzie, K.J. & Ryder, K.S. (2007), Electrodeposition of zinc-tin alloys from deep eutectic solvents based on choline chloride, *J. Electroanal. Chem.*, Vol.599, No.2, January 2007, pp. 288-294, ISSN 0022-0728

Abbott, A.P.; Barron, J.C.; Ryder, K.S. & Wilson, D. (2007), Eutectic-Based Ionic Liquids with Metal-Containing Anions and Cations, *Chem. Eur. J.*, Vol.13, No.22, July 2007, pp. 6495 – 6501, ISSN 1521-3765

Abbott, A.P.; Nandhra, S.; Postlethwaite, S.; Smith, E.L. & Ryder, K. (2007), Electroless deposition of metallic silver from a choline chloride-based ionic liquid: a study using acoustic impedance spectroscopy, SEM and atomic force microscopy, *Phys.Chem.Chem.Phys.*, Vol. 9, No.28, 2007, pp. 3735-3743, ISSN 1463-9076

Abbott, A.P.; Griffith, J.; Nandhra, S.; Connor, C.O.; Postlethwaite, S.; Ryder, K.S. & Smith, E.L. (2008), Sustained electroless deposition of metallic silver from a choline chloride-based ionic liquid, *Surf. Coat. Technol.*, Vol. 202, No.10, February 2008, pp. 2033-2039, ISSN 0257-8972

Abbott, A.P.; Ttaib, K.El; Ryder, K.S. & Smith, E.L. (2008), Electrodeposition of nickel using eutectic based ionic liquids, *Trans.Inst. Met. Finish.*, Vol.86, July 2008, pp. 234-240, ISSN 0020-2967

Abbott, P.; Barron, J. C. & Ryder, K. S. (2009), Electrolytic deposition of Zn coatings from ionic liquids based on choline chloride, *Trans.Inst.Met. Finish.*, Vol. 87 , No.4, July 2009, pp. 201-207, ISSN 0020-2967

Abbott, A.P.; Ttaib, K.El.; Frisch, G.; McKenzie, K. J. & Ryder, K.S. (2009), Electrodeposition of copper composites from deep eutectic solvents based on choline chloride, *Phys. Chem. Chem. Phys.*, Vol.11, No. 21, pp. 4269-4277, ISSN 1463-9076

Abbott, A.P.; Barron, J.C.; Frisch, G.; Ryder, K. S. & Silva, A. F. (2010), The Effect of Additives on Zinc Electrodeposition from Deep Eutectic Solvents, *Electrochim. Acta*, doi:10.1016/j.electacta.2011.02.095 ISSN 0013-4686

Anicai, L; Cojocaru, A.; Florea, A. & Visan, T. (2008), Electrochemical investigation of silver/silver ion couple reversibility in choline chloride-urea based ionic liquid, *Studia Univ.Babes-Bolyai, Chemia*, Vol.53, No.1, 2008, pp. 119-135, ISSN 1224-7154

Benaben, P. & Sottil, B. (2006), Hard Chromium deposition using ionic liquids as a potential substitute to hexavalent chromium plating: First results in European IONMET programme 57-th Annual Meeting of ISE, Edinburgh, 2006, Abstracts, S5-O-3.

Chromium Electroplating: Ion Age Dawns,- Sustainble Technologies Initiative Program, 2004, www.oakdenehollins.co.uk/sti.html

Cojocaru, A.; Costovici, S.; Anicai, L.; Visan, T.; Cojocaru, P.; Magagnin, L. & Cavallotti, P.L. (2009), Tin-Nickel alloy deposition from choline chloride-urea ionic liquids

(Deposizione di una lega Nichel-Stagno da soluzioni ioniche a base di cloruro di colina-urea), *Galvanotecnica e nuove finiture*, No. 3, 2009. pp. 140-146

Cojocaru, A.; Costovici, S.; Anicai, L. & Visan, T. (2009), Studies of cathodic processes during NiSn alloy deposition using choline chloride based ionic liquids, *Metallurgia International*, Vol.14, No.11, pp. 38-46, ISSN 1582-2214

Datta, M. & Landolt, D. (2000), Fundamental aspects and applications of electrochemical microfabrication, *Electrochim. Acta*, Vol.45, No.15-16, May 2000, pp. 2535-2558, ISSN 0013-4686

Endres, F. (2002), Ionic Liquids: Solvents for the Electrodeposition of Metals and Semiconductors, *Chem. Phys. Chem.*, Vol.3, February 2002, pp.144-154, ISSN 1439-7641

Endres, F.; Abbott, A. P. & MacFarlane D.R. (2008), *Electrodeposition from Ionic Liquids*, Wiley-VCH Verlag, GmbH & CO,KgaA, ISBN .978-3-527-31565-9, Weinheim

Florea, A.; Petica, A.; Anicai, L. & Visan, T. (2010), Preliminary Studies Of Silver Coatings Formation From Choline Chloride Based Ionic Liquids, *UPB Sci.Bull*, Series B, Vol.72, No. 2, 2010, pp. 115-126, ISSN 1454-2331

Florea, A.; Anicai, L.; Costovici, S.; Golgovici, F & Visan, T. (2010), Ni and Ni Alloys Coatings Electrodeposited from Choline Chloride Based Ionic Liquids – Electrochemical Synthesis and Characterization, *Surface and Interface Analysis*, Vol.42, No.6-7, June – July 2010, pp. 1271–1275, ISSN 1096-9918

Florea, A. (2010), Metals and conducting polymers electrochemical coatings with controlled properties using ionic liquids media, Ph.D. Thesis, POLITEHNICA University, July 2010

Gurunathan, K.; Vadivel Murugan, A.; Marimuthu, R.; Mulik, U.P. & Amalncrkar, D.P. (1999), Electrochemically synthesised conducting polymeric materials for applications towards technology in electronics, optoelectronics and energy storage devices, *Materials Chemistry and Physics*, Vol.61, No.3, November 1999, pp. 173-191, ISSN 0254-0584

Ji, J. & Cooper, W.C. (1996), Nickel speciation in aqueous chloride solutions, *Electrochim. Acta*, Vol.41, No.9, June 1996, pp. 1549-1560, ISSN 0013-4686

Mukhopadhyay, I.; Aravinda, C.L.; Borissov, D. & Freyland, W. (2005), Electrodeposition of Ti from $TiCl_4$ in the ionic liquid l-methyl-3-butyl-imidazolium bis (trifluoro methyl sulfone) imide at room temperature: study on phase formation by in situ electrochemical scanning tunneling microscopy, *Electrochim. Acta*, Vol.50, No.6, January 2005, pp.1275-1281, ISSN 0013-4686

Wasserscheid, P. & Welton, T. (2007), Ionic Liquids in Synthesis, Weinheim: Wiley-VCH Verlag, ISBN: 9783527312399

Whitehead, A.H.; Pölzler, M. & Gollas, B. (2010), Zinc electrodeposition from a Deep Eutectic System Containing Choline Chloride and Ethylene Glycol, *J. Electrochem. Soc.*, Vol. 157, No.6, pp. D328-D334, ISSN 0013-4651

Zheng, G.; Zheng, L. & Cao, H. (2005), Formation of black nickel in leaching solution containing ammonia and chloride, *Trans. Nonferrous Met. Soc. China*, Vol.15, No.1, February 2005, pp. 165-170, ISSN 1003-6326

Ionic Liquids as Advantageous Solvents for Preparation of Nanostructures

Alexandrina Nan[1] and Jürgen Liebscher[1,2]
[1]National Institute of Research and Development for Isotopic and Molecular Technologies
[2]Institute of Chemistry, Humboldt-University Berlin
[1]Romania,
[2]Germany

1. Introduction

Nanoparticles are defined as solid particles with a size in the range of 10-1000 nm, depending on the applied synthesis method different structural motifs can be obtained: nanoparticles, nanorods, nanowire, nanoclusters or nanospheres. The nanoparticles have drawn great attention across various areas of science and nanotechnology due to their exclusive physicochemical properties, which are attributable to the quantum size effect and single-electron transitions.
The sciences as chemistry, physics, biology, medicine and electrical engineering are becoming more and more connected with nanotechnology and strongly interrelated. (Kumar, 2010; Suh et al., 2009).
When one talks about nanoparticle synthesis, characterization and their properties, it is today obligatory to consider issues related to a tight control over size, shape, crystal structure and morphological purity. Obviously, chemical composition is also critical, but important aspects in the (nano)materials science have been widely demonstrated to correlate with the morphological/structural parameters (Mudring et al., 2009).
First, the ionic liquids were used in electrosynthesis of various metallic nanoparticles, such as palladium (Pd), iridium (Ir) and semiconductor nanoparticles such as stable germanium or Ge nanoclusters (Deshmukh et al., 2001; Dupont et al., 2001; Endres & Abedin, 2002; Endres et al., 2003). Then the preparations of Ti nanowires onto graphite by electroreduction (Mukhopadhyay & Freyland, 2003) were extremely stimulating. In all of these examples the potential for electrochemical reactions and the high polarity of ionic liquids are exploited.
In addition to electrochemical reactions different other techniques are used to fabricate nanostructures, wherein often various solvents are applied. As a recent and very effective technique also ionic liquids are employed in the creation of nanostructured materials, because of their unique properties such as extremely low volatility, wide temperature range in liquid state, good dissolving ability, high thermal stability, excellent microwave absorbing ability, high ionic conductivity, wide electrochemical window, non-flammability, etc (Antonietti et al., 2004; Suh et al., 2004). Many ionic liquids are not only non-flammable but also stable in a broad temperature range from room temperature to 400 ^{0}C. These unique properties can be exploited in an effective preparation of nanostructures with control of size

and size distribution. This aspect is crucial for a number of applications of nanostructures such as in medicine, where nanoparticles should have a size in the range of 2 to 10 nm. The heat stability of ionic liquids can be exploited for example in the preparation of iron oxide at high temperature, in this case the nanoparticle size can be controlled by the temperature of their preparation. Furthermore alloy nanoparticles are obtained at a high temperature, ionic liquids can act as a promising solvent for their preparation.

The present review presents a study of selected exemplary works that outline the progress in research on the application of ionic liquids in the nanoparticle field covering the last ten years period. Diverse types of nanoparticles are presented which were synthesized in ionic liquids, using different techniques. Our aim is to provide a comprehensive overview on the synthesis of different nanoparticles in ionic liquids, and also the possibility to influence the morphology, structure and properties of nanoparticles using ionic liquids. Very recently a review appeared (Patzke et all., 2011) describing the application of ionic liquids in the preparations of meal nanoparticles excluding metal oxides which however are also very important from several points of view.

Ionic liquids are able to offer exceptional properties as media for the synthesis of nanoparticles. The low surface tension of many ionic liquids leads to high nucleation rates and, in consequence, to small particles. The ionic liquid itself can perform as an electronic as well as a steric stabilizer and thus can lower particle growth. As highly structured liquids, ionic liquids have a strong effect on the morphology of the particles formed (Mudring et al., 2009).

Ionic liquids that have an imidazolium ring as cation can interact with many different species, as they provide hydrophobic or hydrophilic regions and a high directional polarizability. This is one of the special qualities of imidazolium ionic liquids that differentiate them from the classical ion aggregates of which ion pairs and ion triplets are widely recognized examples. This structural organization of ionic liquids can be used as "entropic drivers" for spontaneous, well-defined, and extended ordering of nanoscale structures. Indeed, the unique combination of adaptability towards other molecules and phases plus the strong hydrogen-bond-driven structure makes ionic liquids potential key tools in the preparation of a new generation of chemical nanostructures.

The morphology and the properties of the nanoparticles are very well correlated, therefore depending on the size and shape of nanoparticles their properties can change very much. However, the size and shape can be tailored by the synthesis method and certainly at the same time the properties of nanoparticles. That means that by synthesis method one can achieve nanoparticles with required properties for different applications.

Owing to their outstanding properties, ionic liquids have been exploited successfully in the synthesis of different inorganic nanoparticles. Therefore a variety of inorganic nanostructures including metal nanoparticles, metal oxide nanoparticles, metal complexes nanoparticles, alloys etc. have been produced in various room temperature ionic liquids. However, in contrast to their application in organic chemistry, the use of ionic liquids in inorganic synthesis in particular in the synthesis of inorganic nanoparticles is still in its primary stage, and more exploration is necessary to utilize their advantages fully.

The first part of this chapter describes the synthesis of metal nanoparticles in ionic liquids, in the second part we will illustrate the synthesis of some metal oxide nanoparticles in ionic liquids and the last part points out the type of alloys that can be obtained in ionic liquids.

Ionic liquids mostly applied in the preparation of metal, metal-oxide and alloy nanoparticles are shown in Scheme 1.

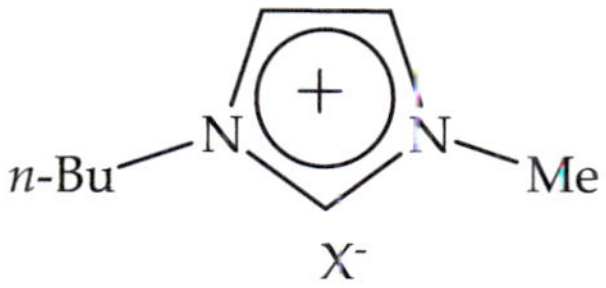

$X = BF_4$
$X = PF_6$
$X = N(SO_2CF_3)_2$
$X = CF_3SO_3$

Scheme 1. Examples of imidazolium ionic liquids used in the preparation of metal nanoparticles.

2. Synthesis of metal nanoparticles in ionic liquids

The optical and electrical properties of metal and semiconductor nanoparticles have been extensively utilized in the creation of advanced materials for use in electronics, sensors, and photonics. Metal nanoparticles have also served as precursors to form nanowires or nanorods. The extraordinary properties of metal nanoparticles largely depend on their morphology and particle size. Accordingly, both the precise control of the particle size and a clear understanding of their chemical phenomena remain a research task. Various synthetic methods for metal nanoparticles have been reported, but most of them were based on the reduction of a metal ion by toxic reducing agents such as $NaBEt_3H$, $LiBEt_3H$, and $NaBH_4$.

To obtain stable nanoparticles consisting exclusively of chemically pure transition metals, without using any stabilizing agent, is still an unsolved topic that remains the subject of contrasting opinions (Ott & Finke, 2007). In the case of physical preparation techniques, traces of hydrocarbons deriving from the vacuum systems, gases present in the deposition environment, or other contaminants that come into contact with the nanoparticles during real applications, always give rise to an outer shell, even of a small thickness, which stabilizes the pure metal core nanoparticles. Even in solution when the stabilizers are not added on purpose, either the solvent itself or dissolved oxygen can be responsible for interactions with the nanoparticles surface, leading to a stabilizing zone that surrounds the particle and has its own composition, this shell being different from that of the pure metal nanoparticle core.

Metal nanoparticles are easily prepared using the following methods:

- controlled decomposition of organometallic compounds in the formal zero oxidation state such as $[Pt_2 (dba)_3]$ (Scheeren et al., 2003), Ru (cod) (cot)] (Silveira et al., 2004), or [Ni $(cod)_2$] (Migowski et al., 2007) (dba=dibenzylideneacetone, cod=1,5-cyclooctadiene, and cot=1,3,5-cyclooctatriene) dispersed in ionic liquids;
- chemical reduction (usually with hydrogen or a hydride source) of transition-metal compounds such as [Pd-$(acac)_2$] (Umpierre et al., 2005), (acac=acetylacetonate), $[Ir(cod)Cl]_2$, $RhCl_3$ (Fonseca et al., 2003), and RuO_2 (Rossi et al., 2004) dispersed in ionic liquids;
- simple transfer of the nanoparticles freshly prepared in water or "classical" organic solvents to the ionic liquids (Zhao et al., 2006).

Alternatively, *in situ* laser radiation may be used to induce the fragmentation of relatively large nanoparticles dispersed in ionic liquids into smaller particles with a narrow size distribution (Gelesky et al., 2005).

The formation of the nanoparticles from the first two methods apparently follows the autocatalytic mechanism (Besson et al., 2005; Besson et al., 2005) that basically involves two steps: nucleation and surface growth. In various cases these ionic liquids colloidal mixtures can be used directly as catalysts or they may be isolated and used as powders in solventless conditions (the substrates/products are per definition the solvent) or re-dispersed in the ionic liquids.

Nevertheless one has to know that metals which are obtainable in nanophases are basically unstable materials difficult to handle. Therefore, storage and using uncovered metal nanoparticles in nanotechnological applications is almost impossible. In aiming to avoid or to slow down the instability of the metal nanoparticles, a large number of investigations have been conducted to determine how the correct stabilization of metal nanoparticles can be achieved. However, it should also be noted that the alternative of a specific stabilizer is currently driving the bibliographic classifications. Many papers are specifically titled in terms of the most important stabilizer selected for the nanoparticles preparation. The number of agents that have been successfully applied to the synthesis of good nanoparticles has undergone a vast increase, thus evidencing the need for a comprehensive examination of this field.

Those species capable of stabilizing a nanosized metal phase can be classified in three classes, according to their mechanisms of action: (i) electrostatic, based on the classical theory of electrostatic colloidal stabilization, (ii) steric; and (iii) electrosteric, which is a combination of the electrostatic and steric modes (Ott & Finke, 2007).

The stabilization of nanoparticles in solution can be achieved by electrostatic and/or steric protection using among others, quaternary ammonium salts that are the most popular and investigated classes of stabilizing agents for soluble metal nanoparticles.

Due to the fact that the metal nanoparticles cannot be handled without a former stabilization, several synthetic routes for producing different metal nanoparticles covered with different stabilizers came out in the literature. This variety of synthetic routes allowed manufacturing a large diversity of metal nanoparticles having different morphology and in particular different properties. Anyway, the assignment of stabilizing the nanoparticles remains a very important issue to solve. For this reason a main part of the published papers describe not only the synthesis of nanoparticles in ionic liquids but also the stabilization of nanoparticles and the influence of the stabilizer molecule on the nanoparticles size, shape and properties.

2.1 Ionic liquid assisted of eighth metal group nanoparticles Ru(0)

Very fine and stable metal nanoparticles as Ru(0) were synthesized in ionic liquids by chemical reduction (Fonseca et al., 2003). The colloidal system metal-nanoparticle/ionic liquids-stabilizer is extraordinarily stable and ligands are not equired. Extraordinarily high turnover numbers are achieved with this system in catalytic hydrogenation.

Few years ago ionic liquids were investigated as suspension media for catalytically active nanoclusters (Ott & Finke, 2007; Calò et al., 2005). The main advantages of synthesizing nanoparticles in ionic liquids resides in the direct possibility of using them as highly recyclable catalysts in several processes, thus providing a low environmental impact (Jacob et al., 2006). The so - called 'green' character of such a synthetic approach is that it does not require conventional solvents; rather, separation of the reaction products can be easily achieved by physical treatments, such as distillation inducing solidification of the ionic liquid solvent, etc. Furthermore, the ionic liquids may itself afford the nanoparticles stabilization, thus greatly facilitating catalyst recycling and improving the total turnover number.

Although the nanoscale Ru(0) particles can be prepared by decomposition of Ru(0) organometallic precursors or reduction of $RuCl_3$ in the presence of stabilizing agents such as polymers or ligands, the development of a more simple and efficient method for the preparation of Ru(0) nanoparticles is still a challenge in nanocatalysis (Rossi et al., 2004).

Quaternary ammonium salts are one of the most popular and investigated classes of stabilizing agents for soluble metal nanoparticles. It is assumed that the stabilization in these cases is essentially due to the positive charge on the metal surface which is ultimately induced by the adsorption of the anions on to the coordinatively unsaturated, electron-deficient, and initially neutral metal surface.

The Ru(0) nanoparticles can be prepared via a very simple method based on H_2 reduction of the commercially available precursor $RuO_2 \cdot 3H_2O$, which avoids the use of an organometallic precursor. An efficient catalyst for hydrogenation of benzene and other benzene derivatives was obtained in this way. In biphasic experiments, the RuO_2 was dispersed in the ionic liquid 1-*n*-butyl-3-methylimidazolium hexafluorophosphate [BMIM][PF_6], 1-butyl-3-methylimidazolium tetrafluoroborate [BMIM][BF_4] or 1-*n*-butyl-3-methylimidazolium trifluoromethane sulphonate [BMIM][CF_3SO_3] prior to the addition of the substrate. The reactor was submitted to vacuum, placed in an oil bath at 75 °C under stirring (700 rpm) and connected to the hydrogen gas reservoir. The reaction was initiated by the admission of H_2 gas of 4 atm. The reaction was monitored by the fall in hydrogen pressure (Rossi & Machado, 2009).

The reduction of [Ru(COD)(2-methylallyl)$_2$] (COD = 1,5-cyclooctadiene) dispersed in various ionic liquids, namely, 1-*n*-butyl-3-methylimidazolium and 1-*n*-decyl-3-methylimidazolium, associated with the N-bis(trifluoromethanesulfonyl)imidates (NTf_2) and the corresponding tetrafluoroborates (BF_4) with hydrogen gas at 50 °C leads to well-dispersed immobilized nanoparticles. Nanoparticles with a smaller mean diameter were obtained in the ionic liquids containing N-bis(trifluoromethanesulfonyl)imidates than that in the tetrafluoroborate analogues, which is less coordinated (Prechtl et al., 2008).

2.2 Ionic liquid assisted of metal nanoparticles: Ir(0), and Rh(0)

Very fine and stable noble-metal nanoparticles of Ir(0) were synthesized in ionic liquids by chemical reduction (Fonseca et al., 2003). By monodispersed Ir(0) nanoparticles with a mean diameter of 2.1 nm were prepared by reduction of [Ir(cod)Cl]$_2$ (cod=1,5-cyclooctadiene) dissolved in 1-butyl-3-methylimidazolium hexafluorophosphate ionic liquid at 75 °C (Dupont et al., 2002) with molecular hydrogen. It is important to note that the synthesis of Ir(0) nanoparticles should be performed in the absence of water (Sweeny & Peters, 2001).

Accordingly, the addition of a ligand constitutes an alternative approach to increase the stability of rhodium nanospecies.

A particularity in the case of Rh(0) colloids, is that the precipitation of nanoparticles has been reported during catalytic applications such as the hydrogenation of substituted arene compounds (Migowski & Dupont, 2007; Léger et al., 2008) . The same paper explains that a very robust stabilization of small nanoparticles with a very narrow size distribution provides almost inactive catalysts for the hydrogenation reactions. In contrast, the nanoparticles stabilized by compounds binding less strongly to the metal surface generate higher catalytic activity.

Recent study described an efficient approach for the preparation of Rh(0) nanoparticles by reduction of $RhCl_3$ with $NaBH_4$ (scheme 2) followed by the stabilized by different isomers of bipyridine (scheme 3), the most used was 2,2′-bipyridine as ligand (Léger et al., 2008). These colloidal suspensions have shown interesting activities and selectivity for the hydrogenation of aromatic compounds in several non-aqueous ionic liquids according to the nature of the anion and cation fragments (scheme 2).

$$RhCl_3 \cdot 3H_2O \xrightarrow[NaBH_4]{THF/IL} Rh(0)/IL \xrightarrow{Bipyridine} Rh(0)/Bipyridine/IL$$

Scheme 2. Synthesis of 2,2′ bipyridine-stabilized rhodium nanoparticles in ionic liquids.

It is important to note that the synthesis of Rh(0) nanoparticles should be performed in the absence of water (Sweeny & Peters 2001).

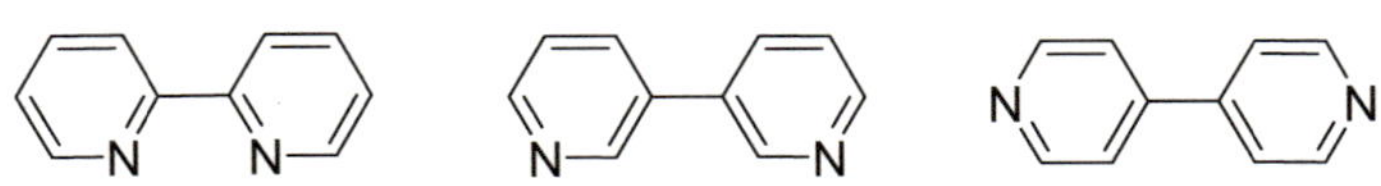

2,2'-Bipyridine 3,3'-Bipyridine 4,4'-Bipyridine

Scheme 3. Bipyridine isomers used in the preparation of Rh-nanoparticles.

A phenomenon, which was studied in the literature, is the role of extra water in ionic liquids. This phenomenon is complex and depends on the supramolecular structure of the ionic liquid. It is assumed that its structure and chemical reactivity is far from that of bulk water, as it is tightly bound and activated in the H-bonding system of the IL. As a result, reactions with water take place quite rapidly in these systems. On the other hand, water cannot function as a solvating ligand here since it is too involved in IL binding. This was deduced, for instance, from the absence of so-called solvent pores and represents a quite singular situation for colloid chemistry and material synthesis.

Gelesky et al. showed that *in situ* laser radiation of relatively large Rh(0) nanoparticles dispersed in 1-*n*-butyl-3-methylimidazolium hexafluorophosphate ionic liquid induces their fragmentation into smaller particles with narrow size distribution. The Rh nanoparticles were prepared by simple hydrogen reduction of $[Rh(cod)\text{-}\mu\text{-}Cl]_2$ (cod=1,5-cyclooctadiene) dispersed in 1-*n*-butyl-3-methylimidazolium hexafluorophosphate at 75 °C. After 60 minutes darkening of the solution was observed indicating the formation of Rh(0) nanoparticles (Gelesky et al., 2005).

Stable ruthenium, rhodium, and iridium metal nanoparticles have been reproducibly obtained by facile, rapid, and energy-saving microwave irradiation under an argon atmosphere from their metal-carbonyl precursors $[M_{(x)}(CO)_{(y)}]$ in the ionic liquid 1-butyl-3-methylimidazolium tetrafluoroborate (Redel et al., 2009; Vollmer et al., 2010). The metal nanoparticles synthesized have a very small (<5 nm) and uniform size and are prepared without any additional stabilizers or capping molecules as long-term stable metal nanoparticles-ionic liquid dispersions. The ruthenium, rhodium, or iridium nanoparticles dispersed in ionic liquids are highly active and easily recyclable catalysts for the biphasic liquid-liquid hydrogenation of cyclohexene to cyclohexane.

The treatment of an orange solution of $[IrCl(cod)]_2$ (cod=1,5-cyclooctadiene) in 1-butyl-3-methylimidazolium hexafluorophosphate with hydrogen at 75 °C for 10 min affords a black solution. This ionic solution promotes the biphasic hydrogenation of various olefins under mild reaction conditions, and the products were isolated almost quantitatively by simple decantation. The catalytic activity of this system is significantly superior to those obtained in biphasic conditions by classical transition-metal catalyst precursors in ionic liquids under similar reaction conditions. A mercury test clearly indicated the presence of Ir(0) particles in the system formed by the reduction of the Ir(I) precursor in 1-butyl-3-methylimidazolium hexafluorophosphate. These nanoparticles could be isolated by centrifugation from the catalytic mixture (Dupont et al., 2002).

2.3 Ionic liquid assisted metal nanoparticles: Ni(0), Pd(0) and Cu(0)

Colloidal suspensions of Ni(0) nanoparticles were obtained by thermal decomposition at 75 °C and 5 bar of molecular hydrogen from bis(-1,5-cyclooctadiene) Ni(0) as organometallic precursor in 1-butyl-3-methylimidazolium ionic liquids with three different anions, BF_4^-, $CF_3SO_3^-$ and $N(Tf)_2^-$ (Cassol et al., 2006).

Different N-donor compounds have already proved their efficiency as protective agents in the case of colloidal suspensions in non-aqueous ionic liquids. As an example, the synthesis of phenanthroline stabilized palladium nanocatalysts and their use for olefin hydrogenation in 1-butyl-3-methylimidazolium hexafluorophosphate media was reported (Huang et al., 2003).

The reduction of $Pd(acac)_2$ (acac=acetylacetonate), dissolved in 1-*n*-butyl-3-methylimidazolium hexafluorophosphat [BMIM][PF_6] or 1-*n*-butyl-3-methylimidazolium tetrafluoroborate[BMIM][BF_4] as ionic liquids, by molecular hydrogen (4 atm) at 75 °C affords stable, nanoscale Pd(0) particles with sizes of 4.9±0.8 nm. In as much as 1,3-butadiene is at least four times more soluble in the 1-*n*-butyl-3-methylimidazolium tetrafluoroborate than butenes, the selective partial hydrogenation could be performed by Pd(0) nanoparticles embedded in the ionic liquid (Umpierre et al., 2005).

Gelesky et al., showed that *in situ* laser radiation of relatively large Pd(0) nanoparticles dispersed in 1-*n*-butyl-3-methylimidazolium hexafluorophosphate ionic liquid induces the fragmentation into smaller particles with narrow size distribution. The Pd nanoparticles were prepared by simple hydrogen reduction of $PdCl_2$ dispersed in 1-*n*-butyl-3-methylimidazolium hexafluorophosphate at 75 °C. After 15 minutes a dark solution was obtained which indicate the formation of Pd(0) nanoparticles (Gelesky et al., 2005).

Reduction of copper precursors was carried out by borohydride (Singh et al., 2008; Singh et al., 2008), by microwave irradiation (Jacob et al., 2006), or even by spontaneous dissolution of macroscopic copper powders (copper bronze) reacting with the counterions of the ionic liquids through an oxidation/reduction multistep process that occurs while using nanoparticles in heterogeneous catalysis (as a part of the overall catalytic cycle). Tetraalkylammonium halides, imidazolium derivatives and other ionic liquids with ammonium moieties (Singh et al., 2008; Singh et al., 2008) have been employed as dispersing media where copper nanoparticles have been generated and successfully used as catalysts.

2.4 Ionic liquid assisted of Ge(0) and Te(0) nanoparticles

Tellurium Te(0) nanorods and nanowires were prepared in ionic liquids applying microwave irradiation. By combining the advantages of both ionic liquids and microwave heating a fast and controlled synthesis of tellurium nanorods and nanowires was obtained. By controlling experimental parameters, it was possible to manufacture either Te nanorods or nanowires (Zhu et al., 2004).

Germanium nanoclusters and quantum dots with dimensions of only a few nanometers have been intensively investigated in the past. Such small Ge(0) clusters show, for example, a photoluminescence which is shifted to higher energies with decreasing particle size, thus quantum size effects are present. Most of such studies were performed under ultrahigh vacuum conditions which would complicate a possible future nanotechnological process. Therefore scientists were seeking for a new method to prepare germanium nanoclusters. Endres et al., developed a synthesis method for germanium nanoclusters with a narrow height distribution by electrodeposition from a diluted solution of $GeCl_4$ in 1-butyl-3-methylimidazolium hexafluorophosphate. Under the reported conditions, the lateral sizes of most of the clusters ranged between 20 and 30 nm while their heights vary from 1 to 10 nm with most of them between 1 and 5 nm (Endres & Abedin, 2002).

2.5 Ionic liquid assisted Au, Ag, Pt nanoparticles

The controlled decomposition of the organometallic Pt(0) precursor $Pt_2(dba)_3$ (dba = dibenzylideneacetone) was reported in 1-*n*-butyl-3-methylimidazolium tetrafluoroborate and 1-*n*-butyl-3-methylimidazolium hexafluorophosphate[BMIM][PF_6] as ionic liquid (Scheeren et al., 2006). Stable and catalytically active Pt(0) nanoparticles with a mean diameter of 2.3 nm (Scheeren et al., 2003) were obtained. Moreover, from transmission electron microscopy (TEM) contrast density fluctuations around the metal nanoparticles in the ionic liquid, and different thicknesses of mass density between the ionic liquid and crystalline structures were observed.

Spherically shaped Ag and Au nanoparticles of 20 nm diameters were synthesized through the reduction of silver nitrate and tetrachloroauric acid, respectively in ionic liquids (Singh et al., 2009). A facile one-pot synthesis of gold nanoparticles took place in an aqueous solution using *N*-(2-hydroxyethyl)-*N*-methylmorpholinium tetrafluoroborate (Kim et al., 2006). Single-crystal gold nanosheets and nanoparticles have been successfully prepared by microwave irradiation process in 1-butyl-3-methylimidazolium tetrafluoroborate (Li et al., 2005). Aggregation-induced colour changes of the gold nanoparticles in an aqueous solution were used as an optical sensor for anions via anion exchange of ionic liquid moiety (Itoh et al., 2004).

Because there is no additional template agent in such syntheses, the synthetic procedures and the related treating processes are very simple. The synthetic route is fast and the size of the resultant nanosheets is very large, and it is favourable to produce gold nanosheets in large scale. Ag and Pt (0) nanoparticles with a very narrow size distribution were prepared in 1-butyl-3-methylimidazolium bis(triflymethyl-sulfonyl) imide process (Wang & Yang 2006), 1-*n*-butyl-3-methylimidazolium tetrafluoroborate and 1-*n*-butyl-3-methylimidazolium hexafluorophosphate ionic liquid (Scheeren et al., 2008) using oleic acid as the major capping agent. In this case the combination of oleic acid with 1-butyl-3-methylimidazolium bis(triflymethyl-sulfonyl)imide led to an automatic separation of colloidal metal nanoparticles from the ionic liquid mixtures through a settling process (Wang & Yang 2006). Gold nanoparticles were also directly synthesized in pure 1-*n*-butyl-3-methylimidazolium tetrafluoroborate ionic liquid via addition of $HAuCl_4$, followed by reduction with a ten times excess of $NaBH_4$. Alternatively, Au nanoparticles were prepared in 1-*n*-butyl-3-

methylimidazolium tetrafluoroborate ionic liquid to which 1-methylimidazole had been added. The presence of 1-methylimidazole at a level of 1.0 mM was found to have a significant effect on the stability of the final Au nanoparticles (Dash & Scott, 2009).

The influence of the temperature on the size distribution of the gold nanoparticles was also studied using small-angle X-ray scattering (SAXS) technique (Hatakeyama et al., 2010). The results showed that the particle size and its distribution were greatly affected by the temperature of the capture ionic liquid. The temperature change caused a drastic change in the viscosity of the ionic liquid and the diffusive velocities of the sputtered Au nanoparticles. In addition to the diffusion factor, the stabilities of the formed Au nanoparticles and the coordination and stabilization by the constituent ions of the ionic liquid, especially the anion, influenced the generating process of Au nanoparticles. Once prepared and stabilized by ions, Au nanoparticles were stably dispersed in the ionic liquid at lower temperatures. This study indicates that one can control the size and dispersion of Au nanoparticles only by regulating the temperature of the capture ionic liquid. In this way very small nanoparticles of less than 1nm diameter can be produced.

A new one phase method for the synthesis of uniform monodisperse crystalline Ag nanoparticles in aqueous systems was developed by using newly synthesized mono and dihydroxylated ionic liquids and cationic surfactants based on 1,3-disubstituted imidazolium cations and halogens anions. The hydroxyl functionalized ionic liquids and hydroxyl functionalized cationic surfactants simultaneously act both as the reducing and protective agent. By changing the carbon chain length, alcohol structure and anion of the hydoxy-functionalized 1,3-imidazolium based ionic liquids and the hydroxyl functionalized cationic surfactants the particle size, uniformity and dispersibility of nanoparticles in aqueous solvents could be controlled (Dorjnamjin et al., 2008).

One-phase synthesis of gold and platinum nanoparticles using new thiol-functionalized ionic liquids is also described in the literature (Kim et al., 2004). Thiol-functionalized ionic liquids as stabilizing agents for gold and platinum nanoparticles were designed to have thiol groups on either the cation or anion and symmetrical or unsymmetrical positions at the cation. The metal nanoparticles formed using thiol-functionalized ionic liquids are crystalline structures with face-centred cubic packing arrangements and have small sizes (the average diameters 3.5, 3.1, and 2.0 nm for Au and 3.2, 2.2, and 2.0 nm for Pt, respectively) and uniform distributions. It is believed that the nanoparticle size and distribution depend on the number and position of thiol groups in the ionic liquid.

3. Synthesis of metal oxide nanoparticles in ionic liquids

Traditionally, synthetic approaches for the production of functional metal oxide materials have involved high-temperature reaction environments with energy-intensive techniques such as laser ablation, ion implantation, chemical vapour deposition, photolithography or thermal decomposition. Therefore, ionic liquids which by definition are salts existing in liquid form typically below 100 °C can be used as thermally stable and non-volatile solvents, which can be beneficial to the reduction of solvent emission in comparison to conventional volatile organic compounds. Several imidazolium-based ionic liquids, including 1-butyl-3-methylimidazolium bis(triflylmethylsulfonyl)imide [BMIM][Tf_2N], are not only non-flammable but also stable in a broad temperature range-from room temperature to over 400 °C.

3.1 Ionic liquid assisted iron oxide nanoparticles

Magnetic iron oxide nanoparticles (Fe_2O_3 and Fe_3O_4) have been extensively studied because of their potential applications in biomedical fields. The fluid dispersions of magnetic iron oxide nanoparticles are known as ferrofluids. These ferrofluids are much used in medicine. A main problem of these magnetic nanoparticles is their aggregation, due to the magnetic attractive forces combined with inherently large surface energies. Therefore, many attempts have been made to prepare magnetic nanoparticle dispersions in the presence of surfactants or polymeric compounds with functional groups as stabilizers.

All these technological and medical applications require that the nanoparticles are superparamagnetic with size smaller than 20 nm and the overall particle size distribution is narrow so that the particles have uniform physical and chemical properties. In the last years, various methods for the preparation of such particles have been developed and optimized. High-quality monodisperse iron oxide nanoparticles with high crystalline and narrow size distribution have been prepared by a high-temperature organic phase decomposition method.

A simple and economical method to achieve stable water-soluble iron oxide nanoparticles is based on dispersions in an ionic liquid containing acrylic acid anion. Distinct particles were synthesized by an organic phase decomposition method. The synthesis of magnetite nanoparticles in our laboratories was carried out only by simple precipitation of ferrous salts in alkaline aqueous solutions at low temperature. It was found that the ionic liquid can be used as solvent and stabilizer to efficiently tune the particle growth and prevent nanoparticle aggregation (Zhai et al., 2009).

The synthesis of iron oxide nanostructures with well-defined shapes, including rods, cubes, and spheres, in 1-butyl-3-methylimidazolium bis(triflylmethyl-sulfonyl) imide ([BMIM][Tf_2N]), was also reported. Surfactants including oleic acid and oleylamine, which are commonly used as surface capping agents for size and shape control in molecular solvents, can be employed for making morphologically well defined nanostructures in this ionic liquid. Iron pentacarbonyl thermally decomposes at elevated temperatures in 1-butyl-3-methylimidazolium bis(triflylmethyl-sulfonyl) imide ionic liquid and subsequently form nanoparticles. Nanorods, nanocubes, and spherical particles were synthesized depending mainly on the reaction temperatures and surfactants (Wang & Yang, 2009). It was also reported that this ionic liquid can be recycled and reused without detrimental effects on size and size distribution of the particles made (Wang et al., 2007). α-Fe_2O_3 with various morphologies has been successfully synthesized *via* an ionic liquid-assisted hydrothermal synthetic method (Cao & Zhu, 2009; Lian et al., 2009). Good photocatalytic activities of α-Fe_2O_3 nanospheres were obtained by microwave–hydrothermy in 1-butyl-3-methylimidazolium tetrafluoroborate ionic. This ionic liquid influences not only the crystal phase but also the morphology of the final product (Cao & Zhu, 2009).

We also prepared iron oxide magnetic nanoparticles using as ionic liquid 1-Butyl-3-methylimidazolium tetrafluoroborate [BMIM][BF_4]. Iron (II) chloride and iron (III) chloride which dissolves in [BMIM][BF_4] are coprecipitated in the presence of potassium hydroxide yielding magnetite. The stabilization of magnetite was realized without further purification with glycolic acid. The coprecipitation reaction of iron(II) and iron(III) in ionic liquid and basic condition is published in a Romanian patent (A. Nan et al., 2010)

3.2 Ionic liquid assisted titanium oxide nanoparticles

Titanium oxide nanostructures have versatile applications, for example, in photocatalysis solar-energy conversion, sensors, and ductile ceramics. The synthesis of derivatives with all kinds of size and shape (spherical particles, nanotubes, and nanorods) has been described in numerous studies. Out of the three main titanium polymorphs (anatase, brookite, rutile), research so far has been centred on the synthesis of anatase nanoparticles. However, recently the generation of nanometer-sized rutile has received growing attention due to its promising potential as a photocatalyst and as an electrode material.

Although the photocatalytic activity of rutile is still unclear, several groups found that rutile phase shows higher photocatalytic activity than anatase, and a lot of experimental evidence supported that the existence of a synergistic effect between the anatase and rutile phases is beneficial for reducing the recombination of photogenerated electrons and holes, which usually results in an enhancement of photocatalytic activity. However, it is known that anatase modification of TiO_2 can be obtained first in all solution synthesis pathways and it is even believed that anatase with particle size of 10-20 nm is a thermodynamically stable modification of TiO_2, when considering the contribution of surface energy.

The synthesis of 2-3 nm sized titanium dioxide nanocrystals and their self-assembly toward mesoporous TiO_2 spheres was achieved in an ionic liquid under mild conditions. The resulting structures combine the convenient handling of larger spheres with a considerable high surface area and narrow pore size distribution and are expected to have potential in solar energy conversion, catalysis, and optoelectronic devices. In a typical synthesis of TiO_2 nanoparticles, 1-butyl-3-methylimidazoliumtetrafluoroborate was used as solvent (Zhou & Antonietti, 2003).

Pure rutile and rutile-anatase nanocomposites have been successfully synthesized by hydrolysis of titanium tetrachloride in hydrochloric acid with the assistance of (1-ethyl-3-methyl-imidazolium bromide). The content of rutile in the composites can be controlled by simply adjusting the amount of (1-ethyl-3-methyl-imidazolium bromide) added to the reaction mixture. According to this approach, the composites with arbitrary content of rutile have been rationally and reproducibly prepared in high yields. This method has the following obvious advantages: a simple process; high yield, reaction can be performed under atmospheric pressure in a glass vessel and because high-pressure and high temperature is unnecessary. Products are highly crystalline and their components are controllable. The method can be developed into a general way to synthesize other metal oxide nanoparticles on a large scale (Zheng et al., 2009).

Highly crystalline, well-defined rutile nanorods were synthesized directly by a modified sol–gel reaction in ionic liquid containing imide moieties. In particular, the crystallization was performed under ambient temperature conditions. Compared to previously reported syntheses of rutile and anatase, the present approach features several peculiarities. In essence, the crystallization starts from an amorphous TiO_2 network, which is stabilized by the ionic liquid and converted to rutile upon removal of the ionic liquids simply by extraction. This mechanism is markedly different from the water-based nucleation of rutile in solution, which requires delicate control of various parameters, especially the pH. It is believed that the nucleation of rutile from amorphous TiO_2, stabilized by the ionic liquid through binding of the imide or amide, has a lower barrier compared to the situation in aqueous solution, in which the crystallization of anatase in the form of nanocrystals is usually regarded as favoured due to its lower interfacial energy. These findings point to a different option for controlling the crystallization of nanostructured oxides in general by the stabilization of an intermediate amorphous oxide, which is then converted into its crystalline counterpart by removal of the stabilizer (Kaper et al., 2007).

Nanocrystalline particles of TiO_2 were synthesized via an alkoxide sol-gel method employing water-immiscible ionic liquids as a solvent medium and later modification with non-ionic surfactant as a pore-templating material. Detailed information on the preparative method, crystallographic and structural properties, and photocatalytic activity of the TiO_2 particles are described (Choi et al., 2006). The possible synthetic route and mechanism of the formation of TiO_2 particles and the advantages of using water-immiscible ILs [BMIM][PF_6] in this sol-gel method are discussed in comparison with the challenges of conventional solvent systems (Choi et al., 2006). In the same study, nanostructured TiO_2 particles with high surface area, controlled porosity, and narrow pore size distribution have been synthesized via a sol-gel method modified with ionic liquids and surfactant molecules. Due to the special characteristics of the [BMIM][PF_6] as well as the role of surfactant as a pore templating material in sol-gel network, highly porous TiO_2 particles with anatase crystalline structure were formed even at low temperatures. The TiO_2 particles were thermally stable and thus resistant to pore collapse and anatase-to-rutile phase transformation during calcinations. The thermal stability induced high photocatalytic activity after heat treatment of TiO_2 particles up to 800 °C. This synthesis method has important implications since obtaining crystal structure at low temperatures and keeping good photocatalytic activity at high temperatures can make it possible to utilize the TiO_2 particles and their preparative method in various applications. This new methodology using ionic liquid and surfactant in combination can also be useful for preparing similar crystal nanostructures of other oxide materials at low temperatures.

3.3 Ionic liquid assisted ruthenium oxide nanoparticles

The reaction of $NaBH_4$ with $RuCl_3$ dissolved in 1-*n*-butyl-3-methylimidazolium hexafluorophosphate [BMIM][PF_6] and 1-*n*-butyl-3-methylimidazolium tetrafluoroborate [BMIM][BF_4] as ionic liquid is a simple and reproducible method for the synthesis of stable ruthenium oxide nanoparticles with a narrow size distribution within 2 - 3 nm. These nanoparticles showed high catalytic activity either in the solventless or liquid-liquid biphasic hydrogenation of olefins and arenes under mild reaction conditions. Nanometric ruthenium oxide confined in a zeolitic framework was recently reported as an efficient catalyst for alcohol oxidation under mild aerobic conditions. The nanoparticles could be re-used in solventless conditions up to 10 times in the hydrogenation of 1-hexene yielding a total turnover number for exposed Ru(0) atoms of 175,000 (Rossi et al., 2004; Rossi et al., 2004).

3.4 Ionic liquid assisted vanadium pentoxide, zinc oxide and cobalt oxide nanoparticles

A methodology for the generation of vanadium pentoxide nanostructures was developed based on ionic liquids serving as both reaction medium and structure providing template at the same time (Kaper et al., 2008).

ZnO nanostructures with different shapes have been synthesized via a hydrothermal route, using zinc acetate and sodium hydroxide as the reactants in the presence of different ionic liquids. The ionic liquids can act as morphology templates for ZnO. The ionic liquids can be considered as potential candidates for constructing novel nano-sized ZnO. The morphology of ZnO changes from rod-like to star-like and flower-like in the presence of different ionic liquids.

Zou and co-workers have used a relatively new technique incorporating ionic liquids to direct the synthesis of the Co_3O_4 nanocrystals (Zou et al., 2008). Ionic liquids have recently attracted attention as stable, highly reactive solvents, with good electrical conductivity, high ionic mobility and good chemical and thermal stabilities. In this synthesis 1-*n*-butyl-3-methylimidazolium hydroxide [BMIM][OH] was dissolved in $Co(NO_3)_3 \cdot 6H_2O$ at room temperature (Scheme 3) and both NaOH and H_2O_2 were added separately to the solution, over a period of 30 min. After a 6 h period of stirring, a black - brown precipitate was collected, purified and dried in the oven at 80 ° C for 10 h. The ionic liquid was collected and reused.

$$Co^{2+} + 2OH^{-} \longrightarrow Co(OH)_2$$

$$3\,Co(OH)_2 + H_2O_2 \xrightarrow[\text{rt., 1h}]{\text{[BMIM][OH]}} Co_3O_4 + 4\,H_2O$$

Scheme 3. Ionic reaction of cobalt precursors in the presence of NaOH and H_2O_2

4. Synthesis of alloy nanoparticles in ionic liquids

Alloy nanoparticles, exhibit different catalytic, magnetic, optical and other chemical and physics properties to those of single metal nanoparticles. They are potentially useful for applications such as chemical and biochemical sensing, catalysis, optoelectronics devices, medical diagnostic and therapeutic.

In the case of preparation of alloy nanoparticles in ionic liquid the literature only a few publications have appeared so far. The properties of alloy nanoparticles do not only depend on the size and shape of nanoparticles but also on their composition.

Although various synthetic methods have been developed for the morphological controls of nanoparticles, rods, and wires of semiconductors, metals, and metal oxides, only limited success has been made in making alloy or intermetallic nanoparticles, such as FePt, CoPt, and AuCu (Lee et al., 2002; Shevchenko et al., 2002; Hyeon, 2003; Manna et al., 2003; Zhang et al., 2003; Milliron et al., 2004; Teng & Yang, 2004).

The formation of one-dimensional nanorods and nanowires has rarely been reported. For technologically important Co-Pt and Fe-Pt systems, shape, crystal phase, and composition are essential for the outstanding catalytic, magnetic, and other properties. For example, the structured $CoPt_3$ and $FePt_3$ have been predicted to have higher electrocatalytic activities than pure Pt in hydrogen oxidation reactions. The face centred tetragonal FePt nanoparticles are the most designable material among the Fe-Pt system for magnetic data storage media applications. Cobalt platinum nanorods were made in freshly dried 1-butyl-3-methylimidazolium bis(trifluoromethylsulfonyl) imide [BMIM]-[Tf_2N] at 350 °C. Platinum acetylacetonate ($Pt(acac)_2$) and cobalt acetylacetonate ($Co(acac)_3$) were used as Pt and Co precursors, respectively. The reaction temperature is usually high in order to obtain crystal phase- and composition-specific alloy nanomaterials (Reiss et al., 2004). These temperatures are typically above the boiling points of most of the conventional solvents. Post synthesis high-temperature treatments often affect the particle size, size distribution, and structural property of the nanoparticles. For this type of reaction, 1-butyl-3-methylimidazolium bis(trifluoromethylsulfonyl) imide was chosen as the solvent because of its thermal stability. The combined high-temperature environment and the unique property of ionic liquid bis(trifluoromethylsulfonyl)imide could attribute to the observed formation of CoPt alloy nanoparticles and nanorods with different compositions and crystal phases. The decrease in melting point of nanoparticles could be relevant for the formation of stable CoPt alloys (Wang & Yang, 2005).

A new methodology which allowed identifying peptide sequences that both specifically bind to the ferromagnetic phase of FePt and control the crystallization of FePt nanoparticles using a modified arrested precipitation technique was also described in the literature (Reiss et al., 2004). They applied a peptide-based synthetic strategy for the synthesis of the technologically important magnetic material FePt, in which biological interactions control the nucleation of nanoparticles that have no isomorphous complement in nature. In addition, this biological strategy has several advantages over more recent chemical methods, including the potential synthesis of the desired high anisotropy ferromagnetic crystalline phase in aqueous conditions under ambient temperature, pressure, and atmosphere. Bacterial amplification of the "organic" virus templates makes the production of both the nucleating peptides and FePt nanoparticles easy and cost-effective.

Au–Ag bimetallic nanoparticles have been fabricated by one-step simple electrochemical deposition method using ionic liquid as green electrolyte (1-butyl-3-methylimidazolium tetrafluoroborate). The indium tin oxide (ITO) surface has been utilized for the fabrication of Au–Ag bimetallic nanoparticles.

The use of a stabilizing agent to prepare alloy nanoparticles in ionic liquids can be avoided by using a technique known as sputter deposition. The technique ejects atoms from a solid surface by the energetic impact of gaseous ions, typically used are Ar^+ ions. This technique was used for the first time to synthesize gold-silver alloy nanoparticles in an ionic liquid (Okazaki et al., 2008). The chemical composition and optical properties of the Au-Ag alloy nanoparticles were easily controlled just by varying the area ratio of the individual pure metal foils in the sputtering targets.

PdAu alloy nanoparticles were also directly synthesized in pure 1-*n*-butyl-3-methylimidazolium hexafluorophosphate as ionic liquid (particle size 6.0 - 2.9 nm) and in the presence of a 1-methylimidazole additive in 1-*n*-butyl-3-methylimidazolium hexafluorophosphate (particle size 3.0 - 0.5 nm). Investigations of the stability of the PdAu alloy nanoparticles revealed aggregation in pure ionic liquid, while stability was observed in the presence of 1.0 mM 1-methylimidazole or other stabilizers. The activity of the nanoparticles synthesized directly in the pure ionic liquid began to drop dramatically within 10 min upon addition of the substrate, while PdAu nanoparticles synthesized in the presence of 1.0 mM 1-methylimidazole showed dramatically improved retention of catalytic activity over time. Finally, PdAu nanoparticles were also synthesized in 1-*n*-butyl-3-methylimidazolium trifluoromethanesulfonate in order to demonstrate the generality of this mode-of-stabilization. PdAu nanoparticles in pure 1-*n*-butyl-3-methylimidazolium trifluoromethanesulfonate solutions precipitated within 1 h in the absence of 1-methylimidazole additives during the hydrogenation reaction, whereas stability for prolonged periods of times was achieved in the presence of 1-methylimidazole, however higher concentrations of 1-methylimidazole (ca. 10–100 mM) were needed to stabilize the nanoparticles (Dash & Scott, 2009).

Nanocrystalline alloys were also possible to obtain by deposition. In this way several aluminium alloys from ionic liquids based on $AlCl_3$ were prepared. Of particular interest are electrochemically made Al-Mn alloys, which are widely used in the automobile industry for lightweight construction. An attempt was made to obtain nanocrystalline Al_xMn_y alloys electrochemically by adding $MnCl_2$ and nicotinic acid to a Lewis acidic ionic liquid. The deposition was performed by using a current density of I=0.5 $mAcm^{-2}$ on glassy carbon.

5. Conclusion

In summary, preparation of metal or metal oxide nanoparticles in ionic liquids is an upcoming method providing several advantages over conventional methods. The metal nanoparticles are often of great interest as catalysts. As an additional effect imidazolium-based ionic liquids also form protective layers at the metal nanoparticles thus providing improved stability. By the formation of such ionic liquid protective layers discrete supramolecular species appear through the loosely bound anionic moieties and/or NHC carbenes together with an oxide layer when present on the metal surface. These loosely surface-bound protective species are easily displaced by other substances present in the media. This on the one hand is responsible to some extend for their catalytic activity, but on the other hand explains their relatively low stability that leads to aggregation/agglomeration and eventually to the bulk metal. Therefore the stability and the catalytic activity of metal nanoparticles in imidazolium ionic liquids are also highly influenced by coordinative strength of the aggregates with the metal surface and the type and nature of the substrates/products.

Moreover, metal nanoparticles dispersed in these fluids are stable and can serve as active catalysts for reactions in multiphase conditions. The catalytic properties (activity and selectivity) of these soluble metal nanoparticles indicate that they possess a pronounced surface like (multi-site) rather than single site like catalytic properties. In other cases the metal nanoparticles are not stable and tend to aggregate or alternatively they serve as simple reservoirs of mononuclear catalytically active species. However, the ionic liquid provides a favourable environment for the formation of metal nanoparticles with, in most cases, a small diameter and size distribution under very mild conditions.

In the hydrogenation of simple alkenes the recovered ionic dispersion can be re-used several times without any significant loss in catalytic activity, as observed with ionic liquid-dispersed Ir(0), Rh(0), Pt(0), and Ru(0) nanoparticles. However, in the case of aromatic compounds and ketones some metal nanoparticles tend to aggregate with loss of their catalytic activity such as Rh(0) in the hydrogenation of benzene and Ir(0) in the reduction of ketones. However, these nanoparticles are more stable in the hydrogenation reactions when dispersed in the ionic liquids than in solventless conditions. It has to be mentioned that aromatics, ketones, and alcohols are much more soluble in the ionic liquids than the alkenes and alkanes.

In some cases, which were described for the synthesis of iron oxide, the different solubility of precursors, reactive intermediates and nanoparticles in ionic liquid helps to regulate the delivery of agents in different phases. This shows that high level morphological control of nanomaterials is feasible using ionic liquids by selecting proper capping agents and reaction conditions, which is an important step forward in using ionic liquids as solvents for controlling the size and shape of nanomaterials.

Another important conclusion which comes out is that the simultaneous sputter deposition of different kinds of pure materials can be applied to prepare alloy nanoparticles containing three or more elements, the physicochemical properties being tunable for desired purposes, such as catalysts for fuel cells.
We expect that ionic liquids will find, in addition to organometallic synthesis, catalysis, and electrochemistry, a fourth area of application—the synthesis of nanostructured solids, either to make nanoobjects (e.g. particles and fiber) or for the design of nanopores and nanochannels in solids. The unique combination of adaptability towards other molecules and phases plus the strong H-bond-driven solvent structure makes ionic liquids potential key tools in the preparation of a new generation of chemical nanostructures.

6. References

Antonietti, M.; Kuang, D. B. Smarsly, B. & Yong, Z. (2004). Ionic liquids for the convenient synthesis of functional nanoparticles and other inorganic nanostructures. *Angewandte Chemie, International Edition*, Vol.43, No.38, (September 2004), pp. 4988–4992, ISSN 1521-3773

Calò, V.; Nacci, A.; Monopoli, A.; Ieva, E. & Cioffi, N. (2005). Copper bronze catalyzed Heck reaction in ionic liquids. *Organic Letters*, Vol.7, No.5, (January 2005), pp. 617 - 620, ISSN 1523-7060

Cao, S. W. & Zhu, Y. J. (2009). Iron oxide hollow spheres: Microwave–hydrothermal ionic liquid preparation, formation mechanism, crystal phase and morphology control and properties. *Acta Materialia*, Vol.57, No.7, (April 2009), pp. 2154–2165, ISSN 1359-6454

Cassol, C. C.; Ebeling, G.; Ferrera, B. & Dupont, J. (2006). A simple and practical method for the preparation and purity determination of halide-free imidazolium ionic liquids. *Advanced Synthesis & Catalysis*, Vol.348, No.1-2, (January 2006), pp. 243-248, ISSN 1521-3765

Choi, H.; Kim, Y. J.; Varma, R. S. & Dionysiou, D. D. (2006). Thermally stable nanocrystalline TiO_2 photocatalysts synthesized via sol-gel methods modified with ionic liquid and surfactant molecules. *Chemistry of Materials*, Vol.18, No.22, (October 2006), pp. 5377-5384, ISSN 0897-4756

Cimpeanu, V.; Kocevar, M.; Parvulescu, V. I. & Leitner, W. (2009). Preparation of rhodium nanoparticles in carbon dioxide induced ionic liquids and their application to selective hydrogenation. *Angewandte Chemie, International Edition*, Vol.48, No.6, (January 2009), pp. 1085-1088, ISSN 1521-3773

Dash, P. & Scott, R. W. J. (2009). 1-Methylimidazole stabilization of gold nanoparticles in imidazolium ionic liquids. *Chemical Communication*, No.7, pp. 812-814, ISSN 1359-7345

Deshmukh, R. R.; Rajagopal, R. & Srinivasan, K. V. (2001). Ultrasound promoted C–C bond formation: Heck reaction at ambient conditions in room temperature ionic liquids. *Chemical Commununication*, No.17, pp 1544–1545, ISSN 1359-7345

Ding, K. L.; Miao, Z. J.; Liu, Z. M.; Zhang, Z. F.; Han, B. X.; An, G. M.; Miao, S. & Xie, Y. (2007). Facile synthesis of high quality TiO_2 nanocrystals in ionic liquid via a microwave-assisted process. *Journal of the American Chemical Society*, Vol.129, No.20, (May 2007), pp. 6362-6363, ISSN 0002-7863

Dorjnamjin, D.; Ariunaa M. & Shim, Y. K. (2008). Synthesis of silver nanoparticles using hydroxyl functionalized ionic liquids and their antimicrobial activity. *International Journal of Molecular Sciences*, Vol. 9, No.5, (May 2008), pp. 807-820, ISSN 1422-0067

Dupont, J.; Fonseca, G. S.; Umpierre, A. P.; Fichtner, P. F. P. & Teixeira, S. R. (2002). Transition-metal nanoparticles in imidazolium ionic liquids: recycable catalysts for biphasic hydrogenation reactions, *Journal of the American Chemical Society*, Vol.124, No.16, (April 2002), pp. 4228–4229, ISSN 0002-7863

Endres, F.; Bukowski, M.; Hempelmann, R. & Natter, H. (2003) Electrodeposition of nanocrystalline metals and alloys from ionic liquids. *Angewandte Chemie, International Edition*, Vol.42, No.29, (July 2003), pp., 3428-3430, ISSN 1521-37732003

Endres, F. & Abedin, S. Z. (2002). Electrodeposition of stable and narrowly dispersed germanium nanoclusters from an ionic liquid, *Chemical Commununication*, No. 8, pp. 892 – 893, pp. 1359-7345, ISSN 1359-7345

Fonseca, G. S.; Umpierre, A. P.; Fichtner, P.; Teixera, F. P. & Dupont, S. R. (2003). The Use of imidazolium ionic liquids for the formation and stabilization of Ir0 and Rh0 nanoparticles: efficient catalysts for the hydrogenation of arenes. *Chemistry A European Journal*, Vol.9, No.14, (July 2003), pp. 3263 – 3269, ISSN 1521-3765

Gelesky, M. A.; Umpierre, A. P.; Machado, G.; Correia, R. R. B.; Magno, W. C.; Morais, J.; Ebeling, G. & Dupont, J. (2005). Laser-induced fragmentation of transition metal nanoparticles in ionic liquids. Journal of the American Chemical Society, Vol.127, No.13, (April 2005), pp. 4588 –4589, ISSN 0002-7863

Hatakeyama, Y.; Takahashi, S. & Nishikawa, K. (2010). Can Temperature Control the Size of Au Nanoparticles Prepared in Ionic Liquids by the Sputter Deposition Technique? *The Journal of Physical Chemistry C*, Vol.114, No.25, (July 2010), pp. 11098-11102, ISSN 1932-7447

Huang, J.; Jiang, T.; Han, B. X.; Gao, H. X.; Chang, Y. H.; Zhao, G. Y. & Wu, W. (2003). Hydrogenation of olefins using ligand-stabilized palladium nanoparticles in an ionic liquid. *Chemical Commununication*, No.14, pp. 1654 – 1655, ISSN 1359-7345

Hyeon, T. (2003). Chemical synthesis of magnetic nanoparticles. *Chemical Communication*, No.8, pp. 927-934, ISSN1359-7345

Itoh, H.; Naka, K. & Chujo, Y. (2004). Synthesis of gold nanoparticles modified with ionic liquid based on the imidazolium cation. *Journal of the American Chemical Society*, Vol.126, No.10, (February 2004), pp. 3026–3027, ISSN 0002-7863

Jacob, D. S.; Genish, I.; Klein, L. & Gedanken , A. (2006) Carbon coated core shell structured copper and nickel nanoparticles synthesized in a ionic liquid. *The Journal of Physical Chemistry B*, Vol.110, No.36, (august 2006), pp. 17711 – 17714, ISSN 1520-6106

Kim, K. S.; Demberelnyamba, D. & Lee, H. (2004). Size-selective synthesis of gold and platinum nanoparticles using novel thiol-functionalized ionic liquids. *Langmuir*, Vol.20, No.3, (December 2003), pp. 556-560, ISSN 0743-7463

Kim, K. S.; Choi, S.; Cha, J. H.; Yeon, S. H. & Lee, H. (2006). Facile one-pot synthesis of gold nanoparticles using alcohol ionic liquids. *Journal of Materials Chemistry*, Vol.16, No.14, (April 2006), pp. 1315-1317, ISSN 0959-9428

Kumar, C. S. S. R. E. (Ed.). (2010). *Nanomaterials for the Life Sciences*, Wiley-VCH, ISBN 9783527610419, Weinheim

Lee, S. M.; Jun, Y. W.; Cho, S. N. & Cheon, J. (2002). Single-crystalline star-shaped nanocrystals and their evolution: programming the geometry of nano-building blocks. *Journal of the American Chemical Society*, Vol.124, No.38, (August 2002), pp. 11244-11245, ISSN 0002-7863

Léger, B.; Nowicki, A. D.; Roucoux, A. & Bourbigou, H. O. (2008). Synthesis of bipyridine-stabilized rhodium nanoparticles in non-aqueous ionic liquids: A new efficient approach for arene hydrogenation with nanocatalysts. *Advanced Synthesis & Catalysis*, Vol.350, No.1, (January 2008), pp. 153 – 159, ISSN 1521-3765

Li, Z. H.; Liu, Z. M.; Zhang, J. L.; Han, B. X.; Du, J. M.; Gao, Y. N. & Jiang, T. (2005). Synthesis of single-crystal gold nanosheets of large size in ionic liquids. *The Journal of Physical Chemistry B*, Vol.109, No.30, (July 2005), pp. 14445-14448, ISSN 1089-5647

Lian, J.; Duan, X.; Ma, J.; Peng, P.; T. Kim, & W. Zheng, (2009). Hematite (α-Fe_2O_3) with various morphologies: ionic liquid-assisted synthesis, formation mechanism, and properties. *ACS Nano, Vol.3, No.11, (November 2009), pp.* 3749–3761, *ISSN* 1936-0851

Li, Z.; Jia, Z.; Luan, Y. & Mu, T. (2008). Ionic liquids for synthesis of inorganic nanomaterials. *Current Opinion in Solid State and Materials Science*, Vol.12, No.1, (January 2008), pp. 1-8, ISSN 1359-0286

Lin, H.; de Oliveira, P.W.; Grobelsek, I.; Haettich, A. & Veith, M. (2010). The synthesis of anatase TiO_2 nanoparticles by solvothermal method using ionic liquid as additive. *Zeitschrift für Anorganische und Allgemeine Chemie*, Vol.636, No.11, (September 2010), pp. 1947–1954, ISSN 1521-3749

Kaper, H.; Endres, F.; Djerdj, I.; Antonietti, M.; Smarsly, B. M.; Maier, J. & Hu, Y. S. (2007). Direct low-temperature synthesis of rutile nanostructures in ionic liquids. Small, Vol.3, No.10, pp. 1753-1763, ISSN 1613-6829

Kaper, H.; Willinger, , M. G.; Djerdj, I.; Gross, S.; Antonietti, M. & Smarsly, B. M. (2008). IL-assisted synthesis of V_2O_5 nanocomposites and VO_2 nanosheets. *Journal of Materials Chemistry*, Vol.18, No.47, (December 2008), pp. 5761–5769, ISSN 0959-9428

Manna, L.; Milliron, D. J.; Meisel, A.; Scher, E. C. & Alivisatos, A. P. (2003). Controlled growth of tetrapod-branched inorganic nanocrystals. *Nature Materials*, Vol.2, No.6, (May 2003), pp. 382-385, ISSN 1476-1122

Migowski, P. & Dupont, J. (2007). Catalytic applications of metal nanoparticles in imidazolium ionic liquids. *Chemistry A European Journal*, Vol.13, No.1, (November 2006), pp. 32-39, ISSN 1521-3765

Migowski, P.; Teixeira, S. R.; Machado, G.; Alves, M. C. M.; Geshev J. & Dupont, J. (2007). Structural and magnetic characterization of Ni nanoparticles synthesized in ionic liquids. Journal of Electron Spectroscopy and Related Phenomena, Vol.156-158, (May 2007), pp. 195-199, ISSN 0368-2048

Milliron, D. J.; Hughes, S. M.; Cui, Y.; Manna, L.; Li, J. B.; Wang, L. W. & Alivisatos, A. P. (2004). Colloidal nanocrystal heterostructures with linear and branched topology. *Nature*, Vol.430, No.6996, (July 2004), pp. 190-195, ISSN 0028-0836

Mudring, A. V.; Alammar, T.; Bäcker, T. & Richter, K. (2009). Nanoparticle synthesis in ionic liquids, In: *Ionic Liquids: From Knowledge to Application*, N. V. Plechkova, R. D.

Rogers, K. R. Seddon (Eds.), pp. 177–188, *ACS Symposium Series*, ISBN: 9780841269972, Washington, USA

Nan, A.; Turcu, R. & Liebscher, J. (2010). Application of the coprecipitation method in ionic liquid for preparation of iron oxide nanoparticles. RO nr. A/01357/2010.

Mukhopadhyay, I. & Freyland, W. (2003). Electrodeposition of Ti nanowires on highly oriented pyrolytic graphite from an ionic liquid at room temperature. *Langmuir*, Vol.19, No.6, (March 2003), pp. 1951–1953, ISSN 0743-7463

Okazaki, K.; Kiyama, T.; Hirahara, K.; Tanaka, N.; Kuwabata, S. & Torimoto, T. (2008). Single-step synthesis of gold–silver alloy nanoparticles in ionic liquids by a sputter deposition technique, *Chemical Communication*, No.6, (February 2006), pp. 691-693, ISSN 1359-7345

Ott, L. S. & Finke, R. G. (2007). Transition - metal nanocluster stabilization for catalysis: a critical review of ranking methods and putative stabilizers. *Coordination Chemistry Reviews* , Vol.251 , No.9-10, (May 2007), pp. 1075 – 100, ISSN 0010-8545

Patzke, G. R.; Zhou, Y.; Kontic, R. & Conrad, F. (2011). Oxide nanomaterials: synthetic developments, mechanistic studies, and technological innovations. *Angewandte Chemie, International Edition*, Vol.50, No.4, (Jamuary 2011), pp. 826-859, ISSN 1521-3773

Prechtl, M. H. G.; Scariot, M.; Scholten, J. D.; Machado, G.; Teixeira, S. R. & Dupont, J. (2008). Nanoscale Ru(0) Particles: Arene Hydrogenation Catalysts in Imidazolium Ionic Liquids. *Inorganic Chemistry*, Vol.47, No.19, pp. 8995–9001, ISSN 0020-1669

Redel, E.; Krämer, J.; Thomann, R. & Janiak, C. (2009). Synthesis of Co, Rh and Ir nanoparticles from metal carbonyls in ionic liquids and their use as biphasic liquid–liquid hydrogenation nanocatalysts for cyclohexene, *Journal of Organometallic Chemistry*, Vol.694, No.7-8, (April 2009), pp. 1069–1075, ISSN 0022-328X

Reiss, B. D.; Mao, C. B.; Solis, D. J.; Ryan, K. S.; Thomson, T. & Belcher, A. M. (2004). Biological routes to metal alloy ferromagnetic nanostructures. *Nano Letters*, Vol.4, No.6, (April 2004), pp. 1127-1132, ISSN 1530-6984

Rossi, L. M.; Machado, G.; Fichtner, P. F. P.; Teixeira, S. R. & Dupont, J. (2004). On the use of ruthenium dioxide in 1-n-butyl-3-methylimidazolium ionic liquids as catalyst precursor for hydrogenation reactions. *Catalysis Letters* , Vol.92, No.3-4 , (February 2004), pp. 149 –155, ISSN 1011-372X

Rossi, L. M.; Dupont, J.; Machado, G.; Fichtnerb, P. F. P.; Radtke, C.; Baumvol I. J. R. & Teixeira, S. R. (2004). Ruthenium dioxide nanoparticles in ionic liquids: synthesis, characterization and catalytic properties in hydrogenation of olefins and arenes. *Journal of the Brazilian Chemical Society*, Vol.15, No.6, (December 2004), pp. 904-910, ISSN 0103 - 5053

Rossi, L. M. & Machado, G. (2009). Ruthenium nanoparticles prepared from ruthenium dioxide precursor: Highly active catalyst for hydrogenation of arenes under mild conditions. *Journal of Molecular Catalysis A: Chemical*, Vol.298, No.1-2, (February 2009), pp. 69–73, ISSN 1381-1169

Scheeren, C. W.; Machado, G.; Dupont, J.; Fichtner, P. F. P. & Texeira, S. R. (2003). Nanoscale Pt(0) particles prepared in imidazolium room temperature ionic liquids: synthesis from an organometallic precursor, characterization, and catalytic properties in hydrogenation reactions. *Inorganic Chemistry*, Vol.42, No.15, (June 2003), pp. 4738-4742, ISSN 0020-1669

Scheeren, C. W.; Machado, G.; Teixeira, S. R.; Morais, J.; Domingos, J. B. & Dupont, J. (2006). Synthesis and characterization of Pt(0) nanoparticles in imidazolium ionic liquids. *Journal of Physical Chemistry B*, Vol.110, No.26, (June 2006), pp. 13011–13020, ISSN 1089-5647

Scheeren, C. W.; Domingos, J. B.; Machado G. & Dupont, J. (2008). Hydrogen reduction of Adams' catalyst in ionic liquids: formation and stabilization of Pt(0) nanoparticles. *The Journal of Physical Chemistry C*, Vol.112, No.42, (September 2008), pp 16463–16469, ISSN 1932-7447

Shevchenko, E. V.; Talapin, D. V.; Rogach, A. L.; Kornowski, A.; Haase, M. & Weller, H. (2002). Colloidal synthesis and self-assembly of $CoPt_3$ nanocrystals. *Journal of the American Chemical Society*, Vol.124, No.38, (September 2002), pp. 11480-11485, ISSN 0002-7863

Silveira, E. T.; Umpierre, A. P.; Rossi, L. M.; Machado, G.; Morais, J.; Soares, G. V.; Baumvol, I. L. R.; Teixeira, S. R.; Fichtner, P. F. P. & Dupont, J. (2004). The partial hydrogenation of benzene to cyclohexene by nanoscale ruthenium catalysts in imidazolium ionic liquids. *Chemistry A European Journal*, Vol.10, No.15, (august 2004), pp. 3734 –3740, ISSN 1521-3765

Singh, P.; Katyal, A.; Kalra, R. & Chandra, R. (2008). Copper nanoparticles in an ionic liquid: an efficient catalyst for the synthesis of bis-(4-hydroxy-2-oxothiazolyl)methanes. *Tetrahedron Letters*, Vol.49, No.4, (January 2008), pp. 727 – 730, ISSN 0040-4039

Singh, P.; Katyal, A.; Kalra, R. & Chandra, R. (2008). Copper nanoparticles in an ionic liquid: an easy and efficient catalyst for the coupling of thiazolidine-2,4-dione, aromatic aldehyde and ammonium acetate. *Catalysis Communications*, Vol.9, pp. 1618 – 1623, ISSN 1566-7367

Singh, P.; Kumari, K.; Katyal, A.; Kalra, R. & R. Chandra, (2009). Synthesis and characterization of silver and gold nanoparticles in ionic liquid. *Spectrochimica Acta Part A: Molecular and Biomolecular Spectroscopy*, Vol.73, No.1, (July 2009), pp.218-220, ISSN 1386-1425

Suh, W. H.; Suh, Y. H. & Stucky, G. D. (2009). Multifunctional nanosystems at the interface of physical and life sciences. *Nano Today*, Vol.4, No.1, (February 2009), pp. 27-36, ISSN 1748-0132

Sweeny, B. K. & D. G. Peters (2001). Cyclic voltammetric study of the catalytic behaviour of nickel(I) salen electrogenerated at a glassy carbon electrode in an ionic liquid (1-butyl-3-methylimidazolium tetrafluoroborate, $BMIM^+BF_4^-$. *Electrochemistry Communication*, Vol.3, No.12, (December 2001), pp. 712-715, ISSN 1388-2481

Tsai, T. H.; Thiagarajan, S. & Chen, S. M. (2010). Ionic liquid assisted one step green synthesis of Au–Ag bimetallic nanoparticles. *Journal of Applied Electrochemistry*, Vol.40, No.3, (March 2010), pp. 493-497, ISSN 0021-891X

Taubert, A. (2004). CuCl Nanoplatelets from an Ionic Liquid-Crystal Precursor A*ngewandte Chemie, International Edition*, Vol.43,No.40, (octomber 2004), pp. 5380-5382, ISSN 1521-3773

Teng, X. W. & Yang, H. (2004). Effects of surfactants and synthetic conditions on the sizes and self-assembly of monodisperse iron oxide nanoparticles. *Journal of Materials Chemistry*, Vol.14, no.4, pp.774-779, ISSN 0959-94282004

Umpierre, A. P.; Machado, G.; Fecher, G. H.; Morais, J. & Dupont, J. (2005). Selective Hydrogenation of 1,3-Butadiene to 1-Butene by Pd(0) Nanoparticles embedded in imidazolium ionic liquids. *Advanced Synthesis & Catalysis*, Vol.347, No.10, (August 2005), pp. 1404 – 1412, ISSN 1521-3765

Vollmer, C.; Redel, E.; Abu-Shandi, K.; Thomann, R.; Manyar, H.; Hardacre, C. & Janiak, C. (2010). Microwave irradiation for the facile synthesis of transition-metal nanoparticles (NPs) in ionic liquids (ILs) from metal–carbonyl precursors and Ru-, Rh-, and Ir-NP/IL dispersions as biphasic liquid–liquid hydrogenation nanocatalysts for cyclohexene. *Chemistry A European Journal*, Vol.16, No.12, (March 2010), pp. 3849-3858, ISSN 1521-3765

Zhai, Y.; Liu, F.; Zhang, Q. & Gao, G. (2009). Synthesis of magnetite nanoparticle aqueous dispersions in an ionic liquid containing acrylic acid anion. *Colloids and Surface A: Physicochemical and Engineering Aspects*, Vol.332, No.2-3, (January 2009), pp. 98-102, ISSN 0927-7757

Zhang, Z. T.; Blom, D. A.; Gai, Z.; Thompson, J. R.; Shen, J. & Dai, S. (2003). High-yield solvothermal formation of magnetic CoPt alloy nanowires. *Journal of the American Chemical Society*, Vol.125, No.5, pp. 7528-7529, ISSN 0002-7863

Zhao, D. B.; Fe, Z. F.; Ang, W. H. & Dyson, P. J. (2006). A strategy for the synthesis of transition-metal nanoparticles and their transfer between liquid phases. *Small*, Vol.2, No.7, (May 2006), pp. 879-883, ISSN 1613-6829

Zheng, W.; Liu, X.; Yan, Z. & Zhu, L. (2009). Ionic liquid-assisted synthesis of large-scale TiO_2 nanoparticles with controllable phase by hydrolysis of $TiCl_4$. *ACS Nano, Vol.3, No.1, (January 2009), pp. 115-122, ISSN* 1936-0851

Zhou, Y. & Antonietti, M. (2003). Synthesis of very small TiO_2 nanocrystals in a room-temperature ionic liquid and their self-assembly toward mesoporous spherical aggregates. *Journal of the American Chemical Society*, Vol.125, No.49, (November 2003), pp. 14960-14961, ISSN 0002-7863

Zhu, Y. J.; Wang, W. W.; Qi, R. J. & Hu, X. L. (2004). Microwave-assisted synthesis of single-crystalline tellurium nanorods and nanowires in ionic liquids. *Angewandte Chemie, International Edition*, Vol.43, No.11, (March 2004), pp. 1410-1414, ISSN 1521-3773

Zou, D.; Xu, C.; Luo, H.; Wang, L. & Ying, T. (2008). Synthesis of Co_3O_4 nanoparticles via an ionic liquid-assisted methodology at room temperature. *Materials Letter*, Vol.62, No.12-13, (April 2008), pp.1976-1978, ISSN 0167-577X

Yavari, I.; Mahjoub, A. R.; Kowsari & Movahedi, , E. M. (2009). Synthesis of ZnO nanostructures with controlled morphology and size in ionic liquids. *Journal of Nanoparticles Research*, Vol.11, No.4, (May 2009), pp. 861-868, ISSN 1388-0764

Wang, W. W. & Zhu, Y. J. (2005). Synthesis of $PbCrO_4$ and Pb_2CrO_5 rods via a microwave-assisted ionic liquid method. *Crystal Growth & Design*, Vol.5, No.2, (January 2005), pp. 505-507, ISSN 1528-7483

Wang, Y. & Yang, H. (2005). Synthesis of CoPt nanorods in ionic liquids. *Journal of the American Chemical Society*, Vol.127, No.15, (March 2005), pp. 5316-5317, ISSN 0002-7863

Wang, Y. & Yang, H. (2006). Oleic acid as the capping agent in the synthesis of noble metal nanoparticles in imidazolium-based ionic liquids. *Chemical Communication*, No.24, pp. 2545-2547, ISSN 1359-7345

Wang, Y.; Maksimuk, S.; Shen, R. & Yang, H. (2007). Synthesis of iron oxide nanoparticles using a freshly-made or recycled imidazolium-based ionic liquid. *Green Chemistry*, Vol.9, No.10, pp. 1051-1056, ISSN 1463-9262

Wang, Y. & Yang, H. (2009). Synthesis of iron oxide nanorods and nanocubes in an imidazolium ionic liquid. *Chemical Engineering Journal*, Vol.147, No.1, (April 2009), pp. 71-78, ISSN 1385-8947

Part 5

Separation Methods

15

Ionic Liquids in Separation Techniques

Jolanta Flieger and Anna Czajkowska-Żelazko
Medical University of Lublin
Poland

1. Introduction

Recently, ionic liquids (ILs) have gained in popularity as unique solvents in different areas of separation techniques. Owing to tunable properties which can be selected by choosing appropriate cationic or anionic constituents, they can be applied in chromatographic and electro-chromatographic methods as mobile phase or stationary phase modifiers or in sample preparations processes as new extraction solvents. Their advantageous properties, from the point of view of separation mechanisms, such as good thermal stability, good extractability for various organic and inorganic ions and the possibility of choosing the optimal viscosity and miscibility with either water or organic solvents make this class of compounds an attractive alternative to classical organic solvents. Furthermore, replacing organic solvents with ILs appears to be much safer for environment regarding the non-flammable and non-volatile nature of these compounds.

This review focuses on applications of ILs in separation techniques: extraction, capillary electrophoresis (CE) and liquid chromatography (LC). So far many studies have been published on the applications of ILs in different separation processes and their number is still growing and attracting great attention. With no doubt, this is connected with huge number of ILs consisting of numerous combinations of various bulky, nonsymmetrical organic cations and inorganic or organic anions. It seems to be obvious that new possibilities of ILs applications will also be discovered in future.

ILs in the form of molten stearates of transition metals were applied firstly in gas chromatography as new liquid stationary phases by Baber et al. (Barber et al., 1959), but since alkylimidazolium-based ILs were introduced for modification of silica capillaries, GC systems have been achieving satisfactory efficiency (Armstrong et al. 1999; Berthod et al., 2001; Anderson & Armstrong, 2003).

In liquid chromatography: either thin layer chromatography (TLC) or high performance liquid chromatography (HPLC), ILs are applied as mobile phase additives as they seem too viscous and non-transparent to UV light to be used as pure solvents. In diluted form, mixed with other low viscosity solvents, they revealed their dual nature responsible for unusual selectivity towards ionic and nonionic analytes. Among the whole group only pyridinium and imidazolium hexafluorophosphates or tetrafluoroborates in reversed-phase chromatographic systems are the most commonly applied for this purpose.

ILs have been used also in capillary electrophoresis (CE) for modification of running buffer and for dynamic coating or covalent attachment to the capillary walls. The main aim of this modification is reverse the electroosmotic flow (EOF) provoked by silanol groups on the inner surface of silica capillaries.

It's worth noting that many reviews devoted to application of ILs in separation techniques systematically appear in the scientific literature (Buszewski & Studzińska, 2008; Marszałł & Kaliszan, 2007; Berthod et al., 2008; Sun & Armstrong, 2010; Han & Row,2010).
This review focuses on experiments conducted by our group (Flieger, 2006; Flieger, 2009; Flieger & Czajkowska-Żelazko 2011a; Flieger & Czajkowska-Żelazko 2011b; Flieger & Czajkowska-Żelazko 2011c) and described in the literature by others.

2. Physicochemical properties

Physicochemical properties and solvation characteristics are useful for choosing appropriate ILs as solvents for extraction or chromatographic methods. Examples for typical cations and anions of ILs used in separation techniques are presented below (sigma-aldrich.com /chemicalsynthesis):

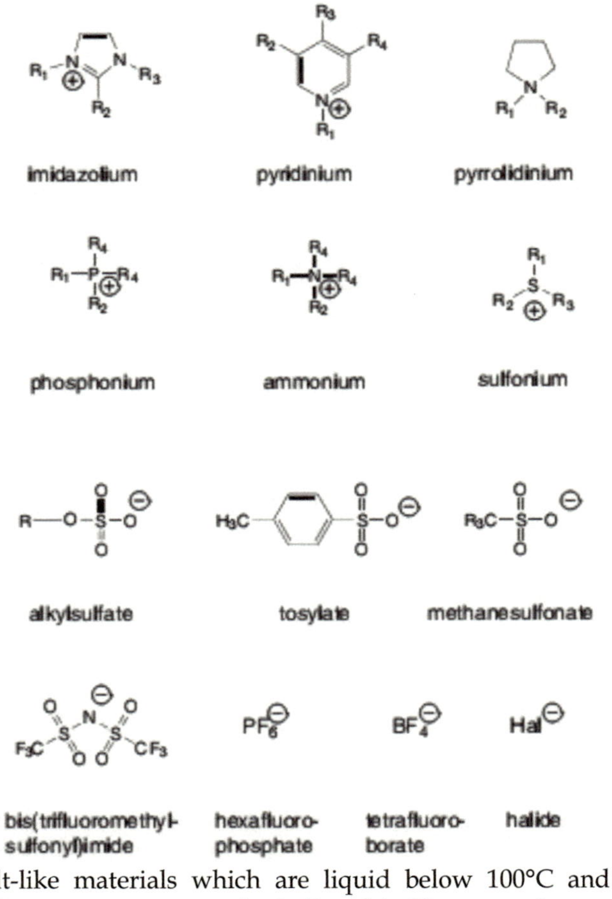

ILs belong to salt-like materials which are liquid below 100°C and even below room temperature (RTIL-room temperature ionic liquids). The most characteristic properties of ILs cover good termal stability over the wide range of temperatures 300-400°C, good electric conductivity enabling their application in electro-migration techniques and low vapor pressure; owing to this they are less air-polluting than common organic solvents. The solvent properties of ionic liquids are also interesting. The densities and viscosities of ILs are generally greater than those of conventional solvents indicating their usefulness in gas chromatography as stationary phase modifiers. On the other hand, they have an extended range of solubility either in polar or non polar solvents, so they could be used in separation techniques including different extraction techniques and as mobile and stationary phase modifiers in liquid chromatography and capillary electrophoresis.

2.1 Solvatochromic parameters

The polarity of ILs can be compared by the use of empirical scales. The most widely applied scale has been based on changes in the charge transfer π-π* absorption band for the betaine dye, 2,6-diphenyl-4-(2,4,6-t triphenylpyridinium-1-yl) phenolate also known as Reichardt's dye (Dimroth et al., 1963). The $E_T(30)$ value describing solvent polarity is calculated on the basis of the following equation:

$$E_T(30)(kcalmol^{-1} = \frac{28591}{\lambda_{max}^{abs}(nm)} \quad (1)$$

Fletcher et al (Flechter et al., 2001) compared the polarity of BMIM PF_6 with other solvents using $E_T(30)$. Obtained $E_T(30)$ value for Reichardt's betaine dye dissolved in BMIM PF_6 and other solvents reveal that polarity for BMIM PF_6 is similar to that of ethanol.

Poole et al. (Poole & Poole, 2010; Poole, 2004) collected other solvatochromic parameters introduced by Kamlet which have been determined for numerous RTILs. According to this approach solvents could be characterized by their polarity/polarizability (π^*) as well as acidity of hydrogen bond expressed as α value and β value reflecting basicity of the hydrogen-bond. As could be seen in Table 1, presented parameters are clearly related to the structure of ILs. To compare the collected parameters with solvents commonly used in analytical chemistry, methanol, water and acetonitrile have been added at the end of the above-mentioned Table 1.

Wakai and coworkers (Wakai et al., 2005) determined the static dielectric constants of 1-alkyl-3-methylimidazolium ionic liquids by microwave dielectric spectroscopy in the megahertz/gigahertz regime. The obtained results classify the ILs as moderately polar solvents. The observed ε-values at 298.15 K fall between 15.2 and 8.8 and decrease with increasing chain length of the alkyl residue of the cation. The anion sequence is trifluoromethylsulfonate > tetrafluoroborate ≈ tetrafluorophosphate. The results indicate markedly lower polarities than the ones found by spectroscopy with polarity-sensitive solvatochromic dyes (Wakai et al., 2005).

2.2 Lipophilicity parameters

Lipophilicity of a molecule is measured by its distribution behavior in a biphasic system either liquid-liquid (partition coefficient in 1-octanol-water) or solid-liquid (retention in RP-TLC or RP-HPLC) systems. According to definition suggested by IUPAC lipophilicity expresses the affinity of a molecule for a lipophilic environment. A reference scale representing lipophilicity appears to be the solute distribution between octanol and water. Berthod & Carda-Broch (Berthod & Carda-Broch) proposed another lipophilicity scale measuring ionic liquid: BMIM PF_6-water distribution constants. Relationship between the obtained values and respective octanol-water coefficients for a series of aromatic compounds differing in acid-base properties revealed that only the neutral compounds or ionizable ones with zwitterionic properties showed similar distribution behavior in the

Ionic liquid	Reichardt's dye		Kamlet-Taft		
	$E_{T(30)}$	E_T^N	π *	α	β
Ethylammonium nitrate	61.6	0.95	1.24	0.85	0.46
n -Propylammonium nitrate	60.6	0.92	1.17	0.88	0.52
Di-n -propylammonium thiocyanate	63.3	1.01	1.16	0.97	0.39
1-Butyl-3-methylimidazolium Hexafluorophosphate	52.3	0.67	1.03	0.63	0.21
1-Butyl-3-methylimidazolium Tetrafluoroborate	52.5	0.67	1.05	0.63	0.38
1-Butyl-3-methylimidazolium Trifluoroacetate	51.1	0.63			
1-Butyl-3-methylimidazolium Trifluoromethanesulfonate	52.3	0.67	1.01	0.63	0.46
1-Butyl-3-methylimidazolium Bis(trifluoromethylsulfonyl)imide	51.5	0.64	0.98	0.62	0.24
1-Butyl-2,3-dimethylimidazolium Bis(trifluoromethylsulfonyl)imide	49.3	0.57	0.99	0.45	0.26
1-Butyl-2,3-dimethylimidazolium Tetrafluoroborate	49.4	0.58	1.08	0.40	0.36
1-Ethyl-3-methylimidazolium Bis(trifluoromethylsulfonyl)imide	52.6	0.68			
1-Ethyl-3-methylimidazolium Tetrafluoroborate	53.7	0.71			
1-Hexyl-3-methylimidazolium Bis(trifluoromethylsulfonyl)imide	51.9	0.65	0.98	0.65	0.25
1-Hexyl-3-methylimidazolium Tetrafluoroborate	53.6	0.71			
1-Hexyl-3-methylimidazolium Trifluoromethanesulfonate	52.5	0.67	0.98	0.67	0.52
1-Octyl-3-methylimidazolium Bis(trifluoromethylsulfonyl)imide	51.0	0.63	0.97	0.60	0.28
1-Octyl-3-methylimidazolium Tetrafluoroborate	52.4	0.67			
1-Octyl-3-methylimidazolium Hexafluorophosphate	50.0	0.60	0.88		
Water	63.1	1.00	1.09	1.17	0.18
Methanol	55.8	0.77	0.60	0.93	0.62
Acetonitrile	51.9	0.65	0.80	0.35	0.38

Table 1. Solvent strength determined by solvatochromism for representative room temperature ionic liquids and traditional organic solvents (Poole & Poole, 2010; Poole, 2004).

compared systems. As can be seen in Fig.1, the amine-containing compounds showed higher affinity to ionic liquid whereas acidic compounds to 1-octanol phase. This result is in consistent with the solvation model proposed by Abraham et al. (Abraham et al., 2003) indicating that BMIM PF_6 has a greater affinity for electron lone pair interactions.

More data for liquid–ionic liquid partition coefficients of different organic compounds were described for the biphasic systems containing alkyl ammonioum ionic liquids: ethylammonium nitrate, n-propylammonium nitrate, or di-n -propylammonium thiocyanate and hexane, toluene, octan-1-ol, or dichloromethane (Poole, 2007; Shetty et al., 1987; Shetty et al., 1990) or aqua biphasic systems formed by 1-butylammonium-3-methylimidazolium hexafluorophosphate and water or heptane (Berthod & Carda-Broch, 2003; Khachatryn et al., 2005; Carda-Broch et al., 2003) or the 1-octyl-3-methylimidazolium hexafluorophosphate–water biphasic system (Liu et al., 2004).

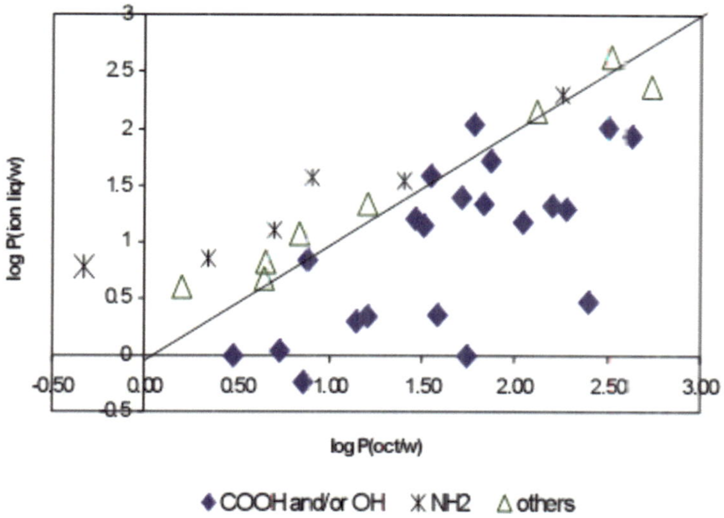

Fig. 1. Ionic liquid/water distribution coefficients compared to the octanol/water values (log scale). Crosses: amino-aromatic compounds, open triangles: neutral compounds or compounds with both acidic and basic substituents, filled-diamond: acidic and/or phenolic compounds (http://www.mariecurie.org/annals/volume3/berthod.pdf).

IL additives have been also applied as a mobile phase modifiers in RP-HPLC systems used for determination of the chromatographic lipophilicity parameters log k_w of ionized forms of basic drugs (Marszałł & Kaliszan, 2007). The correlation of log k_w versus the reference lipophilicity parameters log $P_{o/w}$ was evidently better when ionic liquid EMIM BF_4 was added to the mobile phase at very small concentration (1.5%).
To determine the lipophilicity of ionic liquids, different stationary phases have been tested such as Supelcosil LC-8-DB, Symmetry C8, ACE 5 C18, Symmetry C18 with aqueous acetonitrile buffered at pH 3.55 mobile phases (Molikova et al., 2010a). In this study the log k_w parameters were correlated with log P calculated by the HyperChem software. Good correlations were found for homologous alkylimidazolium cations analyzed by RP HPLC especially in the case of the Symmetry C8 column (R^2 = 0.9966).

3. Liquid-liquid extraction systems

3.1 Aqueous biphasic systems based on ionic liquid

It has been recently demonstrated that hydrophilic ionic liquids induce formation of aqueous biphasic systems (ABS) or aqueous two-phase systems (ATPS) in the presence of inorganic salts with water-structuring properties (Gutowski et al., 2003; Shehong et. al., 2005). According to Flory (Flory, 1953) the ATPS in classical mixtures consisting of two aqueous solution of polymers or a polymer and salt, are formed when enthalpy of interaction between the polymer molecules is bigger than the loss in entropy. It appears that the addition of cosmotropic salt to an aqueous solution of a hydrophilic ionic liquid also causes a salting out effect leading to liquid-liquid demixing and finally to the formation of two water-rich immiscible phases where upper phase contains ionic liquid and lower one concentrated salt. Abraham et al. proved that ATPS could be regarded as novel liquid partitioning systems. (Abraham et al., 2003). Considering that, stability, activity and enantioselectivity of enzymes have all been enhanced in aqueous solution of ionic liquids (Dreyer & Kragl, 2008; Dreyer and Kragl, 2009) and water-rich phases protect biomolecules against denaturation. ATPS have been used mainly in biotechnology for the separation of biomolecules, such as cells, organelles, membrane fractions and proteins (Walter & Johansson, 1994; Albertsson, 1986; Rito-Palomares, 2004; Gunduz &Tolga, 2004; Roobol-Boza et al., 2004; Agasøster, 1998).This method appears to be an attractive alternative to conventional extraction methods such as liquid-liquid extraction and solid-phase extraction, to simultaneously carry out isolation, purification and enrichment of biomolecules (Rito-Palomares, 2004; Gunduz & Tolga, 2004; Roobol-Boza et al., 2004). Apart from proteins like bovine serum albumin, lysozyme, trypsin, myoglobin (Pei et al., 2009, Du et al., 2007; Dreyer & Kragl, 2008; Dreyer and Kragl, 2009) ATPS have been used for the recovery of small organic and inorganic molecules such as metal ions, radiochemicals, dyes, drug molecules (Huddleston et al., 1999; Willauer et al., 2002; Zhu et al., 2001; Bridges & Rogers, 2008). Short chain alcohols (Gutowski et al., 2003), phenol (Chen et al., 2009), testosterone and epitestosterone (He et al., 2005), penicillin G (Liu et al., 2006), opium alkaloids (Li et al., 2005a), L-tryptophan (Neves et al., 2009; Ventura et al., 2009; Louros et al., 2010), antibiotics (Dominguez-Perez et al., 2010), caffeine and nicotine (Freire et al., 2010) have been analysed as partitioning solutes so far.

Biphasic liquid systems could find special applications in countercurrent chromatography (CCC) where both the mobile and stationary phase are immiscible liquids. Berthod and coworkers used biphasic liquid system consisting of water-acetonitrile-BMIM PF_6 for determination of the distribution constants of different organic compounds (Berthod & Carda-Broch, 2004).

3.1.1 ATPS driving forces

Most water-soluble room temperature ionic liquids could be thought of as chaotropic salts. That is why after addition of water-structuring salts (kosmotropic salts) they are salted out and aqueous biphasic systems are formed (Bridges et al., 2007; Zafarani-Moattar & Hamzehzadeh, 2009; Neves et al., 2009; Deng et al., 2007; Pei et al., 2007). Classification of salts is based on the Hofmeister series ordered from kosmotropic to chaotropic accordingly to the ability of salt to protein precipitation. Alternatively to ionic reagents appropriate sugars such as fructose and sucrose could be used as kosmotropic agents known also as water-structuring additives. The Hofmeister series was established in 1888 (Hofmeister, 1888).

Kosmotropic ions **Chaotropic ions**

citrate $^{3-}$ > sulfate^{2-} > phosphate^{2-} > F^- > Cl^- > Br^- > I^- > NO_3^- > ClO_4^-

Strongly hydrated **weakly hydrated**

$N(CH_3)_4^+$ > NH_4^+ > Cs^+ > Rb^+ > K^+ > Na^+ > H^+ > Ca^{2+} > Mg^{2+} > Al^{3+}

weakly hydrated **strongly hydrated**

Originally it was stated that the influence of ions on macromolecules properties was connected with structuring or breaking of bulk water structure. However recent studies have demonstrated that either direct ion-macromolecule interactions or interactions with water molecules in the first hydration shell of the macromolecules govern the Hofmeister effect as well as phase separation in ATP systems (Zhang & Cremer, 2006). Thus, in ATPS systems based on ionic liquid the competition between the ionic liquid and inorganic salt for water molecules appears to be the main driving force responsible for phase separation. It can be confirmed by the thermodynamic approach utilizing the Gibbs free energy of hydration ΔG_{hyd}. Kosmotropic ions with a large negative ΔG_{hyd} possess the higher affinity for water; in this way forcing the flow of water far away from the ionic liquid decreasing its solubility (Bridges et al., 2007). Typically the following inorganic salts have been used as kosmotropes: ammonium, potassium and sodium of phosphates, sulfates, carbonates or citrates whereas ionic liquids chosen for ATPS formation are mainly hydrophilic 1,3-dialkylimidazolium-based water soluble salts.

To characterize the obtained biphasic systems, the phase diagrams determined by the cloud-point titration are usually constructed. The binodal has been described by an empirical model developed by Merchuk and coworkers (Merchuk et al., 1998). The mathematical representation of the curves fitted the best to experimental data was the following:

$$Y = M_1 \exp[(M_2 X^{0.5}) + (M_3 X^3)] \tag{2}$$

Bridges and coworkers (Bridges et al., 2007) studied the BMIM Cl-K_3PO_4 phase diagram. As can be seen in Fig.2 the binodal separates the monophasic region from the two phase region which is located above the binodal. The tie line length measures the relative divergence in two phases. According to Merchuk (Merchuk et al., 1998) the phase ratio can be derived from the inverse lever rule. Composition of biphasic systems at any point of tie line could be determined by the intersecting points of appropriate tie line with the binodal curve.

Some papers described (Li et al., 2005a; Shehong et al., 2005) different salts which were also tested for the formation of ATPS with BMIM Cl (Fig.3). It appeared that ATPS can be formed only by adding appropriate amount of alkali or alkaline salts, such as KOH, NaOH, $K_3 PO_4$, $K_2 HPO_4$, $K_2 CO_3$ or $Na_2 HPO_4$ whereas acidic or neutral salts, such as $KH_2 PO_4$, $(NH_4)_2 SO_4$, NaCl or KCl did not create biphasic system. The abilities of the salts for phase separation can be ordered as follows:

$$K_2 CO_3 \approx K_3 PO_4 \approx K_2 HPO_4 > KOH$$

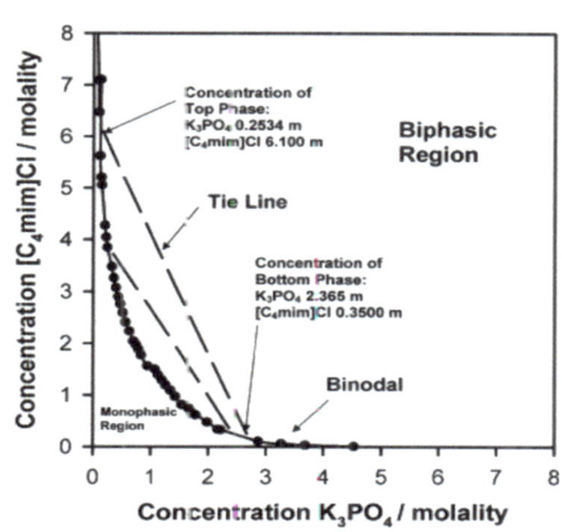

Fig. 2. The phase diagram for [C_4 mim]Cl–K_3PO_4 illustrating the bimodal (—) and the tie lines (---). (Bridges et al., 2007).

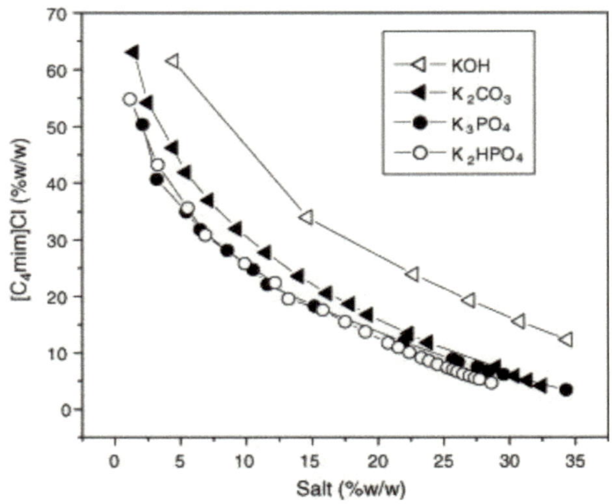

Fig. 3. Phase diagrams for the [C_4 mim]Cl/salt/water systems at 22 °C. In the region above the binodial curve, the system is separated into two phases, below the binodial curve; the system is a homogeneous phase (Shehong et al., 2005).

Several other papers have been also dealing with phase diagrams of IL-based ABS (Pei et al., 2007; Zafarani-Moattar & Hamzehzadeh, 2007; Bridges et al., 2007; Visak et al., 2007; Najdanovic-Visak et al., 2007; Wu et al., 2008a,b,c; Zhang et al., 2007,2008). Research regarding the use of IL-based ATPS has been mostly focused on the influence of inorganic salts using first of all imidazolium-based ILs. Bridges et al (Bridges et al., 2007) determined the ability of different ionic liquids chlorides to ATPS formation using the strongest kosmotropic salt K_3PO_4 (Fig.4). An overall chaotropic order of the studied salts can be arranged as follows: P_{4444}Cl > N_{4444}Cl >> [C_4py]Cl >>[C_4 mmim]Cl≈ [C_4 mim]Cl. This order reflects the increased chaotropicity of ionic liquids due to the differences in the cation structure. Fig.4 shows small, but still visible, quantitative differences in the binodal curves for the phosphonium-based ILs. The ability of the ILs for phase separation can be described by the following order: [P_{4444}]Br > [$P_{i(444)4}$][Tos] > [P_{4441}][$MeSO_4$].
In turn Louros (Louros et al., 2010) studied phosphonium-based ILs which are more effective in promoting ATPS in comparison to the imidazolium-based IL with similar anions (Fig.5).

3.1.2 The recovery of proteins and amino acids

The partitioning behavior of proteins in IL-based ATPS is not clear. Some authors state that the salting-out effect and the electrostatic interactions are responsible for protein extraction into the IL-containing phase (Dreyer & Kragl, 2008). This model is based on published data illustrating the visible influence of the surface charge of the proteins and their molecular weight on the partitioning in ATPS systems. It appears that only bigger proteins with the negative charges on the surface possess higher affinity to upper IL-rich phase. Interactions between negatively charged amino acid residues and the cation of the IL were identified as the main driving forces for the extractions. In contrast to the above idea, another study elaborated on by Pei (Pei et al., 2009) points out hydrophobic interactions as a dominating factor in extraction process. The authors based their conclusion on thermodynamic calculations indicating that, the extraction is controlled by entropy changes, similarly to hydrophobic interactions. Determined TΔS values were always greater than the ΔH at each

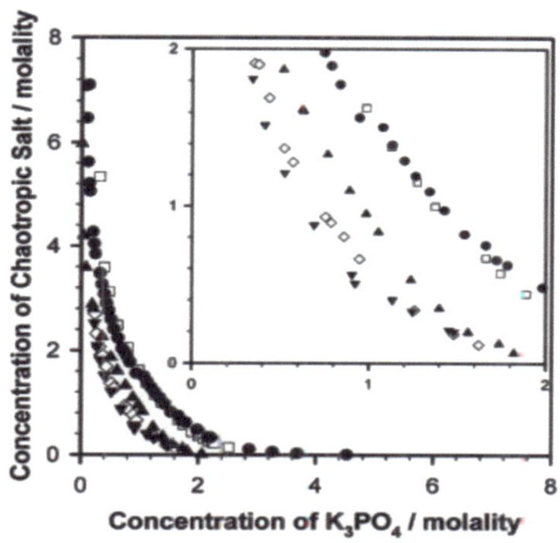

Fig. 4. Phase diagrams of K_3PO_4, with (●)[C_4 mim]Cl,(□)[C_4 mmim]CL(▲)[C_4 py]Cl,(◊)N_{4444} Cl, and(▼)P_{4444} Cl. (Bridges et al., 2007).

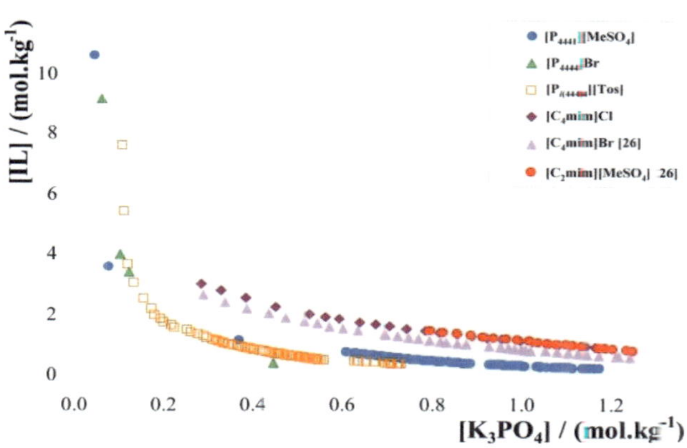

Fig. 5. Phase diagrams for the phosphonium-based ILs and [C_4mim]Cl ternary systems composed by IL + K_3PO_4 + H_2O at 298 K. (Louros et al., 2010).

case of ATPS systems containing ionic liquids differing in hydrophobicity of cation: [C_4mim]Br, [C_6mim]Br, [C_8mim]Br.
A series of dinitrophenylated amino-acids was studied in biphasic systems based on 1-decyl-3-methylimidazolium tetrafluoroborate and 1-hexyl-3-methylimidazolium tetrafluoroborate at slightly basic conditions fixed at 7.4 using sodium phosphate buffer (Rodriguez et al., 2007). Authors observed that amino-acids distribute preferentially to the ionic liquid phase according to the length of their chain.

3.1.3 The recovery of antibiotics, hormones and alkaloids

IL-based ATPS systems have been used for recovery of antibiotics especially the ones produced in fed-batch fermentation process by microorganisms. Lui (Lui et al., 2005, Lui et al., 2006) elaborated on methods of penicillin G extraction from fermentation media with an ATPS based on BMIM BF_4 and BMIM CL and NaH_2PO_4 as an inorganic salt. Improvement of the yield of extraction has been achieved with increasing phosphate concentration in the lower phase and increasing ionic liquid concentration in the upper phase. The maximum extraction yield of above 90% has been reached with 40%kosmotropic salt and 20% of BMIM Cl. In turn Yangyang and coworkers (Yangyang et al., 2009) used BMIM PF_6 simply to support antibiotic extraction by conventional ATPS system containing an imidazolium-terminated polyethylene glycol allowing efficient (96%) extraction of penicillin G. A second biphasic system based on IL and adjusted to slightly basic pH can recover I-PEG which is further extracted into slightly acidic water phase at third step. This method enables recycling not only of the imidazolium-terminated polymer but also of the added ionic liquid.

Despite of the great extractive potential of IL-based ATPS systems for fundamental biomolecules, most studies have been dealing with phase diagrams. Since the ionic liquid containing phase is compatible with liquid chromatography and UV detection, IL-based ATPS appear to be suitable for real samples investigation. As partitioning solutes only a few alkaloids have been analyzed so far (Freire et al., 2010; Li et al., 2005). Lui and coworkers (Lui et al., 2005) demonstrated aqueous two-phase systems (ATPS) based on an ionic liquid, 1-butyl-3-methylimidazolium chloride BMIM Cl/K_2HPO_4 in combination with high-performance liquid chromatography (HPLC) for analysis of the major opium alkaloids: codeine and papaverine in *Pericarpium papaveris* (Fig.6). The recoveries of codeine and papaverine were 90.0–100.2% and 99.3–102.0%, respectively. More recently Freire (Freire et al., 2010) reported strong ability of a series of IL-based ATPS for extraction of caffeine and nicotine in a single-step procedure. The authors focused on testing the concentration of inorganic salts and the usefulness of several ionic liquids differing in the anion and the cation in the aqueous phase and more complex matrixes such as human urine. The results showed that the extraction of both alkaloids significantly improved in human urine samples in comparison to simpler aqueous phases. Authors concluded that the presence of a biological matrix containing additionally neutral salt NaCl and urea with chaotropic properties favored the alkaloid partitioning for the IL phase.

Li studied BMIM Cl-K_2HPO_4 (Li et al., 2005) ATPS system for extraction of two alkaloids - morphine and papaverine used as abused drugs. The maximum extraction efficiency which was about 96% for papaverine and 65% for morphine was obtained for a system consisting of 3% IL and 40% of K_2HPO_4 within the temperature range from 15 to 65°C.

Because the ion exchange between salt and IL can complicate separation procedure, the use of sugars (a, b, c Wu et al. 2008) and aminoacids (Zhang et al., 2007; Dominiguez-Perez et al., 2010) has been suggested to form IL-based ATPS. Perez evaluated the capability of imidazolium-based ILs (BMIM) differing in the type of anion (triflate, dicyanamide, tetrafluoroborate) and amino acids proline and lysine to promote ATPS. The suitability of the proposed ATPS was further checked according to their potential for the extraction of caffeine and antibiotic ciprofloxacin in two forms differing in solubility in water. On the basis of ternary phase diagrams for the examined systems ability of the aminoacids to form ATPS could be expressed as the distance between the origin and the binodal curves. It appeared that more soluble lysine leads to a more pronounced salting-out effect. In turn the hydrogen-bond basicity of the anion composing the IL ($[BF_4^-] > [CF_3\ SO_3^-] > [N(CN)_2^-]$) decides about the ability of [C_4 mim]-based ILs for aqueous phase separation.

Aqueous two-phase systems (ATPS) based on hydrophilic ionic liquid 1-butyl-3-methylimidazolium chloride and K_2HPO_4, coupled with RP-HPLC were proposed for analysis of testosterone (T) and epitestosterone (ET) in human urine (He et al., 2005). The method has been reached satisfactory extraction efficiencies 80–90% and detection limits of 1 ng/mL and linear ranges of 10–500 ng/mL for both compounds (Fig.7).

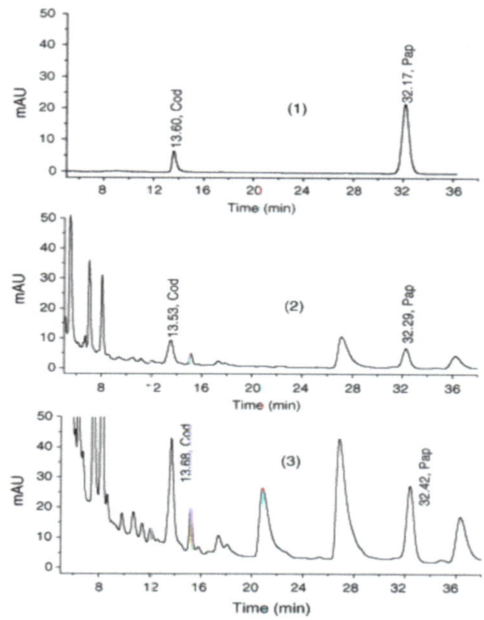

Fig. 6. Chromatogram of (1) standard solution of 10 μg/ml codeine (Cod) and 4 μg/ml papaverine (Pap); (2) methanol LLE of Pericarpium papaveris sample; (3) ATPS extraction of Pericarpium papaveris sample (Li et al., 2005).

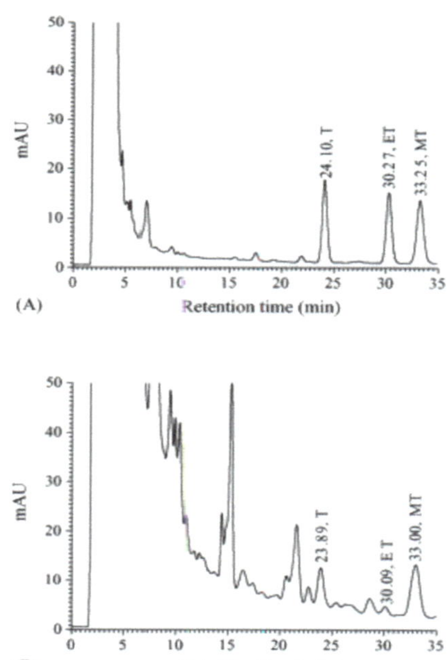

Fig. 7. HPLC chromatograms with UV detection of: (A) SFUS spiked with standard (300.0 ng/mL) and I.S. (333.3 ng/mL); (B) a human urine sample added with I.S. (333.3 ng/mL). Column: Zorbax SB-C_{18}. Mobile phase: 39/61% (v/v), acetonitrile–water. Flow rate: 1.0 mL/min. UV detection at 245 nm. (He et al., 2005).

3.2 Liquid-liquid extraction by hydrophobic ionic liquids

Pure ILs with hydrophobic properties can replace organic solvents in liquid-liquid extraction. Examples of different extraction systems are presented in Table 2. It should be emphasised that the extractability of acids (Matsumoto et al., 2004) and amino acids (Smirnova et al., 2004) could be enhanced by the addition of different extractants such as tri-n-butylphosphate or crown ether respectively to the ionic liquid phase. If analytes exist in ionic form, the extraction occurs by an ion-exchange mechanism involving transfer of appropriate ions to maintain electroneutrality. In this case, stronger hydrogen-bonding interactions will generate tetrafluoroborates owing to the stronger effective charge in comparison to hexafluorophosphates (Fan et al., 2008), taking into account extraction of phenols. In turn increasing the hydrophobicity of the cation provided decreases the extraction efficiency. Ionic liquids are also good extractants of neutral compounds (benzene derivatives) from water (McFarlane et al., 2005). The partition coefficients obtained in IL-water system correlate with that ones determined by classical shake-flask method in n-octanol-water extraction. Addition of n-octanol to ionic liquid phase can significantly improve extraction of carboxylic acids. Besides imidazolium ionic liquids also ionic liquids with quaternary ammonium cations, such as tetrahexylammonium dihexylsulfosuccinate and trioctylmethylammonium salicylate could be successfully applied to extraction of neutral compounds. Egorov (Egorov et al., 2008) showed that the novel RTILs are a more efficient solvent for extraction of phenols and amines than commonly applied for this purpose imidazolium based ILs: $HMImTf_2N$ and $DMImTf_2N$.

Compounds extracted into ionic liquid phase can be recovered by back extraction enabling phase-forming compounds recycling. Such multi-steps extraction procedures have been applied in biotechnology using the ILs in large-scale processes.

Hydrophobic ILs have been considered as a solvent for biocatalytic reactions. It was shown that the reaction of 2-octanone reduction to 2-octanol catalysed by alcohol dehydrogenase is faster when IL BMIM $(CF_3SO_2)_2N$ is present in comparison to organic solvent applied so far (Eckstein et al., 2004). Additional beneficial effects of ILs is reduction of the toxicity towards the cell and possibility of their application in whole-cell biocatalysis (Weuster-Botz, 2007; Pfruender et al,. 2004). This advantageous effect comes from excellent solvent properties of ILs extracting either products or reaction substrates.

RTLs are being considered for selective extraction either in macro or micro scale. The isolation of target compounds from chemical process systems for instance alcohols from fermentation liquors (Chapeaux et al., 2008; Swatloski et al., 2002; Najdanovic-Visak et al., 2002), aromatic compounds from petroleum products (Arce et al., 2007a,b; Arce et al., 2009; Arce et al., 2008), aromatic sulfur-containing compounds (Mochizuki & Sugawara, 2008; Alonso et al., 2007; Eβer et al., 2004) from diesel-fuel represent the use of ILs in large-scale processes.

In turn liquid-phase microextraction techniques such as dispersive, single-drop, hollow-fiber based liquid-phase microextractions and solid-phase with liquid film represent miniaturized systems also utilizing ionic liquids as the extraction solvents. The low solubility, the low vapor pressure, high viscosity of ILs allow the use of long extraction time and eliminate evaporation losses. Using 1-hexyl-3-methylimidazolium tetrafluoroborate in temperature-controlled dispersive liquid-phase microextraction pyrethroid pesticides (Zhou et al., 2008) were isolated from aqueous solution. Headspace liquid-phase microextraction using 1-octyl-3-methylimidazolium hexafluorophosphate was applied for extraction of halomethanes, alkyl aromatic compounds (Aguilera-Herrador et al. 2008a,b, 2009), and chlorophenols (Peng et al., 2007). 1-Butyl-3-methylimidazolium hexafluorophosphate enabled the separation of chlorobenzenes and chloroanilines (Vidal et al., 2007). Amphetamine and methamphetamine were extracted from urine by the use of solid-phase microextraction employing an immobilized liquid film of 1-ethoxyethyl-3-methylimidazolium bis(trifluoromethylsulfonyl)imide on a silicone matrix (He et al.,2009).

4. Determination of ionic liquids

Taking into account widespread application of ionic liquids in the chemical industry, they should also be considered as analytes determined in different matrices, solid or liquids coming from environment. Several techniques such as capillary electrophoresis (Qin et al., 2002; Markuszewski et al., 2004), isotachophoresis (Kosobucki & Buszewski, 2008), high performance liquid chromatography (Stepnowski et al., 2003; Ruiz-Angel & Berthod, 2006; Buszewski et al., 2006; Kowalska & Buszewski, 2006), ion chromatography (Stuff, 1991; Isildak & Covington, 1993; Markowska & Stepnowski, 2008, Molikowa et al., 2010) (Fig.8) have been applied for determination of ILs.

Cations and anions derived from ionic liquids can be analyzed separately or simultaneously. In the first case, two different analytical methods should be applied. Separation of ILs cations can be analyzed on a cation exchange column (Stuff et al., 1991; Molikowa et al., 2010) whereas halides anion by the use of anion exchanger also by ion chromatography (Hao et al., 2008).

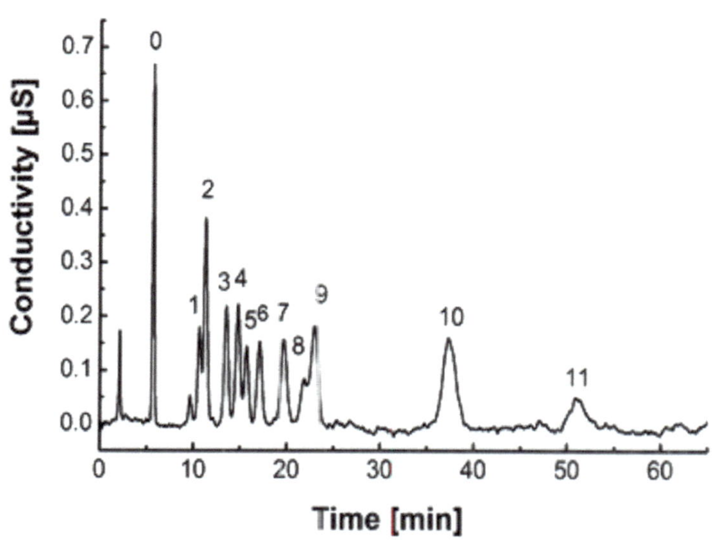

Fig. 8. Chromatogram of separation of ionic liquid cations mixture with isocratic elution for mobile phase consisting of 30% v/v of acetonitrile and: 70% v/v of 5mM MSA (methanesulfonic acid). Key:0–unknown compound,1–EEIM,2–PMIM,3–BMIM,4–BEIM,5–MBPy,6 - AMIM, 7 - BzMIM, 8 - EBzMIM, 9 - HMPy, 10 - HMIM, 11 - OMIM. The cations of IL were analyzed on a Dionex IonPac CS15 column (250mm x 8.5mm) (Molikowa et al., 2010).

Extracted compounds	Ionic liquid	environment	Ref.
acetic,glycolic, propionic,lactic, butyric, pyruvic	The hexafluorophosphates o f1-butyl-,1-hexyl- 1-octyl-3-methylimidazolium containing tris-n-butylphosphate	water	Matsumoto et al., 2004
phenols	1-butyl-3-methylimidazolium hexafluorophosphate	Aqueous solution adjusted to $pH<pK_a$	Khachatryan et al.,2005
phenol, tyrosol, p-hydroxybenzoic acid	1-alkyl-3-methylimidazolium hexafluorophosphates and tetrafluoroborates	Aqueous solution	Vidal et al., 2004
phenol, bisphenol A, pentachlorophenol, 4-octylphenol, 4-nonylphenol	1-methyl-3-alkyl imidazolium hexafluorophosphate 1-methy-3-alkylimidazolium tetrafluoroborate	Aqueous solution pH<7	Fan et al., 2008
amino acids: arginine, glycine, tryptophan, leucine, alanine, lysine, valine	1-butyl-3-methylimidazolium hexafluorophosphate containing the crown ether dicyclohexano-18-crown-6	Aqueous solution at pH 1.5-4.0	Smirnova et al., 2004
amino acids (valine, leucine, tyrosine, phenylalanine, and tryptophan)	1-butyl-3-methylimidazolium hexafluorophosphate, 1-hexyl-3-methylimidazolium hexafluorophosphate and tetrafluoroborate, and 1-octyl-3-methylimmidazolium tetrafluoroborate	Dependence on pH of aqueous solution	Wang et al.,
amoxicillin and ampicillin	1-octyl-3-methylimidazolium tetrafluoroborate	aqueous solution pH = 8	Soto et al., 2005
azo dyes (naphthalene sulfonic acids)	N -butyl-N -methylpyrrolidinium bis(trifluoromethylsulfonyl)imide	aqueous solution	Vijayaraghavan et al., 2006
toluene, cyclohexanone, nonan-1-ol, acetic acid and hexanoic acid	(1-alkyl-3-methylimidazolium bis(trifluoromethylsulfonyl)imide, hexafluorophosphate, trihexyl-tetradecyl phosphonium bis(trifluoromethylsulfonyl) imide, dodecylbenzene sulfonate, trihexyl-tetradecyl phosphonium methanesulfonate	water at various pH, temperature, sodium chloride concentration	McFarlane et al., 2005
phenols and amines	tetrahexyl ammonium dihexylsulfosuccinate, trioctylmethyl ammonium salicylate	aqueous solution	Egorov et al., 2008.
organic compounds	di-n -propylammonium thiocyanate	hexane	Shetty et al., 1990

Table 2. Examples of different liquid-liquid extraction systems based on ILs.

However, the most common method of ILs cations determination appears to be high performance liquid chromatography in reversed-phase mode (Ruiz-Angel & Berthod, 2006; Ruiz-Angel & Berthod, 2008; Stepnowski et al., 2003). Using C-18 bonded phase with organic-aqueous mobile phase, selectivity of separation especially for more polar cations is rather poor. Distorted and even split peaks may appear in the chromatogram depending on the nature and concentration of the injected ILs (Ruiz-Angel & Berthod, 2008). Significant improvement of peak shape and reproducibility of retention factors could be achieved by addition of appropriate acidic buffer, small amount of acid or significant amounts of a salt containing a chaotropic anion to the mobile phase. To separate the mixture of ionic liquids differing with the hydrophobicity of cations, gradient separation is required. Stepnowski and coworkers proposed very efficient chromatographic system with electrospray ionization mass detection for separation of selected 1-alkyl and 1-aryl-3-methylimidazolium-based RTIL cations (Stepnowski et al., 2003). The results obtained on C_8 MetaSil Basic column (Fig.8) were more efficient than that ones performed on C18 column and organic aqueous mobile phase with addition of different salts proposed by Ruiz-Angel and Berthod (Ruiz-Angel & Berthod , 2006). The key matter in determination of ionic liquids appears to be simultaneous analysis of the cations and anions of ILs. This approach involves the use of a mixed bed cation and anion exchange material or cation and anion exchange columns connected in tandem. Markowska & Stepnowski proposed the tandem of columns consisting of silica-based strong anion exchange packing (Phenosphere SAX with quaternary ammonium groups) and strong cation exchange packing (Luna SCX, with sulfonic groups) for simultaneous separation of ionic liquid components in one run followed by conductometric detection (Markowska & Stepnowski, 2008). Separation of a mixture of ionic liquids is presented in Fig.10. The limits of detection (LOD) values obtained by the use of this method were in the range from 0.008 to 0.050 mM; which were similar to the minimum detectable levels determined in the case of capillary electrophoresis ranging from 0.2 to 1.9 mg/L (Soga & Ross, 1999).

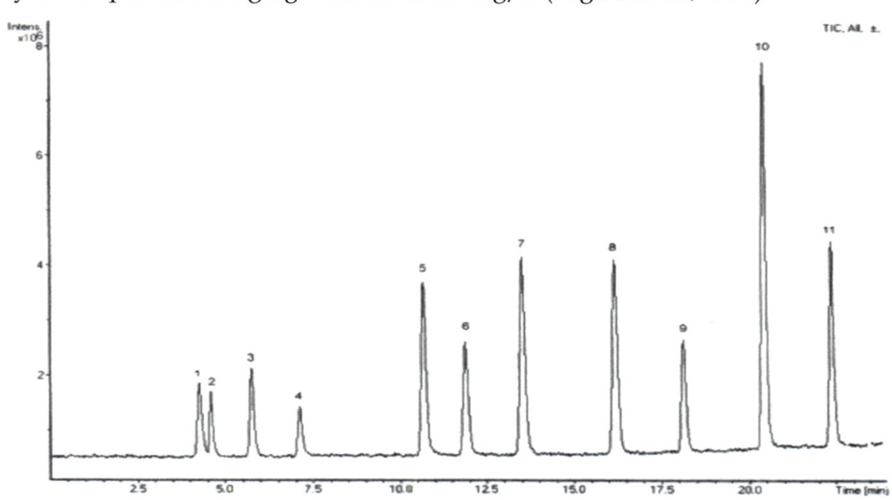

Fig. 9. Gradient separation of the test mixture. Column RP C_8 MetaSil Basic 250×4.6 mm I.D. 5 µm (Varian). Mobile phase A, acetonitrile; B, water (1% acetic acid/20 m M ammonium acetate) 0 min. 10% B, 10 min. 20% B, 25 min. 50% B.1-EEIM Br, 2- PMIM Br, 3- PEIM Br, 4- BMIM Cl, 5- BzMIM BF_4, 6-AMIM Cl, 7- EBzMIM Cl, 8- pMBzMIM Cl, 9-HMIM Cl, 10- HEIM Cl, 11- HpMIM Cl. (Stepnowski et al., 2003).

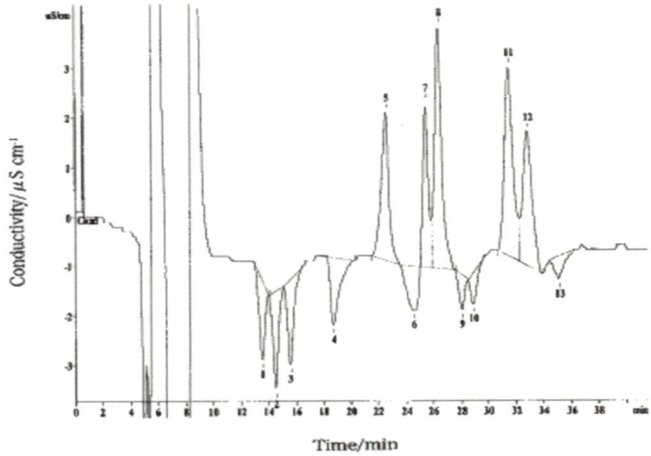

Fig. 10. Separation of a mixture of ionic liquid cations and anions (concentration 0.5 mM each) with use of tandem ion exchange columns. 1, DMIM; 2, NMIM; 3, OMIM; 4, HMIM; 5, Tf_2N^-; 6, AMIM; 7, PF_6^-; 8, BF_4^-; 11, Br^- ; 12, Cl^- ; 13, BMIM; signals 9,10 unidentified. Conductivity detection. Mobile phase: 5 mM potassium hydrogen phthalate/15 mM phthalic acid, 45% (v/v) acetonitrile pH 3.05. Absolute conductivity of mobile phase, 476 µS/cm; flow rate, 1 mL/min; injection volume, 100 µL. Columns in the order: strong anion exchange Phenosphere SAX, strong cation exchange Luna SCX. (Markowska & Stepnowski, 2008).

5. Application of ionic liquids in Liquid chromatography

5.1 Mobile phase modification with ILs

The application of ionic liquids in liquid chromatography has been restricted to mobile phase modification at low percent levels. This limitation is connected with, first of all, high viscosities which are higher than typical chromatographic solvents and secondly with low UV transparency. As mobile phase additives ILs could be advantageous, coating residual hydroxyl groups on the silica surface, modifying the stationary phase or acting as ion-pairing agents. Owing to these additives improvement of peak symmetry, efficiency and separation selectivity could be achieved. Suppression of deleterious silanol effects as a result of incomplete silanization of silica based packings have been described and compared to typical suppressor such as alkylamines used in the past for this purpose (Kaliszan et al., 2004; Marszałł et al., 2006a; Marszałł et al., 2005; Ruiz-Angel et al., 2006). It was found that ionic liquids of the imidazolium tetrafluoroborate class, added to mobile phases at concentrations of 0.5–1.5% (v/v), blocked free silanols exceeding the standard mobile phase additives, like triethylamine, dimethyloctylamine and ammonia either in thin-layer chromatographic separations or high-performance liquid chromatography of strongly basic drugs. To evaluate the silanol-suppresing potency of ILs the dual retention model based on Nahum and Horvàth equation has been applied (Marszałł et al., 2006a):

$$\frac{[A]}{k_0 - k} = \frac{1}{k_2 K_A} + \frac{[A]}{k_2} \qquad (3)$$

where k_0 is the retention factor obtained in the absence of silanol suppressor and k is the retention factor obtained at a certain concentration of silanol suppressor [A]. It was apparent that higher the binding constants K_A were obtained for bromide and chloride derivatives than for tetrafluoroborate and sulfate whereas OMIM Br reduced silanol effect at the lowest concentration.

The ILs additives also significantly improve the asymmetry factor leading to higher column efficiencies especially for basic solutes (Martin-Calero et al., 2009; Tang et al., 2006; Xiao et al., 2004; He et al., 2003; Zhang et al., 2003; Jiang et al., 2008; Flieger & Czajkowska-Żelazko, 2011a,b,c; Flieger, 2009). It is obvious that dual nature of ionic liquids is responsible for observed effects. Whereas the ionic liquid cations undergo absorption onto the C-18 chain interacting with residual silanols, ILs anion depending on its position in the Hofmeister series acts as an ion-pairing or ion-exchanging agents.

Fig.11 shows the peak profile of pindolol analyzed by mobile phase without any ionic liquid additives and with addition of 10 mM BMIM Cl and 10 mM BMIM PF_6

Addition of BMIM Cl as mobile phase modifier provides a decrease of peak tailing and increase of efficiency in connection to retention decrease. Lack of ion-pairing properties of chloride anion and adsorption of ionic liquid cation on the surface provokes repulsion forces between protonated analytes and charged surface responsible for all advantageous effect. Addition of even 10 mM of BMIM PF_6 led to almost two-fold retention time increase as a result of ion-pair creation between chaotropic anion of added IL and protonated basic analyte involving worseness of efficiency and peak symmetry (Flieger & Czajkowska-Żelazko, 2011b).

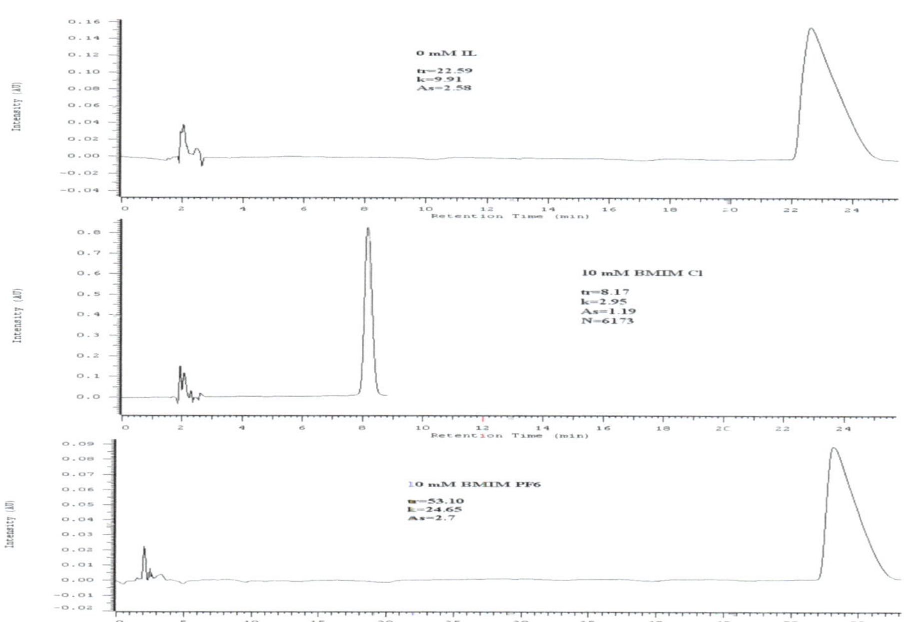

Fig. 11. Peak profile of pindolol obtained on Zorbax SB-Phenyl by the use of 10% ACN / 20 mM phosphate buffer pH = 2.9 mobile phase without any ionic liquid additives (0 mM ILs) and with addition of 10 mM BMIM Cl and 10 mM BMIM PF_6.(Flieger & Czajkowska-Żelazko, 2011b).

5.2 Stationary phases modification with ILs

Recently, ILs covalently bound to silica matrices have been described as stationary phases for liquid chromatography. Imidazolium-based ILs were applied for synthesis of anion exchange stationary phases by Qiu and coworkers (Qiu et al, 2006a,b) enabling to separate inorganic and organic anions. In turn Sun and coworkers using the same modifier synthesized stationary phase attaching IL (1-butyl-3-heptylimidazolium bromide) to silica matrix through an alkyl tether receiving column with properties similar to conventional phenyl stationary phase (Sun et al., 2005). Wang (Wang et al., 2006) modified the surface of 3 μm diameter silica particles with the following derivatives: 1-methyl-3-(trimethoxysilylpropyl) imidazolium bromide and 1-butyl-3-(trimethoxysilylpropyl) imidazolium bromide. It was confirmed by thermogravimetric analysis (TGA) and ^{13}C and ^{29}Si NMR spectroscopies that the ionic liquid moiety was predominantly attached to the silica surface through two siloxane bonds. The obtained columns were used for aromatic carboxylic acids separation under HPLC conditions (Fig.12). The major factors contributing to the retention of analytes on these types of columns appear to be ion-exchange process and the hydrophobic interactions.

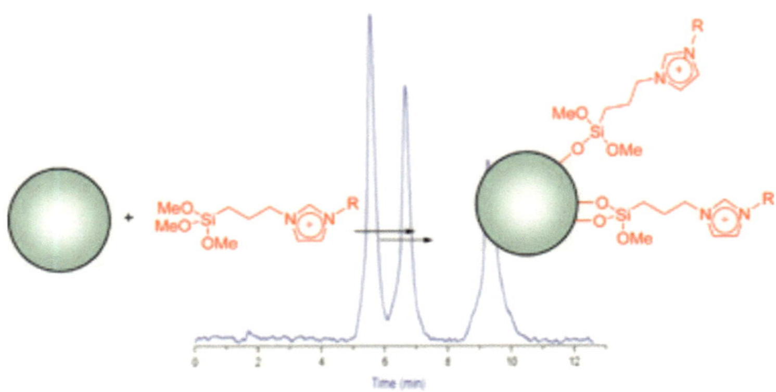

Fig. 12. Chromatogram of amixture of acids eluted using 5 mM phosphate at pH 4.2 in an MPIM-modified silica particle packed column, using UV detection at a wavelength of 214nm. Flow rate is 200 mL/ min. Peaks, from left to right: 3-(p-hydroxyphenyl) propionic acid (10mM), p-hydroxyphenylacetic acid (10mM), and benzoic acid (10mM) (Wang et al., 2006).

6. Application of ILs in capillary electrophoresis

Thanks to their good electrical conductivity, tunable viscosity and solubility, ILs could be applied as additives to electrolytes either aqueous or nonaqueous and as coating reagents of the capillary walls. The most popular approach appears to be application of ILs as running electrolytes. In this case the separation mechanism is complex and covers interactions of free IL ions with analytes in bulk solution and simultaneously dynamically coating the capillaries affecting an electroosmotic flow (EOF) and alleviating the wall adsorption usually occurring during separation of basic proteins by CE.

6.1 Modification the capillary wall with ILs

Ionic liquid (IL) could be covalently bonded onto the silica capillary surface affecting the electroosmotic flow due to the positive charges of the bonded ILs cation. Qin and Li (Qin &

Li 2002) determined sildenafil and its metabolite in human serum by capillary zone electrophoresis-mass spectrometry analysis using covalently bonded PMIM Cl to a silica capillary. Scheme representation of the IL coating procedure is presented on Fig.13 The adsorption of the analytes onto the bare capillary wall was eliminated by the IL coating and the drugs were baseline-separated within 14 min with detection limits of 14 and 17 ng/mL for sildenafil and metabolite respectively.

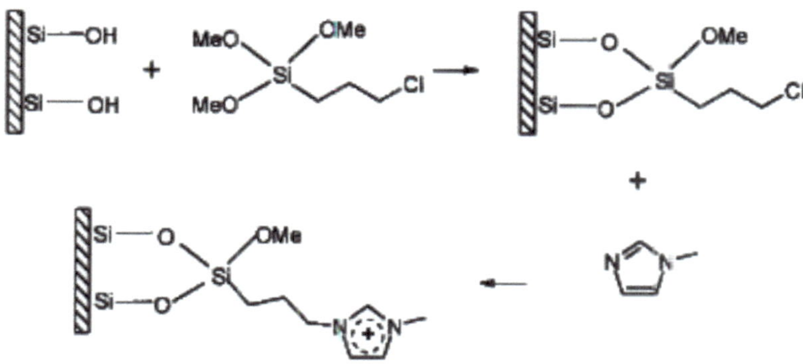

Fig. 13. Scheme representation of the IL coating procedure (Qin & Li, 2002).

6.2 ILs addition to the electrolyte

Use of ILs as background electrolytes in aqueous and non-aqueous electrophoresis is common practice. This approach has been applied for the separation of a variety of organic analytes (Table 3). The separation mechanism covers influence of either cation or anion of IL, however the resolving power is closely related to IL cation. The dynamic coating of the capillary wall by positively charged entire could reverse the electroosmotic flow and improve separations. The association between free ionic additives and ionized analytes could not be neglected. ILs were also used as additives in micellar electrokinetic chromatography (MEKC). Different ILs have been tested as modifiers in this chromatographic technique by Mwongela and coworkers (Mwongela et al., 2003) but only BMIM BF_4 could improve peak efficiency and resolution of tested mixtures of ketones, phenols and enantiomers of binaphtyl derivatives. ILs can be used not only as additives but also as surfactants. This approach requires appropriate length of the alkyl tail of the IL cation enabling micelles formation in aqueous media (Borissowa et al., 2008). The practical utilization of long-chain (C_{12} and C_{14}) alkylimidazolium ionic liquids as a new type of surfactant in MEKC was used to separate neutral analytes-methylresorcinol isomers and benzene derivatives. Newly synthesized chiral ILs as surfactants in MEKC have been described in literature (Rizvi & Shamsi, 2006). Chiral separation of two acidic analytes, (±)-α-bromophenylacetic acid and (±)-2-(2-chlorophenoxy)propanoic acid (±)-(2-PPA) was achieved with both monomers and polymers of undecenoxycarbonyl-L-pyrrolidinol bromide and undecenoxycarbonyl-L-leucinol bromide at 25 mM surfactant concentration at pH 7.50. Authors claim that interaction of the acidic analytes with the cationic headgroup of chiral selectors contribute to observed the chiral recognition. Addition of chiral ILs could also improve enatioseparation using another chiral selector (Francois et al., 2007; Tran & Mejac, 2008). Two new chiral ILs (ethyl- and phenylcholine of bis(trifluoromethylsulfonyl) imide acting synergistically with classical cyclodextrins as chiral selectors, improved the enantioselectivity for 2-arylpropionic acids.

Ionic liquids added to the running electrolyte	Analyzed compounds	References
alkylammonium salts	Water-insoluble dyes	Vaher et al., 2000
tetramethylammonium tetrafluoroborate, 1-alkyl-3-methylimidazolium based ILs	Polyphenols in grape seed extracts	Yanes et al., 2000 Yanes et al., 2001
1-alkyl-3-methylimidazolium(BMIM OMIM)based ionic liquids with different anions: $[PF_6]^-$, $[CH_3COO]^-$. $[CF_3COO]^-$, $[(CF_3SO_2)_2N]^-$	Carboxylic acids and phenols: p-ethylphenol, p-cresol,phenol, hydroquinone , resorcinol, pyrocatechol ,p-chloro-glycinol, pyrogallol,1,3-dihydroxynaphthalene.	Vaher et al., 2002 a,b
1-ethyl-3-methylimidazolium chloride (EMIM Cl) and the hexafluorophosphate salt (EMIM PF_6)	Chlorophenoxy and benzoic acid herbicides	Yu et al., 2005
1-ethyl-3-methylimidazolium tetrafluoroborate tetraethylammonium tetrafluoroborate	Monohalogenated phenols	Cabovska et al.,2003
dimethyldinonylammonium bromide	Carboxylates as copper complexes	Laamanen et al., 2005
1-alkyl-3-methylimidazolium based ionic liquids	Nicotinic, isonicotinic, picolinic acids	Marszałł et al., 2006 b
1-ethyl-3-methylimidazolium ,1-butyl-3-methylimidazolium, 1-pentyl-3-methylimidazolium, 1-heptyl-3-methylimidazolium tetrafluoroborates, 1-ethyl-3-methylimidazolium hexafluorophospha, 1-ethyl-3-methylimidazolium sulfate	Basic proteins: lysozyme, cytochrome c, trypsinogen, α-chymotrypsinogen A	Jiang et al., 2003
1-butyl-3-methylimizolium tetrafluoroborate	Anthraquinone extracts	Qi et al., 2004
propane-1,3-bis(tripropylphosphonium)fluoride	Small organic anions	Krizek et al., 2009
1-ethyl-3-methylimidazoium, 1-butyl-3-methylimidazoium tetrafluoroborates, 1-butyl-3-methylimidazoium hexafluorophosphate	Flavone derivatives	Qi et al., 2006

Table 3. Use of ILs as background electrolytes.

7. Conclusion

The use of ILs in separation techniques has often been found to provide advantageous effects such as higher separation selectivity and efficiency of the systems. Recently developed ILs offer greater separation potential also for enantiomeric compounds. Special attention has been devoted to application of ILs as matrixes in matrix-assisted laser desorption ionization mass spectrometric analysis (MALDI-MS). Armstrong (Armstrong et al., 2001) was the first who noticed that low volatile ILs with good solvent properties could be suitable as liquid MALDI plates. To detect various types of molecules with good sensitivity and a wide detection range, new generation of ILs matrixes has been developed. It appeared that the cation of ILs used for this purpose should have $pk_a > 11$ and proton affinity higher than 930 kJ mol^{-1} (Crank & Armstrong, 2009). Quantification of large biomolecules such as proteins (Tholey et al., 2006) and synthetic polymers (Berthod et al., 2009) as well as low molecular weight compounds such as amino acids, sugars and vitamins (Zabet-Moghaddam et al., 2004) using ILs matrixes is a very promising approach of the last years. Additionally, considering environmental friendly properties of ILs, their separation applications will be further extensively explored.

8. References

Abraham, M.H., Zissimos, A.M., Huddleston, J.C., Willauer, H.D., Rogers, R.D.& Acree, W.E. (2003). Some Novel Liquid Partitioning Systems:⊥ Water−Ionic Liquids and Aqueous Biphasic Systems. *Ind. Eng. Chem. Res.*, Vol. 42 (No.3):413-418.

Agasoster,T.(1998).Aqueous two-phase partitioning sample preparation prior to liquid chromatography of hydrophilic drugs in blood. *J Chromatogr B Biomed Sci Appl* Vol. 716 (No.1–2):293–298.

Aguilera-Herrador, E., Lucena, R., Cardena,s S. & Valcarcel, M. (2009). Ionic liquid-based single drop microextraction and room-temperature gas chromatography for on-site ion mobility spectrometric analysis. *J. Chromatogr. A* Vol. 1216 (No.29):5580-5587.

Aguilera-Herrador, E., Lucena, R., Cardenas, S. & Valcarcel, M. (2008a). Ionic liquid-based single-drop microextraction/gas chromatographic/mass spectrometric determination of benzene, toluene, ethylbenzene and xylene isomers in waters, *J. Chromatogr. A* Vol. 1201 (No.1):106-111.

Aguilera-Herrador, E., Lucena, R., Cardenas, S. & Valcarcel, M., (2008b). Determination of trihalomethanes in waters by ionic liquid-based single drop microextraction/gas chromatographic/mass spectrometry, *J. Chromatogr. A* Vol. 1209 (No.1-2):76-82.

Albertsson, P.-Å. (1986). Partition of Cell Particles and Macromolecules, third ed., Wiley, New York. p. 23

Alonso, L. & Arce, A., Francisco, M., Rodriguez, O., Soto, A. (2007). Gasoline desulfurization using extraction with [C_8mim][BF_4] ionic liquid, *AIChE J.* Vol. 53 (No.12):3108-3115.

Anderson, J.L. & Armstrong, D.W. (2003). High-stability ionic liquids. A new class of stationary phases for gas chromatography, *Anal.Chem.* Vol. 75 (No. 19): 4851-4858.

Arce, A., Earle, M., Rodriguez, H. & Seddon, K.R. (2007). Separation of aromatic hydrocarbons from alkanes using the ionic liquid 1-ethyl-3-methylimidazoliumbis{(trifluoromethyl) sulfonyl}amide, *Green Chem.* Vol. 9 (No.1):70-74.

Arce, A., Earle, M.J., Rodriguez, H. & Seddon, K.R. (2007). Separation of Benzene and Hexane by Solvent Extraction with 1-Alkyl-3-methylimidazolium Bis{(trifluoromethyl)sulfonyl}amide Ionic Liquids: Effect of the Alkyl-Substituent Length, *J. Phys. Chem. B* Vol. 111 (No.18):4732-4736.

Arce, A., Earle, M.J., Rodriguez, H., Seddon, K.R. & Soto, A. (2008). 1-Ethyl-3-methylimidazolium bis{(trifluoromethyl)sulfonyl}amide as solvent for the separation of aromatic and aliphatic hydrocarbons by liquid extraction - extension to C_7- and C_8-fractions. *Green Chem.*, Volume: 10 (No.12):1294-1301.

Arce, A., Earle, M.J., Rodriguez, H., Seddon, K.R. & Soto, A. (2009). Bis{(trifluoromethyl)sulfonyl}amide ionic liquids as solvents for the extraction of aromatic hydrocarbons from their mixtures with alkanes: effect of the nature of the cation, *Green Chem*. Vol. 11 (No.3):365-372.

Armstrong, D.W., He, L. & Liu, Y.S.(1999). Examination of ionic liquids and their interactions with molecules, when used as stationary phase in gas chromatography, *Anal.Chem*. Vol. 71 (No. 17): 3873-3876.

Armstrong, D.W., Zhang, Li-K., He, L. & Gross, M.L. (2001). Ionic Liquids as Matrixes for Matrix-Assisted Laser Desorption/Ionization Mass Spectrometry, *Anal. Chem.* Vol.73 (No.15): 3679–3686.

Barber, D.W., Phillips, C.S.G., Tusa G. F. & Verdin, A. (1959). The chromatography of gases and vapours. Part VI. Use of the stearates of bivalent manganese, cobalt, nickel, copper, and zinc as column liquids in gas chromatography, J. Chem. Soc.18-24.

Berthod, A. & Carda-Broch, S. (2004). Use of the ionic liquid 1-butyl-3-methylimidazolium hexafluorophosphate in countercurrent chromatography, *Anal. Bioanal. Chem.*, Vol.380 (No.1):168-177.

Berthod, A. & Carda-Broch, S.http://www.mariecurie.org/annals/volume3/berthod.pdf

Berthod, A. & Carda-Broch, S. (2003).A New Class of Solvents for CCC: The Room Temperature Ionic Liquids ,*J. Liq. Chromatogr. Rel. Technol.*, Vol. 26 (No.9-10):1493-1508.

Berthod, A., He, L. & Armstrong, D,W. (2001). Ionic liquids as stationary phases solvents for methylated cyclodextrines in gas chromatography, *Chromatographia* Vol. 53 (No.1-2): 63-68.

Berthod, A., Ruiz-Angel, M.J. & Carda-Broch, S. (2008). Ionic liquids in separation techniques, *J.Chromatogr. A* Vol. 1184 (No. 1-2): 6-18.

Berthod, A., Crank, J.A., Rundlett, K.I. & Armstrong, D.W.(2009). Second-generation ionic liquid matrix-assisted laser desorption/ionization matrix for effective mass spectrometric analysis of biodegradable polymers, *Rapid Commun. Mass Spectrom.*Vol.23 (No.21):3409-3422.

Borissova, M., Palk, K. & Koel, M. (2008). Micellar electrophoresis using ionic liquids *J. Chromatogr. A* Vol. 1183 (No.1-2):192-195.

Bridges, N.J.& Rogers, R.D. (2008). Can kosmotropic salt/chaotropic ionic liquid (salt/salt aqueous biphasic systems) be used to remove pertechnetate from complex salt waste, *Separation Science and Technology* Vol.43 (No.5): 1083-1090.

Bridges, N.J., Gutowski, K.E. & Rogers, R.D. (2007). Investigation of aqueous biphasic systems formed from solutions of chaotropic salts with kosmotropic salts (salt–salt ABS), *Green Chem*. Vol. 9 (No.2): 177-183.

Buszewski, B. & Studzińska, S. (2008). A review of ionic liquids in chromatographic and electromigration techniques, *Chromatographia* Vol. 68 (No. 1-2): 1-10.

Buszewski,B., Kowalska,S. & Stepnowski,P. (2006). Influence of stationary phase properties on th separation of ionic liquid cations by RP-HPLC, *J.Sep.Sci*. Vol. 29 (No.8): 1116–1125.

Cabovska, B., Kreishman, G.P., Wassell, D.F. & Stalcup, A.M. (2003). Capillary electrophoretic and nuclear magnetic resonance studies of interactions between halophenols and ionic liquid or tetraalkylammonium cations, *J Chromatogr A* Vol.1007 (No.1-2):179–187.

Carda-Broch, S., Berthod, A. & Armstrong, D.W. (2003). Solvent properties of the 1-butyl-3-methylimidazolium hexafluorophosphate ionic liquid, *Anal. Bioanal. Chem*. Vol. 375 (No.2):191-199.

Chapeaux, A., Sinomi, L.D., Runan, T.S., Stadherr, M.A. & Brennecke, J. (2008). Extraction of alcohols from water with 1-hexyl-3-methylimidazolium bis(trifluoromethylsulfonyl)imide, *Green Chem*. Vol. 10 (No.12):1301-1306.

Chen, Y., Wang, Y., Cheng, Q., Liuand, X. & Zhang, S. (2009). Carbohydrates-tailored phase tunable systems composed of ionic liquids and water, *J.Chem. Thermodyn*. Vol.41 (No.9):1056–1059.

Crank, J.A. & Armstrong, D.W. (2009).Towards a Second Generation of Ionic Liquid Matrices (ILMs) for MALDI-MS of Peptides, Proteins, and Carbohydrates, *J.Am.Soc.Mass Spectrom*. Vol.20 (No.10):1790-1800.

Deng, Y., Chen, J. & Zhang, D.(2007). Phase Diagram Data for Several Salt + Salt Aqueous Biphasic Systems at 298.15 K, *J. Chem. Eng. Data* Vol. 52 (No.4): 1332-1335.

Dimroth, K.; Reichardt, C.; Siepmann, T. & Bohlmann, F. (1963). Überpyridinium-N-phenol-betaine und ihre verwendung zur charakterisierung der polarität von lösungsmitteln. *Justus Liebigs Ann.Chem*. Vol. 661:1-37.

Domínguez-Pérez, M., Tomé,L.I.N., Freire, M.G., Marrucho, I.M., Cabeza, O. & Coutinho, J.A.P. (2010). (Extraction of biomolecules using) aqueous biphasic systems formed by ionic liquids and aminoacids, *Separation and Purification Technol*. Vol.72 (No.1):85-91.

Dreyer, S., Salim,P. & Kragl, U. (2009).Driving forces of protein partitioning in an ionic liquid-based aqueous two-phase system. *Biochem. Eng.J*. 46 (No.2):176-185.

Dreyer, S. & Kragl, U. (2008). Ionic liquids for aqueouis two-phase extraction and stabilization of enzymes. *Biotechnol. Bioeng*. 99 (6): 1416-1424.

Du, Z., Yuand, Y.-L. &, Wang, J.-H. (2007). Extraction of proteins from biological fluids by use of an ionic liquid/aqueous two-phase system, *Chem.–Eur.J*. Vol.13 (No.7):2130–2137.

Eckstein, M., Liese, A. & Kragl, U. (2004). Use of an ionic liquid in a two-phase system to improve an alcohol dehydrogenase catalysed reduction, *Chem.Commun*. Vol.40 (No.9): 1084-1085.

Egorov, V.M., Smirnova, S.V. & Pletnev, I.V. (2008).Highly efficient extraction of phenols and aromatic amines into novel ionic liquids incorporating quaternary ammonium cation, *Sep. Purif. Technol.* Vol. 63 (No.3):710-715.

Eβer, J., Wassercheid, P. & Jess, A. (2004). Deep desulfurization of oil refinery streams by extraction with ionic liquids, *Green Chem.*Vol. 6 (No.7):316-323.

Fan, J., Fan, Y., Pei, Y., Wu, K., Wang, J. & Fan, M.(2008). Solvent extraction of selected endocrine-disrupting phenols using ionic liquids, *Sep. Purif. Technol.* Vol. 61 (No.3):324-331.

Fletcher, K.A., Storey, I. A., Hendricks A. E., Pandey, S. & Pandey, S. (2001). Behavior of the solvatochromic probes Reichardt's dye, pyrene, dansylamide, Nile Red and 1-pyrenecarbaldehyde within the room-temperature ionic liquid bmimPF_6, *Green Chemistry* Vol. 3 (No.5):210–215.

Flieger, J. & Czajkowska-Żelazko, A. (2011a). Ionic liquids in reversed-phase high-performance liquid chromatography of phenolic acids on different stationary phases, *Analytical Letters* in press

Flieger, J.& Czajkowska-Żelazko, A. (2011b). Ionic liquids as mobile phase additives in reversed phase high performance liquid, *J.Liquid Chromatogr. Rel. Technol.* In press.

Flieger, J. & Czajkowska-Żelazko, A. (2011c).Comparison of chaotropic salt and ionic liquid as mobile phase additives in reversed-phase high-performance liquid chromatography of biogenic amines. *J.Sep.Sci.* in press

Flieger, J. (2006). The effect of chaotropic mobile phase additives on the separation of selected alkaloids in reversed-phase high-performance liquid chromatography, *J. Chromatogr. A* Vol. 1113 (No. 1-2): 37-44.

Flieger, J. (2009). Effect of ionic liquids as mobile phase additives on chromatographic parameters of neuroleptic drugs in reversed-phase high-performance liquid chromatography, *Anal..Lett.* Vol. 42 (No. 11): 1632–1649.

Flory, P.J. (1953). Principles of polymer chemistry, ISBN 9780801401343.

Francois, Y., Varenne, A., Juillerat, E., Villemin, D. & Gareil, P. (2007). Evaluation of chiral ionic liquids as additives to cyclodextrins for enantiomeric separations by capillary electrophoresis, *J. Chromatogr. A* Vol. 1155 (No.2):134-141.

Freire, M.G., Neves, C.M.S.S, Marrucho, I.M. & Lopes,J.N.C. (2010). High-performance extraction of alkaloids using aqueous two-phase systems with ionic liquids, *GreenChemistry* Vol.12 (No.10): 1715-1718.

Gunduz, U. & Tolga, A. (2004). Optimization of bovine serum albumin sorption and recovery by hydrogels, *J. Chromatogr. B* Vol.807 (No.1): 13-16.

Gutowski, K.E., Broker, G.A., Willauer, H.D., Huddleston, J.G., Swatloski, R.P., Holbreyand, J.D. & Rogers, R.D. (2003). Controlling the Aqueous Miscibility of Ionic Liquids: Aqueous Biphasic Systems of Water-Miscible Ionic Liquids and Water-Structuring Salts for Recycle, Metathesis, and Separations, *J.Am.Chem.Soc.* Vol.125 (No.22):6632–6633.

Han, D.,& Row, K.H. (2010). Recent Applications of Ionic Liquids in Separation Technology, *Molecules* Vol. 15 (No.4): 2405-2426.

Hao,F.; Haddad, P.R. & Ruther, T. (2008). IC determination of halide impurities in ionic liquids, *Chromatographia* Vol.67 (No.5-6): 495–498.

He, C., Li,S., Liu,H., Li,K., Liu, F. (2005). Extraction of testosterone and epitestosterone in human urine using aqueous two-phase systems of ionic liquid and salt, *J.Chromatogr.A* Vol.1082 (No.2): 143-149.

He, L., Zhang, W., Zhao, L., Liu, X. & Jiang, S. (2003). Effect of 1-alkyl-3-methylimidazolium-based ionic liquids as the eluent on the separation of ephedrines by liquid chromatography. *J.Chromatogr. A* 1007 (No.1-2): 39-45.

He, Y., Pohl, J., Engel, R., Rothman, L. & Thomas, M. (2009). Preparation of ionic liquid based solid-phase microextraction fiber and its application to forensic determination of methamphetamine and amphetamine in human urine, *J. Chromatogr. A* Vol. 1216 (No.24):4824-4830.

Hofmeister, F. (1888). Zur Lehre von der Wirkung der Salze Dritte Mittheilung, *Arch. Exp. Pathol. Pharmakol.*, XXV, (No.1):1-30.

Huddleston, J.G., Willauer, H.D., Griffinand, S.T. & Rogers, R.D. (1999). Aqueous Polymeric Solutions as Environmentally Benign Liquid/Liquid Extraction Media, *Ind.Eng.Chem.Res.* Vol.38 (No.7):2523–2539.

Isildak,I. & Covington, A.K. (1993). Ion-selective electrode potenticmetric detection in ion-chromatography, *Electroanalysis* Vol.5 (No.9-10): 815-824.

Jiang, Q., Qiu, H., Wang, X., Liu, X. & Zhang, S. (2008). Effect of ionic liquids as additives on the separation of bases and amino acids in HPLC. *J.Liquid Chromatogr. Rel.Technol.* 31 (No.10); 1448-1457.

Jiang, T.F., Gu, Y.L., Liang, B., Li, J.B., Shi, Y.P. & Ou, Q.Y. (2003). Dynamically coating the capillary with 1-alkyl-3-methylimidazolium-based ionic liquids for separation of basic proteins by capillary electrophoresis, *Anal Chim Acta* Vol. 479 (No.2):249–254.

Kaliszan, R., Marszałł, M.P., Markuszewski, M.J., Bączek, T. & Pernak, J.(2004) Suppression of deleterious effects of free silanols in liquid chromatography by imidazolium tetrafluoroborate ionic liquids. *J.Chromatogr. A* 1030 (No.1-2): 263-271.

Khachatryn, K.S., Smirnova, S.V., Torocheshnikova, I.L., Shvedene, N.V., Formanovsky, N.V. & Pletnev, I.V.(2005). Solvent extraction and extraction–voltammetric determination of phenols using room temperature ionic liquid, *Anal. Bioanal. Chem.* Vol. 381 (No.2):464-470.

Kosobucki,P. & Buszewski,B. (2008). Isotachophoretic separation of selected imidazolium ionic liquids, *Talanta* Vol. 74 (No.5): 1670–1674.

Kowalska,S. & Buszewski,B. (2006). Effect of stationary phase polarity on the retention of ionic liquid cations in reversed phase liquid chromatography. *J.Sep.Sci.* Vol. 29 (No.17): 2625–2634.

Krizek, T., Breitbach, Z.S., Armstrong, D.W., Tesarova, E. & Coufal, P. (2009). Separation of inorganic and small organic anions by CE using phosphonium based mono and dicationic reagents, *Electrophoresis*, Vol. 30 (No.22):3955-3963.

Laamanen, P-L., Busi, S., Lahtinen, M. & Matilainen, R. (2005). A new ionic liquid dimethyldinonylammonium bromide as a flow modifier for the simultaneous determination of eight carboxylates by capillary electrophoresis, *J Chromatogr A* Vol.1095 (No.1-2):164–171.

Li, S., He, C., Liu, H., Liand, K. & Liu, F. (2005a). Ionic liquid-based aqueous two-phase system, a sample pretreatment procedure prior to high-performance liquid chromatography of opium alkaloids, *J.Chromatogr. B* Vol.826 (No.1-2):58–62.

Li, S.H., He, C.Y., Liu, H.W., Li K.A. & Liu, F. (2005b). Ionic liquid-salt aqueous two-phase system, a novel system for the extraction of abused drugs, *Chinese Chemical Letters* Vol.16 (No.8): 1074-1076.

Liu, J.-F., Chi, Y.G., Peng, J.-F., Jiang, G.-B. & Jonsson, J.A. (2004). Ionic Liquids/Water Distribution Ratios of Some Polycyclic Aromatic Hydrocarbons, *J. Chem. Eng. Data* Vol. 49 (No.5):1422-1424.

Liu, Q., Jiang, Y., Li, W., Hu, X., Xia, H., Liu, H. & Yang, P. (2006). Partitioning Behavior of Penicillin G in Aqueous Two Phase System Formed by Ionic Liquids and Phosphate, *Sep.Sci. Technol.* Vol.41 (No.12):2849–2858.

Louros, C.L.S., Cláudio, A.F.M., Neves,C.M.S.S., Freire,M.G., Marrucho,I.M., Pauly,J. & Coutinho, J.A.P. (2010).Extraction of Biomolecules Using Phosphonium-Based Ionic Liquids + K_3PO_4 Aqueous Biphasic Systems, *Int J Mol Sci.* Vol. 11(No.4): 1777–1791.

Lui, Q., Xuesheng, H., Wang, Y., Yan,g P., Xia, H. & Yu, J. (2005). Extraction og penicillin G by aqueous two-phase system of [BMIM] BF_4/NaH_2PO_4, *Chin. Sci. Bull.* Vol.50 (No.15): 1582-1585.

Lui, Q., Yu, J., Li, W.L., Hu, X.S., Xia, H.S., Liu, H.Z. & Yang, P. (2006). Partitioning behavior of penicillin G in aqueous two phase system formed by ionic liquids and phosphate, *Sep.Sci. Technol.* Vol. 41 (No.12): 2849-2858.

Markowska, A. & Stepnowski, P. (2008). Simultaneous determination of ionic liquid cations and anions using ion chromatography with tandem ion exchange columns: a preliminary assessment, *Analytical Sciences*, 24 (No.10):1359-1361.

Markuszewski,M.; Stepnowski,P. & Marszałł, M.P.(2004). Capillary electrophoretic separation of cationic constituents of imidazolium ionic liquids, *Electrophoresis* Vol. 25 (No.20): 3450–3454.

Marszałł, M.P.& Kaliszan, R., (2007). Application of ionic liquids In liquid chromatography, *Critical Reviews in Analytical Chemistry* Vol. 37 (No.14): 127-140.

Marszałł, M.P., Bączek, T. & Kalszan, R. (2006a). Evaluation of the silanol-suppressing potency of ionic liquids. *J.Sep.Sci.* Vol.29 (No.8): 1138-1145.

Marszałł, M.P., Markuszewski, M.J. & Kaliszan, R. (2006b). Separation of nicotinic acid and its structural isomers using 1-ethyl-3-methylimidazolium ionic liquid as a buffer additive by capillary electrophoresis, *J Pharm Biomed Anal* Vol.41 (No.1):329–332.

Marszałł, M.P., Bączek, T. & Kaliszan, R. (2005). Reduction of silanophilic interactions In liquid chromatography with the use of ionic liquids. *Anal.Chim.Acta* Vol.547 (No.2): 172-178.

Martin-Calero, A., Pino, V., Ayala, J.H., Gonzalez, V. & Alfonso, A.M. (2009). Ionic liquids as mobile phase additives in high-performance liquid chromatography with electrochemical detection: Application to the determination of heterocyclic aromatic amines in meat-based infant foods. *Talanta* Vol.79 (No.3): 590-597.

Matsumoto, M., Mochiduki, K., Fukunishi, K. & Kondo, K. (2004). Extraction of organic acids using imidazolium-based ionic liquids and their toxicity to Lactobacillus rhamnosus, *Sep. Purif. Technol.* Vol.40 (No.1):97-101.

McFarlane, J., Ridenour, W.B., Luo, H., Hunt, R.D., DePaoli, D.W. & Ren, R.X. (2005). Room Temperature Ionic Liquids for Separating Organics from Produced Water ,*Sep. Sci. Technol.* Vol.40 (No.6):1245-1265.

Merchuk, J.C., Andrews, B.A. & Asenjo, J.A. (1998). Aqueous two-phase systems for protein separation: Studies on phase inversion, *J.Chromatogr.B Biomed.Appl.* Vol.711 (No.1-2):285-293.

Mochizuki, Y. & Sugawara, K. (2008). Removal of Organic Sulfur from Hydrocarbon Resources Using Ionic Liquids, *Energy Fuels* Vol. 22 (No.5):3303-3307.

Molíková, M., Markuszewski, M.J., Kaliszan, R. & Jandera, P. (2010a). Chromatographic behaviour of ionic liquid cations in view of quantitative structure-retention relationship, *J.Chromatogr. A* Vol.1217 (No.8): 1305-1312.

Molikovaa, M., Studzinska,S., Kosobucki,P., Jandera,P. & Buszewski,B. (2010b). Determination of imidazolium and pyridinium ionic liquid cations by ion chromatography, *J.Liquid Chromatography and Related Technologies* Vol. 33 (No.2): 225-238.

Mwongela, S.M., Numan, A., Gill, N.L., Agbaria, R.A. & Warner, I.M. (2003). Separation of Achiral and Chiral Analytes Using Polymeric Surfactants with Ionic Liquids as Modifiers in Micellar Electrokinetic Chromatography, *Anal. Chem.*Vol. 75 (No.22):6089-6096.

Najdanovic-Visak, V., Canongia Lopes, J.N., Visak, Z.P., Trindade, J. & Rebelo, L.P.N. (2007).Salting-out in Aqueous Solutions of Ionic Liquids and K_3PO_4: Aqueous Biphasic Systems and Salt Precipitation, Int. J. Mol. Sci. Vol.8 (No.8):736–748.

Najdanovic-Visak, V., Esperanca, J.M.S.S., Rechelo, L.P.N., Nunes da Fonte, M., Guedes, J.R., Seddon, K.R. & Szydlowski, J. (2002).Phase behaviour of room temperature ionic liquid solutions: an unusually large co-solvent effect in (water + ethanol), *Phys. Chem.* Vol. 4 (No.10): 1701-1703.

Neves, C.M.S.S., Ventura, S.P.M., Freire, M.G., Marrucho, I.M. & Coutinho, A.P. (2009). Evaluation of Cation Influence on the Formation and Extraction Capability of Ionic-Liquid-Based Aqueous Biphasic Systems, *J.Phys.Chem.B* Vol.113 (No.15):5194–5199.

Pei, Y., Wang, J., Liu, L., Wu, K. & Zhao, Y. (2007).Liquid−Liquid Equilibria of Aqueous Biphasic Systems Containing Selected Imidazolium Ionic Liquids and Salts. *J. Chem. Eng. Data* Vol. 52 (No.5): 2026-2031.

Pei, Y., Wang, J., Wu, K., Xuan, X. & Lu, X. (2009). Ionic liquid-based aqueous two-phase extraction of selected proteins, *Sep.Purif.Technol.* Vol.64 (No.3):288–295.

Peng, J.-F., Liu, J.-F., Hu, X.-L. & Jiang, G.-B. (2007). Direct determination of chlorophenols in environmental water samples by hollow fiber supported ionic liquid membrane extraction coupled with high-performance liquid chromatography *J. Chromatogr. A* Vol. 1139 (No.2): 165-170.

Pfruender, H., Amidjojo, M., Kragl, U. & Weuster-Botz, D. (2004). Efficient whole-cell biotransformation in a biphasic ionic liquid/water system. *Angew.Chem.Int.Ed.* Vol.43 (No.34):4529-4531.

Poole, C. F. & Poole, S. K. (2010). Extraction of organic compounds with room temperature ionic liquids, *J.Chromatogr. A* 1217 (No.16): 2268-2286.

Poole, C.F. (2004). Chromatographic and spectroscopic methods for the determination of solvent properties of room temperature ionic liquids, *J. Chromatogr. A* Vol. 1037 (No.1-2):49-82.

Poole, C.F. (2007). Applications of ionic liquids in extraction, chromatography, and electrophoresis, *Adv. Chromatogr.* Vol. 45:89-124.

Qi, S., Cui, S., Chen, X. & Hu, Z. (2004). Rapid and sensitive determination of anthraquinones in Chinese herb using 1-butyl-3-methylimidazolium-based ionic liquid with beta.-cyclodextrin as modifier in capillary zone electrophoresis, *J Chromatogr A* Vol 1059 (No.1-2):191–198.

Qi, S., Li, Y., Deng, Y., Chen, Y., Chen, X. & Hu, Z. (2006) Simultaneous determination of bioactive flavone derivatives in Chinese herb extraction by capillary electrophoresis used different electrolyte systems—Borate and ionic liquids, *J. Chromatogr. A* Vol. 1109 (No.2):300-306.

Qin, W. & Li, S.F.Y. (2002).An ionic liquid coating for determination of sildenafil and UK-103,320 in human serum by capillary zone electrophoresis-ion trap mass spectrometry, *Electrophoresis*, Vol. 23 (No.24):4110-4116.

Qin,W.; Wei,H. & Li,S.F.Y. (2002). Separation of ionic liquid cations and related imidazole derivatives by α-cyclodextrin modified capillary zone electrophoresis, *Analyst* Vol.127 (No.4):490–493.

Qiu, H., Jiang, S. & Liu, X. (2006a). N-Methylimidazolium anion-exchange stationary phase for high-performance liquid chromatography, *J. Chromatogr. A* Vol.1103 (No.2):265-270.

Qiu, H., Jiang, S., Liu, X. & Zhao, L. (2006b).Novel imidazolium stationary phase for high-performance liquid chromatography, *J. Chromatogr. A* Vol. 1116 (No.1-2):46-50.

Rito-Palomares, M. (2004). Practical application of aqueous two-phase partition to process development for the recovery of biological products, *J. Chromatogr. B* Vol. 807 (No.1): 3-11.

Rizvi, S.A.A. & Shamsi, S.A. (2006).Synthesis, Characterization, and Application of Chiral Ionic Liquids and Their Polymers in Micellar Electrokinetic Chromatography, *Anal. Chem.* Vol. 78 (No.19):7061-7069.

Rodriguez, O., Madeira, P.P. & Macedo, E.A. (2007). Amino-acid recovery using ionic liquids: Partitioning in water+ionic liquid systems. Proceedings of European Congress of Chemical Engineering (ECCE-6) Copenhagen, 16-20 September 2007.

Roobol-Boza, M., Dolby, V., Doverskog, M., Barrefelt, A., Lindqvist, F., Oppermann, U.C., Van Alstine, K.K. & Tjerneld, F., (2004). Membrane protein isolation by in situ solubilization, partitioning and affinity adsorption in aqueous two-phase systems: Purification of the human type 1 11β-hydroxysteroid dehydrogenase, *J. Chromatogr. A* Vol. 1043 (No.2):217-223.

Ruiz-Angel M.J., Carda-Broch, S. & Berthod, A (2006). Ionic liquids versus triethylamine as mobile phase additives in the analysis of β-blockers. *J.Chromatogr. A* 1119 (No.1-2): 202-208.

Ruiz-Angel, M.J. & Berthod, A. (2008). Reversed-phase liquid chromatography analysis of alkyl-imidazolium ionic liquids: II. Effects of different added salts and stationary phase influence, *J.Chromatogr. A* Vol.1189 (No.1-2): 476-482.

Ruiz-Angel,M.J. & Berthod,A. (2006). Reversed-phase liquid chromatography analysis of alkyl-imidazolium ionic liquids, *J.Chromatogr.A* Vol. 1113 (No.1-2): 101–108.

Shehong, L., Chiyang, H., Huwei, L., Kean, L. & Feng, L. (2005). Ionic liquid-based aqueous two-phase system, a sample pretreatment procedure prior to high-performance liquid chromatography of opium alkaloids. *J.Chromatogr.* 826 (1-2): 58-62.

Shetty, P.H., Poole, S.K. & Poole, C.F. (1990). Applications of ethylammonium and propylammonium nitrate solvents in liquid-liquid extraction and chromatography, *Anal. Chim. Acta* Vol. 236:51-62.

Shetty, P.H., Youngberg, P.J., Kersten, B.R. & Poole, C.F. (1987). Solvent properties of liquid organic salts used as mobile phases in microcolumn reversed-phase liquid chromatography, *J. Chromatogr.A* Vol.411:61-79.

Smirnova, S.V., Torocheshnikova, I.I., Formanovsky, A.A. & Pletnev, I.V. (2004). Solvent extraction of amino acids into a room temperature ionic liquid with dicyclohexano-18-crown-6, *Anal. Bioanal. Chem.* Vol. 378 (No.5):1369-1375.

Soga, T. & Ross, G.A. (1999). Simultaneous determination of inorganic anions, organic acids and metal cations by capillary electrophoresis , *J. Chromatogr., A* Vol. 834 (No.1-2): 65-71.

Soto, A., Arce, A. & Khoshkbarchi, M.K. (2005). Partitioning of antibiotics in a two-liquid phase system formed by water and a room temperature ionic liquid, *Sep. Purif. Technol.*, Vol. 44 (No.3):242-246.

Stepnowski,P.; Muller,A.; Behrend,P.; Ranke,J. ;Hoffmann,J. & Jastorff,B. (2003). Reverse phase liquid chromatographic method for the determination of selected room temperature ionic liquids cations, *J.Chromatogr.A* Vol. 993(No.1-2) 173–178.

Stuff,J.R. (1991).Separation of cations in buffered 1-methyl-3-ethylimidazolium chloride-aluminum chloride ionic liquids by ion chromatography, *J.Chromatogr.A* Vol. 547: 484–487.

Sun, Y., Cabovska, B., Evans, C.E., Ridgway, T.H. & Stalcup, A.M. (2005). Retention characteristics of a new butylimidazolium-based stationary phase, *Anal. Bioanal. Chem.* Vol. 382 (No.3):728-734.

Sun,P., Armstrong, D.W. (2010). Ionic liquids in analytical chemistry, *Anal.Chim.Acta* Vol.661 (No. 1): 1-16.

Swatloski, R.P., Visser, A.E., Reichert, W.M., Broker, G.A., Farina, L.M., Holbrey, J.D. & Rogers, R.D. (2002). On the solubilization of water with ethanol in hydrophobic hexafluorophosphate ionic liquids, *Green Chem.* Vol. 4 (No.2):81-87.

Tang,F., Tao, L., Luo, X., Ding, L., Guo, M., Nie, Li. & Yao, S. (2006). Determination of octopamine, synephrine, and tyramine in Citrus herbs by ionic liquid improved 'green' chromatography. *J.Chromatogr. A* 1125 (No.2): 182-188.

Tholey, A, Zabet-Mghaddam, M. & Heinzle, E. (2006). Quantification of Peptides for the Monitoring of Protease-Catalyzed Reactions by Matrix-Assisted Laser Desorption/Ionization Mass Spectrometry Using Ionic Liquid Matrixes, *Anal.Chem.* Vol.78 (No.1):291-297.

Tran, C.D. & Mejac, I. (2008). Chiral ionic liquids for enantioseparation of pharmaceutical products by capillary electrophoresis, *J. Chromatogr. A* Vol. 1204 (No.2):204-209.

Vaher, M., Koel, M. & Kaljurand, M. (2002a). Ionic liquids as electrolytes for nonaqueous capillary electrophoresis, *Electrophoresis* Vol.23 (No.2)::426–430.

Vaher, M., Koel, M., Kaljurand, M. (2002b). Application of 1-alkyl-3-methylimidazolium-based ionic liquids in non-aqueous capillary electrophoresis, *J Chromatogr A* Vol. 979 (No.1-2):27–32.

Vaher, M., Koel, M. & Kaljurand, M. (2000). Non-aqueous capillary electrophoresis in acetonitrile using ionic-liquid buffer electrolytes, *Chromatographia* Vol.53 (No.1):S302–S306.

Ventura, S.P.M., Neves, C.M.S.S., Freire, M.G., Marrucho, I.M., Oliveiraand, J. & Coutinho, J.A.P. (2009). Evaluation of Anion Influence on the Formation and Extraction Capacity of Ionic-Liquid-Based Aqueous Biphasic Systems, *J.Phys.Chem.*B Vol.113 (No.27): 9304–9310.

Vidal, L., Psillakis, E., Domini, C.E., Grane, N. & Marken, F. (2007). An ionic liquid as a solvent for headspace single drop microextraction of chlorobenzenes from water samples, *Anal. Chim. Acta* Vol. 548 (No.1): 189-195.

Vidal, S.T.M., Correla, M.J.N., Marques, M.M., Ismael, M.R. & Reis, M.T.A. (2004). Studies on the use of ionic liquids as potential extractants of phenolic compounds and metal ions, *Sep. Sci. Technol.*, Vol. 39 (No.9): 2155-2169.

Vijayaraghavan, R., Vedaraman N., Surianarayanan, M. & MacFarlane, D.R. (2006). Extraction and recovery of azo dyes into an ionic liquid, *Talanta* Vol. 69 (No.5):1059-1062.

Visak, Z.P., Canongia Lopes, J.N. & Rebelo, L.P.N.(2007). Ionic Liquids in Polyethylene Glycol Aqueous Solutions: Salting-in and Salting-out Effects, Monatsh. Chem. Vol.138:1153–1157.

Wakai, C., Oleinikova,A., Ott,M. & Weingärtner, H. (2005). How Polar Are Ionic Liquids? Determination of the Static Dielectric Constant of an Imidazolium-based Ionic Liquid by Microwave Dielectric Spectroscopy, *J. Phys. Chem. B* Vol.*109* (No.36):17028–17030

Walter, H. & Johansson, G., (Ed.) (1994). Methods of Enzymology, Vol. 228 Academic Press, New York

Wang, Q., Baker, G.A., Baker, S.N. & Colón, L.A. (2006). Surface confined ionic liquid as a stationary phase for HPLC,*Analyst* Vol. 131 (No.9):1000-1005.

Weuster-Botz, D. (2007). Process intensification of whole-cell biocatalysis with ionic liquids. *Chem.Rec.*Vol. 7:334-340.

Willauer, H.D., Huddleston, J.G. & Rogers, R.D. (2002). Solute Partitioning in Aqueous Biphasic Systems Composed of Polyethylene Glycol and Salt: The Partitioning of Small Neutral Organic Species, *Ind.Eng.Chem.Res.* Vol.41 (No.7):1892–1904.

Wu, B., Zhang, Y. & Wan,g H.(2008a). Phase Behavior for Ternary Systems Composed of Ionic Liquid + Saccharides + Water, *J. Phys. Chem. B* Vol. *112* (No.20):6426–6429.

Wu, B., Zhang, Y., Wang, H. & Yang, L.(2008b). Temperature Dependence of Phase Behavior for Ternary Systems Composed of Ionic Liquid + Sucrose + Water, *J. Phys. Chem. B* Vol. *112* (No.41): 13163–13165.

Wu, B., Zhang, Y.M. & Wang, H.P.(2008c). Aqueous Biphasic Systems of Hydrophilic Ionic Liquids + Sucrose for Separation, *J. Chem. Eng. Data Vol.*53 (No.4):983–985.

Xiao, X., Zhao, L., Liu, X. & Jiang, S. (2004). Ionic liquids as additives In high performance liquid chromatography. Analysis of amines and the interaction mechanism of ionic liquids. *Anal.Chim.Acta* 519 (No.2): 207-211.

Yanes, E.G., Gratz, S.R. & Stalcup, A.M. (2000). Tetraethylammonium tetrafluoroborate: A novel electrolyte with unique role in the capillary electrophoretic separation of polyphenols found in grape seed extracts. *Analyst* Vol 125 (No. 11):1919-1923.

Yanes, E.G., Gratz, S.R., Baldwin, M.J., Robison, S.E. & Stalcup, A.M. (2001).Capillary electrophoretic application of 1-alkyl-3-methylimidazolium-based ionic liquids, *Anal.Chem*. Vol. 73 (No.15): 3838-3844.

Yangyang, J., Hansong, X., Jiang, Y., Chen, G. & Huizhou, L. (2009). Hydrophobic ionic liquids-assisted polymer recovery during penicillin extraction in aqueous two-phase system, *Chem.Eng. J*. Vol. 147 (No.1):22-26.

Yu, L., Qin, W. & Li, S.F.Y. (2005). Ionic liquids as additives for separation of benzoic acid and chlorophenoxy acid herbicides by capillary electrophoresis, *Anal Chim Acta* Vol.547 (No.2):165–171.

Zabet-Moghaddam, M., Heinzle, E. & Tholey, A. (2004). Qualitative and quantitative analysis of low molecular weight compounds by ultraviolet matrix-assisted laser desorption/ionization mass spectrometry using ionic liquid matrice, *Rapid Commun. Mass Spectrom.* Vol 18 (No.2): 141-148.

Zafarani-Moattar, M.T. & Hamzehzadeh, S. (2009).Phase Diagrams for the Aqueous Two-Phase Ternary System Containing the Ionic Liquid 1-Butyl-3-methylimidazolium Bromide and Tri-potassium Citrate at T = (278.15, 298.15, and 318.15) K, *J. Chem. Eng. Data* Vol. 54 (No.3): 833-841.

Zafarani-Moattar, M.T. & Hamzehzadeh, S. (2007). Liquid−Liquid Equilibria of Aqueous Two-Phase Systems Containing 1-Butyl-3-methylimidazolium Bromide and Potassium Phosphate or Dipotassium Hydrogen Phosphate at 298.15 K, J. Chem. Eng. Data *Vol.* 52 (No.5):1686–1692.

Zhang, W., He, L., Gu, Y., Liu, X. & Jiang, S. (2003). Effect of ionic liquids as mobile phase additives on retention of catecholamines in reversed-phase high-prformance liquid chromatography. *Anal.Lett.* 36 (No.4): 827-838.

Zhang, Y. & Cremer, P.S. (2006). Interactions between macromolecules and ions: the Hofmeister series, *Current opinion in Chemical Biology* Vol. 10 (No.6): 658-663.

Zhang, Y., Wang, H. & Yang, L.(2008). Temperature Dependence of Phase Behavior for Ternary Systems Composed of Ionic Liquid + Sucrose + Water, J. Phys. Chem. *B Vol.*112 (No.41):13163–13165.

Zhang, Y., Zhang, S., Chen, Y. & Zhang, J.(2007). Aqueous Biphasic Systems Composed of Ionic Liquid and Fructose, Fluid Phase Equilib. Vol.257 (No.2):173–176.

Zhou, Q., Bai, H., Xie, G. & Xiao, J. (2008). Temperature-controlled ionic liquid dispersive liquid phase micro-extraction, *J. Chromatogr. A* Vol. 1177 (No.1):43-49.

Zhu, Z., Guan, Y. & .Li, M. (2001). Applications of Aqueous Two-phase Systems in Extraction, *J.Chem.Ind.Eng.* Vol.52:1039–1048.

16

Ionic Liquid-Mediated Liquid-Liquid Extraction

Qilong Ren, Qiwei Yang, Yan Yan, Huabin Xing,
Zongbi Bao, Baogen Su and Yiwen Yang
Department of Chemical and Biological Engineering,
Zhejiang University
China

1. Introduction

Liquid-liquid extraction is an important kind of separation method that is based on the distribution of chemicals between two different liquid phases. Compared to other separation methods, liquid-liquid extraction often has unique advantages for the separation of chemicals that have high or similar boiling points, with relatively large capacity and low consumption of material and energy (Treybal, 1951). However, the kinds of extractants that could be used for liquid-liquid extraction process are relatively small at present, so only limited separation efficiency could be achieved for the separation of many mixtures, especially those having similar structures. Besides, the volatility of existing extractants or extraction solvents not only can bring contamination to the environment, but also can lead to difficulty on the recovery of extractant and subsequent purification of products when the solutes are also volatile. These problems have limited the application of liquid-liquid extraction to more separation processes.

Using ionic liquids (ILs) as extractants may be a prospective solution to the above problems. The physicochemical properties of IL could be designed and adjusted task-specifically, so large separation selectivity may be achieved for various mixtures that need to be separated (Han & Armstrong, 2007). Moreover, the cohesive energy of ILs is always very large, so ILs are easy to form various immiscible liquid-liquid biphasic systems with other solvents (Marcus, 2010). Besides, the vapor pressures of ILs are extremely low, so ILs could be regarded as green extractants and the separation of ILs with volatile solutes could also be simplified (Welton, 1999). These characteristics have made ILs appropriate to be used as extractants in liquid-liquid extractions.

Liquid-liquid extraction using IL as extractant has been studied by many researchers in recent years since the partition of substituted-benzene derivatives between water and a hydrophobic IL 1-butyl-3-methylimidazolium hexafluorophosphate ($[C_4mim]PF_6$) was initially reported (Huddleston et al., 1998). These studies include the removal of sulfides and nitrides from diesel and gasoline (Holbrey et al., 2008; Xie et al., 2008), the separation of aromatics from aliphatics (Meindersma & de Haan, 2008), the removal of pollutants from water (Egorov et al., 2008; Vijayaraghavan et al., 2006), the isolation of biological substances from aqueous mixtures (Wang et al., 2007), the extraction of glycerol from biodiesel (Abbott et al., 2007), the extractive essential oil terpenless (Arce et al., 2006), and so on. From those works, ILs have been revealed to have the strong ability of interacting with organic

molecules through various mechanisms (e.g., π-π, dispersion, ionic exchange, hydrogen bonding). Moreover, these interactions can be finely adjusted by the change of IL's cation or anion task-specifically, thus often bringing on elevated separation efficiency compared with traditional solvents. IL-mediated liquid-liquid extraction has been studied not only at a lab scale but also a pilot scale, and has exhibited a great prospect for the industrial application.
In this chapter, three subtopics will be introduced subsequently: liquid-liquid equilibrium of IL-molecular solvent mixtures; applications of IL-mediated liquid-liquid extraction; investigation of IL-solute interaction by quantum chemistry method. The extraction of metal ions, sulfides and nitrides using ILs as extractant are not be included in this chapter because they are introduced in other chapters of this book.

2. Liquid-liquid equilibrium of ionic liquid-molecular solvent mixtures

Liquid-liquid extraction is a kind of separation technology based on immiscible liquid-liquid biphasic systems, so the liquid-liquid equilibrium of the mixtures consisted of IL with traditional molecular solvent has to be investigated. Currently, the liquid-liquid equilibrium of both binary and ternary IL-solvent mixtures has been reported.

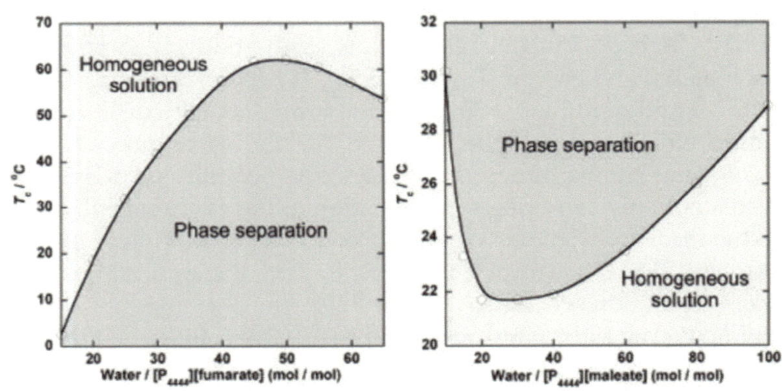

Fig. 1. Phase transition temperature Tc for ionic liquids/water mixtures at different water content. Left: tetra-*n*-butylphosphonium fumarate-water mixtures, UCST. Right: tetra-*n*-butylphosphonium maleate-water mixtures, LCST. Reprinted from reference which is downloaded from www.rsc.org (Fukaya et al., 2007).

As for the binary mixtures, liquid-liquid equilibrium has been reported for the mixtures of IL plus water, aliphatic alcohol, alkane, alkene, halogenated hydrocarbon, etc. (Crosthwaite et al., 2004; Domanska et al., 2006; Shiflett et al., 2009). Because ILs often have a relatively large polarity, the interaction between ILs and weak-polar solvents are often not strong and liquid-liquid biphasic systems with low immiscibility can be formed between ILs and those solvents. Different types of solubility curve have been found for IL-solvent binary mixtures. Most of the IL-solvent mixtures have a solubility curve with an ultimate critical solubility temperature (UCST). IL-aliphatic alcohol mixtures are typical examples that possess UCST-type solubility curves (Crosthwaite et al., 2004). In the immiscible region, the content of IL in the alcohol-rich phase is quite low. However, the content of alcohol in the IL-rich phase is much higher and could not be ignored. Due to different intermolecular interactions such as hydrogen-bonding, π-π and Van der Waals, the phase equilibrium can be effected by various structural factors, including the alkyl side chain length of IL's cation, the kind of IL's anion, the length of alcohol, the branching of alcohol, and so on. Solubility curves with a lower critical solution temperature (LCST) were found for several IL-solvent mixtures. Ohno et al. found LCST in the mixtures of water plus amino acid-functionalized ILs those possess a carboxyl group in the anion (Fukumoto & Ohno, 2007), as well as the mixtures of water plus tetra-*n*-butylphosphonium maleate (Fukaya et al., 2007, Figure 1). The mixtures of 1-alkyl-3-methylimidazolium bis(trifluoromethylsulfonyl)amide ($[C_n mim]NTf_2$) plus aromatics may have different phase diagrams depending on the length of alkyl chain in the cation (Lachwa et al., 2006), whose solubility curve could change from LCST type to UCST type with the increase of chain length.

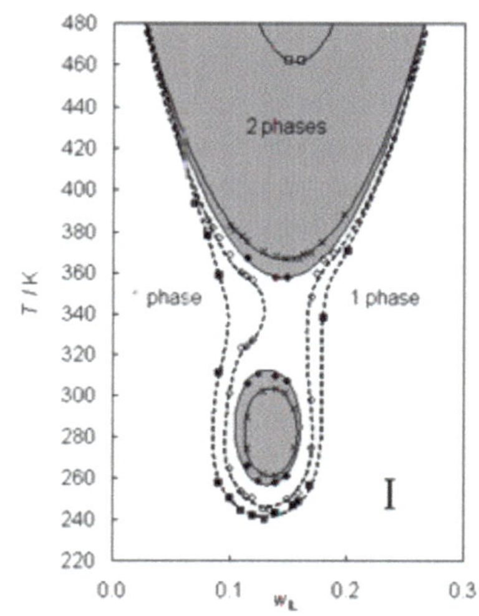

Fig. 2. Phase equilibria of $[C_n mim]NTf_2$-$CHCl_3$ mixtures; w_{IL} is weight fraction of IL. (□) n = 5.000; (×) n = 4.337; (●) n = 4.330; (○) n = 4.320; (■) n = 4.300. The shaded areas depict the demixing regions for n = 4.330. Reprinted from reference which is downloaded from pubs.acs.org (Lachwa et al., 2005).

Liquid-liquid equilibrium of IL-molecular solvent ternary mixtures has also been determined. Those mixtures are usually consisted of one IL and two molecular solvents, such as IL + aromatic + aliphatic, IL + ether + alcohol, IL + alkane + alkene, and so on (Arce et al., 2004; Letcher & Deenadayalu, 2003). In fact, most of those works are also complete liquid-liquid extraction studies themselves, where distribution coefficients and selectivity can be obtained directly from the triangle phase diagrams. Besides, Swatloski et al. studied the phase equilibrium of $[C_4mim]PF_6$ + ethanol + water ternary mixture (Swatloski et al., 2002). Although $[C_4mim]PF_6$ could not be fully miscible with either ethanol or water, it could be fully miscible with the mixture of ethanol + water when the composition of the ethanol-water mixture was within certain range. With the increase of water content in the $[C_4mim]PF_6$ + ethanol + water mixture, the UCST of mixture decreased first and then increased (Najdanovic-Visak et al., 2002). Therefore, through adjusting the content of water and ethanol in the $[C_4mim]PF_6$ + ethanol + water system, different kinds of liquid-liquid biphasic systems including fully miscible, partially miscible and fully immiscible systems could be obtained. Phase diagrams containing a closed loop at low temperature and a high-temperature immiscible region were found in the ternary or so-called *quasi*-binary mixtures of $[C_nmim]NTf_2$ plus chloroform (Figure 2), where n was not a integer due to the mixing of two different ILs that had different *n* values (Lachwa et al., 2005). The closed loop was accounted the result of a delicate balance between enthalpic and entropic contributions to the Gibbs energy of a mixture.

3. Applications of ionic liquid-mediated liquid-liquid extraction

3.1 Extraction of organic compounds from aqueous phase

Extensive studies have been conducted for the extraction of organic compounds from aqueous phase with ILs, depending on the affinity between hydrophobic ILs and organic solutes. The extraction mechanism includes ion exchange, hydrogen bond, Van der Waals interaction, and so on.

Khachatrya et al. reported the extraction of phenolic compounds from aqueous solution into $[C_4mim]PF_6$, in which the solutes could be extracted nearly quantitatively at $pH<pK_a$ (Khachatryan et al., 2005). The relatively large distribution coefficient of phenolate anions indicated the ion exchange mechanism in the extraction that when phenolate anion entered into the IL phase, an equal amount of hexafluorophosphate anion was transferred to water.

The performance of a neutral IL *N*-butyl-*N*-methyl pyrrolidinium bis(trifluoromethanesulfonyl)amide as extractant was studied by Vijayaraghavan et al. for the removal of azo dyes from aqueous solutions (Vijayaraghavan et al., 2006). During the extraction, the azo dye went into the organic phase in its ionic form, and the distribution coefficient of the dye between IL and water was about 2.0. Repeated extractions (two to three times) with fresh IL proceeded to a removal fraction about 95% of the dyes from aqueous phase into the IL phase. The ion exchange mechanism was suggested again by Li et al. when investigating the extraction process of acid dyes from aqueous solution into $[C_4mim]PF_6$ (Li et al., 2007), in which the solvated part of IL played an important role as counter-ions for the anions of acidic dyes. Fan et al. investigated the suitability of imidazolium-based ILs, 1-methyl-3-alkylimidazolium hexafluorophosphate ($[C_nmim]PF_6$, n = 4, 6, 8) and 1-methy-3-alkylimidazolium tetrafluoroborate ($[C_nmim]BF_4$, n = 6, 8), as a substitute for volatile organic solvents in the liquid–liquid extraction of selected phenols from aqueous solutions (Fan et al., 2006). The selected phenols included phenol, bisphenol A, pentachlorophenol, 4-octylphenol and 4-nonylphenol. A deep analysis of experimental results suggested the existence of strong hydrogen-bonding interaction between the anions of ILs and the phenols, which contributed to the high extraction efficiency of ILs for the phenols. As a result, the value of distribution ratio for phenol into $[C_nmim]BF_4$ was about 10 times as high as that into dichloromethane under the same conditions. The effect of aqueous phase pH on the partitioning of an indicator dye, thymol blue, was studied by Visser et al (Visser et al., 2000). A remarkable dependence of distribution ratio between $[C_nmim]PF_6$ and water on the pH value was revealed, suggesting a possible approach of separating ILs and extract after extraction (Figure 3).

Besides the extraction of organic pollutes from aqueous solutions, ILs can be also applied to the extraction of target products in biological conversion and separation process. For example, several ILs were reported as extractants to separate antibiotics and proteins from aqueous solutions (Cull et al., 2000; Tzeng et al., 2008).

Aqueous two-phase systems (ATPS) composed of salt solution and ILs can find its utilization in the extraction of certain organic compounds. For example, ATPS formed by

$[C_4mim]BF_4$ and KH_2PO_4 could be used to extract bovine serum albumin (Deng & Guo, 2006). High extraction efficiency could be achieved under the following conditions: the concentration of KH_2PO_4 was 80 g/L, the concentration of $[C_4mim]BF_4$ was in the range of 160-240 mL/L, the concentration of BSA was 30-50 mg/L, the acidity of the solution was pH = 4-8. Also, ATPS formed by $[C_4mim]BF_4$ and KH_2PO_4 exhibited high extraction efficiency of separation of penicillin G (Liu et al., 2005).

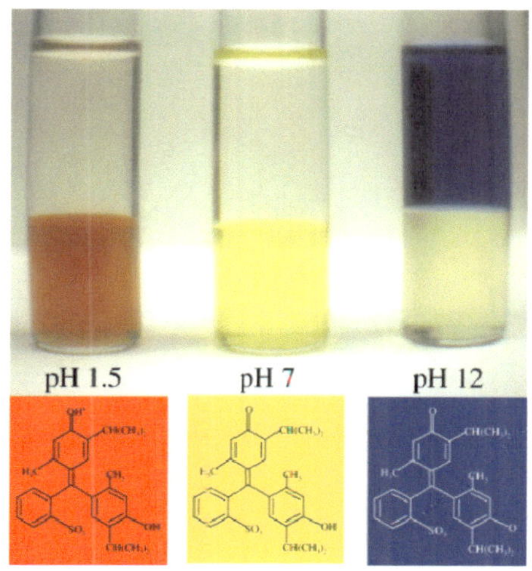

Fig. 3. Distribution of thymol blue in $[C_4mim]PF_6$-water biphasic system at different pH. Reprinted from reference which is downloaded from www.rsc.org (Visser et al., 2000).

3.2 Separation of mixtures with close boiling points and azeotropes

Extractive distillation serves as a useful solution for the separation of azeotropic mixtures and mixtures consisted of components which have small relative volatilities. The frequently-used extractants refer to solid salts and organic liquids. While the solid salts can corrode and block the distillation towers and the pipes, the volatility of organic liquids leads to larger solvent and energy consumption. ILs are liquid organic salts with negligible vapor pressure, so they can be used to overcome these shortages. Pereiro et al. studied the performance of 1-hexyl-3-methylimidazolium hexafluorophosphate ($[C_6mim]PF_6$) and 1-butyl-3-methylimidazolium methyl sulfate $[C_4mim]CH_3SO_4$ in the removal of heptane from its azeotropic mixture with ethanol (Pereiro et al., 2006; Pereiro & Rodriguez, 2008). Hu et al. studied ILs with tetrafluoroborate anion as solvent to separate ethyl acetate from its azeotropic mixture with ethanol (Hu et al., 2008), which included ILs 1-ethyl-3-methylimidazolium tetrafluoroborate ($[C_2mim]BF_4$), 1-(2-hydroxyethyl)-3-methylimidazolium tetrafluoroborate ($[C_2OHmim]BF_4$), 1-ethyl-2,3-dimethylimidazolium tetrafluoroborate ($[C_2dmim]BF_4$) and 1-(2-hydroxyethyl)-2,3-dimethylimidazolium tetrafluoroborate ($[C_2OHdmim]BF_4$). It was found that the separation efficiency could be influenced by the hydrogen bond basicity of anions, the acidity of hydrogen at the C2 position of imidazolium ring, and the alkyl chain length of imidazolium cation (Figure 4). Seiler et al. showed the excellent performance of $[C_2mim]BF_4$ as entrainer for the separation of ethanol/water mixture. Compared to a conventional entrainer 1,2-ethanediol, $[C_2mim]BF_4$ could dramatically reduce the energy consumption of the separation process.

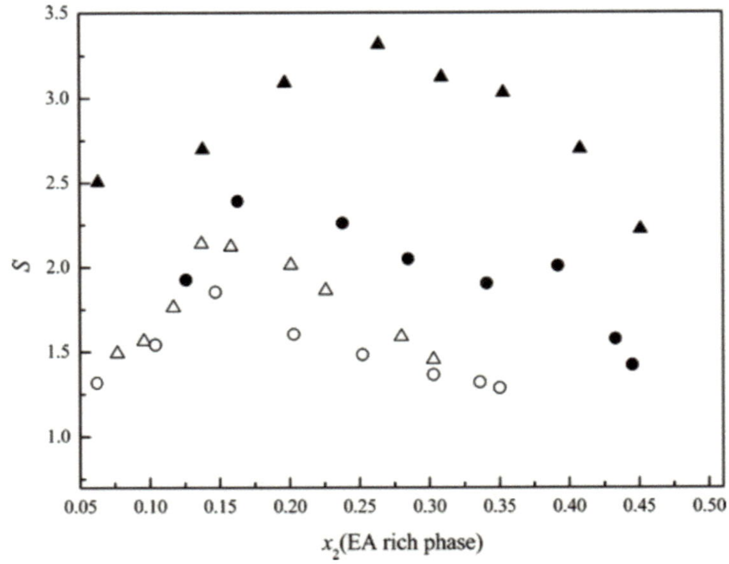

Fig. 4. Extraction selectivity of ethanol and ethyl acetate with different ILs. ▲, [C_2OHmim]BF_4; •,[$C_2OHdmim$]BF_4; △, [C_2mim]BF_4; ○, [C_2dmim]BF_4. x_2 is the mole fraction of ethanol. Reprinted from reference which is downloaded from pubs.acs.org (Hu et al., 2008).

Since the boiling points of aromatic hydrocarbons (benzene, toluene, ethylbenzene, xylene) and aliphatic hydrocarbons with four to ten carbons are really close (some even form azeotropic mixtures), it's of great difficulty to separate them from each other. For now, the common separation method in industry is liquid-liquid extraction with organic solvents such as *N*-methylpyrrollidone and sulfolane as extractant (Chen et al., 2000; Krishna et al., 1987), which often have low separation selectivity and high energy consumption. Nevertheless, the unique structure of ILs qualifies them as an alternative extractant for separation of aromatic and aliphatic hydrocarbons. On the one hand, one or more unsaturated functional groups in the structure of ILs entitle them to form π-π interaction with aromatic hydrocarbons, leading to a peculiar affinity of aromatic hydrocarbons for ILs. On the other hand, the relatively strong polarity of ILs induces a small solubility of alkanes. Meindersma et al. have given their efforts to the extraction of aromatics from aromatic/aliphatic mixtures in recent years (Meindersma et al., 2005; Meindersma et al., 2006; Meindersma et al., 2006; Meindersma & de Haan, 2008; Onink et al., 2010). They selected 4-methyl-*N*-butylpyridinium tetrafluoroborate ([mebupy]BF_4), 1-methyl-3-methylimidazolium methylsulfate ([mmim]CH_3SO_4) and several other ILs as novel extractants to separate toluene/heptane mixtures. The result showed that the separation selectivity when using IL as extractant could be 1.5 to 2.5 times higher than that when using sulfolane and the increase of polarity of ILs (i.e., the reduction of alkyl length on the cation, introduction of polar functional groups such as hydroxyl group) could improve the selectivity (Figure 5). What's more, since ILs are nearly non-volatile, the regeneration of ILs could be achieved by simple flash evaporation, which dramatically reduced the energy consumption compared to the distillation of sulfolane. Pilot research in a rotary disk column has been conducted and inspiring result was obtained.

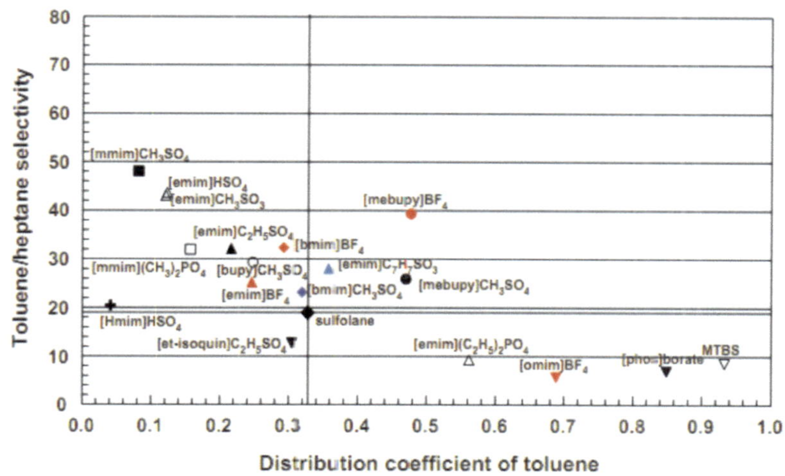

Fig. 5. Tolune/heptane separation with ionic liquids or sulfolane. Reprinted from reference which is downloaded from www.elsevier.com (Meindersma et al., 2005).

3.3 Separation of structurally related natural compounds

There is a growing interest in obtaining organic products from natural resources in recent years. Many of these natural products have certain physiologically activities, which makes them as preferable components of medicine, health products and cosmetics. However, a common problem existing in obtaining chemicals from natural resources is that the product of interest often appears in a mixture with various structurally related compounds (Chen & Payne, 2001), so the effective separation of them is necessary and challenging (Sticher, 2008). For now, the widely-used methods for separating structurally related natural compounds are adsorption methods which base on the distribution equilibrium of chemicals between a solid adsorbent and a liquid solvent. Nevertheless, these methods have some shortages. First, since the active chemical structures are just located at the surface of solid adsorbent, their apparent density is very small, leading to a low throughput capacity and a large solvent consumption and energy consumption. Second, since the adsorbents always possess a lot of minute pores but the natural compounds often have a relatively large molecular size, the diffusion and transmission of compounds are difficult, which can improve the consumption of solvent and energy again. Although new approaches such as simulated moving bed chromatography and supercritical fluid chromatography could improve separation efficiency to some extent, they are not widely-used because of their requirement for special equipment (Rajendran et al., 2009; Taylor, 2009). So, novel separation methods besides adsorption are still desired.

Liquid-liquid extraction using ILs as extractant have been applied for the separation of structurally related natural compounds. Arce et al. studied the separation of limonene and linalool by extraction, using organic solvents or ILs as extractant (Arce et al., 2006). The separation of linalool from limonene is necessary for the production of citrus essential oil, and the common method is distillation. However, essential oils can be damaged by distillation, so liquid extraction may be a good alternative. Six types of organic solvents, and one type of IL were used as extractant in extraction, including 2-butene-1,4-diol, diethylene glycol, 1,2-propanediol, 1,3-propanediol, 2-aminoethanol, ethylene glycol, and 1-ethyl-3-methylimidazolium methanesulfonate ([C_2mim]OMs). Based on the measured distribution ratio and selectivity (Figure 6), extraction process simulations were conducted for using these solvents. As shown by the results, although the IL did not get the highest linalool recovery, it provided the highest purity. Li et al. studied the extraction of α-tocopherol by IL-modified SBA-15 adsorbent (Li et al., 2008). Although the process seems not liquid-liquid extraction but solid-phase extraction from the macroscopic view, the mechanism is still based on the distribution of chemicals between IL and organic solvent. Very high adsorption selectivity for α-tocopherol was achieved by using 1-ethyl-3-methylimidazolium aminoacetate ([C_2mim]Gly)/SBA-15 as adsorbent, in which [C_2mim]Gly was a kind of amino acid-functionalized IL possessing a glycine anion. The separation selectivity of α-tocopherol to the major interference component glyceryl triundecanoate was 10.5 and the concentration of α-tocopherol was elevated from 15.7% in the feedstock to 73.0% after stripping by diethyl ether in the first cycle.

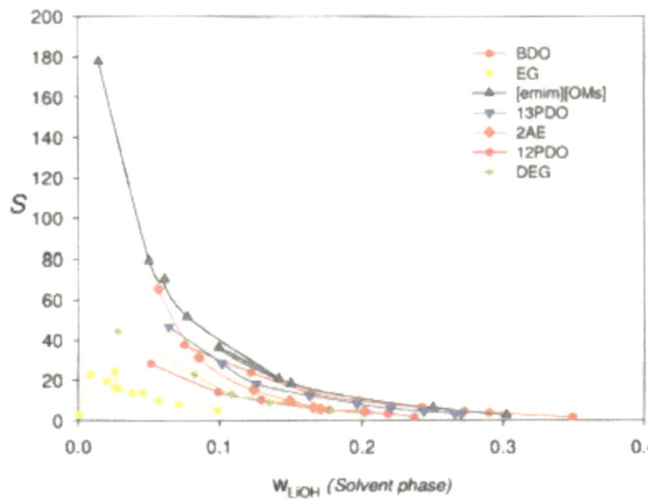

Fig. 6. Selectivity S for different solvents to extract linalool from its mixtures with limonene at 298.15 K. Reprinted from reference which is downloaded from onlinelibrary.wiley.com (Arce et al., 2006).

The separation of tocopherol homologues by IL-mediated liquid-liquid extraction have been investigated by Yang et al. in recent years and will be introduced in more detail hereinafter. Tocopherols are the main compositions of natural Vitamin E and play an important role in human health due to their antioxidative capacity and ability to act as free radical scanveger (Packer, 1984). Tocopherols are widely contained in many vegetables and fruits, and can be extracted as a mixture of four different homologues, viz. α-, β-, γ- and δ-tocopherol. These different tocopherols possess a same basic structure that consists of a chromanol head and an alkyl side chain, while differing in the number and position of methyl groups on the chromanol head (Figure 7). Since α-tocopherol has been reported to possess the highest biological activity, there is a strong need to separate it from the other forms of tocopherol (Hosomi et al., 1997). However, the separation of mixed tocopherol homologues has been very challenging because of their high structural similarity. The available methods for this problem mainly include various chromatographic technologies, such as high performance liquid chromatography (Ng et al., 2004; Pyka & Sliwoik, 2001), capillary electrochromatography

(Chaisuwan et al., 2008), microemulsion electrokinetic chromatography (Chang et al., 2006), nano-liquid chromatography (Fanali et al., 2002), supercritical fluid chromatography (Jiang et al., 2003) and low pressure column chromatography (Wan et al., 2008). Although these technologies could give effective separation for the tocopherol homologues, they are almost only applied at a laboratory scale on account of the low throughput capacity and large consumption of absorbents and solvents or the requirement for special equipment.

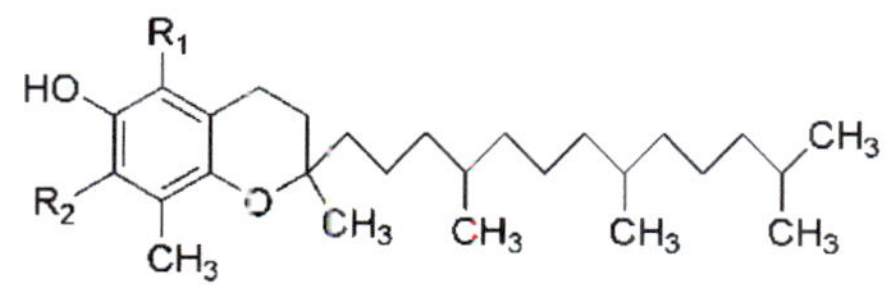

Compound	R_1	R_2
α-Tocopherol	$-CH_3$	$-CH_3$
β-Tocopherol	$-CH_3$	$-H$
γ-Tocopherol	$-H$	$-CH_3$
δ-Tocopherol	$-H$	$-H$

Fig. 7. The structures of tocopherols. Reprinted from reference which is downloaded from pubs.acs.org (Yang et al., 2009).

As an alternative to the chromatographic methods, Yang et al. carried out the liquid-liquid extraction of mixed tocopherol homologues with ILs (Yang et al., 2009). Since the four tocopherol homologues are likely to have different hydrogen bond acidities due to their structural difference in the number and position of methyl groups on the chromanol head, they may be effectively separated with each other by certain ILs via hydrogen bonding interaction. So, 1-butyl-3-methylimidazolium chloride ($[C_4mim]Cl$) was selected as a model extractant for the separation of tocopherol homologues, as it had a relatively strong ability of interacting with hydrogen bond donor molecules (Armstrong et al., 1999; Xie et al., 2008). (Armstrong et al., 1999; Xie et al., 2008) (Armstrong et al., 1999; Xie et al., 2008) (Armstrong et al., 1999; Xie et al., 2008) (Armstrong et al., 1999; Xie et al., 2008) (Armstrong et al., 1999; Xie et al., 2008) $[C_4mim]Cl$ was diluted with methanol before use as it was highly viscous or even solid near room temperature. Non polar solvent, hexane, was selected as the feedstock solvent for mixed tocopherols to form a biphasic system with IL-methanol solution. As the result presented in Table 1, significant difference among the distribution coefficients of each tocopherol homologue was found, which decreased as the order δ-tocopherol > β&γ-tocopherol > α-tocopherol (β and γ-tocopherols were considered as one combined component, because their structure and property were too similar to be distinguished in HPLC analysis). Especially, α-tocopherol exhibited a much lower value of distribution coefficient than the other forms of tocopherols. While most of α-tocopherol remained in the hexane phase, the other forms were much easier to distribute into the extract phase, resulting in large selectivities of α-tocopherol toward the other tocopherols. For example, while the mole ratio of $[C_4mim]Cl$ to methanol was 1:1.3 and temperature was 303.15 K, the selectivity of δ-tocopherol to α-tocopherol was as large as 21.3 and the selectivity of β&γ-tocopherol to α-tocopherol was also higher than 8. Therefore, it is reasonable to consider that α-tocopherol could be selectively obtained from the mixture of various homologues in high purity by multi-stage extractions. And then, it was also showed in Table 1 that the selectivities of δ-tocopherol to α-tocopherol and β&γ-tocopherol to α-tocopherol tended to increase along with the increase of IL's content in the IL-methanol solution, thus suggesting a key role of [C_4mim]Cl in the extractive separation. Since the differences among the tocopherol homologues just lie in the different structural environment around the hydroxyl group, it appears that the selective separation probably results from the hydrogen bonding interaction between the hydroxyl groups and the chloride anion as expected. The methyl group which exists around the hydroxyl group of tocopherol could reduce tocopherol's hydrogen bond acidity and weaken the affinity of tocopherol for [C_4mim]Cl, consequently resulting in a smaller distribution coefficient of α-tocopherol between IL phase and hexane phase.

Mole ratio of [C_4mim]Cl to methanol	Distribution coefficient			Selectivity		
	δ	β&γ	α	δ/α	β&γ/α	δ/β&γ
1:1.3	2.34	0.95	0.11	21.3	8.6	2.5
1:2.7	2.58	1.18	0.16	16.1	7.4	2.2
1:5.4	1.94	1.06	0.24	8.1	4.4	1.8
Pure methanol[a]	1.01	0.94	0.85	1.2	1.1	1.1
Pure [C_4mim]Cl[b]	0.23	0.09	0.01	18.8	7.7	2.4

Table 1. Extractive separation of mixed tocopherols using [C_4mim]Cl at 303.15 K. Initial concentration of tocopherol in hexane (mg/ml): δ 0.54, β&γ 0.53, α 0.11. Volume ratio of two phases: 1:1. a) methanol amount was adjusted to reach an equilibrium phase volume ratio about 1:1. b) carried out at 328.15 K, mixed for 18 h. Reprinted from reference which is downloaded from pubs.acs.org (Yang et al., 2009).

To clarify the hydrogen-bonding interaction mechanism proposed above, the authors investigated the extraction when the structures of IL and tocopherol were modified. First, to demonstrate the role of phenolic hydroxyl group in extraction, distributions of mixed tocopherol acetates in [C_4mim]Cl-methanol-hexane biphasic system were determined and the results were used for a comparison (Table 2). Since the hydroxyl groups on chromanol ring had been converted to ester groups, all of the tocopherol acetates were poor hydrogen bond donors and thus were expected to show weak affinity for [C_4mim]Cl. Well, the experimental results showed in Table 2 accord with this expectation. The distribution coefficients of tocopherol acetates are much smaller than those of tocopherols, as well as sharply decreased selectivities.

Then, to demonstrate the role of IL's anion in extraction, two other kinds of ILs with different anions were investigated for the extractive separation of tocopherol homologues. Consequently, the distribution coefficients of tocopherols in biphasic system varied with the order [C_4mim]BF_4 < [C_4mim]CF_3SO_3 < [C_4mim]Cl under the same conditions ([C_4mim]CF_3SO_3 refers to 1-butyl-3-methylimidazolium trifluoromethanesulfonate), and the selectivities of tocopherol homologues had the same order (Table 3). This order is consistent with the order of these ILs' hydrogen bond basicity strength (Anderson et al., 2002). Combined with the experimental results about tocopherol acetates, it was revealed that the hydrogen-bonding interaction between IL's anion and phenolic hydroxyl group played an important role on the separation selectivity.

Solutes	Distribution coefficient			Selectivity		
	δ	$\beta\&\gamma$	α	δ/α	$\beta\&\gamma/\alpha$	$\delta/\beta\&\gamma$
Tocopherols	2.10	0.89	0.11	19.1	8.1	2.4
Tocopherol acetates	0.002	0.002	0.002	1.0	1.0	1.0

Table 2. Extractive separation of mixed tocopherol acetates and mixed tocopherols using $[C_4mim]Cl$ at 303.15 K. Initial concentration of tocopherol in hexane (mg/ml): δ 2.31, $\beta\&\gamma$ 2.27, α 0.46. Initial concentration of tocopherol acetates in hexane (mg/ml): δ 2.30, $\beta\&\gamma$ 2.25, α 0.46. Volume ratio of two phases: 1:1. Mole ratio of $[C_4mim]Cl$ to methanol: 1:1.3. Reprinted from reference which is downloaded from pubs.acs.org (Yang et al., 2009).

IL	Distribution coefficient			Selectivity		
	δ	$\beta\&\gamma$	α	δ/α	$\beta\&\gamma/\alpha$	$\delta/\beta\&\gamma$
$[C_4mim]BF_4$	0.02	0.01	0.003	6.7	3.3	2.0
$[C_4mim]CF_3SO_3$	0.31	0.15	0.04	7.8	3.8	2.1
$[C_4mim]Cl$	2.34	0.95	0.11	21.3	8.6	2.5

Table 3. Extractive separation of mixed tocopherols using different ILs at 303.15 K. Initial concentration of tocopherol in hexane (mg/ml): δ 0.54, $\beta\&\gamma$ 0.53, α 0.11. Volume ratio of two phases: 1:1. Mole ratio of IL to methanol: 1:1.3. Reprinted from reference which is downloaded from pubs.acs.org (Yang et al., 2009).

The effect of temperature and concentration of tocopherols on extraction was investigated. As the temperature increased, both the distribution coefficients and selectivities tended to decrease. As the concentration of tocopherols increased, the distribution coefficients decreased slowly at the beginning but much more sharply after a certain concentration, where the equilibrium concentration of tocopherols in extract phase started to achieve saturation.

3.4 Regeneration of ionic liquids

The method of regenerating ILs varies according to diverse properties of different extracts. As for the solutes with low boiling points and good thermostability such as aromatics, short-chain aliphatic compounds and so on, distillation is suggested due to negligible vapor pressure of ILs. When it comes to the solutes which are less volatile or sensitive to heat, for example, aromatic sulfur compounds and natural bioactive products, ILs can be recovered by stripping with solvents. With respect to the solutes which can dissociate in water, such as phenolic compounds and organic acids, ILs can be regenerated by stripping with aqueous solution whose pH value is in a reasonable range. Moreover, supercritical CO_2, which is well-known as another variety of green solvent beside IL, can be used to separate the solutes from ILs (Blanchard & Brennecke, 2001).

4. Ionic liquid-solute interaction investigated by quantum chemistry method

Quantum chemistry based method have attracted more and more attention in separation science to get the understanding of underlying mechanism for experimental results and provide reasonable predictions before further experiments. The extractant-solute interaction is one of the most important factors that decide the distribution coefficient and separation selectivity in liquid-liquid extraction, so the investigation of IL-solute interaction by quantum chemistry method is worth carrying out to improve the research and application on IL-based liquid-liquid extraction.

The interaction between thiophene and ILs [C_4mim]PF_6 and [C_4mim]BF_4 were investigated by *ab initio* calculations (Zhou et al., 2008), as a lot of attention has been paid to the extraction of sufur-containing compound from fuel and [C_4mim]PF_6 showed a larger capacity of thiophene than [C_4mim]BF_4. The geometry optimization and electronic charge analysis were carried out at MP2/6-31G level, and energy calculation was finished at MP2/6-311G(d,p) level. Various interacting systems, including thiophene-PF_6^-, thiophene-BF_4^-, thiophene-[C_4mim]$^+$ and neat IL, were analyzed. Columbic attraction was found to dominate the interaction between thiophene and IL. Two factors mainly contributed to the larger capacity of thiophene in [C_4mim]PF_6 than [C_4mim]BF_4, of which one was the relatively stronger electron donation of phosphor atom to fluorine atoms in PF_6^- and the other was the lower compactness degree of [C_4mim]PF_6 compared to that of [C_4mim]BF_4.

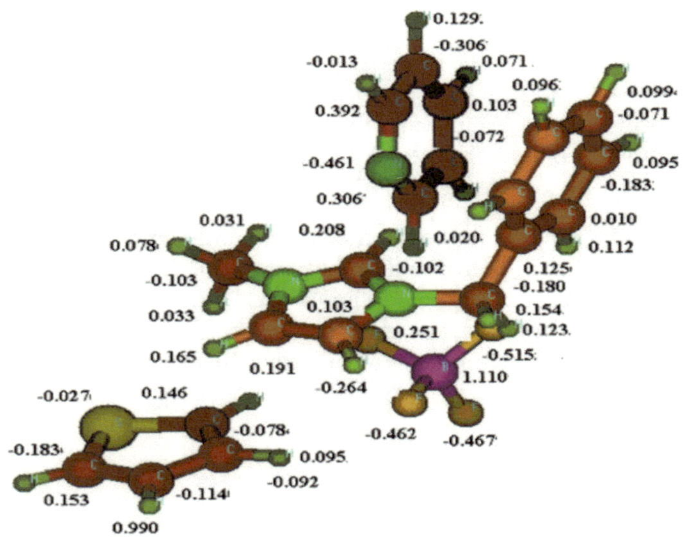

Fig. 8. Optimized geometries and partial charges for [Bemim]BF_4–thiophene–pyridine. Reprinted from reference which is downloaded from www.elsevier.com (Anantharaj & Banerjee, 2010).

The simultaneous interaction of IL with thiophene and pyridine was studied by density functional theory (Figure 8), due to the requirement of simultaneous removal of aromatic sulfur-compounds and nitrogen-compounds from diesel oil (Anantharaj & Banerjee, 2010). Calculations were carried out to determine several global scalar properties such as HOMO/LUMO energies, HOMO-LUMO energy gap, chemical hardness, chemical potential, electronegativity, global hardness, global softness and electrophilicity index for clusters containing ILs with thiophene and pyridine, providing a prediction that the simultaneous removal followed an overall order 1-benzyl-3-Methylimidazolium tetrafluoroborate ([Bemim]BF_4) > 1-butyl-1-methylpyrrolidinium tetrafluoroborate ([C_4mpyr]BF_4) > 1-butyl-4-methylpyridinium hexafluorophosphate ([C_4mpy]PF_6) > 1-butyl-4-methylpyridinium tetrafluoroborate ([C_4mpy]BF_4) > 1-butyl-1-methylpyrrolidinium hexafluorophosphate

([C_4mpyr]PF_6) based on the combination of various scalar parameters. This trend was in accordance with that of infinite dilution activity coefficients of pyridine plus thiophene in ILs predicted by quantum chemical based conductor like screening model-real solvents (COSMO-RS) method. The CH-π interaction between the CH group in the imidazolium cation and the aromatic ring in the sulfur-compound was considered as the main interaction between imidazolium ILs and sulfur-compounds (Anantharaj & Banerjee, 2011).

Investigation was also conducted for the extraction of another common sulfur-compound, benzothiophene, from diesel oil (Varma et al., 2011). Both extraction experiments in IL-hexane biphasic systems and theoretical calculations with COSMO-RS method were performed. Based on the activity coefficients of component calculated by COSMO-RS method, the tie-lines for benzothiophene-IL-hexane ternary systems could be predicted with a good agreement with experimental results (Figure 9). The root mean square deviations were 4.36% and 7.87% for 1-ethyl-3-methylimidazolium ethylsulphate ([C_2mim]$C_2H_5SO_4$) and 1-ethyl-3-methylimidazolium acetate ([C_2mim]CH_3COO) based system, respectively. Besides, it was notable that the authors also utilized the COSMO-RS method for the prediction of quaternary liquid-liquid biphasic systems containing two ILs, benzothiophene and hexane.

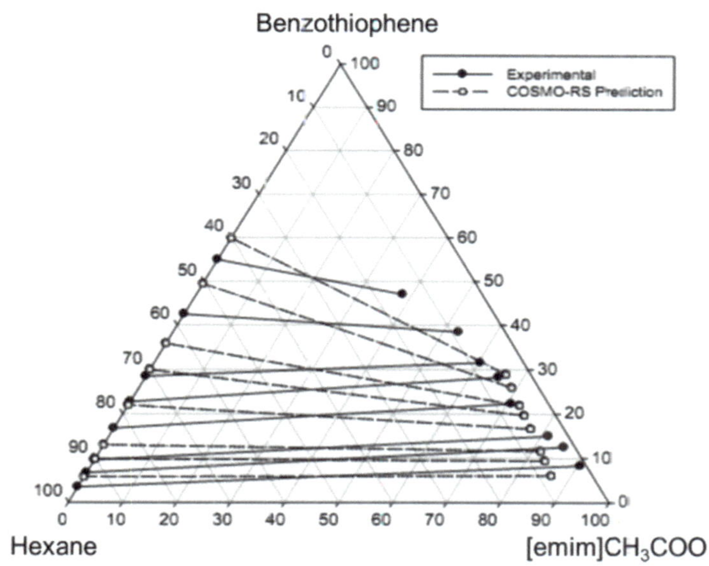

Fig. 9. Experimental vs. COSMO-RS predictions for the composition tie lines of the system [emim]CH_3COO–benzothiophene–n-hexane at 308.15K. Reprinted from reference which is downloaded from www.elsevier.com (Varma et al., 2011).

5. Conclusion

IL-mediated liquid-liquid extraction is receiving more and more attention now. In general, ILs can be regarded as "green" alternatives to common organic solvents, because their negligible volatility can reduce the air pollution and facilitate the regeneration from volatile components. Nevertheless, they can also be superior to common solvents in many cases not due to the nonvolatile features but to some other benefits such as the designable physicochemical properties against specific tasks and unusual liquid-liquid equilibrium

with common solvents. Overall, favorable distribution coefficient and separation selectivity, together with less energy consumption and less volatile pollution, may be achieved in liquid-liquid extraction using IL as extraction solvent.
It should be noted that the environmental impact of IL should be evaluated more carefully. Despite the non-volatility, some amount of ILs may move into water and soil inevitably during the extraction process, especially for a large-scale application in industry. More research is needed on the leach of IL to environment, enrichment of IL in environment, toxicity of IL and degradability or recovery of IL. Dynamics and transport properties in IL-based extraction require more attention, which is of special significance due to the large viscosity of ILs compared with common solvents. Recent work on the hydrodynamic behavior analysis of a rotating disc contactor for aromatics extraction with [mebupy]BF_4 by an experimental and numerical analysis has been reported (Onink et al., 2010). More research should be carried out on the diffusion of components, dispersion of biphasic systems, interfacial transport, process intensification, and so on. Moreover, quantum chemistry and molecular dynamic simulation methods are expected for a further development to improve the design of new ILs and novel applications of current ILs. As the variety of IL is considered a very huge number, it seems unreasonable to develop IL-mediated applications just by numerous experiments.

6. Acknowledgement

The authors are grateful to the financial supports of National Natural Science Foundation of China (20936005, 20806066 and 21076175), Zhejiang Provincial Natural Science Foundation of China (Y4080167) and Ministry of Science and Technology of the People's Republic of China (2009QNA4030).

7. References

Abbott, A. P., Cullis, P. M., Gibson, M. J., Harris, R. C. & Raven, E. (2007). Extraction of Glycerol from Biodiesel into a Eutectic Based Ionic Liquid. *Green Chemistry*, Vol. 9, No. 8, pp. 868-872, ISSN 1463-9262

Anantharaj, R. & Banerjee, T. (2010). Evaluation and Comparison of Global Scalar Properties for the Simultaneous Interaction of Ionic Liquids with Thiophene and Pyridine. *Fluid Phase Equilibria*, Vol. 293, No. 1, pp. 22-31, ISSN 0378-3812

Anantharaj, R. & Banerjee, T. (2011). Quantum Chemical Studies on the Simultaneous Interaction of Thiophene and Pyridine with Ionic Liquid. *AIChE Journal*, Vol. 57, No. 3, pp. 749-764, ISSN 0001-1541

Anderson, J. L., Ding, J., Welton, T. & Armstrong, D. W. (2002). Characterizing Ionic Liquids on the Basis of Multiple Solvation Interactions. *Journal of the American Chemical Society*, Vol. 124, No. 47, pp. 14247-14254, ISSN 0002-7863

Arce, A., Marchiaro, A., Rodriguez, O. & Soto, A. (2006). Essential Oil Terpenless by Extraction Using Organic Solvents or Ionic Liquids. *AIChE Journal*, Vol. 52, No. 6, pp. 2089-2097, ISSN 0001-1541

Arce, A., Rodriguez, O. & Soto, A. (2004). Experimental Determination of Liquid-Liquid Equilibrium Using Ionic Liquids: Tert-Amyl Ethyl Ether Plus Ethanol Plus 1-Octyl-3-Methylimidazolium Chloride System at 298.15 K. *Journal of Chemical and Engineering Data*, Vol. 49, No. 3, pp. 514-517, ISSN 0021-9568

Armstrong, D. W., He, L. F. & Liu, Y. S. (1999). Examination of Ionic Liquids and their Interaction with Molecules, When Used as Stationary Phases in Gas Chromatography. *Analytical Chemistry*, Vol. 71, No. 17, pp. 3873-3876, ISSN 0003-2700

Blanchard, L. A. & Brennecke, J. F. (2001). Recovery of Organic Products from Ionic Liquids Using Supercritical Carbon Dioxide. *Industrial & Engineering Chemistry Research*, Vol. 40, No. 1, pp. 287-292, ISSN 0888-5885

Chaisuwan, P., Nacapricha, D., Wilairat, P., Jiang, Z. & Smith, N. W. (2008). Separation of Alpha-, Beta-, Gamma-, Delta-Tocopherols and Alpha-Tocopherol Acetate on a Pentaerythritol Diacrylate Monostearate-Ethylene Dimethacrylate Monolith by Capillary Electrochromatography. *Electrophoresis*, Vol. 29, No. 11, pp. 2301-2309, ISSN 0173-0835

Chang, L. C., Chang, H. T. & Sun, S. W. (2006). Cyclodextrin-Modified Microemulsion Electrokinetic Chromatography for Separation of Alpha-, Gamma-, Delta-Tocopherol and Alpha-Tocopherol Acetate. *Journal of Chromatography A*, Vol. 1110, No. 1-2, pp. 227-234, ISSN 0021-9673

Chen, J., Duan, L. P., Mi, J. G., Fei, W. Y. & Li, Z. C. (2000). Liquid-Liquid Equilibria of Multi-Component Systems Including *n*-Hexane, *n*-Octane, Benzene, Toluene, Xylene and Sulfolane at 298.15 K and Atmospheric Pressure. *Fluid Phase Equilibria*, Vol. 173, No. 1, pp. 109-119, ISSN 0378-3812

Chen, T. H. & Payne, G. F. (2001). Separation of Alpha- and Delta-Tocopherols Due to an Attenuation of Hydrogen Bonding. *Industrial & Engineering Chemistry Research*, Vol. 40, No. 15, pp. 3413-3417, ISSN 0888-5885

Crosthwaite, J. M., Aki, S., Maginn, E. J. & Brennecke, J. F. (2004). Liquid Phase Behavior of Imidazolium-Based Ionic Liquids with Alcohols. *Journal of Physical Chemistry B*, Vol. 108, No. 16, pp. 5113-5119, ISSN 1520-6106

Cull, S. G., Holbrey, J. D., Vargas-Mora, V., Seddon, K. R. & Lye, G. J. (2000). Room-Temperature Ionic Liquids as Replacements for Organic Solvents in Multiphase Bioprocess Operations. *Biotechnology and Bioengineering*, Vol. 69, No. 2, pp. 227-233, ISSN 0006-3592

Deng, F. Z. & Guo, D. F. (2006). Extraction Separation of Bovine Serum Albumin in Ionic Liquid Aqueous Two-Phase System. *Chinese Journal of Analytical Chemistry*, Vol. 34, No. 10, pp. 1451-1453, ISSN 0253-3820

Domanska, U., Pobudkowska, A. & Eckert, F. (2006). Liquid-Liquid Equilibria in the Binary Systems (1,3-Dimethylimidazolium, or 1-Butyl-3-Methylimidazolium Methylsulfate Plus Hydrocarbons). *Green Chemistry*, Vol. 8, No. 3, pp. 268-276, ISSN 1463-9262

Egorov, V. M., Smirnova, S. V. & Pletnev, I. V. (2008). Highly Efficient Extraction of Phenols and Aromatic Amines into Novel Ionic Liquids Incorporating Quaternary Ammonium Cation. *Separation and Purification Technology*, Vol. 63, No. 3, pp. 710-715, ISSN 1383-5866

Fan, J., Fan, Y. C., Wang, J. J. & Cui, F. L. (2006). Extraction of Aminobenzene Sulfonic Acids with Room-Temperature Ionic Liquids. *Acta Chimica Sinica*, Vol. 64, No. 14, pp. 1495-1499, ISSN 0567-7351

Fanali, S., Catarcini, P., Quaglia, M. G., Camera, E., Rinaldi, M. & Picardo, M. (2002). Separation of Delta-, Gamma- And Alpha-Tocopherols by Cec. *Journal of Pharmaceutical and Biomedical Analysis*, Vol. 29, No. 6, pp. 973-979, ISSN 0731-7085

Fukaya, Y., Sekikawa, K., Murata, K., Nakamura, N. & Ohno, H. (2007). Miscibility and Phase Behavior of Water-Dicarboxylic Acid Type Ionic Liquid Mixed Systems. *Chemical Communications*, No. 29, pp. 3089-3091, ISSN 1359-7345

Fukumoto, K. & Ohno, H. (2007). LCST-Type Phase Changes of a Mixture of Water and Ionic Liquids Derived from Amino Acids. *Angewandte Chemie-International Edition*, Vol. 46, No. 11, pp. 1852-1855, ISSN 1433-7851

Han, X. & Armstrong, D. W. (2007). Ionic Liquids in Separations. *Accounts of Chemical Research*, Vol. 40, No. 11, pp. 1079-1086, ISSN 0001-4842

Holbrey, J. D., Lopez-Martin, I., Rothenberg, G., Seddon, K. R., Silvero, G. & Zheng, X. (2008). Desulfurisation of Oils Using Ionic Liquids: Selection of Cationic and Anionic Components to Enhance Extraction Efficiency. *Green Chemistry*, Vol. 10, No. 1, pp. 87-92, ISSN 1463-9262

Hosomi, A., Arita, M., Sato, Y., Kiyose, C., Ueda, T., Igarashi, O., Arai, H. & Inoue, K. (1997). Affinity for Alpha-Tocopherol Transfer Protein as a Determinant of the Biological Activities of Vitamin E Analogs. *Febs Letters*, Vol. 409, No. 1, pp. 105-108, ISSN 0014-5793

Hu, X. S., Li, Y. X., Cui, D. & Chen, B. H. (2008). Separation of Ethyl Acetate and Ethanol by Room Temperature Ionic Liquids with the Tetrafluoroborate Anion. *Journal of Chemical and Engineering Data*, Vol. 53, No. 2, pp. 427-433, ISSN 0021-9568

Huddleston, J. G., Willauer, H. D., Swatloski, R. P., Visser, A. E. & Rogers, R. D. (1998). Room Temperature Ionic Liquids as Novel Media for 'Clean' Liquid-Liquid Extraction. *Chemical Communications*, Vol. 16, No. 16, pp. 1765-1766, ISSN 1359-7345

Jiang, C. W., Ren, Q. L. & Wu, P. D. (2003). Study on Retention Factor and Resolution of Tocopherols by Supercritical Fluid Chromatography. *Journal of Chromatography A*, Vol. 1005, No. 1-2, pp. 155-164, ISSN 0021-9673

Khachatryan, K. S., Smirnova, S. V., Torocheshnikova, I. I., Shvedene, N. V., Formanovsky, A. A. & Pletnev, I. V. (2005). Solvent Extraction and Extraction-Voltammetric Determination of Phenols Using Room Temperature Ionic Liquid. *Analytical and Bioanalytical Chemistry*, Vol. 381, No. 2, pp. 464-470, ISSN 1618-2642

Krishna, R., Goswami, A. N., Nanoti, S. M., Rawat, B. S., Khanna, M. K. & Dobhal, J. (1987). Extraction of Aromatics from 63-69-Degrees-C Naphtha Fraction for Food Grade Hexane Production Using Sulfolane and NMP as Solvents. *Indian Journal of Technology*, Vol. 25, No. 12, pp. 602-606, ISSN 0019-5669

Lachwa, J., Szydlowski, J., Makowska, A., Seddon, K. R., Esperanca, J., Guedes, H. & Rebelo, L. (2006). Changing from an Unusual High-Temperature Demixing to a UCST-Type in Mixtures of 1-Alkyl-3-Methylimidazolium Bis{(Trifluoromethyl)Sulfonyl}Amide and Arenes. *Green Chemistry*, Vol. 8, No. 3, pp. 262-267, ISSN 1463-9262

Lachwa, J., Szydlowski, J., Najdanovic-Visak, V., Rebelo, L., Seddon, K. R., Da Ponte, M. N., Esperanca, J. & Guedes, H. (2005). Evidence for Lower Critical Solution Behavior in Ionic Liquid Solutions. *Journal of the American Chemical Society*, Vol. 127, No. 18, pp. 6542-6543, ISSN 0002-7863

Letcher, T. M. & Deenadayalu, N. (2003). Ternary Liquid-Liquid Equilibria for Mixtures of 1-Methyl-3-Octyl-Imidazolium Chloride Plus Benzene Plus an Alkane at T-298.2 K and 1 Atm. *Journal of Chemical Thermodynamics*, Vol. 35, No. 1, pp. 67-76, ISSN 0021-9614

Li, C. P., Xin, B. P., Xu, W. G. & Zhang, Q. (2007). Study on the Extraction of Dyes into a Room-Temperature Ionic Liquid and their Mechanisms. *Journal of Chemical Technology and Biotechnology*, Vol. 82, No. 2, pp. 196-204, ISSN 0268-2575

Li, M., Pham, P. J., Jr. Pittman, C. U. & Li, T. (2008). Selective Solid-Phase Extraction of α-Tocopherol by Functionalized Ionic Liquid-Modified Mesoporous Sba-15 Adsorbent. *Analytical Sciences*, Vol. 24, No. 10, pp. 1245-1250, ISSN 0910-6340

Liu, Q. F., Hu, X. S., Wang, Y. H., Yang, P., Xia, H. S., Yu, J. & Liu, H. Z. (2005). Extraction of Penicillin G by Aqueous Two-Phase System of [Bmim]BF_4/NaH_2PO_4. *Chinese Science Bulletin*, Vol. 50, No. 15, pp. 1582-1585, ISSN 1001-6538

Marcus, Y. (2010). The Cohesive Energy of Molten Salts and its Density. *Journal of Chemical Thermodynamics*, Vol. 42, No. 1, pp. 60-64, ISSN 0021-9614

Meindersma, G. W. & de Haan, A. B. (2008). Conceptual Process Design for Aromatic/Aliphatic Separation with Ionic Liquids. *Chemical Engineering Research & Design*, Vol. 86, No. 7A, pp. 745-752, ISSN 0263-8762

Meindersma, G. W., Podt, A. & de Haan, A. B. (2005). Selection of Ionic Liquids for the Extraction of Aromatic Hydrocarbons from Aromatic/Aliphatic Mixtures. *Fuel Processing Technology*, Vol. 87, No. 1, pp. 59-70, ISSN 0378-3820

Meindersma, G. W., Podt, A. & de Haan, A. B. (2006). Ternary Liquid-Liquid Equilibria for Mixtures of Toluene Plus *n*-Heptane Plus an Ionic Liquid. *Fluid Phase Equilibria*, Vol. 247, No. 1-2, pp. 158-168, ISSN 0378-3812

Meindersma, G. W., Podt, A., Klaren, M. B. & De Haan, A. B. (2006). Separation of Aromatic and Aliphatic Hydrocarbons with Ionic Liquids. *Chemical Engineering Communications*, Vol. 193, No. 11, pp. 1384-1396, ISSN 0098-6445

Najdanovic-Visak, V., Esperanca, J., Rebelo, L., Da Ponte, M. N., Guedes, H., Seddon, K. R. & Szydlowski, J. (2002). Phase Behaviour of Room Temperature Ionic Liquid Solutions: An Unusually Large Co-Solvent Effect in (Water Plus Ethanol). *Physical Chemistry Chemical Physics*, Vol. 4, No. 10, pp. 1701-1703, ISSN 1463-9076

Ng, M. H., Choo, Y. M., Ma, A. N., Cheng, H. C. & Hashim, M. A. (2004). Separation of Vitamin E (Tocopherol, Tocotrienol, and Tocomonoenol) in Palm Oil. *Lipids*, Vol. 39, No. 10, pp. 1031-1035, ISSN 0024-4201

Onink, F., Drumm, C., Meindersma, G. W., Bart, H. J. & de Haan, A. B. (2010). Hydrodynamic Behavior Analysis of a Rotating Disc Contactor for Aromatics Extraction with 4-Methyl-Butyl-Pyridinium Center Dot BF_4 by CFD. *Chemical Engineering Journal*, Vol. 160, No. 2, pp. 511-521, ISSN 1385-8947

Packer, L. (1984). Protective Role of Vitamin E in Biological Systems. *American Journal of Clinical Nutrition*, Vol. 53, No. pp. 1050-1055, ISSN 0002-9165

Pereiro, A. B. & Rodriguez, A. (2008). Azeotrope-Breaking Using [Bmim] [$MeSO_4$] Ionic Liquid in an Extraction Column. *Separation and Purification Technology*, Vol. 62, No. 3, pp. 733-738, ISSN 1383-5866

Pereiro, A. B., Tojo, E., Rodriguez, A., Canosa, J. & Tojo, J. (2006). HmimPF_6 Ionic Liquid that Separates the Azeotropic Mixture Ethanol Plus Heptane. *Green Chemistry*, Vol. 8, No. 3, pp. 307-310, ISSN 1463-9262

Pyka, A. & Sliwoik, J. (2001). Chromatographic Separation of Tocopherols. *Journal of Chromatography A*, Vol. 935, No. 1-2, pp. 71-76, ISSN 0021-9673

Rajendran, A., Paredes, G. & Mazzotti, M. (2009). Simulated Moving Bed Chromatography for the Separation of Enantiomers. *Journal of Chromatography A*. Vol. 1216, No. 4, pp. 709-738, ISSN 0021-9673

Shiflett, M. B., Niehaus, A. & Yokozeki, A. (2009). Liquid-Liquid Equilibria in Binary Mixtures Containing Chlorobenzene, Bromobenzene, and Iodobenzene with Ionic

Liquid 1-Ethyl-3-Methylimidazolim Bis(Trifluoromethylsulfonyl)Imide. *Journal of Chemical and Engineering Data*, Vol. 54, No. 7, pp. 2090-2094, ISSN 0021-9568

Sticher, O. (2008). Natural Product Isolation. *Natural Product Reports*, Vol. 25, No. 3, pp. 517-554, ISSN 0265-0568

Swatloski, R. P., Visser, A. E., Reichert, W. M., Broker, G. A., Farina, L. M., Holbrey, J. D. & Rogers, R. D. (2002). On the Solubilization of Water with Ethanol in Hydrophobic Hexafluorophosphate Ionic Liquids. *Green Chemistry*, Vol. 4, No. 2, pp. 81-87, ISSN 1463-9262

Taylor, L. T. (2009). Supercritical Fluid Chromatography for the 21St Century. *Journal of Supercritical Fluids*, Vol. 47, No. 3Sp. Iss. SI, pp. 566-573, ISSN 0896-8446

Treybal, R. E. (1951). *Liquid Extraction*, McGraw-Hill Book Co., ISBN 1443724742, New York; Toronto; London

Tzeng, Y. P., Shen, C. W. & Yu, T. (2008). Liquid-Liquid Extraction of Lysozyme Using a Dye-Modified Ionic Liquid. *Journal of Chromatography A*, Vol. 1193, No. 1-2, pp. 1-6, ISSN 0021-9673

Varma, N. R., Ramalingam, A. & Banerjee, T. (2011). Experiments, Correlations and COSMO-RS Predictions for the Extraction of Benzothiophene from *n*-Hexane Using Imidazolium-Based Ionic Liquids. *Chemical Engineering Journal*, Vol. 166, No. 1, pp. 30-39, ISSN 1385-8947

Vijayaraghavan, R., Vedaraman, N., Surianarayanan, M. & MacFarlane, D. R. (2006). Extraction and Recovery of Azo Dyes into an Ionic Liquid. *Talanta*, Vol. 69, No. 5, pp. 1059-1062, ISSN 0039-9140

Visser, A. E., Swatloski, R. P. & Rogers, R. D. (2000). PH-Dependent Partitioning in Room Temperature Ionic Liquids Provides a Link to Traditional Solvent Extraction Behavior. *Green Chemistry*, Vol. 2, No. 1, pp. 1-4, ISSN 1463-9262

Wan, J. C., Zhang, W. N., Jiang, B., Guo, Y. H. & Hu, C. R. (2008). Separation of Individual Tocopherols from Soybean Distillate by Low Pressure Column Chromatography. *Journal of the American Oil Chemists Society*, Vol. 85, No. 4, pp. 331-338, ISSN 0003-021X

Wang, J. H., Cheng, D. H., Chen, X. W., Du, Z. & Fang, Z. L. (2007). Direct Extraction of Double-Stranded DNA into Ionic Liquid 1-Butyl-3-Methylimidazolium Hexafluorophosphate and its Quantification. *Analytical Chemistry*, Vol. 79, No. 2, pp. 620-625, ISSN 0003-2700

Welton, T. (1999). Room-Temperature Ionic Liquids. Solvents for Synthesis and Catalysis. *Chemical Reviews*, Vol. 99, No. 8, pp. 2071-2083, ISSN 0009-2265

Xie, L. L., Favre-Reguillon, A., Wang, X. X., Fu, X., Pellet-Rostaing, E., Toussaint, G., Geantet, C., Vrinat, M. & Lemaire, M. (2008). Selective Extraction of Neutral Nitrogen Compounds Found in Diesel Feed by 1-Butyl-3-Methyl-Imidazolium Chloride. *Green Chemistry*, Vol. 10, No. 5, pp. 524-531, ISSN 1463-9262

Yang, Q. W., Xing, H. B., Cao, Y. F., Su, B. G., Yang, Y. W. & Ren, Q. L. (2009). Selective Separation of Tocopherol Homologues by Liquid-Liquid Extraction Using Ionic Liquids. *Industrial & Engineering Chemistry Research*, Vol. 48, No. 13, pp. 6417-6422, ISSN 0888-5885

Zhou, J. X., Mao, J. B. & Zhang, S. G. (2008). *Ab Initio* Calculations of the Interaction between Thiophene and Ionic Liquids. *Fuel Processing Technology*, Vol. 89, No. 12, pp. 1456-1460, ISSN 0378-3820

17

Liquid Matrices in MALDI-MS

Yuko Fukuyama
Koichi Tanaka Mass Spectrometry Research Laboratory,
Shimadzu Corporation
Japan

1. Introduction

Matrix-assisted laser desorption/ionization mass spectrometry (MALDI-MS) (Karas, M. & Hillenkamp, F., 1988; Tanaka, K. et al., 1988), in addition to elecrospray ionization mass spectrometry (ESI-MS) (Fenn J. B. et al., 1989) have developed into practical analytical tools in proteomics and glycomics as both have higher throughput and sensitivity than previous mass spectrometric techniques in this area. In MALDI-MS, one benefit is the detection of mainly singly charged ions, whereas ions are detected in a multiply charged state in ESI-MS. This benefit enables easy interpretation of the mass spectra acquired and thus MALDI-MS finds particular use in mixture analysis. On the other hand, a weak point of MALDI is difficulty in selecting appropriate matrices and preparation methods for each sample. Therefore development of an analyte-specific, polarity independent matrix is still required.

In MALDI the matrix is essential for ionizing the analytes that are then detected as ion peaks in the mass spectrometer. Thus the general structure of matrices has a benzene ring for absorbing N_2 laser energy and hydroxyl groups to have high affinity with analytes such as peptides, proteins and carbohydrates. In addition, the matrices have some functional groups to work as a proton donor and/or a proton acceptor to help with ionizing analytes. So far, many matrices have been reported for biopolymers and synthetic polymers. The most common matrices for biopolymers are α-cyano-4-hydroxycinnamic acid (CHCA) for peptides, 3,5-dimethoxy-4-hydroxycinnamic acid (sinapic acid or sinapinic acid, SA) for proteins and 2,5-dihydroxybenzoic acid (DHB) for carbohydrates (Fig. 1). Typically, the matrices are dissolved in solvent and then mixed with analytes solution on MALDI sample plate. After volatilization of the solvent, solid crystals containing analytes and matrix are formed on the plate. The crystals are irradiated by the laser in mass spectrometer to ionize the analytes. On the other hand, a "liquid matrix" forms a liquid droplet containing analytes and matrix on the plate. We show here the properties of liquid matrices and differences from conventional solid matrices.

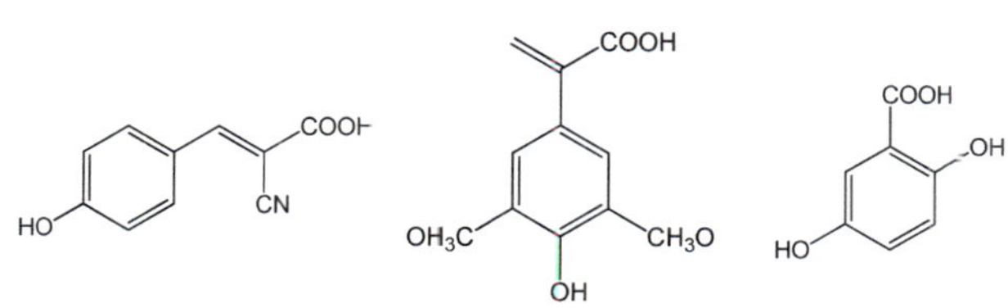

Fig. 1. Conventional MALDI matrices: CHCA, SA and DHB.

2. Conventional MALDI matrices

CHCA is the most common matrix for peptides and classified as a "hot" matrix which induced fragmentation. Typically, CHCA solution is prepared by dissolving CHCA in water/acetonitrile (1:1 v/v) with 0.1% trifluoroacetic acid (TFA) at 10 mg/mL. 0.5 μL of the CHCA solution is mixed on the plate with 0.5 μL of peptide solution dissolved in 0.1% TFA water. After volatilization of the solvent, solid crystals containing the analyte and CHCA are formed. As the analyte is co-crystallized with an excess of solid matrix, the crystal state depends on the matrix species. CHCA is known to make a thin layer of many small granular white crystals. Analyte ions are uniformly detected across the co-crystal spot when the laser irradiates the co-crystal surface. However, as a result of the nature of the thin co-crystal surface, upon irradiation with the laser the intensity of the ions from any given point in the co-crystal spot begins to decrease as the laser penetrates the thin layer. Thus, CHCA have been useful for automatic analyses using a raster function that continually moves the laser to a fresh position on the co-crystal surface.

On the other hand, DHB is the most common matrix for carbohydrates and classified as "cool" matrix compared with CHCA. For example, DHB solution is prepared by dissolving DHB in 50% aqueous acetonitrile at 10 mg/mL. 0.5 μL of the DHB solution was mixed on the plate with 0.5 μL of carbohydrate solution dissolved in water. After volatilization of the solvent, solid crystals containing the analytes and DHB are formed. DHB is known to make needle-shaped crystals. Analyte ions are detected from only a few small areas called "sweet spot" or "hot spot". This has made the application of DHB difficult because of long measurement time to find the "sweet spot" and poor reproducibility of the results.

3. Liquid matrices

Ionic liquid matrices (ILMs) introduced by Armstrong et al. were reported to have not only the property to make a homogeneous spot surface of analyte-matrix mixture but also the suitable properties for ionization of analytes (Anderson, J. L. et al., 2002; Armstrong, D. W. et al., 1999, 2001; Carda-Broch, S. et al., 2003). The essential point is that the ILMs consist of a conventional solid MALDI matrix, *e.g.*, CHCA, DHB or SA and an organic base, *e.g.*, tributylamine, pyridine or 1-methylimidazole which enables a relative state of 'liquidity' under vacuum conditions (Tholey, A. & Heinzle, E., 2006). The constituent solid matrices probably contribute to the ionization process. Several ILMs have been described and increased sensitivity analyses at the fmol or amol level have been reported for peptides and carbohydrates (Bungert, D. et al., 2004; Cramer, R. & Corless, S., 2005; Crank, J. A. & Armstrong, D. W., 2009; Fukuyama, Y. et al., 2008a; Kaneshiro, K. et al., 2011; Laremore, T. N. et al., 2006, 2007; Mank, M. et al., 2004).

Our group reported high sensitivity analyses of oligosaccharides using an optimized 1,1,3,3-tetramethylguanidine (TMG) salt of *p*-coumaric acid (G_3CA) and a TMG salt of CHCA (G_2CHCA) as liquid matrices (Fig. 2) (Fukuyama, Y. et al., 2008a). G_3CA was made by mixing *p*-coumaric acid (CA) with TMG at a 1:3 molar ratio in methanol. After evaporation of the methanol, they were dissolved in methanol at 9 mg/mL to be used as matrix solution. The matrix solution was mixed with analyte solution at a 1:1 ratio (v/v). 0.5 μL of the analyte-matrix mixture solution was spotted on a MALDI plate. After evaporating solvent, a small liquid droplet of analyte-matrix mixture was remained on the plate (Fig. 2) (Fukuyama, Y. et al., 2008a). These small droplets were irradiated by UV laser light and analyzed in the mass spectrometer.

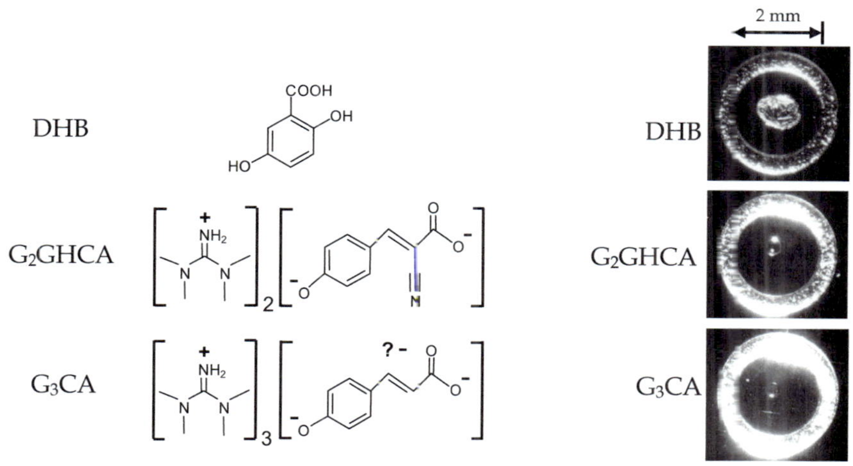

Fig. 2. Structure of matrices (left), and stereoscopic microscope photographs of analyte-matrix mixtures on a mirror-polished stainless-steel plate (right). The location of the third negative charge on the CA moiety in G_3CA has not been clarified.

3.1 Homogeneity

Fig. 3 is ion intensity distributions on analyte-matrix surfaces of ILMs (G_2CHCA and G_3CA) and DHB for $[M-Na]^-$ of Analyte-1 (Table 1, Fig. 4) using an incorporated automatic analytical function (Fukuyama, Y. et al., 2008a). The ion $[M-Na]^-$ was uniformly detected across the sample spot when using the ILMs whereas it was detected in only a few small areas called "sweet-spots" using DHB (Fig. 3). G_2CHCA gave near-perfect uniform distribution (Fig. 3). It was noted that analyses using the ILMs were carried out easily and rapidly when compared to DHB.

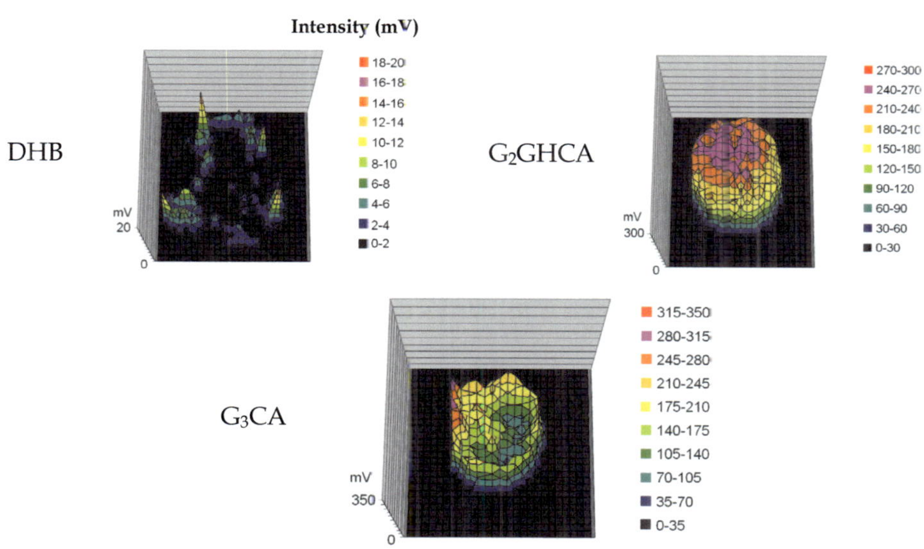

Fig. 3. Ion intensity distributions of analyte-matrix surface (810 μm × 810 μm) using DHB and ILMs (G_2CHCA and G_3CA) for $[M-Na]^-$ of Analyte-1 (in Table 1).

Analyte-#	oligosaccharides	n^a	FW
1	Neocarratetraose-41,3-di-O-sulfate (2Na$^+$)	2	834.6
2	Neocarrahexaose-41,3,5-tri-O-sulfate (3Na$^+$)	3	1242.9
3	Neocarrahexaose-24,41,3,5-tetra-sulfate (4Na$^+$)	4	1345.0
4	Neocarradodecaose-41,3,5,7,9,11-hexa-sulfate (6Na$^+$)	6	2467.9

[a] *n* corresponds to the number of repeating units in the oligosaccharides (see Fig. 4).

Table 1. Carrageenan oligosaccharides.

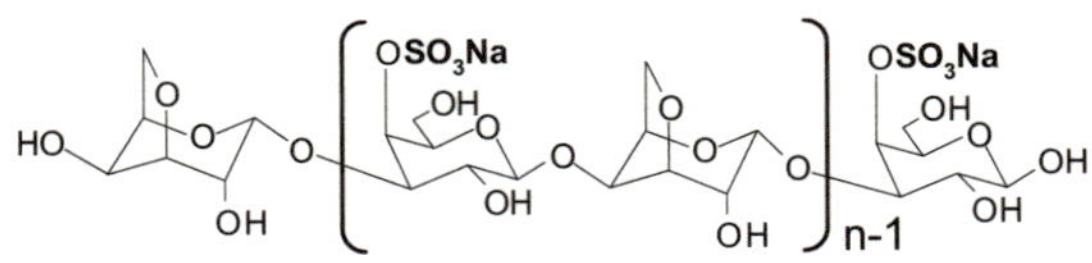

Fig. 4. Structure of carrageenan oligosaccharides. *n* corresponds to the number of repeating units in the oligosaccharides (see Table 1).

3.2 High sensitivity analyses of carbohydrates

Our group demonstrated the highly sensitive detection of sulfated/sialylated/neutral oligosaccharide molecules (Table 1 and 2, Fig. 4) using G_3CA and G_2CHCA (Fukuyama, Y. et al., 2008a). As a result, all oligosaccharides were detected with high sensitivity (*e.g.* 1 fmol) using the ILMs, especially using G_3CA, in both positive and negative ion extraction modes (Table 3 and 4) (Fukuyama, Y. et al., 2008a).

Analyte-#	oligosaccharides	M.W.
5	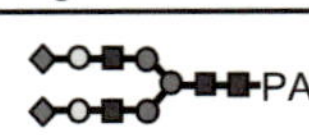	2302.1
6	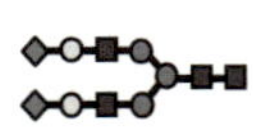	2224.0
7	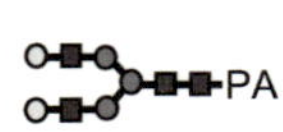	1719.4
8	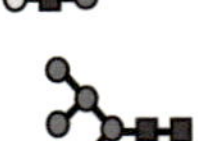	1235.1
9		1865.8
10	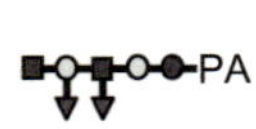	1281.3

GlcNAc: ■ ; Galactose: O ; Glucose: ● ; Mannose: ● ;
Fucose: ▼ ; NeuAc: ◆ ; PA: Aminopyridine

Table 2. Sialylated and neutral oligosaccharides.

Analyte-#	Positive: [M+Na]$^+$			Negative: [M-Na]$^-$		
	DHB	G_2CHCA	G_3CA	DHB	G_2CHCA	G_3CA
1 (2S)	1 p	100 f	1 f	100 f	10 f	1 f
2 (3S)	1 p	10 f	5 f	100 f	1 f	1 f
3 (4S)	ND	100 f	10 f	1 p	10 f	1 f
4 (6S)	ND	10 f	1 f	ND	1 f	1 f

[a]The highest sensitivity (mol/well) is shown for each analysis when 1 pmol – 1 fmol/well analytes were analyzed using MALDI-QIT-TOF mass spectrometer. ND denotes that analyte molecular ions are not detected.

Table 3. Detection limits of sulfated oligosaccharide molecules (Analyte-1 – 4 in Table 1) [a]

Analyte-#	Positive: [M+Na]$^+$			Negative: [M-Na]$^-$		
	DHB	G_2CHCA	G_3CA	DHB	G_2CHCA	G_3CA
5	1 p	10 f	10 f	50 f	1 f	1 f
6	ND	100 f	100 f	1 p	100 f	100 f
7	100 f	10 f	10 f	100 f	25 f	10 f
8	100 f	10 f	10 f	ND	ND	ND
9	10 f	1 f	10 f	10 f	100 f	1 f
10	10 f	1 f	1 f	10 f	100 f	10 f

[a] The highest sensitivity (mol/well) is shown for each analysis when 1 pmol – 1 fmol/well analytes were analyzed using MALDI-QIT-TOF mass spectrometer. ND denotes that analyte molecular ions are not detected.

Table 4. Detection limits of sialylated and neutral oligosaccharide molecules (Analyte-5 – 10 in Table 2).[a]

Furthermore, 3-aminoquinoline/CHCA (3-AQ/CHCA) reported by Kumar et al. (Kolli, V. S. K. et al, 1996) is one of most widely used liquid matrices (Fig. 5). It was reported to work for highly sensitive analyses of peptides and proteins, additionally for oligosaccharides (Fukuyama, Y. et al., 2009). Several tens of attomole of oligosaccharides were detected with this matrix (Table 5) (Fukuyama, Y. et al., 2009).

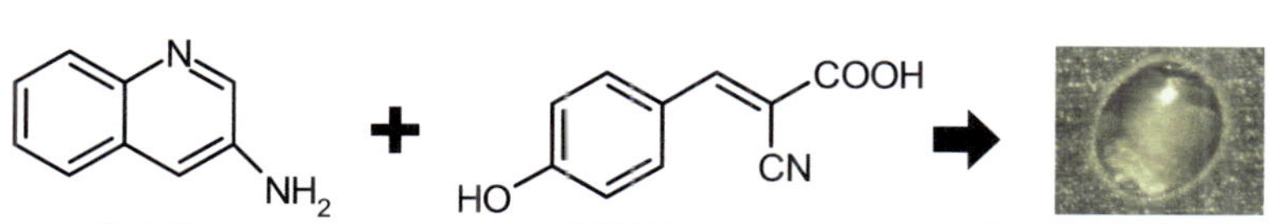

3-AQ, CHCA, 3-AQ/CHCA

Fig. 5. Liquid matrix: 3-AQ/CHCA.

				oligosaccharides	
				I	II
MALDI-QIT-TOFMS		pos	DHB	50f	10f
			3-AQ/CHCA	50a	100a
		neg	DHB	100f	100f
			3-AQ/CHCA	1f	5f
MALDI-TOFMS	lin	pos	CHCA	1p	50f
			DHB	500f	50f
			3-AQ/CHCA	500a	50a
		neg	CHCA	500f	100f
			DHB	50f	50f
			3-AQ/CHCA	5f	500a

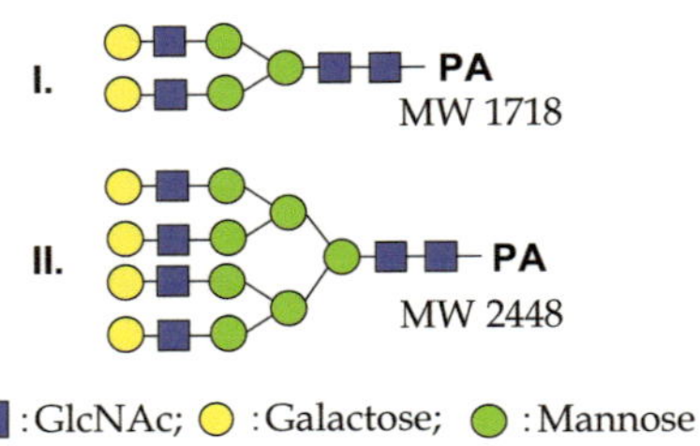

[a] The highest sensitivity (mol/well) is shown for each analysis when 10 amol - 1 pmol/well analytes were analyzed. ND denotes that analyte molecular ions are not detected.

Table 5. Detection limits of PA-labeled oligosaccharide molecules with 3-AQ/CHCA and conventional solid matrices.[a]

Additionally, a labeling function of this matrix allowed the analyses of oligosaccharides on attomole level (see section 3.5) (Kaneshiro, K. et al., 2011).

3.3 Suppression of dissociation of acidic groups acids and acidic sugars

In sulfated and sialylated oligosaccharide analyses using MALDI-MS, dissociation of sulfated groups and sialic acids are observed and thus intact molecular ions are difficult to detect. Fig. 6 and Fig. 7 show results of evaluation of sulfated (Table 1, Fig. 4) and sialylated (Table 2) oligosaccharides using DHB and optimized G_2CHCA and G_3CA (Fukuyama, Y. et al., 2008a). As a result, dissociation of sulfate groups and sialic acids were suppressed using the liquid matrices in both positive and negative ion extraction modes.

3.4 Durability of ionization

One concern in automatic MALDI analyses is amount of time dedicated to the raster scanning function during automated analyses on crystalline solid matrices such as CHCA. We have reported comparison of an optimized liquid matrix CHCA butylamine salt (CHCAB) and CHCA for uniformity and durability of ionization (Fig. 8) (Fukuyama, Y. et al., 2010). Homogeneous property was confirmed using CHCA, however durability of ionization was not so long (<100 laser shots) (Fig. 8). On the other hand, peptide was ionized from whole area of analyte-matirx mixture on plate using CHCAB (Fig. 8). All these peaks showed long ionizing durability (>500 laser shots). The homogeneous property and long durability of ionization using the liquid matrix enabled irradiation on just 1-3 discrete positions with a laser for MS and MS/MS (Fukuyama, Y. et al., 2010). As a result, it was confirmed that automatic analytical time became shorter than that required using CHCA (one-third or one-quarter of the total analytical time) (Fukuyama, Y. et al., 2010).

3.5 Reaction in liquid (digestion, separation and labeling)

In addition to oligosaccharide analysis, optimized liquid matrices were applied to glycopeptide analysis. Ribonuclease B (RNase B) digests containing several peptides and glycopeptides were analyzed using DHB and G_3CA (Fukuyama, Y. et al., 2008a). The results

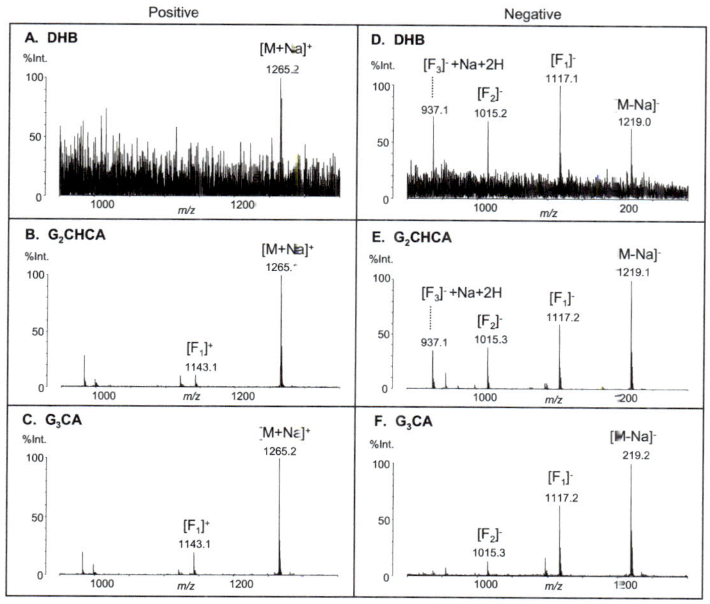

Fig. 6. Positive and negative ion mass spectra of Analyte-2 (100 fmol/well) with DHB (A and D), G_2CHCA (B and E) and G_3CA (C and F) using MALDI-QIT-TOF mass spectrometer. $[Fn]^+ = [M+Na-nSO_3Na+nH]^+$. $[Fn]^- = [M-Na-nSO_3Na+nH]^-$.

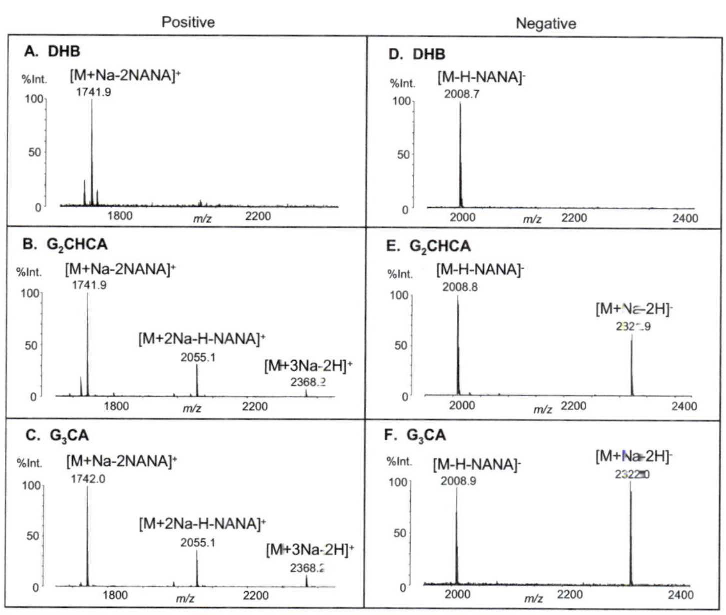

Fig. 7. Positive and negative ion mass spectra of Analyte-5 (100 fmol/well) with DHB (A and D), G_2CHCA (B and E) and G_3CA (C and F) using MALDI-QIT-TOF mass spectrometer. Dissociation of N-acetylneuraminic acid (NANA; sialic acid) was suppressed using G_2CHCA or G_3CA (B, C, E and F).

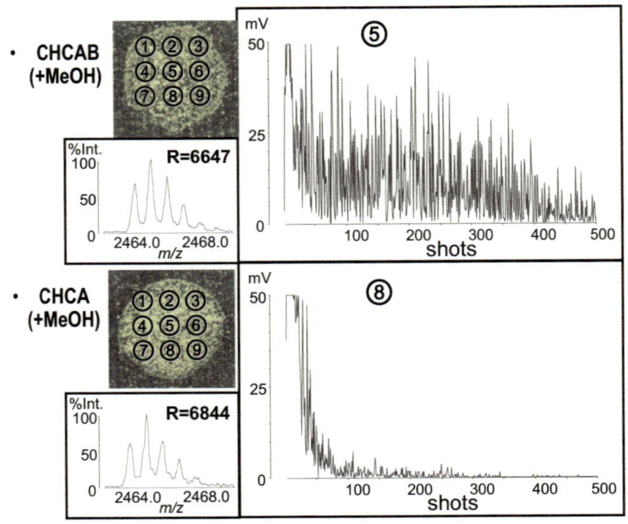

Fig. 8. Photos of analyte-matrix mixtures using optimized CHCAB and CHCA for ACTH18-39, and ionizing durability graphs of ⑤ for CHCAB and ⑧ for CHCA by manual evaluation using MALDI-TOF mass spectrometer. 500 shots of laser was irradiated to each of ①~⑨ on the photos. Almost same results were obtained for ①~⑨. CHCAB and CHCA had homegeneous property especially by adding MeOH.

shows that a series of five glycopeptides ions were preferentially detected with both positive and negative ion extractions using ILMs (Figure 9B, C, D, F, G and H) whereas they were detected only with positive ion extraction using DHB (Figure 9A and E) (Fukuyama, Y. et al., 2008a). On this basis, it may be possible to confirm the presence or absence of glycopeptides in a mixture by comparing the negative ion mass spectra obtained using the liquid matrix and DHB.

Furthermore, we confirmed on-target digestion of neutral glycoproteins: RNase B and asialofetuin and acidic glycoprotein: fetuin using liquid matrix GCA or G_3CA (Fukuyama, Y. et al., 2008b). Glycoprotein was dissolved in water and mixed with the liquid matrices and dithiothreitol (DTT). The mixture was dropped on a stainless-steel plate for reduction at 60°C for 1h. Then, trypsin was added on the mixture for digestion at 37°C for 3hs. Finally obtained mixtures were analyzed with MALDI-QIT-TOF mass spectrometer. Glycopeptide ions of RNase B digests using liquid matrix were preferentially detected in both positive and negative ion modes (Fukuyama, Y. et al., 2008b). However, glycopeptides of asialofetuin were detected with low S/N ratio and dissociation of sialic acids was observed for fetuin (Fukuyama, Y. et al., 2008b). Although the on-target digestion requires further development, these results suggest that liquid matrices can work not only as a matrix, but also a reaction medium for rapid digestion and MS analyses.

On the other hand, on-target desalting within 3-AQ/CHCA droplet and selective detection of glycopeptides ions was reported (Sekiya, S. et al., 2008). Sekiya et al. found that 3-AQ/CHCA has a property to concentrate hydrophilic compounds on a small surface area of a matrix droplet (Sekiya, S. et al., 2008). Then it was confirmed by analyzing glycoprotein digests that glycopeptides were ionized from the center hydrophilic small area on 3-AQ/CHCA droplet whereas peptides were detected from the outer area on it (Fig. 10) (Sekiya, S. et al., 2008).

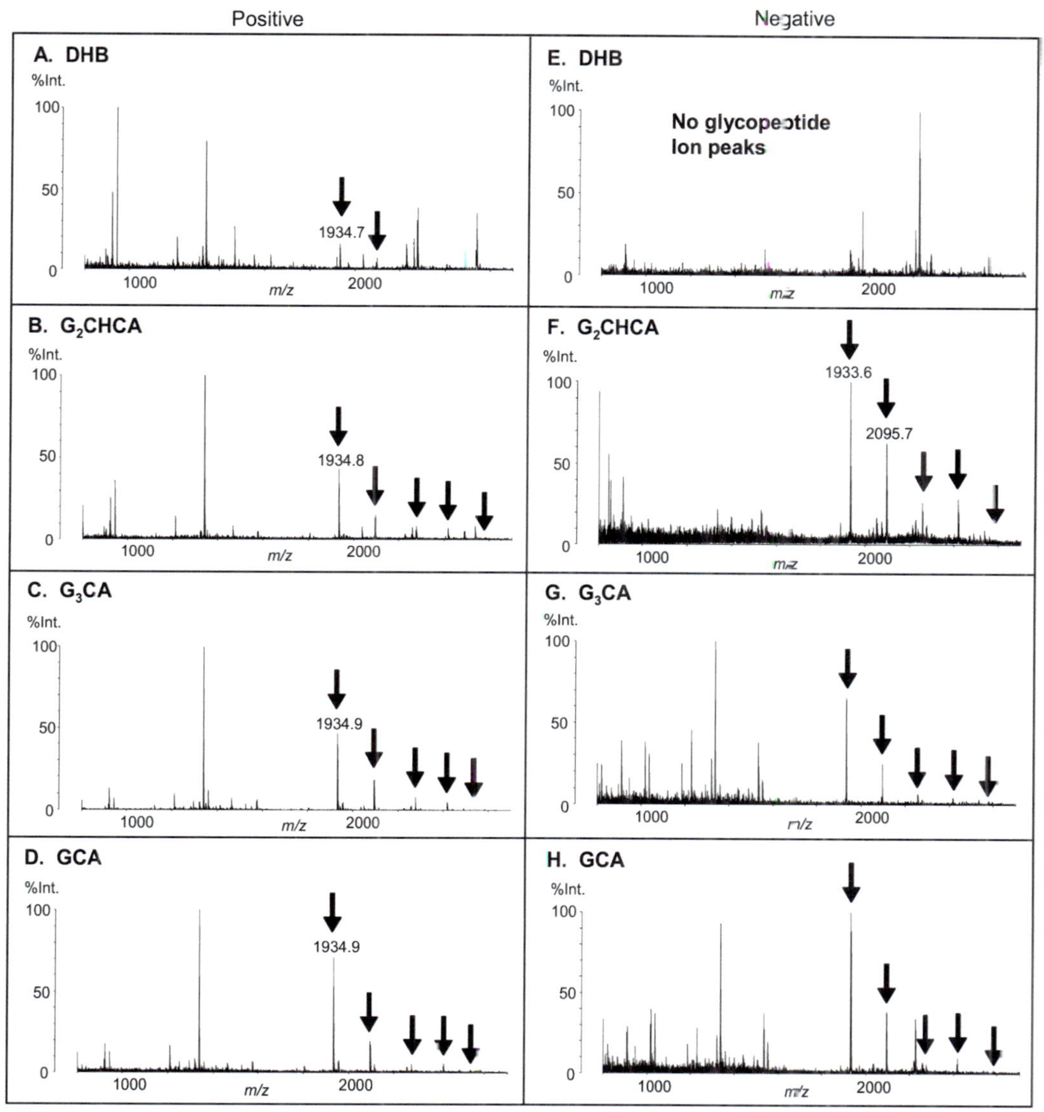

Fig. 9. Positive and negative ion mass spectra of RNase B digests (*ca.* 1 pmol/well) with DHB (A and E), G_2CHCA (B and F), G_3CA (C and G) and GCA (D and H) using MALDI-QIT-TOF mass spectrometer. Arrowed peaks are derived from glycopeptide ions.

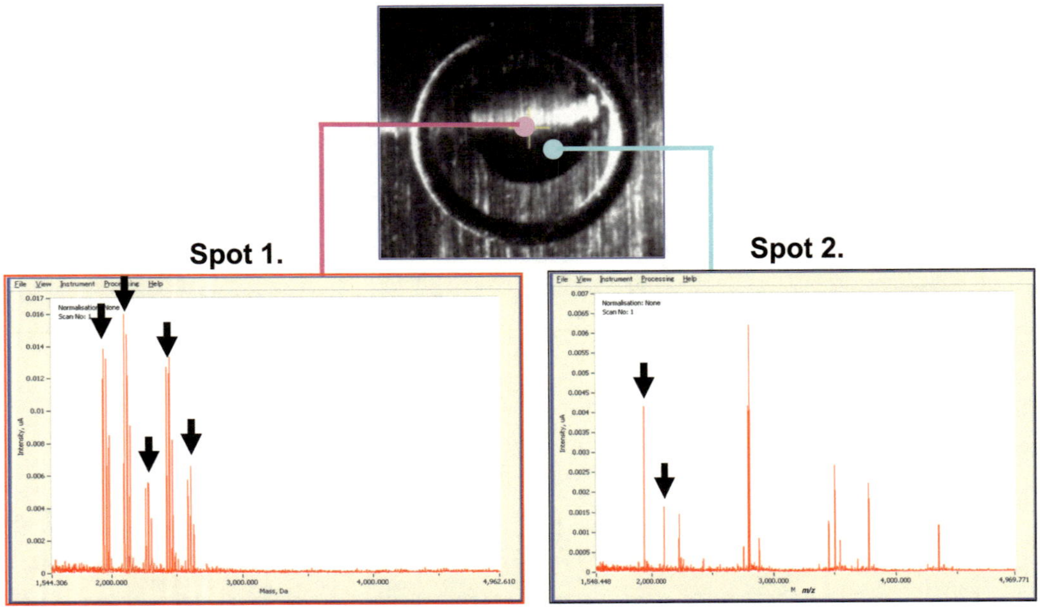

Fig. 10. Mass spectrum of RNase B Lys-C digests using 3-AQ/CHCA at a spot 1 and spot 2, respectively. Arrowed peaks are derived from glycopeptide ions.

Addionally, Kaneshiro et al. reported a high sensitive AQ-labeling method of glycans on a MALDI target using 3-AQ/CHCA (Kaneshiro, K. et al., 2011). Glycans were detected as AQ-labeled molecular ions on attomolle level (Table 6 and 7) (Kaneshiro, K. et al., 2011).

structure	name	M.W.
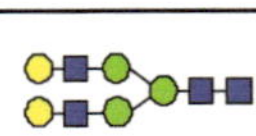	NA2	1,641
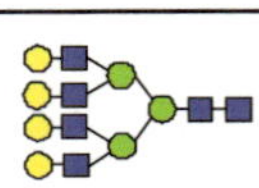	NA4	2,372
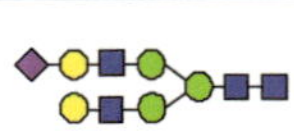	A1	1,933
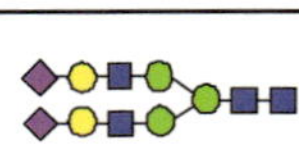	A2	2,224

■: GlcNAc ○: Galactose
●: Mannose ◆: Sialic Acid

Table 6. *N*-linked glycan standards.

	positive ion mode		negative ion mode	
	DHB	3-AQ/CHCA	DHB	3-AQ/CHCA
NA2	10 fmol	100 amol	ND	50 amol
NA4	5 fmol	50 amol	ND	50 amol
A1	100 fmol	500 amol	100 fmol	10 amol
A2	100 fmol	500 amol	100 fmol	50 amol

[a] 1 amol-1 pmol/well of analytes were analyzed using MALDI-QIT-TOF mass spectrometer. ND denotes that analyte molecular ions were not detected.

Table 7. Detection limits of *N*-linked glycan with DHB and 3-AQ/CHCA.[a]

3.6 Quantitative capability

Quantitative analysis is challenging area in MALDI. One of the reasons is inhomogeneous crystal surface using solid matrices. We confirmed that normalized ion peak intensities of peptides have a high correlation with loaded amounts at 1fmol-100 pmol range using 3-AQ/CHCA (Fig. 11) (Fukuyama, Y. et al., 2009).

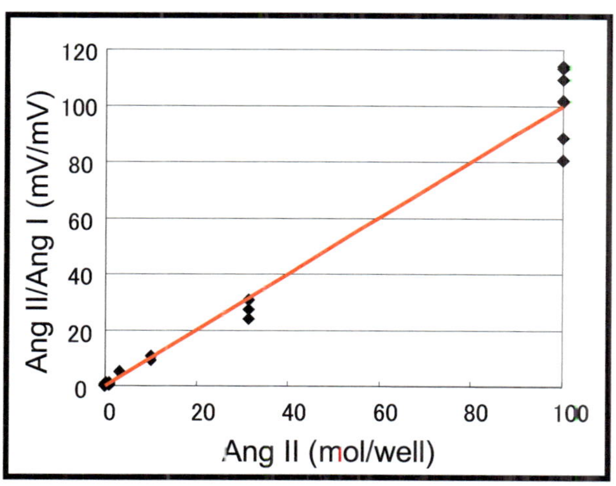

Fig. 11. Dynamic range for absolute quantification of angiotensin II (Ang II) (1 fmol - 100 pmol) using angiotensin I (Ang I) as an internal standard with 3-AQ/CHCA. Ion peak intensity (mV) of the Ang II was normalized by that of the Ang I. MALDI-QIT-TOFMS was used.

4. Conclusion

Liquid matrices are a relatively new area in MALDI. However they have great potential to overcome issues of conventional solid matrices. This chapter indicates that with liquid matrices it is possible to improve sensitivity and homogeneity, suppress dissociation of labile sites like acidic groups and acidic sugars of carbohydrates, increase the durability of ionization, enable quantitative analyses and on-target reactions such as digestion, separation and labeling in MALDI. A rapid and highly sensitive analytical method for any analytes is a difficult but attractive goal and liquid matrix research may open a new insight into this.

5. Acknowledgment

The author thanks Shimadzu Corporation. The author also thanks Koichi Tanaka, Shinichi Iwamoto, Helen Montgomery and staff of Koichi Tanaka Mass Spectrometry Research Laboratory for thoughtful consideration and helpful comments, and Sadanori Sekiya and Kaoru Kaneshiro for providing figures for this chapter.

6. References

Anderson, J. L.; Ding, J.; Welton, T. & Armstrong, D. W. (2002). Characterizing Ionic Liquids on the Basis of Multiple Solvation Interaction. *J. Am. Chem. Soc.*, Vol.124, No.47, pp. 14247-14254

Armstrong, D. W.; He, L. & Liu, Y.-S. (1999). Examination of Ionic Liquids and Their Interaction with Molecules, When Used as Stationary Phases in Gas Cheomatography. *Anal. Chem.*, Vol.71, No.17, pp. 3873-3876

Armstrong, D. W.; Zhang, L.-K.; He, L. & Gross, M. L. (2001). Ionic Liquids as Matrixes for Matrix-assisted Laser Desorption/Ionization Mass Spectrometry. *Anal. Chem.*, Vol.73, No.15, pp. 3679-3686

Bungert, D.; Bastian, S.; Heckmann-Pohl, D. M.; Giffhorn, F.; Heinzle, E. & Tholey, A. (2004). Screening of Sugar Converting Enzymes using Quantitative MALDI-TOF Mass Spectrometry. *Biotechnology Letters*, Vol.26, pp. 1025-1030

Carda-Broch, S.; Berthod, A. & Armstrong, D. W. (2003). Ionic Matrices for Matrix-assisted Laser Desorption/Ionization Time-of-flight Detection of DNA Oligomers. *Rapid Commun. Mass Spectrom.*, Vol.17, No.6, pp. 553-560

Cramer, R. & Corless, S. (2005). Liquid Ultraviolet Matrix-assisted Laser Desorption/Ionization-Mass Spectrometry for Automated Proteomic Analysis. *Proteomics*, Vol.5, pp. 360-370

Crank, J. A. & Armstrong, D. W. (2009). Towards a Second Generation of Ionic Liquid Matrices (ILMs) for MALDI-MS of Peptides, Proteins, and Carbohydrates. *J. Am. Soc. Mass Spectrom.*, Vol.20, pp. 1790-1800

Fenn, J. B.; Mann, M.; Meng, C. K.; Wong, S. F. & Whitehouse, C. M. (1989). Electrospray Ionization for Mass Spectrometry of Large Biomolecules. *Science*, Vol.246, No.4926, pp. 64-71

Fukuyama, Y.; Nakaya, S.; Yamazaki, Y. & Tanaka, K. (2008). Ionic Liquid Matrixes Optimized for MALDI-MS of Sulfated/Sialylated/Neutral Oligosaccharides and Glycopeptides. *Anal. Chem.*, Vol.80, No.6, pp. 2171-2179

Fukuyama, Y.; Toyama A. & Tanaka, K. (2008). On-target digestion using liquid matrixes for glycopeptide analysis. *Poster (proceedings) of* 2008 56th *the American Society for Mass Spectrometry (ASMS) conference on Mass Spectrometry*, THP382, Denver, Colorado, USA, Jun 2-6, 2008

Fukuyama, Y.; Kaneshiro, K.; Taniguchi, K.; Sekiya S.; Iwamoto, S. & Tanaka, K. (2009). High versatility and quantitative capability at femto mol level of the liquid matrix 3-aminoquinoline/CHCA in MALDI mass spectrometry. *Poster (proceedings) of* 2009 57th *the American Society for Mass Spectrometry (ASMS) conference on Mass Spectrometry*, MP569, Philadelphia, Pennsylvania, USA, May 31-June 4, 2009

Fukuyama, Y.; Iwamoto, S.; Funakoshi, N.; Taniguchi, K. & Tanaka, K. (2010). Rapid automated off-line LC-MALDI analyses using a liquid matrix optimized for uniformity and durability of ionization. *Poster (proceedings) of* 2010 58th *the American Society for Mass Spectrometry (ASMS) conference on Mass Spectrometry*, WP601, Salt Lake City, Utah, USA, May 23-27, 2010

Tanaka, K.; Waki, H.; Ido, Y.; Akita, S.; Yoshida, Y. & Yoshida, T. (1988). Protein and Polymer Analysis up to m/z 100,000 by Laser Ionization Time-of-flight Mass Spectrometry. *Rapid Commun. Mass Spectrom.*, Vol.2, No.8, pp 151-153

Kaneshiro, K.; Fukuyama, Y.; Iwamoto S.; Sekiya, S. & Tanaka, K. (2011). Highly sensitive MALDI analyses of glycans by a new AQ-labeling method using 3-AQ/CHCA liquid matrix. *Anal. Chem.*, (Submitted)

Karas, M. & Hillenkamp, F. (1988). Laser Desorption Ionization of Proteins with Molecular Mass Exceeding 10,000 Daltons. *Anal. Chem.*, Vol.60, No.20, pp. 2299-2301

Kolli, V. S. K. & Orlando, R. (1996). A New Matrix for Matrix-assisted Laser Desorption /Ionization on Magnetic Sector Instruments with Point Detectors. *Rapid Commun. Mass Spectrom.*, Vol.10, No.8, pp. 923-926

Laremore, T. N.; Murugesan, S.; Park, T.-J.; Avci, F. Y.; Zagorevski, D. V. & Linhardt, R. J. (2006). Matrix-assisted Laser Desorption/Ionization Mass Spectrometric Analysis of Uncomplexed Highly Sulfated Oligosaccharides Using Ionic Liquid Matrices. *Anal. Chem.*, Vol.78, No.6, pp. 1774-1779

Laremore, T. N.; Zhang, F. & Linhardt, R. J. (2007). Ionic Liquid Matrix for Direct UV-MALDI-TOF-MS Analysis of Dermatan Sulfate and Chondroitin Sulfate Oligosaccharides. *Anal. Chem.*, Vol.79, No.4, pp. 1604-1610

Mank, M.; Stahl, B. & Boehm, G. (2004). 2,5-Dihydroxybenzoic Acid Butylamine and Other Ionic Liquid Matrixes for Enhanced MALDI-MS Analysis of Biomolecules. *Anal. Chem.*, Vol.76, No.10, pp. 2938-2950

Sekiya, S.; Taniguchi, K. & Tanaka, K. (2008). Desalting within sample/ liquid matrix droplet and selective detection of sample ion species using liquid matrix. *Poster (proceedings) of* 2008 56th *the Annual Conference on Mass Spectrometry*, 2P-11, Tsukuba, Japan, May 14-16, 2008

Tholey, A. & Heinzle, E. (2006). Ionic (Liquid) Matrices for Matrix-assisted Laser Desorption/Ionization Mass Spectrometry-Applications and perspectives. *Anal. Bioanal. Chem.*, Vol.386, pp. 24-37

18

Ionic Liquids in Separation of Metal Ions from Aqueous Solutions

Magdalena Regel-Rosocka and Maciej Wisniewski
Poznan University of Technology, Institute of Chemical Engineering and Technology
Poland

1. Introduction

In 1992 the information on the first, stable in water and air, room temperature ionic liquid was published. Since then the number of publications about ionic liquids (ILs) has been rapidly growing. Only in the year 2005 alone more than one thousand articles concerning ILs' synthesis, analysis and applications appeared. Various fields of ILs application are presented in Figure 1.

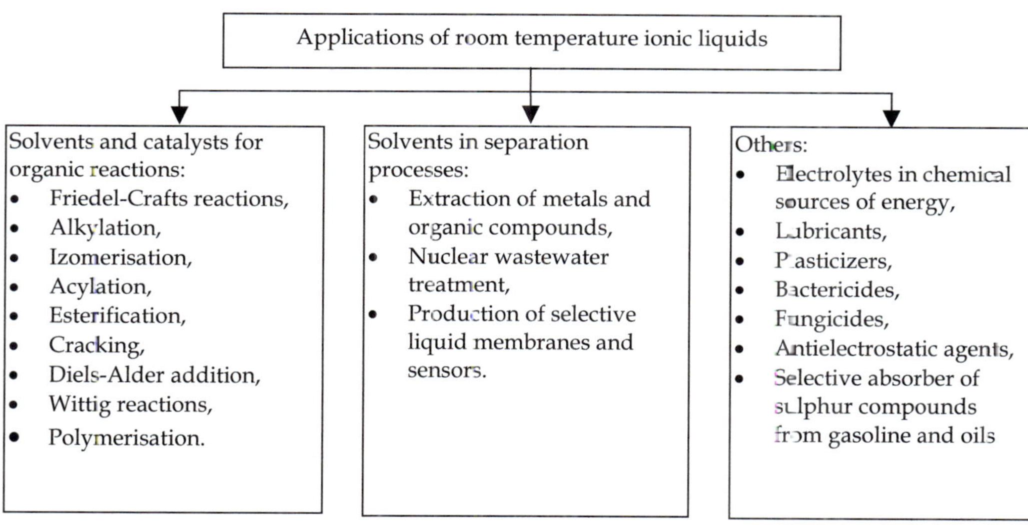

Fig. 1. Scheme of ILs' applications (Adams, 2002; Holbrey & Seddon, 1999; Kosmulski et al., 2002; Pernak, 2003; Seddon et al., 2000).

Ionic liquids (ILs) have become widely used as solvents for organic reactions; however, recently they are more frequently used for separation of metal ions both in extraction, membrane and adsorption systems. In this paper their current applications as solvents and carriers in liquid-liquid extraction of metal ions are discussed and possible extraction mechanisms in ILs are considered in the light of their further use and prospective development.

2. General information on ILs

ILs are salts that exist in liquid form at temperatures below 100°C (Holbrey & Rogers, 2002) and their molecules are composed of a large asymmetric cation, e.g., 1-alkyl-3-methylimidazolium, N-alkylpyridinium, and an organic or inorganic anion. The most common cations and anions used for ILs synthesis are presented in Table 1. Depending on the cation and anion used, the properties of IL such as viscosity, cloud point and solubility in water can vary. The hydrophobicity of IL is determined by the anion. Most of the ionic liquids containing chloride anions are miscible with water, while those composed of [PF_6^-] or [Tf_2N^-] are very hydrophobic. A great advantage of ILs is the facility of anion replacement that enables designing compounds with properties required for particular application. These tuned salts are called task specific ionic liquids (TSIL). Moreover, their non-measurable volatility, non-flammability and thermal stability even up to 350°C make them very attractive for industrial use. They are expected to replace volatile organic compounds (especially halogenated organic solvents) in some systems and applications.

Cation	Anion	Hydrophobicity of anion
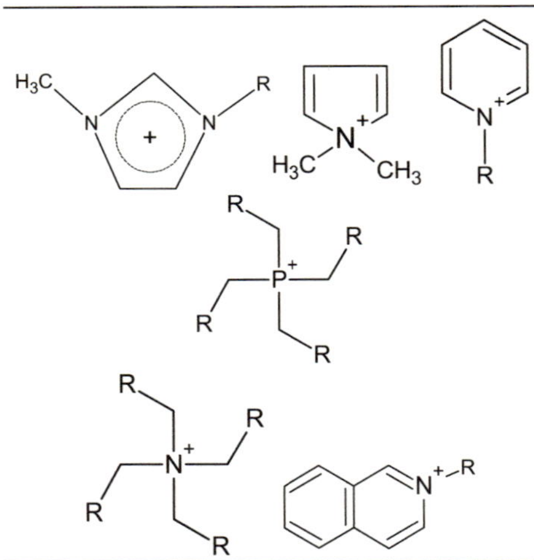	NO_3^-, Cl^-, Br^-, $CuCl_2^-$, ClO_4^- ,BF_4^-, PF_6^-, $CF_3SO_3^-$, $N(SO_2CF_3)_2^-$, $N(SO_2CF_2CF_3)_2^-$	↓

Table 1. Composition of ILs (Bradaric et al., 2003a, 2003b; Cocalia et al., 2005a; Del Sesto et al., 2005; Han & Armstrong, 2007; Holbrey & Rogers, 2002; Hunddleston et al., 1998; Pernak et al., 2005; Visser et al., 2001a; Visser et al., 2002a; Visser et al., 2003).

The wide range of their applications includes also those in extraction processes. Imidazolium ILs are solvents of main interest and have been comprehensively described. They are well defined and their synthesis is well established. Although they are employed in several extraction systems, only a few of them with successful and efficient stripping are described. Stripping from loaded organic phase containing IL is difficult because of strong interactions among ions.

Some ammonium and phosphonium ILs have been also applied in extraction processes, and are considered as prospective solvents and carriers in separation techniques (Bradaric et al., 2003a, 2003b; Del Sesto et al., 2005; Pernak et al., 2005).

3. Liquid-liquid extraction

3.1 Imidazolium ILs

3.1.1 Extraction systems

Ionic liquids have become widely used as solvents for organic reactions, however their use as solvents in extraction systems seems much promising. Separation with imidazolium ILs is best described and broadly investigated. In most cases ILs replace typical solvent extraction diluents. The most frequently applied are 1-alkyl-3-methylimidazolium hexafluorophosphate [$C_n$mim][PF_6], tetrafluoroborate [$C_n$mim][BF_4] and bis(trifluoromethyl-sulphonyl)imide [$C_n$mim][$N(SO_2CF_3)_2$] (anion also abbreviated as [Tf_2N]) (Table 2).
As shown in Table 2, alkali metals, UO_2^{2+}, Cs^+, Sr^{2+} and lanthanides are most frequently extracted with imidazolium ILs. In most cases, extraction of metal ions into the hydrophobic ionic liquid phase is insignificant because metal cations are strongly hydrated in the aqueous phase and affinity of the IL phase to the aqueous one is too small. Thus, an extractant or ligand must be applied, which is a substance that, when diluted in IL, forms complexes with metal ions increasing their hydrophobicity and facilitating their transport to the IL phase. Examples of such ligands are the following macrocyclic compounds: pyridinecalix-4-arene, 18-crown-6 ether (18C6) or dicyclohexano-18-crown-6 (DCH18C6), industrial extractants - TBP (tributyl phosphate), CMPO (octyl(phenyl)-N,N-diisobutylcarbamoylmethylphosphine oxide), PAN (1-(2-pyridylazo)-2-naphthol), TAN (1-(2-thiazolyl)-2-naphthol) or neutral - TODGA (N,N,N',N'-tetra(octyl)diglycolamide) or DEHEHP (di(2-ethylhexyl)2-ethylhexyl phosphonate) (Table 2). Significant improvement in extraction efficiency of metal cations has been achieved when [$C_n$mim][Tf_2N] replaces molecular solvents, such as chloroform, dodecane or 1-octanol, in pyridinecalix-4-arene (extraction of Ag^+) (Shimojo & Goto, 2004), in CMPO – extraction of Ce^{3+}, Eu^{3+}, Y^{3+} (Nakashima et al., 2003, 2005) – in PAN and TAN (extraction of Hg^{2+}) (Visser et al., 2001b) by [$C_n$mim][PF_6], and in crown ethers 18C6 and DCH18C6 (extraction of alkali metals and Sr^{2+}) (Dai et al., 1999; Dietz & Stepinski, 2005; Jensen et al., 2002; Luo et al., 2004a; Stepinski et al., 2005). However, not always the presence of ILs in the organic phase instead of conventional solvents increases the partitioning of the species to be extracted. For example, uranium extraction into dodecane is more efficient than into [$C_n$mim][Tf_2N] (Dietz & Stepinski, 2008).
It is reported that Cs^+ extraction with crown ethers or calixarene tends to increase with shortening of the alkyl chain in IL cation (Luo et al., 2004b). However, a compromise should be made between the extraction efficiency and the solubility of ligand in IL. In other words, the shorter the alkyl chain the lower the solubility of a hydrophobic calixarene in the IL phase.
Furthermore, sodium extraction with DCH18C6 increases in the presence of [$C_n$mim][Tf_2N] compared with 1-octanol (Table 3). However, it is noted that the partitioning of Na^+ is strongly affected by stereochemistry of the crown ether applied, which was not observed in conventional solvents (Dietz et al., 2008). In general, the mechanism of extraction can be tuned by changing the isomeric form of the extractant. The presence of trans isomers of DCH18C6 make the neutral complex extraction dominate over the ion-exchange mechanism. This mechanism is more environmentally friendly because of no release of IL to the aqueous phase.
Papaiconomou et al. (2008) investigated extraction of Cu^{2+}, Hg^{2+}, Ag^+ and Pd^{2+} from aqueous chloride solutions at pH 7 with ten imidazolium, pyridinium, piperidinium and pyrrolidinium ionic liquids comprising typical anions, i.e., [BF_4^-], [Tf_2N^-], trifluoromethyl

IL	n	Ligand or solvent	Extracted species	Ref.
$[C_n mim] [PF_6]$	4	Dithizone	Pb^{2+}, Cu^{2+}, Hg^{2+}, Zn^{2+}, Cd^{2+}	Wei et al., 2003
		TBP	UO_2^{2+} in NO_3^-	Giridhar et al., 2005
		CMPO/TBP	Pu^{4+}, Th^{4+}, Am^{3+}, UO_2^{2+} in the presence of less than 1 M HNO_3	Visser & Rogers, 2003; Visser et al., 2003
		CMPO	Ce^{3+}, Eu^{3+}, Y^{3+} in NO_3^-	Nakashima et al., 2003, 2005
	4, 6	PAN and TAN	Hg^{2+}, Na^+, Ni^{2+}, Co^{2+}, Cd^{2+}, Fe^{3+}; in the presence of KSCN, NaOCN, NaCN, NaX (X = halide); pH = 1; 7; 14	Visser et al., 2001b
	4, 6, 8	18C6, DCH18C6, Dtb18C6	Na^+, Cs^+, Sr^{2+} in HCl, $NaNO_3$, HNO_3 or Na_3citrate	Visser et al., 2000
	4-9	DC18C6	K^+, Na^+, Rb^+, Cs^+, Li^+	Chun et al., 2001
			Mg^{2+}, Ca^{2+}, Sr^{2+}, Ba^{2+} Pb^{2+}	Bartsch et al., 2002
	4, 6, 8	$^tBu[1]CH_2Py$, $^tBu[4]CH_2Py$	Ag^+ in the presence of Cu^{2+}, Zn^{2+}, Co^{2+}, Ni^{2+}	Shimojo & Goto, 2004
	4, 8	-	Zn^{2+}, Fe^{3+} in 1 M HCl	Perez de los Rios et al., 2010
	8	-	Ce^{4+}, Th^{4+}, RE in HF and HNO_3	Zuo et al., 2008
	8	DEHEHP	Ce^{4+}, Th^{4+}, RE in HF and HNO_3	Zuo et al., 2009
$[C_n mim][Tf_2N]$	5-10	DCH18C6	Na^+ in NO_3^-	Dietz & Stepinski 2005
	2, 4, 6, 8	N-alkylaza-18-crown-6 ether	Na^+, K^+, Cs^+, Sr^{2+} in NO_3^- or Cl^-; $pH_i \approx 6$	Luo et al., 2004a
		BOBCalixC6	Cs^+ K^+, Na^+, Sr^{2+} in NO_3^-	Luo et al. 2004b
		18C6, DC18C6	Sr^{2+} in NO_3^-, Cl^-, SO_4^{2-}	Dietz & Dzielawa, 2001; Jensen et al. 2002
	2, 5, 6, 8, 10	DC18C6/TBP	Sr^{2+} in NO_3^-	Stepinski et al.,

IL	n	Ligand or solvent	Extracted species	Ref.
		DC18C6/DAAP		2005
	5, 8,10	TBP	UO_2^{2+} in HNO_3	Dietz & Stepinski, 2008
	8	CMPO	UO_2^{2+} in less than 1 M HNO_3	Visser et al., 2003
	10	DEHPA, Cyanex 272	UO_2^{2+}, Am^{3+}, Nd^{3+}, Eu^{3+}	Cocalia et al., 2005b
	4	Htta	Ln^{3+}	Jensen et al., 2003
		CMPO	Ce^{3+}, Eu^{3+}, Y^{3+} in NO_3^-	Nakashima et al., 2005
	2, 4, 6	TODGA	La^{3+}, Eu^{3+}, Lu^{3+} in NO_3^-	Shimojo et al., 2008
	4, 8	-	Zn^{2+}, Fe^{3+} in 1 M HCl	Perez de los Rios et al., 2010
[$C_n$mim][BETI]	2, 4	DCH18C6	Cs^+, Sr^{2+} in Cl^-	Luo et al., 2006b
[1-R^1-2-R^2-3-mim][Tf_2N] or [1-R^1-2-R^2-3-mim][PF_6]	R^1 = ethyl, propyl, butyl R^2 = H or methyl	DCH18C6	Sr^{2+} in NO_3^-	Dai et al., 1999
[diC_2hist18C6][Br] [diC_2hist18C6][Tf_2N] [$C_n$mim18C6][Br] [$C_n$mim18C6][Tf_2N]	4	-	Cs^+, Sr^{2+} in Cl^-	Luo et al., 2006a
On the base of [$C_n$mim][PF_6] thioether, thiourea and urea derivatives	4	-	Hg^{2+}, Cd^{2+}, pH = 1 or 7	Visser et al., 2001c; Visser et al., 2002b
[$C_n$mim][PF_6], [C_2OC_2mim][PF_6], [$(C_2OC_2)_2$im][PF_6] or [$C_n$mim][BF_4]	4, 6, 8	-	Cu^{2+}, Zn^{2+}, Cr ions in the presence of NaCl	Vidal et al., 2004; Correia et al., 2005

BOBCalixC6 calix[4]arene-bis(tert-octylbenzo-crown-6); Htta (1-(2-thienyl)-4,4,4-trifluoro-1,3-butanedione); BETI bis(perfluoroethanesulphonyl)imide; [C_2OC_2 mim] 1-(2-ethoxyethyl)-3-methylimidazolium; [(C_2OC_2)2im] 1,3-bis(2-ethoxyethyl)imidazolium; [diC_2hist18C6] N-(N,N-diethyl)-histamonium aza-18-crown-6 ether; [C_4mim18C6] N-(3-butylimidazolium propyl)aza-18-crown-6 ether

Table 2. Imidazolium ionic liquids in selected extraction systems.

sulphonate [TfO-] or nonafluorobutyl sulphonate [NfO-]. Only mercury has been efficiently extracted (D > 24) with imidazolium and pyridinium ionic liquids, while the other metal cations are not transferred to the IL phase.
Moreover, good extraction abilities of imidazolium ILs with [NfO-] have been confirmed in the studies on Li^+, Na^+, Cs^+, Ca^{2+}, Sr^{2+} and La^{3+} extraction (Kozonoi & Ikeda, 2007). According to the authors the metal ions with greater charge are more easily transferred to the [$C_n$mim][NfO] phase.
Another point of view is represented by Wei et al. (2003a, 2003b) and Domanska & Rekawek (2009). They propose to complex metal ions in the aqueous phase with dithizone, 8-hydroxyquinoline or 1-(2-pyridylazo)-2-naphthol and next, to extract such metal complexes with imidazolium ILs. The [$C_n$eim] ILs (Domanska & Rekawek, 2009) show better extraction efficiency of Ag^+ and Pb^{2+} than chloroform, however, it decreases with increasing alkyl chain length in the cation and with increasing hydrophobicity of an anion (i.e., [Tf_2N^-], [PF_6^-]). The extraction with [C_4mim][PF_6] (Wei et al. 2003a, 2003b) of various metal ions (e.g., Ag^+, Cu^{2+}, Pb^{2+}, Cd^{2+} and Zn^{2+}) is strongly dependent on pH and allows selective separation of Cu^{2+} from Pb^{2+} and Zn^{2+} at pH 2 and from Cd^{2+} at pH 1.9. Furthermore, Ag^+ is selectively separated from Pb^{2+} also at pH 1.9. The dependence on pH is advantageous for stripping because metal ion can be stripped from the loaded IL phase by dissociation of metal-dithizone complex with acid solution. 0.1 M HNO_3 is used to regenerate IL and the reproducibility of extraction is confirmed in five cycles of recycling (extraction-stripping).
Not always an additional ligand is necessary to extract metal ions. For example, Zn^{2+} and Fe^{3+} can be transported directly to [C_8mim][BF_4] phase. The extraction tends to increase in the following order: [Tf_2N^-]<[PF_6^-]<[BF_4^-] that corresponds to decreasing hydrophobicity of the anions studied (Perez de los Rios et al., 2010). In parallel, the same authors observed increasing extraction efficiency with lengthening of the alkyl chain of imidazolium cation contrary to Cs^+ extraction studied by Luo et al. (2004b). Similarly, efficient extraction of Ce^{4+} from HNO_3 is shown in pure [C_8mim][PF_6] (Zuo et al., 2008). However, its application for the recovery of Ce^{4+} from bastnasite leaching liquor, containing Th^{4+} and rare earth metals (RE), is limited by the presence of F^- that negatively affects extraction efficiency. Because of this, neutral extractant (DEHEHP) has been added to [C_8mim][PF_6] to overcome this problem and the extraction efficiency of metal ions is compared to that in the traditional DEHEHP-heptane system (Table 3) (Zuo et al., 2009). The selectivity of extraction in both IL and heptane systems can be ordered as follows: $Ce^{4+} > Th^{4+} > RE^{3+}$.
The higher capacity for Ce^{4+} of DEHEHP-[C_8mim]PF_6 than of DEHEHP-heptane indicates that both DEHEHP and [C_8mim]PF_6 may act as extractants. The mechanism of extraction is presented in section 3.2.2.
Further, ILs dedicated to very specific extractions have been synthesized and called task specific ionic liquids (TSILs). Elimination of the use of additional extractant or ligand from the organic phase can be pointed out as a consequence of imidazolium cation modification with other compounds. However, according to Abbot et al. (2011), the term TSIL should be changed into 'functionalised ILs', as now most of ILs are designed and synthesised for a dedicated application.
N-(3-butylimidazolium propyl)aza-18-crown-6 ether bis[(trifluoromethyl)sulphonyl]imide [C_4mim18C6][Tf_2N] illustrates TSIL with the IL-cation structure modified with aza-crown ether via covalent bonds (Luo et al., 2006a). Another type of TSIL, based on thioether, thiourea and urea derivatives, are involved as carriers in Hg^{2+} and Cd^{2+} liquid-liquid extraction (Visser et al., 2001c, 2002b). Correspondingly, ILs composed of a functional

IL system	D or E	Conventional solvent system	D	Ref.
0.1 M in$[C_4mim][Tf_2N]$ $[diC_2hist18C6][Tf_2N]$ $[C_4mim18C6][Tf_2N]$	D_{Cs}=14.5 D_{Sr}=447 D_{Cs}=23.9 D_{Sr}=213	0.1 M in $[C_4mim][Tf_2N]$ N-octylaza18C6 DCH18C6	Cs^+=25.7 D_{Sr}=1070 Cs^+=380 D_{Sr}=935	Luo et al., 2006a; Luo et al., 2006b
8 mM BOBCalixC6 in $[C_4mim][Tf_2N]$	D_{Cs}=576 D_K=8.4 D_{Sr}=0	0.01 M BOBCalixC6 in 1,2-dichloroethane	D_{Cs} negligible	Luo et al., 2004b; Haverlock et al., 2000
0.15 M DCH18C6 in $[C_2mim][Tf_2N]$ $[C_4mim][PF_6]$	 D_{Sr}=1100 D_{Sr}=2.4	0.15 M DCH18C6 in toluene chloroform	 D_{Sr}=0.76 D_{Sr}=0.77	Dai et al., 1999
0.1 M DCH18C6 in $[C_5mim]$ $[Tf_2N]$ $[C_{10}mim][Tf_2N]$	 D_{Sr} ~8 D_{Sr} ~4	0.1 M DCH18C6 in 1-octanol	D_{Sr}=1	Dietz et al., 2003
0.1 M DCH18C6 in $[C_5mim]$ $[Tf_2N]$ $[C_{10}mim][Tf_2N]$	in 1 M HNO_3 D_{Na}~0.11 D_{Na} ~0.1	0.1 M DCH18C6 in 1-octanol	in 1 M HNO_3 D_{Na} ~0.06	Dietz & Stepinski, 2005
0.1 M CMPO in $[C_4mim][PF_6]$	D_{UO2}=1000	0.1 M CMPO in dodecane	D_{UO2}=100	Visser & Rogers,
0.6 mM TODGA in $[C_2mim][Tf_2N]$	D_{La} =100	5 mM TODGA in isooctane	D_{La}=0.01	Shimojo et al., 2008
0.34 M DEHEHP in $[C_8mim][PF_6]$	$E_{Ce(IV)}$=99% $E_{Ce(III)}$=2% E_{Th}=49%	0.34 M DEHEHP in heptane	$E_{Ce(IV)}$=96% $E_{Ce(III)}$=2% E_{Th}=30%	Zuo et al., 2009

Table 3. Comparison of selected distribution coefficients of metal ions between IL or conventional solvent and the aqueous phase.

disulphide group bonded to piperidinium or pyrrolidinium cation show high selectivity toward Hg^{2+} or Cu^{2+} extraction. Moreover, ionic liquids containing a functional nitrile group bonded to pyridinium or piperidinium cation are found to extract completely Ag^+ and Pd^{2+} from the aqueous phase (Papaiconomou et al., 2008). It can be concluded that the type of functional group, cation ring and anion strongly influence the extraction abilities and selectivity.

Nevertheless, limited quantities of these tuned compounds should be emphasised as the main drawback of TSIL use for extraction. For this reason they could be applied for removal of metal ions in very low concentrations or immobilised at any support. Although metal ions can be stripped easily, it seems that aza-crown modified TSILs are less efficient in Cs^+ and Sr^{2+} extraction than the IL phase mixed with DCH18C6 or N-octylaza 18C6 as extractants (Table 3). High Cs^+ extraction is observed for BOBCalixC6 in $[C_4mim][Tf_2N]$ (Luo et al., 2004b). What is important, the selectivity of Cs^+ extraction over Na^+ and Sr^{2+} is higher than in conventional organic solvents, while Cs^+ extraction over K^+ is lower in

$[C_4mim][Tf_2N]$. The selectivity of Sr^{2+}/Cs^+ can be tuned by the choice of IL anion. Sr^{2+} extraction prevails over that of Cs^+ with increasing anion hydrophobicity for $[C_4mim]$ cation (Luo et al., 2006b).

The synergism observed in the presence of ILs is attributed to ion-recognition capabilities of complexing ligands, unique ionic solvation environment and ion-exchange capabilities of ILs (Luo et al., 2006a).

3.2 Mechanism of extraction

The differences in extraction efficiency between IL and conventional solvent systems are significant, as presented in Table 3. Generally, in IL systems an increase in distribution ratios and selectivity is observed. It has been indicated that cation-exchange is a mechanism of extraction with imidazolium ILs (Table 4) and emphasized that a mechanism different from that in conventional solvents, affects the extraction efficiency. It can be attributed to a change in coordination environment of extracted species, in other words, to a change in the number of ligands coordinating with metal ions. An example is La^{3+} extraction with TODGA in isooctane and in $[C_2mim][Tf_2N]$ (Shimojo et al., 2008), in which a change in extraction mechanism from ion-pair extraction accompanied by anions (in isooctane) into cation-exchange (in IL) is observed (eq. 1). The same observations have been made by Shen et al. (2011) for UO_2^{2+} extraction with other diglycolamides (TBDA – N,N,N',N'-tetrabutyl-3-oxapentanediamide and MBDA – N,N,N',N'-dimethyldibutyl-3-oxapentanediamide). It can be concluded that the change in extraction mechanism depends on the acid (in this case HNO_3) concentration. At low acidity ion exchange is favoured, while at high acid content neutral uranyl-diamide nitrato complexes are extracted. For example, for La^{3+} extraction with $[C_nmim][NfO]$ (without additional ligands) cation exchange between two phases with the partial transfer of non-charged species has been proposed (Kozonoi & Ikeda, 2007).

Reaction		Ref.
$Ln^{3+}_{(w)} + 3TODGA_{(o)} + 3[C_nmim^+]_{(o)} \Leftrightarrow LnTODGA^{3+}_{3(o)} + 3[C_nmim^+]_{(o)}$	(1)	Shimojo et al., 2008
$Ag^{+}_{(w)} + {}^tBu[4]CH_2Py_{(o)} + [C_8mim^+]_{(o)} \Leftrightarrow Ag\,{}^tBu[4]CH_2Py^{+}_{(o)} + [C_8mim^+]_{(w)}$	(2)	Shimojo & Goto, 2004
$Sr^{2+}_{(w)} + CE_{(o)} + 2[C_5mim^+]_{(o)} \Leftrightarrow Sr(CE)^{2+}_{(o)} + 2[C_5mim^+]_{(w)}$	(3)	Dietz & Dzielawa, 2001; Jensen et al., 2002
$M^{3+}_{(w)} + 3CMPO_{(o)} + 3[C_4mim^+]_{(o)} \Leftrightarrow M(CMPO)^{3+}_{3(o)} + 3[C_4mim^+]_{(w)}$	(4)	Nakashima et al., 2005
$Ln^{3+}_{(w)} + Htta_{(o)} + [C_4mim][Tf_2N]_{(o)} \Leftrightarrow [C_4mim][Ln(tta)_4]_{(o)} + 4H^{+}_{(w)} + [Tf_2N^-]_{(}$	(5)	Jensen et al., 2003
$Ce^{4+}_{(w)} + 6NO^{-}_{3(w)} + 2[C_8mim][PF_4]_{(o)} \Leftrightarrow [C_8mim]_2[Ce(NO_3)_6]_{(o)} + 2[PF_6{}^-]_{(w)}$	(6)	Zuo et al., 2008
$Ce^{4+}_{(w)} + 4NO^{-}_{3(w)} + HF_{(w)} + DEHEHP_{(o)} \Leftrightarrow Ce(HF)(NO_3)_4 \cdot DEHEHP_{(o)}$	(7)	Zuo et al., 2009

${}^tBu[4]CH_2Py$ - pyridinecalix-4-arene; CE – crown ether; (w) and (o) denote aqueous and organic phase.

Table 4. Reactions of metal cations with imidazolium Ils.

Equations (2)-(4) can be combined into one equation of extraction according to cation-exchange mechanism, where L is a ligand complexing metal cation in the organic phase:

$$M^{m+}_{(w)} + mL_{(o)} + m[C_n mim^+]_{(o)} \Leftrightarrow ML^{m+}_{m(o)} + m[C_n mim^+]_{(w)} \quad (8)$$

Metal cation binds the ligand contained in the organic phase, and then it is exchanged for imidazolium cation [$C_n mim^+$] of the ionic liquid. When metal cation becomes, as a result of extraction, a part of ionic liquid and is strongly bound in the organic phase, its stripping is very difficult. Additionally, the loss of IL cation to the aqueous phase is not advantageous for high cost and environmental impact (Dietz, 2006). However, the change in the Sr^{2-} and Cs^+ extraction mechanism from cation-exchange to extraction of neutral complexes, observed with increasing alkyl chain in imidazolium cation (Dietz et al., 2003), causes a reduction in IL loss to the aqueous phase. Alternatively, to overcome the loss of IL cation Luo et al. (2004b) have proposed addition of organophilic species ($NaBPh_4$) to control a transfer of imidazolium cations to the aqueous phase. Its addition decreases the loss of IL cation by 24%.
The knowledge of extraction mechanism with participation of ILs has been continuously extended. Recently, Dietz & Stepinski (2005) have reported a complex extraction process of Na^+ from nitrate solution with crown ether in [$C_n$mim][Tf_2N], described by a combination of three processes:

- sodium nitrato-crown ether complex partitioning:

$$Na \cdot DCH18C6^+ + NO_3^- \rightarrow (NaNO_3 \cdot DCH18C6)_{(o)} \quad (9)$$

- exchange of the 1:1 sodium-crown ether complex for the IL cation

$$Na \cdot DCH18C6^+ + [C_n mim][Tf_2N]_{(o)} \rightarrow [NaDCH18C6][Tf_2N]_{(o)} + C_n mim^+_{(w)} \quad (10)$$

- crown ether mediated Na^+/H_2O exchange at high acidity:

$$Na^+ + (H_3O \cdot DCH18C6^+)_{(o)} \rightarrow (Na \cdot DCH18C6^+)_{(o)} + H_3O^+ \quad (11)$$

This complex three-path-mechanism opens the opportunities to design IL based extraction systems to improve selectivity of conventional organic phases.
Recycling of the crown ether-IL phase loaded with metal ions can proceed by simple pH change. For example, at low pH protonated aza crown releases 98% Cs^+ and Sr^{2+}, and can be reused for extraction after deprotonation with base (Luo et al., 2004a).
In contrast, the anion exchange mechanism is proposed for lanthanide extraction (Ln^{3+}) in the system with Htta in [C_4mim][Tf_2N] (Jensen et al., 2003). As a result of complex $Ln(tta)_4^-$ formation in the organic phase, four protons are released, and simultaneously [Tf_2N^-] anion is transferred to the aqueous phase according to reaction (5) presented in Table 4. A similar extraction mechanism is proposed for pure [C_8mim][PF_6], while with DEHEHP in [C_8mim][PF_6] the neutral complexing extraction mechanism, eq. (7), is proposed (Zuo et al., 2008, 2009). The latter mechanism seems to be more advantageous as no IL is lost to the aqueous phase.
As a result, stripping in these systems is easier than for the cation-exchange mechanism, because it is enough to use an aqueous solution of $H(Tf_2N)$ or KPF_6 to reverse the

equilibrium of reactions (5-7). At the same time pure cerium as CeF_3 nano-particles or $Ce_2(SO_4)_3$ solutions are obtained as products after regeneration of the [C_8mim][PF_6] phase.
However, it is proven by Rickert et al. (2007) that the presence of certain solutes (e.g., crown ethers) in a hydrophobic IL, even in the absence of metal ion in the extraction system, can significantly increase the solubility of the ionic liquid in acidic aqueous media. Thus, it is still an open question whether they are environmentally friendly and can replace traditional organic solvents.
Apart from small amounts of synthesised ILs and their solubility in the aqueous phase, there is one more issue that must be indicated as limiting the ILs use for liquid-liquid extraction, which is their hydrolysis in the contact with an acidic solution (Swatloski et al., 2003). It is particularly risky to use [PF_6^-] containing ILs because the decomposition reaction of the anion leads to a toxic and corrosive product HF. Additionally, during acidic reactions gas HF may be released. Thus, Swatloski et al. (2003) propose to consider the list of non-toxic pharmaceutically acceptable anions when designing ILs as solvents for extraction.

3.3 Ammonium ILs

3.3.1 Extraction systems

Some ionic liquid extraction systems operate very efficiently without a ligand complexing metal ions. An example of such liquids is methyltrioctylammonium salt (Aliquat 336), a reagent used for extraction for many years.
Several Aliquat 336 [A336] derived ILs have been already described in literature. Some of them can be applied for separation of metal ions; however, methyltrioctylammonium thiosalicylate [A336][TS] seems to be the most efficient among those studied.
Extraction of Pd^{2+} with [A336][Cl] and [A336][NO_3] is efficient, however, [A336][Cl] provides stronger binding abilities (Giridhar et al., 2006). Direct electrodeposition is proposed as an interesting and effective method of metal recovery from the IL phase due to the wide electrochemical window of ILs and their ion exchange abilities.
[A336][TS], thiocyanate [SCN] and methionate [Met] are used for preconcentration of UO_2^{2+} (Srncik et al., 2009) and [A336][TS], benzoate [BA] and hexanoate [Hex] for removal of cadmium(II) from natural river matrix (Kogelnig et al., 2008) by extraction. The extraction efficiency from river water decreases in the following order of ILs used: [A336][TS] (>99.9%) > [A336][Hex] (14%) > [A336][BA] (11%). It is suggested that the application of [A336][TS] may be considered as more sustainable than that of systems with volatile solvents because the risk related to volatile and flammable solvents is eliminated. Pure [A336][Cl] allows almost complete removal of Zn^{2+} and Fe^{3+} from 1 M HCl (Perez de los Rios et al., 2010), which is in agreement with numerous previous studies on [A336][Cl] solutions in typical VOCs (Kejun et al., 2004; Sato et al., 2010; Wionczyk, 2009).
Aliquat 336 and quaternary phosphonium cation [QP] with 2-(methylthio)benzoate [MTBA] and [TS] have been applied as extracting agents for Pt^{4+} from chloride solution. The extraction efficiency tends to diminish with decreasing viscosity of IL in the following order: [QP][MTBA] (100%) > [A336][TS] (85%) > [QP][TS] (76%) > [A336][MTBA] (40%) (Stojanovic et al., 2010). Thus, it proves that not only 'functionality appended' to the anion but also physico-chemical properties of IL play an important role in extraction of metal ions.

3.3.2 Mechanism of extraction

Egorov et al. (2010) have used methyltrioctylammonium salicylate [A336][Sal] for extraction of Fe^{3+}, Cu^{2+}, Ni^{2+}, Mn^{2+} from various solutions of their salts: sulphate, nitrate, chloride,

chloride, respectively. [A336][Sal] efficiently extracts Fe^{3+} and Cu^{2+} (99 and 89%). The mechanism proposed for the extraction of Fe^{3+} may be represented by equation (12) given in Table 5.

Reaction		Ref.
$Fe^{3+}_{(w)} + 2[A336][HSal]_{(o)} + HSO^{-}_{4(w)} \Leftrightarrow [A336][FeSal_2]_{(o)} + [A336][HSO_4]_{(o)} + 2H^{+}_{(w)}$	(12)	Egorov et al., 2010
$M^{2+}_{(w)} + [A336][HSal]_{(o)} + HSO^{-}_{4(w)} \Leftrightarrow [MSal]_{(o)} + [A336][HSO_4]_{(o)} + H^{+}_{(w)}$	(13)	Egorov et al., 2010
$Co^{2+}_{(w)} + [A336][CA-12]_{(o)} + SO^{2-}_{4(w)} \Leftrightarrow CoSO_4 \cdot [A336][CA-12]_{(o)}$	(14)	Sun et al., 2010
$2Ln^{3+}_{(w)} + 2[A336][Bis]_{(o)} + 6NO^{-}_{3(w)} \Leftrightarrow [A336][Ln(NO_3)_3Bis]_{(o)} + [A336][NO_3]_{(o)} + [Ln(NO_3)_2Bis]_{(o)}$	(15)	Belova et al., 2010

Table 5. Reactions of metal ions with ammonium ILs.

For divalent metals (M^{2+}) the mechanism is similar, however, in this case a formation of metal ion salicylate in the organic phase is expected (Table 5, eq. (13)). In each reaction, protons are released to the aqueous phase while no transfer of IL is observed. Separation factors of Fe^{3+} over other metals such as Co^{2+}, Ni^{2+}, Mn^{2+}, Cr^{3+} and Zn^{2+} are high and in all cases exceed 10. It means that Fe^{3+} can be selectively separated from such solutions.

0.02 M [A336][CA-12] (sec-octylphenoxy acetate) mixed with toluene has been studied for Co^{2+} and Ni^{2+} separation from sulphate solutions (Sun et al., 2010). The inner synergistic effect, defined as synergistic effect coming from cation and anion of a bifunctional IL [A336][CA-12] is observed, and the ion association mechanism (Table 5, eq. (14)) is proposed for the reaction of Co^{2+} extraction from sulphate liquor. Easy stripping with diluted H_2SO_4 may confirm rather weak binding of Co^{2+} in this IL phase.

Again methyltrioctylammonium based IL is proposed for La^{3+} and Y^{3+} extraction from nitric acid solution (Belova et al., 2010). After analysis of extraction constants for various extraction reactions of [A336][Bis] (bis(2,4,4-trimethylpentyl)phosphinate) and lanthanides (Ln^{3+}), these authors have suggested that several species are formed in the organic phase according to eq. (15) (Table 5). The ratio between them depends on their stability and is difficult to evaluate.

3.4 Phosphonium ILs

Besides imidazolium and ammonium ILs, also phosphonium ionic liquids are considered to be prospective for substance separation (Bradaric et al., 2003b), and have been proposed as solvents for crown ether extractants mixed with organic solvents and as carriers in impregnated resins (Campos et al., 2008a, 2008b; Gallardo et al., 2008; Guibal et al., 2008).

In fact, many phosphonium ILs used for separation are based on widely known and applied Cytec Industries extractants such as trialkylphosphine oxides or Cyanex 272; however there is still little information about their use in extraction processes. Some commercial phosphonium ionic liquids (Cyphos series) studied in separation processes are presented in Table 6.

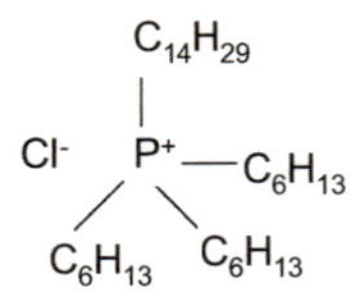 Cyphos IL 101, *Trihexyl(tetradecyl)phosphonium chloride* **[QP][Cl]**	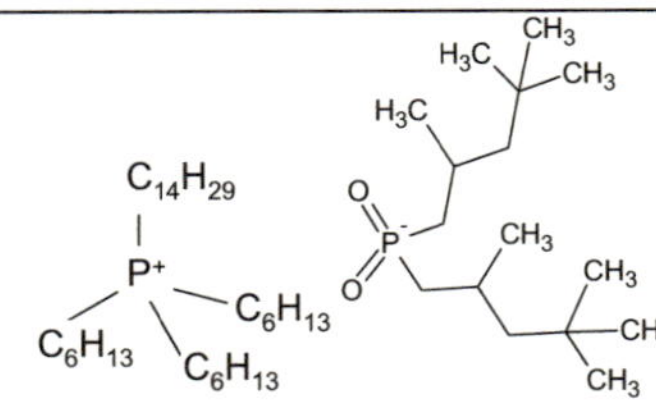 Cyphos IL 104, *Trihexyl(tetradecyl)phosphonium bis(2,4,4-trimethylpentyl)phosphinate* **[QP][Bis]**
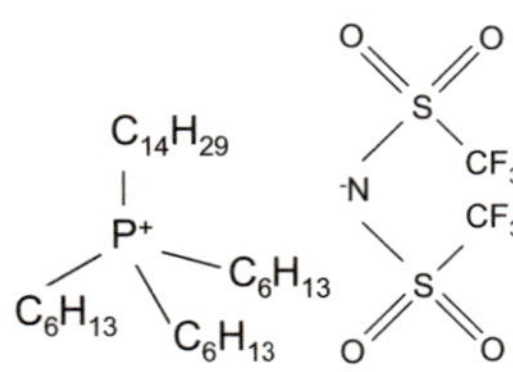 Cyphos IL 109, *Trihexyl(tetradecyl)phosphonium bis(trifluoromethylsulphonyl)imide* **[QP][Tf$_2$N]**	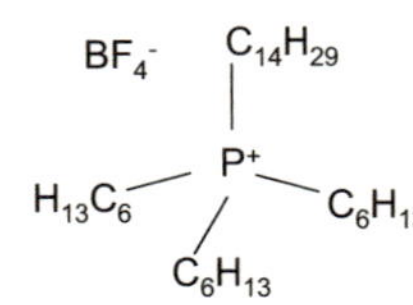 Cyphos IL 111, *Trihexyl(tetradecyl)phosphonium tetrafluoroborate* **[QP][BF$_4$]**

Table 6. Structures of exemplary phosphonium ILs used in extraction systems.

Turanov et al. (2008) have studied partitioning of lanthanide chlorides between HCl and organic solutions of neutral organophosphorous compounds and IL – e.g., butyldiphenyl-phosphonium hexafluorophosphate and bis(trifluoromethylsulphonyl)imide. However, the presence of phosphonium ILs does not affect lanthanide extraction as strongly as imidazolium ILs, i.e., [C$_4$mim][PF$_6$] and [C$_4$mim][Tf$_2$N].

Efficient extraction of Fe^{3+} from 6 M HCl with [QP][Cl] dissolved in chloroform is reported by Kogelnig et al. (2010). At the same time Ni^{2+} is not extracted which enabled Fe^{3+} to be separated from the other metal ions. The spectroscopic data support the assumption that tetrachloroferrate is formed in the organic phase according to the equation (16) (Table 7).

Reaction		Ref.
$FeCl^{-}_{4(w)} + [QP][Cl]_{(o)} \Leftrightarrow [QP][FeCl]_{(o)} + Cl^{-}_{(w)}$	(16)	Kogelnig et al., 2010
$Na^{+}_{(w)} + TcO^{-}_{4(w)} + CE_{(o)} \Leftrightarrow Na \cdot CE^{+}TcO^{-}_{4(o)}$	(17)	Stepinski et al., 2010

Table 7. Reactions of metal ions with phosphonium Ils.

Recently, phosphonium ILs have been reported as solvents for DCH18C6 to extract TcO_4^- and ReO_4^- from NaOH and/or NH_4OH solutions (Stepinski et al., 2010). The use of [QP][Tf$_2$N] ensures the highest distribution of TcO_4^- both in the presence and absence of a crown ether and prefers extraction of TcO_4^- over ReO_4^-. The most important is that, unlike imidazolium ILs, [QP][Tf$_2$N] extracts without IL loss (negligible amounts of [Tf$_2$N$^-$]

determined) to the aqueous phase. Owing to this, it has been suggested by Stepinski et al. that 'these solvents may provide the basis for improved approaches to the extraction and recovery of a variety of anions'. The dominant extraction mechanism is the ion pair transfer according to eq. (17) included in Table 7. The enhancement in TcO_4^- extraction with crown ether, compared with conventional solvents, is attributed to improvement in the solvation properties of IL.

The authors of the chapter have studied Cyphos ILs as extractants for regeneration of spent pickling solutions from hot-dip galvanizing plants (Marszalkowska et al., 2010; Nowak et al., 2010; Regel-Rosocka et al. 2006, 2007; Regel-Rosocka, 2009, 2010).

Among the studied extractants trihexyl(tetradecyl)phosphonium chloride (Cyphos IL 101), trihexyl(tetradecyl)phosphonium bis(2,4,4-trimethylpentyl)phosphinate (Cyphos IL 104), trihexyl(tetradecyl)phosphonium bis(trifluoromethylsulphonyl)imide (Cyphos IL 109), and trihexyl(tetradecyl)phosphonium tetrafluoroborate (Cyphos IL 111) in mixture with toluene have been investigated as reagents to extract Zn^{2+}, Fe^{3+} or Fe^{2+} from chloride media. Toluene has been applied to overcome some drawbacks caused by the high viscosity of ILs. In some cases alkylene carbonates (propylene or butylene carbonate) as novel diluents in extraction of Zn^{2+} and Fe^{3+} have been used replacing toluene. In addition, TBP has been used to modify the organic phase properties.

3.4.1 Extraction of Zn^{2+}

The removal of toxic Zn^{2+} is an important issue in the area of disposal of spent pickling solutions (SPS) from hot-dip galvanizing plants. As a result of pickling, Zn^{2+} concentration in spent solutions increases even up to 130 g/dm^3, iron content to 100 g/dm^3, HCl to 10% (Maass & Peissker, 1998). Taking into account economic, environmental and technical advantages and disadvantages of various regeneration methods, solvent extraction seems to be a reasonable solution for hot-dip galvanizing plants.

Among various investigated extractants, phosphonium ILs have been selected as potentially effective organic phases that are more stable than imidazolium ILs. Model aqueous feeds contained various amounts of Zn^{2+}, 1.8% (0.58 M) HCl and a constant concentration of chloride anions (5 M).

[QP][Cl] and [QP][Bis] extract quickly and almost completely most chlorocomplexes of the metals studied (extraction efficiency near 100%). They are effective and prospective extractants. Salts with more hydrophobic anions [BF_4^-] extract about 60% of Zn^{2+}, while [QP][Tf_2N], as having the most hydrophobic anion, poorly extracts metal ions. The metal ion transfer to the organic phase is influenced by the type of acid, electrolyte and IL concentration. The affinity of [QP][Tf_2N] phase to the aqueous phase is very low and the transfer of Zn^{2+} species is difficult. Water content in the organic phase after extraction confirms changes in the hydrophilicity of the IL phase, which increases in the following order: [QP][Tf_2N] in toluene < [QP][Tf_2N] in TBP < [QP][Cl] in toluene < [QP][Cl] in butylene carbonate. When toluene is replaced with butylene carbonate no decrease in Zn^{2+} extraction is observed and this system is more environmentally friendly.

Extraction efficiency of the phosphonium ILs decreases with increasing hydrophobicity of the anion in the following sequence (Nowak et al., 2010): [QP][Cl] > [QP][Br] > [QP][Bis] > [QP][BF_4]> [QP][PF_6]> [QP][Tf_2N]. The dependence is also observed for the extraction with pure ILs, as it is mentioned for Zn^{2+} and Fe^{3+} extraction with imidazolium salts (Perez de los Rios et al., 2010). Due to low extraction of Zn^{2+} [QP][Tf_2N] cannot be considered as an effective extractant in the system studied (Figs. 2 and 3).

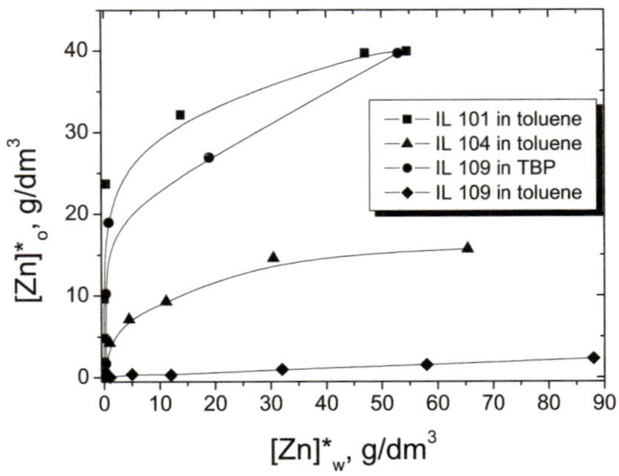

Fig. 2. Isotherms of Zn^{2+} extraction with various phosphonium IL/solvent mixtures.

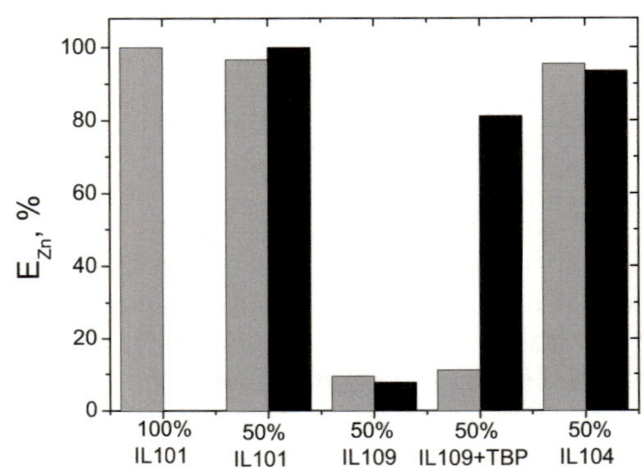

Fig. 3. Percentage extraction of Zn^{2+} with various phosphonium ILs from feed containing 0.58 M HCl (■) and without HCl (■) (Regel-Rosocka et al., 2006).

[QP][Cl] is selected as the most effective extractant among the phosphonium ILs studied (Figs. 2 and 3). Extraction equilibrium is achieved in 5 minutes. It transfers more than 95% of Zn^{2+}, and up to 80% Fe^{2+} from the individual metal ion solutions. Twofold molar excess of the extractant over Zn^{2+} is necessary for efficient extraction (100%). Moreover, Zn^{2+} extraction is preferred over Fe^{2+} when both are present in a mixture. The kinetics of both Zn^{2+} and Fe^{2+} extraction is very fast, and can be successfully applied to separate Zn^{2+} from Fe^{2+} when Zn^{2+} exceeds Fe^{2+} content in the feed. The presence of HCl in the feed enhances Zn^{2+} extraction (Fig. 3). The following reactions of Zn^{2+} extraction mechanism are proposed:

$$\mathrm{ZnCl}_{2(\mathrm{w})} + 2[QP][Cl]_{(o)} \Leftrightarrow [QP]_2[ZnCl_4^-]_{(o)} \tag{18}$$

$$\mathrm{ZnCl}_{4(\mathrm{w})}^{2-} + 2[QP][Cl]_{(o)} \Leftrightarrow [QP]_2[ZnCl_4]_{(o)} + 2\mathrm{Cl}_{(\mathrm{w})}^- \tag{19}$$

Moreover, the studies on stripping and regeneration of IL phase have revealed that sulphuric acid is the best stripping solution from among those studied. The ability to reuse the [QP][Cl]/toluene mixture in several cycles of Zn^{2+} extraction-stripping has been proven. However, Zn^{2+} recovery from the organic phase needs three steps (Regel-Rosocka, 2009).

3.4.2 Extraction of Pd^{2+}

Solvent extraction is proposed as a suitable method for PGMs (platinum group metals) recovery from low concentrated sources. Several extractants have been studied and proposed, e.g., hydroxyoximes, alkyl derivatives of 8-hydroxyquinoline, dialkyl sulphides, hydrophobic amines and quaternary ammonium salts, derivatives of pyridine and pyridinecarboxamides. However, the problems of a slow extraction rate or low metal separation remain to be solved. Looking for extractants permitting possibly fastest extraction of Pd^{2+}, two phosphonium ionic liquids, i.e., [QP][Cl] and [QP][Bis], have been used (Cieszynska et al., 2007; Cieszynska, 2010; Cieszynska & Wisniewski, 2010, 2011). To overcome problems caused by the high viscosity of ILs, the mixtures with toluene have been used for extraction, similarly as for extraction of Zn^{2+}.

Extraction of Pd^{2+} with [QP][Cl] and [QP][Bis] is very fast and the equilibrium is achieved after 5 minutes (Fig. 4) contrary to the case when dialkyl sulphides are used in conventional PGMs extraction. The increase in HCl concentration affects Pd^{2+} extraction, which decreases from more than 90 to near 50% for 0.1 and 3 M HCl, respectively (Fig. 5). Moreover, spontaneous transfer of Pd^{2+} to the organic phase, controlled by diffusion, is observed that suggests great mobility of the interface system.

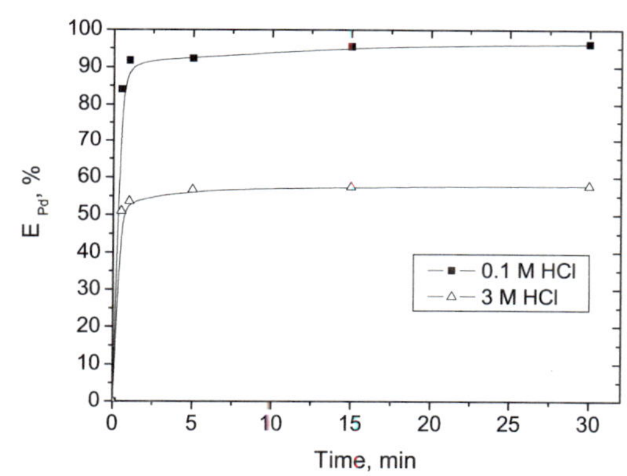

Fig. 4. Effect of contact time on Pd^{2+} extraction with [QP][Cl]: (feed: 5 mM Pd^{2+}, 0.1 or 0.3 M HCl; organic: 5 mM [QP][Cl] in toluene) (Cieszynska & Wisniewski, 2010).

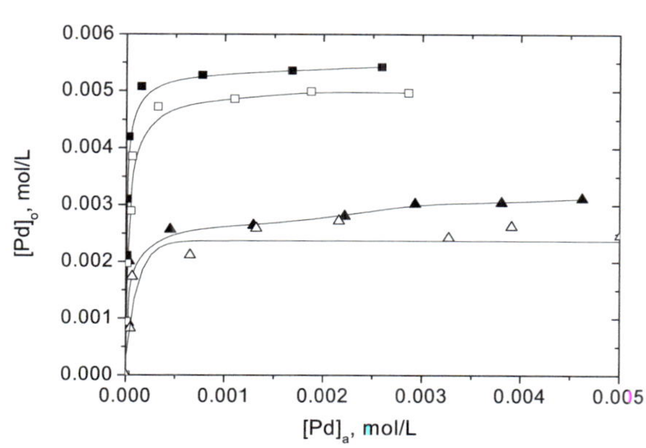

Fig. 5. The isotherms of Pd^{2+} extraction from 0.1 (■,□) and 3 M HCl (▲, △) with [QP][Cl] (■, ▲) and [QP][Bis] (□, △); (feed: 1 - 8 mM Pd^{2+}; organic: 5 mM [QP][Cl] in toluene) (Cieszynska, 2010; Cieszynska & Wisniewski, 2011).

As a result of our investigation, the ion exchange mechanism has been proposed to describe the extraction of Pd^{2+} from 0.1 and 3 M HCl:

[QP][Cl]

$$\mathrm{PdCl}^{2-}_{4(w)} + [QP][Cl]_{(o)} \Leftrightarrow [QP]_2[PdCl_3]_{(o)} + 2\mathrm{Cl}^{-}_{(w)} \tag{20}$$

$$\mathrm{PdCl}^{2-}_{4(w)} + 2[QP][Cl]_{(o)} \Leftrightarrow [QP]_2[PdCl_4]_{(o)} + 2\mathrm{Cl}^{-}_{(w)} \tag{21}$$

[QP][Bis]

$$\mathrm{H}^{+}_{(w)} + \mathrm{PdCl}^{2-}_{4(w)} + [QP][Bis]_{(o)} \Leftrightarrow [QP][PdCl_3]_{(o)} + HBis_{(o)} + \mathrm{Cl}^{-}_{(w)} \tag{22}$$

$$2\mathrm{H}^{+}_{(w)} + \mathrm{PdCl}^{2-}_{4(w)} + 2[QP][Bis]_{(o)} \Leftrightarrow [QP]_2[PdCl_4]_{(o)} + 2HBis_{(o)} \tag{23}$$

On this basis, $[QP][PdCl_3]$ and $[QP]_2[PdCl_4]$ are proposed as ion pairs formed in the organic phase by Pd^{2+}. Consequently, the ionic bonds in the IL phase are very strong, which is a drawback when stripping the metal ions from the loaded IL with water. Different stripping phases have been examined (Table 8). Ammonia solution is found to be an effective one, resulting in almost 100% Pd^{2+} stripped from the loaded IL in one step.

Stripping solution	S_{Pd} (%)
0.5 M H_2SO_4	1.3
0.5 M HCl	10.3
1 M HCl	22.0
3 M HCl	38.6
0.5 M NH_4OH	97.1

Table 8. Percentage stripping (S_{Pd}) of Pd^{2+} stripping from loaded [QP][Cl] with various stripping solutions (Cieszynska, 2010; Cieszynska & Wisniewski, 2010).

Regeneration of spent IL phases and their reuse in extraction is an important economical and ecological issue. Recycling of 0.005 M [QP][Cl], loaded with Pd^{2+}, was tested in 5 cycles of extraction (one step) and stripping (two steps with 0.5 M NH_4OH). After this procedure percentage extraction of Pd^{2+} does not change and remains 98 and 55% for the feed containing 0.1 and 3 M HCl, respectively (Cieszynska, 2010; Cieszynska & Wisniewski, 2011). [QP][Cl] is a stable extractant that can be regenerated and used again for efficient extraction of Pd^{2+}.

To sum up, the presented extraction data indicate that [QP][Cl] and [QP][Bis] can be considered as active extractants for the extraction of metal chlorocomplexes.

3.4.3 Extraction of Co^{2+}

Extraction of Co^{2+} can be given as another example to confirm the usefulness of phosphonium ILs in liquid-liquid extraction. Selective separation of Co^{2+} and Ni^{2+} in hydrometallurgical processes has been investigated for many years. Their chemical affinity means that their behaviour is similar which makes problems with their separation. Both Ni^{2+} and Co^{2+} form divalent ions hydrated by six water molecules in diluted aqueous solutions. The ratio of water exchange for Co^{2+} is much greater than for Ni^{2+}, therefore the complexing ion is more easily formed with Co^{2+} than Ni^{2+}. Moreover, Co^{2+} in the divalent state in high electrolyte concentrations tends to exhibit a tetrahedral rather than an octahedral configuration, giving it six coordinated sites.

As Cyanex 272 is a well known extractant for Co^{2+} ions, Cyphos IL 104, that is a derivative of Cyanex 272, has been applied for cobalt chloride solutions. Efficiency of extraction with both [QP][Cl] and [QP][Bis] is much higher for Co^{2+} than Ni^{2+}.

Aqueous phase	[QP][Cl]			[QP][Bis]		
	$D_{Co(II)}$	$D_{Ni(II)}$	$S_{Co(II)/Ni(II)}$	$D_{Co(II)}$	$D_{Ni(II)}$	$S_{Co(II)/Ni(II)}$
Ni + Co; without HCl	0.15	0.05	3.00	20.4	0.29	70.4
Ni + Co; 0.1 M HCl	0.15	0.05	3.00	0.18	0.03	6.00
Ni + Co; 0.5 M HCl	0.18	0.19	0.95	0.19	0.00	0.00

Table 9. Distribution coefficients of metal ion between organic and aqueous phase and the selectivity of Co^{2+} over Ni^{2+}

An increase in HCl concentration in the feed aqueous phases containing metal ions, negatively influences the extraction efficiency and selectivity (shown in Table 9). The stripping efficiency of Co^{2+} decreases in the following order of stripping phases: 2 M H_2SO_4 (97%) > 0.5 M HCl (96%) > 0.25 M H_2SO_4 (84%) > 4 M HCl (61%) > H_2O (39%), and for Ni^{2+}: 2 M H_2SO_4 (65%) > 0.25 M H_2SO_4 (55%) > distilled water (39%), while HCl does not strip Ni^{2+} ions from the loaded organic phase. On this basis sulphuric acid is selected as the most efficient stripping phase.
Spectrophotometric analysis allows the coordination of Co^{2+} complexes formed in the presence of chlorides to be determined and indicates a change from an octahedral complex in the aqueous phase into tetrahedral one, existing in the organic phase (Ma et al., 2008). [QP][Bis] is very selective for Co^{2+} over Ni^{2+} in the aqueous solutions without HCl ($E_{Co(II)}$ = 95% at pH 6). The same transformation of metal complex coordination is observed during extraction from a mixture of metal ions.
The proposed equations describing Co^{2+} extraction mechanism are as follows:

[QP][Cl]

$$CoCl_{2(w)} + [QP][Cl]_{(o)} \Leftrightarrow [QP][CoCl_3]_{(o)} \tag{24}$$

$$H^+_{(w)} + CoCl^-_{3(w)} + [QP][Cl]_{(o)} \Leftrightarrow [QP][CoCl_3]_{(o)} + H^+_{(w)} + Cl^-_{(w)} \tag{25}$$

$$2H^+_{(w)} + CoCl^{2-}_{4(w)} + 2[QP][Cl]_{(o)} \Leftrightarrow [QP]_2[CoCl_4]_{(o)} + 2H^+_{(w)} + 2Cl^-_{(w)} \tag{26}$$

[QP][Bis]

$$H^+_{(w)} + CoCl^{2-}_{4(w)} + [QP][Bis]_{(o)} \Leftrightarrow [QP][CoCl_3]_{(o)} + HBis_{(o)} + Cl^-_{(w)} \tag{27}$$

$$H^+_{(w)} + CoCl^{2-}_{4(w)} + [QP][Bis]_{(o)} \Leftrightarrow [QP]_2[CoCl_4]_{(o)} - HBis_{(o)} \tag{28}$$

Nevertheless, further studies should be carried out on the mechanism of Co^{2+} extraction with [QP][Bis]. It is known that $CoCl_4^{2-}$ is not stable in the aqueous phase, so the extraction could proceed in another way.

4. Conclusions

Ionic liquids, considered as 'green solvents', have been studied as potential solvents or carriers of metal ions in liquid-liquid extraction. Although recently their 'greenness' is questioned because of possible hydrolysis with formation of toxic HF or partial loss to the aqueous phase, they are still interesting and important compounds in metal processing. Further, mutual solubility of imidazolium ILs and aqueous phase, i.e., problems with extraction but also (when the solubility is too high) with loss of ILs must be pointed out as their limitations. Hence, a compromise between IL hydrophobicity and its extraction power must be achieved. Their potential lies in improvement of extraction efficiency, vast possibilities to design 'functionalised ILs', significant reduction in the volume of the low concentration aqueous streams and process intensification.

Looking at literature data, it is obvious that imidazolium ILs are the best described and applied for a variety of metal ion systems. However, research on extraction with ammonium and phosphonium ILs has been developed in the last five years, and indicates successful application of some of them for separation of metal ions. There are still plenty of research to be pursued in the field of separation with phosphonium and new ammonium ILs, to describe their extraction behaviour, mechanism of metal ion, water and other species transfer to the organic phase. Additionally, their stability and regeneration in separation processes, particularly those operating in strongly acidic or basic solutions, should be investigated.

The mechanism of extraction with ILs in most cases differs from that in conventional solvents, and seems to be more complex. ILs prefer extraction of charged species and, as a result, most metal ions are transported to the IL phase according to the cation or anion-exchange mechanism. It can be attributed to the unique ionic solvation environment and the ion-exchange capabilities of ILs that influence their specific extraction behaviour.

Finally, regeneration of ILs is recently highlighted as an important issue affecting their greenness. Not only stripping with various solutions but pH change and even electrowinning of metal ions are proposed. Nonetheless, it seems that ILs will be rather applied for special separation processes, but not on a large scale.

5. Acknowledgment

We thank Anna Cieszynska for an agreement to present some results from her doctoral thesis. We thank Cytec Industries Inc. for providing us with free samples of Cyphos ILs. This work was supported by the grant No. 32/067/11/DS.

6. References

Abbott, A.P.; Frisch, G.; Hartley, J. & Ryder, K.S. (2011). Processing of Metals and Metal Oxides Using Ionic Liquids. *Green Chem.*, Vol.13, pp. 471-481

Adams, C.J. (2002). Neoteric Solvents: an Examination of Their Industrial Attractiveness, In : *Ionic Liquids. Industrial Applications for Green Chemistry*, R.D. Rogers, K.R. Seddon (Eds.), Vol.818, 15-29, ACS, ISBN13: 978-08-41237-89-6, Washington

Bartsch, R.A.; Chun, S. & Dzyuba, S.V. (2002). Ionic Liquids as Novel Diluents for Solvent Extraction of Metal Salts by Crown Ethers, In : *Ionic Liquids. Industrial Applications for Green Chemistry*, R.D. Rogers, K.R. Seddon (Eds.), Vol.818, 58-68, ACS, ISBN13: 978-08-41237-89-6, Washington

Belova, V.V.; Voshkin, A.A.; Egorova, N.S. & A. I. Kholkin (2010). Extraction of Rare Earth Metals from Nitrate Solutions with a Binary Extractant Based on Cyanex 272. *Russ. J. Inorg. Chem.*, Vol.55, No.4, pp. 629-633

Bradaric, C.J.; Downard, A.; Kennedy, C.; Robertson, A.J. & Zhou, Y. (2003a). Phosphonium Ionic Liquids. *The Strem Chemiker*, Vol. XX, No.1, pp. 2-11

Bradaric, C.J.; Downard, A.; Kennedy, C.; Robertson, A.J. & Zhou, Y. (2003b). Industrial Preparation of Phosphonium Ionic Liquids. *Green Chem.*, Vol. 5, pp. 143-152

Campos, K.; Domingo, R.; Vincent, T.; Ruiz, M.; Sastre, A.M. & Guibal, E. (2008a). Bismuth Recovery from Acidic Solutions Using Cyphos IL-101 Immobilized in a Composite Biopolymer Matrix. *Water Res.*, Vol.42, No.14, pp. 4019-4031

Campos, K.; Vincent, T.; Bunio, P.; Trochimczuk, A. & Guibal, E. (2008b). Gold Recovery from HCl Solutions Using Cyphos IL-101 (a Quaternary Phosphonium Ionic Liquid) Immobilized in Biopolymer Capsules. *Solvent Extr. Ion Exch.*, Vol.26, No.5, pp. 570 – 601

Chun, S.; Dzyuba, S.V. & Bartsch, R.A. (2001). Influence of Structural Variation in Room-Temperature Ionic Liquids on the Selectivity and Efficiency of Competitive Alkali Metal Salt Extraction by a Crown Ether. *Anal. Chem.*, Vol.73, pp. 3737-3741

Cieszynska, A.; Regel-Rosocka, M. & Wisniewski, M. (2007). Extraction of Palladium(II) Ions from Chloride Solutions with Phosphonium Ionic Liquid Cyphos IL101. *Pol. J. Chem. Technol.*, Vol.9, No.2, pp. 99-101

Cieszynska, A. & Wisniewski, M. (2010). Extraction of Palladium(II) from Chloride Solutions with Cyphos IL 101/Toluene Mixtures as Novel Extractant. *Sep. Purif. Technol.*, Vol.73, pp. 202-207

Cieszynska, A. (2010). Extraction of the Palladium(II) in the Presence of Selected Metals from Aqueous Chloride Solutions with Phosphonium Extractants. *Doctoral thesis*, Poznan (in Polish).

Cieszynska, A. & Wisniewski, M. (2011). Selective Extraction of Palladium(II) from Hydrochloric Acid Solutions with Phosphonium Extractants. *Sep. Purif. Technol.* (after revision).

Cocalia, V.A.; Holbrey, J.D.; Gutowski, K.E.; Bridges, N.J. & Rogers, R.D. (2005a). Separations of Metal Ions Using Ionic Liquids: the Challenges of Multiple Mechanisms. *Proceedings of International Solvent Extraction Conference*, ISBN 7-900692-02-9, Beijing, September 2005

Cocalia, V.A.; Jensen, M.P.; Holbrey, J.D.; Spear, S.K.; Stepinski, D. & Rogers, R.D. (2005b) Identical Extraction Behaviour and Coordination of Trivalent or Hexavalent f-Element Cations Using Ionic Liquid and Molecular Solvents. *Dalton Trans.*, Vol.2005, 1966-1971

Correia, M.J.N.; Marques, M.M.; Ismael, M.R. & Reis, M.T.A. (2005). Ionic Liquids as Extractants of Phenolic Compounds and Metal Ions: Influence of the Type of Ionic Liquid. *Proceedings of International Solvent Extraction Conference*, ISBN 7-900692-02-9, Beijing, September 2005

Dai, S.; Ju, Y.H. & Barnes, C.E. (1999). Solvent Extraction of Strontium Nitrate by a Crown Ether Using Room-Temperature Ionic Liquids. *J. Chem. Soc., Dalton Trans.*, Vol.1999, pp. 1201-1202

Del Sesto, R.E.; Corley, C.; Robertson A. & Wilkes, J.S. (2005). Tetraalkylphosphonium-Based Ionic Liquids. *J. Organometall. Chem.*, Vol.690, pp. 2536-2542

Dietz, M.L. & Dzielawa, J.A. (2001). Ion-Exchange as a Mode of Cation Transfer into Room-Temperature Ionic Liquids Containing Crown Ethers: Implications for the "Greenness" of Ionic Liquids as Diluents in liquid-Liquid Extraction. *Chem. Commun.*, Vol.2001, pp. 2124-2125

Dietz, M.L.; Dzielawa, J.A.; Laszak, I.; Young, B.A. & Jensen, M.P. (2003). Influence of Solvent Structural Variations on the Mechanism of Facilitated Ion Transfer into Room-Temperature Ionic Liquids. *Green Chem.*, Vol.5, No.6, pp. 682-685

Dietz, M.L. & Stepinski, D.C., (2005). A Ternary Mechanism for the Facilitated Transfer of Metal Ions into Room-Temperature Ionic Liquids (RTILs): Implications for the "Greenness" of RTILs as Extraction Solvents. *Green Chem.*, Vol.7, pp. 747-750

Dietz, M.L. (2006). Ionic Liquids as Extraction Solvents: Where do We Stand? *Sep. Purif. Technol.*, Vol.41, pp. 2047-2063

Dietz, M.L.; Jakab, S.; Yamato, K. & Bartsch, R.A. (2008). Stereochemical Effects on the Mode of Facilitated Ion Transfer into Room-Temperature Ionic Liquids. *Green Chem.*, Vol.10, pp. 174-176

Dietz, M.L. & Stepinski, D.C., (2008). Anion Concentration-Dependent Partitioning Mechanism in the Extraction of Uranium into Room-Temperature Ionic Liquids. *Talanta*, Vol.75, pp. 598-603

Domanska, U. & Rekawek A. (2009). Extraction of Metal Ions from Aqueous Solutions Using Imidazolium Based Ionic Liquids. *J. Solution Chem.*, Vol.38, pp. 739–751

Egorov, V.M.; Djigailo, D.I.; Momotenko, D.S.; Chernyshov, D.V.; Torocheshnikova, I.I.; Smirnova, S.V. & Pletnev, I.V. (2010). Task-Specific Ionic Liquid Trioctylmethylammonium Salicylate as Extraction Solvent for Transition Metal Ions. *Talanta*, Vol.80, No.3, pp. 1177-1182

Gallardo, V.; Navarro, R.; Saucedo, I.; Ávila, M. & Guibal, E. (2008). Zinc(II) Extraction from Hydrochloric Acid Solutions Using Amberlite XAD-7 Impregnated with Cyphos IL 101 (Tetradecyl(trihexyl)phosphonium Chloride). *Sep. Sci. Technol.*, Vol.43, No.9-10, pp. 2434 - 2459

Giridhar, P.; Venkatesan, K.A.; Srinivasan, T.G. & Vasudewa Rao, P.R. (2005). Extraction of Uranium(VI) from Nitric Acid Medium by 1.1 M Tri-n-butylphosphonate in Ionic Liquid Diluent. *J. Radioanalytical Nuc. Chem.*, Vol.265, pp. 31-38

Giridhar, P.; Venkatesan, K.A.; Srinivasan, T.G. & Vasudeva Rao, P.R. (2006). Extraction of Fission Palladium by Aliquat 336 and Electrochemical Studies on Direct Recovery from Ionic Liquid Phase. *Hydrometallurgy*, Vol.81, pp. 30-39

Guibal, E.; Gavilan, Campos, K.; Bunio, P.; Vincent, T. & Trochimczuk, A. (2008). CYPHOS IL 101 (Tetradecyl(trihexyl)phosphonium Chloride) Immobilized in Biopolymer Capsules for Hg(II) Recovery from HCl Solutions. *Sep. Sci. Technol.*, Vol.43, No.9-10, pp. 2406 - 2433

Han, X. & Armstrong, D.W. (2007). Ionic Liquids in Separations. *Acc. Chem. Res.*, Vol.40, pp. 1079-1086

Haverlock, T.J.; Bonnesen, P.V.; Sachleben, R.A. & Moyer, B.A. (2000). Analysis of Equilibria in the Extraction of Cesium Nitrate by Calix[4]arene-bis(t-octylbenzo-crown-6) in 1,2-Dichloroethane. *J. Inclusion Phenom. Macrocyclic Chem.*, Vol.36, pp. 21-37

Holbrey, J.D. & Seddon, K.R. (1999). Ionic Liquids. *Clean Products and Processes*, Vol.1, pp. 223–236

Holbrey, J.D. & Rogers, R.D. (2002). Green Chemistry and Ionic Liquids: Synergies and Ionies. In : *Ionic Liquids. Industrial Applications for Green Chemistry*, R.D. Rogers, K.R. Seddon (Eds.), Vol.818, 2-14, ACS, ISBN13: 978-08-41237-89-6, Washington

Hunddleston, J.G.; Willauer, H.D.; Swatloski, R.P.; Visser, A.E. & Rogers, R.D. (1998). Room Temperature Ionic Liquids as Novel Media for "Clean" Liquid-Liquid Extraction. *Chem. Commun.*, Vol.1998, pp. 1765-1766

Jensen, M.P.; Dzielawa, J.A.; Rickert P. & Dietz, M.L. (2002). EXAFS Investigations of the Mechanism of Facilitated Ion Transfer into a Room-Temperature Ionic Liquid. *J. Am. Chem. Soc.*, Vol.124, pp. 10664-10665

Jensen, M.P.; Neuefeind, J.; Beitz, J.V.; Skanthakumar, S. & Soderholm, L. (2003). Mechanisms of Metal Ion Transfer into Room-Temperature Ionic Liquids: the Role of Anion Exchange. *J. Am. Chem. Soc.*, Vol.125, pp. 15466-15473

Kejun, L.; Yen, W.T.; Shibayama, A.; Miyazaki, T. & Fujita T. (2004). Gold Extraction from Thiosulfate Solution Using Trioctylmethylammonium Chloride. *Hydrometallurgy*. Vol. 73, pp. 41–53

Kogelnig, D.; Stojanovic, A.; Galanski, M.; Groessl, M.; Jirsa, F.; Krachler, R. & Keppler, B.K (2008). Greener Synthesis of New Ammonium Ionic Liquids and Their Potential as Extracting Agents. *Tetrahedron Lett.*, Vol.49, No.17, pp. 2782-2785

Kogelnig, D.; Stojanovic, A.; Jirsa, F.; Körner, W.; Krachler, R. & Keppler, B.K. (2010). Transport and Separation of Iron(III) from Nickel(II) with the Ionic Liquid Trihexyl(tetradecyl)phosphonium Chloride. *Sep. Purif. Technol.*, Vol.72, No.1, pp. 56-60

Kosmulski, M.; Marczewska-Boczkowska, K. & Saneluta, C. (2002). Low Temperature Ionic Liquids – a Laboratory Curiosity or a Technological Revolution? *Przem. Chem.*, Vol.81, pp. 106-110 (in Polish)

Kozonoi, N. & Ikeda, Y. (2007). Extraction Mechanism of Metal Ion from Aqueous Solution to the Hydrophobic Ionic Liquid, 1-Butyl-3-methylimidazolium Nonafluorobutanesulfonate. *Monatshefte fur Chemie*, Vol.138, pp. 1145–1151

Luo, H.; Dai, S. & Bonnesen, P.V. (2004a). Solvent Extraction of Sr^{2+} and Cs^{+} Based on Room-Temperature Ionic Liquids Containing Monoaza-Substituted Crown Ethers. *Anal. Chem.*, Vol.76, pp. 2773-2779

Luo, H.; Dai, S.; Bonnesen, P.V.; Buchanan III, A.C.; Holbrey, J.D.; Bridges, N.J. & Rogers, R.D. (2004b). Extraction of Cesium Ions from Aqueous Solutions Using Calix[4]arene-bis(tert-octylbenzo-crown-6) in Ionic Liquids. *Anal. Chem.*, Vol.76, pp. 3078-3083

Luo, H.; Dai, S.; Bonnesen, P.V. & Buchanan III, A.C. (2006a). Separation of Fission Products Based on Ionic Liquids : Task-specific Ionic Liquids Containing an Aza-crown Ether Fragment. *J. Alloy Compd.*, Vol.418, 195-199

Luo, H.; Dai, S.; Bonnesen, P.V., Haverlock T.J., Moyer B.A & Buchanan III, A.C. (2006b). A Striking Effect of Ionic-Liquid Anions in the Extraction of Sr^{2+} and Cs^{+} by Dicyclohexano-18-Crown-6. *Solvent Extr. Ion Exch.*, Vol.24, pp. 19-31

Ma, H.; Wan, C. & Zewail, A.H. (2008). Dynamics of Ligand Substitution in Labile Cobalt Complexes Resolved by Ultrafast T-jump. *Proceedings of the National Academy of Sciences of the USA*, Vol.105, No.35, pp. 12754–12757

Maass, P. & Peissker, P. (1998). *Hot-dip Galvanizing*. Agencja Wydawnicza Placet, ISBN 978-83-85428-28-3, Warsaw (in Polish)

Marszalkowska, B.; Regel-Rosocka, M.; Nowak, Ł. & Wisniewski, M. (2010). Quaternary Phosphonium Salts as Effective Extractants of Zinc(II) and Iron(III) Ions from Acidic Pickling Solutions. *Pol. J. Chem. Technol.*, Vol.12, No.4, pp. 1-5

Nakashima, K.; Kubota, F.; Maruyama, T. & Goto, M. (2003). Ionic Liquids as a Novel Solvent for Lanthanide Extraction. *Anal. Sci.*, Vol.19, pp. 1097-1098

Nakashima, K.; Kubota, F.; Maruyama, T. & Goto, M. (2005). Feasibility of Ionic Liquids as Alternative Separation Media for Industrial Solvent Extraction Processes. *Ind. Eng. Chem. Res.*, Vol.44, pp. 4368-4372

Nowak, L.; Regel-Rosocka, M.; Marszałkowska, B. & Wisniewski, M. (2010). Removal of Zn(II) from Chloride Acidic Solutions with Hydrophobic Quaternary Salts. *Pol. J. Chem. Technol.*, Vol.12, No.3, pp. 24-28

Papaiconomou, N.; Lee, J.M.; Salminen, J.; von Stosch, M. & Prausnitz, J.M. (2008). Selective Extraction of Copper, Mercury, Silver, and Palladium Ions from Water Using Hydrophobic Ionic Liquids. *Ind. Eng. Chem. Res.*, Vol.47, pp. 5080-5086

Perez de los Rios, A.; Hernandez-Fernandez, F.J.; Sanchez-Segado, S.; Lozano, L.J.; Moreno, J.I. & Godinez, C. (2010). Selective Separation of Zn(II) and Fe(III) from Acidic Media Using Ionic Liquids as Sole Extraction Agents. *Chem. Eng. Transactions*, Vol.21, pp. 625-630, ISBN 978-88-95608-05-1

Pernak, J. (2003). Ionic Liquids. Compounds for the 21st Century. *Przem. Chem.*, Vol. 82, pp. 521-524 (in Polish)

Pernak, J.; Stefaniak, F. & Weglewski, J. (2005). Phosphonium Acesulfamate Based Ionic Liquids. *Eur. J. Org. Chem.*, Vol.2005, No.4, pp. 650-652

Regel-Rosocka, M.; Cieszyńska K. & Wiśniewski M. (2006). Extraction of Zinc(II) with Selected Phosphonium Ionic Liquids, *Przem. Chem.*, Vol.85, pp. 651-654 (in Polish)

Regel-Rosocka, M.; Cieszynska, A. & Wisniewski, M. (2007). Methods of Regeneration of Spent Pickling Solutions from Steel Treatment Plants. *Pol. J. Chem. Technol.*, Vol.2, pp. 42-45

Regel-Rosocka, M. (2009). Extractive Removal of Zinc(II) from Chloride Liquors with Phosphonium Ionic Liquids/Toluene Mixtures as Novel Extractants, *Sep. Purif. Technol.*, Vol.66, pp. 19-24

Regel-Rosocka, M. (2010). A Review on Methods of Regeneration of Spent Pickling Solutions from Steel Processing. *J. Hazard. Mater.*, Vol.177, No.1-3, pp. 57-69

Rickert, P.G.; Stepinski, D.C.; Rausch, D.J.; Bergeron, R.M.; Jakab, S. & Dietz, M.L. (2007). Solute-Induced Dissolution of Hydrophobic Ionic Liquids in Water. *Talanta*, Vol.72, pp. 315-320

Sato, T.; Takayanagi, T. & Sato, K. (2010). Liquid-Liquid Extraction of Palladium(II) from Hydrochloric Acid Solutions by High Molecular Weight Amines. *Solvent Extr. Res. Dev., Japan*, Vol.17, pp. 95-110

Seddon, K.R.; Stark, A. & Torres, M.J. (2000). Influence of Chloride, Water, and Organic Solvents on the Physical Properties of Ionic Liquids. *Pure Appl. Chem.*, Vol.72, pp. 2275-2287

Shen, Y.; Tan, X.; Wang, L. & Wu, W. (2011). Extraction of the Uranyl Ion from the Aqueous Phase into Anionic Liquid by Diglycolamide. *Sep. Purif. Technol.*, doi: 10.1016/j.seppur.2011.01.042, in press.

Shimojo, K. & Goto, M. (2004). Solvent Extraction and Stripping of Silver Ions in Room-Temperature Ionic Liquids Containing Calixarenes. *Anal. Chem.*, Vol.76, pp. 5039-5044

Shimojo, K.; Kurahashi, K. & Naganawa, H. (2008). Extraction Behavior of Lanthanides Using a Diglycolamide Derivative TODGA in Ionic Liquids. *Dalton Trans.*, No.37, pp. 5083-5088

Srncik, M.; Kogelnig, D.; Stojanovic, A.; Körner, W.; Krachler, R. & Wallner, G. (2009). Uranium Extraction from Aqueous Solutions by Ionic Liquids. *Appl. Radiat. Isotopes*, Vol.67, No.12, pp. 2146-2149

Stepinski, D.C.; Jensen, M.P.; Dzielawa, J.A. & Dietz, M.L. (2005). Synergistic Effects in the Facilitated Transfer of Metal Ions into Room-Temperature Ionic Liquids. *Green Chem.*, Vol.7, pp. 151-158

Stepinski, D.C.; Vandegrift, G.F.; Shkrob, I.A.; Wishart, J.F.; Kerr, K.; Dietz, M.L.; Qadah, D.T.D. & Garvey, S.L. (2010). Extraction of Tetra-Oxo Anions into a Hydrophobic, Ionic Liquid-Based Solvent without Concomitant Ion Exchange. *Ind. Eng. Chem. Res.*, Vol.49, pp. 5863–5868

Stojanovic, A.; Kogelnig, D.; Fischer, L.; Hann, S.; Galanski, M.; Groessl, M.; Krachler, R. & Keppler, B. K. (2010). Phosphonium and Ammonium Ionic Liquids with Aromatic Anions: Synthesis, Properties, and Platinum Extraction. *Aust. J. Chem.*, Vol.63, pp. 511–524

Sun, X.; Ji, Y.; Zhang, L.; Chen, J. & Li, D. (2010). Separation of Cobalt and Nickel Using Inner Synergistic Extraction from Bifunctional Ionic Liquid Extractant (Bif-ILE). J. *Hazard. Mater.*, Vol.182, pp. 447-452

Swatloski, R.P.; Holbrey J.D. & Rogers, R.D. (2003). Ionic Liquids Are Not Always Green: Hydrolysis of 1-Butyl-3-methylimidazolium Hexafluorophosphate. *Green Chem.*, Vol.5, pp. 361–363

Turanov, A.N.; Karndashev, V.K. & Baulin, V.E. (2008). Effect of Ionic Liquids on the Extraction of Rare-earth Elements by Bidentate Neutral Organophosphorus Compounds from Chloride Solutions. *Russ. J. Inorg. Chem.*, Vol.53, pp. 970-975

Vidal, S.T.M.; Correia, M.J.N.; Marques, M.M.; Ismael, M.R. & Reis, M.T.A. (2004). Studies on the Use of Ionic Liquids as Potential Extractants of Phenolic Compounds and Metal Ions. *Sep. Sci. Technol.*, Vol.39, pp. 2155-2169

Visser, A.E.; Swatloski, R.P.; Reichert, W.M.; Griffin, S.T. & Rogers, R.D. (2000). Traditional Extractants in Nontraditional Solvents: Groups 1 and 2 Extraction by Crown Ethers in Room-Temperature Ionic Liquids. *Ind. Eng. Chem. Res.*, Vol 39, pp. 3596-3604

Visser, A.E.; Holbrey, J.D. & Rogers, R.D. (2001a). Hydrophobic Ionic Liquids Incorporating N-alkylisoquinilinium Cations and Their Utilization in Liquid-Liquid Separations. *Chem. Commun.*, pp. 2484-2485

Visser, A.E.; Swatloski, R.P.; Griffin, S.T.; Hartman, D.H. & Rogers, R.D. (2001b). Liquid/Liquid Extraction of Metal Ions in Room Temperature Ionic Liquids. *Sep. Sci. Technol.*, Vol.36, pp. 785-804

Visser, A.E.; Swatloski, R.P.; Reichert, W.M.; Mayton, R.; Sheff, S.; Wierzbicki, A.; Davis, J.H. & Rogers, R.D. (2001c). Task-specific Ionic Liquids for the Extraction of Metal Ions from Aqueous Solutions. *Chem. Commun.*, Vol.2001, pp. 135-136

Visser, A.E.; Holbrey, J.D. & Rogers, R.D. (2002a). Room Temperature Ionic Liquids as Alternatives to Traditional Organic Solvents in Solvent Extraction. *Proceedings of Solvent Extraction Conference*, ISBN 1-919783-24-5, Cape Town, March 2002

Visser, A.E.; Swatloski, R.P.; Reichert, W.M.; Mayton, R.; Sheff, S.; Wierzbicki, A.; Davis, Jr., J.H. & Rogers, R.D. (2002b). Task-Specific Ionic Liquids Incorporating Novel Cations for the Coordination and Extraction of Hg^{2+} and Cd^{2+}: Synthesis, Characterization, and Extraction Studies. *Environ. Sci. Technol.*, Vol.36, pp. 2523-2529

Visser, A.E.; Jensen, M.P.; Laszak, I.; Nash, K.L.; Choppin, G.R. & Rogers, R.D. (2003). Uranyl Coordination Environment in Hydrophobic Ionic Liquids: An in Situ Investigation. *Inorg. Chem.*, Vol.42, pp. 2197-2199

Visser, A.E. & Rogers, R.D. (2003). Room-Temperature Ionic Liquids: New Solvents for f-element Separations and Associated Solution Chemistry. *J. Solid State Chem.*, Vol.171, pp. 109-113

Wei, G.T.; Yang, Z. & Chen, C.J. (2003a). Room Temperature Ionic Liquid as a Novel Medium for Liquid/Liquid Extraction of Metal Ions. *Anal. Chim. Acta*, Vol.488, pp. 183-192

Wei, G.T.; Chen, C.J. & Yang, Z. (2003b). Studies on Liquid-Liquid Extraction of Copper Ion with Room Temperature Ionic Liquid. *Journal of Chinese Chemical Society*, Vol.50, pp 1123-1130

Wionczyk, B. (2009). Effect of Temperature on the Extraction of Chromium(III) from Alkaline Aqueous Solutions with Trioctylmethylammonium Chloride (Aliquat 336). *Solvent Extr. Ion Exch.*, Vol.27, No.3, pp. 423 - 446

Zuo, Y.; Liu, Y.; Chen, J. & Li, D.Q. (2008). The Separation of Cerium(IV) from Nitric Acid Solutions Containing Thorium(IV) and Lanthanides(III) Using Pure $[C_8mim]PF_6$ as Extracting Phase. *Ind. Eng. Chem. Res.*, Vol.47, pp. 2349-2355

Zuo, Y.; Liu, Y.; Chena, J. & Lia, D.Q. (2009). Extraction and Recovery of Cerium(IV) along with Fluorine(I) from Bastnasite Leaching Liquor by DEHEHP in $[C_8mim]PF_6$. *J. Chem. Technol. Biotechnol.*, Vol.84, pp. 949–956

Part 6

Miscellaneous Applications

19

Inner Filter Effect in the Fluorescence Emission Spectra of Room Temperature Ionic Liquids with-β-Carotene

Krzysztof Pawlak[1], Andrzej Skrzypczak[2] and Grazyna E. Bialek-Bylka[1]
[1]*Institute of Physics, Faculty of Technical Physics,*
[2]*Faculty of Chemical Technology,*
Poznan University of Technology,
Poland

1. Introduction

Steady-state fluorescence spectroscopy is highly sensitive an analytical fluorescence detecting tool limited to diluted samples. The serious problem of quantitative analysis is dominant in the samples with optical density higher than 0.05 (10 mm cell) detected in the right angle cell geometry, because of the two kinds of inner filter effects IFEs (Lakowicz, 2006; Li & Hu, 2007; Kao et al., 1998). The primary inner filter effect (PIFE) - defined as the decrease in the intensity of the excitation beam at the point of observation because of the chromophore optical absorption in the excitation region. In the fluorescence emission experiment, an apparent decrease in the fluorescence emission quantum yield and/or distortion of band shape as a result of re-absorption of emitted radiation is observed. Finally, some molecules of the investigated samples will be excited by the less intense light. For example the excitation light intensity (I_o) at center of 10 mm cuvette is (0.88 I_o) for (CD = 0.1). The secondary inner filter effect (SIFE) takes place when the fluorescence intensity decreases as a result of the chromophore absorption in the emission region. More serious is PIFE than SIFE because excitation wavelengths are always shorter than emission wavelength, especially in organic samples. Generally, the relationship between fluorescence intensity (I_F) and fluorophore concentration (c) is logarithmic (Guilbault, 1990) - $I_F = I_o \phi_F (1-e^{-\varepsilon l c})$, where ($I_o$) is the incident light intensity, (ϕ_F) is the fluorescence quantum efficiency, (ε) is the molar extinction coefficient, (l) is the cell path length. In the concentrated sample the excitation radiation is absorbed significantly by the fluorophore or other chromophores in the cell and the (PIFE) plays a significant role.

For dilute solutions (absorbance <0.01) fluorescence is uniformly distributed. The above equation is reduced to the linear form: $I_F = k I_o \phi_F \varepsilon l c$ with approximately 1% error, where (k) is proportionality constant. For the fluorescence experiment it is always important to establish either by calculation or by measurement the concentration value (C_{max}) at which a plot of fluorescence emission against concentration becomes nonlinear. This value can be calculated from equation: $C_{max} = 0.05/\varepsilon l$, where: symbols ($\varepsilon$) and (l) have the same meaning as in the other equations. From the nonlinear calibration curves (fluorescence intensity *versus* fluorophore concentrations) the information of the presence of the IFEs in concentrated samples can be found. The fluorescence data (calibration curve) of high molecule concentrations (10 mm cuvette) can be corrected for IFE (Lakowicz, 1983): $F_{corr} = F_{obs}$ Antilog$(A_{ex} + A_{em})/2$ where: (F_{obs}) and (F_{corr}) - the observed and corrected fluorescence intensity values, respectively and (A_{ex}) and (A_{em}) - the absorbance values at the excitation and the emission wavelengths, respectively. If calibration curve is not perfectly linear, it means that the molar extinction coefficient (ε) depends on the refractive index (n) of the solvent: $n\,\varepsilon / (n^2 + 2)^{1/2}$.

In order to be able to measure the IFE - less fluorescence spectra of RTILs alone and with non-polar (very low fluorescence yield in conventional solvent like hexane) β-carotene molecules in RTILs, a special quartz Suprasil μ-cuvette has been used. Short pass cells are designed in order to reduce the PIFE along a particular (usually excitation) axis (Koel, 2008; Wasserscheid & Welton, 2008).

According to our knowledge the IFE study of strongly fluorescing RTILs alone and with β-carotene has not been undertaken yet, but they are very important data for fluorescence result discussion. Generally, both the fluorescence emission or excitation spectra can be measured, but in our project we concentrated on the measurement of the emission spectra, because of the less subject to IFE than in fluorescence excitation spectra. The less dependence on IFE fluorescence excitation spectra will be discuss in order to explain 'red edge effect'.

2. Room temperature ionic liquids - synthesis and purification

The following designed structures of RTILs were synthesized and carefully purified: 1-methyl-3-octyloxymethylimidazolium tetrafluoroborate (*IL1*), 1-methyl-3-hexyloxymethylimidazolium bis(trifluoromethylsulfonyl)imide (*IL2*), 1-methyl-3-octyloxymethylimidazolium bis(trifluoromethylsulfonyl)imide (*IL3*) and 1-methyl-3-octylimidazolium tetrafluoroborate (*IL4*). The (*IL2*) and (*IL3*) RTILs differ in the cationic part $[C_6H_{13}OCH_2\text{-}C_1Im]^+$, $[C_8H_{17}OCH_2\text{-}C_1Im]^+$, respectively. According to the literature (Koel, 2008; Wasserscheid & Welton, 2008) and our measurements viscosity of tested compounds increased due to the cations: $[C_6H_{13}OCH_2\text{-}C_1Im]^+$, $[C_8H_{17}OCH_2\text{-}C_1Im]^+$ and $[C_8\text{-}C_1Im]^+$. The (*IL3*) and (*IL1*) RTILs have the same cationic part: $[C_8H_{17}OCH_2\text{-}C_1Im]^+$ but different anions $[(CF_3SO_2)_2N]^-$ and $[BF_4]^-$, respectively. The (*IL1*) and (*IL4*) RTILs having the same anion $[BF_4]^-$ differ in the cationic part: $[C_8H_{17}OCH_2\text{-}C_1Im]^+$ and $[C_8\text{-}C_1Im]^+$, respectively.

The ionic liquid 1-methyl-3-octylimidazolium tetrafluoroborate was purchased from the Sigma-Aldrich and carefully purified.

2.1 Preparation of testing ILs

Preparation of 1-methyl-3-octyloxymethylimidazolium and 1-methyl-3-hexyloxymethyl-imidazolium chlorides (Fig. 1.). To anhydrous solution of 0.1 mol 1-methylimidazole in acetonitrile, 0.105 mol of chloromethyloctyl ether or chloromethylhexyl ether was added. The reaction was carried out for 1h and the product was purified by extraction with heptane in 343K. The final product was hygroscopic compound with the yield 92.5% for 1-methyl-3-octyloxymethylimidazolium chloride and 93.2% for 1-methyl-3-hexyloxymethyl-imidazolium chloride.

Preparation of 1-methyl-3-octyloxymethylimidazolium tetrafluoroborate (*IL1*) (Fig. 2) (Pernak et al., 2001) and 1-methyl-3-octyloxymethylimidazolium bis(trifluoromethylsulfonyl)imide (*IL3*), and 1-methyl-3-hexyloxymethylimidazolium bis(trifluoromethylsulfonyl)imide (*IL 2*).

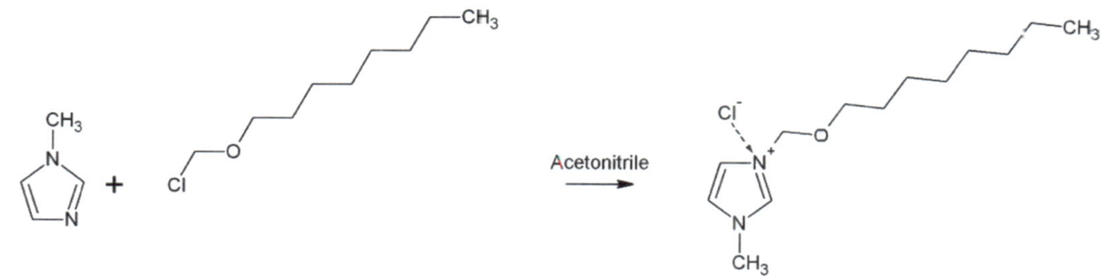

Fig. 1. Preparation of 1-methyl-3-octyloxymethylimidazolium chloride.

To 1-methyl-3-octyloxymethylimidazolium chloride or 1-methyl-3-hexyloxymethyl-imidazolium chloride (0.09 mol) dissolved in 30 ml methanol, sodium tetrafluoroborate (0.095 mol) or lithium bis(trifluoromethylsulfonyl)imide was added. The reaction was completed by heating (at 328K) on a water bath with stirring (24 h). After removing solvent in vacuum, product was once again dissolved in anhydrous acetone in order to filter small amounts of NaCl and excess of $NaBF_4$ or $Li[(CF_3SO_2)_2N]$. Products were a colorless liquids with the yield of 98%.

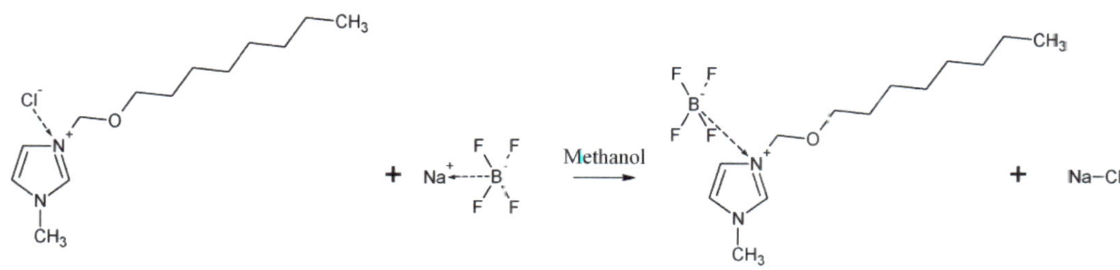

Fig. 2. Preparation of 1-methyl-3-octyloxymethylimidazolium tetrafluoroborate (IL1).

2.2 Structures of synthesized and tested ionic liquids

(*IL1*), $[C_8H_{17}OCH_2\text{-}C_1Im][BF_4]$, 1-methyl-3-octyloxymethylimidazolium tetrafluoroborate:

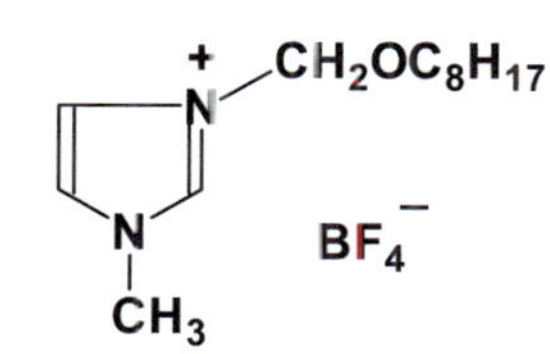

(*IL2*), $[C_6H_{13}OCH_2\text{-}C_1Im][(CF_3SO_2)_2N]$, 1-methyl-3-hexyloxymethylimidazolium bis(trifluoromethylsulfonyl)imide:

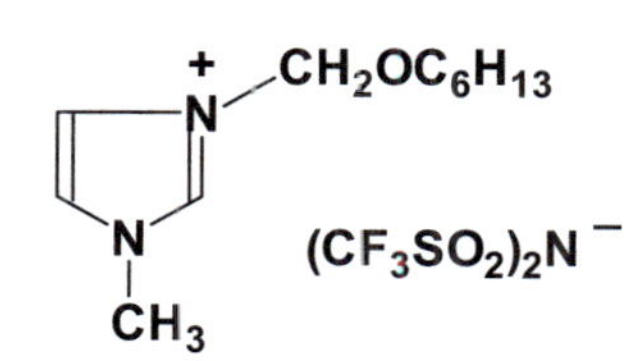

(*IL3*), $[C_8H_{17}OCH_2\text{-}C_1Im][(CF_3SO_2)_2N]$, 1-methyl-3-octyloxymethylimidazolium bis(trifluoromethylsulfonyl)imide:

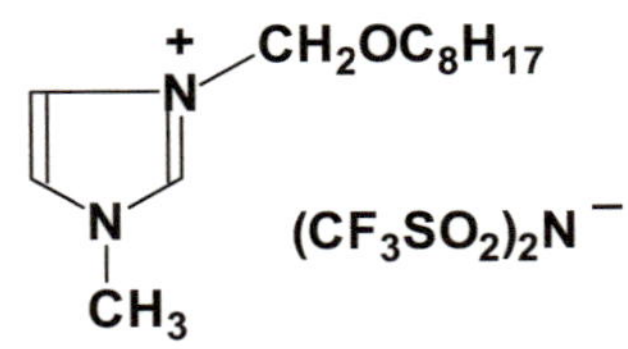

(*IL4*), $[C_8\text{-}C_1Im][BF_4]$, 1-methyl-3-octylimidazolium tetrafluoroborate:

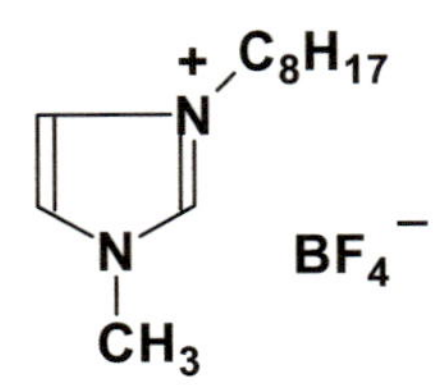

2.3 Purification of testing ILs

It is well known that ILs should be colorless, but frequently they are not. Chromophores which are present in ionic liquids are the products of side reactions occurring at high temperatures (higher than 75^{0}C) especially in the alkylation reactions when ILs halides are made (Earle et al. 2007). Our attempt to have "spectroscopic grade" ILs was done in four ways (Gordon et al., 2003):

1. purification of started materials - all substrates were distilled just before reactions,
2. control temperature of quaternisation reactions – temp. below 50^{0}C,
3. controlling conditions of anion exchange,
4. cleaning of ILs – passing dichloromethane solution of IL (3:1) through the special column, similar to the one used in conventional chromatography. Column was packed with charcoal and Celite and pretreated with dichloromethane.

2.4 Experimental system and spectroscopic methods

The aim of this chapter also is the introduction of the photo physical study of model system (β-carotene in RTILs) mimicking very well spectroscopic data of this photosynthetic pigment *in situ* in the photosynthetic system. Unique fluorescence properties of β-carotene in room temperature ionic liquids (RTIL) (Bialek-Bylka et al., 2007) and new (β-carotene) electronic states ($3A_g^-$ and $1B_u^-$) of both important in photosynthesis isomers: all-*trans* and 15-*cis* in ionic liquid with dimethylformamide (DMF) were found (Bialek-Bylka et al., 2008) and also a sensor detecting the methoxy group in the cation part of imidazolium ionic liquid was developed by us (Pawlak et al., 2009). In order to determine a detailed answer to the unique fluorescence properties, around 100 times higher fluorescence yield of β-carotene in (*IL1*) (1-methyl-3-octyloxymethylimidazolium tetrafluoroborate) than in standard solvent *n*-hexane, the designed structures of RTILs were synthesized and carefully purified: 1-methyl-3-octyloxymethylimidazolium tetrafluoroborate (*IL1*), 1-methyl-3-hexyloxymethylimidazolium bis(trifluoromethylsulfonyl)imide (*IL2*), 1-methyl-3-octyloxymethylimidazolium bis(trifluoromethylsulfonyl)imide (*IL3*), 1-methyl-3-

octylimidazolium tetrafluoroborate (*IL4*). The differences in the structures of the cationic and anionic part of the RTILs investigated by us are shown in the Scheme 1. In this scheme: the first (*IL2*) and second (*IL3*) RTILs differ in the cationic part $[C_6H_{13}OCH_2\text{-}C_1Im]^+$, $[C_8H_{17}OCH_2\text{-}C_1Im]^+$, respectively. According to the literature (Koel, 2008; Wasserscheid & Welton, 2008) and our measurements viscosity of tested compounds increased due to the cations: $[C_6H_{13}OCH_2\text{-}C_1Im]^+$, $[C_8H_{17}OCH_2\text{-}C_1Im]^+$ and $[C_8\text{-}C_1Im]^+$. The second (*IL3*) and third (*IL1*) RTILs have the same cationic part: $[C_8H_{17}OCH_2\text{-}C_1Im]^+$. Finally third (*IL1*) and fourth (*IL4*) RTILs differ again in the cationic part: $[C_8H_{17}OCH_2\text{-}C_1Im]^+$ and $[C_8\text{-}C_1Im]^+$, respectively, but having the same anion $[BF_4]^-$.

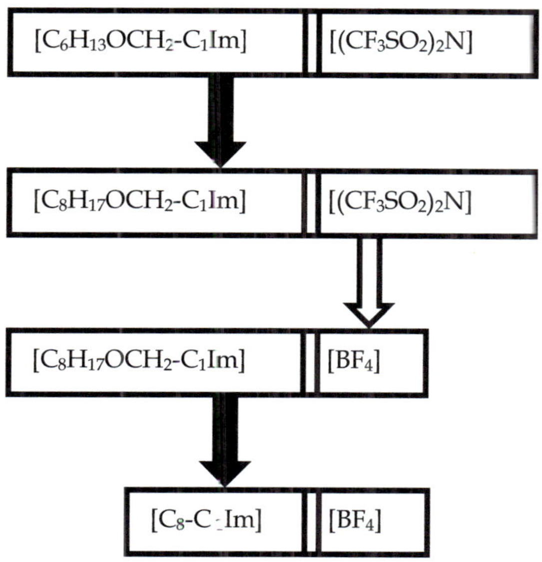

Scheme 1. The differences in the structure of the cationic and anionic part of the investigated RTILs: $[C_8H_{17}OCH_2\text{-}C_1Im][BF_4]$, (1-methyl-3-octyloxymethylimidazolium tetrafluoroborate) (*IL1*); $[C_6H_{13}OCH_2\text{-}C_1Im][(CF_3SO_2)_2N]$, (1-methyl-3-hexyloxymethylimidazolium bis(trifluoromethylsulfonyl) imide) (*IL2*); $[C_8H_{17}OCH_2\text{-}C_1Im][(CF_3SO_2)_2N]$, (1-methyl-3-octyloxymethylimidazolium bis(trifluoromethylsulfonyl)imide) (*IL3*). $[C_8\text{-}C_1Im][BF_4]$, (1-methyl-3-octylimidazolium tetrafluoroborate) (*IL4*).

In order to be able to measure the IFE - less fluorescence spectra of RTILs alone and with non-polar and almost non-fluorescing β-carotene molecules in RTILs, a special quartz Suprasil μ-cuvette (1 mm x 10 mm) was used. Short pass cell is used in order to reduce the PIFE along a particular (usually excitation) axis (Koel, 2008; Wasserscheid & Welton, 2008). Generally, the fluorescence emission or excitation spectra can be measured. For our project we concentrated on measurement of the emission spectra because they are less subject to IFE than fluorescence excitation spectra.
The ultraviolet (UV) and visible (Vis) absorption and fluorescence spectra were measured with UV-VIS Lambda 20 Perkin-Elmer Spectrophotometer and LS-55 Perkin-Elmer fluorescence spectrometer, respectively. Both fluorescence quartz cuvettes: a standard square 10 mm x 10 mm and a short path (μ-cuvette) 1 mm x 10 mm, were used. The right angle fluorescence measurements geometry is shown in the Fig. 3.

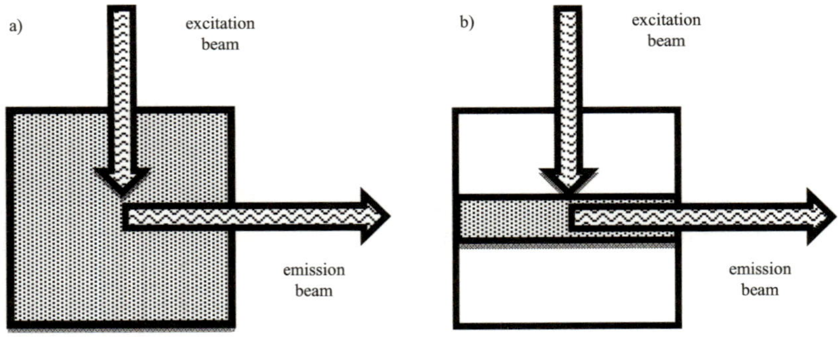

Fig. 3. The right angle cell geometry of the fluorescence measurements: a) standard cuvette 10 mm x 10 mm, b) μ- cuvette 1 mm x 10 mm.

Fluorescence quantum yield was calculated according to the formula (Albani, 2007):

$$Q_x = Q_r \frac{A_r}{A_x} \frac{\int_0^\infty F_x d\lambda}{\int_0^\infty F_r d\lambda} \frac{n_x^2}{n_r^2} \quad (1)$$

where: (Q_x) and (Q_r) are the fluorescence quantum yields of measured sample (x) and reference (r) (Coumarin 6-$C_{20}H_{18}N_2O_2S$; 98%; Sigma-Aldrich), respectively; $\int_0^\infty F d\lambda$ are areas under the fluorescence spectra of the sample (F_x) and the reference (F_r), respectively; (A_x) and (A_r) are the optical densities of the sample and the reference solution at the excitation wavelength, respectively; and (n_x) and (n_r) are the values of the refractive index for the sample and reference, respectively.

3. Results and discussion

In this chapter the attempt in order to explain the unusual high fluorescence quantum yield of β-carotene in RTILs has been undertaken. The inner-filter effect in fluorescence spectra has been observed in a conventional right angle cell. In order to minimize this effect a short pass cell (μ-cuvette) was applied. The right angle geometry is the most common in the spectroscopy, because at right an angle the elastic light scattering signal is the least intense. Also the cell walls directly illuminated by the excitation beam are not directly viewed, so the contribution of fluorescence cell wall distortion is reduced (Ingle & Crouch, 1988). Application of the RTILs as solvents for photo physical studies of molecules in RTILs (in our case photosynthetic pigment all-*trans* β-carotene) depends on transparency of solvents in the visible region of absorption.

Figure 4 shows the absorption spectra of synthesized ionic liquids. These ionic liquids are characterized by a spectral purity in the range above 350 nm. Comparison of these spectra and literature data (Paul & Samanta, 2006) has shown that our ionic liquid purification technique is very efficient.

The figure 5 presents normalized absorption spectra of all-*trans* β-carotene in the following RTILs: *(IL1)*, *(IL2)*, *(IL3)*, *(IL4)*.

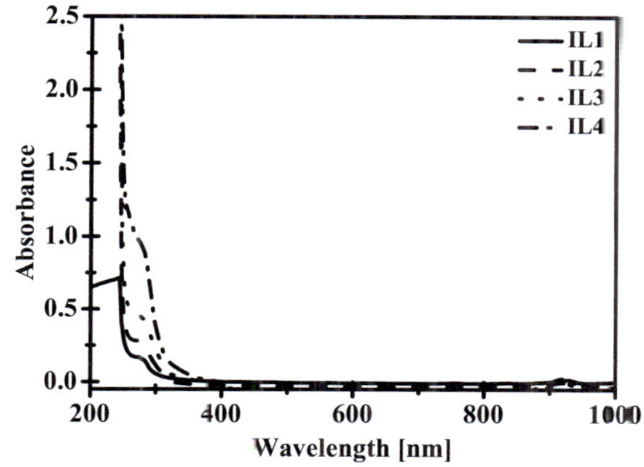

Fig. 4. The absorption spectra of *(IL1)* $[C_8H_{17}OCH_2\text{-}C_1Im][BF_4]$; *(IL2)* $[C_6H_{13}OCH_2\text{-}C_1Im][(CF_3SO_2)_2N]$; *(IL3)* $[C_8H_{17}OCH_2\text{-}C_1Im][(CF_3SO_2)_2N]$; *(IL4)* $[C_8\text{-}C_1Im][BF_4]$ in μ-cuvette.

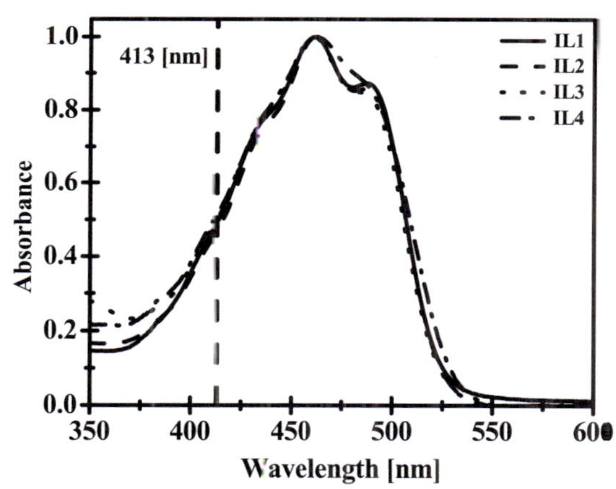

Fig. 5. Normalized absorption spectra of all-*trans* β-carotene in *(IL1)* $[C_8H_{17}OCH_2\text{-}C_1Im][BF_4]$; *(IL2)* $[C_6H_{13}OCH_2\text{-}C_1Im][(CF_3SO_2)_2N]$; *(IL3)* $[C_8H_{17}OCH_2\text{-}C_1Im][(CF_3SO_2)_2N]$; *(IL4)* $[C_8\text{-}C_1Im][BF_4]$; T = 298K; 413 nm is the excitation wavelength used for the fluorescence emission spectra.

The fluorescence emission spectra of all-*trans* β-carotene in RTILs: 1-methyl-3-octyloxymethylimidazolium tetrafluoroborate *(IL1)*, 1-methyl-3-hexyloxymethylimidazolium bis(trifluoromethylsulfonyl)imide *(IL2)*, 1-methyl-3-octyloxymethylimidazolium bis(trifluoromethylsulfonyl)imide *(IL3)* and 1-methyl-3-octylimidazolium tetrafluoroborate *(IL4)* in standard cuvette and in μ-cuvette are shown in Figure 6 (A-D).

The comparison of the RTIL fluorescence spectra measured in both cells (regular and micro) shows the so called 'red tail' (Paul & Samanta, 2006) especially in a case of (*IL2*) and (*IL3*). This effect can be eliminated to high extend in all samples except (*IL4*) where practically

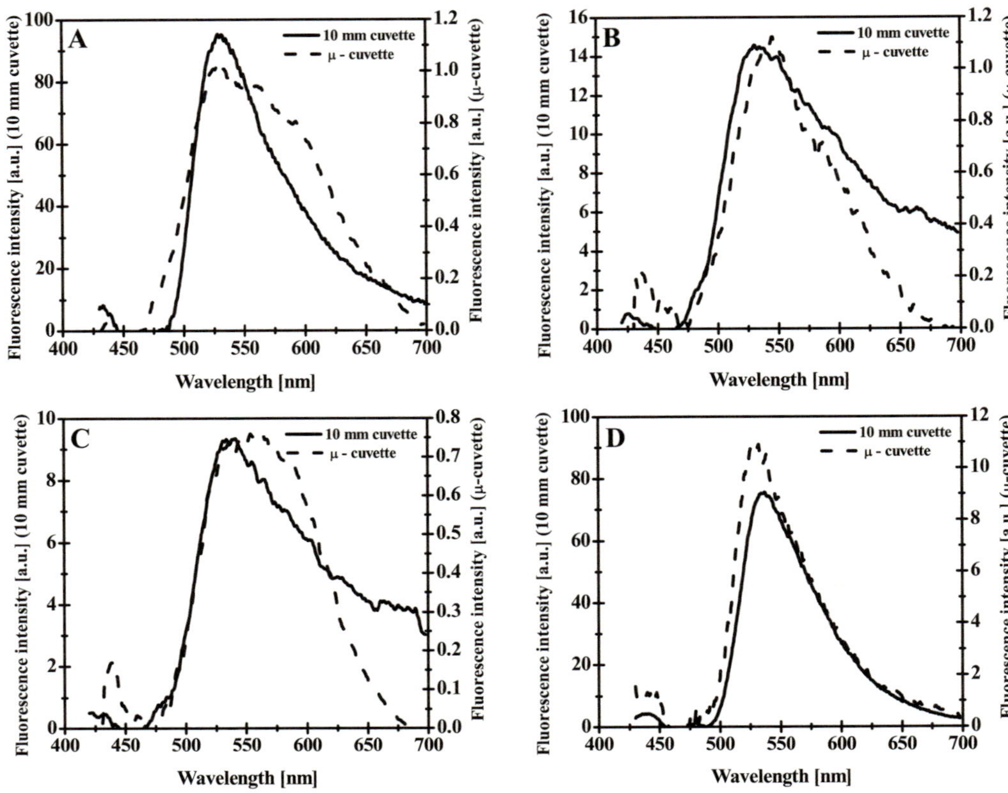

Fig. 6. The fluorescence emission spectra of all-*trans* β-carotene (5 μM) in: (A)- (*IL1*), $[C_8H_{17}OCH_2\text{-}C_1Im][BF_4]$; (B)- (*IL2*), $[C_6H_{13}OCH_2\text{-}C_1Im][(CF_3SO_2)_2N]$; (C)- (*IL3*), $[C_8H_{17}OCH_2\text{-}C_1Im][(CF_3SO_2)_2N]$; (D)- (*IL4*), $[C_8\text{-}C_1Im][BF_4]$ in the standard cuvette and in the μ-cuvette; $\lambda_{ex} = 413$ [nm].

such effect does not exist. The maxima of the fluorescence emission spectra of β-carotene in the investigated RTILs are at around 530 nm. Quenching of the fluorescence intensity of RTILs by β-carotene is not linear with path length of the cell (changing from regular -10 mm to micro cell -1 mm).

The fluorescence emission spectra of the ionic liquids: (*IL1*), 1-methyl-3-octyloxymethylimidazolium tetrafluoroborate; (*IL2*), 1-methyl-3-hexyloxymethylimidazolium bis(trifluoromethylsulfonyl)imide; (*IL3*), 1-methyl-3-octyloxymethylimidazolium bis(trifluoromethylsulfonyl)imide and (*IL4*), 1-methyl-3-octylimidazolium tetrafluoroborate in the standard and μ-cuvette are shown in Figure 7 (A-D).

The values of fluorescence intensities of RTIL spectra differ considerably with RTIL structure. For investigated samples these values in [a. u.] are: 341.0, 45.0, 26.6, 942.6 in 10 mm standard cuvette and 5.4, 7.4, 4.6, 203.9 in μ–cuvette. There is different 'red tail' effect in the fluorescence emission spectra of RTILs. The most pronounce effect is in a case of (*IL4*) but in a sample of β-carotene in (*IL4*) practically there is no such effect. The fluorescence intensity of RTILs is quenched by β-carotene. This effect one can see by comparison of the

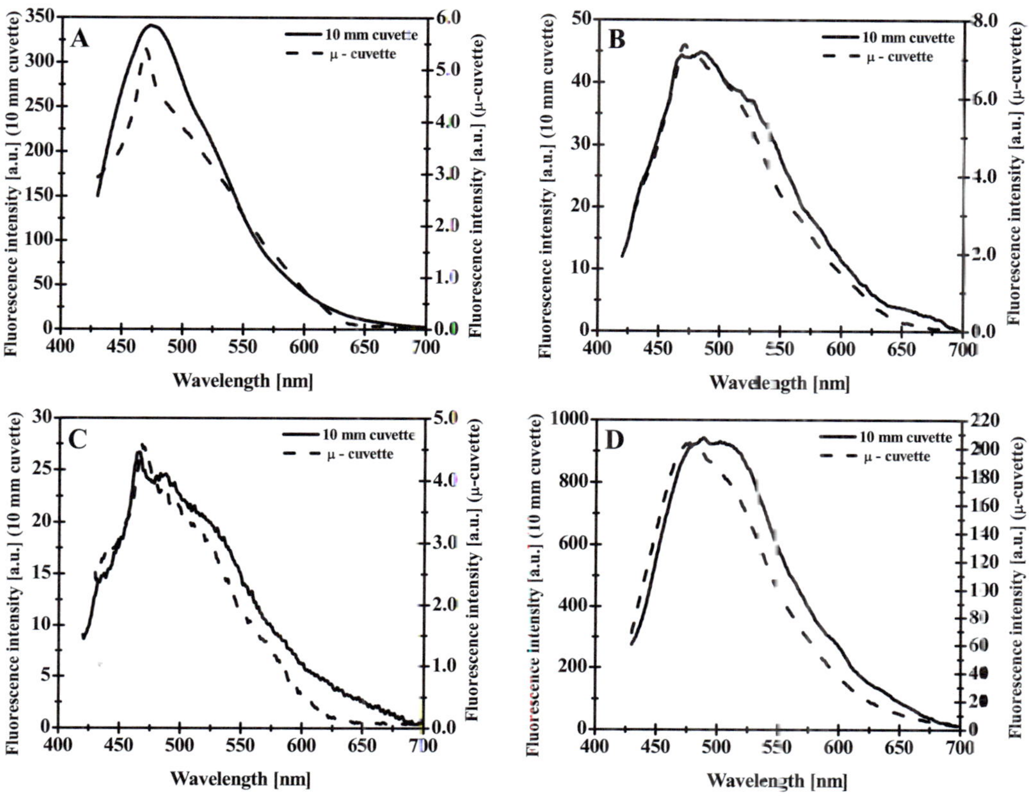

Fig. 7. The fluorescence emission spectra of: (A)- (*IL1*) $[C_8H_{17}OCH_2\text{-}C_1Im][BF_4]$; (B)- (*IL2*) $[C_6H_{13}OCH_2\text{-}C_1Im][(CF_3SO_2)_2N]$; (C)- (*IL3*) $[C_8H_{17}OCH_2\text{-}C_1Im][(CF_3SO_2)_2N]$; (D)- (*IL4*) $[C_8\text{-}C_1Im][BF_4]$ in the standard cuvette and μ-cuvette; λ_{ex} = 413 [nm].

set of Figure 6 (A-D) and Figure 7 (A-D). For spectroscopic details see Table 1. The β-carotene quenching of RTIL fluorescence is at the same level in a case of (*IL2*) and (*IL3*). The highest effect of β-carotene fluorescence quenching of RTIL in 10 mm cell is for (*IL4*) but the lowest for (*IL1*).

The analysis of all RTIL fluorescence data (Table 1) shows that the half band width $\lambda_{1/2}$ is constant within the experimental error ± 2 nm in a case of measurement in micro cell. But this values ($\lambda_{1/2}$) are bigger for β-carotene in RTILs with longer chain as we can see from comparison of (IL4) and (IL1); and also (IL2) and (IL3). The wavelength at the fluorescence maximum (λ_{max}) in the micro cell measurement is 530 nm for β-carotene in (IL1) and (IL4) (both with an anion $[BF_4]^-$). But for (IL2) and (IL3) (with an anion $[(CF_3SO_2)_2N]^-$) the (λ_{max}) is shifted towards longer wavelengths 543 nm and 554 nm in a case of shorter (IL2) chain, and longer (IL3) chain, respectively. In a case of fluorescence measurement in standard cuvette such relation is not observed.

The all-*trans* β-carotene fluorescence yield (ϕ_F = 1.96 ± 0.03) in ionic liquid (1-methyl-3-octyloxymethylimidazolium tetrafluoroborate) according to our previous study (Bialek-Bylka, 2007) is around hundred times higher than in standard solvent (*n*-hexane). The excitation wavelength at 413 nm was selected in order to excite both RTILs alone and β-

		standard or μ- cuvette	IL1	IL2	IL3	IL4
β-carotene in RTILs	λ_{max} ± 1 [nm]	Standard	528	530	534	537
		Micro	530	543	554	530
		$\Delta\lambda_{max}$	-2	-13	-20	7
	F_{max} ± 0.1	Standard	95.3	14.5	9.3	75.5
		Micro	1.1	1.1	0.7	10.9
	$\lambda_{1/2}$ ± 2 [nm]	Standard	80	126	127	73
		Micro	123	96	117	68
		$\Delta\lambda_{1/2}$	-43	30	10	5
RTILs	λ_{max} ± 1 [nm]	Stand.	473	483	464	488
		Micro	468	470	467	479
		$\Delta\lambda_{max}$	5	13	-3	-9
	F_{max} ± 0.1	Standard	341.0	45.0	26.6	942.6
		Micro	5.4	7.4	4.6	203.9
	$\lambda_{1/2}$ ± 2 [nm]	Standard	103	129	125	119
		Micro	107	107	109	110
		$\Delta\lambda_{1/2}$	-4	22	16	9
	F_{max} RTIL/ β-car	Standard	3.6	3.1	2.9	12.5
	F_{max} RTIL/ β-car	Micro	4.9	6.7	6.6	18.7

(*IL1*) $[C_8H_{17}OCH_2\text{-}C_1Im][BF_4]$ (*IL2*) $[C_6H_{13}OCH_2\text{-}C_1Im][(CF_3SO_2)_2N]$
(*IL3*) $[C_8H_{17}OCH_2\text{-}C_1Im][(CF_3SO_2)_2N]$ (*IL4*) $[C_8\text{-}C_1Im][BF_4]$

Table 1. The spectroscopic parameters of RTILs and β-carotene in RTILs.

carotene in RTILs. The quantum yield (ϕ_F) of β-carotene in RTILs is much lower than (ϕ_F) of only RTILs. The values of quantum yields (ϕ_F) of β-carotene in RTILs and RTILs alone are listed in Table 2.

Since the fluorescence efficiency of RTILs is sensitive to the excitation wavelength, the fluorescence quantum yield is also dependent on the excitation wavelength. The fluorescence quantum yield of RTILs and β-carotene in RTILs (Table 2) was estimated for excitation wavelength 413 nm at room temperature; as reference sample Coumarin was used.

The polarity of the most imidazolium ionic liquids lies between that of acetonitrile and methanol and is comparable to the short chain alcohols (Mandal et al., 2006).

RTILs	ϕ_F (%) RTILs	ϕ_F (%) β-carotene in RTILs	n ± 0.001
(IL1) $[C_8H_{17}OCH_2\text{-}C_1Im][BF_4]$	1.009 ± 0.005	0.034 ± 0.002	1.433
(IL2) $[C_6H_{13}OCH_2\text{-}C_1Im][(CF_3SO_2)_2N]$	2.628 ± 0.027	0.03 ± 0.01	1.433
(IL3) $[C_8H_{17}OCH_2\text{-}C_1Im][(CF_3SO_2)_2N]$	3.044 ± 0.015	0.026 ± 0.003	1.431
(IL4) $[C_8\text{-}C_1Im][BF_4]$	6.26 ± 0.63	0.26 ± 0.05	1.432

Table 2. The fluorescence quantum yield (ϕ_F) and refractive index (n) of RTILs.

The ionic liquid viscosity change effect was obtained by the bipolar solvent (DMF). Influence of the ionic liquid viscosity has been detected by the fluorescence emission and absorption spectra. Important information is that the new electronic states of β-carotene were found in the mixture of RTIL and DMF (Bialek-Bylka, 2008).

In the Figure 8 (B1) the fluorescence maxima of different concentrations of RTIL in DMF are at shorter wavelength at around 475 nm but bands around 525 nm belong to the fluorescence of all-*trans* β-carotene in RTIL. The λ_{ex} at 413 nm was selected in order to give an opportunity for excitation of both the all-*trans* β-carotene in RTILs and only the RTILs.

The ionic liquid fluorescence signal of RTILs is quenched by β-carotene. The Figure 8 (A1) shows the fluorescence spectra (λ_{ex} = 413 nm) of binary solvent (*IL4*, $[C_8\text{-}C_1Im][BF_4]$ + DMF) with different concentration of RTIL. The 'pure' (corrected for fluorescence intensity of ionic liquids) spectra of all-*trans* β-carotene in binary solvent with different concentration of RTIL in DMF; λ_{ex} = 413 nm; concentration of β-carotene 5 μmol are shown in Figure 8 (C1). The spectra (Fig. 8 (A1)) can be resolved into two Gaussian components: short and long wavelength. In the literature (Paul & Samanta, 2006) also two components were reported in the case of imidazolium based RTILs. Authors assigned the short wavelength component to unassociated imidazolium ion and the long wavelength component to various associated forms of the imidazolium ion. These associated species (stacked dimers and oligomers of the imidazolium ions) are responsible (Paul & Samanta, 2006) for the unique absorption and fluorescence of the imidazolium RTILs.

Figure 8 (C1) shows the highest intensity of the fluorescence of β-carotene in a case of pure (100%) RTIL (*IL4*). The fluorescence intensities of only RTIL (Figure 8 (A1)) and β-carotene in RTIL (Figure 8 (B1)) are different (for example for 100% ionic liquid: 221 [a. u.] and 63 [a. u.], respectively). There exists quenching of RTIL fluorescence by β-carotene molecule in this system.

The fluorescence emission intensity of (*IL4*) (Figure 9 A) and all-*trans* β-carotene in (*IL4*), $[C_8\text{-}C_1Im][BF_4]$ increases nonlinearly with concentration of RTIL (Figure 9 B). The three linear phases are present in both Figures 9 A and B.

The explanation of the parameters of the curves: a, b and c of Figures 9 (A-D) is in Table 3.

Absorption measurement sensitivity of the spectrometer is limited by the ability of the instrument to discriminate between the two nearly equal signals. Detection limit for even the most favorable cases rarely exceeds 10^{-8} moles. But fluorometric instruments are limited only by the intensity of exciting light and the ability to detect low light levels. Under the ideal conditions the concentrations of 10^{-12} moles can be measured.

The decrease of the fluorescence intensity by interaction of the excited state of the fluorophore with its surroundings is known as quenching (not random process). The intermolecular electronic energy transfer is the possible way of RTIL fluorescence quenching:

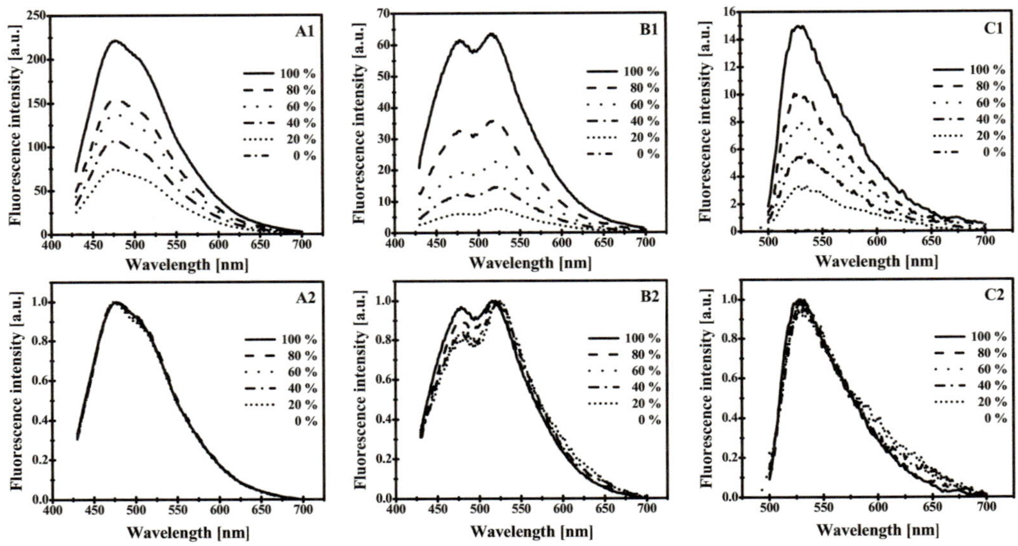

Fig. 8. The fluorescence emission spectra (A1) of binary solvent (A2 – normalized curves) (*IL4*), $[C_8\text{-}C_1Im][BF_4]$ + DMF); the fluorescence emission spectra (B1) of all-*trans* β-carotene (B2 – normalized curves) 5 μmol in binary solvent and (C1) the 'pure' (corrected for fluorescence of ionic liquid) (C2 – normalized curves) fluorescence emission spectra (measured in the 10% step) of all-*trans* β-carotene in binary solvent mixture with different concentrations of RTIL (DMF signal is at the zero level), respectively; λ_{ex} = 413 nm; μ-cuvette.

RTIL* + β-carotene → RTIL + β-carotene*; where an excited molecule (RTIL*) transfers excitation energy to a quencher molecule β-carotene, causing deexcitation of RTIL and forming an excited quencher molecule, β-carotene*. If β-carotene* is a fluorescent species, its fluorescence, called sensitized fluorescence, can be detected. This is the phenomenon allowing observation of fluorescence from a molecule like β-carotene that may be difficult to excite directly because of the optically forbidden (for symmetry reason) singlet electronic states. The all-*trans* β-carotene in standard solvent, for example hexane, fluorescence quantum yield of electronic state S_2 is 2 x 10^{-4} (Sherve et al., 1991; Andersson, 1992) and of S_1 state is 4 x 10^{-6} (Wasielewski, 1986, 1989).

Figure	9A		9B		9C		9D	
	Slope	Slope [deg]	Slope	Slope [deg]	Slope	Slope [deg]	Slope	Slope [deg]
a	0.91	42	0.65	33	0.33	18	0.38	21
b	0.46	25	0.47	25	0.65	33	0.67	34
c	0.83	40	0.96	44	2.25	66	2.07	64

Table 3. The values of the slope for fitting linear dependence of fluorescence intensity on the percentage content of ionic liquid in binary solvent mixture ($[C_8\text{-}C_1Im][BF_4]$ + DMF).

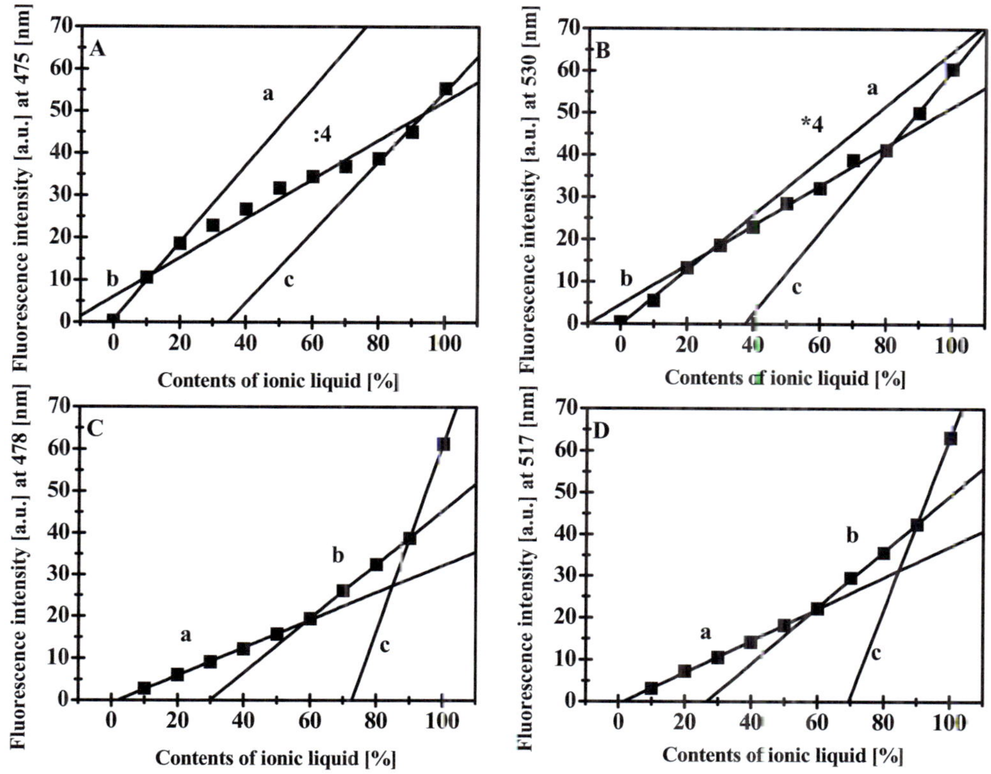

Fig. 9. The fluorescence emission intensity (A) of binary solvent (*IL4*), ([C_8-C_1Im][BF_4] + DMF) with different concentration of RTIL; (B) the 'pure' (corrected for fluorescence of ionic liquid) fluorescence emission spectra of all-*trans* β-carotene in binary solvent mixture (*IL4*), ([C_8-C_1Im][BF_4] + DMF). The fluorescence emission intensity (not corrected) of all-*trans* β-carotene in binary solvent for maxima: short wavelength (C) and long wavelength (D) with different concentration of RTIL. For all Figures. A, B, C, D measurements were done in the steps of 10% within concentration range of (0% - 100%) of RTIL; λ_{ex} = 413 nm; μ-cuvette.

Inner filter effects are often responsible for distorted emission spectra and nonlinear calibration curves between fluorescence intensity and fluorophore concentration. In a case of concentrated samples, in order to minimize or correct for IFE either instrumental or mathematical corrections can be done (Kao et al., 1998). Instrumental corrections are based on the observation that fluorescence intensity and fluorophore concentration are linear at low absorbance. In principle, any method that can lower the absorbance will reduce IFEs. Sample dilution (Guilbault, 1990; Lakowicz, 1983) is the most popular approach, but it may cause changes: in conformation, bonding, solvation, and the degree of association. Also other chemical events may alter absorption-fluorescence processes and, there by, introduce large unknown errors (Holland, 1977).

IFEs can distort fluorescence emission spectral profiles. Serious IFE will be indicated by changes in the shape of fluorescence spectra and position of max intensities in the fluorescence measurements of the samples in standard and μ-cuvette. In a case of experiment in μ-cuvette, the PIFE is decreased because the excitation beam passes only through the 1 mm path instead of 10 mm. The SIFE is not negligible because optical density of the absorber (β-carotene) in emission wavelengths is not low. If samples absorb at both wavelengths: the fluorescence excitation and emission, the correction of the observed fluorescence for (IFE) is necessary (Tucker et al., 1992). This is the case of β-carotene in RTILs, the fluorescence excitation wavelength is at 413 nm and maximum of fluorescence emission at around 530 nm in standard cell.

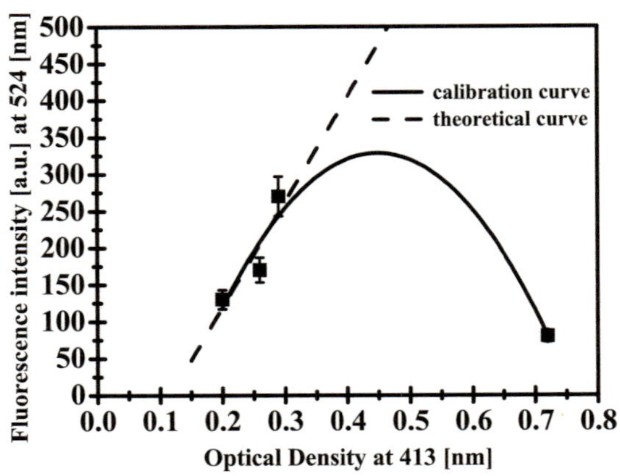

Fig. 10. The calibration curve of β-carotene in RTIL (*IL1*), 1-methyl-3-octyloxymethylimidazolium tetrafluoroborate) in 10 mm cell (right angle cell configuration).

If the calibration curve (Fig. 10) of the sample in 10 mm cuvette is perfectly linear this means that the (ε) molar extinction coefficient dependence on the (n) refractive index of the solution is negligible (Skoog & Leary, 1992). This is the case in our study where only up to OD = 0.2 the calibration curve is linear within experimental error.
The optical density (OD) of samples: β-carotene in RTILs and only RTILs at excitation and emission wavelengths are listed in Table 4.

	OD			
RTILs	β-carotene in RTIL		RTIL	
	413 [nm]	525 [nm]	413 [nm]	525 [nm]
IL1	0.28	0.04	0.00	0.00
IL2	0.26	0.04	0.00	0.00
IL3	0.25	0.06	0.00	0.00
IL4	0.10	0.04	0.63	0.06

Table 4. The sample optical density (OD) at excitation and emission wavelengths.

With temperature increased (in the physiological range: 283K and 313K, the fluorescence emission intensity decreased in both samples β-carotene in RTIL and also RTIL alone. Figures 11 (A-D) and 12 A & B show fluorescence emission spectra of RTIL and β-carotene in RTIL as a function of temperature.

The RTIL's fluorescence and absorption spectra changes induced by the temperature are correlated with the ionic liquid viscosity changes. The room temperature viscosity measurements of investigated RTILs are listed in Table 5.

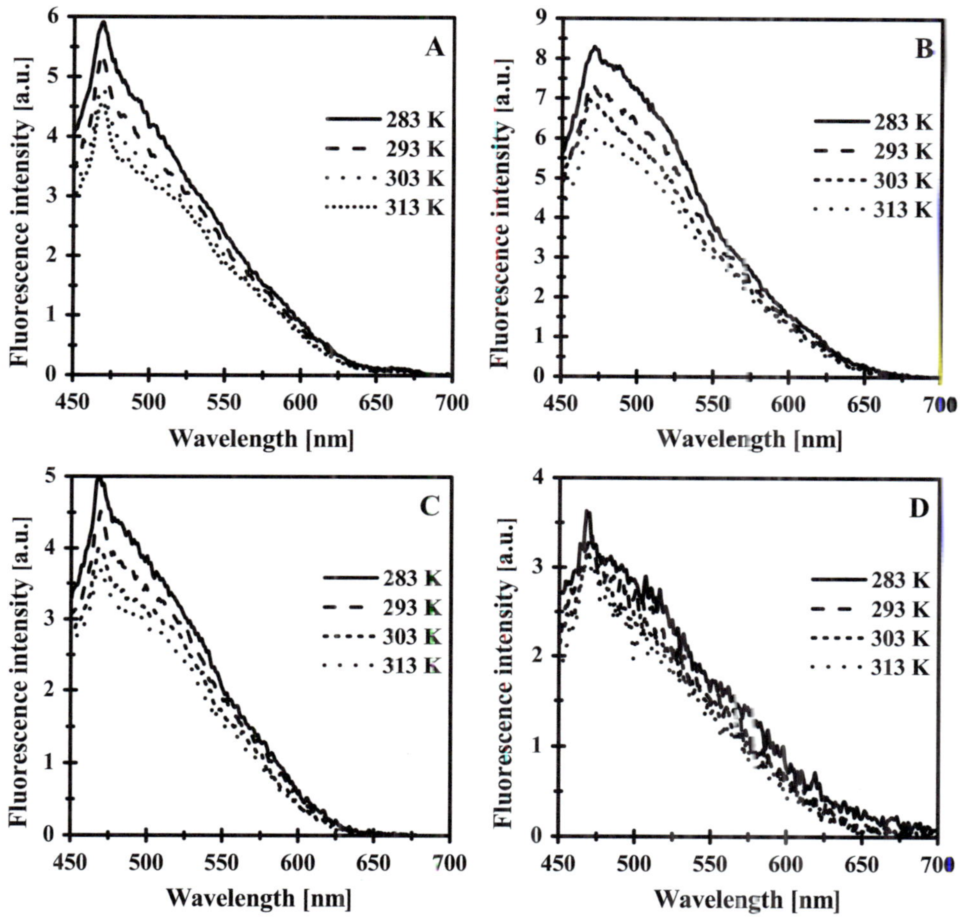

Fig. 11. Fluorescence emission spectra of: (A)- (*IL1*) [$C_8H_{17}OCH_2$-C_1Im][BF_4]; (B)- (*IL2*) [$C_6H_{13}OCH_2$-C_1Im][$(CF_3SO_2)_2N$]; (C)- (*IL3*) [$C_8H_{17}OCH_2$-C_1Im][$(CF_3SO_2)_2N$]; (D)- (*IL4*) [C_8-C_1Im][BF_4] (μ-cuvette) as a function of temperature.

RTILs	Viscosity [cP]	Reference data
[$C_8H_{17}OCH_2$-C_1Im] [BF_4]	574	
[$C_8H_{17}OCH_2$-C_1Im] [$(CF_3SO_2)_2N$]	105	
[$C_6H_{13}OCH_2$-C_1Im] [$(CF_3SO_2)_2N$]	87	
[C_8-C_1Im] [BF_4]	420	439 (Seddon et al., 2002)

Table 5. The viscosity (error 2%) of RTILs.

The fluorescence quantum yield of all-*trans* β-carotene in the investigated ionic liquids are shown in Figure 12.

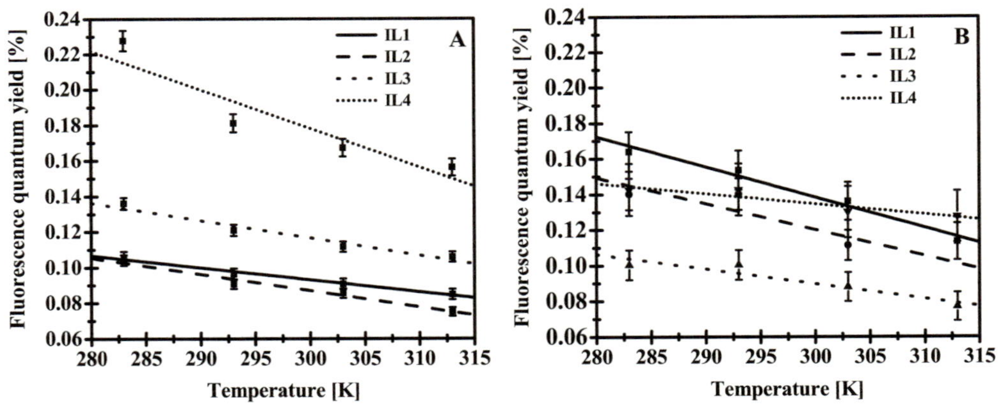

Fig. 12. The fluorescence quantum yield of all-*trans* β-carotene in: *IL1* - $[C_8H_{17}OCH_2\text{-}C_1Im][BF_4]$; *IL2* - $[C_6H_{13}OCH_2\text{-}C_1Im][(CF_3SO_2)_2N]$; *IL3* - $[C_8H_{17}OCH_2\text{-}C_1Im][(CF_3SO_2)_2N]$; *IL4* - $[C_8\text{-}C_1Im][BF_4]$: (A) - before correction for fluorescence ionic liquid; (B)- 'pure' all-*trans* β-carotene spectra corrected for fluorescence of ionic liquid.

In order to check fluorescent impurities of the sample the fluorescence emission spectra with different excitation wavelengths ought to be measured. The fluorescence maxima of the RTILs investigated by us based on imidazolium are strongly dependent on the excitation wavelengths similarly as was found by other authors (Paul & Samanta, 2006). In the present study we focus on four RTILs with different combination of cations or anions (*see* Scheme 1). Special care was taken for the purification of the synthesized ionic liquids.
The dependence of fluorescence emission spectra of all-*trans* β-carotene on excitation wavelength is presented in Figure 13.
There exist two sets of data (Figures 14 A and B): one with max around 400 nm and second with max shifted toward longer wavelength with increasing excitation wavelength in the range between 350 nm and 450 nm. The RTILs based on imidazolium are not as transparent as commonly has been believed. Also an excitation wavelength dependent fluorescence effect known as 'red edge effect' is observed. The various energetically different associated species in RTILs are responsible for this 'red tail'.

4. Conclusion

The $[BF_4]^-$ and $[(CF_3SO_2)_2N]^-$ based imidazolium ionic liquids have non-negligible absorption in the UV-VIS region with the absorption red tail extending far into the visible region. Such effect is also observed in very carefully purified RTILs. Inner filter effects are responsible for distorted emission spectra and nonlinear calibration curves between fluorescence intensity and fluorophore concentration. In a case of samples like RTILs in order to minimize IFE the μ-cell ought to be used. Other methods may cause changes in the nanostructure of ions and influence the intensity and shape of the fluorescence spectra.

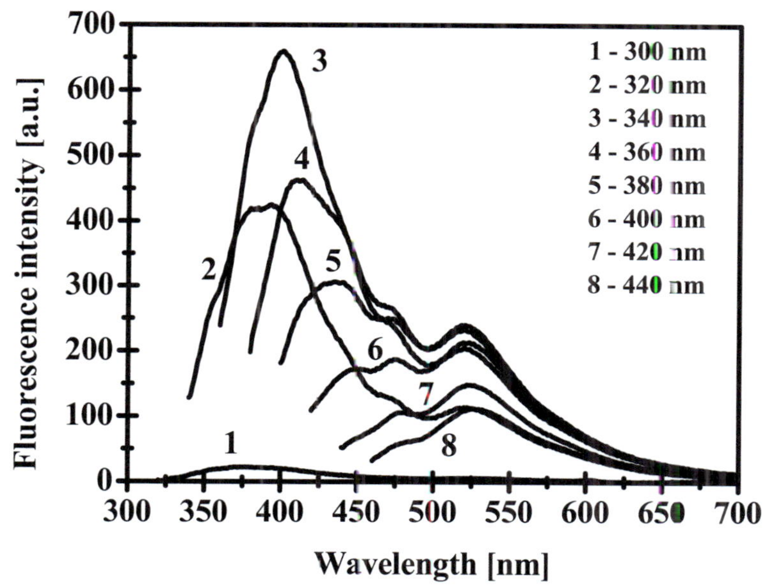

Fig. 13. Fluorescence emission spectra of all-*trans* β-carotene; c = 5 [μM] in (*IL1*), $C_8H_{17}OCH_2$-C_1Im][BF_4]; 10 mm cuvette; as a function of excitation wavelength.

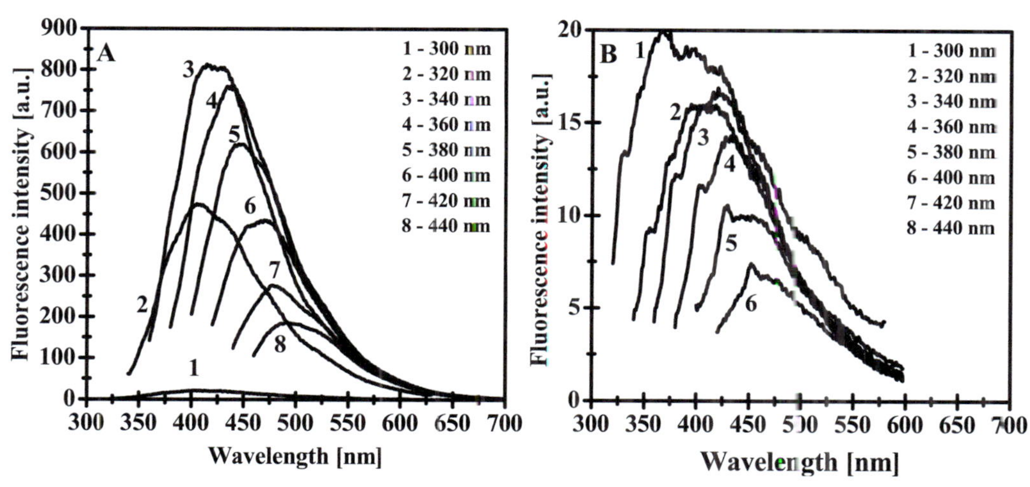

Fig. 14. Fluorescence emission spectra of (*IL1*), [$C_8H_{17}OCH_2$-C_1Im][BF_4] in: (A) -10 mm standard cuvette and (B) - μ–cuvette, as a function of excitation wavelength.

5. Acknowledgements

Acknowledgement is made to project DS. 62-176/2011.
ISBN 978-1-4051-3891-8

6. References

Albani, J.R. (2007). *Principles and Applications of Fluorescence Spectroscopy*, Wiley-Blackwell, ISBN 978-1-4051-3891-8, Malaysia.

Andersson, P.O., Gillbro, T.A., Asato, E., & Liu, R.S.H. (1992). Dual singlet state emission in a series of mini-carotenes. *Journal of Luminescence*, Vol. 51, No. 1-3, pp. 11- 20, ISSN 0022-2313.

Bialek-Bylka, G.E., Pawlak, K., Jazurek, B., Skrzypczak, A., & Koyama, Y. (2007). Spectroscopic properties and temperature induced electronic configuration changes of all- *trans* and 15- *cis* β-carotenes in ionic liquids. *Photosynthetica*, Vol. 45, No. 2, pp. 161-166, ISSN 0300-3604.

Bialek-Bylka, G.E., Kakitani, Y., Li, Ch., Koyama, Y., Kuki, M., Yamano, Y., & Nagae, H. (2008). Excitation followed by stimulated-emission from diabatic levels in all-*trans* and 15-*cis*-β-carotenes: Effects of molecular symmetry and solvent polarity. *Chemical Physics Letters*, Vol. 454, No. 4-6, pp. 367-373, ISSN 0009-2614.

Earle, M.J., Gordon, Ch.M., Pletchkova, N.V., Seddon, K.R., & Welton, Th. (2007). Decolorization of Ionic Liquids for Spectroscopy. *Analytical Chemistry*, Vol. 79, No. 2, pp. 758-764, ISSN 0003-2700.

Gordon, C.M., McLean, A.J., Muldoon, M.J., & Dunkin I.R. (2003). Diffusion-Controlled Reactions in Room Temperature Ionic Liquids, *Proceedings of Ionic Liquids as Green Solvents: Progress and Prospects-224th American Chemical Society National Meeting*, ISBN 9780841219618, Boston, August 2002.

Guilbault, G.G. (1990). *Practical fluorescence* (2nd. edition), Mercel Dekker Inc., ISBN 0824783506, New York.

Holland, J.F., Teets, R.E., Kelly, P.M., & Timnick, A. (1977). Correction of right-angle fluorescence measurements for the absorption of excitation radiation. *Analytical Chemistry*, Vol. 49, No. 6, pp. 706-710, ISSN 0003-2700.

Ingle, J.D., & Crouch, S.R. *Spectrochemical Analysis*, Prentice-Hall Inc., ISBN 0-13-826876-2, Upper Saddle River.

Kao, S., Asanov, A.N., & Oldham, P.B. (1998) A Comparison of Fluorescence Inner Filter Effects for Different Cell Configurations. *Instrumentation Science & Technology*, Vol. 26, No. 4, pp. 375-387, ISSN 1073-9149.

Koel, M. (2008). *Ionic Liquids in Chemical Analysis*, CRC Press, ISBN 1-4200-4646-2, Boca Raton.

Lakowicz, J.R. (1983). *Principles of fluorescence spectroscopy* (1st edition), Plenum Press, ISBN 0306412853, New York.

Lakowicz, J.R. (2006). *Principles of fluorescence spectroscopy* (3rd edition), Springer, ISBN 0387312781, Singapore.

Li, H., & Hu, Y. (2007). Spectroscopic investigation of inner filter effect by magnolol solutions. *Spectrochimica Acta Part A: Molecular and Biomolecular Spectroscopy*, Vol. 68 No. 5, pp. 1263-1268, ISSN 1386-1425.

Lidke, D., Nagy, P., Barisas, B., Heintzmann, R., Post, J., Lidke, K., Clayton, A., Arndt-Jovin, D., & Jovin, T. (2003). Imaging molecular interactions in cells by dynamic and static fluorescence anisotropy (rFLIM and emFRET). *Biochemical society transactions*, Vol. 31, No. 5, pp. 1020-1027, ISSN 0300-5127.

Mandal, P.K., Saha, S., Karmakar, R., & Samanta, A. (2006). Solvation dynamics in room temperature ionic liquids: Dynamic Stokes shift studies of fluorescence of dipolar molecules. *Current Science*, Vol. 90, No. 3, pp. 301-310, ISSN 0011-3891.

Moss, T., & Leblanc, B. (2009). *DNA-Protein Interactions: Principles and Protocol* (3rd edition), Humana Press, ISBN 1603270140, Dordrecht Heidelberg London New York.

Pawlak, K., Skrzypczak, A., & Bialek-Bylka, G.E. (2009). Reichardt's dye as sensor material detected oxymethyl group in the cation part of imidazolium ionic liquid structure. *Journal of Engineering and Applied Sciences*, Vol 4, No. 7, pp. 71-74, ISSN 1819-6608.

Paul, A., & Samanta, A. (2006). Optical absorption and fluorescence studies on imidazolium ionic liquids comprising the bis(trifluoromethane - sulphonyl)imide anion. *Journal of Chemical Sciences* Vol. 118, No. 4, pp. 335-340, ISSN 0974-3626.

Pernak, J., Czepukowicz, A., & Pozniak, R. (2001). New Ionic Liquids and Their Antielectrostatic Properties. *Industrial & Engineering Chemistry Research*, Vol. 40, No. 11, pp. 2379-2383, ISSN 0888-5885.

Seddon, K., Stark, A., & Torres, M.J. (2002). Viscosity and Density of 1-Alkyl-3-methylimidazolium Ionic Liquids, *Proceedings of Clean Solvents: Alternative Media for Chemical Reactions and Processing*, ISBN 9780841219182, Washington DC, August 2000.

Shreve, A.P., Trautman, J.K., Owens, T.G., & Albrecht, A. C. (1991). A femtosecond study of electronic state dynamics of fucoxanthin and implications for photosynthetic carotenoid-to-chlorophyll energy transfer mechanisms. *Chemical Physics*, Vol. 154, No. 1, pp. 171-178, ISSN 0301-0104.

Skoog, D. A., & Leary, J. J. (1992). *Principles of Instrumental Analysis* (5th edition), Saunders College Publishing, ISBN 0030020786, Madrid.

Tucker, S.A., Amszi, V.L., & Acree, W. E. (1992). Primary and secondary inner filtering. Effect of $K_2Cr_2O_7$ on fluorescence emission intensities of quinine sulfate. *Journal of Chemical Education*, Vol. 69, No. 1, pp. A8, ISSN 0021-9584.

Wasielewski, M.R., & Kispert, L.D. (1986). Direct measurement of the lowest excited singlet state lifetime of all-*trans*-β-carotene and related carotenoids. *Chemical Physics Letters*, Vol. 128, No. 3, pp. 238-243, ISSN 0009-2614.

Wasielewski, M.R., Johnson, D.G., Bradford, E. G., & Kispert, L.D. (1989). Temperature dependence of the lowest excited singlet-state lifetime of all-*trans*-β-carotene and fully deuterated all-*trans*-β-carotene. *The Journal of Chemical Physics* Vol. 91, No.11, pp. 6691-6697, ISSN 0021-9606.

Wasserscheid, P., & Welton, T. (2008). *Ionic liquids in synthesis* (2nd edition), Vch Verlagsgesellschaft Mbh, ISBN 978-3-527-31239-9, Weinheim.

20

Ionic Liquid Lubricant with Ammonium Salts for Magnetic Media

Hirofumi Kondo
Sony Chemical & Information Device
Japan

1. Introduction

Magnetic recording systems have been responsible for the widespread and inexpensive recording of sound, video, and information processing. Despite the availability of other means of storing data, such as optical recordings and semiconductor devices, magnetic recording media have the advantages of low cost, stable storage, a relatively higher data transfer rate, a relative short seek time, and high volumetric storage capacity.

1.1 Hard disk drive systems (HDDs)

In current hard disk drive (HDD) systems (Fig.1 (a)), a rigid disk is rotated by a spindle motor at a speed of 10000 revolutions per minute (rpm). Information is written and read by a magnetic head with a tiny electric current attached to the end of the slider (Fig. 1(b)). The physical spacing between the magnetic sensors and the disk is down to almost 10 nm in recent systems, and it will be necessary to be within 5-7 nm for areal densities in the $Tbin^{-2}$ range. (Wood et al.,2002) The read/write magnetic heads are mounted in the slider and travel across the data zone during the reading and writing operations.

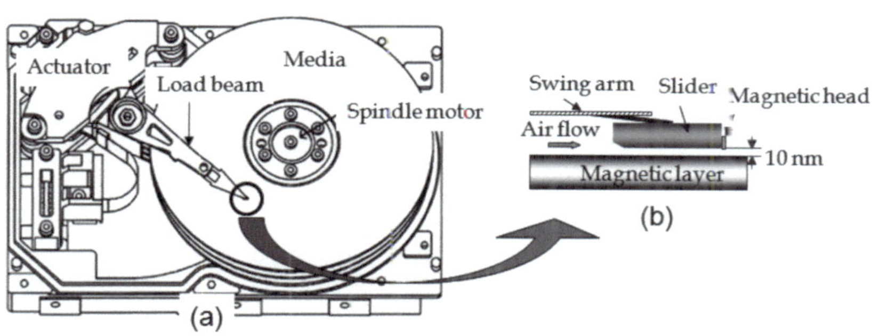

Fig. 1. Schematic drawing of a hard disk drive (a)(Kato et al., 2000) and a head-media interface (b)

However, when the drive stops, this head assembled device rests in the landing zone which is typically textured in order to reduce wear during contact-start-stop (CSS) operations. Most drives require that the static and kinetic friction forces at the head-media-interface (HMI) remain low under extreme environmental conditions and after the required number of CSS that is usually 10,000 or greater. (Jhon et al., 2001)

The rigid disk consists of an Al-Mg alloy or glass substrate, undercoat layer, a magnetic multilayer, and a carbon overcoat as illustrated in Fig. 2(a). (Maesaka and Ohmori, 2002) Nowadays, the magnetic media are perpendicular media, which consist of a Co-Cr based film. A carbon overcoat is used to enhance the wear and corrosion resistance. Finally a molecular thin lubricant, which is the topic of this chapter, is coated on the carbon layer to further reduce both the wear and friction between the HMI.

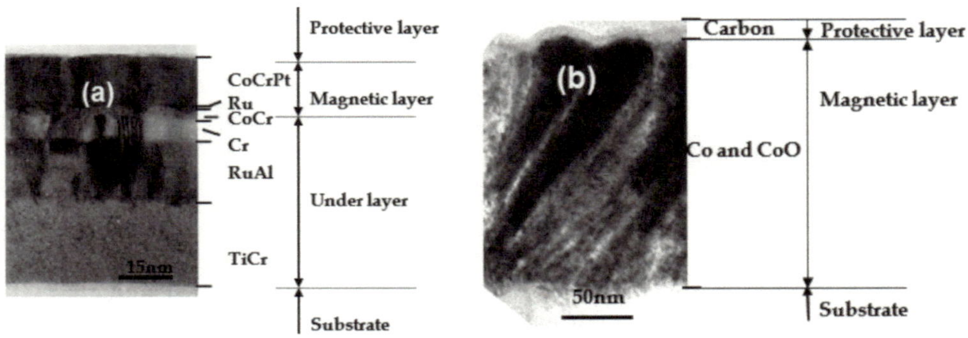

Fig. 2. Cross-sectional TEM image of magnetic layer for rigid disk (a) and magnetic tape (b).

1.2 Tape drive systems

Magnetic tape media are divided into two categories (Onodera et al., 1996a); i.e., particulate media where magnetic particles are dispersed in a polymer matrix with some additives and coated onto the polyethylene terephthalate (PET) substrate, and thin film media in which monolithic magnetic thin films are deposited onto the substrate in a vacuum, which is discussed in this chapter. For magnetic tape helical scan systems, the tape is driven by a pinch roller and a tension (0.2 - 0.5 N) is applied.

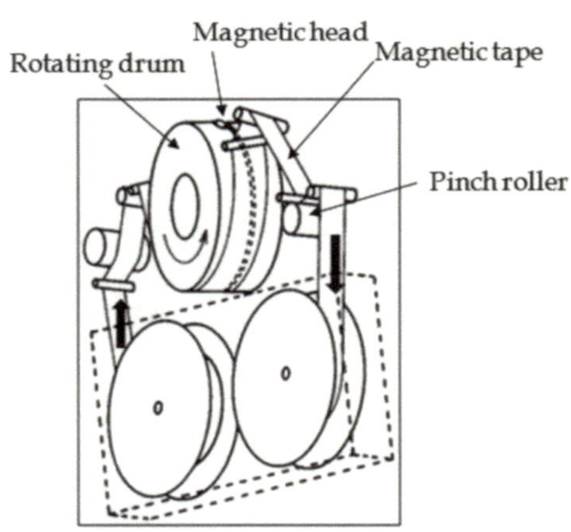

Fig. 3. Tape path for helical scan tape systems.

A video recorder uses a rotating head configuration as shown in Fig. 3. (Shiraishi et al., 1978) Much higher data transfer rates can be achieved using this rotating head drive system. (Kanota et al., 1990) The tape used is a PET substrate with an evaporated film of Co which is typically 100 nm thick. The magnetic layer fabricated with oxygen gas has obliquely aligned fine particles (Fig. 2(b)), (Ito et al., 2002) which leads to higher electromagnetic characteristics. (Kaneda, 1997) The carbon overcoat and the lubricant layer are deposited onto the magnetic layer in a similar way to the rigid disks. (Onodera et al., 1996b)

1.3 Demand for lubricant of thin film magnetic media

In conventional magnetic recording, thin film media typically have their surfaces lubricated to reduce friction and wear resulting from contacts between the read/write magnetic head and media surface. In practice, to avoid adhesion related problems, lubrication has to be achieved with a molecularly thin lubricant film. (Homola, 1996) However, the main challenge in selecting the best lubricant for a magnetic media surface is finding a material which provides wear protection while the media surface is exposed to various environmental situations. It is important that the lubricant remain on the media surface over the life of the file without being subject to desorption, spin-off, or chemical degradation. This problem has become more difficult with the advent of a very smooth thin film surface. because thin film media do not have a mechanism for lubricant replenishment. (Klaus & Bhusha, 1985;Lin & Wu, 1990) Furthermore, lubricant adhesion to the overcoat surface is often insufficient to prevent lubricant depletion that eventually results in accelerated wear.
The presence of a mobile lubricant film, an excess lubricant, is often deemed necessary to replenish itself after sliding events. Increasing the amount of a lubricant enhances the durability, but exceeding the surface roughness of the media generally leads to adhesion related problem, such as deleterious stiction. (Bhushan, 1996; Sato et al., 1999) This frictional problem cannot be solved by using the conventional functional perfluoropolyether (PFPE) as the lubricant. In order to reduce this trade-off, novel lubricants must be designed and synthesized for the smoother surface magnetic thin film media. The very large body of patents relating to the lubricity of PFPEs on thin film magnetic media shows the importance of this problem to manufacturers. (Gunsel et al., 2003; Liu et al., 2002; Ishida et al., 2005; Kondo & Uchimi, 1992; Marchon et al., 2006a)
New types of PFPE lubricants whose chemical structure are summarized in Table 1 have been reported to enhance the performance and reliability. Z-DOL has hydroxyl groups at both chain ends, which has been widely used for the rigid disk application. With the additional functional hydroxyl groups in the middle of the PFPE backbone chain, Z-tetraol multi-dentate (ZTMD) can achieve a reduced clearance, while still achieving an overall drive reliability. (Chiba et al., 2005; Marchon et al., 2006b) A20H was invented to improve the tribological performance, which can effectively block the Lewis acid sites for degradation of the PFPE chain. (Waltman et al., 2003) The backbone chain and functional group of Mono are different from the above PFPE. Mono effectively reduced the head-disk adhesive interaction during near-contact operations. (Sakae et al., 2003) Sometimes the additives, such as X1-P, are used to enhance the reliability. (Jiaa& Liu, 1999) The linear or branched PEPE physically deposited on the diamond-like carbon surface with a covalent linkage formed by thermal decomposition of the peroxide improves the stability. (Navarrini et al, 2010)
Topical lubrication of solid lubricants has not been successful because the solid lubricating layer is often found to interfere with the sensitive magnetic transducing process, and because most solid lubricants have a poor wear resistance, they tend to wear away in the tracks under the head and generate debris. (Klaus & Bhushan, 1985) Liquid lubricants have the advantage that they will creep across the surface to replenish a portion of the layer which has been removed by abrasion or head wear. However, because of their mobility, liquid lubricants may suffer the disadvantage of spinning off from the disk surfaces during operation, especially at higher operating temperatures. These lubricants may also slowly

evaporate with time at the high temperatures, thereby reducing their protection. The use of higher viscosity, low-volatility liquid lubricants may help to decrease the evaporation rate and prolong their life.

	$X-CF_2\,(OCF_2CF_2)_n\,(OCF_2)_m\,OCF_2-X$ (0.5<n/m<1)
Z	X = - OCF_3
Z-DOl	X = - CH_2OH
Z-DIAC	X = - COOH
Z-Tetraol	X = - $CH_2\,OCH_2CHCH_2\,OH$ OH
AM2001	X = - CH_2OCH_2 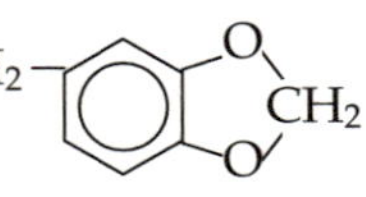
A20H	$HOCH_2CF_2\,(OCF_2CF_2)_n\,(OCF_2)_m OCF_2CH_2O$ 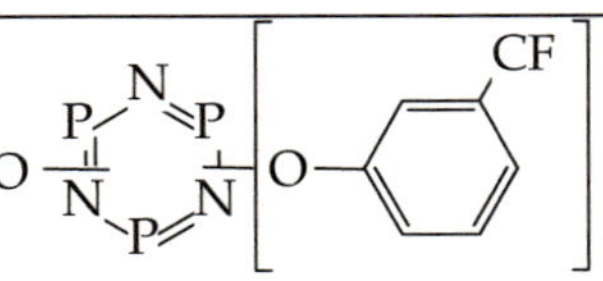
Mono	$F-(CF_2CF_2CF_2O)_l - CF_2CF_2CH_2\,-N\,(C_3H_7)_2$

Table 1. Molecular structure of commercial PFPE lubricants.

Based on these understandings of lubrication mechanisms, the demand for lubricants for magnetic thin film media with low friction and wear can be summarized as follows: (Klaus & Bhushan, 1985)

1. Low vapor pressure (volatility)
2. Low surface tension (replenish surface with lubricant)
3. Some affinity to the surface (with polar end group)
4. Good thermal and oxidative stability to prevent degradation during storage and use
5. Chemical inertness to met als, glasses and plastics and nondeposit formings to insure no damage to heads, guides, devices, etc.
6. No toxicity and flammability (to prevent degradation in the presence of water)
7. Good boundary lubricant (forms easily sheared chemical film during asperity contact)
8. Solubility in organic solvents

Over the past decade, ionic liquids have received a great deal of attention as a class of green solvents with a wide range of potential applications including organic and inorganic synthesis, (Sheldon, 2001) energy storage devices, (Ohno, 2005) separations, (Anderson et al., 2003; Visser et al., 2001) and catalysis. (Binnemans, 2005; Gordon, 2001; Welton, 1999) The term ionic liquid is broadly used to describe a large class of low melting fused salts that are liquids below 100°C. The most notable characteristics of many ionic liquids are their low vapor pressure, non-flammability, thermal stability, wide liquid range, and solvating properties for diverse substances. Limited results from very recent studies have shown the potential for using ionic liquids as a new class of lubricants. Friction and wear reductions have been reported on met allic and ceramic surfaces lubricated by selected ionic liquids compared to the conventional hydrocarbon lubricants.(Liu et al., 2002; Ye et al., 2001) Some fluoroalkyl-substituted immidazolium and a new series of imidazolium cation-based ionic liquids have been successfully prepared.(Singh et al., 2002; Branco et al., 2002) Alkylimidazolium tetrafluoroborates and hexafluoro-phosphate exhibit good tribological performances for steel, aluminum, copper, single crystal SiO_2 and Si, and sialon (Si-Al-O-N) ceramics as compared to phosphazene (X-1P) and PFPE.(Liu et al., 2002; Mu et al., 2004; Mu et al., 2005; Ye et al., 2001; Ye et al., 2002) Ammonium-based ionic liquids provide friction reduction from elastohydrodynamic to boundary lubrication regimes compared to the fully-formulated base oil. (Qu et al., 2006) Ionic liquids have also been studied to determine their effectiveness as additives for base oil and water and the chemical and tribochemical reactions have been evaluated to understand the lubrication mechanisms. (Iglesias et al., 2004; Omotowa et al., 2004; Philips & Rabinski, 2004; Reich et al., 2003)

Protic ionic liquids are a subset of ionic liquids formed by the stoichiometric (equimolar) combination of a Bronsted acid with a Bronsted base.(Greaves et al., 2006; Greaves & Drummond, 2008; Xu & Angel, 2003;Yoshizawa et al., 2003) Relevant investigations into the molecular interactions of carboxylic acids and amines were conducted by Kohler et al., and the complexes of acid and amine with the molecular ratio of 1:1 can be found. (Kohler, 1981a, 1981b) Ionic liquids, which possess an octadecyl ammonium salt with pentadecafluoro octanate, significantly reduce the friction compared to the corresponding amide and Z-DOL. (Kondo, 1987a,b, 1997; Kondo et al., 1989, 1990, 1993a,b, 1994a,b) The modified PFPEs having the same hydrophilic group have also been synthesized and also show better frictional properties, which have been used as a lubricant for magnetic thin film media for a long time. (Kondo et al., 1996, 1994; Kondo, 2008) In this chapter, a series of ionic lubricants having the same hydrophilic group stated above are deposited on the magnetic thin film media and the effect of their molecular structures on the frictional properties is systematically investigated.

The lubricant is required to be very thin on the order of a mono-molecular layer. Therefore, the frictional properties depend not only on the molecular structure, but also on the microscopic structure of the lubricant film.(Seki & Kondo, 1991; Kondo & Seki, 1993a; Kondo et al., 1996; Kondo, 2008) Microscopic coverage of this alkylammonium-based protic ion liquid film on the medium surface is also examined using FTIR and X-ray photoelectron spectroscopy (XPS) and related to the spectra to the frictional properties.

2. Materials

Three types of lubricants which possess both the perfluoroalkyl group and hydrocarbon long chain, that is an ester, amide, and carboxylic acid amine salt, were synthesized by the following Scheme 1 in Fig.4. The ester and the amide were prepared by the addition of carboxylic acid chloride to the hexane solution of the corresponding alcohol and amine in the presence of a base agent. The perfluorocarboxylic acid and the carboxylic acid ammonium salts are prepared by warming the mixture of the perfluorocarboxylic acid and the amine to 80°C until the complete dissolution was obtained (Scheme 2).(Kondo et al., 1989, 1994)

Three different types of PFPEs, which possess a carboxylic acid group as the end group, were used as the raw materials. K-lube is a homopolymer of perfluoro-isopropylene oxide, and D-lube is a homopolymer of perfluoro-n-propylene oxide. K-lube and D-lube have one end group. Z-lube is a random copolymer of the perfluoro oxymethylene and oxyethylene oxide monomers, which have two identical end groups. The average molecular weight of the PFPEs is about 2000.

The ammonium salts with PFPE carboxylate lubricants were synthesized according to Scheme 2 and Scheme 3 in Fig. 4 in the same manner. (Kondo & Kaneda, 1994; Kondo, 2008) The chemical structure is determined by its infrared spectra: 3200-2800cm^{-1}($N^{+}H_3$

stretching), 2918 cm^{-1} and 2958 cm^{-1} (CH_2 stretching), 1674 cm^{-1} (CO stretching), 1280-1110 cm^{-1} (CF stretching). The CO stretching moved from 1800 cm^{-1} to 1674 cm^{-1}, and the N^+H_3 stretching at 3200-2800 cm^{-1} appears, thus identifying the ammonium salt with a carboxylate structure. The synthesized lubricants are summarized in Table 2.

Scheme 1

$$R_f\text{-COCl} + R_1\text{-XH} \longrightarrow R_f - COXR_1$$

$X = O, NH$

$R_f = F\text{-}(CF_2)_m\text{-} \quad m = 7, 9$

$R_1 = C_nH_{2n+1,\ 2n-1}$

Scheme 2

$$R_f\text{-COOH} + R_1\text{-N}(R_2)_2 \longrightarrow R_f\text{-COO}^-N^+H\ (R_2)_2R_1$$

$R_f = F\text{-}(CF_2)_m\text{-} \quad m = 7, 9$

$R_f = C_nH_{2n+1,\ 2n-1,\ 2n-5}$

$R_f = F\text{-}(CF_2CF_2CF_2O)_n\text{-}CF_2CF_2\text{-}$ for **D-lube**

$R_f = F\text{-}(CF_2CF(CF_3)O)_n\text{-}CF_2\text{-}$ for **K-lube**

$R_1, R_2 = CH_3, C_nH_{2n+1,\ 2n-1,\ 2n-5}$

Scheme 3

$$HOCO\text{-}R_f\text{-}COOH + R_1\text{-N}(R_2)_2 \longrightarrow (R_2)_2R_1N^+HO^-COR_f\text{-}COO^-NH^+(R_2)_2R_1$$

$R_f = \text{-}CF_2O\text{-}(CF_2O)_m\text{-}(CF_2CF_2O)_n\text{-}CF_2O\ \text{-}CF_2\text{-}$ for **Z-lube**

$R_1, R_2 = CH_3, C_nH_{2n+1,\ 2n-1,\ 2n-5}, C_6H_5$

Fig. 4. Synthetic scheme for the new ionic liquid lubricant and the reference compound

3. Friction properties of novel lubricants

3.1 Friction measurement apparatus

The apparatus shown in Fig.5 was used to measure the CSS friction characteristics of the rigid disks. Friction at the head slider was measured by a strain gauge for each CSS operation during the starting of the spindle motor with a 10g load at 25°C, 50% relative humidity.

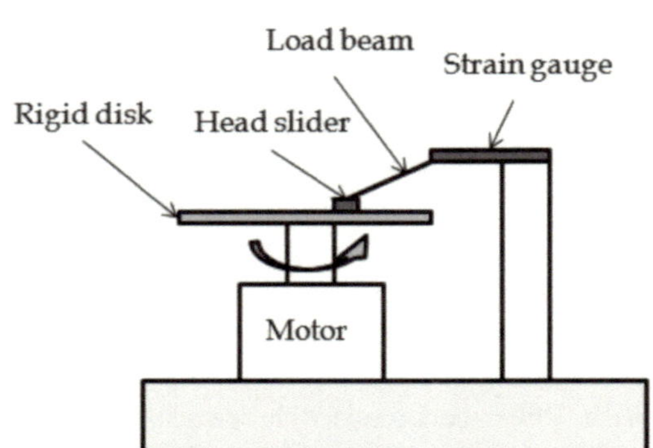

Fig. 5. Friction measuring apparatus for rigid disk.

R_F-CO-Y-R

Number	R_F	Y	R	mp /°C
1	C_7F_{15}	O	$C_{18}H_{37}$	29
2	C_7F_{15}	O	$C_{18}H_{31}$	<20
3	C_7F_{15}	NH	$C_{18}H_{37}$	91
4	C_7F_{15}	NH	$C_{18}H_{31}$	48
5	C_7F_{15}	$O^-H_3N^+$	$C_{18}H_{37}$	55
6	C_7F_{15}	$O^-H_3N^+$	$C_{18}H_{31}$	<20
7	C_9F_{19}	$O^-H_3N^+$	$C_{12}H_{25}$	61
8	C_9F_{19}	$O^-H_3N^+$	$C_{14}H_{29}$	65
9	C_9F_{19}	$O^-H_3N^+$	$C_{18}H_{37}$	71
10	C_9F_{19}	$O^-H_3N^+$	$C_{24}H_{49}$	85
11	$-CF_2O-(CF_2O)_m-(CF_2CF_2O)_n-CF_2O-$	$O^-H_3N^+$	$C_{18}H_{37}$	38-40
12	$-CF_2O-(CF_2O)_m-(CF_2CF_2O)_n-CF_2O-$	$O^-H_3N^+$	$C_{18}H_{35}$	< 30
13	$-CF_2O-(CF_2O)_m-(CF_2CF_2O)_n-CF_2O-$	$O^-H_3N^+$	$C_{18}H_{31}$	< 30
14	$-CF_2O-(CF_2O)_m-(CF_2CF_2O)_n-CF_2O-$	$O^-H_2N(CH_3)^+$	$C_{18}H_{37}$	79-82
15	$-CF_2O-(CF_2O)_m-(CF_2CF_2O)_n-CF_2O-$	$O^-HN(CH_3)_2{}^+$	$C_{18}H_{37}$	55-57
16	$-CF_2O-(CF_2O)_m-(CF_2CF_2O)_n-CF_2O-$	$O^-H_2N(C_{18}H_{37})^+$	$C_{18}H_{37}$	51-55
17	$-CF_2O-(CF_2O)_m-(CF_2CF_2O)_n-CF_2O-$	$O^-H_2N(C_6H_5)^+$	$C_{18}H_{37}$	31-33
18	$F-(CF_2CF_2CF_2O)_n-CF_2CF_2-$	$O^-H_3N^+$	$C_{18}H_{37}$	< 30
19	CF_3 $F-(CF_2CFO)_n-CF_2-$	$O^-H_3N^+$	$C_{18}H_{37}$	

Table 2. Molecular structure and melting point of ionic liquid lubricants.

A schematic diagram of the friction measurement apparatus for the magnetic tapes is shown in Fig.6. The coefficients of kinetic friction are measured for 8-mm wide tapes sliding around a quadrant of a 4-mm diameter polished stainless steel (SUS 304) cylinder. The friction coefficient was calculated from the change in the sliding of the tension (T_1) exerted by a 20-g weight (T_2) hanging from the tape sliding on the cylinder. A 50-mm section of the tape is made to slide against the cylinder at a speed of 5 mms^{-1} in a reciprocating motion at 25°C, 60% relative humidity.

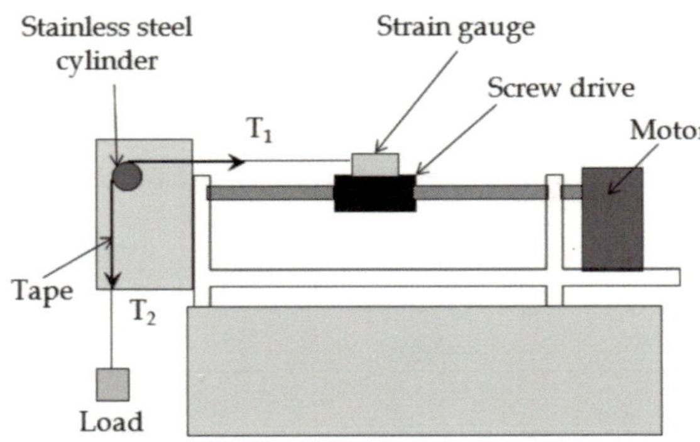

Fig. 6. Schematic diagram of the friction measurement apparatus for ME tapes.

3.2 Frictional performance of the newly synthesized ionic liquid lubricant with ammonium salt for magnetic media

The new lubricants exhibited a good CSS friction properties for the rigid disks compared to the PFPE (Z-DOL) as shown in Fig.7. The relationship between the CSS durability and the molecular structure of the lubricant in terms of the polar group, chain length, and chain symmetry was investigated.

The frictional characteristic of the carboxylic acid ammonium salt coated on the magnetic tape by dip-coating is shown in Fig.8. The friction coefficient is shown as a function of the number of cycles of reciprocating motion over the cylinder. The frictional characteristic of a conventional PFPE (Z-DOL) is shown for comparison. For the ammonium salt, the friction coefficient is low and remains at 0.18 even after 100 cycles of reciprocating motions, but it is more than 0.30 for the PFPE.

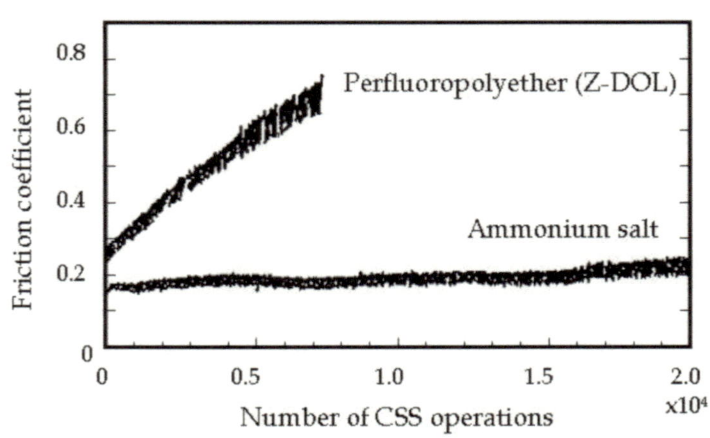

Fig. 7. Frictional coefficient of disk coated with lubricant (10) versus number of CSS operations. The functional PFPE (Z-DOL) shown for comparison.

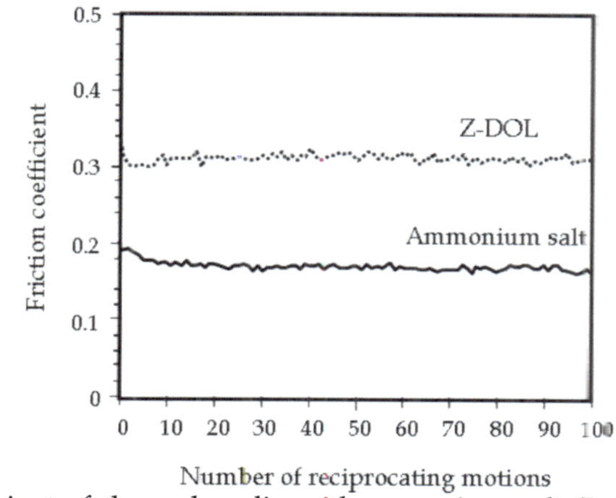

Fig. 8. Friction coefficient of the carboxylic acid ammonium salt (Lubricant 9) and Z-DOL during friction test at 25°C and relative humidity of 60%.

4. Effect of molecular structure on friction

4.1 Hydrophilic group

4.1.1 Effect of hydrophilic group on CSS friction of HDD

The CSS friction properties of three types of lubricants, i.e., ester (Lubricant 1), amide (Lubricant 3), and carboxylic acid ammonium salt (Lubricant 5) are shown in Fig.9. These friction measurements of the synthesized lubricants revealed that the ester and the amide are far less durable than the comparable salt type. For the ester and amide lubricants, the friction coefficients (μ) were around 0.25 for the first ten CSS operations, but rose with the increasing number of CSS operations (n). Especially, for the ester lubricant, μ steeply increased after 20 operations, and the carbon protective layer got damaged when the μ became over 0.90 at n = 138. For the amide lubricant, the slope of $\Delta\mu$ per Δn decreased compared to the ester, however, a wear scar occurred at n = 3279. For the carboxylic acid ammonium salt lubricant, the μ value remained nearly constant at around 0.25 throughout the 10^4 CSS operations and the medium was scarcely damaged. The low initial value of μ, 0.2-0.3, indicates that there is sufficient lubricant film to protect the rubbing surface.

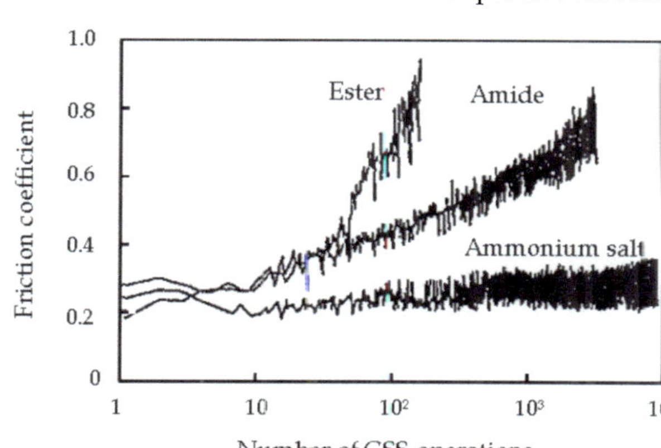

Fig. 9. Friction coefficient of disk coated with ester lubricant (Lubricant 1), amide (Lubricant 3), and carboxylic acid ammonium salt (Lubricant 5) versus number of CSS operations.

4.1.2 Frictional tests for magnetic tapes

Fig. 10 shows the friction coefficient variation vs. the reciprocating cycles for the salt and for the amide. For the salt, the friction coefficient value remained low and steady at approximately 0.23 throughout the 100 reciprocating cycles. The friction coefficient with the amide increased with the number of cycles.

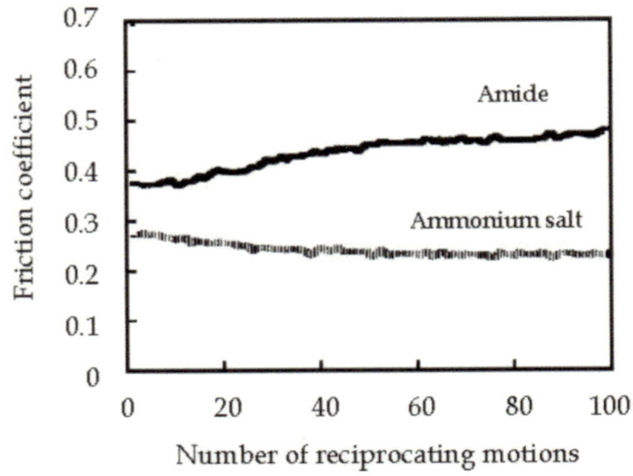

Fig. 10. Comparison of frictional properties of ammonium carboxylate ionic liquid and the corresponding amide during a friction test.

4.2 Hydrophobic group

The hydrocarbon chain length of the salt type lubricant were changed and the CSS durability measured (Fig. 11). As the number of carbon atoms in the amine molecule (n) increased, μ decreased until a nearly constant value of 0.24 was attained. The shorter homologues showed an increase in μ as a result of the breakdown of their film. (Beltzer & Jahanmir, 1987)

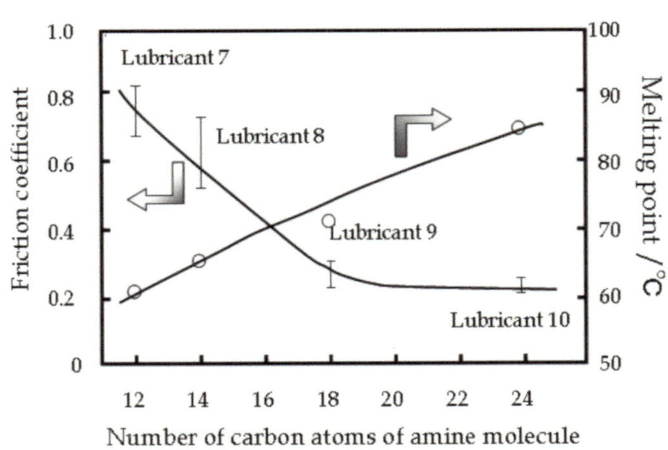

Fig. 11. Friction coefficients of the salt type lubricants after 20,000 CSS operations.

For the given polar group, one of the key properties required for friction reduction is a high intermolecular cohesion energy (due to dispersive or van der Waal's interactions) between the hydrocarbon chains. Melting of the lubricant materials by heating involves disruption of the dispersive interactions between the hydrocarbon chains, (Markley, 1968) therefore, the melting point of the lubricant should be related to the dispersive interactions of the hydrocarbon chain. (Beltzer & Jahanmir, 1987) The melting point becomes higher with the increasing hydrocarbon chain length and μ decreases. Not only the polar group but also the chain length due to dispersive interactions determined the durability.

4.3 Frictional properties of modified PFPEs on magnetic tapes

4.3.1 Frictional properties of new lubricant

The good viscosity characteristics, low melting point, low surface energy, low volatility, and good thermal stability of PFPE are among the important criteria for selecting a lubricant. By changing the perfluorocarboxylic acid with PFPE carboxylic acid, that is Z-lube in Scheme 3 and D- and K-lubes in Scheme 2 of Fig. 4, the modified PFPEs are expected to have lower melting point and improved thermal properties compared to the corresponding perfluorocarboxylic acid homologue.

The frictional characteristics of the three types of modified PFPEs of the ammonium salt with carboxylate are shown in Fig.12. For the ammonium salt, it is low and approximately 0.17 even after 100 cycles of reciprocating motion, and is not dependent on the chain structure of the PFPE. However, it is over 0.30 for the conventional PFPE, and other types of end groups, such as the other hydroxyl(Z-Tetraol) and piperonyl(AM2001), have a similar frictional coefficient of approximately equal to or greater than 0.30 (data not shown).

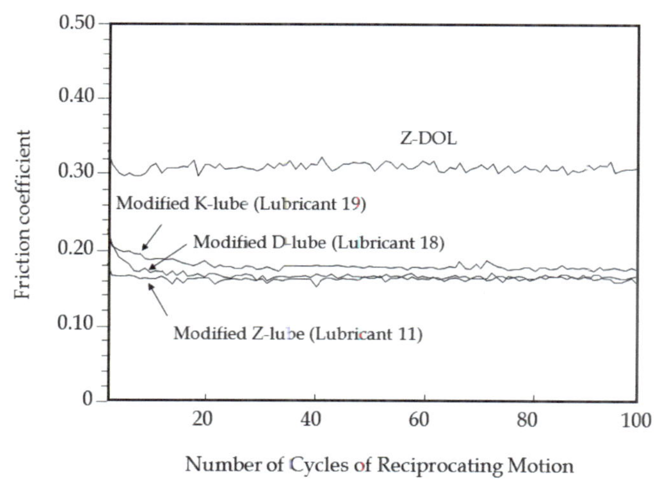

Fig. 12. Frictional properties of the three types of the modified PFPEs using stearyl amine The conventional Z-DOL is shown for comparison.

Fig. 13 shows the dynamic friction coefficient during the reciprocating operation for each PFPE. The amplitude of the saw tooth pattern in the friction curve is significantly high for the Z-DOL tapes; these fluctuations in sliding resulted from the stick-slip process and are associated with squeal and chatter. However, the dynamic friction coefficient was relatively constant and the stick-slip phenomenon is only slightly observed for the ammonium salt tapes (Lubricant 11).

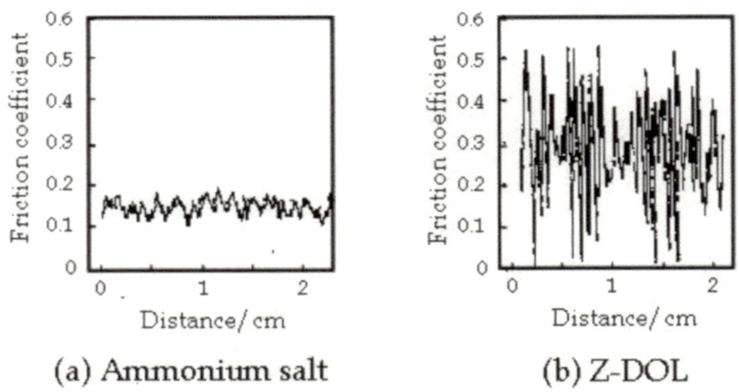

Fig. 13. Change in dynamic friction coefficient during the reciprocating operation for the different PFPE.

4.3.2 Effect of molecular structure of amine

Secondly, the molecular structure of the amine is changed using stearyl amine derivatives in order to fix the chain length of the longest substituent. The frictional results of the PFPE ammonium salts after 100 reciprocating motions and their melting points are summarized in Table 3. The introduction of a double bond into the hydrocarbon long chain, an oleyl with one double bond and a linoleyl with three, also makes the friction coefficient increase.

Lubricant number	Structure of lubricant	Friction coefficient	Melting point /°C
11 (stearyl)	$C_{18}H_{37}NH_2$	0.17	38-40
12 (oleyl)	$C_{18}H_{35}NH_2$	0.20	<30
13 (linolenyl)	$C_{18}H_{31}NH_2$	0.25	<30
14 (methyl stearyl)	$C_{18}H_{37}NHCH_3$	0.19	79-82
15 (dimethyl stearyl)	$C_{18}H_{37}N(CH_3)_2$	0.20	55-57
16 (distearyl)	$(C_{18}H_{37})_2NH$	0.21	51-55
17 (phenyl stearyl)	$C_{18}H_{37}NHC_6H_5$	0.30	31-33

Table 3. Fiction coefficients and melting points of the lubricants with the molecular structure of the amine having 18 carbon derivatives.

By replacing the hydrogen atoms of the amine group with alkyl and phenyl groups, a different series of salts would be obtained. This trend strongly suggests that replacing the hydrogen atom of the amine with an electron donating alkyl group increases the electrostatic interactions (including possible hydrogen bonding) between the cation and anion in the salts, which, in turn, raises the melting point. The alkyl substituted ammonium salt, e.g., by methyl and stearyl groups, have higher melting points than the non-substituted stearyl amine salt, nevertheless, the friction coefficient was somewhat inferior to the stearyl amine. Exchanging the amine of the salts for the larger phenyl substituent generally produced a further decrease in the melting point. As a general trend, for each anion, the salts with cations of lower symmetry show a lower melting point than those with cations of higher symmetry. Also, adding the bulky phenyl group significantly increased the friction. These magnitudes of increase in the friction suggest a blocking effect by the large

substituent attached to the amine nitrogen, which hinders adsorption of the polar group on the media surface. Steric hindrance by the polar group caused the high friction.
The cohesive energy density is normally lower for the fluorocarbon and ether group compared to the hydrocarbon. Cong et al. indicated that the film strength controlled by intermolecular attractive forces is an important factor that affects the frictional properties of the monolayers, which are associated with a higher load-capacity. (Cong et al., 2000) The attractive forces between the fluorocarbon chains are lower than those between the hydrocarbon chains and the presence of oxygen atoms in the PFPE backbone decreases the intermolecular attractive force. For the ammonium salts of the PFPEs, the reason for the lower friction might be the incorporation of a hydrocarbon into the molecules. Therefore, it is expedient to increase the dispersive interaction by introduction of a hydrocarbon chain into a PFPE molecule without steric hindrance between the polar group of the lubricant and media surface. To be precise, a saturated straight long chain ammonium salt is the best selection.
The use of a conventional PFPE is limited by the solvent. However, since the modified PFPEs contain a hydrocarbon moiety and ammonium salt moiety, it is soluble in alcohols, and other conventional fluorinated solvents, which makes its practical use convenient. Fig. 14 shows the consequences for friction of the stearyl ammonium salt of Z-lube by changing to a thinner dip-coating. In this case, ethanol, n-hexane-20%wt.ethanol, and a fluorinated solvent were used. It is evident that the frictional properties are independent of the dip-coating solvent.

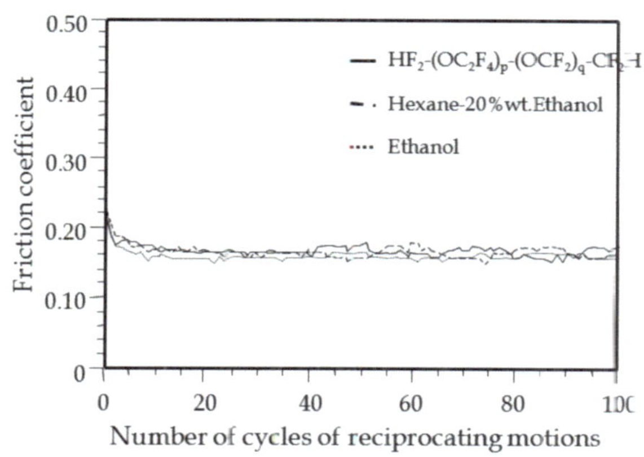

Fig. 14. Influence of changing to a thinner lubricant on the friction coefficient.

5. Langmuir-Blodgett (LB) films of the salt type lubricant

In order to elucidate the molecular level structure of the spontaneously adsorbed layers, a comparative structural study with films prepared by the Langmuir-Blodgett (LB) method is useful. Langmuir films (L films, i.e., monolayer on the water surface) and Langmuir-Blodgett films (LB films) of the lubricants were prepared and studied using a Jpyce-Loebl.

5.1 Basic properties of the L films and LB films

The ammonium salt with perfluorocarbxylate (Lubricant 5), and the corresponding ester (Lubricant 1) and the amide (Lubricant 3) are compared. The salt forms a much more stable monolayer on the water surface than the other two ester and amide lubricants. Fig.15 shows the area decay curve of the L films of the salt and the amide on the water surface when they are held at a surface pressure of 25mNm^{-1}. The curve for the ester is unstable such that it collapsed as it was compressed. The area decay over 1 hour was less than 5% for the salt, whereas for the amide it was 45%, indicating a good stability for the salt monolayer. This result suggests a good balance between the polar and hydrophobic nature of the salt molecule, which is a necessary condition for producing a stable monolayer on the water surface.

Furthermore, the isotherm of the salt shown in Fig.16 suggests that the molecules of the salt were closely packed in the L film. From Fig. 16, the area per molecule at 25 mNm^{-1} is about 0.6nm^2. The areas occupied by an alkyl chain and a perfluoroalkyl chain are about 0.2 nm^2 and 0.4 nm^2, respectively, when their chain axes are perpendicular to the water surface. Therefore, it is suggested that the alkyl chains and perfluoroalkyl chains of the salt molecules are highly ordered and closely packed with their chain axes perpendicular to the water surface in the L film.

The closely packed monolayer is transferred onto the surface of the magnetic layer at 25 mNm^{-1} with a dipping speed of 5 mmmin^{-1} upon withdrawal from the water. The deposition trace suggested that there is an even deposition and the transfer ratio is about 0.75. Therefore, in the LB film monolayer, the area occupied by one salt molecule is calculated to be about 0.8 nm^2. The difference in occupied area per molecule indicates that the chain leans slightly to the surface normal. The tilt of the chains in the LB monolayer is induced by the deposition process.

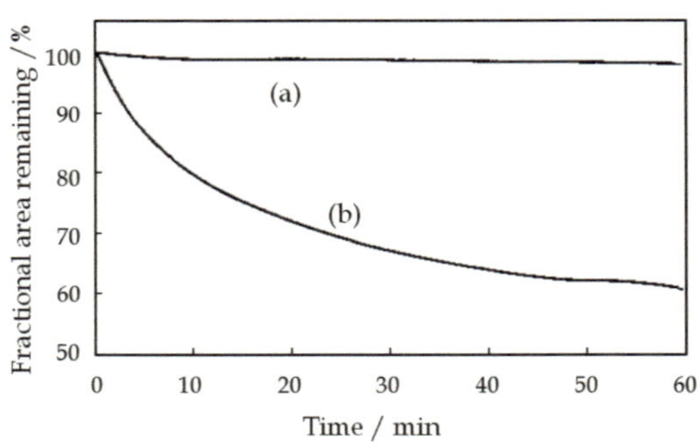

Fig. 15. Area decay curve of the L films of the salt (a) and the amide (b) at a surface pressure of 25mNm^{-1}.

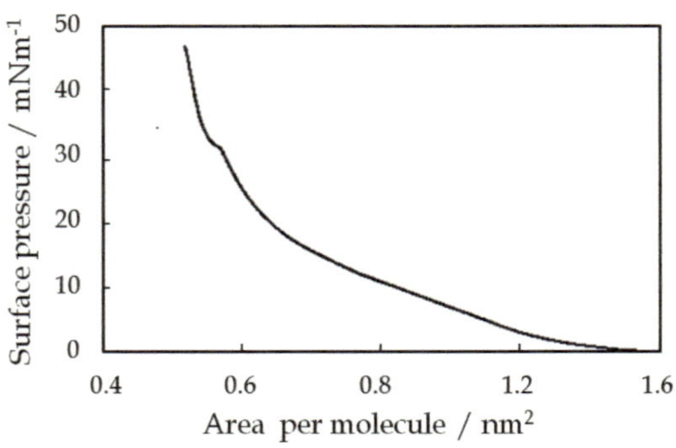

Fig. 16. Isotherm of the salt.

5.2 FTIR study for LB films

The friction properties may be enhanced by a microscopically smooth coverage of the lubricant film over the media surface. In the FTIR-RAS spectra, the components of the vibrational moments of the chemical bonds, which are parallel to the substrate normal, selectively appear. Therefore, the FTIR-RAS spectra are useful for investigating the molecular orientation of a lubricant film on a substrate.

The salt-type lubricant films, which are prepared by the LB method and the spontaneous adsorption from the lubricant solution, are compared in Fig.17. The bands in the 3000-2800 cm^{-1} and 1370-1100 cm^{-1} regions are assigned to the CH and the CF stretching vibrations, respectively. The band at around 1674cm^{-1} is assigned to the COO^- anti-symmetric vibration. The absolute intensities of these bands are very similar in both spectra, which suggest that the adsorbed layer of the salt is in fact a monolayer and that the degree of orientation and packing of the alkyl and perfluoroalkyl chains closely resembles that in the LB monolayer Therefore, the area occupied by one salt molecule in the adsorbed layer is approximately 0.8nm^2.

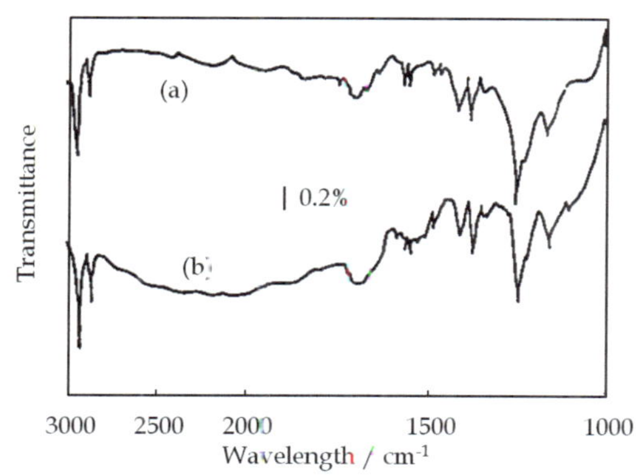

Fig. 17. FTIR reflection adsorption spectra of the novel lubricant films prepared by (a) spontaneous adsorption method and (b) LB method on the surface of the magnetic layer.

6. Spectroscopy of lubricant film

6.1 FTIR study for lubricant film of different hydrophilic groups

Figs. 18 and 19 compare the RAS of the lubricant film and transmission spectra of the bulk material for the salt and the amide, respectively. The mode assignment and peak positions for the RAS and for the transmission spectra are summarized in Table 4.

For the salt, the differences in the spectra between the RAS of the adsorbed lubricant film on the substrate (a) and the transmission spectra of the bulk materials (c) are described as follows:

1. In the RAS spectrum of the adsorbed film, the bond assigned to the COO^- asymmetric stretching vibration at 1674 cm^{-1} is much weaker than in the spectrum of the bulk material.
2. The relative intensity in wavenumber of the CF_2 stretching vibrations in the 1250-1140 cm^{-1} region is changed and shifted to a higher frequency in the RAS spectrum.
3. The band assigned to the CH_2 stretching vibration in the RAS spectrum in the vicinity of 2900 cm^{-1} is also shifted to a higher frequency than that in the spectrum of the bulk.

The weakness of the band at 1674 cm^{-1} suggests that asymmetric stretching of the COO^- (hydrophilic) group is almost parallel to the substrate surface in the adsorbed film. The changes in the relative intensity and the higher frequency shift in the 1250 to 1140 cm^{-1} region are similar to those reported for a monolayer of perfluorocarboxylic acid. (Chau & Porter, 1990) These changes have been attributed to adsorption of the molecules with preferential orientation and perfluoroalkyl chains tilted in the adsorbed layer.

The higher frequency shift of the CH_2 stretching vibrations is often observed when the alkyl chains are in a liquid phase or in a solution, where the cohesive interaction between them is weak. (Asher & Levin 1977) A probable cause of the higher frequency shift in the RAS spectrum is that the perfluoroalkyl chains hinder a cohesive interaction between the alkyl chains in the monolayer.

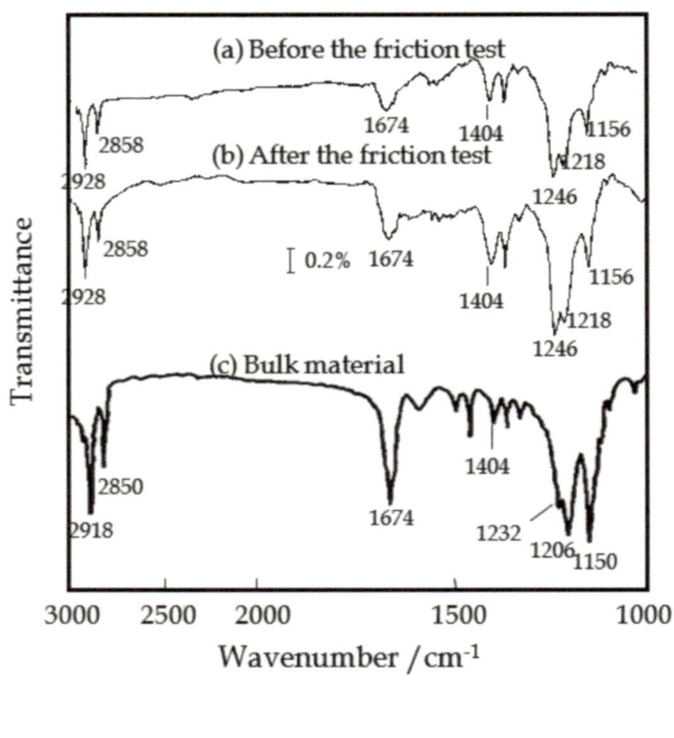

Fig. 18. FTIR spectra of ammonium, C_9F_{19} -$COO^-H_3N^+$-$C_{18}F_{37}$. (a)RAS spectra of adsorbed film; (b)RAS spectra of adsorbed film after friction test; (c) Transmission spectrum of bulk material.

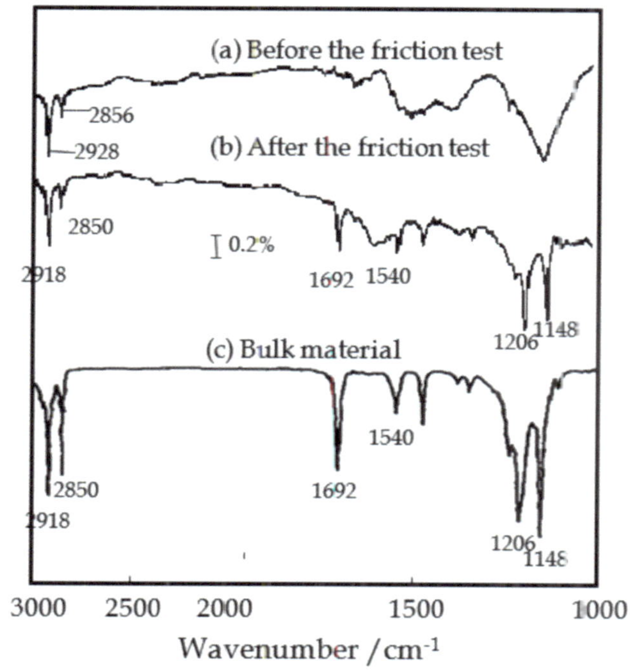

Fig. 19. FTIR spectra of amide, C_9F_{19} -CONH-$C_{18}F_{37}$. (a)RAS spectra of adsorbed film; (b)RAS spectra of adsorbed film after friction test; (c)Transmission spectrum of bulk material.

For the amide, the spectral pattern of the film in the region of the CF_2 and CF_3 stretching vibrations, 1250 to 1140 cm^{-1}, is different from that of the salt. However, the decrease in intensity of the stretching vibrations for C=O at 1692 cm^{-1} and the high frequency shift of the CH_2 stretching mode behaves similar to the carbonyl stretch and high frequency CH_2 shift for the salt.

The FTIR-RAS spectra of the lubricant film after the friction test are also shown in Figs. 18(b) and 19(b). The absolute intensity of the CH_2 stretching region is similar in the spectra both before and after the friction test, indicating that sliding causes no substantial change in the surface concentration of the alkyl groups. For the salt, the spectra both before and after the test are similar, therefore, the rubbing motion caused no change in the microscopic structure of the lubricant film. On the other hand, there are several differences in the RAS spectrum of the adsorbed film of the amide before and after the test. These results are summarized as follows:

1. The bands at approximately 1692 cm^{-1} and 1540 cm^{-1}, which are assigned to the C=O stretching vibration and the NH bending vibration, respectively, appear after the test.
2. The pattern of the CF_2 and CF_3 stretching vibrations at 1250 cm^{-1} and 1140 cm^{-1} changes.
3. The band of the CH_2 antisymmetrical and symmetrical stretching vibrations in the 3000 to 2800 cm^{-1} region shifts to a lower frequency.

Thus, the spectrum of the adsorbed amide film approachs the spectrum of the bulk material with sliding. This suggests rearrangement of the amide monolayer by the sliding process, perhaps into three-dimensional crystals or amorphous piles.

These results show that the layers of the amide were composed of orientated molecules, although they produced a higher friction coefficient than those of the salt. A possible cause

for the higher friction coefficient is that a layer structure with a high degree of molecular orientation is less stable for the amide compared to the salt. This is demonstrated by the decay curve on the water surface in Fig.15.
The FTIR-RAS spectra showed that the polar group of both lubricants interacted with the magnetic surface before the friction test. For the ammonium salt, the friction coefficient is low and constant throughout the 100 reciprocating cycles (Fig. 10). From a spectroscopic point of view, before and after the spectra of the friction test are similar, suggesting no change in the structure of the lubricant film. In contrast, for the amide, the friction increases with the number of cycles, thus the adsorbed lubricant was changed into bulk-like aggregates, which brings about the bare contact and leads to a higher friction.

Peak position /cm^{-1}				Vibration mode
Amide		Ammonium salt		
Bulk material	Film	Balk material	Film	
2918	2928	2918	2928	CH_2 asymmetric stretching
2850	2856	2850	2858	CH_2 symmetric stretching
		1674	1674	COO^- asymmetric stretching
		1404	1404	COO^- symmetric stretching
1692				CO stretching
1232		1232	1246	CF_2 asymmetric stretching (E_1 symmetry)
1206		1206	1218	CF_2 asymmetric stretching (A_2 symmetry) + CF_3 stretching
1148		1150	1156	CF_2 symmetric stretching (E_1 symmetry)

Table 4. Mode assignment and peak positions for the ammonium salt and amide in KBr and as adsorbed on a magnetic layer.

6.2 Angle resolved X-ray photoelectron spectroscopy (ARXPS) study for microscopic coverage of modified PFPE

The uniformity of the monolayer level lubricant on the magnetic thin film media has been investigated using ARXPS with the finding that the PFPE is a discontinuous film. (Kondo et al., 1996;Moulder et al., 1986) Kimachi et al. proposed an island model to describe the coverage of conventional PFPEs on the magnetic recording media. (Kimachi et al., 1987) On the other hand, the modified PFPE has a good lubricity compared to the conventional ones, which implies a different film formation. In order to gain insight into this effect, microscopic coverage of the lubricant on the surface is investigated by ARXPS. The Z-lubes with and without modification are deposited under the same conditions.
A lubricant layer completely covers the surface with a constant thickness of d in the uniform model. In the island model, the surface is discretely covered with lubricant islands having the average thickness of d (Fig. 20). It is assumed in these models that the surface is flat and that elemental atoms are homogeneously distributed in the lubricant layer. (Fadley, 1974) Based on these assumptions, the photo-electron intensity ratio, I_{lub}/I_{sub} is expressed as a function of the take-off angle, φ, and the lubricant coverage ratio, Θ, on the surface of the

magnetic layer. I_{lub} is the intensity of the photoelectron detected from the lubricant layer and, I_{sub} is the intensity from the under layer.

In Fig. 21, I_{lub}/I_{sub} is plotted as a function of sin φ, and the calculated model curves of Θ using the island model are also shown. Because the I_F of the modified PFPE is somewhat low, I_{lub}/I_{sub} is normalized in such a way that the I_{lub}/I_{sub} at 90 degrees is 1.

The coverage of the modified PFPE is greater than the conventional lube, despite the fact that the amount of lube on the surface is somewhat lower. The higher coverage reduces the dry contacts at the rubbing surface and minimizes friction.

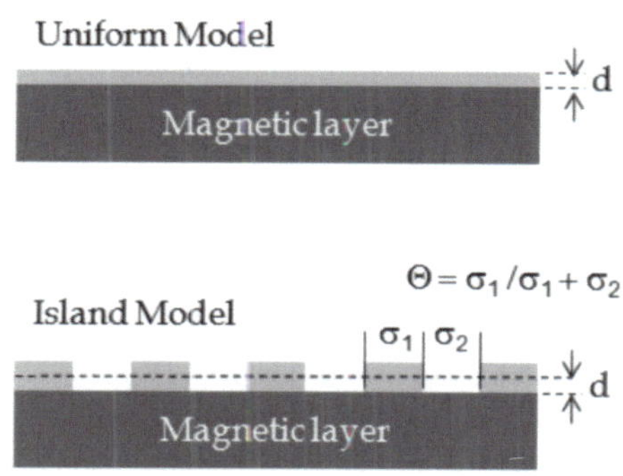

Fig. 20. Lubricant coverage models. Θ is the lubricant coverage ratio in island model and d is the average thickness.

It can be considered that two factors determine low friction, namely, better coverage and a strong interaction between the lubricant and the media surface. Better coverage is ascribed to the balance of the hydrophobic and hydrophilic properties of the lubricant. It is well-balanced when the hydrocarbon chain is introduced. The polar group of the ammonium salt without steric hindrance has strong interactions at the surface.

6.3 Heat of the preferential adsorption

These three types of lubricants, the salt (lubricant 5), the amide (Lubricant 3), and the ester (Lubricant 1), have the same hydrophobic groups, but have different friction coefficients, which are related to the polar group of the lubricant molecule. A plausible explanation is that the salt-type lubricant leads to greater adhesion than the ester and amide. As the surface of the sputtered carbon protective layer is classified as rather polar because it contains 5 and 7atom% hydroxyl and carbonyl groups, respectively, a source of the attractive force at the surface may be the interactions between these polar groups. (Kondo & Nishida, 2007) In order to gain insight into this effect on the friction properties, the dielectric loss measurements are conducted.

The relative dielectric constant (ε) of the lubricants is measured at 3MHz and summarized in Table 6. The relative dielectric constant, which is a parameter of the dipole moment, is 2.01 and 2.36 for the ester and the amide, respectively, but 2.66 for the salt-type. The higher the dielectric constant, the more strongly the lubricant seems to adsorb on the medium.

The formation of the lubricant film is a spontaneous process caused by a decrease in the free energy of the solid surfaces and lubricant molecule adsorption. The heat of adsorption of lubricants on a rubbing surface can be taken as a measure of the strength of attachment of the lubricant molecules to the surfaces and are also shown in Table 6.

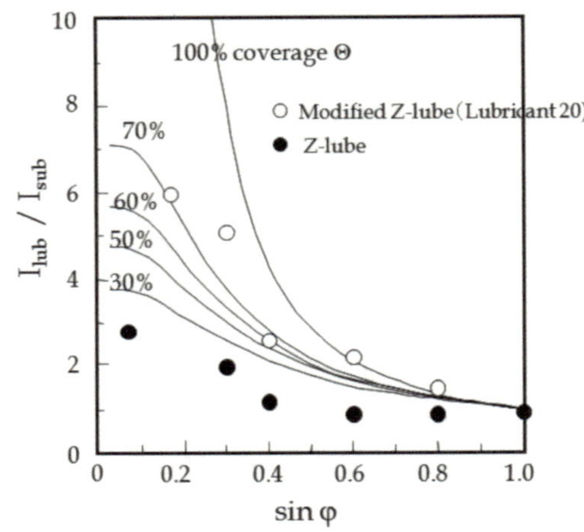

Fig. 21. Coverage of the media surface with lubricant by ARXPS and the calculated curves using island model shown for comparison.

For the ester and the amide type, the heat of adsorption is small, and the heat of desorption is partially observed. However, for the salt-type, the heat of adsorption was very high compared to the ester and amide, and the heat of desorption could not be detected. These results confirm that the ester and amide were held by weak interactions on the carbon surface, whereas irreversible adsorption had taken place for the salt-type. The high heat of adsorption of the salt-type on the carbon surface results from these strong interactions and accounts for its low and steady friction. The low heat of adsorption for the ester and the amide, by contrast, apparently produces a film which is easily disrupted by sliding to give a rising friction coefficient with sliding.

Structure and lubricant number	Dielectric constant**	Heat of adsorption /erg cm^{-2}	Heat of desorption /erg cm^{-2}
(2)$C_7F_{15}CO\ O\ C_{18}H_{31}$	2.01	2.1	1.0
(4)$C_7F_{15}CO\ NH\ C_{18}H_{31}$ *	2.36	2.7	1.9
(6)$C_7F_{15}\ O^-H_3N^+C_{18}H_{31}$	2.66	18.9	Not detected

* Stearyl amide is slightly soluble in the solvent, therefore, the double bonded lubricants are used.
**Dielectric constant is measured at 3 MHz.

Table 6. Heat of adsorption and desorption of the lubricants on the surface.

7. Conclusions

The frictional properties of newly synthesized ionic liquid lubricants for magnetic media have been investigated. A novel ionic liquid, which has an ammonium salt with a carboxylate as a hydrophilic group, has a lower frictional coefficient than the other conventional PFPE. It can be considered that two factors determined the low friction, namely, better coverage and strong interaction between the lubricant and the media surface. These ionic liquid lubricants invented around 1987 (Haga & Kondo, 1987;Kondo, 1987, 1992; Kondo & Haga, 1987; Kondo & Uchimi, 1992) have been used for magnetic tapes of 8mm video, digital video cassette, the AIT system, and the broadcast application for about a quarter century because of their good lubricity and also from an environmental point of view.

The effects of the molecular structure of the modified PFPE on the frictional properties are summarized as follows:

1. Stronger adsorption due to the adhesive interaction of the polar group, thus the novel carboxylic acid ammonium salt has a lower and more stable friction coefficient.
2. Sufficient length and symmetry of the hydrocarbon chain cause the extensive cohesive interactions, and these dispersive interactions compensate for the friction reduction.
3. On the contrary, for the olefinic lubricant, the friction coefficient increased with the increasing number of reciprocating cycles because of the weaker cohesive interactions.
4. The ammonium salt polar group can be introduced to the commercial PFPE, and the frictional coefficient is independent of the molecular structure of the PFPE backbone polymer and of the thinner dip-coating, which makes practical use convenient without any environmental problems.

From a microscopic point of view, the lubricant film coverage was also investigated as follows:

1. A layer of the carboxylic acid ammonium salt lubricant film prepared by the spontaneous adsorption is highly ordered and closely packed as in the LB monolayer.
2. The polar COO^- groups of the salt are adsorbed almost parallel to the surface, and the sliding scarcely changes the salt-type lubricant film, while the arrangement of the amide film into a bulk phase occurs due to the sliding process.
3. The modified PFPE uniformly covers the magnetic surfaces; this is why it minimizes the friction.

8. References

Anderson, J.L., & Armstrong, D.W. (2003) High-Stability Ionic Liquids. A New Class of Stationary Phases for Gas Chromatography. *Anal. Chem.* Vol.75, pp.4851-4858, ISSN: 0003-2700

Asher, I., & Levin I.W., (1977). Effect of temperature and molecular interactions on the vibrational infrared spectra of phospholipids vehicles, *Biochim. Biophys. Acta,* Vol.468, No. 1, (Jul. 1977), pp.63-72, ISSN: 0005-2736

Beltzer, M., & Jahanmir, S., (1987). Role of Dispersion Interactions Between Hydrocarbon Chains in Boundary Lubrication, *Triblogy Trans.* Vol.30, No.1, pp.47-54 , ISSN: 1547-397X

Bhushan, B., (1996), *Tribology and Mechanics of Magnetic Storage Devices,* (2nd Edition), Springer-Verlag, ISBN:0-387-96926-8, 3-540-96926-8, NewYork

Binnemans, K., (2005). Ionic liquid crystals, *Chem. Rev.,* Vol.105, No.11, (Nov. 2005), pp. 4148-4204, ISSN: 0009-2665

Branco, L.C., Rosa, J.N., Moura Ramos, J.J., & Afonso, A.M., (2002). Preparation and characterization of new room temperature ionic liquids, *Chem. A-Eur. J.,* Vol. 8, No. 16, (Aug. 2002), pp.3671-3677, ISSN:0947-6539

Chau, L-K., & Porter, M.D., (1990). Composition and structure of spontaneously adsorbed monolayers of *n*-perfluorocarboxylic acids on silver, *Chem. Phys. Lett.,* Vol.167, No.3, (Mar. 1990), pp.198-204, ISSN: 0009-2614

Chiba, H., Yamasaka, E., Tokairin, T., Oshikubo, Y., & Watanabe, K., (2005). Synthesis of multi-functional PFPE lubricant and its tribological characteristics, *Proc. 3rd World Tribol. Congr.*, Washington DC, U.S.A., Sept. 2005, Paper WTC2005-63165

Clark, R.J.H, (1988). *Spectroscopy of Surface,* John Wiley and Sons, ISBN-13: 978-0471918950, New York

Cong, P., Igari, T., & Mori, S. (2000). Effects of film characteristics on frictional properties of carboxylic acid monolayers, *Tribology Letters*, Vol. 9, No. 3-4, (Jan. 2001), pp.175-179, ISSN: 1023-8883

Fadley, C.S., (1974). Instrumentation for surface studies:XPS angular distributions, *J. Electron Spectroscopy Related Phenomena*, Vol. 5, No.1, pp.725-754, ISSN: 0368-2048

Gordon, C. M. (2001) New developments in catalysis using ionic liquids. *Appl. Catal. Gen.* Vol.222, No. 1-2, (Dec. 2001), pp.101-117, ISSN: 0926-860X

Greaves, T.L., Weerawardene, A., Fong, C., Krodkiewska, I., & Drummond, C.J., (2006). Protic Ionic Liquids: Solvents with Tunable Phase Behavior and Physicochemical Properties. *Phys. Chem. B.* Vol. 110, No. 45, (Oct. 2006)22479-22487, ISSN: 1520-6106

Greaves, T.L., & Drummond C.J., (2008). Protic ionic liquids: Properties and applications, *Chem. Rev.*, Vol.108, No.1, (Dec. 2007), pp.206-237, ISSN: 00092665

Gunsel, S., Venier, C., & Chiu, I-C., (2003). Lubricant for magnetic recording medium and use thereof, *U.S. Patent* 6667284

Haga, S., & Kondo, H., (1987). Magnetic recording medium, *Japanese patent* 2590482

Homola, A.M., (1996) Lubrication Issues in Magnetic Disk Storage Devices. *IEEE Trans. Magn.* Vol. 32, No. 3, pp. 1812-1818 (1996), ISSN: 0018-9464

Iglesias, P., Bermudez, M.D., Carrion, F.J., & Martinez-Nicolas, G., (2004). Friction and wear of aluminium-steel contacts lubricated with ordered fluids-neutral and ionic liquid crystals as oil additives, Wear, Vol. 256, No. 3-4 , (Feb. 2004), pp.386-392, ISSN: 0043-1648

Ishida, M., Nakakawaji, T., Ito, Y., Matsumoto, H., Tani, H., & Ishihara, H., (2005). Lubricant, magnetic disk and magnetic disk apparatus, *U.S. Patent* 6869536

Ito, T., Iwasaki, Y., Tachikawa, H., Murakami, Y., & Shindo, D., (2002) Microstructure of a Co–CoO obliquely evaporated magnetic tape, *J. Appl. Phys.*, vol. 91, no.7, pp.4468-4473, ISSN: 0021-8979

Jhon, M.S., & Choi, H.J, (2001) Lubricants in future data storage technology, *J. Ind. Eng. Chem.*, vol.7, no. 5, pp.263-275 ISSN:1226-086X

Jiaa, C.L., & Liu, Y.M., (1999). Tribological evaluation and analysis of the head/disk interface with perfluoropolyether and X1-P phosphazene mixed lubricants, *Tribology Lett.* Vol.7, No.1 pp.11-16, ISSN:1023-8883

Kaneda, Y., (1997). Tribology of met al-evaporated tape for high density magnetic recording, *IEEE Trans. Magn.*, Vol. 33, No. 2, pp. 1058-1068, ISSN:0018-9464

Kanota, K., Inoue, H., Uetake, A., Kawaguchi, M., Chiba, K., & Kubota, Y., (1990). A high density recording technology for digital VCRs, *IEEE Trans., Consumer Electronics*, Vol. 36, No. 3, pp. 540-545, ISSN: 0098-3063

Kato, Y., Umeda, H., & Fujimoto, H., (2000), Past and Future Developments of Aluminum Substrates for High DensityMagnetic Disks, *KOBE STEEL ENGINEERING REPORTS*, Vol. 50, No. 3, December, pp.50-53

Kimachi, K., Yoshimura, F., Hoshino, M., & Terada, A. (1987). Uniformity Quantification of Lubricant Layer on Magnetic Recording Media. IEEE Trans. Magn., Vol.23, No.5, (Sep. 1987), pp.2392-2394, ISSN:0018-9464

Klaus, E.E., & Bhushan, B.: Lubricants in Magnetic Media - a Review. *Tribology and Mechanics of Magnetic Storage System*, IV SP-19, 7-15 (1985)

Kohler, F., Atrops, H., Kalall, H., Liebermann, E., Wilhelm, E., Ratkovics, F., & Salamon, T., (1981a). Molecular Interactions in Mixtures of Carboxylic acids with amines. 1. Melting Curves and Viscosities, *J. Phys. Chem.* Vol. 85, No. 17, (Aug. 1981), pp. 2520-2524, ISSN: 0022-3654

Kohler, F., Gopal, R., Goetze, G., Atrops, H., Demiriz, M.A., Liebermann, E., Wilhelm, E., Ratkovics, F., & Palagyl, B., (1981b). Molecular Interactions in Mixtures of Carboxylic acids with amines. 2. Volumetric, Conductimetric, and NMR Properties, *J. Phys. Chem.* Vol. 85, No. 17, (Aug. 1981), pp.2524-2529, ISSN: 0022-3654

Kondo, H., (1987a). Magnetic recording medium, *Japanese patent* 2581090

Kondo, H., & Haga, S., (1987b). Magnetic recording medium, *Japanese patent* 2629725

Kondo, H., Seto, J., Ozawa, & K. Haga, S., (1989). Novel Lubricant for Magnetic Thin Film Media. *J. Magn. Soc. Jpn.* Vol. 13, Suppl.,No.S1, pp. 213-218

Kondo, H., Seki, A., Watanabe, H., & Seto, J., (1990). Frictional Properties of Novel Lubricants for Magnetic Thin Film Media," *IEEE Trans. Magn.* Vol.26, No. 5, (Sep. 1990), pp.2691-2693, , ISSN:0018-9464

Kondo, H., & Uchimi, T., (1992a). Novel Perfluoropolyether Derivatives Lubricants and Magnetic Recording Medium using the same, *US Patent* 5453539

Kondo, H., (1992b). Magnetic recording medium, *US Patent* 5741593.

Kondo, H., & Seki, A., (1993a). Relation between the microscopic structure of the lubricant film and the frictional properties, *J. Japanese Society of Tribology*, Vol. 38, No. 1, (Feb. 1992) pp. 40-45, ISSN: 1045-7828

Kondo, H, Seki, A., & Kita, A. (1993b). The effect of molecular structure and microscopic coverage of lubricants on the frictional properties, *J. Surface Sci. Soc. Jpn*, Vol. 14, No. 6, (Aug. 1993) pp. 331-335, ISSN: 1881-4743

Kondo, H., Seki, A., & Kita, A., (1994a). Comparison of an Amide and Amine Salt as Friction Modifiers for a Magnetic Thin Film Medium. *Tribology Trans.* Vol.37, No. 1, (Jan. 1994), pp. 99-105, ISSN: 0569-8197

Kondo, H., & Kaneda, Y., (1994b). Development of Modified Perfluoropolyether Tape. *Proceedings of the 4th International Tribology Conference- AUSTRIB '94*, pp. 415-420, Perth, Australia, 5-8 December 1994

Kondo, H., Hisamichi, Y., & Kamei, T., (1996). Lubrication of Modified Perfluoropolyether on Magnetic media, *J. Magnetism and Magnetic Materials*, Vol. 155, No.1-3, (Mar. 1996), pp.332-334, ISSN: 0304-3853

Kondo, H. (1997) Effect of double bonds on friction in the boundary lubrication of magnetic thin film media, *Wear*, Vol. 202, No.2, (Jan. 1997) pp.149-153, ISSN: 0043-1648

Kondo, H, & Nishida, Y., (2007) Quantitative analysis of surface functional groups on the amorphous carbon in magnetic media with XPS preceeded by chemical derivatization, *Bull. Chem. Soc. Jpn*, Vol. 80, No. 7, pp. 1405-1412, ISSN: 0009-2673

Kondo, H., (2008) Protic Ionic Liquids with Ammonium Salts as Lubricants for Magnetic Thin Film Media, *Tribology Lett.* Vol. 31, No. 3, (Aug. 2008) pp.211-218, ISSN: 1023-8883

Lin, J., & Wu, A.W., (1990). Lubricants for Magnetic Rigid Disks, *Proceedings of International Tribology Conference*, pp.599-604, Nagoya, Japan, Oct.29-Nov.1, 1990

Liu, J., Stirniman, M. J., & Gui, J., (2005). Lubricant for thin film storage media, *U.S. Patent* 6916531

Liu, W., Ye, C., Gong, Q., Wang, H., & Wang, P., (2002). Tribological perfprmance of room-temperature ionic liquids as lubricant, *Tribology Letters*, Vol. 13, No. 2, (Aug. 2002) pp.81-85, ISSN:1023-8883

Maesaka, A., & Ohmori, H., (2002) Transmission electron microscopy analysis of lattice strain in epitaxial Co-Pd multilayers, *IEEE Trans. Magn.* vol. 38, no.5, Sep. pp.2676-2678, ISSN:0018-9464

Marchionni, G., Ajroldi, G., Righetti, M. C., & Pezzin, G., (1993). Molecular Interactions in Perfluorinated and Hydrogenated Compounds, *Macromolecules* Vol. 26, No. 7, (Mar. 1993), pp. 1751-1757, ISSN: 0024-9297

Marchon, B., Dai, Q., Hendriks, F., & Nayak, U. V., (2006a). Disk drive having reduced variation of disk surface lubricant layer thickness, *U.S. Patent* 7002768

Marchon, B., Guo, X.-C., Karis, T., Deng, H., Dai, Q., Burns, J., & Watlman, R, (2006b). Fomblin multidentate lubricants for ultra-low magnetic spacing, *IEEE Tran. Magn*, vol.42, no.10, pp.2504-2506, ISSN:0018-9464

Markley, K.S., (1968). Fatty acids and their chemistry, properties, production, and uses, *Part I* (2nd Ed.), John Wiley & Sons Inc, ISBN-13: 978-0470571248, (New York)

Moulder, J.F., Hammond, J.S., & Smith, K.L., (1986). Using Angle Resolved ESCA to Characterize Winchester Disks. *Applied Surface Sci.*, Vol. 25, No. 4, (May 1986), pp. 446-454, ISSN: 0169-4332

Mu, Z., Liu, W., Zhang, S., & Zhou, F., (2004). Functional room-temperature ionic liquids as lubricants for an aluminium-on-steel system, *Chem. Letters*, Vol. 33, No. 5, (Apr. 2004) pp.524-426, ISSN: 0366-7022

Mu., Z., Zhou, F., Zhang, S., Liang, Y., & Liu, W., (2005). Effect of the functional groups in ionic liquid molecules on the friction and wear behavior of aluminium alloy in lubricated aluminium-on-steel contact, *Tribology International*, Vol.38, pp. 725-731, No. 8, (Aug. 2005), ISSN: 0301-679X

Navarrini,W., Bianchi, C. L., Magagnin, L., Nobili, L., Carignano, G., Metrangolo, P., Resnati, G., Sansotera, M., (2010). Low surface energy coatings covalently bonded on diamond-like carbon films, *Diamond and Related Materials,*Vol. 19, No. 4, (Apr. 2010), pp. 336-341, ISSN: 0925-9635

Ohno, H.: Electrochemical Aspects of Ionic Liquids, Wiley-Interscience, ISBN-13: 978-0471648512 , Hoboken, NJ. (2005)

Omotowa, B.A., Philips, B.S., Zabinski, J.S., & Shreeve, J.M., (2004). Phosphazene-based ionic liquids: synthesis, temperature-dependent viscosity, and effect as additives in water lubrication of silicon nitride ceramics, *Inorganic Chem.*, Vol. 43, No.17, (Jul. 2004), pp.5466-5471, ISSN: 0020-1669

Onodera, S., Kondo, H., & Kawana, T., (1996a) Materials for magnetic tape media. *MRS Bulletin,* Vol. XXI, No. 9, (Sep. 1996) pp.35-41, ISSN: 0883-7694

Onodera, S., Takeda, T., & Kawana, T. (1996b) The archival stability of met al evaporated tape for consumer digital VCRs, *J. Appl. Phys.* Vol.79, No.8, pp.4875-4877, (Apr. 1996), ISSN: 0021-8979

Overney, R.M., Meyer, E., Frommer, J., Gruntherodt, H.J., Fujihira, M., Takano, H., & Goto, Y., (1994). Force Microscopy Study of Friction and Elastic Compliance of Phase-Separated Organic Thin Films, *Langmuir,* Vol. 10, No. 4, (Apr. 1994), pp.1281-1286, ISSN: 0743-7463

Paserba, K., Shukla, N., Gellman, A.J., Gui, J., & Marchon, B., (1999). Bonding of Ethers and Alcohols to a-CN$_x$ Films, *Langmuir,* Vol. 15, No. 5, (Feb. 1999), pp.1709-1715, ISSN: 0743-7463

Philips, B.S., & Zabinski, J.S., (2004). Ionic liquid lubrication effects on ceramoics in water environment, *Tribology Letters,* Vol.17, No. 3, (Feb. 2004), pp.533-541, ISSN:1023-8883

Qu, J., Truhan, J.J., Dai, S., Luo, H., & Blau, P.J., (2006). Ionic liquids with ammonium cations as lubricants or additives, *Tribology Letters,* Vol. 22, No. 3, (June 2006), pp.207-214, ISSN:1023-8883

Reich, R.A., Stewart, P.A., Bohaychick, J., & Urbanski, J.A., (2003). Base oil properties of inonic liquids, *J. Soc. Tribologists and Lubrication Engineers,* pp.16-21 ISSN: 0024-7154

Sakae, Y., Wakabayashi, A., Li, L., & Kasai, P.H., (2003). Effect of molecular structure of PFPE lubricant on interaction at HDI in near-contact operation, *IEEE Trans. Magn.,* Vol.42, No.10, pp.2501-2503, ISSN:0018-9464

Sato, S., Arisaka, Y., & Matsumura, S., (1999). Surface design of Aramid film for future ME tapes, *IEEE Trans. Magn.,* Vol. 35, No. 5, pp.2760-2762, ISSN: 0018-9464

Seki, A. & Kondo, H., (1991). FTIR Reflection Absorption Spevtra of Novel Lubricant Layer on the Magnetic Thin Film Media, *J. Magn. Soc. Jpn,* Vol. 15, No. Suupl., S2. pp.745-749

Sheldon, R., (2001) Catalytic reactions in ionic liquids. *Chem. Commun* Vol. 23, pp.2399-2403, ISSN: 1359-7345

Shiraishi, & Y. Hirota, A. (1978). Magnetic recording at video cassette recorder for home use, *IEEE Trans. Magn.* Vol.14, No.5, pp.318-320, ISSN : 0018-9464

Shukla, N., & Gellman, A. J., (2000). The Interaction of CF_3CH_2OH and $(CF_3CF_2)_2O$ with Amorphous Carbon Films, *Langmuir,* Vol.16, No.16, (Jul. 2000), pp. 6562, ISSN: 0743-7463

Singh, R.P., Manandhar, S., & Shreeve, J.M., (2002). New dense fluoroalkyl-substituted imidazolium ionic liquids, *Tetrahedron Lett.,* Vol.43, (Dec. 2002), pp.9497-9499, ISSN: 0040-4039

Visser, A.E., Swatloski, R.P., Reichert, W.M., Davis Jr., J.H., Rogers, R.D., Mayton, R., Sheff, S., & Wierzbicki, A., (2001) Task-specific ionic liquids for the extraction of met al ions from aqueous solutions. *Chem. Commum.* vol.23 pp.135-136, ISSN: 1359-7345

Waltman, R.J., Pocker, D.J., Deng, H., Kobayashi, N., Fujii, Y., Akada, T., Hirosaka, K, & Tyndall, G.W.,(2003). Investigation of a new cyclotriphoshazene-terminated perfluoropolyether lubricant: Properties and interactions with a carbon surface, *Chem. Mater.,* Vol.15, No.12, pp.2362-2375, ISSN: 0897-4756

Welton, T., (1999). Room-temperature ionic liquids. Solvents for Synthesis and Catalysis, *Chem. Rev.,* Vol.99, No. 8, (July 1999), pp.2071-2083, ISSN: 0009-2665

Wood, R. W., Miles, J., & Olson, T., (2002). Recording technologies for terabit per squire inch systems. *IEEE Trans. Magn.,* Vol. 38, No.4, pp.1711-1718, ISSN: 0018-9464

Xu, X., & Angel, C.A., (2003). Solvent-Free Electrolytes with Acueous Solution-Like Conductivities. *Science* Vol. 302, Issue 5644, (Oct. 2003), pp.422-425, ISSN: 0036-8075

Ye, C., Liu, W., Chen, Y., & Yu, L. (2001). Room temperature ionic liquids: a novel versatile lubricant. *Chem. Commum.*, No. 21, pp.2244-2245, ISSN: 1359-7345

Ye, C., Liu, W., Chen, Y., & Ou, Z., (2002). Tribological behavior of dry-sialon ceramics sliding against Si3N4 under lubrication of fluorine-containing oils, *Wear*, Vol. 253, No. 5-6, (Sep. 2002), pp.579-584, ISSN: 0043-1648

Yoshizawa, M., Xu, W., & Angel, C.A., (2003). Ionic Liquids by Proton Transfer: Vapor Pressure, Conductivity, and the Relevance of ΔpK_a from Aqueous Solutions, *J Am. Chem. Soc.* Vol. 125, No.50, (Nov. 2003), pp.15411-15419, ISSN: 0002-7863

21

Application of Ionic Liquids to Space Propulsion

Fahrat Kamal, Batonneau Yann, Brahmi Rachid and Kappenstein Charles
University of Poitiers, Laboratory of Catalysis, LACCO, CNRS UMR-6503, Poitiers France

1. Introduction

1.1 Propellants and ignition systems

A propellant is an energetic compound or a mixture in gas, liquid or solid state, which undergoes a rapid and predictable combustion resulting in a large volume of hot gas used to thrust a projectile (rocket, missile, launcher). Liquid propellants are subdivided into monopropellants and bipropellants. Monopropellants are liquids which decompose in the absence of external oxygen. They have comparatively low energy content and display lower performance by comparison with bipropellants. They need an ignition system and are used in small missiles and engines which require low thrust. Hydrazine is currently the most widely used monopropellant for orbit and attitude control of satellites or for roll control of launcher (Schmidt, 2001). Bipropellants consist of two components, a fuel (or reducer) and an oxidizer, which are stored in separate tanks and are injected simultaneously into a combustion chamber where they come into contact and burn. They display higher energetic content and higher thrust and are used on the most demanding missions. Current fuels include hydrogen, methanol, kerosene, hydrazine, monomethylhydrazine (MMH), unsymmetrical dimethylhydrazine (UDMH). Common oxidizers are nitric acid and dinitrogen tetroxide (NTO).

The ignition systems can be sorted into three types:

- Spontaneous reaction between oxidizer and fuel as it is the case for the MMH-NTO mixture ($CH_3N_2H_3$ - N_2O_4); such mixtures are called *hypergolic*. The reaction can be restarted several times and enable launcher upper stage to reach precisely the expected satellite orbit.
- Energy supply using thermal, electrical or photochemical procedures. It is applied to non-hypergolic bipropellants (i.e. hydrogen–oxygen) and to monopropellants. The main drawback is the ability to enable restarts which depend on the ignition mode.
- Catalytic ignition of bipropellant or catalytic decomposition of monopropellant, using an adequate catalyst. This is a very simple and convenient ignition system, which enables numerous restarts. The major hurdle is the long term stability of the catalyst which suffers very severe conditions: high pressure and temperature increases in very short time.

1.2 Propulsion parameters and propulsion types

The critical parameters of space propulsion engines are:

- The thrust F (unit N) given by the 2nd Newton's law: $F = (dm/dt).V_e$, where dm/dt (unit kg s^{-1}) is the mass flow rate of the exhaust propellant and V_e (unit m s^{-1}), the average velocity of the exhaust gas;
- The specific impulse I_{sp} (unit s) and the volumetric impulse I_{vol} (unit s kg L^{-1}) which characterize the propellant, and are defined by the relation: $I_{sp} = V_e/g = F/(dm/dt).g$ (g is the gravitational constant, 9.807 m s^{-2}) and $I_{vol} = I_{sp}{}^*\rho$ where ρ is the density of the propellant;
- The efficiency η given in percentage and corresponding to the ratio (power associated with propulsive force)/(power input); it depends on the propulsion type and system.

From the kinetic theory of gases, the exit velocity is proportional to $(T_c/M)^{1/2}$ where M is the average molecular weight of gas mixture (in kg mol^{-1}) and T_c the gas temperature in the chamber (in K). Therefore, to increase the thrust, we have to increase the chamber temperature and/or decrease the average molar weight of the exit gas. This can be obtained using more energetic propellant, adding electric energy through heating elements or using an electric field and charged particles.

The simplest and cheapest propulsion system is the *cold gas system* where a gas under pressure is released through a remote-controlled valve (Fig. 1). Chemical propulsion uses (i) *liquid monopropellant system* with catalytic decomposition of an energetic liquid (Fig. 1); (ii) *bipropellant system* that provides higher performances for launchers; and (iii) *solid propellant system* best represented by the huge boosters of launchers, that contain mainly ammonium perchlorate (Fig. 2). For electric propulsion systems, three families have been developed, depending on the force used: (i) Coulomb force for *electrostatic system,* i.e. an electric field is applied to accelerate ions or charged droplets, (ii) electromagnetic field to generate plasma for *electrothermal systems,* and (iii) Lorentz force for *electromagnetic systems.* Concerning micropropulsion applications, the first family was developed as ion electrospray (Lozano, 2011). The characteristics of the propulsion types are compared in Table 1.

	Cold gas	Chemical monopropellant	Chemical bipropellant	Electric Ion electrospray
Isp	Low (60 s)	Medium (200 s)	Medium (320 s)	High (5000 s)
Miniaturization	Yes	Yes	No	Yes
Safe materials	Yes	No	No	Yes
Power needed	No	No	No	Yes
Simplicity	Yes	No	No	Yes

Table 1. Comparison of the propulsion types; adapted from (Lozano, 2011).

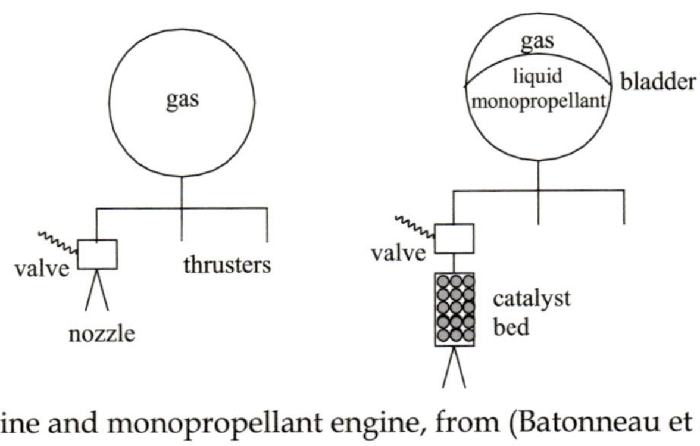

Fig. 1. Cold gas engine and monopropellant engine, from (Batonneau et al., 2008).

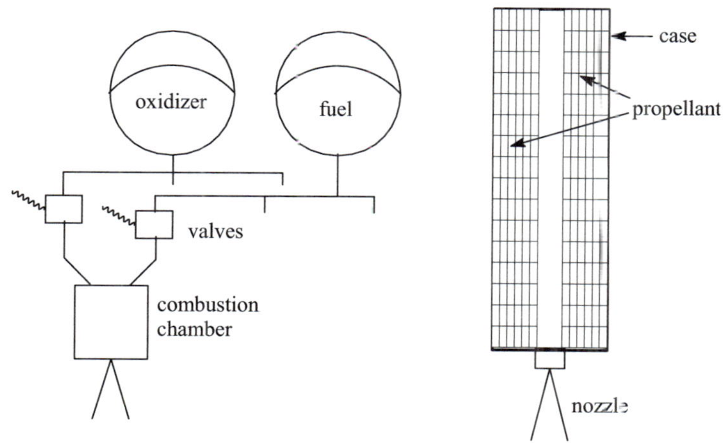

Fig. 2. Bipropellant system and solid motor, from (Batonneau et al., 2008).

1.3 Oxygen balance

When using propellants containing an oxidizer and a fuel in the same chemical (as in NH_4NO_3) or in two independent species, the best performances are obtained when the oxidizer to fuel ratio is close to the value corresponding to a complete combustion giving nitrogen, steam, carbon dioxide and the most stable metal oxides (if metals are present) as sole products. In this case, the oxygen balance is zero. The oxygen balance corresponds to the mass percentage of released or consumed oxygen by the theoretical decomposition of the oxidizer or the fuel. Oxidizers present a positive oxygen balance whereas fuels display negative oxygen balance. Recently, it was proposed to take the formation of CO instead of CO_2 at high temperatures, into account to calculate the oxygen balance (Jones et al., 2006). In the case of monopropellants composed of single energetic species, the oxygen balance is an important parameter linked to the performance.

1.4 Replacement of hydrazine and hydrazine derivatives: toward "green propellants"

Liquid hydrazine as monopropellant and its derivatives (MMH and UDMH) for bipropellant systems have been the most used propellants for space propulsion since the fifties. But they display several drawbacks that lead to increasing handling and storage costs: (i) high crystallization point (1.4 °C for hydrazine); (ii) poisonous and carcinogenic properties; (iii) high vapor pressure at room temperature; and (iv) sensitivity to adiabatic compression (risk of detonation). These drawbacks have prompted, in the last years, the search for possible "green monopropellants" which would be cheaper, and less toxic. Different substitutes have been proposed and can be sorted into three main types: energetic ionic liquids, hydrogen peroxide, and nitrous oxide (Batonneau et al., 2008).

1.5 Previous review papers

Energetic ionic liquids have been previously studied in the late seventies as LGP (Liquid Gun Propellants), but due to development problems, they have been finally discarded for this peculiar application (Klingenberg et al., 1997). More recently, they have been proposed as space monopropellants to replace hydrazine and numerous research papers appeared in this field. A general review focusing on the catalytic decomposition of propellants, including ionic liquids, is presented in (Batonneau et al., 2008). Two recent reviews concerning new energetic ILs appeared in (Smiglak et al., 2007) and (Singh et al., 2006), whereas (Chiu & Dressler, 2007) presented a review paper entitled *Ionic liquids for space propulsion* focusing mainly on ion electrospray electric propulsion (vide infra).

2. Monopropellants

2.1 HAN, ADN and HNF-based propellants

The first family of EILs (energetic ionic liquids) used as monopropellant component is represented by ionic compounds used; they are used in concentrated aqueous solutions, in order to control the performance and the maximum reached temperature in steady state flow. The propellant contains an IL as an ionic oxidizer associated to a fuel (or reductant) to have a zero oxygen balance. The most promising oxidizers are (the indicated temperatures correspond to the melting point and the percentage to the oxygen balance):

HAN, hydroxylammonium nitrate	$[NH_3OH]^+[NO_3]^-$	44 °C	33.3 %
ADN, ammonium dinitramide	$[NH_4]^+[N(NO_2)_2]^-$	92 °C	25.8 %
HNF, hydrazinium nitroformate	$[N_2H_5]^+[C(NO_2)_3]^-$	118 °C	13.1 %
AN, ammonium nitrate	$[NH_4]^+[NO_3]^-$	170 °C	20.0 %
HN, hydrazinium nitrate	$[N_2H_5]^+[NO_3]^-$	70 °C	8.4 %

Different ionic or molecular fuels have been associated to these oxidizers:

TEAN, tris(ethanol)ammonium nitrate	$[NH(C_2H_4OH)_3]^+[NO_3]^-$	81 °C	-105.6 %
AA, ammonium azide	$[NH_4]^+[N_3]^-$	160 °C	-53.3 %
HA, hydrazinium azide	$[N_2H_5]^+[N_3]^-$	75 °C	-53.3 %
HEHN, 2-hydroxyethylhydrazinium nitrate	$[HO\text{-}C_2H_4\text{-}N_2H_4]^+[NO_3]^-$	57 °C	-51.8 %
Methanol, ethanol, glycerol, glycine, urea			

Kinetics of the thermal decomposition of ADN, HAN, HNF and HAN-TEAN-H_2O mixture, have been intensively studied in the past, but the results concerning catalytic ignition remain scarce. The major merit of the catalytic decomposition versus the thermal decomposition is to strongly decrease the ignition temperature needed to trigger the decomposition and, at the same time, to increase the decomposition rate. This is obtained by a complete change in the reaction pathway, with lower activation energies due to the surface active centers of the catalyst. However, this introduces limitations coming from the thermal stability of the catalyst bed as the decomposition temperature rises. Electric ignition has also been used with success for HAN-water solution (Meng et al., 2009), HAN-TEAN-water propellant (Risha et al., 2007) and ADN-based monopropellant (Wingborg et al., 2005). The effects of electric current, voltage, volume, temperature and propellant concentration on the ignition delay time were investigated. For the HAN-TEAN-water mixture, the time delay to peak power decayed exponentially from 160 s to 2-3 s with an increase in the input voltage from 7 to 12 V; beyond 12 V, the time delay dependency became less significant and appeared to remain constant. Resonant laser ignition was also demonstrated for HAN and HEHN mixtures which were successfully and reliably ignited with single laser pulses in the IR, whereas UV did not lead to ignition (Alfano et al., 2009).

2.1.1 Results on HAN-based propellants

HAN-based aqueous blends have been investigated mainly by US (Meinhardt et al., 1999; Wucherer et al., 2000), French (Courthéoux et al., 2002) and Japanese groups (Katsumi et al.,

2010). The percentage of water controls the maximum adiabatic temperature, and several mixtures have been proposed, depending on the adiabatic temperature that can be reached in steady state conditions: low temperature propellants (up to about 1300 °C) or high temperature propellants (up to about 1800 °C); examples of weight composition are given below:

- 55 % HAN + 12 % CH_3OH + 33 % H_2O 1090 °C
- 59 % HAN + 26 % glycine + 15 % H_2O 1720 °C

Another example is given by HAN-TEAN-H_2O mixture, previously known as LGP (liquid gun propellant) (Klingenberg et al., 1997; Hisatsune et al., 2004). The mass composition HAN 60.8 %-TEAN 19.2 %-H_2O 20 % corresponds to the balanced equation:

$$7\ NH_3OHNO_3 + NH(C_2H_4OH)_3NO_3 \rightarrow 22\ H_2O(g) + 8\ N_2(g) + 6\ CO_2(g)$$

Different catalysts for HAN decomposition have been previously patented (Schmidt & Gavin, 1996): unsupported platinum group metals (wire mesh and sponge) and supported catalysts (32 % Ir, 10 % Pt, 12 % Rh) were proposed to start ignition between 80 and 120 °C. Decomposition of HAN-glycine-water mixtures on Shell 405 (Ir/Al_2O_3 catalyst developed for the decomposition of hydrazine) and other home made catalysts was studied for application in small thrusters (Meinhardt et al., 1999).

Catalysts prepared more recently by a French team are able to decompose HAN catalytically at room temperature with an ignition delay of around 1 s. Successive injections of the propellant (80 wt.-% HAN, 40 °C) decompose easily on such catalysts, thus displaying a very good catalytic activity (Courthéoux et al., 2005a). Fig. 3 shows the mass spectrum of the gas phase during 5 successive injections of 100 µL aliquots on a catalyst preheated at 85 °C (Farhat et al., 2007). The catalysts consist of 10 wt.-% platinum supported on a silicon-doped alumina prepared using sol-gel process (Nguéfack et al., 2003). The impact of drying procedure (xerogel Vs. aerogel) and active phase introduction way (impregnation or one-step addition before sol formation) on catalytic activity are compared (Courthéoux et al., 2004, 2005), showing that aerogel based catalysts display always a better activity and stability. The thermal and catalytic decomposition using the optimized catalyst were followed for different HAN concentrations (20, 40 and 79 wt.-% HAN) (Courthéoux et al., 2006).

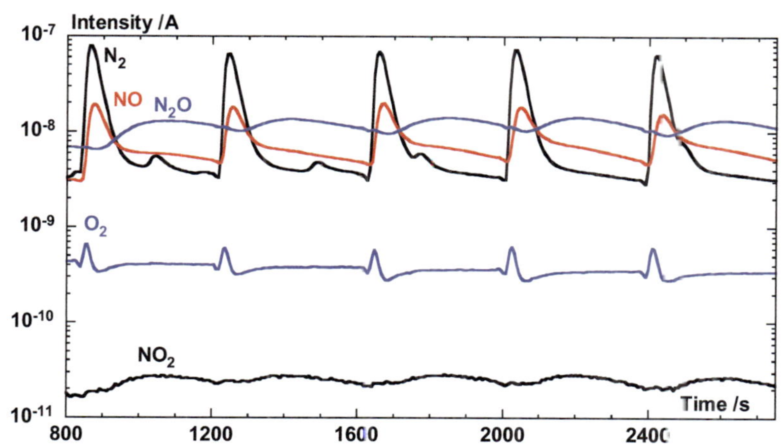

Fig. 3. Catalytic decomposition of HAN(80 wt.-%)-water mixture at 85 °C. Mass spectrometer data versus time after successive HAN injections (Farhat et al., 2007).

The analysis of the products during the decomposition of HAN-based propellants was performed using a dynamic reactor with on-line mass spectrometer analysis of the gas phase and acid-base titration of the trapped aqueous solution, associated with Raman spectroscopy (Amariei et al., 2007; Farhat et al., 2007). For thermal as well as catalytic decomposition, the gaseous primary products are nitrogen N_2 (major, thermodynamic product) and nitrogen oxide NO (medium, kinetic product) (Fig. 4), whereas nitric acid is the major product in the trapped solution, associated with ammonium nitrate (Farhat et al., 2007). The amount of oxygen species (second expected thermodynamic product) is about two orders of magnitude lower than nitrogen and corresponds to oxygen impurities introduced during the injection.

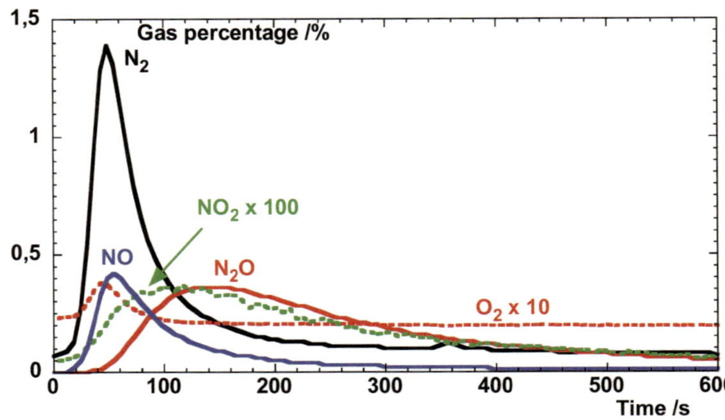

Fig. 4. Thermal decomposition of HAN (80 wt.-%)-water mixture at 200 °C. Gas percentage versus time after the HAN injection (100 μL).

Secondary products are nitrous oxide N_2O (medium) and nitrogen dioxide NO_2 (traces). The formation of NO_2 traces can be the result of NO oxidation by oxygen impurities. The rise of NO_2 follows the drop of oxygen. The concentration difference between oxygen and NO_2 (about one order of magnitude) is related to the solubility of nitrogen dioxide in water.
The simultaneous formation of nitrogen, nitrogen oxide and nitric oxide (primary products) are described by parallel competitive redox reactions between hydroxylammonium and nitrate ions; then, the delayed formation of N_2O is the result of the decomposition of intermediate ammonium nitrate:

$$6\,NH_3OHNO_3(aq) = 3\,N_2(g) + 2\,NO(g) + 10\,H_2O(g) + 4\,HNO_3(g)$$

$$6\,NH_3OHNO_3(aq) = 2\,N_2(g) + 2\,NH_4NO_3(s\text{ or }l) + 6\,H_2O(g) + 4\,HNO_3(g)$$

$$2\,NH_4NO_3(s) = 2\,N_2O(g) + 4\,H_2O(g)$$

These results have been supplemented by a study of the decomposition of ammonium nitrate (AN) (Farhat et al., 2009a). The heating of small amounts of aqueous AN (50-55 wt-%) shows only the quantitative endothermic vaporization of water and AN, with no evidence of thermal decomposition even at high temperature. In the presence of various mono- and bimetallic catalysts, we can observe a true decomposition, but only Pt-based

catalysts were able to trigger the decomposition at low temperature (210 °C). The addition of Cu or Zn to platinum leads to an improvement of the catalytic activity and a reduction of the decomposition temperature (close to the melting point 170 °C). The product analysis reveals the presence of the same gaseous and condensed products as observed with HAN e.g. major nitrogen and nitric acid.
The influence of the fuel (methanol and glycerol) added in stoichiometric proportions, or in excess, to HAN, ADN, or HNF was evaluated (Amariei et al., 2005). Fig. 5 displays the thermogravimetric results obtained for binary (no fuel) and ternary HAN-based propellants. Obviously, the presence of the fuel increases the ignition temperature for both thermal and catalytic decomposition, disclosing an inhibiting behavior.

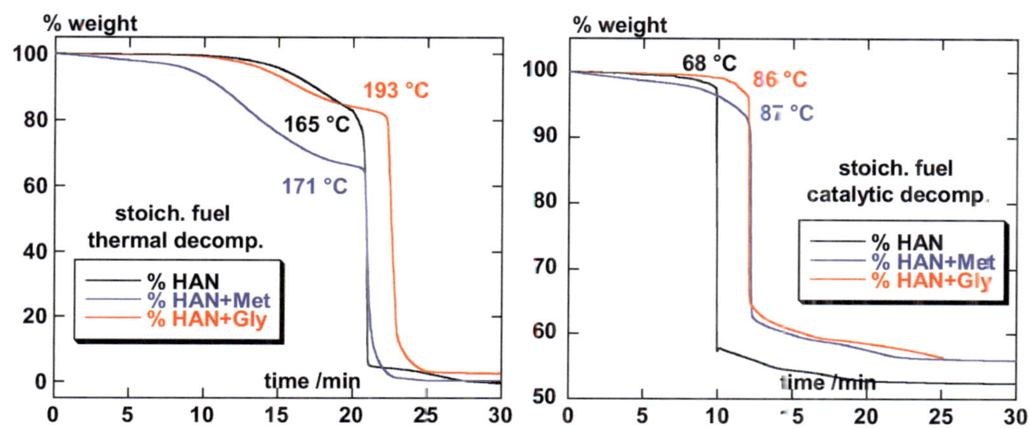

Fig. 5. Comparison of thermal and catalytic decomposition of HAN-based propellants.
"Met" = methanol, "Gly" = glycine (Amariei et al., 2005).

To overcome this problem, azides have been proposed as ionic fuels (Farhat et al., 2009b). But the first results were disappointing: for binary ammonium azide (AA)-water mixture, no azide decomposition occurred even in the presence of the catalyst and only endothermic processes corresponding to water and azide vaporization have been recorded. For ternary mixtures HAN-azide-water, no redox reaction has been observed; when using AA, the predominant step is linked to acido-basic reactions. In the case of ammonium nitrate-azide-water ternary mixtures, no reaction between reactants has been observed, but the presence of azide compounds influences strongly the decomposition of ammonium nitrate and the onset temperature decreases enough to enter the solid domain of AN.
Using strand burner technique, it was showed that for binary HAN-water solutions (50-95 wt.-%), the combustion mode is very complex as the linear burning rate increases up to 80 % and decreases from 80 to 95 %. The role of the two-phase region is very important and the boiling of water caused by superheat is responsible for the high burning rate (Katsumi et al., 2010). The combustion mechanism of HAN-based propellant containing ammonium nitrate and methanol shows that the instability of the liquid interface may trigger a sudden increase in the burning rate to a violently high region; methanol was found to reduce the bubble growth rate in the solution. S405 catalyst (which replaces Shell 405 catalyst) showed the best performance (Katsumi et al., 2008, Matsuo et al., 2008). Strand burner investigation of HAN

60 %-glycine 14 %-water 26 % showed that the burning rate exhibited four burning rate regimes for pressures ranging from 1.5 to 18.2 MPa. The temperatures of combustion products were found to be near the water boiling point at pressures lower than 8.8 MPa and under at pressures larger than 8.8 MPa. The observed slope breaks in burning rates are shown to be associated with the reaction mechanism changes between adjacent pressure regimes. Major species detected are nitrogen, nitric oxide, carbon dioxide, and formaldehyde (Chang & Kuo, 2002).

2.1.2 Results on ADN-based propellants

Aqueous ammonium dinitramide (ADN) based mixtures, using as a fuel, glycerol (defining the LMP 101 propellant) or methanol (LMP 103), have been developed by Swedish groups (Wingborg, 2004; Wingborg et al., 2006; Groenland et al. 2004; Anflo & Möllerberg, 2009) with promising results. A new thruster with propellant LMP-103-S (ADN-methanol-water-ammonia) is currently under flight demonstration on-board the Prisma satellite (ESA, 2010). This group has patented catalysts for ADN decomposition and tested them on mixtures comprising Ca 65 %ADN + 10 %CH_3OH + 25 % H_2O and stabilized with weak base (urea and hexamine). Catalysts are supported Pt, Rh or Ir onto hexaaluminates ($LaAl_{11}O_{18}$) which are prepared by a combined sol-gel and microemulsion technique. The so-obtained catalysts are heat resistant above 1000 °C (Groenland et al., 2002). Such catalysts have been investigated and developed in combination with the preparation of other ADN-based ternary mixtures for the propulsion of rocket engines. ADN-based mixtures have been also tested using catalysts developed for HAN (Kappenstein et al., 2004; Farhat et al., 2008). Fig. 6 compares thermal and catalytic decomposition of aqueous ADN (50 wt.-%) and we can observe a full decomposition at lower temperature in the presence of the catalyst. The products are similar to the ones observed during the decomposition of HAN and again ammonium nitrate was found as an intermediate product.

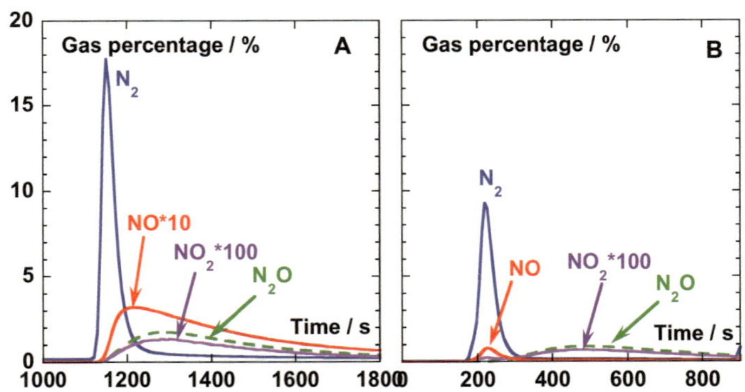

Fig. 6. Gas analysis after thermal (A, 320 °C) and catalytic (B, 250 °C) decomposition of ADN.

2.1.3 Results on hydrazinium-based propellants (HNF, HEHN)

Hydrazinium NitroFormate (HNF), was mainly developed as a substitute for ammonium perchlorate for solid propulsion, but has been also proposed as a more energetic oxidizer than HAN for monopropellant, when combined with reductants (Fick et al., 2001;

Slettenhaar et al., 2003). 44 wt.-% HNF in water (maximum solubility is about 50 wt.-% at 20 °C) was catalytically decomposed with the catalyst Pt/Al_2O_3Si previously described for HAN (Farhat et al., 2008). The gaseous products formed during the catalytic decomposition are major N_2, medium NO, N_2O and CO_2, minor CO and traces of NO_2. Raman analysis of the trapped solution evidenced the complete disappearance of hydrazinium cation, the formation of nitric acid and ammonium, whereas nitroformate anion remains in the solution. Pt, Rh and Pd catalysts supported on doped alumina were used for HNF decomposition (Courthéoux et al., 2004b).

The use of 2-hydroxyethylhydrazinium nitrate (HEHN) was proposed recently as hydrazine substitute (Shamshina et al., 2010). HEHN is more stable than HAN and displays a glass transition temperature of -57 °C. The catalytic decomposition in contact with iridium supported on alumina (Shell 405) was demonstrated when the catalyst is preheated at 150 °C, but with the formation of soot most probably linked to a negative oxygen balance:

$$[HO\text{-}C_2H_4\text{-}N_2H_4]^+[NO_3]^- + 2.25\ O_2 = 2\ CO_2 + 1.5\ N_2 + 4.5\ H_2O$$

2.2 New energetic ionic liquids for space propulsion

Energetic ionic liquids (EILs) are now numerous and fully described in several papers (Singh et al., 2006; Smiglak et al., 2007). New EILs have been recently prepared and characterized, mainly by the research groups of R. D. Rogers in Alabama and J. M. Shreeve in Idaho. The required properties of a possible propellant are: melting point < -40 °C, density > 1.4 g cm^{-3}, surface tension < 100 dyne cm^{-1}, viscosity as low as possible, long term stability; the development of an IL toolbar is proposed (Smiglak et al., 2007). Imidazolium, triazolium, and tetrazolium families, with different N- and C-substituents (alkyl, amino, azido…) are presented and discussed (Fig. 7). Azolates are surprisingly stable (Smiglak et al., 2010) and the most promising anion for liquid propellant is 3,5-dinitro-1,2,4-triazolate (Fig. 7, species 7). But few EILs have been experimentally investigated as monopropellants. The compound 4-amino-1-methyl-1,2,4-triazolium nitrate (2, Fig.7) displays a glass transition point of -55 °C, shows low sensitivity against impact and friction, and a burn rates with low pressure dependence (Schaller et al., 2009). The decomposition of triazolium (2, Fig. 7) (Chowdhury & Thynell, 2010a) and tetrazolium-based EILs (3, Fig. 7) has been followed by confined rapid thermolysis using FTIR spectroscopy and time-of-flight mass spectroscopy to identify the products. The primary formation of nitrogen and methylisocyanide was evidenced (Chowdhury et al., 2009a). For triazolium EILs, the processes governing the decomposition were found to be autocatalytic in nature and the catalytic species were strong acids generated by the initial decomposition step, with an activation energy of 167-188 kJ mol^{-1} (Chowdhury & Thynell, 2010b).

As an oxygen balance close to zero remains a strong requirement for new propellants, the search for such compounds has been fruitful and new EILs have been proposed (Christe & Drake, 2003; Jones et al., 2006; Tao et al., 2008) and some examples are presented in Fig. 8. Table 2 presents the corresponding oxygen balance values for the formation of CO_2 and CO. As an example, the oxygen balance equation for 1 (Fig. 8), based on CO formation is:

$$2\ [N_4C_5H_{11}]^+[Al(NO_3)_4]^- = Al_2O_3 + 8\ N_2 + 11\ H_2O + 10\ CO$$

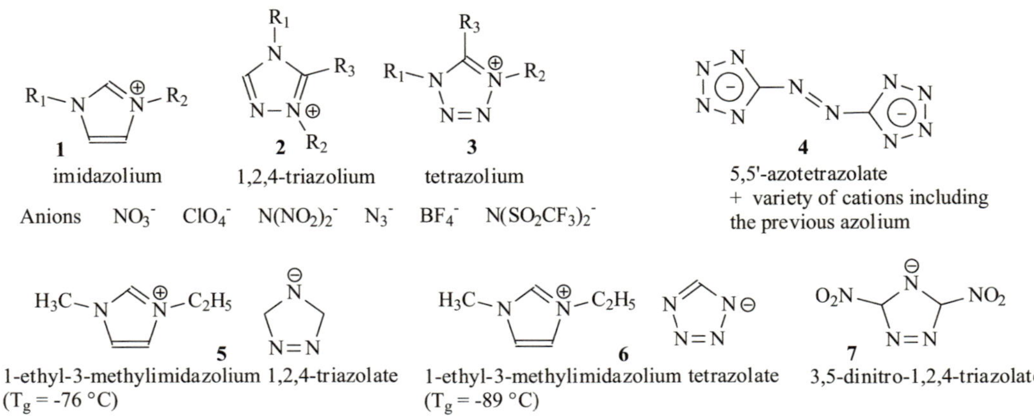

Fig. 7. Representative EILs based on rich nitrogen anions and cations.

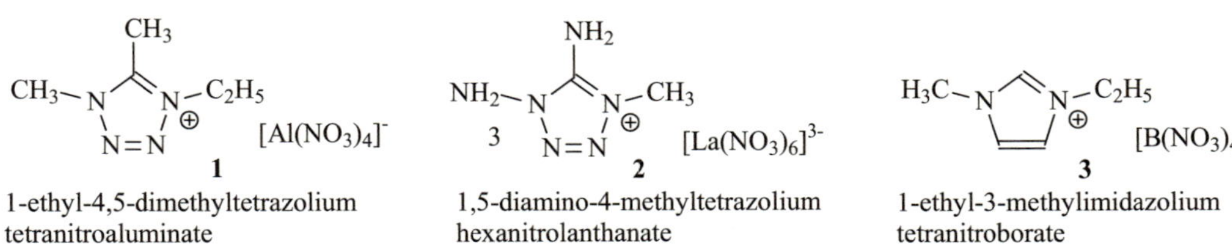

Fig. 8. New EILs with reduced oxygen balance.

Sample (Fig. 8)	Formula	Tm/Tg	OB, CO_2	OB, CO
1	$[N_4C_5H_{11}]^+[Al(NO_3)_4$	- 46 °C	-39.8 %	0 %
2	$3[N_6C_2H_7]^+[La(NO_3)_6]^-$	88 °C	-22.4 %	0 %
3	$[N_2C_6H_{11}]^+[B(NO_3)_4]^-$	-25 °C	-60.5 %	-8.6 %

Table 2. Examples of ILs with low oxygen balance (OB).

3. Hypergolic bipropellants

3.1 New hypergolic IL fuels for bipropellant systems

Hypergolic fuel-oxidizer systems hypergolic systems have been widely used in bipropellants engines, mainly for upper stage of launchers because they afford restart of the engine, leading to a better accuracy of the trajectory and finally the satellite orbit. The most used fuels are hydrazine, monomethylhydrazine (MMH) or dimethylhydrazine (UDMH), with oxidizing agents such as white fuming nitric acid (WFNA, 100 % HNO_3), red fuming nitric acid (RFNA, HNO_3 plus NO_2) or nitrogen tetroxide (NTO, N_2O_4). The main drawback of these systems is linked to the toxicity of hydrazines associated with high vapor pressures, which increase strongly the handling and storage costs. The need for non-toxic "green" hydrazine substitutes appeared recently and new hypergolic liquid fuels must present low vapor pressures, high energy densities, low viscosities to enhance ease of mixing, as well as short ignition delays to preclude accumulation of fuel and oxidizer that could lead to an explosion. EILs are the candidates of choice for this application and have been proposed by different research groups in the US.

The first family or ILs fuels exhibiting hypergolic activity toward WFNA was reported by the research group at AFRL Edwards. It corresponds to dicyanamide-based ILs (Chambreau et al., 2008); it must be quoted that the same group prepared also azide-based ILs which show high reactivity with nitric acid, but do not ignite (Schneider et al., 2008a). Then other IL families have been more recently proposed, mainly by the Idaho group, like nitrocyanamide-based (He et al., 2010), or dicyanoborate-based fuels (Zhang & Shreeve, 2011).

3.2 Composition and properties of hypergolic IL fuels

The composition of the different hypergolic IL families is presented in Fig. 9 and some of the properties in Table 3. The borate and cyanoborate ILs are water sensitive, whereas the DCB-based ILs remain water-stable. From Table 3, we can sort the three hypergolic families for ignition delay and viscosity that decrease in the order: NCA > DCA > DCB. Table 3 shows that the DCB-based ILs present much shorter ID than the DCA or NCA, associated with these ILs meet all the important criteria to be selected as hypergolic hydrazine substitutes, with ignition delay close to the ID of the current MMH-NTO bipropellant (2 ms).

cation	anion	T_m/T_g /°C	T_d /°C	Density ρ /g cm^{-3}	Viscosity η /mPa s	ID /ms	Ref.
1	DCB	< -80	307	0.96	17.3	28	a
1	DCA					47	b
1	NCA	-90	256	1.13	57	81	c
2	DCB	< -80	222	0.91	39.4	6	a
2	DCA	20	263	1.01	113.9	45	d
2	NCA	9	286	1.11	119.5	228	d
3	DCB	< -80	189	0.93	35.0	4	a
3	DCA		199	1.05	78.6	30	d
3	NCA		208	1.16	84.9	130	d
4	DCB	< -80	266	0.99	12.4	8	a
4	DCA	-85	207		42	43	b
4	NCA	-91	220	1.11	44	46	c
5	DCB	< -80	252	0.96	19.8	18	a
6	DCB	< -80	203	1.00	13.5	6	a
7	DCB	< -80	220	0.99	29.9	32	a
8	DCB	< -80	217	1.03	21.0	6	a

T_m/T_g: phase-transition temperature, T_d: decomposition temperature (onset), ρ: density (25 °C), η: viscosity (25 °C), ID: ignition delay time with WFNA. Ref: [a] (Zhang & Shreeve, 2011), [b] (Schneider et al., 2008b), [c] (He et al., 2010), [d] (Zhang et al., 2010).

Table 3. Properties of the dicyanoborate-based (DCB) ILs and the corresponding dicyanamide- (DCA) and nitrocyanamide- (NCA) based Ils.

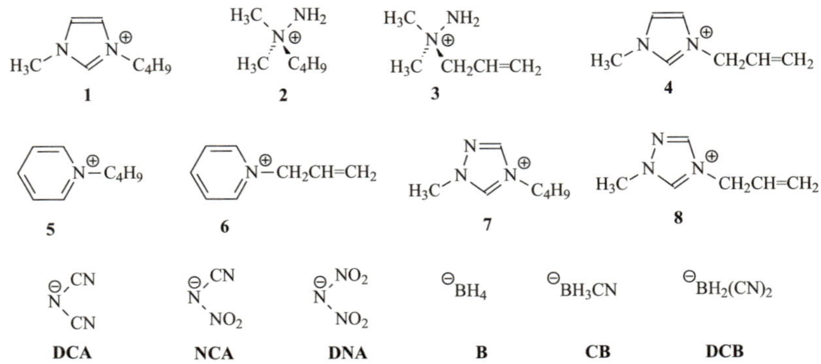

Fig. 9. Constituents of the new hypergolic IL families. DCA = dicyanamide, NCA = nitrocyanamide, DNA = dinitramide, B = tetrahydroborate, CB = cyanoborate, DCB = dicyanoborate.

3.3 Reactivity of hypergolic IL fuels

One example of hypergolic drop test followed by high speed camera is given in Fig. 10 (cation **1** + DCB). The fuel sample is dropped into a beaker containing an excess of WFNA. The ignition delay is defined as the elapsed time between the fuel-oxidizer contact and the appearance of a luminous flame.

New azide-functionalized IL associated with DCA or NCA have been demonstrated to be hypergolic, whereas azide-based ILs display non hypergolic behavior (Joo et al., 2010).

The hypergolic reaction of sodium dicyanamide with nitric acid was followed to investigate the ignition and reaction mechanism (Litzinger & Iyer, 2011). Gas phase products were analyzed by mass spectrometry, and the proposed balanced equation is:

$$Na^+N(CN)_2^- + 4\ HNO_3 \rightarrow 2\ CO_2 + 3\ N_2O + 2\ H_2O + NaNO_3$$

The major products are CO_2 and N_2O. Traces of NO_2, HNO_3 and carbamic acid H_2NCOOH (i.e. HNCO + H_2O) have been detected. No nitrogen, ammonia and hydrogen cyanide has been evidenced.

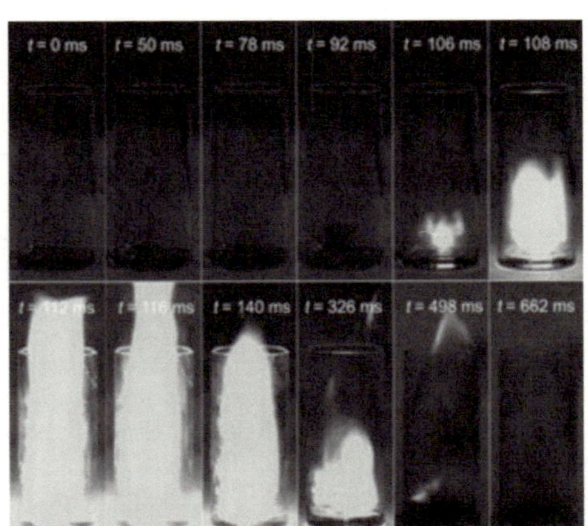

Fig. 10. Ignition delay test shown with a series of high-speed camera photos. A droplet of IL contacting WFNA (Zhang & Shreeve, 2011).

The reaction of 1-ethyl-3-methyl imidazolium dicyanamide (EmimDCA) with WFNA was followed by IR spectroscopy and the presence of CO_2, N_2O, H_2O and isocyanic acid HNCO was evidenced whereas HCN, HONO, NO and CO do not form (Chowdhury at al. 2009b). The use of labeled nitric acid ($H^{15}NO_3$) leads to the formation of $^{14}N^{15}NO$ and $H^{14}NCO$, proving that the NO functionality of nitrous oxide derived from nitric acid and the isocyanic acid nitrogen atom comes from DCA anion only.
Propellant gels can bring safety and insensitivity to tactical systems fueled by storable hypergolic propellants. The unique rheological behavior of gel phase can curtail leakage from damaged tanks and can reduce the volatility of the propellants. Drop ignition tests have been followed used gelled fuel - liquid oxidizer or liquid fuel - gelled oxidizer combination. The ignition of liquid and gelled methyl ethyl imidazolium DCA have been demonstrated (Coil, 2010).

4. Electric propulsion

Among the different propulsion options for small satellites (from micro to nano), ion electrospray propulsion system (iEPS) is a good candidate for high precision applications (Table 1) and with the possibility of miniaturization. Ions are emitted from a conducting liquid in contact with a metallic tip (tungsten, nickel) and accelerated by a static electric field (Fig. 11, left). The tip can be externally wetted in the case of a bulk metal or internally in the case of a porous metal. The liquid is a liquid metal (e.g. indium) or an ionic liquid. The advantages of ILs over liquid metals are: (i) very low vapor pressure, (ii) low surface tension, (iii) possibility to emit positive and negative ions; they are thus described as "a plasma in a bottle" (Lozano, 2011). To scale up the thrust, iEPS can be arranged in the form of arrays of densely packed porous metal emitters; the delivery of the IL propellant uses the passive capillary force (Fig. 11, right). The meniscus of the propellant, when under the influence of an electric field exceeding a critical value will deform into a cone (known as Taylor cone) where the surface tension is balanced by the electrostatic force. An alternating polarity of 1 Hz suppresses the electro-chemical effects.
If the electric field

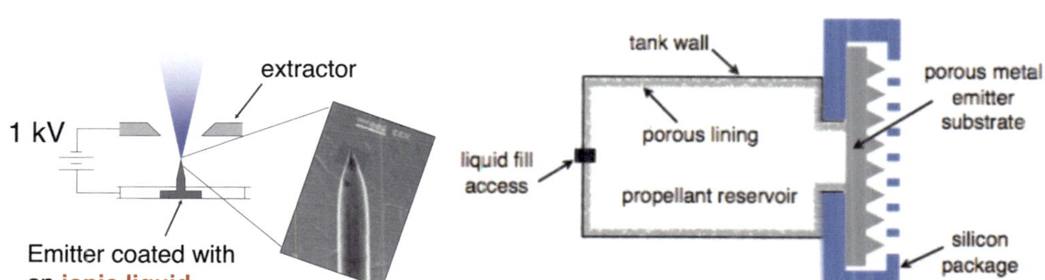

Fig. 11. Scheme of an ion electrospray propulsion system (left); propellant storage and delivery for an array of emitters (right) (Lozano, 2011).

Ionic liquids used as propellants for iEPS do not need to be energetic as it was the case previously, but they must satisfy the requirement to change the classical operation mode from a droplet dominated emission mode (as it is the case for colloid thrusters) to a pure ion emission regime. ILs classically used for iEPS comprise EMIM-BF4, EMIM-Im and EMIM-Beti (Fig. 12) (Daily, 2008; Ticknor et al., 2010; Legge & Lozano, 2011).

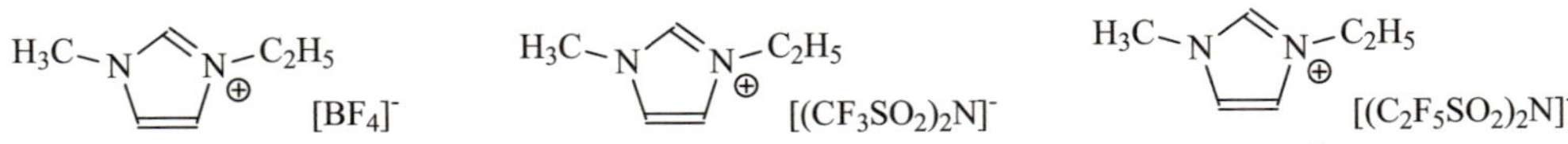

1 EMIM-BF4
1-ethyl-3-methylimidazolium tetranfluoroborate

2 EMIM-Im
1-ethyl-3-methylimidazolium bis(trifluoromethylsulfonyl)imide

3 EMIM-Beti
1-ethyl-3-methylimidazolium bis(perfluoroethylsulfonyl)imide

Fig. 12. Some ILs used for electrospray propulsion.

5. Other propulsion applications

5.1 Hybrid engines

To improve the performances of a monopropellant engine and reduce the risk linked to the simultaneous presence of an oxidizer and a fuel for the long term storage, the oxidizer can be use alone and after decomposition, the hot gaseous oxygen can burn a solid fuel grain in a hybrid solid/liquid engine (Fig. 13). The proposed IL oxidizers can be concentrated aqueous solution of HAN (hydroxylammonium nitrate) or HAN plus AN (ammonium nitrate) and the fuel polyethylene or HTPB (hydroxy-terminated polybutadiene) (Biddle & Sutton, 1985; Ramohalli & Dowler, 1995).

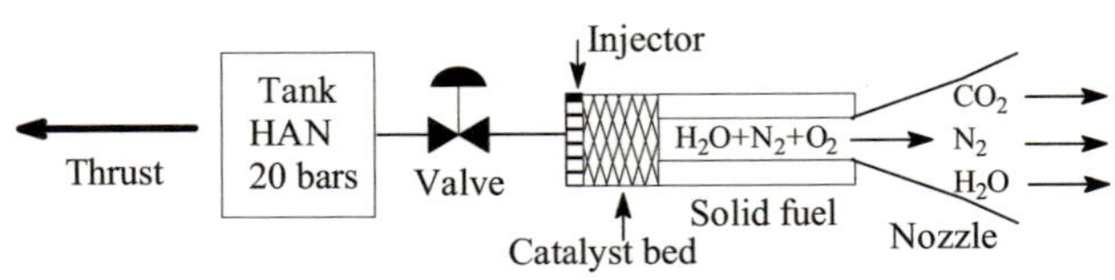

Fig. 13. Scheme of a HAN-based hybrid engine.

Despite the interest of such energetic IL-based hybrid engine, the current development replaced EIL by hydrogen peroxide; this is most probably due to the formation of major nitric acid (kinetic product) instead of expected oxygen (thermodynamic product) during thermal or catalytic decomposition of HAN:

$$5\,[NH_3OH]^+[NO_3]^-(l) \rightarrow 3\,N_2(g) + 8\,H_2O(g) + 4\,HNO_3(g)$$

$$[NH_3OH]^+[NO_3]^-(l) \rightarrow N_2(g) + 2\,H_2O(g) + O_2(g)$$

5.2 Solid propellant stabilizers

Double base (DB) propellants are composed primarily of nitroglycerin (NG) and nitrocellulose (NC). The decomposition reaction is complex, depends on the temperature, but also on the storage history when exposed to conditions in excess of the qualified limit; this last point needs the use of inert stabilizers with a performance decrement (specific impulse, burn rate, energy release…). One solution is to introduce energetic stabilizers that do not carry a performance penalty and ILs are proposed for the next stabilizer generation because they offer the possibility to choose ions that provide task-specific performance: one ion has stabilization capability (generally linked to an affinity for NO_x species) whereas the counterion displays an energetic nature. The feasibility of creating such ILs has been recently demonstrated from NDPA (nitrodiphenylamine) and choline (Fig. 14) (Forton et al., 2010). The use of simpler tetra-alkyl ammonium cations was not successful.

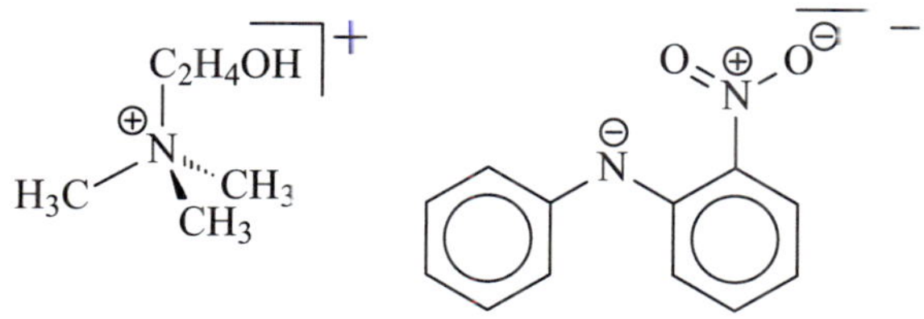

Fig. 14. Example of double base propellant stabilizer: cholinium nitrodiphenylamide.

6. Conclusion

We have seen that different ionic liquids are proposed for the different propulsion systems (monopropellants, hypergolic bipropellants, ion electrospray, hybrid engines). The current and future challenges that can be drawn from this review are summarized as follows: (i) to find new and safe EILs whose thermal or catalytic decomposition avoid the formation of nitric acid as a primary product; (ii) to develop new catalysts able to decompose EILs at low temperature and stable at high temperature; (iii) to find new fuels associated to IL oxidizers but with no inhibiting properties for the catalysts; and (iv) to develop cheaper and safer synthetic method to prepare ILs.
Currently, ionic liquids are part of the GRASP project (GReen Advanced Space Propulsion) of the seven framework program of the European Community (GRASP, FP7).

7. Acknowledgment

We want to thank the French Space Agency (CNES, Centre National d'Etudes Spatiales), the European Space Agency (ESA), the European Community through the 7th Framework Program (FP7, GRASP Project), ONERA company, SAFRAN-SME company, Roxel company, SNECMA-SAFRAN company for constant interest and support for these activities.

8. References

Alfano, A. J., Mills, J. D. & Vaghjiani, G. L. (2009). Resonant laser ignition study of HAN-HEHN propellant mixture, *Combust. Sci. Technol.*, Vol. 181, No. 6, pp. 902-913, ISSN 0010-2202

Amariei, D., Courthéoux, L., Rossignol, S., Batonneau, Y., Kappenstein, C., Ford, M. & Pillet, N. (2005). Influence of Fuel on Thermal and Catalytic Decompositions of Ionic Liquid Monopropellants. *AIAA paper* 2005-3980, ISSN: 0146-3705. Available from http://www.aiaa.org/content.cfm?pageid=320

Amariei, D., Courthéoux, L., Rossignol, S. & Kappenstein, C. (2007). Catalytic and thermal decomposition of ionic liquid monopropellants using a dynamic reactor, *Chem. Eng. Process.*, Vol. 46, No. 2, pp. 165-174

Anflo, K. & Möllerberg, R. (2009). Flight demonstration of new thruster and green propellant technology on the PRISMA satellite, *Acta Astronaut*, Vol.65, No.9-10, pp. 1238-1249, ISSN 0094-5765

Batonneau, Y.; Kappenstein, C. & Keim, W. (2008). Catalytic decomposition of energetic compounds: gas generator, propulsion. In: *Handbook of Heterogeneous Catalysis*, Ertl,

G.; Knözinger, H.; Schüth, F. & Weitkamp J., Eds., 2nd edition, Vol. 5, Chapter 12.7. pp 2647-2680, VCh-Wiley, ISBN 978-3-527-31241-2, Weinheim, Germany

Biddle, R. A. & Sutton, E. S. (1985). Highly soluble, non-hazardous hydroxylammonium salt solutions for use in hybrid rocket motors, *US Patent* N° 4527389, July 9, 1985

Chambreau, S. D., Schneider, S., Rosander, M., Hawkins, T., Gallegos, C. J., Pastewait, M. F., & Vaghjiani, G. L. (2008). Fourier Transform Infrared Studies in Hypergolic Ignition of Ionic Liquids, *J. Phys. Chem. A*, Vol. 112, No. 34, pp. 7816-7824, ISSN 0022-3654

Chang, Y.-P. & Kuo, K. K. (2002). Assessment of combustion characteristics and mechanism of hydroxylammonium nitrate-based liquid monopropellant, *J. Propul. Power*, Vol. 18, No. 5, pp. 1076-1085, ISSN 0748-4658

Chiu, Y. H. & Dressler, R. A. (2007). Ionic liquids for space propulsion, in *ACS Symp. Ser., No. Ionic Liquids IV*, Vol. 975, pp. 138-160

Chowdhury, A., Thynell, S. T. & Lin, P. (2009a). Confined rapid thermolysis/FTIR/ToF studies of tetrazolium-based energetic ionic liquids, *Thermochim. Acta*, Vol. 485, No. 1-2, pp. 1-13, ISSN 0040-6031

Chowdhury, A., Thynell, S. T. & Wang, S. (2009b). Ignition Behavior of Novel Hypergolic Materials, *AIAA Paper* 2009-5352, ISSN: 0146-3705

Chowdhury, A. & Thynell, S. T. (2010a). Confined rapid thermolysis/FTIR/ToF studies of methyl-amino-triazolium-based energetic ionic liquids, *Thermochim. Acta*, Vol. 505, No. 1-2, pp. 33-40, ISSN 0040-6031

Chowdhury, A. & Thynell, S. T. (2010b). Kinetics of Decomposition of Energetic Ionic Liquids, *Propell. Explos. Pyrot.*, Vol. 35, No. 6, pp. 572-581, ISSN 0721-3115

Christe, K. O. & Drake, G. W. (2003). Energetic ionic liquids, Patent US N° 2003-681502

Coil, M. A. (2010). Hypergolic Ignition of a Gelled Ionic Liquid Fuel, *AIAA Paper* 2010-6901, ISSN: 0146-3705

Courthéoux, L.; Eloirdi, R.; Rossignol, S.; Kappenstein, C.; Duprez, D. & Pillet, N. (2002). Catalytic decomposition of HAN-water binary mixtures, *AIAA Paper* 2002-4027, ISSN: 0146-3705

Courthéoux, L., Rossignol, S., Kappenstein, C., Duprez, D. & Pillet, N. (2003). Improvement of catalyst for the decomposition of HAN-based monopropellant - comparison between aerogel and xerogel, *AIAA Papers*, 2003-4645, ISSN: 0146-3705

Courthéoux, L., Popa, F., Gautron, E., Rossignol, S. & Kappenstein, C. (2004a). Platinum supported on doped alumina catalysts for propulsion applications. Xerogels versus aerogels, *J. Non-Cryst. Solids*, Vol. 350, pp. 113-119, ISSN 0022-3093

Courthéoux, L., Amariei, D., Rossignol, S., Kappenstein, C., Pillet, N. & Ford, M. (2004b). Thermal and catalytic decomposition of HNF and HAN-based propellants, *European Space Agency*, SP-557, ISBN 978-92-909286-6-9, pp. 106-111

Courthéoux, L., Amariei, D., Rossignol, S. & Kappenstein, C. (2005a). Facile catalytic decomposition at low temperature of energetic ionic liquid as hydrazine substitute, *Eur. J. Inorg. Chem.*, No. 12, pp. 2293-2295, ISSN 1434-1948

Courthéoux, L.; Gautron, E.; Rossignol, S. & Kappenstein, C. (2005b). Transformation of platinum supported on silicon-doped alumina during the catalytic decomposition of energetic ionic liquid, *J. Catal.*, Vol .232, No.1, pp. 10-18, ISSN: 00219517

Courthéoux, L., Amariei, D., Rossignol, S. & Kappenstein, C. (2006). Thermal and catalytic decomposition of HNF and HAN liquid ionic as propellants, *Appl. Catal. B-Environ.*, Vol. 62, No. 3-4, pp. 217-225, ISSN 0926-3373

Daily, J. W. (2008), Molecular dynamics simulation of ion emission from nanodroplets of ionic liquids, *J. Propul. Power*, Vol. 24, No. 5, pp. 981-986, ISSN 0748-4658

ESA. 2010. 'Green' satellite fuel designed to make space safer, in *ESA news*, available from: http://www.esa.int/esaCP/SEMPJQ9KF6G_index_0.html

Farhat, K., Batonneau, Y.; Florea, O. & Kappenstein, C. (2006). Preparation and use of ammonium and sodium azide as a fuel additive to ionic oxidizer. European Space Agency, SP-635, ISBN 978-92-909294-6-8

Farhat, K., Amariei, D.; Batonneau, Y.; Kappenstein, C. & Ford, M. (2007). Reaction balance of thermal and catalytic decomposition of nitrogen-based ionic monopropellants, *AIAA Paper* 2007-5642, ISSN: 0146-3705

Farhat, K., Batonneau, Y. & Kappenstein, C. (2008). Thermal and Catalytic Decomposition of AN-, ADN and HNF-Based Ionic Monopropellants, *AIAA Paper* 2008-4938, ISSN 0146-3705

Farhat, K., Cong, W., Batonneau, Y. & Kappenstein, C. (2009a). Improvement of catalytic decomposition of ammonium nitrate with new bimetallic catalysts, *AIAA Paper* 2009-4963, ISSN 0146-3705

Farhat, K., Batonneau, Y. & Kappenstein, C. (2009b). Azide-based fuel additives for ionic monopropellants, *AIAA Paper* 2009-4876, ISSN 0146-3705

Fick, M.; Schiebener, P.; Moerel, J. L. P. A.; van den Berg, R. P.; Sanders, H. M. & Welland-Veltmans, W. H. M. (2001). Industrial benefits of applying HNF in monopropellant satellite propulsion, *European Space Agency*, SP-484, pp. 138-147

Forton, M. S.; Sims, J. D.; Askins, R. E.; Stevenson, W. H.; Shamshina, J.; Smiglak, M.; Rogers, R. D.; and Barrow, R. (2010). An Ionic Liquid-Based Next Generation Double Base Propellant Stabilizer, *AIAA Paper* 2010-6587, ISSN 0146-3705

Grasp, FP7. GReen Advanced Space Propulsion. 7th Framework Program of the European Community. Available from https://www.grasp-fp7.eu/grasp/

Grönland, T.-A., Westerberg, B., Bergman, G., Anflo, K., Brandt, J., Lyckfeldt, O., Agrell, J., Ersson, A., Jaeras, S., Buotonnet, M. & Wingborg, N. (2002). Reactor and hexaaluminate-based catalysts for thermal decomposition of ammonium dinitramide single-base monopropellants, Patent N° 2002095207

Grönland, T. A., Anflo, K., Bergman, G. & Nedar, R. (2004). ADN-based propulsion for spacecraft-Key requirements and experimental verification, *AIAA Paper* 2004-4145, ISSN 0146-3705

He, L., Tao, G.-H., Parrish, D. A. & Shreeve, J. n. M. (2010). Nitrocyanamide-Based Ionic Liquids and Their Potential Applications as Hypergolic Fuels, *Chem.-Eur. J.*, Vol. 16, No. 19, pp. 5736-5743, ISSN 0947-6539

Hisatsune, K.; Izumi, J.; Tsutaya, H. & Furukawa, K. (2004). Development of HAN-based liquid propellant thruster. *Proceeding of 2nd Intl Conf. on Green Propellants for Space*

Propulsion, European Space Agency, ESA SP-557, pp. 153-157, ISBN 978-92-909286-6-9

Jones, C. B., Haiges, R., Schroer, T. & Christe, K. O. (2006). Oxygen-balanced energetic ionic liquid, *Angew. Chem., Int. Ed.*, Vol. 45, No. 30, pp. 4981-4984, ISSN 1433-7851

Joo, Y.-H., Gao, H., Zhang, Y. & Shreeve, J. n. M. (2010). Inorganic or Organic Azide-Containing Hypergolic Ionic Liquids, *Inorg. Chem.*, Vol. 49, No. 7, pp. 3282-3288, ISSN 0020-1669

Kappenstein, C., Batonneau, Y., Perianu, E.-A. & Wingborg, N. (2004). Non toxic ionic liquids as hydrazine substitutes. Comparison of physico-chemical properties and evaluation of ADN and HAN, *European Space Agency, [Special Publication]*, Vol. SP-557, pp. 112-117, ISBN 978-92-909286-6-9

Katsumi, T., Kodama, H., Shibamoto, H., Nakatsuka, J., Hasegawa, K., Kobayashi, K., Ogawa, H., Tsuboi, N., Sawai, S. & Hori, K. (2008). Combustion characteristics of HAN-based liquid monopropellant, *International Journal of Energetic Materials & Chemical Propulsion*, Vol. 7, No. 2, pp. 123-137, ISSN 2150-766X

Katsumi, T; Matsuda, R.; Inoue, T.; Tsuboi, N.; Ogawa, H.; Sawai, S. & Hori, K. (2010). Combustion characteristics of hydroxylammonium nitrate aqueous solutions. *Int. J. Energetic Materials Chem. Prop.*, Vol.9, No.3, pp. 219-231, ISSN 2150-766X

Klingenberg, G.; Knapton, J. D.; Morrison, W. F. & Wren, G. P. Eds. (1998). *Liquid Propellant Gun Technology*. Progress in Astronautics and Aeronautics Series, V-175, AIAA (American Institute of Aeronautics & Astronautics), ISBN-10 1-56347-196-5, Reston, Virginia, USA

Legge, R. S., Jr. & Lozano, P. C. (2011). Electrospray propulsion based on emitters microfabricated in porous metals, *J. Propul. Power*, Vol. 27, No. 2, pp. 485-495, ISSN 0748-4658

Litzinger, T. & Iyer, S. (2011). Hypergolic Reaction of Dicyanamide-Based Fuels with Nitric Acid, *Energ. Fuels,* Vol. 25, No. 1, pp. 72-76, ISSN 0887-0624

Lozano, P. (2011). Miniaturizing Electric Propulsion: the ion Electrospray Propulsion System (iEPS), Desktop Delta-V Workshop, Brown University, February 17, available from www.engin.brown.edu/ddv/ppt/Paulo_Lozano.ppt

Matsuo, T., Mishima, H., Hisatsune, K., Katsumi, T., Sawai, S. & Hori, K. (2008). Development of HAN-based liquid propellant thruster,*International Journal of Energetic Materials & Chemical Propulsion*, Vol. 7, No. 2, pp. 139-152, ISSN 2150-766X

Meinhardt, D.; Christofferson, S.; Wucherer, E. & Reed, B. (1999). Performance and life testing of small HAN thrusters. *AIAA Paper* 1999-2881, ISSN: 0146-3705.

Meng, H., Khare, P., Risha, G. A., Yetter, R. A. & Yang, V. (2009). Decomposition and Ignition of HAN-Based Monopropellants by Electrolysis, *AIAA Paper* 2009-451, ISSN 0146-3705

Nguéfack, M., Popa, A. F., Rossignol, S. & Kappenstein, C. (2003). Preparation of alumina through a sol-gel process. Synthesis, characterization, thermal evolution and model of intermediate boehmite, Phys. Chem. Chem. Phys., Vol. 5, No. 19, pp. 4279-4289, ISSN 1463-9076

Ramohalli, K. & Dowler, W. (1995). Helping HAN for hybrid rockets, *Aerospace America*, Vol. 33, No. 1, pp. 20-21, ISSN 0740-722X

Risha, G. A., Yetter, R. A. & Yang, V. (2007). Electrolytic-induced decomposition and ignition of HAN-based liquid monopropellants," *International Journal of Energetic Materials & Chemical Propulsion*, Vol. 6, No. 5, pp. 575-588, ISSN 2150-766X

Schaller, U., Weiser, V., Roth, E., Keicher, T. & Krause, H. (2009). Synthesis and first combustion studies of a nitrate based ionic liquid, *40th International Annual Conference of ICT*, Energetic Materials, pp. 91/1-91/9

Schmidt, E. W. & Gavin, D. F. (1996). Catalytic decomposition of hydroxylammonium nitrate-based monopropellants, US Patent N° 5485722

Schmidt, E. W. (2001). *Hydrazine and its derivatives. Preparation, Properties, Applications.* (2nd edition), John Wiley & Sons, Inc., ISBN 978-0-471-41553-4, New York

Schneider, S., Hawkins, T., Rosander, M., Mills, J., Vaghjiani, G. & Chambreau, S. (2008a). Liquid Azide Salts and Their Reactions with Common Oxidizers IRFNA and N2O4. *Inorg. Chem.*, Vol. 47, No. 13, pp. 6082-6089, ISSN 0020-1669

Schneider, S., Hawkins, T., Rosander, M., Vaghjiani, G., Chambreau, S. & Drake, G. (2008b) Ionic Liquids as Hypergolic Fuels, *Energ. Fuels,* Vol. 22, No. 4, pp. 2871-2872, ISSN 0887-0624

Shamshina, J. L., Smiglak, M., Drab, D. M., Parker, T. G., Dykes, H. W. H., Jr., Di Salvo, R., Reich, A. J. & Rogers, R. D.(2010). Catalytic ignition of ionic liquids for propellant applications, *Chem. Commun.*, Vol. 46, No. 47, pp. 8965-8967, ISSN 1359-7345

Singh, R. P., Verma, R. D., Meshri, D. T. & Shreeve, J. n. M. (2006). Energetic nitrogen-rich salts and ionic liquids, *Angew. Chem., Int. Ed.*, Vol. 45, No. 22, pp. 3584-3601, ISSN 1433-7851

Slettenhaar, B., Zevenbergen, J. F., Pasman, H. J., Maree, A. G. M. & Moerel, J. L. P. A. (2003). Study on catalytic ignition of HNF based non toxic monopropellants, *AIAA Paper* 2003-4920, ISSN 0146-3705

Smiglak, M., Metlen, A. & Rogers, R. D. (2007). The Second Evolution of Ionic Liquids: From Solvents and Separations to Advanced Materials-Energetic Examples from the Ionic Liquid Cookbook, *Accounts Chem. Res.*, Vol. 40, No. 11, pp. 1182-1192, ISSN 0001-4842

Smiglak, M., Hines, C. C., Wilson, T. B., Singh, S., Vincek, A. S., Kirichenko, K., Katritzky, A. R. & Rogers, R. D. (2010). Ionic Liquids Based on Azolate Anions, *Chem.-Eur. J.*, Vol. 16, No. 5, pp. 1572-1584, ISSN 0947-6539

Tao, G.-H., Huang, Y., Boatz, J. A. & Shreeve, J. M. (2008). Energetic ionic liquids based on lanthanide nitrate complex anions, *Chem.--Eur. J.*, Vol. 14, No. 35, pp. 11167-11173, ISSN 0947-6539

Ticknor, B. W., Anderson, J. K., Fritz, B. A. & Chiu, Y.-H. (2010). Effect of Aspect Ratio on the Wettability and Electrospray Properties of Porous Tungsten Emitters with the Ionic Liquid [Emim][Im], *AIAA-Paper* 2010-6618, ISSN 0146-3705

Wingborg, N., Eldsäter, C. & Henrik, S. (2004). Formulation and characterization of ADN-based liquid monopropellants, *European Space Agency*, SP-557, ISBN 978-92-909286-6-9, pp. 101-106

Wingborg, N., Larsson, A., Elfsberg, M. & Appelgren, P. (2005). Characterization and Ignition of ADN-Based Liquid Monopropellants, *AIAA Paper* 2005-4468, ISSN 0146-3705

Wingborg, N. (2006). Ammonium Dinitramide-Water: Interaction and Properties, *J. Chem. Eng. Data*, Vol. 51, No. 5, pp. 1582-1586, ISSN 0021-9568

Wucherer, E. J., Christofferson, S. & Reed, B. (2000). Assessment of high performance HAN-monopropellants, *AIAA Paper* 2000-3872, ISSN 0146-3705

Zhang, Y., Gao, H., Guo, Y., Joo, Y.-H. & Shreeve, J. M. (2010). Hypergolic N,N-Dimethylhydrazinium Ionic Liquids, *Chem.-Eur. J.*, Vol. 16, No. 10, pp. 3114-3120, ISSN 0947-6539

Zhang, Y., and Shreeve, J. M. (2011). Dicyanoborate-Based Ionic Liquids as Hypergolic Fluids, *Angew. Chem., Int. Ed.*, Vol. 50, No. 4, pp. 935-937, ISSN 1433-7851

22

The Latent Application of Ionic Liquids in Absorption Refrigeration

Shiqiang Liang, Wei Chen, Keyong Cheng, Yongxian Guo and Xiaohong Gui
Institute of Engineering Thermophysics, Chinese Academy of Sciences, Beijing 100190, China

1. Introduction

The absorption refrigeration technology, which went through more than 100 years, has attracted much attention all over the world, for the reason that it is environmental friendly and could make use of the low-grade energy, which refers to the ignored energy embedded in the exhaust steam of low pressure and low temperature. The absorption refrigeration is widely used in many fields, such as military, air conditioning, electric power, steelmaking, chemical industry, drugs manufacture and so on. In this section, the absorption refrigeration cycle and its working pairs are the key contents that will be introduced to the readers.

1.1 Principles

Absorption refrigeration uses a source of heat to provide the energy needed to drive the cooling process. The liquid refrigerant evaporates in a low partial pressure environment, thus extracting heat from its surroundings, and the absorbent absorbs the gaseous refrigerant to reduce its partial pressure in the evaporator and allowing more liquid to evaporate. The refrigerant-laden liquid is heated to boil refrigerant vapor out of the absorbent solution and compress the refrigerant vapor to a higher pressure, then it is condensed through a heat exchanger to replenish the supply of liquid refrigerant in the evaporator.

1.2 Absorption refrigeration cycle

The single effect absorption refrigeration cycle is shown in Figure 1, which is the most elementary one. As mentioned above, in absorption refrigeration system, an absorber, generator, pump and recuperative heat exchanger replace the compressor in the actual vapor-compression refrigeration systems. The cycle begins when high pressure liquid refrigerant from the condenser passes into the evaporator through an expansion valve, which reduces the pressure of the refrigerant to the low pressure existing in the evaporator. The liquid refrigerant vaporizes in the evaporator by absorbing heat from the material needing to be cooled and the resulting low-pressure vapor migrates to the absorber, where the vapor is absorbed by the solution coming from the generator, called strong solution.

Here we consider refrigerant as solvent. The resulting low concentration solution, called weak solution, is pumped to the generator, where the refrigerant is boiled off. The remaining strong solution flows back to the absorber and, thus, completes the cycle [1].

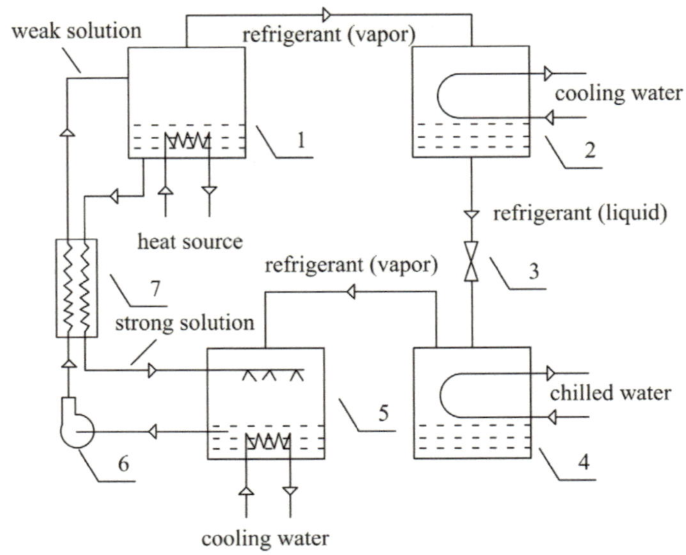

Fig. 1. Schematic diagram of single effect absorption refrigeration: 1- Generator, 2- Condenser, 3- Expansion valve, 4- Evaporator, 5- Absorber, 6- Solution pump, 7- Heat exchanger.

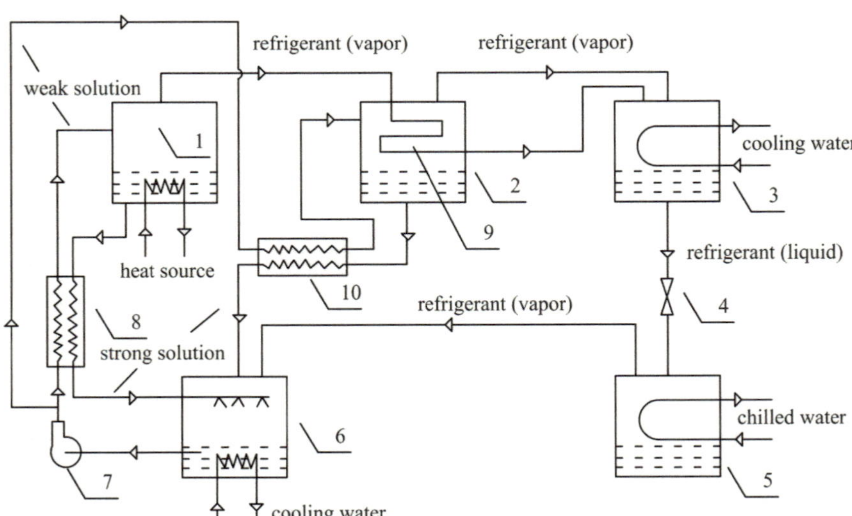

Fig. 2. Schematic diagram of double effect absorption refrigeration: 1- High pressure generator, 2- Low pressure generator, 3- Condenser, 4- Expansion valve, 5- Evaporator, 6- Absorber, 7- Solution pump, 8- High-temperature heat exchanger, 9- Condensing heat exchanger, 10- Low-temperature heat exchanger.

The another common cycle in the actual application, called double-effect absorption refrigeration cycle, is shown in Figure 2, which is mainly composed of a high pressure generator, low pressure generator, absorber, condenser and evaporator. Compared with the single one, the double effect absorption refrigeration system has two generators. The solution in high pressure generator, indirectly heated by the heat source, emits high pressure refrigerant vapor and forms strong solution. The solution in low pressure generator is heated by high pressure refrigerant vapor from the high pressure generator. The solution is concentrated in some degree and passes into high pressure generator. The energy is used twice, directly in the high pressure generator and indirectly in the low one. The double one is more complex and costly than the single one, but its efficiency is higher than the latter.

Besides above, there are some other absorption refrigeration cycles in order to make use of the energy more efficiently and completely, for example, the triple-effect absorption refrigeration cycle, the GAX cycle and the absorption/compression refrigeration cycle.

1.3 Working pairs

Many working fluids are suggested to be used as working pairs for absorption refrigeration in literature, but there is no ideal absorbent-refrigerant pair by now. Currently, the binary systems of NH_3-H_2O and LiBr-H_2O are well known as working fluid pairs to be applied in absorption refrigerators, but they both present advantages and disadvantages. The advantage for refrigerant NH_3 is that it can evaporate at lower temperatures (i.e. from -10 to 0°C) compared to H_2O (i.e. from 4 to 10°C). Therefore, for refrigeration, the NH_3-H_2O cycle is used. However, the NH_3/H_2O system is high-pressure and explosive, and the refrigerant NH_3 is poisonous, and its solution is alkaline and corrosive. The disparity in boiling point between NH_3 and H_2O is not large, which makes it necessary to utilize the distillation equipment. The coefficient of the performance for the H_2O-LiBr system is much higher than that for the NH_3-H_2O system. The only disadvantage is that H_2O-LiBr solution is corrosive to metal and easily crystallized, in addition, the working temperature and pressure of the H_2O-LiBr system are too low.

The refrigerant for the application being investigated should have the following properties: high latent heat of vaporization and low saturation pressures at normal operating temperature. Ammonia, water, methanol, and fluorocarbon refrigerants are at the top of the choice list. The important considerations influencing the choice of a suitable absorbent are: higher boiling point than refrigerant, strong ability to absorb the refrigerant, high thermal and chemical stability, low mass flow rate and heat capacity, non-poisonous, non-corrosive, non-flammable and so on. Ionic liquids are organic salts with a melting point below some arbitrary temperature, such as 100°C. Comparing with frequently-used solvents, ionic liquids exhibit distinctive properties, such as negligible vapor pressure, low combustibility, excellent thermal stability, wide liquid regions, and favorable solvating properties for a range of polar and non-polar compounds. If used as absorbents in absorption refrigeration system, their good solvating properties will make them useful in absorption of large amount of refrigerant under low temperature conditions to yield good COP, and their involatile will ensure them not contaminating with refrigerant stream when desorbed. Therefore, in recent years, RTILs are regarded as the potential candidates of absorbent in absorption refrigeration system [2, 3].

In the following sections, we'd like to introduce to readers a number of studies on the application of ionic liquids in absorption refrigeration system. Whether ionic liquids can be used as refrigeration absorbent, there are two issues needed to be focused on. Firstly, we should make sure whether the physicochemical properties of ionic liquid and refrigerant's binary solution can meet the requests of absorption refrigerator's working pairs. Secondly, we should make sure whether the selected ionic liquid working pairs have competitive advantages in terms of performance compared with those traditional ones.

2. Physicochemical properties of ionic liquid and refrigerant's binary solution

2.1 Vapor liquid equilibrium (VLE)

VLE Properties of binary system containing ionic liquids are the most important factor to determine whether the binary solution is suitable for absorption refrigeration system. Ionic liquids can reduce the saturated vapor pressure of refrigerants with different magnitudes, which is associated with the kinds of ionic liquids and refrigerants.

Saturated vapor pressures of $[BMIm]BF_4$ + 2,2,2-trifluoroethanol (TFE) and [BMIm]Br + TFE mixtures were measured by K.S. Kim et al. [4] using the boiling point method in the concentration range of 40.0 ~ 90.0 mass% of ionic liquids and in the temperature range of 298.2 K ~ 323.2 K. The data were correlated with an Antoine-type equation. The average absolute deviations between experimental and calculated values were 0.6% and 0.4% for the $[BMIm]BF_4$ + TFE and the [BMIm]Br + TFE system, respectively. As shown in Figure 3, the [BMIm]Br + TFE system was found to be more favorable as working pairs in absorption heat pumps or chillers than the $[BMIm]BF_4$ + TFE from the results of VLE.

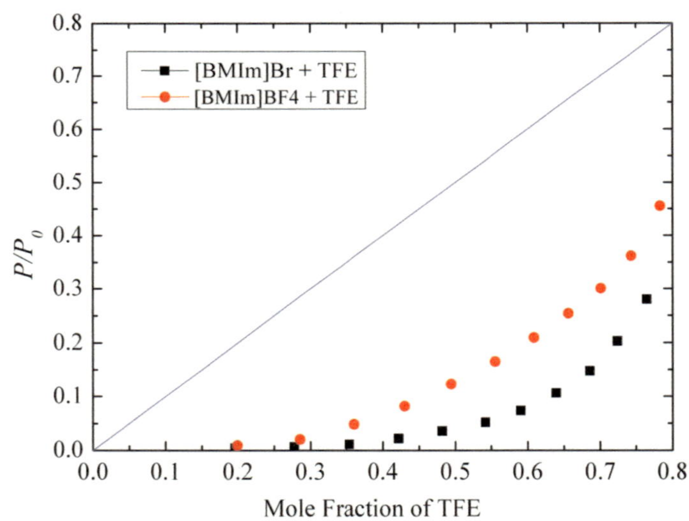

Fig. 3. The comparison of the saturated vapor pressures between the [BMIm]Br + TFE and $[BMIm]BF_4$ + TFE systems at 313.2 K.

J. Zhao et al. [5] measured the vapor pressure data for nine binary systems containing water, methanol or ethanol with the ionic liquids 1-methyl-3-methylimidazolium dimethylphosphate ([MMIm]DMP), 1-ethyl-3- methylimidazolium diethylphosphate ([EMIm]DEP) and 1-butyl-3-methylimidazo- lium dibutylphosphate ([BMIm]DBP) and one ternary system ethanol-water- [MMIm]DMP at varying temperature and ionic liquid mass percent ranging from 10% to 70% by a quasi-static method. The vapor pressure data of the binary systems were correlated by the NRTL model. Similar work by J.F. Wang et al. [6] was conducted in 2007. As shown in Figure 4, based on the vapor pressure depression of binary systems interpolated at ionic liquid mole fraction 5% in the temperature range from 280 K to 370 K, the effect of ionic liquids on the vapor pressure lowering follows the order [MMIm]DMP > [EMIm]DEP > [BMIm]DBP for water, and [BMIm]DBP > [EMIm]DEP > [MMIm]DMP for methanol and ethanol.

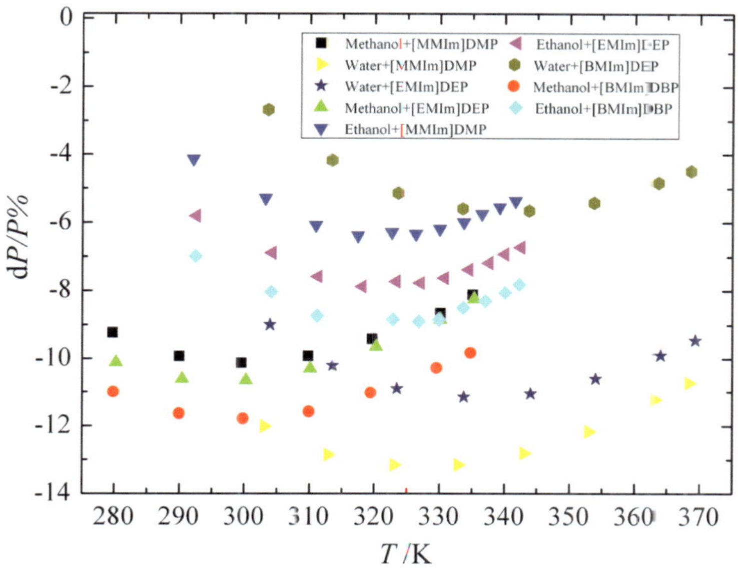

Fig. 4. Vapor pressure lowering ($dP/P\%$) of water, methanol and ethanol caused by different ILs at mole fraction of 0.05.

Solubilities of ammonia in ionic liquids, 1-ethyl-3-methylimidazolium acetate ([EMIm]Ac), 1-ethyl-3-methylimidazolium thiocyanate ([EMIm]SCN), 1-ethyl-3-methylimidazolium ethylsulfate ([EMIm]$EtOSO_3$), and N,N-dimethy- lethanolammonium acetate ([DMEA]Ac) were measured for the first time by A.Yokozeki and M.B. Shiflett [7] in 2007. Six mixture compositions of each binary system were involved from about 30 to 85 mole% of ammonia. Pressure-temperature-composition (*P-T-x*) data were claimed at isothermal conditions of 283, 298, 323, 348, and 373 K. The observed solubility of ammonia in ionic liquids is very high, and all cases show negative deviations from ideal solution behavior. Experimental *P-T-x* data were successfully correlated with the equation-of-state (EOS) model [8]. The experimental data and fitting results are shown in Figure 5 ~ 8. The opportunity for the absorption cycle application using the ammonia-RTIL system, replacing the traditional ammonia-water system, has been discussed [7].

S.P. Verevkin et al. [9] studied the vapor-liquid equilibria (VLE) of binary mixtures containing methanol, ethanol, 1-propanol and benzene in the ionic liquid [BMIm]NTf2 by using a static method. VLE measurements were carried out over the whole concentration range at four different temperatures in the range from 298.15 K to 313.15 K. Activity coefficients γi of these solvents in the ionic liquid have been determined from the VLE data and are described formally by using the NRTL equation.

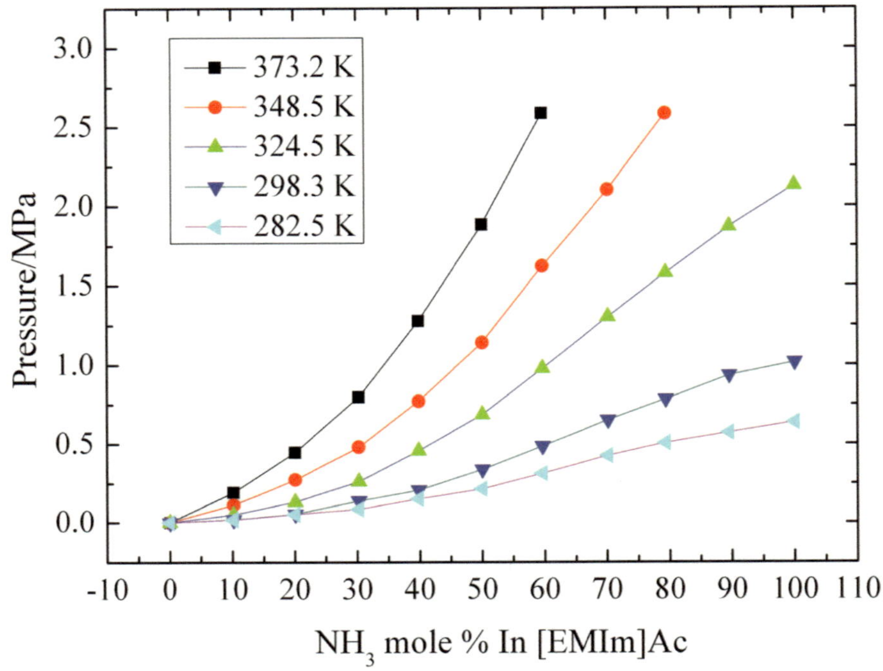

Fig. 5. *P-T-x* phase diagram of NH_3 + [EMIm]Ac mixtures [7].

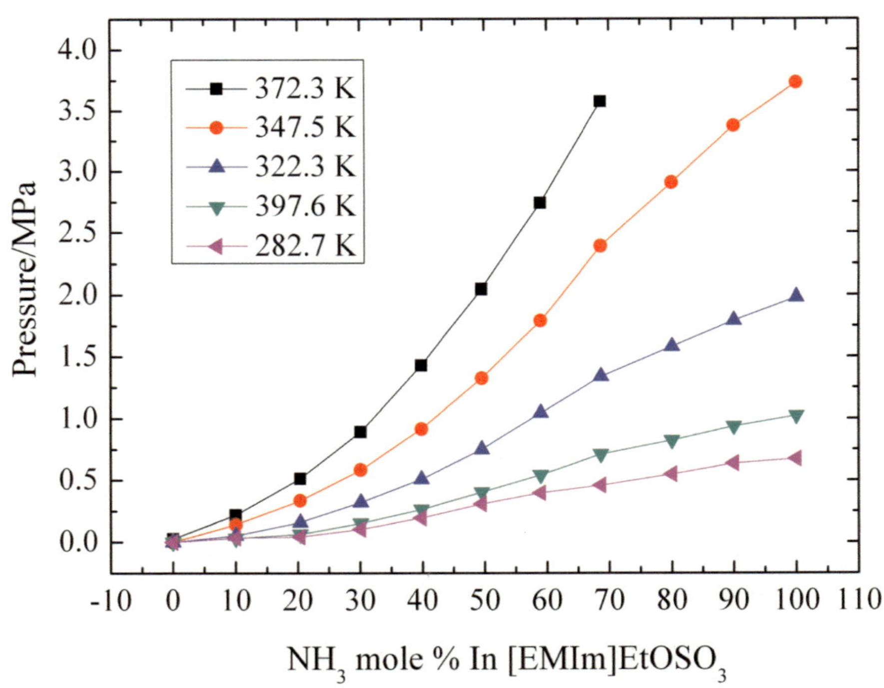

Fig. 6. *P-T-x* phase diagram of NH_3 + [EMIm]EtOSO$_3$ mixtures [7].

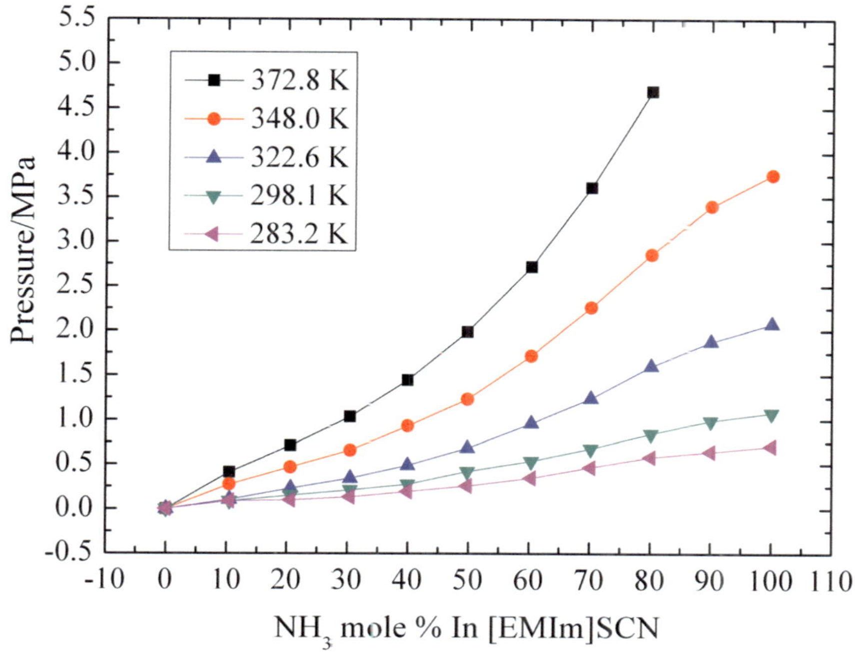

Fig. 7. *P-T-x* phase diagram of NH_3 + [EMIm]SCN mixtures [7].

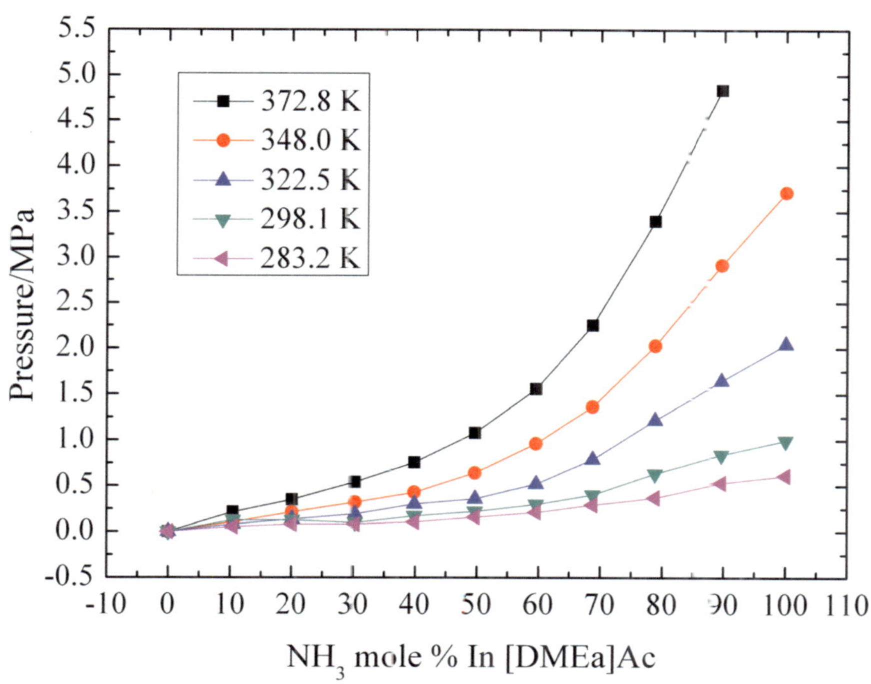

Fig. 8. *P-T-x* phase diagram of NH_3 + [DMEA]Ac mixtures[7].

Liang et al. [10] measured the saturated vapor pressure of the [MMIm]DMP-methanol solutions at 30°C ~ 90°C on condition that the mole fraction of [MMIm]DMP is 12.6%, 29.9%, 39.5%, 41.8%, 44.2% and 47.1%, respectively. The experimental data were correlated with the NRTL model, and the model parameters and the correlation deviation were calculated. The selected NRTL model with five parameters is applicable to the medium to high concentration zones, and the correlation deviation is 0.0159. With this NRTL model, the saturated vapor pressure of the [MMIm]DMP-methanol solutions when the mole fraction of [MMIm]DMP is 17.8% and 30.0% is predicted. As the low concentration solutions, the relationships between the saturated vapor pressure of the medium to high concentrations and the temperature are similar to the pure solvent, which obey the Antoine equation. The saturated vapor pressure of the [MMIm]DMP-methanol solutions when the mole fraction of [MMIm]DMP is 12.6% and 30.0% is also calculated by the NRTL model with three parameters by J. Zhao [5]. All the above *T-P-x* data are shown in Figure 9.

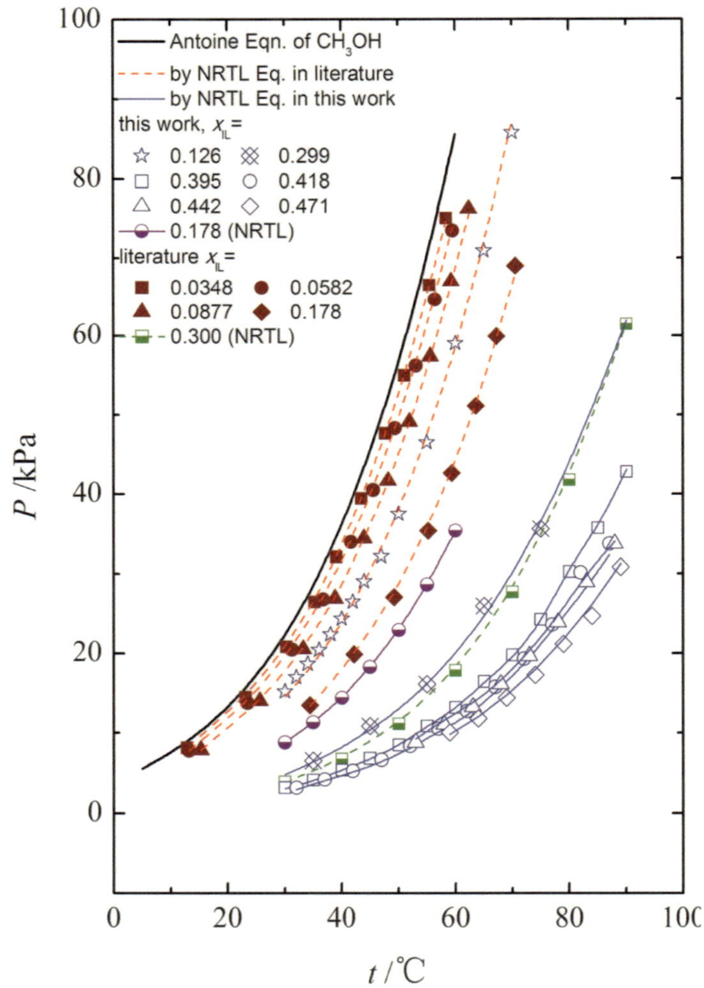

Fig. 9. Saturated vapor pressure of [MMIm]DMP/CH_3OH solutions.

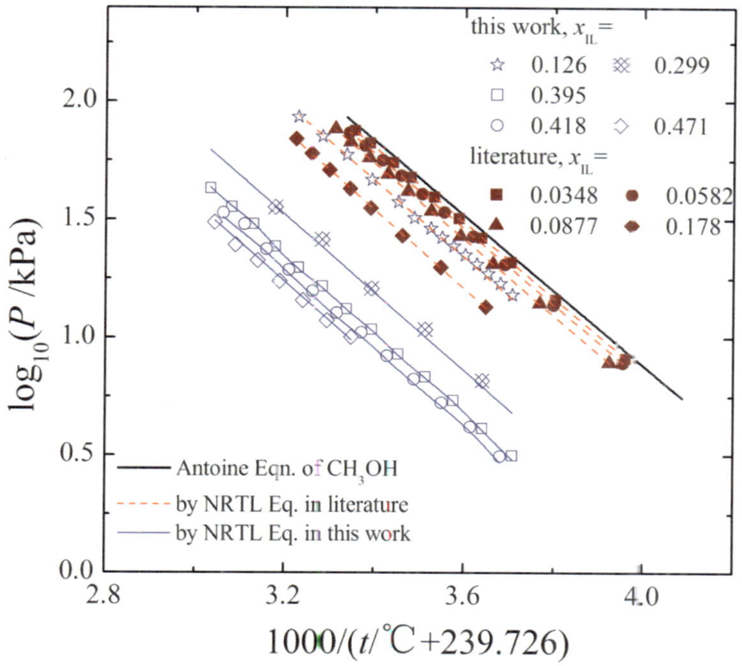

Fig. 10. Relationship of log *P*.vs.1000/(*t*+*C*).

As shown in Figure 10, with the increase of the mole fraction of [MMIm]DMP, the saturated vapor pressure was significantly reduced. The saturated vapor pressure of the [MMIm]DMP-methanol solution at normal temperatures is lower than that of pure methanol at 5°C. This means that the normal temperatures solution can absorbs the low-temperature methanol vapor, which makes the [MMIm]DMP-methanol solution suitable for absorption refrigeration as the working pair.

The solution of working pair can absorb the refrigerant vapor when the saturated vapor pressure of solution is lower than that of the pure refrigerant at the evaporating temperature. As shown in table 1, when the evaporation temperature and the cooling water temperature are 10°C and 30°C, respectively, the lowest mole fraction of [MMIm]DMP in solution is 20.8%. When the evaporation temperature and the cooling water temperature are 5°C and 40°C, respectively, the lowest mole fraction of [MMIm]DMP in solution is 39.3%.

Evaporation Temperature /°C	Vapor Pressure /kPa (Refrigerant)	Minimum Concentration	Vapor Pressure /kPa (Solution)	Minimum Concentration	Vapor Pressure /kPa (Solution)
		Cooling Water Temperature /°C			
		30		40	
5	5.570	0.270	≤5.569	0.393	≤5.561
7	6.284	0.245	≤6.252	0.362	≤6.262
10	7.504	0.208	≤7.500	0.317	≤7.504

Table 1. Minimum concentration (mole fraction) of absorption solution.

The refrigerant can be separated from the solution of working pair when the saturated vapor pressure of solution is higher than that of the pure refrigerant at the condensing temperature. As shown in Table 2, when the regeneration temperature is 90°C, the mole fraction of [MMIm]DMP in the concentrated solution is relatively high. When the regeneration temperature and condensing temperature is 90°C and 40°C, the largest mole fraction of [MMIm]DMP in the concentrated solution is 44.5%. From the above two tables, there is a large optimization space between the lower and higher limit of mole fraction of [MMIm]DMP. Just from the perspective of vapor liquid equilibrium, the [MMIm]DMP-methanol system can be used as working pair of absorption refrigeration with a large adjustable redundancy.

Regeneration Temperature /°C	Maximum Concentration	Vapor Pressure /kPa	Maximum Concentration	Vapor Pressure /kPa
	Cooling Water Temperature /°C (Condensing pressure of CH_3OH /kPa)			
	30 (22.135)		40 (35.876)	
80	0.480	≥22.153	0.349	≥35.936
85	0.531	≥22.201	0.397	≥35.925
90	0.580	≥22.240	0.445	≥35.996

Table 2. Maximum concentration (mole fraction) of absorption solution.

Liang et al. [11] measured the saturated vapor pressure of the [BMIm]Cl-methanol solutions at 30°C~80°C on condition that the mole fraction of [BMIm]Cl is 19.2%, 21.6%, 31.6%, 43.8% and 51.7%, respectively. The experimental data were correlated with the NRTL model, and the model parameters and the correlation deviation were calculated. The data of experiment and prediction by the NRTL model is shown in Figure 11.

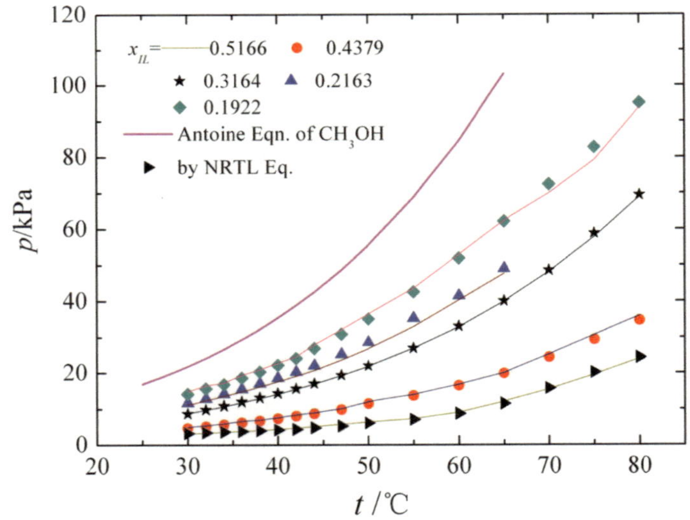

Fig. 11. Saturated vapor pressure of [BMIm]Cl/CH_3OH solutions.

As shown in Table 3, when the evaporation temperature and the cooling water temperature are 10°C and 30°C, respectively, the lowest mole fraction of [BMIm]Cl in solution is 34.3%. When the evaporation temperature and the cooling water temperature are 5°C and 40°C, respectively, the lowest mole fraction of [BMIm]Cl in solution is 49.0%.

Evaporation Temperature /°C	Vapor Pressure /kPa (Refrigerant)	Minimum Concentration	Vapor Pressure /kPa (Solution)	Minimum Concentration	Vapor Pressure /kPa (Solution)
		Cooling Water Temperature /°C			
		30		40	
5	5.570	0.408	≤5.309	0.490	≤5.485
7	6.284	0.381	≤6.086	0.472	≤6.049
10	7.504	0.343	≤7.271	0.439	≤7.221

Table 3. Minimum concentration (mole fraction) of absorption solution.

As shown in Table 4, when the temperature of heat resource is 80°C, the mole fraction of [BMIm]Cl in the concentrated solution is relatively high. When the temperature of heat resource and condensing temperature is 80°C and 40°C, the largest mole fraction of [MMIm]DMP in the concentrated solution is 42.0%. From the above two tables, there is a large optimization space between the lower and higher limit of mole fraction of [BMIm]Cl. Just from the perspective of vapor liquid equilibrium, the [BMIm]Cl-methanol system can be used as working pair of absorption refrigeration with a large adjustable redundancy.

Regeneration Temperature /°C	Maximum Concentration	Vapor Pressure /kPa	Maximum Concentration	Vapor Pressure /kPa
	Cooling Water Temperature /°C (Condensing pressure of CH_3OH /kPa)			
	30 (22.135)		40 (35.876)	
70	0.439	≥24.113	0.354	≥37.101
75	0.472	≥24.014	0.394	≥36.456
80	0.512	≥24.050	0.420	≥37.291

Table 4. Maximum concentration of absorption solution.

2.2 Viscosity

Viscosity is an important physical parameter of the binary system containing ionic liquids, and it can largely influence the application of ionic liquids in absorption refrigeration. The viscosities of binary system containing ionic liquids and water, methanol, TFE has been reported in the literature [12~16]. Viscosities of the binary system are reducing exponentially when the mole fraction of refrigerant and temperature increase. The excess logarithmic viscosities of the binary system in the whole composition range are all positive. Liang et al. [17] measured viscosities of binary systems [MMIm]DMP-methanol, [BMIm]Cl-methanol by the capillary tube method at temperatures (293.15 K to 353.15 K) when the mole fraction of ionic liquids is 20%, 40%, 60%, 80%, 100%, respectively. The experimental data are correlated with the Arrhenius-like equation, Redlich-Kister equation and Seddon equation. The experimental data and correlate results are shown in Figure 12 and Figure 13.

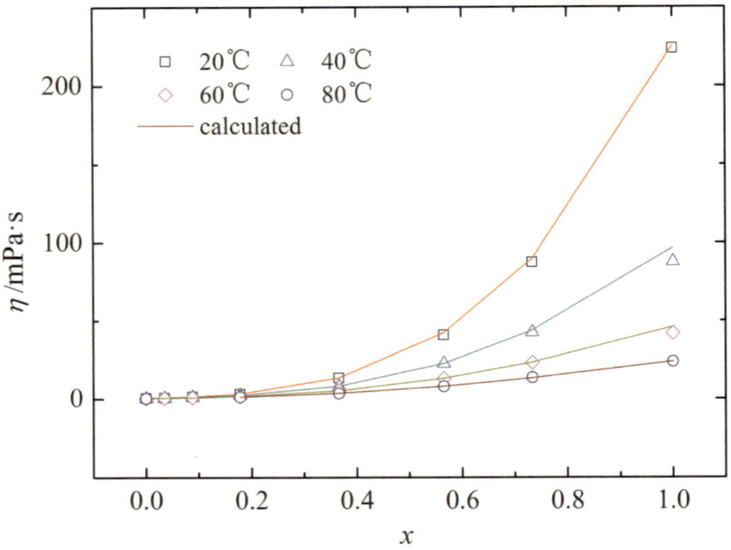

Fig. 12. Viscosities of binary systems [MMIm]DMP-methanol.

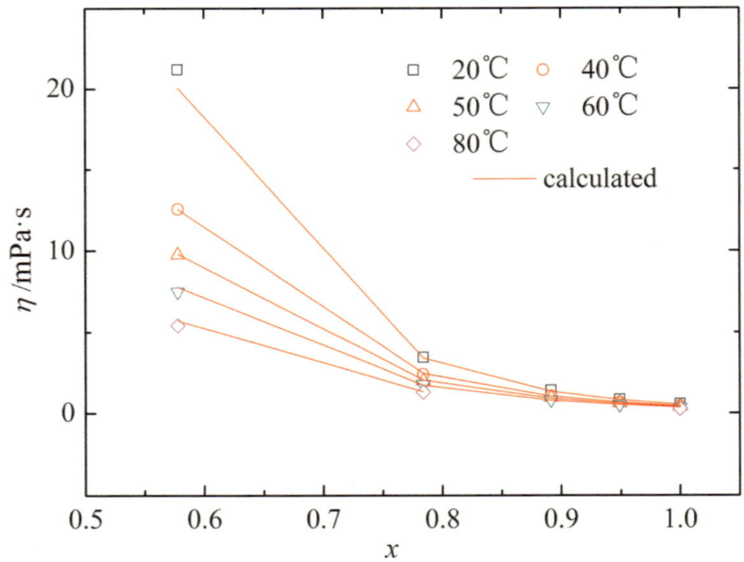

Fig. 13. Viscosities of binary systems [BMIm]Cl-methanol.

As shown in Figure 12 and Figure 13, viscosities of binary systems [MMIm]DMP- methanol, [BMIm]Cl-methanol reduce rapidly with the increase of temperature, especially in the high concentration region. Viscosities of [MMIm]DMP and [BMIm]Cl are very large, but the methanol can reduce their viscosities to a great extent. Viscosity of [MMIm]DMP at 293.15 K is 223.6736 mPa·s, while viscosities of [MMIm]DMP-methanol is 87.5429 mPa·s on condition that the mole fraction of [BMIm]DMP is 73.3%. [BMIm]Cl is solid at 293.15 K, while viscosities of [mmim]Cl-methanol is 20.1732 mPa·s on condition that the mole fraction of [BMIm]Cl is 42.2%.

2.3 Heat capacity

Heat capacity is a physical quantity which is the important bridge connecting the macroscopic observable thermodynamics quantity and the microscopic molecular structure. On the basis of heat capacity, the *h-ω* diagram is mapped, from which we can analyze COP of the whole absorption refrigeration cycle.

Fredlake [18] measured the heat capacities of thirteen imidazolium based ionic liquids at temperature (298.15 K ~ 323.15 K). Heat capacities for nine ionic liquids have been determined with the "three-step" method using two different differential scanning calorimeters (DSC) by Anja et al. [19], the measurements cover a temperature range from 315 K to 425 K. Waliszewski et al. [20] measured the heat capacities of ionic liquids: [EMIm]BF_4, [EMIm]NTf_2, [BMIm]BF_4 and [MPPy]NTf_2 at temperature (283.15 K ~ 358.15 K). Z. H. Zhang et al. [21] measured the molar heat capacities of the room temperature ionic liquid 1-ethyl-3-methylimidazolium ethyl sulfate (EMIES) by an adiabatic calorimeter at temperature (78 K ~ 390 K). The results show that the heat capacities of ionic liquids ranges from 1.2 to 1.9 J·g-1·K-1, which is smaller than that of water, methanol and ammonia.

Tao et al. [22] measured the heat capacity of the binary system [EMIm]BF_4-water at temperature (298.15 K ~ 343.15 K) when the mass fraction change from 25% to 100%. The heat capacity falls when the mass fraction of [EMIm]BF_4 increases. Rebelo et al. [23] measured the heat capacity of [BMIm]BF_4 at temperature (278.15 K ~ 333.15 K), the capacity ranges from 355 to 385 J·mol^{-1}·K^{-1}. They also measured excess heat capacity of binary system of [BMIm]BF_4-water. The results show that excess heat capacity is positive with the mole fraction ranging from 0 to 56.4%, while in the high concentration region excess heat capacity is negative. This indicates that the addition of water augments the heat capacity of ionic liquids.

Liang et al. [17] measured the heat capacities of binary system [MMIm]DMP-methanol and [BMIm]Cl-methanol at temperature (30 K ~ 80 K) with various of mole fractions. The experimental data are correlated with the following equation:

$$C_P = A + B \cdot t$$

where C_P is heat capacity, t is temperature, $A = A_1 \cdot \omega_2 + A_2 \cdot \omega + A_3$, $B = B_1 \cdot \omega_2 + B_2 \cdot \omega + B_3$. The fitting parameters and average absolute deviation is shown in Table 5. The experimental data and correlate results are shown in Figure 14 and Figure 15.

Binary System	A_1	A_2	A_3	B_1	B_2	B_3	ARD
[MMIm]DMP-methanol	-0.9057	0.8936	1.6894	0.0251	-0.0496	0.0273	0.0060
[BMIm]Cl-methanol	-0.7312	0.6737	1.6857	0.0200	-0.0315	0.0166	0.0019

Table 5. Fitting parameters.

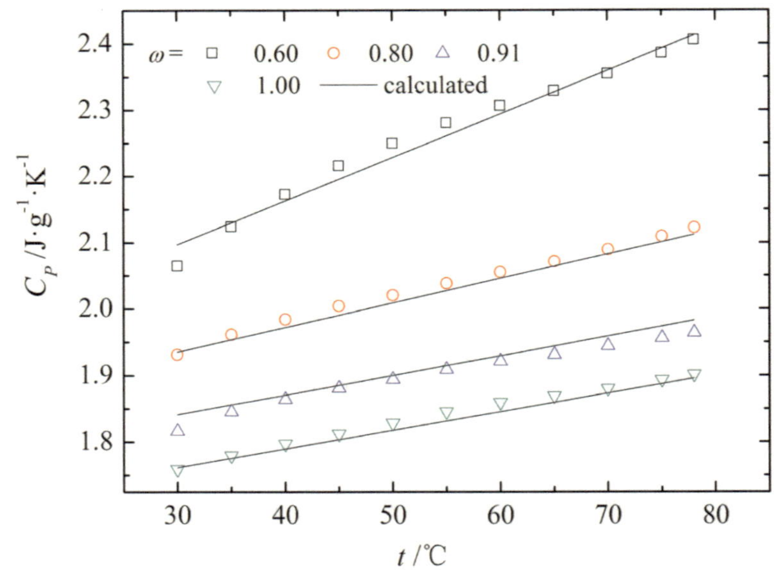

Fig. 14. Heat capacities of binary systems [MMIm]DMP-methanol.

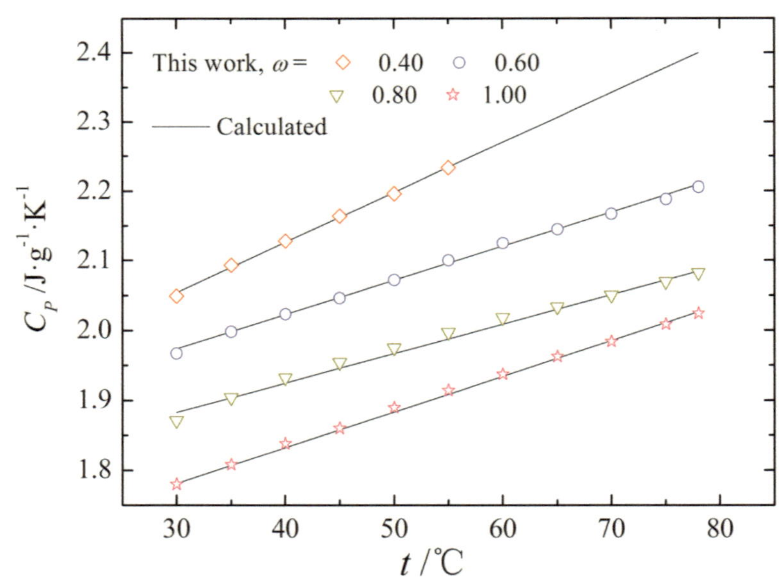

Fig. 15. Heat capacities of binary systems [BMIm]Cl-methanol.

The heat capacity of traditional working pair LiBr-H_2O is very low, which is only 2 J·g^{-1}·K^{-1} at 373.15 K with the mass fraction of 55%. The heat capacities of binary system [MMIm]DMP-methanol and [BMIm]Cl-methanol on the same condition is close to that of LiBr-H_2O solution. This means that the evaporating of methanol out of solution requires only a small amount of heat, which is favorable for enhancing COP of absorption refrigeration cycle.

2.4 Summary

This section investigates the application potential in binary system containing ionic liquids and refrigerant as working pairs for absorption refrigeration from three aspects: vapor liquid equilibrium, viscosity and heat capacity. The saturated vapor pressure and heat capacity in plenty of these binary systems are relatively low. The density and viscosity in these systems are moderate. Compared with the traditional working pairs, the new ones have their own advantages. They are non-corrosive and non-crystalline. Since ionic liquids are non-evaporative, this new kind of absorption refrigeration system is no need to set up distillation equipment. The physical and chemical properties in binary system containing ionic liquids and frequently-used refrigerants are still not perfect, and the studies on the application of ionic liquids in absorption refrigeration are still inadequate. With the increasing of the quantity of ionic liquids and the deepening of studies on their physical and chemical properties, ionic liquid absorption refrigeration must become an important part of refrigeration in near future.

3. Theoretical cycle and efficiency analysis of novel absorption refrigeration system using ionic liquids

3.1 Single-effect absorption refrigeration system

A. Yokozeki [7, 8] measured the solubility, vapor-liquid equilibrium of binary solutions containing ionic liquids, and computed the specific heat capacity at constant pressure, enthalpy, Gibbs energy, and entropy based on the EOS equation. According to the results conducted, various parameters of ideal single-effect absorption refrigeration system were computed. The quality circulation rate f, solution concentration in generator X_g, solution concentration in absorber X_a, and COP are shown in Table 6 when the temperatures of the generator, absorber, condenser, and evaporator are 373 K, 313 K, 303 K, and 283 K respectively. The results present that when ionic liquids are used as working pairs with NH_3, Freon, water, and CO_2 respectively, COP of absorption refrigeration systems are higher than that systems using NH_3/H_2O or H_2O/LiBr.

Wang et al. [24] used TFE-[BMIm]Br in double-effect parallel absorption refrigeration system, and the effect of the effectiveness of solution heat exchanger η on COP was analyzed. In the Figure 16, ω_1, ω_2, ω_3, ω_4 stand for COP in four kinds of process respectively. The results show that COP increases markedly with the improvement of η. Therefore, it is important to use high effect solution heat exchanger for double-effect parallel absorption refrigeration. Considering COP of the system and actual heat transfer performance of the heat exchanger, η is chosen to be 0.9. The effects of evaporating temperature, absorbing temperature, high pressure generating temperature on COP, solution circulation ratio, and operating pressure of the system are presented.

Binary System	f	$x_g/\%$	$x_a/\%$	COP	Literatures
NH_3+[BMIm]PF_6	17.27	94.5	89.0	0.575	[7]
NH_3+[BMIm]BF_4	12.98	95.7	88.3	0.557	[7]
NH_3+[HMIm]Cl	14.26	93.9	87.3	0.525	[7]
NH_3+[EMIm]Tf_2N ,	24.57	96.3	92.4	0.589	[7]
NH_3+[EMIm]AC	12.55	92.3	85.0	0.573	[7]
NH_3+[EMIm]$EtOSO_3$	17.55	85.2	89.8	0.485	[7]
NH_3+[EMIm]SCN	12.42	92.7	85.3	0.557	[7]
NH_3+ H_2O	2.54	59.5	36.1	0.646	[26]
R22+[BMIm]PF_6	5.12	89.70	72.20	0.319	[27]
R32+[BMIm]PF_6	7.35	90.40	78.10	0.385	[27]
R32+[BMIm]BF_4	6.41	90.20	76.10	0.330	[27]
R134+[BMIm]PF_6	4.38	88.80	68.50	0.348	[27]
R134a+[BMIm]PF_6	10.66	92.40	83.70	0.254	[27]
CO_2+[BMIm]PF_6	25.76	88.54	85.12	0.008	[29]
H_2O+[BMIm]BF_4	18.20	98.56	93.14	0.525	[28]
H_2O+LiBr	4.08	66.30	50.00	0.833	[26]

Table 6. Parameters of single-effect absorption refrigeration system.

From Figure 17, it can be concluded that double-effect parallel absorption refrigeration using TFE-[BMIm]Br is more suitable for air conditioning. However, COP is relatively low when evaporation temperature is below zero. As shown in Figure 18, COP increases with the high pressure generating temperature. Considering the heat stability of TFE-[BMIm]Br, it is not suitable for the operation when the high pressure generating temperature is more than 500 K.

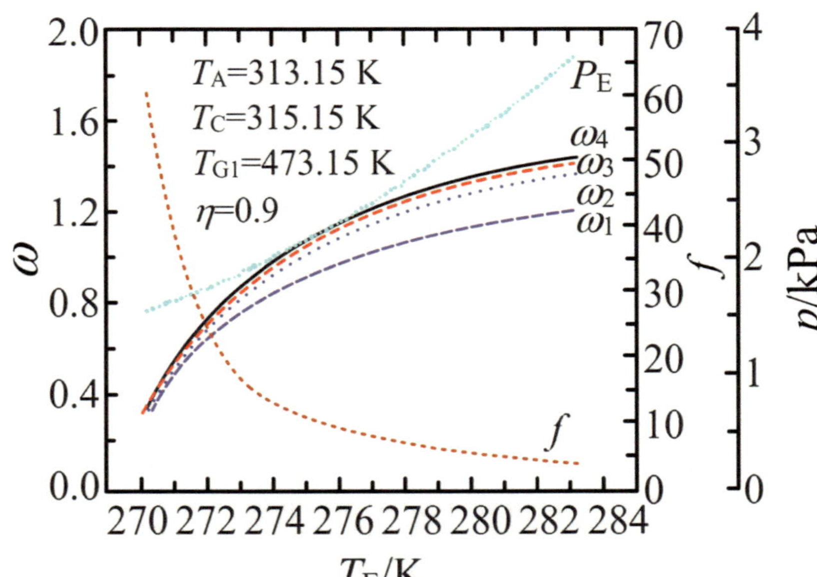

Fig. 16. The effect of evaporation temperature on performances.

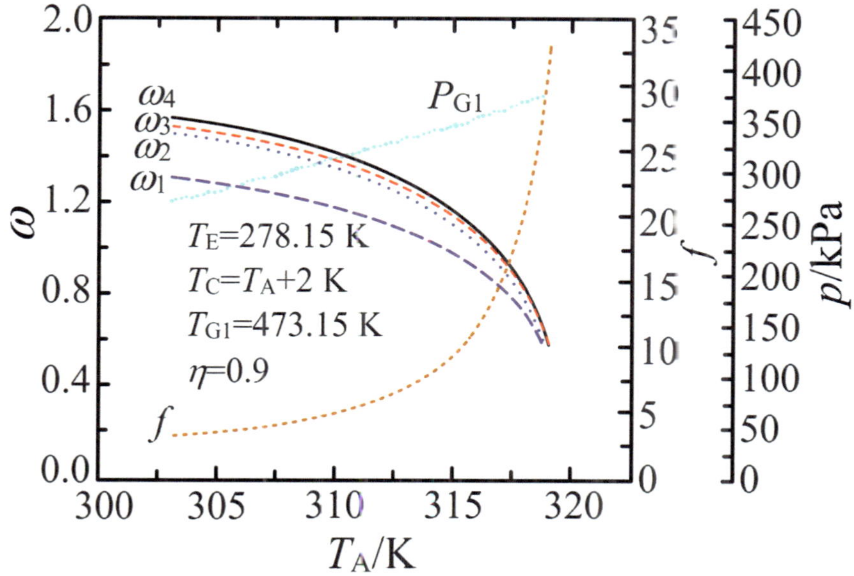

Fig. 17. The effect of absorption temperature on performances.

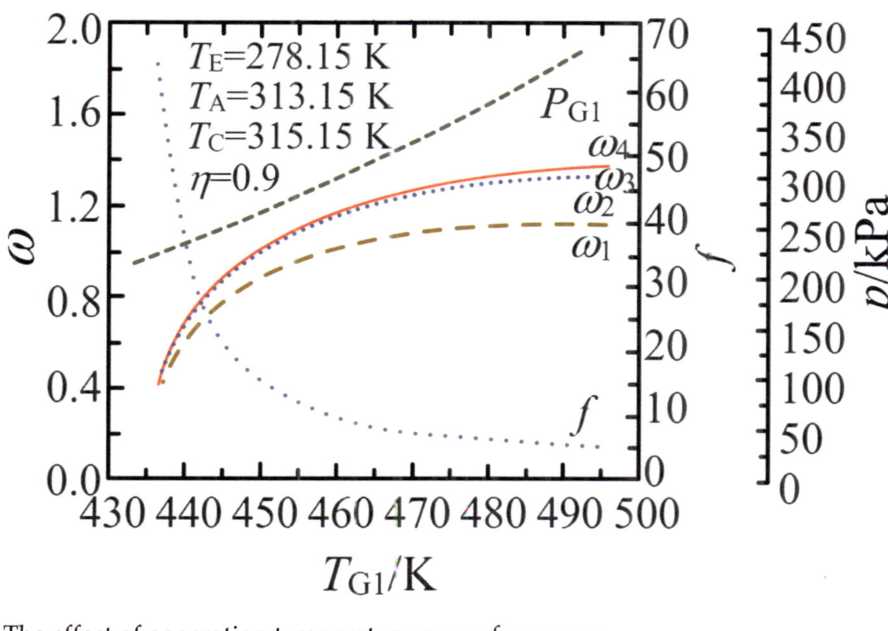

Fig. 18. The effect of generation temperature on performances.

Liang et al. [25] used [MMIm]DMP-CH_3OH to analyze the effectiveness of absorption refrigeration system. In order to analyze the effect of various temperatures on the effectiveness of the system, different values were set, and loads of heat equipment, solution concentration, circulation ratio, gas-emission scope, and COP were computed.

As shown in Figure 19, for single-effect absorption refrigeration system, COP increases with the improvement of heat source temperature and evaporating temperature. When heat source and evaporating temperature are 120°C and 10°C respectively, COP gets 0.895. However, it is not suitable for evaporating temperature to be too low. For example, if evaporating temperature is 5°C, and heat source temperature is below 80 °C, solution concentration will not meet the requirements of solution circulation.

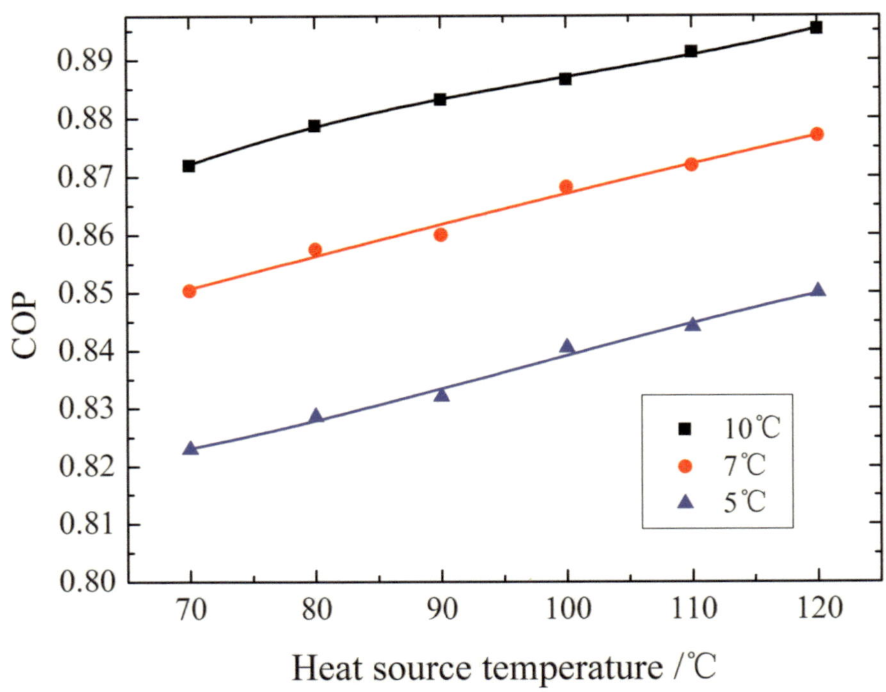

Fig. 19. The effect of heat resource temperature on COP.

Figure 20 and 21 show the variation tendency of gas-emission scope, circulation ratio with heat resource temperature and evaporating temperature. With the improvement of heat source temperature and evaporating temperature, gas-emission scope increases, while circulation ratio decreases. The smaller the gas-emission scope, the bigger the circulation ratio, and the more the heat absorbed by solution liquid, the lower COP. When the evaporating temperatures are 5°C, 7°C, and 10°C respectively, the heat source temperatures should be above 60°C, 70°C and 80°C respectively. Otherwise, the gas-emission scopes will be too small and the circulation ratios will be too big, which will lead to lower effectiveness of the system, huge equipment, and high operating cost. Figure 22, 23 and 24 show the changes of COP, gas-emission scope, and circulation ratio with heat source temperatures at various evaporation temperatures. COP of the system increases with the reduction of the following parameters: condensing temperature, absorbing temperature and circulation ratio, while increases with the improvement of gas-emission scope.

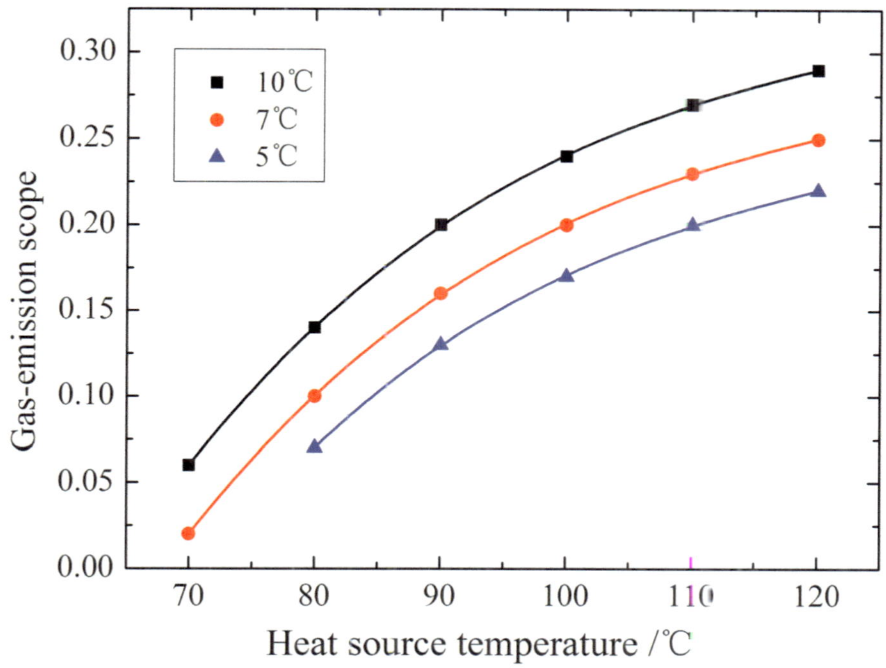

Fig. 20. The effect of heat resource temperature on gas-emission scope.

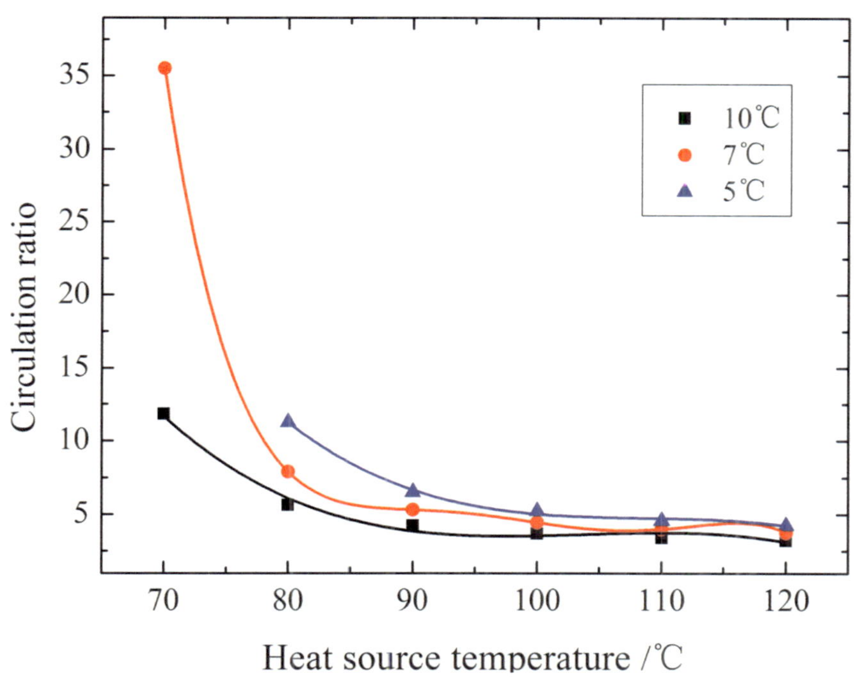

Fig. 21. The effect of heat resource temperature on circulation ratio.

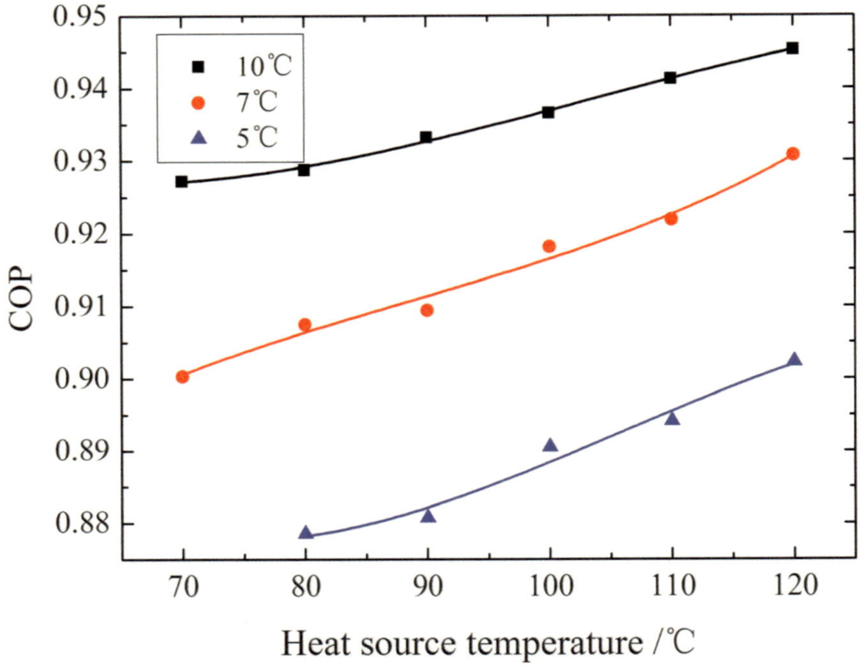

Fig. 22. The effect of heat resource temperature on COP.

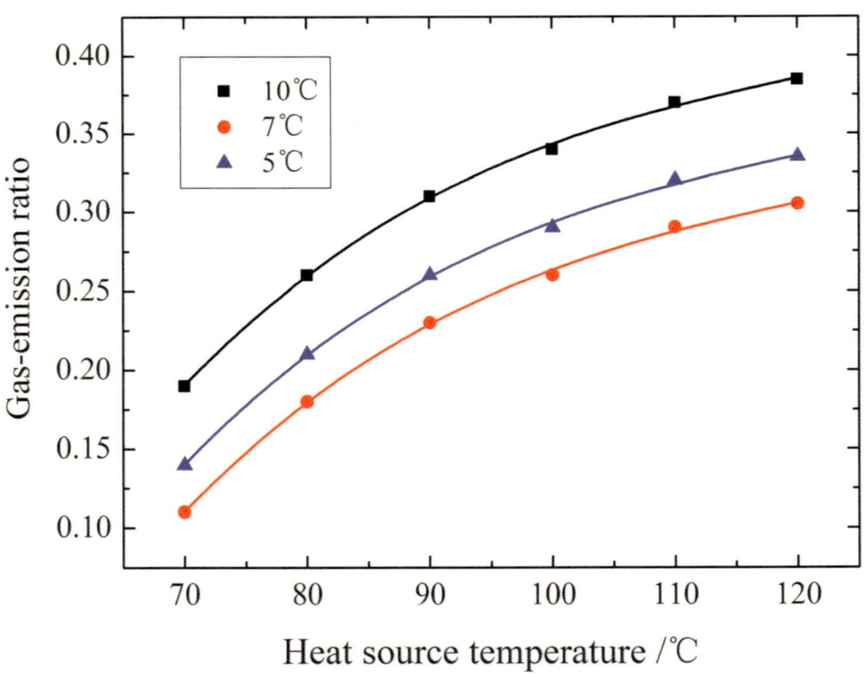

Fig. 23. The effect of heat resource temperature on gas-emission scope.

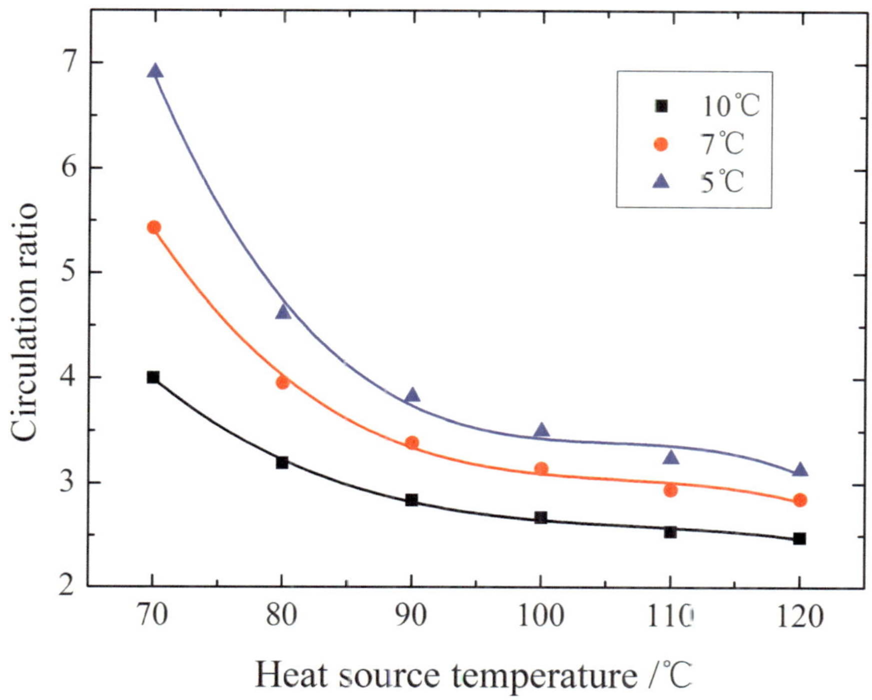

Fig. 24. The effect of heat resource temperature on circulation ratio.

In order to compare the difference between the system using [MMIm]DMP-CH_3OH and that using the traditional working pairs, such as NH_3/H_2O, H_2O/LiBr, under the same working conditions, various parameters were analyzed, such as condensing pressure, evaporating pressure, circulation ratio, and COP. The results are shown in Table 7, the system using [MMIm]DMP-CH_3OH has the following advantages:

1. COP of the system using [MMIm]DMP-CH_3OH is higher than that using NH_3/H_2O and H_2O/LiBr.
2. Unit quality refrigerating capacity is higher than that using NH_3/H_2O, but lower than that using H_2O/LiBr.
3. Circulation ratio is higher than that using NH_3/H_2O, which is helpful to significantly reduce the volume of the equipment.
4. The system using [MMIm]DMP-CH_3OH reduces the requirements of the generating pressure and condensing pressure. The generating and condensing pressures for [MMIm]DMP-CH_3OH are far lower than that for NH_3/H_2O,while a little higher than that H_2O/LiBr. So maintaining and operating of the system are favorable.
5. The mass fraction of [MMIm]DMP-CH_3OH solution is in the range of 65% ~ 89%, and the gas-emission scope can get 0.24, which is far higher than that using H_2O/LiBr, and helpful to reduce the volume of the equipment.

Binary System	P_{con}, P_g /kPa	P_{eva}, P_a /kPa	f	x_g Mass%	x_a Mass%	Q_e /kW	COP
CH_3OH/[MMIm]DMP	35.43	7.4113	3.71	89	65	1242	0.886
NH_3/H_2O	1548	615	2.54	59.5	36.1	1112	0.646
H_2O/LiBr	7.38	1.23	4.08	66.3	50	2502	0.833

Table 7. Comparison of COP in the system with different working pairs.

3.2 Double-effect absorption refrigeration system

Single-effect absorption refrigeration system, with simple structure, convenient operation, and low heat source requirement, is suitable for the situation that the heat source temperature and operating pressure is not too high. However, while the heat source temperature is high, there are several drawbacks of the single system, such as low COP and low utilization rate of energy. In order to utilize energy more efficiently and fully, some other systems are developed, such as double-effect system, three-effect system and multiple-effect system.

Liang et al. analyzed the effectiveness and equipment load of double-effect absorption refrigeration system using [MMIm]DMP-CH_3OH under various heat source temperature and evaporating temperature. Different generating temperatures, condensing temperatures, absorbing temperatures, and evaporating temperatures were set. The loads of heat equipment, solution concentration, circulation ratio, gas-emission scope, and COP were computed. The results are shown in Figure 25, 26 and 27.

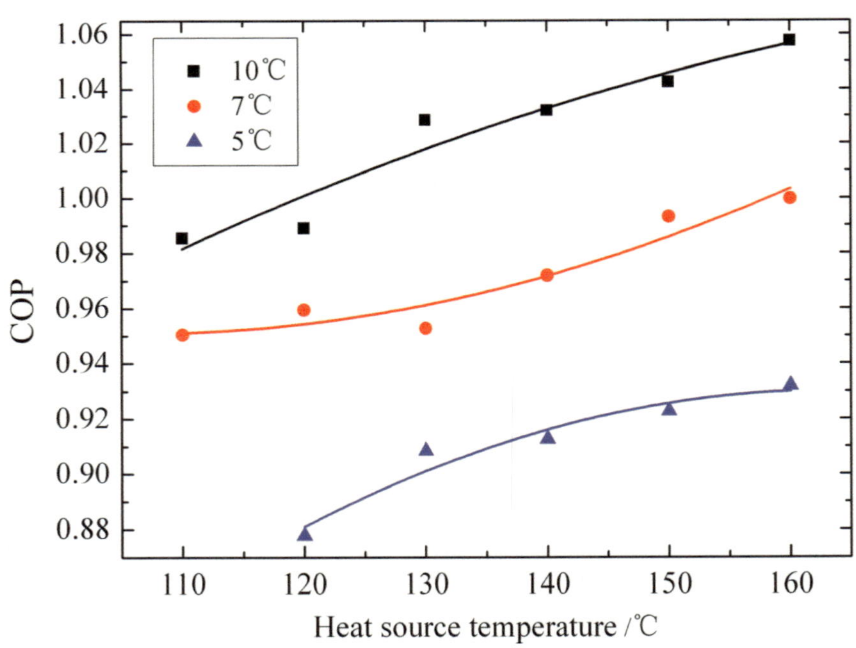

Fig. 25. The effect of heat resource temperature on COP.

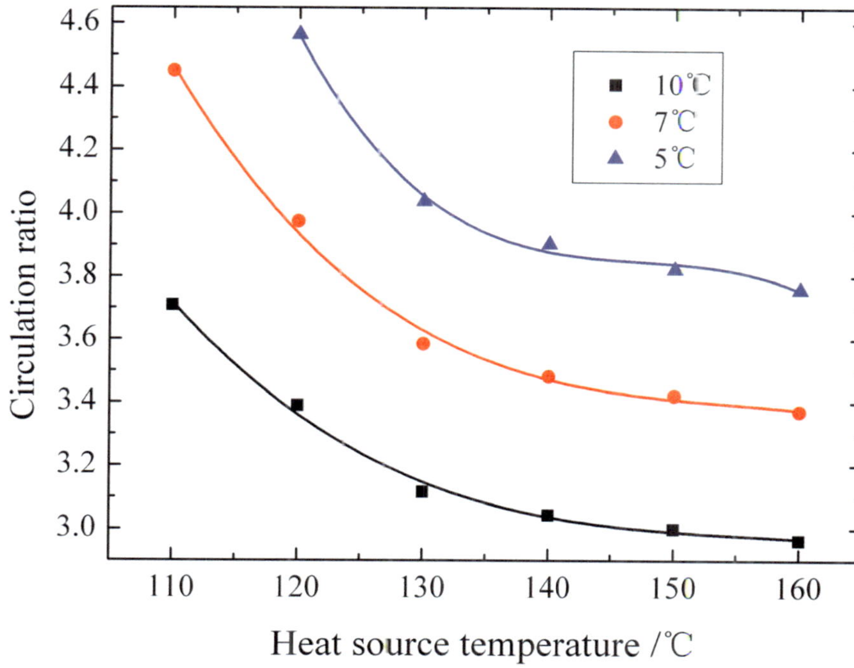

Fig. 26. The effect of heat resource temperature on gas-emission scope.

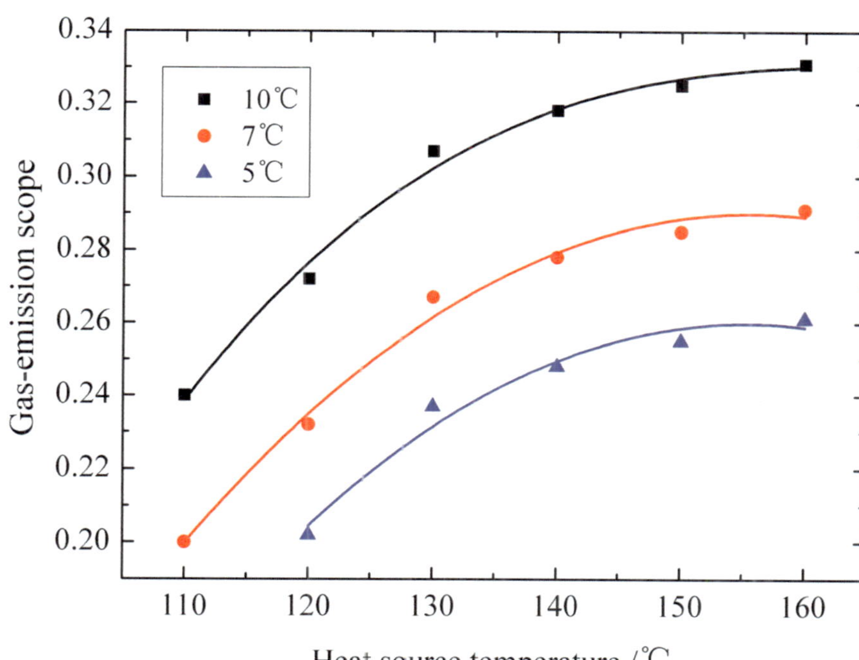

Fig. 27. The effect of heat resource temperature on circulation ratio.

Single-effect absorption refrigeration system, with simple structure, convenient operation, and low heat source requirement, is suitable for the situation that the heat source temperature and operating pressure is not too high. However, while the heat source temperature is high, there are several drawbacks of the single system, such as low COP and low utilization rate of energy. In order to utilize energy more efficiently and fully, some other systems are developed, such as double-effect system, three-effect system and multiple-effect system.

Liang et al. analyzed the effectiveness and equipment load of double-effect absorption refrigeration system using [MMIm]DMP-CH_3OH under various heat source temperature and evaporating temperature. Different generating temperatures, condensing temperatures, absorbing temperatures, and evaporating temperatures were set. The loads of heat equipment, solution concentration, circulation ratio, gas-emission scope, and COP were computed. The results are shown in Figure 25, 26 and 27.

Figure 25 shows that COP of the system increases with the improvement of the heat source temperature and evaporating temperature. For example, COP gets 1.06 when the heat source temperature and evaporating temperatures are 160°C and 10°C respectively. However, the evaporating temperature should not be too low. When the evaporating temperature is 5°C, and heat source temperature is under 120°C, the solution concentration of [MMIm]DMP-CH_3OH will be too low, the circulation ratio will be too high, and the double-effect system will lose its predominance in comparing with the single-effect one.

Figure 26 and Figure 27 show the variation tendency of gas-emission scope and circulation ratio with heat source temperature and evaporating temperature for double-effect absorption refrigeration system. With the improvement of heat source temperature and evaporating temperature, gas-emission scope increases, while circulation ratio decreases. The smaller the gas-emission scope, the bigger the circulation ratio, and the more the heat absorbed by solution liquid, the lower COP. When the evaporating temperatures are 5°C, 7°C, and 10°C respectively, the heat source temperatures should be above 120°C, 110°C and 100°C respectively. Otherwise, gas-emission scope is too small and circulation ratio is too big, so that compared with single-effect absorption refrigeration system, double-effect absorption refrigeration system will not have any advantage. Furthermore, the structure double-effect absorption refrigeration system is huger, and the cost is higher.

In order to compare the difference between the system using [MMIm]DMP-CH_3OH and that using the traditional working pairs, such as NH_3/H_2O, $H_2O/LiBr$, under the same working conditions, various parameters were analyzed, such as condensing pressure, evaporating pressure, circulation ratio, and COP. The results are shown in Table 8, the system using [MMIm]DMP-CH_3OH has the following advantages:

1. COP of double-effect absorption refrigeration system using [MMIm]DMP-CH_3OH is almost the same with that of the system using $H_2O/LiBr$.
2. Since methanol is chosen as refrigerant, evaporation latent heat of methanol is smaller than that of water, and unit quality refrigerating capacity of methanol is smaller than that of $H_2O/LiBr$.
3. Circulation ratio of the system using [MMIm]DMP-CH_3OH is smaller, and gas-emission scope is larger. Compared with that of the system using $H_2O/LiBr$, it is helpful to simplify the structure of the equipment.
4. The requirements of operating pressure, condensing pressure, and vacuum for the system using [MMIm]DMP-CH_3OH are lower, which benefits the operation and the maintaining of the system.

Binary System	Deflation Ratio	ω_a	ω_{r1}	ω_{r2}	Circulation Ratio	Q_e /kW	COP
CH_3OH/[MMIm]DMP	0.255	0.72	0.957	0.975	3.82	1242	0.923
H_2O/LiBr	0.05	56.93	60	61.93	12.38	2502	0.930

Table 8. Comparison of COP in the system with different working pairs.

3.3 Summary

Based on the physical and chemical properties of binary system containing ionic liquids and refrigerant, A. Yokozeki et al. from American, J.Z. Wang et al. and S.Q. Liang et al. from China calculated some fundamental parameters of refrigeration cycle by the EOS model, including several key parameters, such as COP, circulation ratio and gas-emission scope. They also analyzed the flux characteristics of some fundamental parameters with different heat source temperature, condensing temperature, evaporating temperature and absorption temperature. The pressure of the whole cycle is moderate, and there is no distillation equipment, no corrosion and no crystallization in the new type of absorption refrigeration. Compared with the double effect cycle, the single effect cycle is more economical and practical when this new working pairs containing ionic liquids is used. Since the working pairs used by different researchers are not the same, COP of different cycles is different from each other. Further research will focus on the practical applications of the new type absorption refrigeration.

4. Conclusions and outlook

Based on the low-grade heat source, absorption refrigeration, which owns many advantages such as simple and quiet, along with energy-saving and environmental protection, has huge spaces for development. It is revealed that the binary system, containing ionic liquids and refrigerants as working pairs of absorption refrigeration, has large application potential through the researches on the physical and chemical properties. The new type ionic liquid absorption refrigeration can overcome same defects belonging to the traditional type, such as corrosion, crystallization and requirement of distillation equipment. However, the large-scaled industrial application of the new technology is still restricted by several factors as follows. Firstly, COP of some working pairs containing ionic liquids is still not high, requiring that better working pairs with high COP should be screened out. Secondly, the researches are still at the academic stage in defect of the studies on practical design and structure optimization. The experimental units are all theoretical simulation ones, and until now the suitable absorption refrigeration units in practice have not been produced. In addition, the manufacturing costs for the new type of absorption refrigeration are high, for the reason that the price of ionic liquids is high, which prevents the market from popularizing this technology. Overall, the ionic liquid absorption refrigeration has a huge space for development and good market prospect, and it will certainly bring innovative promotion and ground-breaking progress for the absorption refrigeration technology.

5. Acknowledgments

The authors wish to express their thanks to Master Jie Zhao for his helpful advice and selfless assistance. This work is financially supported by the National High Technology Research and Development (863) Program (Grant No.2007AA05Z259).

6. References

[1] Z Crepinsek, D Goricenec, J Krope. Comparison of the performances of absorption refrigeration cycles. *Faculty of Chemistry and Chemical Engineering*, 2009, 3 (4): 65-76

[2] Q Yang, Y Huang, E C Luo, et al. Analysis on thermodynamic performance of absorption refrigeration cycle utilizing transition-critical CO_2-[BMIm]PF_6 as working fluid. *Cryogenics*, 2009, 3(169): 5-10

[3] S Mihir, P Samuel. Using carbon dioxide and ionic liquid for absorption refrigeration. *7th IIR Gustav Lorentzen Conference on Natural Working Fluids*, 2006

[4] K S Kim, B K Shin, H Lee, et al. Refractive index and heat capacity of 1-butyl- 3-methylimidazolium bromide and 1-butyl-3-methylimidazolium tetrafluoroborat, and vapor pressure of binary systems for 1-butyl-3-methylimi-dzoliumbromide+trifluoroe- thanol and 1-butyl-3-methylim-idazolium tetrafluoro borate+trifluoroethanol. *Fluid Phase Equilibria*, 2004, 218 (2): 215-220

[5] J Zhao, X C Jiang, C X Li, et al. Vapor pressure measurement for binary and ternary systems containing a phosphoric ionic liquid. *Fluid Phase Equilibria*, 2006, 247: 190-198

[6] J F Wang, C X Li, Z H Wang. Vapor pressure measurement for water, methanol, ethanol and their binary mixtures in the presence of an ionic liquid 1-ethyl-3-methylimidazolium dimethylphosphate. *Fluid Phase Equilibria*, 2007, 255:186-192

[7] A Yokozeki, B Mark. Vapor–liquid equilibria of ammonia + ionic liquid mixtures. *Applied Energy*, 2007, 84: 1258-1273

[8] A Yokozeki. A simple theoretical model for condensed-phase equilibria based on a hard-sphere equation-of-state and its applications. *Applied Energy*, 2005, 81: 306-321

[9] S P Verevkin, J Safarov, E Bich. Thermodynamic properties of mixtures containing ionic liquids Vapor pressures and activity coefficients of n-alcohols and benzene in binary mixtures with 1-methyl-3-butyl-imidazoliumbis- (trifluoromethyl-sulfonyl) imide. *Fluid Phase Equilibria*, 2005, 236(3): 222-228

[10] J Zhao, S Q Liang, J Chen, et al. VLE for the High Concentration [MMIm]DMP-Methanol Solutions. *Chemical Engineering*, 2010, 38 (3): 52-56

[11] J Zhao, S Q Liang, L Wang, et al. Potential analysis of [BMIm] Cl-CH_3OH as working fluid for absorption refrigeration. *CIESC Journal*, 2009, 60(12): 2957-2962

[12] K R Seddon, A Stark, M J Torres. Influence of chloride, water, and organic solvents on the physical properties of ionic liquids. *Pure and Applied Chemistry*, 2000, 72 (12): 2275-2287

[13] H Xu, D Zhao, P Xu, et al. Conductivity and viscosity of 1-allyl-3-methylimi- dazolium chloride + water and + ethanol from 293.13 K to 333.15 K. *Journal of Chemical and Engineering Data*, 2005, 50 (1):133-135

[14] C Comminges, R Barhdadi, M Laurent, et al. Determination of viscosity, ionic conductivity, and diffusion coefficients in some binary systems: ionic liquid + molecular solvents. *Journal of Chemical and Engineering Data*, 2006, 51 (2): 680-685

[15] B Mokhtarani, A S Hamid, R Mortaheb, et al. Density and viscosity of pyridinium-based ionic liquids and their binary mixtures with water at several temperatures. *Journal of Chemical Thermodynamics*, 2009, 41(1): 323-329

[16] W W Liu, T Y Zhao, Y M Zhang, et al. The Physical Properties of Aqueous Solutions of the Ionic Liquid [BMIm]BF_4. *Journal of Solution Chemistry*, 2006, 35 (4): 1337-1346

[17] S Q Liang, J Zhao, Y X Guo, et al. Experimental Study on the Thermal Properties of Ionic Liquids [MMIm]DMP-Methanol Solution. *Journal of Engineering Thermophysics*, 2011, 32(3): 441-444

[18] C P Fredlake, J M CrosthWaite, D G Hert, et al. Thermophysical Properties of Imidazolium-Based Ionic Liquids. *Journal of Chemical and Engineering Data*, 2004, 49:954-964

[19] A Diedrichs, J Gmehling. Measurement of heat capacities of ionic liquids by differential scanning calorimetry. *Fluid Phase Equilibria*, 2006, 244:68-77

[20] D Waliszewski, I Stepniak, H Piekarski et al. Heat capacities of ionic liquids and their heats of solution in molecular liquids. *Thermochimica Acta*, 2005, 433: 149-152

[21] Z H Zhang, Z C Tan, L X Sun. Thermodynamic investigation of room temperature ionic liquid: The heat capacity and standard enthalpy of formation of EMIES. *Thermochimica Acta*, 2006, 447: 141-146

[22] T Tian, D X Zheng, et al. Determination of the heat capacity of the ionic liquid [EMIm]BF_4 and its aqueous solutions. *Journal of Beijing University of Chemical Technology*, 2008, 35(3): 27-30

[23] L P N Rebelo, V Najdanvic-Visakm, Z P Visak, et al. A detail thermodynamic analysis of [BMIm]BF_4+water as a case study to model ionic liquid aqueous solutions. *Green Chemistry*, 2004, 6: 369-381

[24] J Z Wang, D X Zheng. Performance analysis of absorption cooling cycle utilizing TFE-[BMIm]Br as working fluid. *Journal of Engineering Thermophysics*, 2008, 29 (11): 1813-1816

[25] S Q Liang, J Zhao, L Wang, et al. Absorption refrigeration cycle utilizing a new working pair of ionic liquid type. *Journal of Engineering Thermophysics*, 2010, 31(10): 1627-1630

[26] A Yokozeki. Theoretical performances of various refrigerant-absorbent pairs in a vapor-absorption refrigeration cycle by the use of equations of state. *Applied Energy*, 2005, 80: 383-399

[27] B S Mark, A Yokozeki. Absorption cycle using ionic liquids as working fluids. US2006/0197053 A1, 2006

[28] B S Mark, A Yokozeki. Absorption cycle using ionic liquids and water as working fluids. WO2007/070607 A2, 2007
[29] W H Cai, M Sen, S Paolucci. Dynamic modeling of an absorption refrigeration. *Proceedings of IMECE2007[C], Seattle, Washington, USA: ASME*, 2007, 1-9

23

Optofluidic Compound Lenses Made with Ionic Liquids

Sergio Calixto[1], Martha Rosete-Aguilar[2], Francisco J. Sanchez-Marin[1], Olga L. Torres-Rocha[1], E. Militza Martinez Prado[1] and Margarita Calixto-Solano[3]

[1]Centro de Investigaciones en Óptica, León, Gto.;
[2]Centro de Ciencias Aplicadas y Desarrollo Tecnológico, Universidad Nacional Autónoma de México;
[3]Instituto Tecnologico y de Estudios Superiores de Monterrey, Monterrey, Mexico

1. Introduction

Traditional optical materials comprise glasses, metals, plastics and crystals to mention but a few. The fabrication of optical elements with these materials gives passive elements. However, actual needs demand for dynamical elements with surfaces that can be modulated. To achieve this function several methods have been developed. For example deformable mirrors are used in astronomy to improve the image of stars. Other methods use fluids to achieve the deformation. They are called optofluidic methods. Among the optofluidic elements are the membrane liquid lenses (Erickson et al., 2008; Kuiper & Hendriks, 2004; Werber & Zappe, 2005; Ren & Wu, 2007; Sai et al., 2008; Zeng & Jiang, 2008; Calixto et al, 2009; Yu et al, 2009; Hu et al., 2009; Xu et al., 2010, Zhang et al., 2011). Thus, the optical liquid that composes the lenses is a key element. Due to the good characteristics of ionic liquids, described in section 2, we began a study to find the optical features of some ionic liquids. We hope that the results of the study could be taken as a basis to use the ionic liquids in some optical applications.
Optical liquids should have a specific refractive index and dispersion, thermal stability, low toxicity, transparency for the intended working wavelengths, low cost and compatibility with the material that encloses them. In the past we have suggested briefly the use of ionic liquids as the liquid that composes optofluidic varifocal lenses (Calixto et al, 2009). Now in this chapter a more in-depth optical study of ionic liquids is presented. The paper is organized as follows: Section 2 describes briefly ionic liquids; Section 3 exposes the results of an optical and a thermal study made to some ionic liquids; as an example of an application of ionic liquids we describe in Section 4 an optical design that was developed considering the optical properties of the studied ionic liquids. Following sections describe the efforts to fabricate a compound lens. In section 5 is described the fabrication of cells, that behave as lenses, and contain the ionic liquids. These cells have a flat glass on one end and a flexible membrane on the other side. Section 6 shows the results of a mechanical study made to the membranes. This study shows the profile of the membranes when they are under pressure or vacuum. Section 7 shows the results of an optical study made to cells to find the relation between the pressure applied to the cell and the radii of the membranes. Section 8 describes

the experimental behavior when a compound two-cell lens is used to form images. This compound lens changes its focal length by changing the pressure in each cell. Finally in Section 9 we conclude.

2. Ionic liquids

The development of ionic liquids goes back to 1914. First research efforts dealt with the synthesis of ethylammonium nitrate (Wasserscheid & Keim, 2000). Later the first ionic liquid with chloroaluminate ions was developed in 1948 by Hurley and Wier, at the Rice Institute in Texas, as bath solutions for electroplating aluminum. However, these systems were not studied further until the late 1970s when the groups of Osteryoung and Wilkes rediscovered them (Wasserscheid & Keim, 2000). In the early 1980s the groups of Seddon and Hussey began to use chloroaluminate melts as nonaqueous, polar solvents for the investigation of transition metal complexes (Wilkes, 2002). Spectroscopic and complex chemistry experiments followed. It is specially to Seddon`s work that ionic liquids became more familiar to broad public (Freemantle, 2007).
The use of ionic liquids as solvents for homogeneous transition metal catalyst was described in 1990 by Chauvin et al. and by Wilkes et al. (Wasserscheid & Keim, 2000). The concept of ionic liquid received an impulse by the work of Wilke`s group when they described in 1992 the synthesis of systems with significantly enhanced stability against hydrolysis (Wasserscheid & Keim, 2000).
Ionic Liquids (ILs), containing essentially only ions, have been considered as a group of environmentally friendly solvents with unique properties, such as negligible vapor pressure, thermal and chemical stability, recyclability , and nonflammability.
Ionic Liquids have also attracted great attention among scientists because they are versatile in terms cations, anions, and their combinations, which make their properties designable according to different requirements. An ionic liquid is a salt with a melting temperature below the boiling point of water. Most salts identified in the literature as ionic liquids are liquid at room temperature, and often to substantially lower temperatures. Ionic liquids posses a high degree of asymmetry that frustrates packing and thus inhibits crystallization (Wilkes, 2002).

2.1 Ionic liquid synthesis

The initial step in the synthesis of ionic liquids is the quaternization of an amine or phosphane to form the cation. The most important reported cation types are show in Figure 1. Salts with different anions are obtained by the quaternization reaction depending on the alkylation reagent, Figure 2. Melting points under 100 °C can be obtained for a series of cation/anion combinations with this method (Table 1).

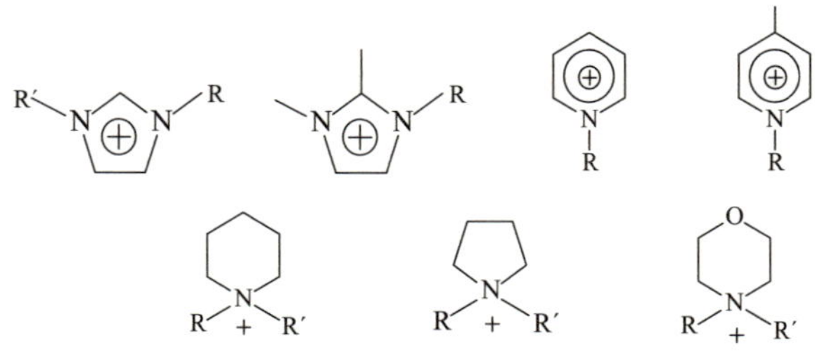

Fig. 1. Important types of cations in ionic liquids. R and R´ are lineal chains: ethyl, propyl, butyl, pentyl, hexyl.

Cl^- Br^- PF_6^- SbF_6^- BF_4^- OTf^- $MeSO_4^-$

NO_3^- $N(CN)_2^-$ SCN^- $Co(CO)_4^-$

Fig. 2. Different anions that configure the ionic liquids.

Ionic liquid	Alkylation reagent	m p. [°C]
[EMIM]CF_3SO_3[(a)]	methyl triflate	-9
[BMIM]CF_3SO_3[(b)]	methyl triflate	16
[Ph_3POc]OTs[(c)]	OcOTs	70-71
[Bu_3NMe]OTs	MeOTs	62
[BMIM]Cl	chlorobutane	65-69

[(a)] EMIM = 1-ethyl-3-methylimidazolium; CF_3SO_3 = triflate anion; [(b)] BMIM = 1-*n*-butyl-3-methylimidazolium; [(c)] Oc = octyl; ; Ts = $H_3CC_6H_4$-SO_2 (tosyl).

Table 1. Examples of ionic liquids that can be formed by direct quaternization.

It should be noted that the synthesis of highly pure, binary ionic liquids places particular demands on the preparative work. The purity of the system is essential for many solvent applications and for the characterization of their physical and chemical properties (Wasserscheid & Keim, 2000).

2.2 Important properties

Density. In general ionic liquids have a larger density value than the density of water. For example, [BMIM][PF_6] has a density of 1.37 g/cm^3 at 20°C. The density of ionic liquids decreases when the N-alkyl chain length increases.

Viscosity. The viscosity of ionic liquids is essentially determined by their tendency to form hydrogen bonding and by the strength of their Van Der Waals interactions. Ionic liquids are generally more viscous than molecular solvents. The Ionic liquid´s coefficients of viscosity at room temperature are between 10 and 1000 centipoises. The viscosity increases when the cation size and alkil chain length increases. An example of this phenomenon is shown in Table 2.

Melting point. A key criterion for the evaluation of ILs is the meltng point. This point for some ILs could be in the range from -90 °C to 200 °C.

Themal stability. Some ionic liquids used as solvents are so stable that the upper temperature limit is not a problem for their employment.

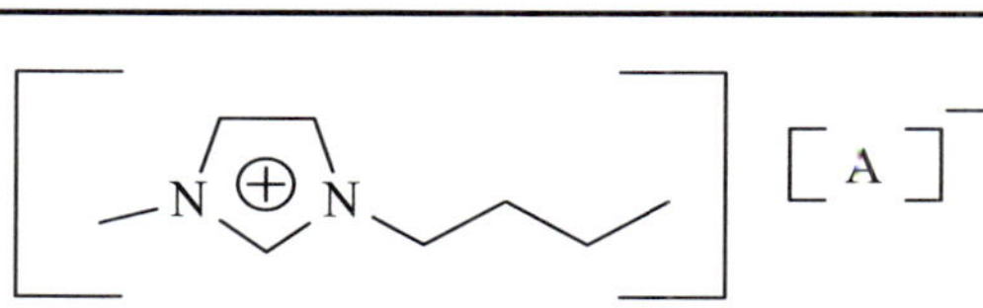

Anion [A]$^-$	η [cP]
$CF_3SO_3^-$	90
n-$C_4F_9SO_3^-$	373
CF_3COO^-	73
n-$C_3F_7COO^-$	182
$(CF_3SO_2)_2N^-$	52

Table 2. Dynamic viscosities η of various 1-*n*-butyl-3-methylimidazolium (BMIM) salts at 20 °C.

Solubility. Due to their ionic nature and to their organic composition ILs are able to dissolve organic and inorganic compounds and sometimes even polymeric materials. This feature makes ILs attractive, not only as solvents for chemical processes, but also for the separation and extraction of materials from solutions and mixtures.
Toxicity. Although ILs impact in air pollution is insignificant due to their very low vapour pressure, their solubility (being of ionic nature) in water is high (Pham et al., 2010). It is necessary to be aware of the potential environmental risks of ILs (Romero, 2008; Wasserscheid & Keim, 2000).

2.3 Potential applications of ionic liquids

Ionic liquids are versatile liquids that have recently been used in researchs concerned with macromolecules. The main reports on ionic liquids and polymers concern its use as polymerization solvents. Also, ionic liquids are currently being used as solvent or additives for macromolecules (Ueki & Watanabe, 2008).
Liquids Separations. Since the choice of cation and anion provides an opportunity to tailor the solubility of various compounds in the ionic liquids, these new compounds should have applications in the selective separation of liquids. This concept was explored by Fadeev and Meagher (Brennecke & Maginn, 2001) who used ionic liquids to separate alcohols from a fermentation broth. In the same field, Rogers and coworkers (Freemantle, 2007) explored the use of ionic liquids/aqueous biphasic systems with added chelating agents to perform metal extractions (Wasserscheid & Keim, 2000).
Lubricants. The characteristics of high thermal stability and large liquid range makes ionic liquids good candidates for lubricants in high-temperature and/or low pressure applications.
Green Chemistry. Recently, ionic liquids have been extensively evaluated as environmentally friendly or "green" alternatives to conventional organic solvents for a broad range of organic synthetic applications. In addition, ionic liquids have been used as catalysts, in organic synthesis, in compositions for stabilizing and/or isolating nucleic acids in or from micro-organisms, as process aids for the synthesis of polynucleotides, as lubricants and for the preparation and stabilization of nanoparticles. Furthermore, after the announcement of the first industrial process involving ionic liquids by BASF (BASIL process) in 2003, the potential of ionic liquids for new chemical processes and technologies is beginning to be recognized (Weyershausen & Lehmann, 2004).
Optical applications. Application of ionic liquids in photophysical studies depends on its transparency in the optical region. Behavior of some imidazolium ionic liquids concerning its UV -visible absorption and fluorescence has been studied (Aniruddha & Anunay, 2006). An important characteristic of these materials is that some of them possess liquid crystal properties, which indicate that these compounds may present large nonlinear optical responses (Souza et al., 2008; Aniruddha & Anunay, 2006). Researchers at some laboratories are using ILs as optical immersion fluids for examining inclusions in gems and minerals (Freemantle, 2007). Besides this application, ILs have been suggested as the base for an extremely large diameter spinning liquid mirror telescope (Borra et al. 2007). Research groups have suggested the use of ionic liquids in optofluidic varifocal compound lenses (Calixto et al. 2009).
More recently, the influence of ionic liquids in photopolymerizable holographic materials has been investigated (Lin et al., 2011). Ionic liquids structure has an important effect on the characteristics of polymeric materials. An extensive investigation of the influence of ionic

liquids on photopolymerizable holographic materials is of interest from both fundamental and application aspects.

3. Optical and thermal characterization of ionic liquids

3.1 Optical studies

From the optical point of view conventional optical materials, like glasses or plastics, are well characterized. By looking at the manufacturer catalogs for optical glass and plastics (www.us.schot.com; www.ohara.corp.com/index.html; Mills, 1986; Simonds, 1964) it is possible to find the properties of tens of glasses and plastics. Among those properties are the chemical, thermal, mechanical (micro hardness) and optical. Chemical properties include climate, staining, acids and alkali resistance. Thermal properties include viscosity and thermal expansion. Optical properties include refractive index and dispersion, secundary spectrum and internal transmittance, to mention but a few.

The refractive index (n) of the materials is a function of the wavelength of light (λ).

Refractive indices are given in the manufacturer´s data sheets for each type of glass. Refractive indices of optical materials are measured at some specific wavelengths (Malacara & Thompson, 2001). For example at the yellow sodium D line (589.2 nm), blue hydrogen F line (486.1 nm), red hydrogen C line (656.2 nm) and yellow helium d line (587.5 nm). The difference in refractive index n_F-n_C is described as the "principal dispersion".

Another quantity that determines the chromatic dispersion is the Abbe value for the D line $V_D = (n_D - 1)/(n_F - n_C)$. A diagram relating the refractive index n_D vs. Abbe value is given by the manufacturers (www.schott.com; www.oharacorp.com/index.html). Each point in the diagram represents a glass. Glasses are somewhat arbitrarily divided in two groups, the *"crown"* glasses and the *"flint"* glasses. Crowns have an Abbe-value of 50 or more, flints 50 or less (Smith, 1966).

Next we show the results of a characterization optical study of some ionic liquids. Ionic Liquids considered were [EMIM][BF_4], [EMIM][TA], [BMIM][TA], [BMIM][BF_4], [BMIM][TF_2N], [EMIM][TF_2N], [EMIM][CF_3COO], [BMIM][CF_3COO], [BMIM][PF_6], [BMIM][NO_3], [BuPy][BF_4], [MePyrr][HSO_4], [BuPy][TF_2N]. Their fabrication method can be found in reference (Calixto et al, 2009).

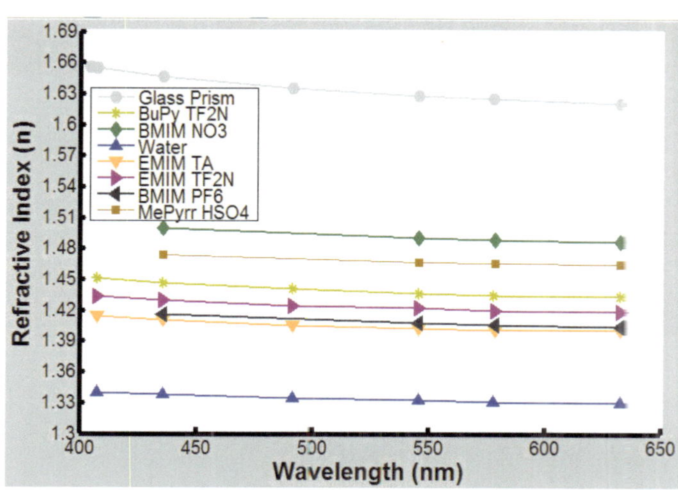

Fig. 3. Behavior of refractive index vs. Wavelenght.

For some of these liquids we found the behavior of the refractive index (n) as a function of some wavelengths (λ). To do this we used a spectroscope. We have not tested all the ionic liquids mentioned above because some of them had close refractive indices. Plot in Fig. 3 shows the results. Also in Fig. 3 is shown the behavior for distilled water, and a glass.
To interpolate refractive index values between measured points we used the Sellmeier or Cauchy equations (Smith, 1966). n_D, n_F and n_C values were calculated with these equations and are shown, together with Abbe values, in the upper part of Table 3. Lower part of Table 3 show values of some glasses taken from data sheets (www.us.schott.com; www.cargille.com).

Material	n_D (λ= 0.5893 microns)	n_F (λ= 0.4861 microns)	n_C (λ= 0.6563 microns)	$n_F - n_C$	V-value
Water	1.3309	1.3348	1.3292	0.0056	58.77
EMIM [TA]	1.4009	1.4051	1.3999	0.0051	78.08
BMIM [PF_6]	1.4059	1.4112	1.4040	0.0072	56.29
EMIM [TF_2N]	1.4200	1.4250	1.4183	0.0067	62.52
BuPy [TF_2N]	1.4346	1.4409	1.4324	0.0085	51.15
MePyrr [HSO_4]	1.4655	1.4696	1.4645	0.0051	91.30
BMIM [NO_3]	1.4881	1.4943	1.4856	0.0086	56.68
Glass Prism	1.6242	1.6363	1.6194	0.0169	36.91
BK7 [15]	1.5158	1.5224	1.51432	0.008054	64.17
PK1 [15]	1.50378	1.5090	1.50146	0.007528	66.92
FK1 [15]	1.47069	1.4755	1.46853	0.006990	67.34
BaFN10 [15]	1.67003	1.68001	1.66579	0.01422	47.11
BaK4 [15]	1.56883	1.57590	1.56576	0.01014	56.13
Immersion Oil (Type B) [21]	1.515	--	--	--	43.2

Table 3. Optical characteristics of some materials. In the upper part are shown data obtained through the experiment. Data in the lower part was obtained from references (www.us.schott.com; www.cargille.com).

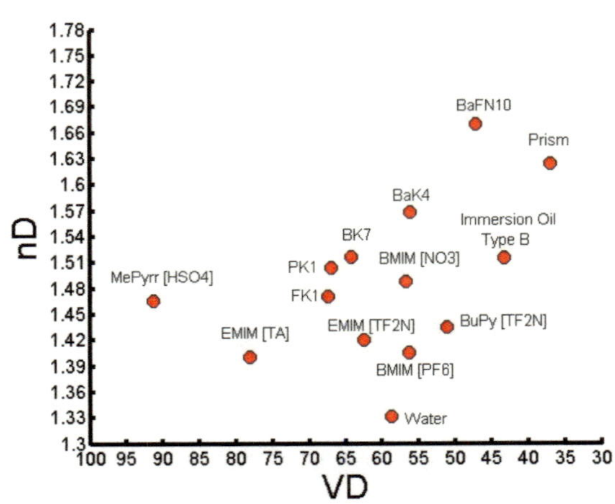

Fig. 4. Abbe Diagram n_D-v_D.

In Fig. 4 is shown the n_D – Abbe value diagram for the materials displayed in Table 3. Here we can notice that most of the ionic liquids have a lower refractive index than the glasses show. Notice that the ionic liquid MePyrr[HSO₄] shows a high Abbe number (91.3) and a refractive index of 1.465. These characteristics were useful when the design of multichamber lenses was done (Section 4).

3.2 Thermal study

It is common that optical elements are used in environments where temperature changes. We investigated the refractive index behavior of some ionic liquids when temperatures above room temperature were reached. The temperature ranged from about 18 °C to about 95 °C. Ionic liquids tested were MePyrrHSO₄, EMIM BF₄ and EMIM[TA]. As a reference we also tested water. Results can be seen in Figure 5.

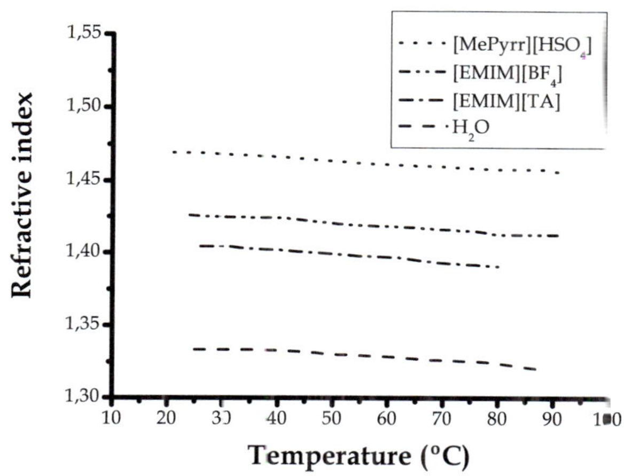

Fig. 5. Behavior of refractve index as a function of temperature for some ionic liquids and water.

4. Optical design

As an example of application of ionic liquids we developed the design of an achromatic varifocal multichamber lens. This lens was first designed by Reichelt and Zappe (Reichelet and Zappe, 2007). Liquids suggested in this reference were commercial (www.cargille.com). An f/3.6 achromatic lens with a positive focal length of 14.4 mm (Fig. 6a), and a negative focal length of -14.4mm (Fig. 6b), was designed by Reichelt and Zappe. The design consists of three chambers (1), (2) and (3). Each chamber is separated by a 400 μm thin, low-dispersive silica glass plate which is also used for the front and back covers. It is supposed that a thin membrane, within each chamber, separates two liquids or a liquid and air. By changing the pressure in the chamber the membrane profile changes. Chamber (1) contains a crown-like liquid in air, (2) contains the same crown-like liquid and a flint-like liquid and (3) contains the same flint-like liquid in air. The refractive indices, dispersions and Abbe-values are given respectively by: $n_1 = 1.3250$, $\Delta n_1 = 0.003788$, $V_1 = 85.8$ for the crown-like liquid and $n_2 = 1.5000$, $\Delta n_2 = 0.0120$ and $V_2 = 41.7$ for the flint-like liquid.

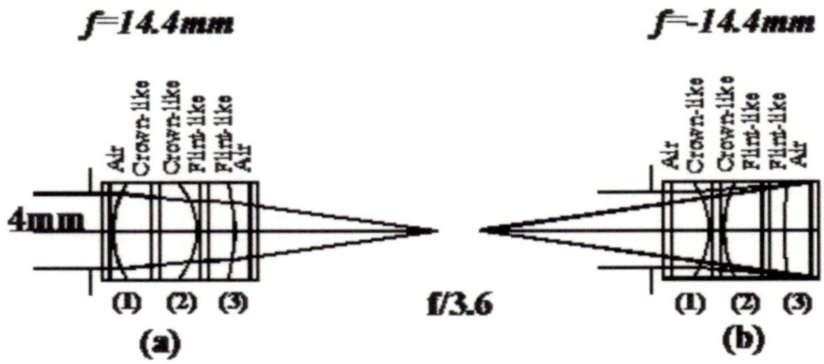

Fig. 6. Diagrams for the lens designed by Reichelet and Zappe. a) Focal dstance is positive, b) focal distance is negative.

In our case by looking at the diagram of n_D-Abbe value (Fig. 4) we found that the combination of ionic liquids MePyrr[HS04] and BMIM[NO3] was able to provide a similar achromatic lens having the same f-number, f/3.6, and working with the same focal length interval between -14.4mm and 14.4mm. A second combination that can provide an achromatic lens is given by the pair EMIM[TA] and BuPy[TF2N] ionic liquids. This lens is achromatic within the focal interval -14.4mm and 14.4mm but working with a larger f-number given by f/4.8. In each combination the first liquid works as a crown-like and the second as a flint-like liquid. The layouts for the MePyrr[HS04] and BMIM[NO3] combination (EMIM[TA] and BuPy[TF2N] combination) are shown in Figures 7a (7c) and 7b (7d) for positive and negative focal length respectively.

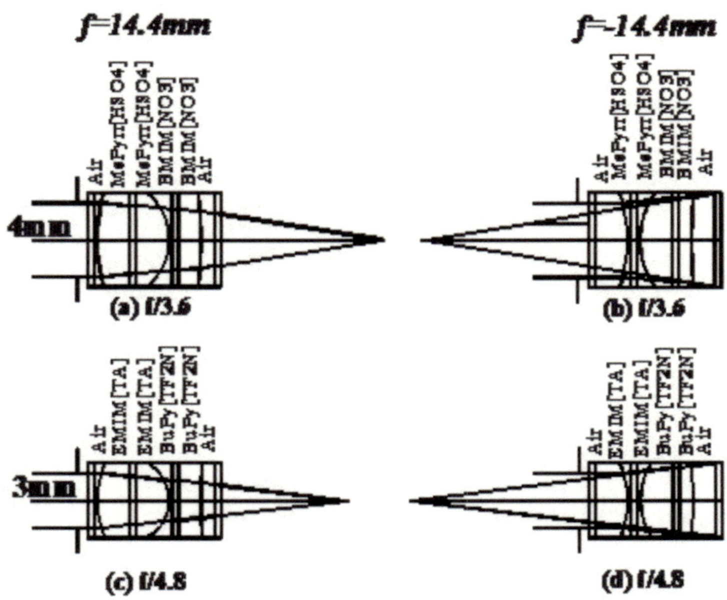

Fig. 7. Layouts for (a,b) the lens with liquids MePyrr[HS04] and BMIM[NO3] and (c,d) the lens with liquids EMIM[TA] and BuPy[TF2N].

The transverse ray aberrations for the achromatic lens designed by Reichelt and Zappe, shown in the first row for a positive and negative focal length, and for the lenses with ionic liquids are plotted in Fig. 8 for comparision. The transverse ray aberration for the MePyrr[HS04] and BMIM[NO3] and for the EMIM[TA] and BuPy[Tf2N] ionic liquids are shown in the second and third row respectively for positive and negative focal lengths. The transverse ray aberration is plotted as a function of the aperture of the lens for three wavelengths: 589.3nm, 486.1nm and 656.3nm. The distance between parallel lines is given by the diameter of the airy disk diffraction pattern.

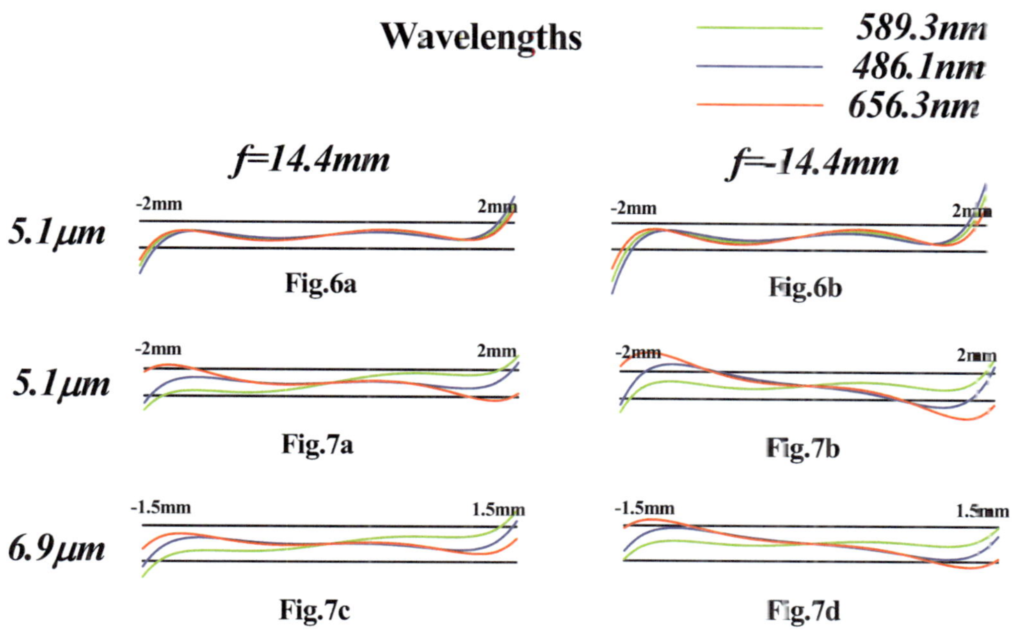

Fig. 8. Transverse ray aberrations.

As can be seen, the ionic liquids also give a nearly diffraction limited achromatic lens almost as good as the achromatic lens designed by Reichelt and Zappe, although the latter gives a better color correction, providing a practically apochromatic lens.

5. Liquid cell fabrication

To implement the ideas exposed in the preceding optical design section (4) we followed the next method. Each cell was made with a clear square acrylic block (11mm X 11mm). Thickness of each block was 4.6 mm. A hole with a diameter of 4.2 mm was drilled in the block center. This hole was used as the liquid chamber. Two small holes, 90° apart, were drilled on the sides of the block. In each hole a syringe needle was glued. Through one needle the liquid was fed into the chamber and through the second needle air was drained. Ends of the acrylic block were covered one with a thin flat glass plate (150 μm thick), the other with a thin silicone membrane. Figure 9 shows side and plain diagrams of the cell. Figure 10 is a photograph of a fabricated cell.

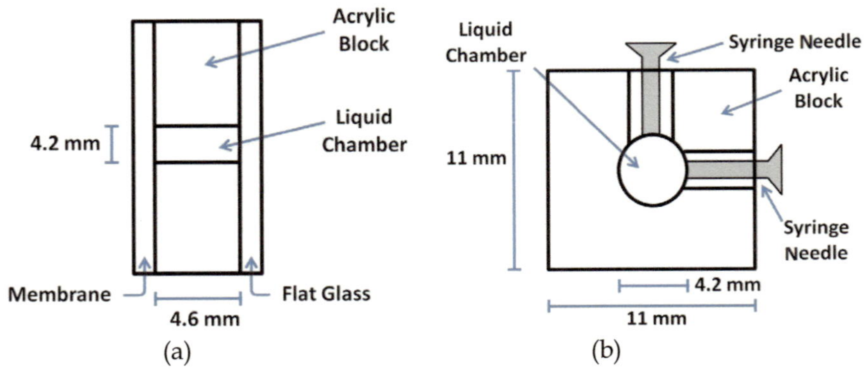

Fig. 9. Side (a) and plain (b) diagramas of a cell. Figure not to sacale.

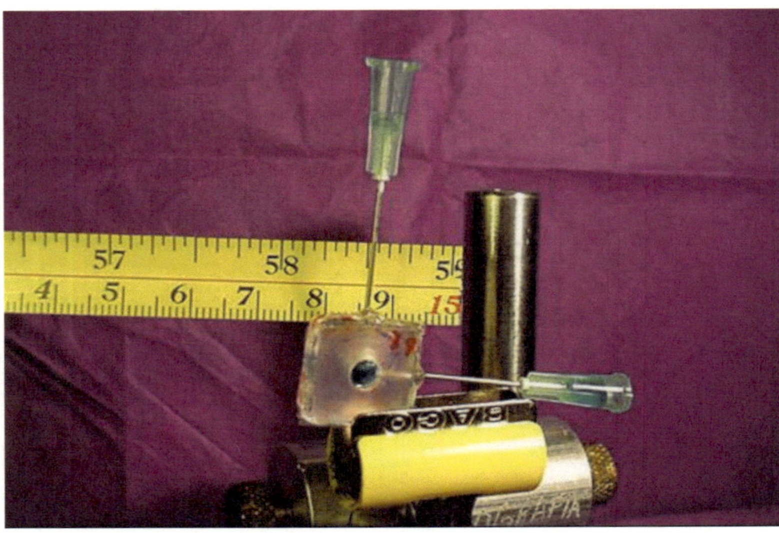

Fig. 10. Photograph of a fabricated cell

Silicone (PDMS) was chosen as the material for the membrane because it is flexible and transparent. Membranes were made by the spinning method. A small amount of silicone and curing agent (10:1) was placed over a flat glass plate that was fixed to the shaft of an electric motor. Then it was spun. Later the plate was leveled and left overnight. Membrane was peeled off the glass and glued to the acrylic block. Membrane thickness ranged between 150 μm to about 220 μm.

Membrane roughness is an important parameter in the cell fabrication process. Membranes with large roughness will scatter light and the lens will give blurred images. To avoid large roughness we followed the next method. It is known that when a mixture of silicone and curing agent is placed in contact with a surface the silicone copies perfectly the surface relief. Thus, we chose a polished glass plate for the spinning time. To find out the glass plate roughness we used an Atomic Force Microscope (AFM). In Figure 11a we can see an image (100 μm x 100 μm) of the glass plate surface. Here we notice that the surface roughness in an area of about 20 μm x 18 μm is 1.15 nm rms. In Fig. 11b is shown a portion (100 μm x 100 μm) of the silicone surface. The roughness in an area of 16 μm x 14 μm is 0.97 nm rms. This value is close to that shown by the glass plate. With these roughness values we can say that light will be poorly scattered by the membrane surfaces. In Fig.11c is shown a perspective view of some part of the membrane surface.

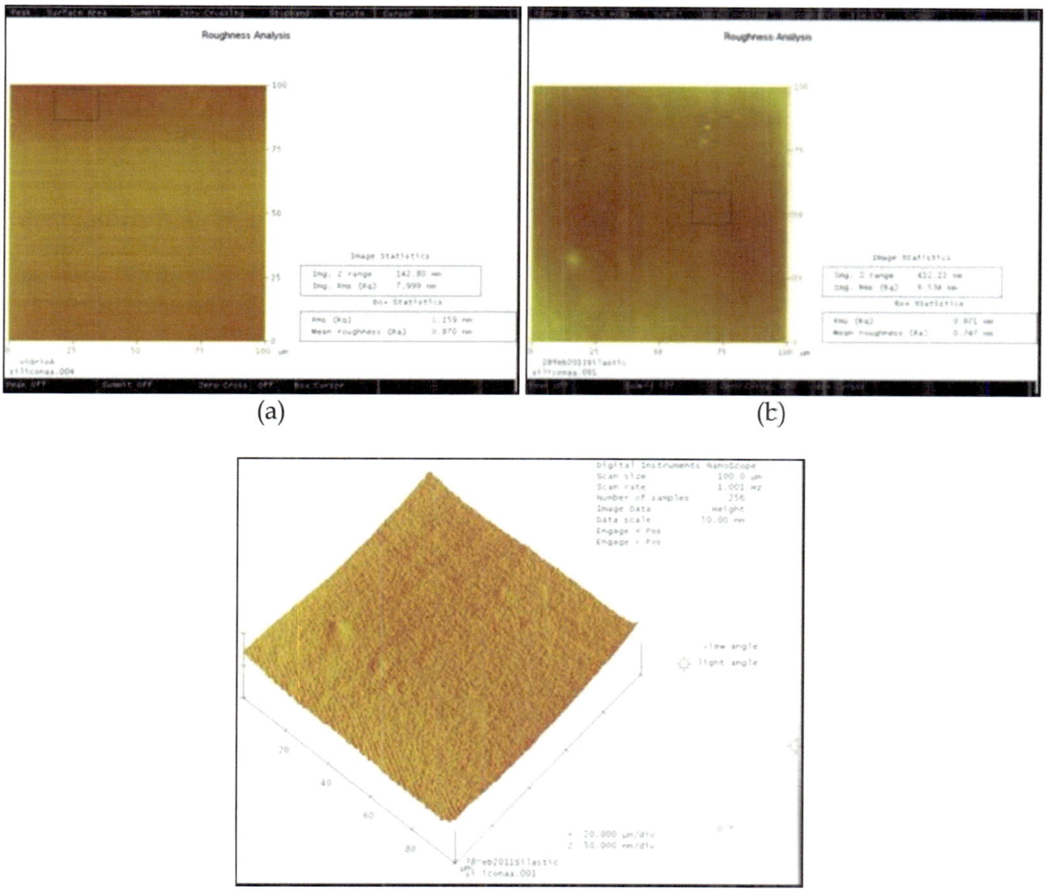

(a) (b)

(c)

Fig. 11. Images of surfaces given by an AFM. (a) glass plate (b) silicone membrane and (c) Perspective view of a silicone surface.

6. Membranes profiles studies

6.1 Profiles studied with a surface analyzer

Membrane profile changes, caused by positive or negative pressure in the cell, were investigated with a surface analyzer. This instrument has a sharp needle that scans the surface under test. The result is a graph relating the height of the profile as a function of a linear coordinate. In Fig. 12a are shown a series of plots given by the surface analyzer. They show the profiles of a membrane when positive pressures were given. It is noticed that as pressure increases the radius of curvature of the surface decreases. In Fig. 12b are shown the membrane profile curves when the chamber was evacuated.

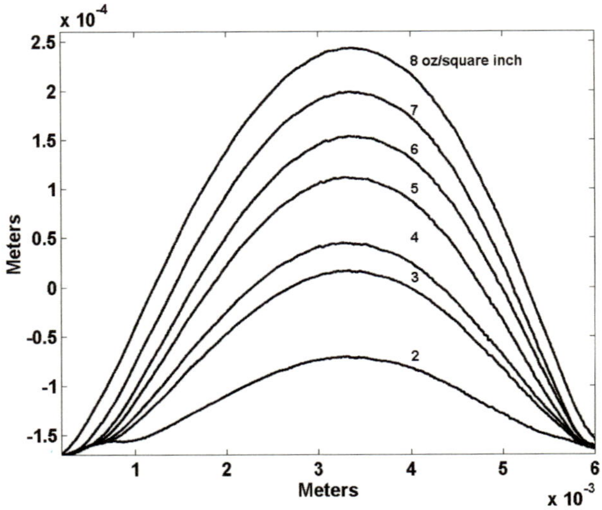

(a)

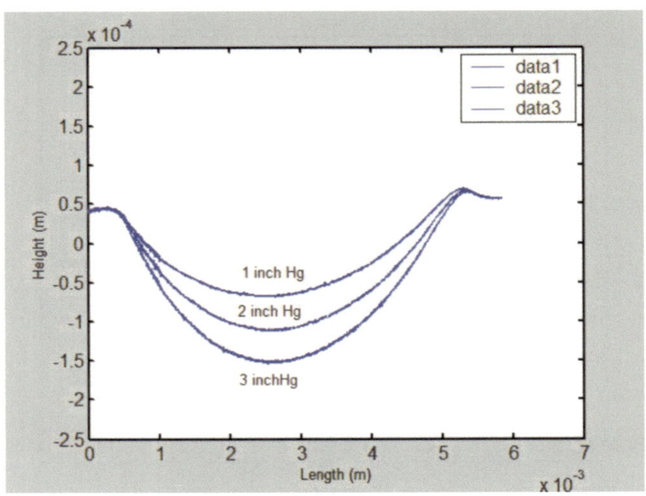

(b)

Fig. 12. Cell membranes reliefs measured with a surface analyzer. Parameter is the pressure. In (a) positive pressure was applied. In (b) negative pressure (vacuum) was applied.

6.2 Comparison of membrane profiles with a theoretical circle

In section 4 we have shown that optical designs predict that good liquid lenses can be made with ionic liquids. These designs suppose that the profile of the membranes in the cell is spherical. To find out how close this is to the profile of the membranes, measured with the surface analyzer, to a segment of a circle we followed the method described in the next paragraph.

To determine the deviation of the central part of the profiles from a theoretical circle, such part was extracted from the profiles obtained with the surface analyzer. To do that, given the small dimensions of the lenses and the need to determine the global maxima and minima of the profiles, it was found convenient to filter the profiles with a Gaussian filter to eliminate quantization noise. Then, the maximum (for positive pressures) or minimum (for negative pressures) of each profile was determined and used as a reference to extract the central part of each profile and two other points (one to the left and another to the right) that were used to calculate the equations of the circumferences that contained those three points of each profile. Once the equation of each circumference was obtained, the corresponding radius was used to determine de deviation of each point of the central part of the profiles from the corresponding theoretical circle. The results of this procedure are shown in figures 13a to 13e. As can be seen, in all cases, for positive and negative pressures, the central parts of the profiles were practically perfect circles. For positive pressures, the maximum deviation from a perfect circle was approximately 0.6 %, while for negative pressures the maximum deviation was close to 0.1 %.

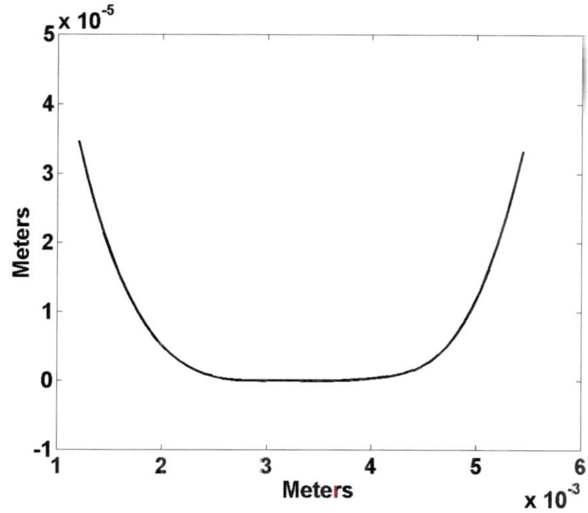

Fig. 13. (a)

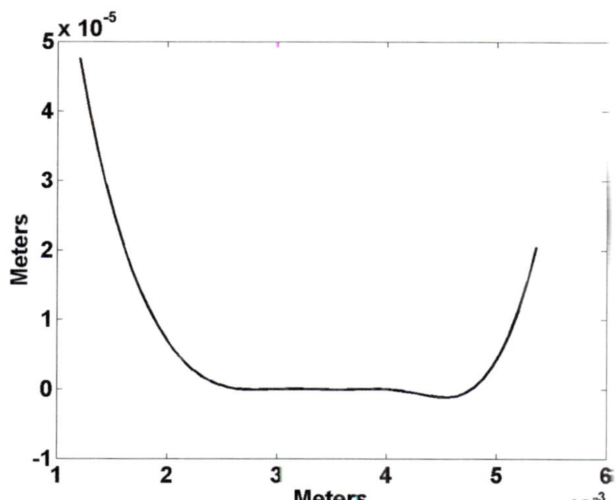

Fig. 13. (b)

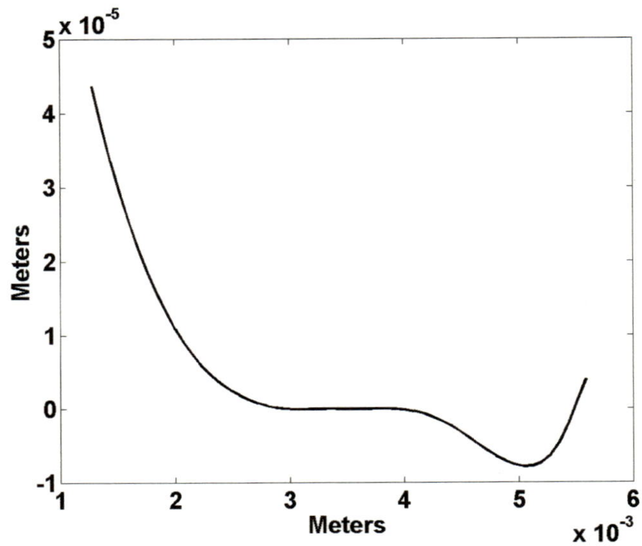

Fig. 13. (c)

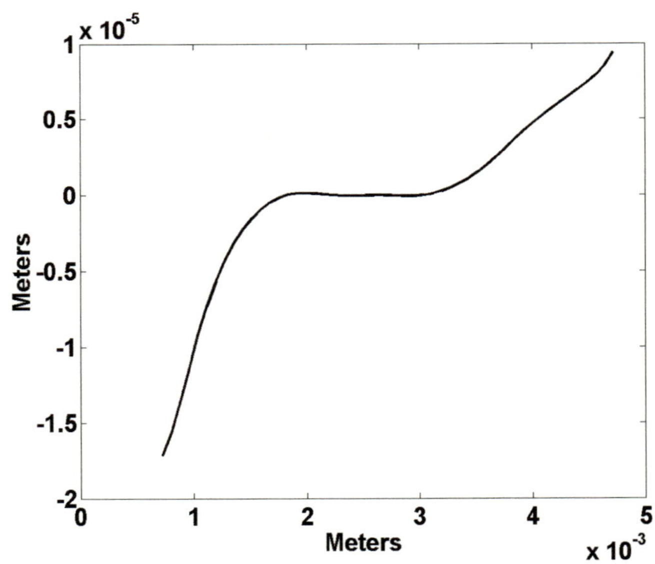

Fig. 13. (d)

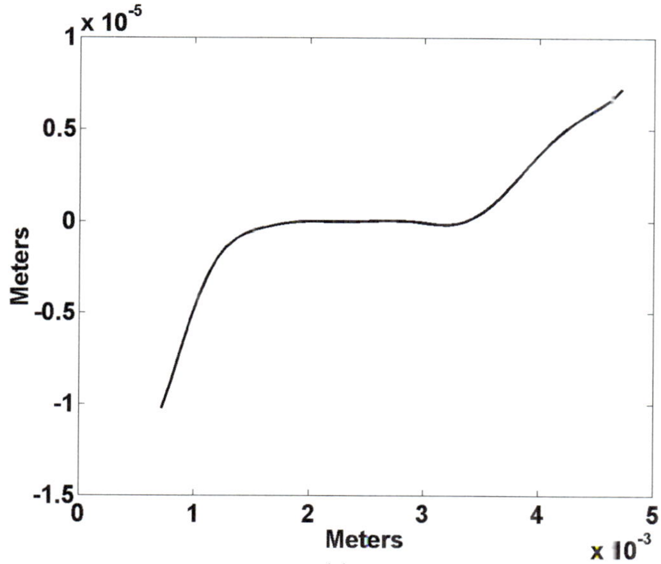

(e)

Fig. 13. Deviation of membrane profile with reference to the best-fitting circle. Membrane pressures were the following: (a) 3 inch Hg, (b) 1 inch Hg, (c) 2 oz/inch^2, (d) 5 oz/inch^2, (e) 8 oz/inch^2.

7. Optofluidic lens optical test

7.1 Lens image formation

In the preceding section (6) we have seen that cells membranes bend under pressure. Thus, the cell becomes a lens. The lens membrane can present positive or negative radius of curvature depending whether positive or negative pressure is applied. In Fig. 14a is shown a photograph of the image given by a lens when a positive pressure of 4 oz/inch^2 was given. Object was a chart printed in a paper placed at about 2 m from the lens. Fig. 14b shows a photograph of the image when a pressure of 2 inch Hg (vacuum) was given.

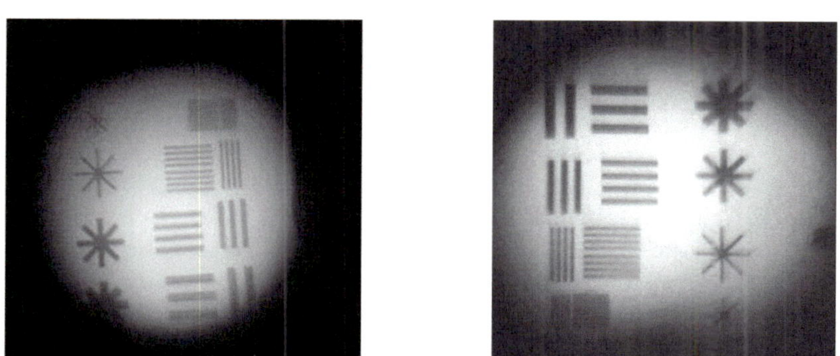
(a) (b)

Fig. 14. Images given by a cell when a positive pressure was given, (a), and when negative pressure was given, (b).

7.2 Behavior of membrane radius with pressure

To find the behavior of the radius of curvature of the lenses as a function of pressure (positive or negative) the following method was used. An object consisting of some vertical and horizontal bars was placed about 3 m from the cell. Then pressure was fed into the cell by steps. For each pressure, image given by the lens was found with a microscope. Image position with reference to the lens was measured. If we suppose that the object was far from the lens this distance can be taken as the back focal length of the lens for that pressure. By means of the next formula the membrane radius of curvature was found: $1/f = (n-1)\,(1/R_1 + 1/R_2 - (n-1)d/R_1 R_2)$, here R_1 and R_2 are the radius of curvature of the glass plate and the membrane, n is the refractive index of the liquid in the cell and d its thickness. Because the glass plate is flat, $R_2 = \infty$. Results of the experiments for a given cell are shown in Fig.15. Here we notice the relation of the pressure in the cell with the focal distance or the membrane radius.

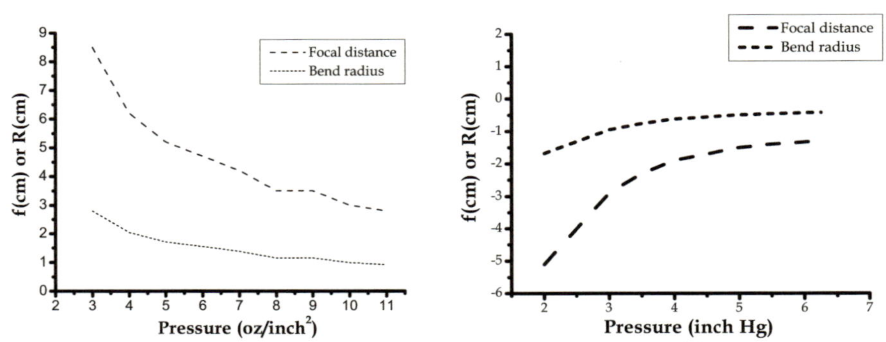

(a) (b)

Fig. 15. Behavior of lens focal distance or radius of curvature as a function of pressure. In (a) positive pressures were given. In (b) negative pressures were given.

8. The compound lens

In the preceding sections we have described the fabrication and testing of cells that could compose a compound lens. To test the performance of ionic liquids a doublet formed by two chambers (cells) was fabricated and tested experimentally. The two cells were built containing different ionic liquids: MePyrr[HSO4] and EMIM [TF2N] for the first and second cells, respectively, see table 3. Each cell is formed by a thin membrane and a thin flat glass. Membrane profile changes with pressure. The two cells are placed (assembled) together by joining the two flat glasses and adding a thin film of immersion oil between the two glasses to decrease internal reflections. The final layout of the compound lens is shown in Fig. 16.

There are some methods to test optical elements (Malacara, 1978). For example those based on the interference of light, formation of images and others. Among the formation of images methods is the star test. Here a small illuminated pinhole is used as object. Its image is formed by the optical system under test and studied with a microscope. Although it is difficult to have quantitative results using this test, for qualitative results requirements are more relaxed. To test the assembled lens we placed an illuminated pinhole (dia. 500 μm) about 3 m away from the compound lens. Figure 16 shows a diagram of the experimental configuration. Light from the pinhole travels until it reaches the first membrane (A). Then it is refracted and crosses the cell where the MePyrrl ionic liquid is. Then it reaches the first flat glass, the thin immersion liquid film and the second flat glass. After this it enters the second cell, filled with EMIM[TF2] ionic liquid and reaches the second membrane (B). Finally it travels to the place where an image is formed. This image is studied with a microscope. Curvature of each membrane can be changed by means of the pressure, positive or negative, given to each cell.

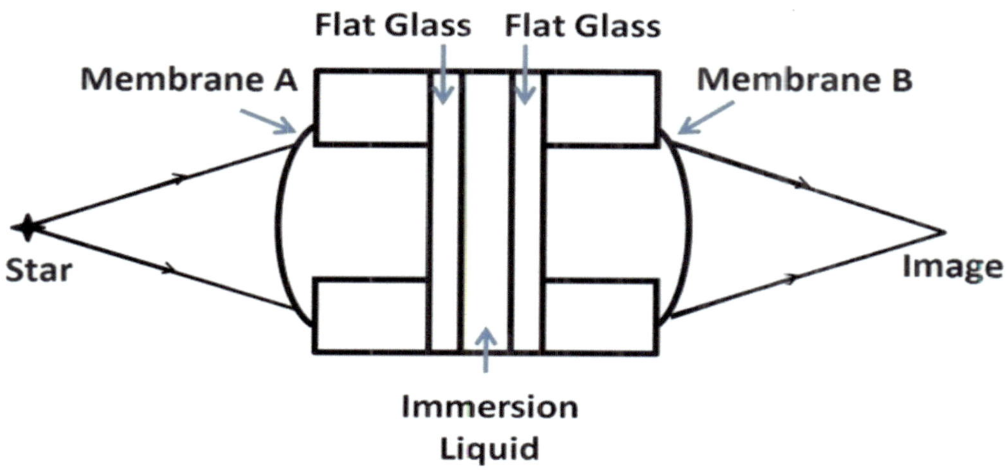

Fig. 16. Diagram showing a compound lens forming an image. Figure not to scale.

In a given experiment we gave to the first cell a pressure of 1 psi. Then we changed the pressure of the second cell by the following values: 3 inch Hg and 2 inch Hg, these values made the second surface (B) of the membrane concave. To have a convex surface we gave the following values: 3 oz/inch^2, 5 oz/inch^2, 7 oz/inch^2. Every time a pressure was given a photograph of the image of the star was taken. These photographs are shown in Figure 17.
Here we can notice that in photographs a) and b) the star image is blurred meaning that aberrations were present. However, for photographs c-f the star image is more rounded meaning aberrations were small.
The star image performance depends on the quality of the lens and the position of the star with respect to the optical axis. By looking at the image of the star shown in figure 17a, we can see that there is astigmatism. This astigmatism may arise from a non spherical surface when changing the pressure of the membrane or may arise from a star which is not located at the optical axis of the doublet. We have seen in section 6.1 that the membranes profile is close to a spherical surface when pressure is applied. Thus, we could assume that the star is not located in the optical axis. Let us assume that the star is located at 10° semifield of view. We know that the first cell had a pressure of 1 psi. Thus it had a radius R= 9 mm. Radius of the second membrane was varied between – 4 mm and – 20 mm for positive pressure, and between 4 mm and 20 mm for negative pressures. Seidel aberration wavefront (Kidger, 2002) was calculated and its beavior is shown in Figure 18. Fig 18a shows behavior when R is in the range from -4 mm to -20 mm. Fig. 18b when R goes from 4 mm to 20 mm.
From Figure 18 we can observe that the aberration decreases when the radius of curvature of the second membrane increases or tends to infinity. In Figure 18(a) we can see that the best image is obtained for a pressure of 1psi in the first cell and 0 inch Hg for the second cell. This last value means the membrane is a plane surface. For negative pressure, Figure 18(b),

the aberration difference between R=9mm and R=20mm is approximately 0.027mm which is larger than the aberration difference for positive pressure between R=-9mm and R=-20mm, given approximately by 0.002mm, see Figure 18(a). This result qualitatively agrees with the star images shown in Figure 17 where a larger image degradation is observed when negative pressure is applied to the second cell.

(a) (b) (c)

(d) (e) (f)

Fig. 17. Photographs showing the images of a pinhole (star), placed at 3 m, given by a two-cell compound lens under different negative and positive pressure. First cell had a pressure of 1 oz/inch^2. Second cell had different pressures: (a) 3 inch Hg, (b) 2 inch Hg, (c) zero pressure, (d) 3 oz/inch^2 , (e) 5 oz/inch^2, (f) 7 oz/inch^2.

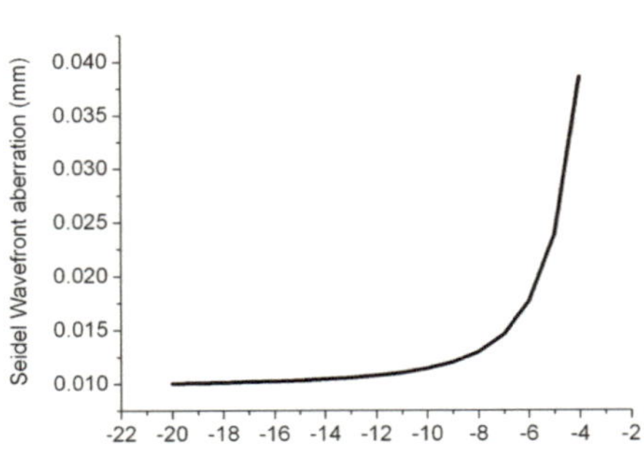

(a)

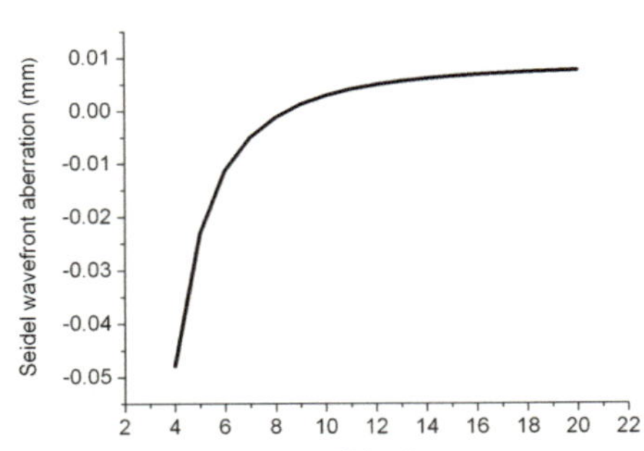

(b)

Fig. 18. Wavefront aberration vs. Radii of curvature.

The two-cell compound lens was also tested to find the possibility to form images of nearby objects. An USAF test chart was placed about 2 cm away from the cell. Then a pressure of 1 psi was given to the first cell and a pressure of 11 oz/inch^2 was given to the second cell. Image was studied with a microscope and is seen in Fig. 19. We can notice that element 3 in group 5 of the test chart can be seen. This means that about 40 lp/mm can be resolved.

Fig. 19. Image of an USAF test target placed about 2 cm from the membranes cell. Element 3 group 5 can be resolved (40 lp/mm).

9. Conclusions and comments

Some ionic liquids have been studied with a spectroscope and a plot relating n_D vs Abbe value has been found. This plot could be used as a basis to choose some ionic liquids for an optical application.

In section 2 we have mentioned that there is a large number of combinations of cations and anions that could produce limitless different ionic liquids. Thus, ideally we can adapt a specific set of ionic liquids to an optofluidic optical system that has a specific task. However, more chemical and optical studies of ionic liquids should be done. With these new studies we will have more points in the n_D – Abbe value diagram and the optical designer will have more freedom to design an optical system.

We have described the characterization studies made to the membranes. Parameters like roughness and profiles under pressure have been exposed. A study of the membrane profiles tell us that their deviation from a theoretical circle is small. Also it was shown that cells were able to form images. Besides this a two-cell compound lens was assembled. This compound lens gave images with different sizes when the pressure on one cell was changed.

In section 4, we have shown an optical design study of a 3 cell compound lens. Liquids considered were ionic liquids. From the theoretical point of view (diffraction limited behavior) optical behavior of this three-cell lens is good. However, the fabrication of such a three cell compound lens needs the development of more accurate fabrication methods. Probably with MEMS technology the three cell compound lens could be made. With our low-technology method, comprising acrylic cells, just a two-cell compound lens with diameters of some millimeters can be done.

10. Acknowledgments

We thank Antonio Martinez Richa and Karla A. Barrera Rivera for supplying the ionic liquids. Also we thank Octavio Pompa Carrera for helpful discussions.

11. References

Aniruddha P. & Anunay S. (2006) Optical absorption and fluorescence studies on imidazolium ionic liquids comprising the bis(trifluoromethanesulphonyl)imide anion. *J. Chem Sci.*, vol. 118, No.4, (July 2006), pp. 335-340, ISSN 0253-4134

Borra E., Seddiki O., Angel R., Eisenstein D., Hickson P., Seddon K.R. & Worden S.P. (2007). Deposition of metal films on an ionic liquid as a basis for a lunar telescope. *Nature,* Vol. 447, pp. 979-981, ISSN 0028-0836

Brennecke J.F. & Maginn E.J. (2001). Ionic Liquids: Innovative Fluids for Chemical Processing. *AIChE Journal*, vol. 47, No.11, pp. 2384-2389, ISSN 0001-1541

Calixto S., Sanchez-Morales M.E., Sánchez-Marin F.J., Rosete–Aguilar M., Martínez–Richa A., & Barrera–Rivera K.A. (2009). Optofluidic Variable Focus Lenses. *Appl. Opt.*, Vol. 48, pp. 2308–2314, ISSN 0003-6935

Cargille Laboratories, Specifications of Cargille *Optical Liquids*, www.cargille.com

Erickson D., Yang C. & Psaltis D. (2008). Optofluidics emerges from the laboratory, *Photonics spectra,* Vol. 42, pp. 74-79, ISSN 0731-1230

Freemantle M. (2007). New frontiers for ionic liquids. *Chemical & Engineering News,* Science and Technology Vol 85, pp. 23-26 ISSN 0009-2347

Hu S., Ren H., Lin Y.J., Moharam M.G.J., Wu S.T., & Tabiryan N. (2009). Adaptive liquid lens actuated by photo-polymer. *Opt. Express,* Vol. 17, pp. 17590–17595, ISSN 1094-4087

Kidger M.J.(2002), "Fundamental Optical Design", SPIE-The International Society for Optical Engineering.

Kuiper S. & Hendriks B.H.W. (2004). Variable focus liquid lens for miniature cameras. *Appl. Phy. Lett.,* Vol. 85, pp. 1128–1130 ISSN 0003-6951

Lin, H. Oliveira P. & Veith M. (2011). Application of ionic liquids in photopolymerizable holografic materials. *Optical Materials*, 33, (January 2011), pp. 759-762, ISSN 0925-3467.

Malacara, D. (1978) *Optical Shop testing*. Wiley, ISBN 0-471-01973-9, NewYork.

Malacara D. & Thompson B.J. (2001). *Handbook of Optical Engineering*, Marcel Dekker, ISBN: 0-8247-9960-7, New York.

Mills N.J. (1986). *Plastics Microstructure. Properties and Application*, Edward Arnol

Pham T.P.T., Cho C.W. & Yun Y.S. (2010). Environmental fate toxicity of ionic liquids: A review. *Water research*, Vol. 44, pp. 352-372, ISSN 0043-1354

Reichelt S. & Zappe H. (2007). Design of spherically corrected, achromatic variable–focus liquid lenses. *Opt. Express*, Vol. 15, pp. 14146–14154, ISSN 1094-4087

Ren H. & Wu S-T. (2007). Variable focus liquid lens. *Opt. Express*, Vol 15, pp. 5931- 5936, , ISSN 1094-4087

Romero A. (2008) Ionic liquids to ambient temperature: a new one means for the chemical reactions. *Rev. R. Acad. Cienc. Exact. Fís. Nat.*, 102, No.1, pp. 79-90, ISSN 1578-7303

Sai F.T., Cho S.H., Lo Y.H.H & Vasko B. (2008). Miniaturized universal device using fluidic lens. *Opt. Lett.*, Vol. 33, pp. 291–293, ISSN 0146-9592

Simonds H.R. & Church J.M. (1964). *Concise Guide to Plastics*, Reinhold Publishing Corporation

Smith W, (1966) *Modern Optical Engineering*, McGraw Hill, ISBN 07-053690-x, New York.

Souza R. F. Alencar M., Meneghetti M. & . Hickmann J. (2008). Nonlocal optical nonlinearity of ionic liquids. *J. Phys. Condens, Matter*, 20, (March 2008), pp. 1-5, ISSN 0953-8984

Ueki T. & Watanabe M. (2008). Macromolecules in Ionic Liquids: Progress, Challenges, and Opportunities. *Macromolecules*, 41, No.11, (June 2008), pp. 3739-3749, ISSN 0024-9297

Wasserscheid P. & Keim W. (2000). Ionic Liquids-New Solutions for Transition Metal Catalysis. *Angew. Chem. Int. Ed.*, 39, pp. 3772-3789, ISSN 1433-7351

Werber A. & Zappe H. (2005). Tunable micofluidic lens. *Applied Optics*, Vol. 44, pp. 3238-3245, ISSN 0003-6935

Weyershausen B. & Lehmann K. (2004). Industrial application of ionic liquids as performance additives. *Green. Chem.*, Vol. 7, (November 2004), pp. 15–19, ISSN 1463-9262

Wilkes J. S. (2002). A short history of ionic liquids-from molten salts to neoteric solvents. *Green Chemistry*, 4, (March 2002), pp. 73-80, ISSN 1463-9262

Xu S., Liu Y., Ren H. & Wu S.T. (2010). A novel adaptive mechanical–wetting lens for visible and near infrared. *Opt. Express*, Vol. 18, pp. 12430–12435, ISSN 1094-4087

Yu H.B., Zhou G.Y., Chau F.S., Lee F.W., Wang S.H., & Leung H.M. (2009). A liquid filled tunable double–focus microlens. *Opt. Express*, Vol. 17, pp. 4782–4790, ISSN 1094-4087

Zeng X.F. & Jiang H. (2008). Tunable liquid microlens actuated by infrared light- responsive hidrogel. *Appl. Phy. Lett.*, Vol 93, pp. 151101, ISSN 0003-6951

Zhang, W, Aljasen K., Zappe H., Seifert A, (2011) Completely integrated, thermo-pneumatically tunable microlens, *Optics Express*, Vol. 3, 247 – 2358, ISSN 1094-4087.

www.us.schott.com
www.oharacorp.com/index.html